Environmental Issues and Waste Management Technologies in the Ceramic and Nuclear Industries III

Related titles published by the American Ceramic Society:

Corrosion of Materials by Molten Glass (Ceramic Transactions Volume 78)
Edited by George Pecoraro, James C. Marra, and John T. Wenzel
© 1996, ISBN 1-57498-019-X

Enviornmental Issues and Waste Management Technologies in the Ceramic and Nuclear Industries II (Ceramic Transactions Volume 72)
Edited by Vijay Jain and David K. Peeler
© 1997, ISBN 1-57498-023-8

Environmental Issues and Waste Management Technologies in the Ceramic and Nuclear Industries (Ceramic Transactions Volume 61)
Edited by Vijay Jain and Ron Palmer
© 1995, ISBN 1-57498-004-1

Environmental and Waste Management Issues in the Ceramic Industry II (Ceramic Transactions Volume 45)
Edited by Dennis Bickford, Steven Bates, Vijay Jain, and Gary Smith
© 1994, ISBN 0-944904-79-3

For information on ordering titles published by the American Ceramic Society, or to request a publications catalog, please contact our Customer Service Department at 614-794-5890 (phone), 614-794-5892 (fax), <customersrvc@acers.org> (e-mail), or write to Customer Service Department, 735 Ceramic Place, Westerville, OH 43081. Visit our on-line book catalog and web site at <www.acers.org>.

Ceramic Transactions
Volume 87

Environmental Issues and Waste Management Technologies in the Ceramic and Nuclear Industries III

Edited by
David K. Peeler
Westinghouse Savannah River Company/
Savannah River Technology Center

James C. Marra
Westinghouse Savannah River Company/
Savannah River Technology Center

Published by
The American Ceramic Society
735 Ceramic Place
Westerville, Ohio 43081

The Joint Fall Meeting of the Basic Science, Cements, Glass & Optical Materials, Electronics, Nuclear & Environmental divisions and the Electronic Division of the Japanese Ceramic Society, held in San Antonio, Texas, October 30–November 2, 1996; the International Symposium on Waste Management Technologies in Ceramic and Nuclear Industry, and the Environmental Issues and Ceramic Manufacturing Symposium, both held at the 99th Annual Meeting of The American Ceramic Society in Cincinnati, Ohio, May 4–7, 1998.

For information on ordering titles published by the American Ceramic Society, or to request a publications catalog, please call 614-794-5890.

Printed in the United States of America.

1 2 3 4–01 00 99 98

ISSN 1042-1122
ISBN 1-57498-035-1

Contents

ENVIRONMENTAL ISSUES IN CERAMIC MANUFACTURING
1997 Annual Meeting
Cincinnati, OH

Environmental Regulations and the Ceramic Industry

Air Emissions and Treatment of Secondary Wastes

WASTE MANAGEMENT TECHNOLOGIES IN THE CERAMIC AND NUCLEAR INDUSTRY
1997 Annual Meeting
Cincinnati, OH

Historical Aspects of Radioactive Materials

Actinide Stabilization I

Waste Form Property Measurement and Modeling

High-Level Waste Processing and Vitrification

Hazardous Waste Stabilization and Alternative Waste Forms

Leaching, Durability and Performance Assessment of Waste Forms

Actinide Stabilization II

Low-Level Radioactive and Mixed Waste Vitrification

Radioactive Waste Stabilization

Preface

Glass and ceramic waste forms are standard treatment means for hazardous and radioactive waste disposal. With continued growth of scientific and technological interest in this field, there has been a major effort to seek solutions to existing environmental problems. These efforts have advanced waste form development issues, aided the development of immobilization technologies, and increased the understanding of product performance for disposal purposes. Along with the technical issues of waste form development, environmental concerns are becoming significant in the ceramic manufacturing industry. Industries are focusing on "green ceramics", performing lifecycle analysis, and adopting "environmental stewardship" programs to manufacture environmentally friendly products.

The Nuclear and Environmental Technology (NET) Division of the American Ceramic Society recently sponsored three symposia focusing on these issues. The first was included in the 1996 Joint Fall Meeting held in San Antonio, October 30–November 2. A symposium on environmental challenges for ceramic manufacturing, applications, and waste management was sponsored by the NET Division in conjunction with the Society's Basic Science, Glass & Optical Materials Division and the Japanese Ceramic Society's Electronic Division. Two symposia in conjunction with the 99th Annual Meeting of the American Ceramic Society held in Cincinnati, Ohio, May 4–7, 1997 were sponsored by the NET, Basic Science, Engineering Ceramics, Glass & Optical Materials, and the Whitewares Divisions of the American Ceramic Society. These two symposia focused on environmental issues in ceramic manufacturing and waste management technologies in the ceramic and nuclear industries.

The fifth in this series, the proceedings you hold is an expansion of the established series on the management of nuclear waste. Earlier proceedings include Ceramic Transactions volumes 39, 45, 61, and 72. This volume contains manuscripts from both the 1996 Fall Meeting and the 1997 Annual Meeting of the American Ceramic Society.

The symposium organizers are grateful to all who contributed to the success of these symposia. First, we wish to thank the authors and reviewers who gave freely of their time and talents to ensure quality publications. We would also like to acknowledge the contributions of the session chairs: Denny Bickford, Greg Chandler, Xiangdong Feng, Vijay Jain, Linda Jones, John Marra, Ron Palmer, Gary Smith, Pete Smith, Arun Wagh, and Wayne Young. In addition to chairing their sessions, these individuals were instrumental in organizing the technical review of each manuscript published in this proceeding.

David K. Peeler
James C. Marra

CERAMICS IN THE ENVIRONMENT
1996 Fall Meeting
San Antonio, TX

Waste Management and
Material Performance in
Severe Environments

DEVELOPMENT OF A REMOTE BUSHING SYSTEM FOR AMERICIUM-CURIUM VITRIFICATION

R.F. Schumacher, W.G. Ramsey, F.M. Johnson, T.M. Jones, D.H. Miller, and B.J. Hardy
Westinghouse Savannah River Company
Savannah River Technical Center
Aiken, SC 29808

ABSTRACT

The Savannah River Site (SRS) and the Savannah River Technology Center (SRTC) are combining their existing experience in handling highly radioactive, special nuclear materials with commercial glass fiberization technology in order to assemble a small vitrification system for radioactive actinide solutions. The vitrification system or "bushing", is fabricated from platinum-rhodium alloy and is based on early marble remelt fiberization technology. Advantages of this unique system include its relatively small size, reliable operation, geometrical safety (nuclear criticality), and high temperature capability. The bushing design should be capable of vitrifying a number of the actinide nuclear materials, including solutions of americium/curium, neptunium, and possibly plutonium. State of the art, mathematical and oil model studies are being combined with basic engineering evaluations to verify and improve the thermal and mechanical design concepts.

INTRODUCTION

A solution containing the isotopes of americium-243 and curium-244 (Am/Cm) is stored in the F-Canyon, Separation Area at the Savannah River Site. This material was accumulated from processing plutonium targets during the 1970's. These materials are the only concentrated kilogram sources of these isotopes in the United States and have an extremely high intrinsic value for the nation. This is particularly true if the material can be further enhanced by transformation at the Oak Ridge National Heavy Element and Advanced Neutron Facility in Oak Ridge, Tennessee.

Alternatively, the disposal of this material to the SRS Tank Farm as a waste would create a number of problems and concerns due to the very high level of radioactivity involved. Transportation of the radioactive Am/Cm in liquid form to Oak Ridge is not considered acceptable under present standards. An acceptable solution is to vitrify this material in available space in the Separation Canyon prior to transporting to Oak Ridge in sealed containers within shipping casks. The vitrified form would be relatively safe to transport and the material would then be redissolved in acid at Oak Ridge. The Am/Cm would then undergo nuclear transformation to californium-252 (Cf) and other useful transplutonium isotopes.

After reviewing a number of available vitrification technologies which could be employed in the canyon environment, a technology was selected which has been employed in the glass fiber industry to remelt glass marbles and draw glass fibers. This technology involves the direct resistance heating of the platinum bushing as described in a later section of this report and in greater detail in a number of published works. [1-2]

Am/Cm PROCESS DESCRIPTION

At the present time, the Am/Cm is dissolved along with other lanthanides in 12,000 liters of 4 normal nitric acid and stored in a stainless steel tank in the Canyon. There are about 10 kg of Americium-243 emitting about 1,800 Curies and 3 kg of Curium-244 emitting about 210,000 Curies. This material is maintained inside a shielded cell behind heavy walls of concrete and glass. The thermal energy generated from the nuclear activity amounts to about 7 kw/hr total. During processing, oxalic acid will be added to the nitric acid solution and the actinides and lanthanides will be precipitated as oxalates. The remaining aqueous fraction will be removed and the lanthanide-actinide oxalates will be redissolved in nitric acid. The Am/Cm solution will be evaporated and steam stripped down to approximately 100 grams/liter solids in 1 to 3 molar nitric acid solution. The solution will be stored in a stainless steel tank until it can be fed along with a special glass frit to the proposed vitrification system. The glass frit and nitric acid solution will be fed in two separate streams and melted together to form a lanthanide borosilicate glass. A compositional description of an initial frit and the glass surrogate are presented as Table 1. Glass research is being undertaken to try to replace the lead and barium oxide constituents.

Complete recovery of the Am and Cm surrogates (La and Nd) from the glass has been demonstrated by a nitric acid extraction technique. While the glass can be attacked by concentrated acid it is several orders of magnitude more durable in water than the required specification for high level waste glass. [3-5]

Table 1
Surrogate Frit and Glass Formulation - Am/Cm

Oxide Composition	Frit (Wt%)	Am/CmGlass (Wt%)
SiO_2	44.8	26.3
B_2O_3	5.6	3.3
BaO	6.1	3.6
Al_2O_3	23.7	14.1
La_2O_3	1.5	16.5
Nd_2O_3	- - -	17.6
CeO_2	- - -	7.3
PbO	18.4	11.0

REMELT FIBERIZATION TECHNOLOGY

The technology selected for the vitrification was derived from the glass marble-remelt fiberization process, which was the primary method of continuous fiberization until the mid 1960's. Specialty glass fibers with diameters less than 6 microns are still manufactured by this type of process. The small platinum alloy melters or "bushings" are heated by passing large electrical currents (thousands of amps) through the end terminals and into the platinum sides and heater screens in the bushing.

The design of a typical marble remelt bushing is complicated by the number of functions the bushing must complete. These functions are to receive and melt glass marbles, condition the glass melt for fiberization, and distribute the glass melt evenly across the fiberization section while maintaining an even temperature. Two factors are employed to adjust and maximize this process. First the position of the water cooled connectors on the end terminals can significantly affect the temperature of the bushing ends. Also the location and thickness of the platinum alloys between the end terminals can influence the thermal energy distribution. The amount of heat generated in a bushing is a function of the current path and the electrical resistance (thickness) of the metal sheets. If the bushing is constructed of continuous sheets between the terminals it can be generalized that the amount of heat produced in a particular area of the bushing is directly proportional to the thickness of the alloy in that region. In order to provide an abundance of heat for the melting of the glass marbles an additional heating screen is included in the top section of the typical marble bushing design as shown in Figure 1.

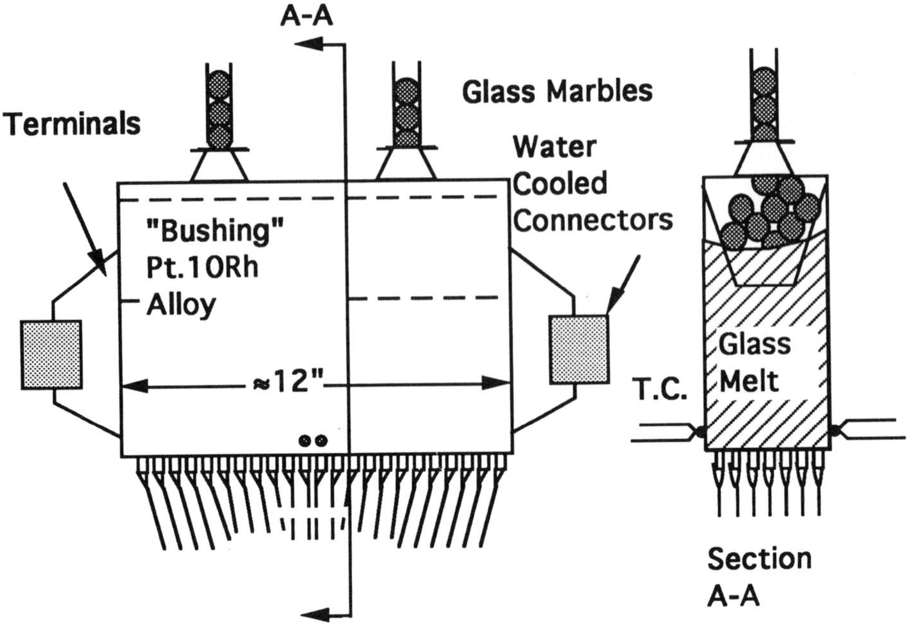

Figure 1. Typical Marble Remelt, Fiberization Bushing.

The typical operating parameters for a 408 fiber marble remelt bushing are presented in Table 2. [2] These parameters would be increased or decreased for other types of bushings.

Table 2
Operating Parameters for Typical Marble Remelt Bushing

Number of Fibers/Bushing	408
Available Power	25 to 35 KVA
Nominal Power	18 KVA
Nominal Voltage Drop	2-6 Volts
Throughput	15 kg/hr
Melt Energy	5 KVA
Platinum Alloy	4 kg. Pt10Rh

Since the bushing alloy is an excellent conductor of electricity it requires a great deal of current to generate thermal energy. This high current density requires heavy air or water cooled bus bar systems to carry the high currents from the step-down transformer to the bushing.

A very simple schematic of a typical remelt bushing process is shown as Figure 2. The temperature and power control systems can include over temperature control, ramping, and loss of power restrictions. Because the glass fiber industry requires a large number of these systems (hundreds/plant), the cost of a single control system is relatively low.

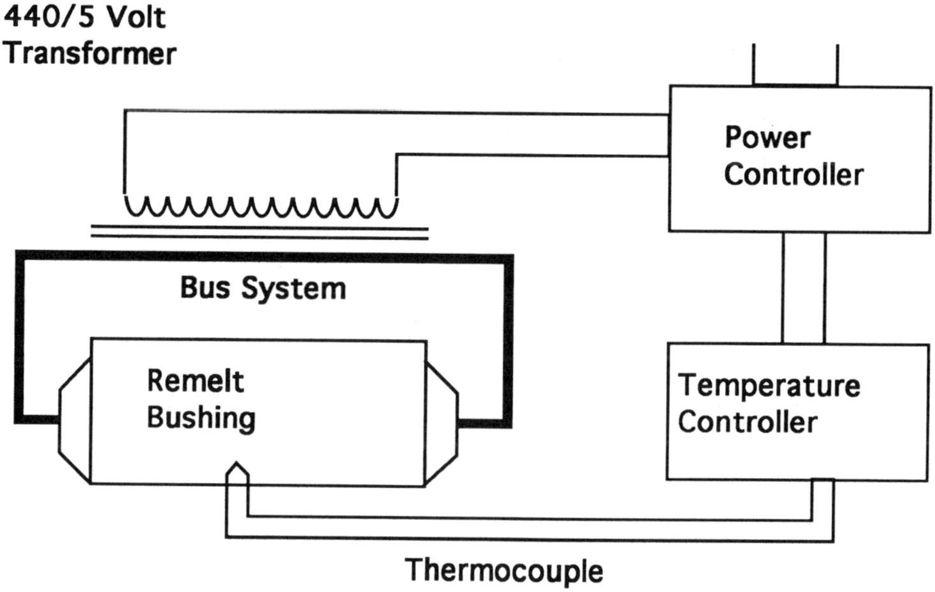

Figure 2. Simple Schematic of a Typical Marble Remelt Bushing.

Am/Cm BUSHING DEVELOPMENT

The melter for the vitrification of the Am/Cm solution was initially based on the marble remelt bushing technology. These types of melters offer several distinct advantages for the vitrification of actinides:

- Bushings can be cooled to room temperature and restarted as required,
- The bushings are chemically inert due to the use of platinum-rhodium alloys,
- The typical commercial bushing has a geometrically favorable design, similar to the slab tanks used in the nuclear materials industry, and
- The bushing melter is relatively small and simple to operate.

A number of design requirements were established for the Am/Cm project. The bushing melter is fed with two streams, one containing the ground frit and the other the nitric acid Am/Cm solution. The nitric acid feed contacting the hot glass was expected to be highly corrosive and generate a highly corrosive, acidic off-gas. It was required that the melter would produce about 300 kg of Am/Cm containing glass over a reasonable period of time. The average melt rate should be between 1 to 2 kg/hr. and a melt temperature approaching 1450°C. should be possible. The exiting glass flow would be controllable by remote means. The melter would be capable of being drained to an empty condition. It was expected that the complete melter portion of the system would fit on a single "process rack" which is normally used to support equipment in the canyons. The melter and in cell support systems had to survive the radiation field from the Am/Cm glass.

Some of the requirements for processing the Am/Cm were quite different from the conventional marble-remelt fiberization bushing technology. The liquid feed was a major departure and has caused considerable concern throughout the development program. Partially compensating the liquid feed issue is the fact that the glass quality does not have to be as high as the glass quality for fiberization. The first bushing designed for the Am/Cm program was similar to the marble remelt bushings. A second heating screen was installed near the bottom to further heat the glass prior to exiting. The tip section for fiberization was replaced with a solid sloping bottom and a single drain tube at the central axis. The inside diameter of the drain tube was 0.125 inch and its length about 0.75 inch. An air jet was installed at the bottom of the bushing, aimed at the drain tube, to stop the glass flow. A platinum wire wound resistance heater was placed around the drain tube to help start the glass flow. The bushing was fabricated and set into a high alumina castable refractory.

A fairly large Inconel-690 off-gas plenum was developed which matched the bushing. A ceramic separator was placed between the two parts to maintain the electrical isolation of the plenum. A small viewing port was also included in the plenum, but is not shown in Figure 3 for the sake of simplicity.

The first bushing was operated and demonstrated the proof of concept. However, the operation over an extended period of time indicated several problem areas and the need for a number of design modifications:

- The air/resistance heat, flow control was successful in starting and stopping glass flow, however a fairly large number of droplets preceded and followed the stream flow and produced an unacceptable number of glass threads. These threads would tend to increase the spread of contamination.

- In order to minimize the number of glass threads formed during draining it was decided to drain at a continuous flow stream of about 7 to 8 kg/hr. This would require that the bushing be operated in a batch mode i.e., feeding to build level over a number of hours then draining glass for a number of minutes. This cycle would then be repeated.

- The large open area at the top of the bushing allowed excessive heat to exit the bushing to the plenum and limited the feed melting rate.

- The lower screen was an impediment to the melting operation. During evaporation and melting of the liquid/frit feed, when more power was required to enhance the melting rate, the bottom melt temperature exceeded the temperature limit setting for the system.

- The expansion of the alloy against the castable ceramic created warpage of the metal which eventually introduced a small tear in the sidewall.

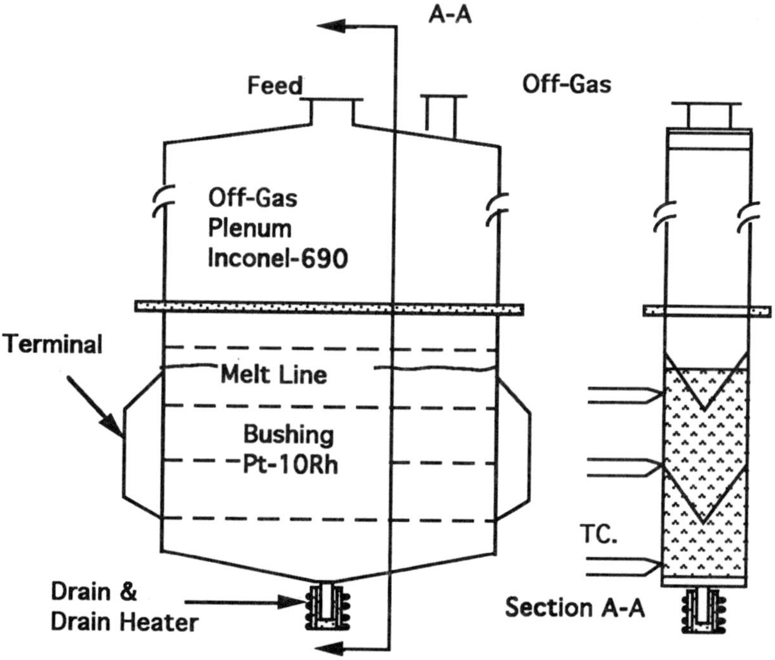

Figure 3. Simplified Schematic, First Am/Cm Bushing.

During this period of time additional commercial glass fiber and specialty glass producers were contacted and extensive discussions were held. It was decided that using geometry and the terminal clamps to control the heat distribution in the bushing would require numerous trial and error designs. A more expeditious route would be to install a two zone power supply and provide two pair of terminal connectors.

Oil modeling of the bushing and drain was established to investigate convection within the bushing and glass flow through the drain tube. In order to obtain more control in starting and stopping the flow a drain tube about six inches long with an inside diameter of about 0.200 inch was selected. Oil studies confirmed the ouput rate calculated using Poiseuille's equation:

$$F = K\, r^4\, h / l\, \mu,$$

where F is the rate of flow of glass from the drain tube, r is the radius of the tube, h is the height of the glass above the drain tube, l is the length of the tube, and μ is the viscosity of the glass in the tube. [2]

Additional mathematical computer modeling of heat transfer and fluid flow were initiated. An ABAQUS ™ Finite Element Computer program was utilized to map the current density in the two zone bushing. Results from this code indicated that the two zones should apportion power independently. In other words, during melting the power could be concentrated in the upper zone, while during pouring some power in the lower zone reduced the temperature gradient and helped initiate pouring. A CFX-4™ heat transfer and fluid flow program was used to study flow patterns in the bushing and flow control in the orifice tube. These mathematical computer models provided a great deal of support and guidance to what had previously been an "intuitive" design effort. A number of potential replacements for the "V" trough heater screens were considered and eventually the V screens were replaced with vertical heater strips which would permit the liquid feed to fall onto the glass surface at the expected melt levels. The vertical strips also reduced the stresses on the side-walls. The over-melt space was redesigned to reduce heat loss and improve evaporation of the liquid. A simple schematic of the second bushing is shown in Figure 3.

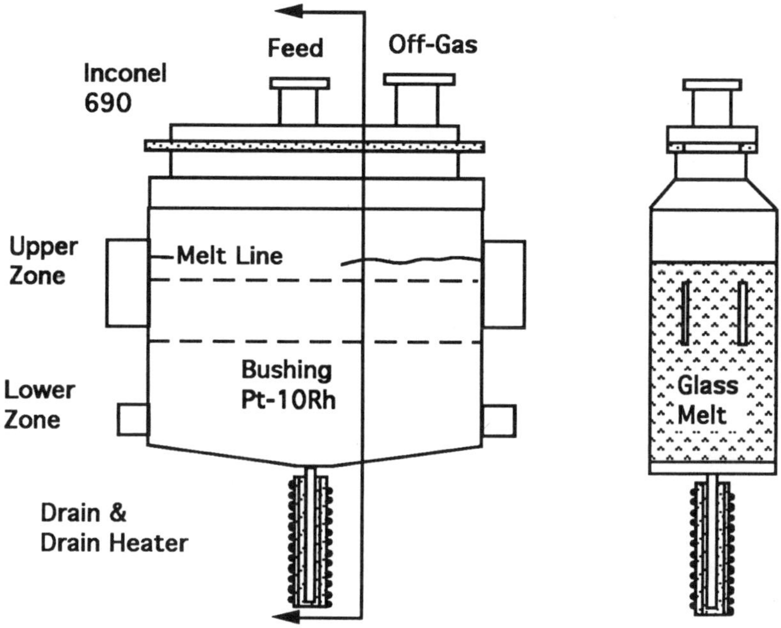

Figure 3. The Second Am/Cm Bushing Design.

The second bushing has now been evaluated and all of the systems appear to have operated as expected. Some problems still remain in obtaining the desired melt rate. The two zone control appears to operate as planned although the upper power level is presently somewhat limited. Efforts are under way to establish a separate bushing dedicated to flow control and to investigate the potential of inductively heating the drain.

CONCLUSIONS

Commercial marble remelt-fiberization technology was combined with existing experience in remote handling of special nuclear materials to provide equipment suitable for the vitrification of residual highly radioactive nuclear materials. Development of this process is continuing, with emphasis in the areas of melt rate enhancement, two zone power control, and remote glass flow control.

ACKNOWLEDGMENTS

The information contained in this article was developed during the course of work under Contract No. DE-AC09-89SR18035 with the U.S. Department of Energy. The authors also wish to acknowledge the GAFTECH licensing group of GAF Corp, Nashville,Tennessee; Ferro Corp. Specialty Glass Group, Cleveland, Ohio, Rathmann Associates, Corning, New York; and the numerous others in the glass fiber and specialty glass industries who contributed suggestions and advice to this effort.

REFERENCES

1. F.V. Tooley, "The Handbook of Glass Manufacture Vol.II," pg 715-737, Books for Industry Inc., New York, 1974.

2. K.L. Lowenstein, "The Manufacturing Technology of Continuous Glass Fibres," Second Edition, pg. 118-168, Elsevier, New York, 1983.

3. W.G. Ramsey, N.E. Bibler, & T.F. Meaker, "Compositions and Durabilities of Glasses for Immobilization of Plutonium and Uranium," Waste Management '95, Record 23828-23907, WM Symposia, Inc., Tuscon, AZ, (1995).

4. N.E. Bibler, W.G. Ramsey, T.F. Meaker, & J.M. Pareizs, "Durabilities and Microstructures of Radioacative Glasses for Immobilization of Excess Actinides at the Savannah River Site," Materials Research Society Symposia Proceedings, Vol 233, pg. 333-336, Materials Research society, Pittsburgh, PA (1995).

5. C.M. Jantzen, N.E. Bibler, & D.C. Beam, "Characterization of the Defense Waste Processing Facility (DWPF) Environmental Assessment (EA) Glass Standard Reference Material (U)," WSRC-TR-92-346. Westinghouse Savannah River Co. Aiken, SC (1992).

ENVIRONMENTALLY BENIGN PROCESSING OF ALUMINUM-BASED CERAMIC MATERIALS: WATER SOLUBLE ALUMOXANES

Rhonda L. Callender,[a] C. Jeff Harlan,[a] Daniel L. Callahan,[b] Mark R. Wiesner,[c] Ron L Cook,[d] and Andrew R. Barron,[a,b]*
[a] Department of Chemistry, Rice University, Houston, Texas 77005
[b] Department of Mechanical Engineering and Materials Science, Rice University, Houston, Texas 77005
[c] Department of Environmental Science and Engineering, Rice University, Houston, Texas 77005
[d] TDA Research, Inc., 12345 West 52nd Ave Wheat Ridge, Colorado, 80033

ABSTRACT

The objective of our research is the development of an environmentally-benign process for the fabrication of alumina-based ceramics. We have designed an alternative synthetic pathway to alumina ceramics which does not utilize toxic reagents or volatile organic chemicals (VOCs); the aqueous synthesis of water soluble carboxylate-alumoxane precursors from inexpensive boehmite feedstock. The synthesis and characterization of these new alumoxanes is reported along with their conversion to alumina. Physical and spectroscopic characterization of the alumoxane will be discussed. The potential environmental impact of the new alumoxane methodology as compared to traditional approaches will be discussed.

INTRODUCTION

With a combined annual production of over 30 million tons, aluminum oxides are undoubtedly among the most industrially important chemicals. [1] Their uses include: precursors for the production of aluminum metal, catalysts and absorbents; structural ceramic materials; reinforcing agents for plastics and rubbers, antacids and binders for the pharmaceutical industry; and as low dielectric loss insulators in the electronics industry.

Traditional ceramic processing involves three basic steps generally referred to as powder-processing, shape-forming, and densification, often with a final mechanical finishing step.[2] Although several steps may be energy intensive, the most direct environmental impact arises from the shape-forming process where various binders, solvents, and other potentially toxic agents are added to form and stabilize a solid ("green") body. The component chemicals include: alumina (27 %), 1,1,1-trichloroethylene/ethanol (solvents, 58 %), menhaden oil (deflocculent, 1.8 %), polyvinyl butyrol (Binder, 4.4 %), and polyethylene glycol/octyl phthalate (plasticizer, 8.8 %). All solvents and additives must be

*Author to whom correspondence should be addressed.

removed as gaseous products via evaporation or pyrolysis.[3] The replacement of 1,1,1 Trichloroethylene (TCE) as a solvent in the traditional ceramic process must be regarded as a high priority for limiting environmental pollution. Due to its wide spread use as a solvent in industrial processes, TCE has become one of the most commonly found contaminants in ground waters and surface waters.[4] Concentrations range from parts per billion to hundreds of milligrams per liter. The United States Environmental Protection Agency (USEPA) included TCE on its 1991 list of 17 high-priority toxic chemicals targeted for source reduction.[5] The 1988 releases of TCE reported under the voluntary right to know provisions of Superfund Amendments and Reauthorization Act (SARA) totaled to 190.5 million pounds. The plasticizers, binders, and alcohols used in the process present a number of potential environmental impacts associated with the release of combustion products during firing of the ceramics, and the need to recycle or discharge alcohols which, in the case of discharge to waterways, may exert high biological oxygen demands in the receiving communities.[6]

Whereas the traditional sintering process is used primarily for the manufacture of dense parts, the solution-gelation (sol-gel) process has been applied industrially primarily for the production of porous materials and coatings. Solution-gelation involves a four stage process: dispersion; gelation; drying; firing. A stable liquid dispersion or "sol" of the colloidal ceramic precursor is formed in a solvent with appropriate additives. The principal environmental results arising from the sol-gel process are those associated with use of strong acids, plasticizers, binders, and solvents. Depending on the firing conditions, variable amounts of organic materials such as binders and plasticizers may be released as combustion products. NO_x's are also produced from the use of nitric acid or nitrate slats. Moreover, acids and solvents must be recycled or disposed of.

Carboxylate-Alumoxanes

The aluminum based sol-gels formed during the hydrolysis of aluminum compounds belong to a general class called alumoxanes.* Alumoxanes were first reported in 1958 by Andrianov,[7] however, have since been prepared with a wide variety of substituents on aluminum. Work from our laboratory has shown that hydrolytically stable alumoxanes are three dimensional cage compounds,[8,9] with an Al-O core analogous to that found in the mineral boehmite, $[Al(O)(OH)]_n$. Based on this knowledge we posed the question: *Can alumoxanes be prepared directly from the mineral boehmite?* Such a "top-down" approach represented a departure from the traditional synthetic methodologies. However, subsequent studies showed that carboxylate-alumoxanes, similar to those formed by sol-gel routes could indeed be prepared, Eq. 1.[10] The physical properties including the solubility of these carboxylate-alumoxane are dependent on the carboxylic acid.[11]

$$[Al(O)OH)]_n \xrightarrow{HO_2CR} [Al(O)_x(OH)_y(O_2CR)_z]_n \qquad (1)$$

* The term alumoxane is commonly used to describe a species containing an oxo (O^{2-}) bridge binding (at least) two aluminum atoms, i.e., Al-O-Al.

We have reported that carboxylate-alumoxanes with significant solubility and stability in water may be prepared by the use of polyether substitution, e.g., methoxyethoxyethoxyacetic acid or MEEA-H (**I**).[12]

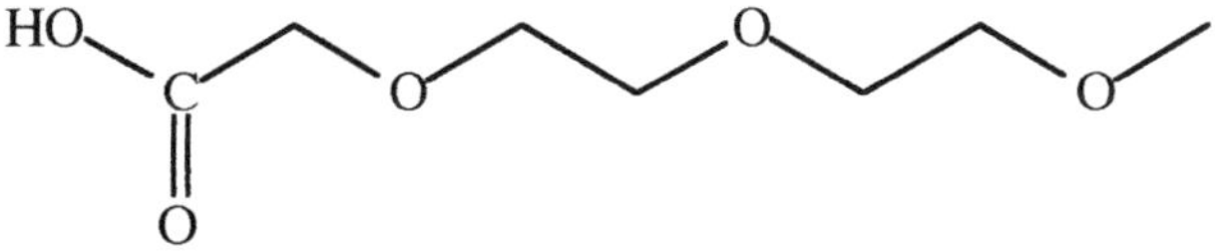

(**I**)

However, as reported the MEEA-alumoxane had two draw-backs with regard to its application as an environmentally benign pre-ceramic system. First, the original samples were prepared in xylene nullifying their environmentally benign nature. Second, although MEEA-alumoxane was converted to alumina upon pyrolysis the ceramic yield was relatively low (*ca.* 20 %). Thus, it is desirable to synthesize water soluble carboxylate-alumoxanes using water as the reaction medium and to employ smaller (low molecular weight) organic substituents to maximize the ceramic yield. The results of this study are presented herein.

RESULTS AND DISCUSSION

Synthesis and Characterization of Water Soluble Alumoxanes

For the present study four different carboxylate alumoxanes were prepared; from acetic acid [HO_2CCH_3], methoxyacetic acid [$HO_2CCH_2OCH_3$], methoxyethoxyacetic acid [$HO_2CCH_2OCH_2CH_2OCH_3$], and methoxyethoxy-ethoxyacetic acid [$HO_2CCH_2(OCH_2CH_2)_2OCH_3$]. For simplicity throughout this paper the following acronyms are used for the alumoxanes thus formed: acetate-alumoxane (A-A), methoxyacetate-alumoxane (MA-A), methoxyethoxyacetate-alumoxane (MEA-A), methoxyethoxyethoxyacetate-alumoxane (MEEA-A).

Pseudoboehmite and the appropriate carboxylic acid were refluxed in water. The clear solution obtained was centrifuged to remove any residual insoluble particles and decanted. The water was removed *in vacuo* to give the carboxylate-alumoxane. In order to remove any residual unreacted carboxylic acid the carboxylate-alumoxane may be dissolved in ethanol and precipitated as a white powder via the addition of diethyl ether. However, this was not found to be necessary for all samples. Redissolving the powder in water and drying (50 °C) yields a clear glassy material suitable for storage. No change in composition or solubility is observed over several months under ambient conditions.

The physical appearance, solubilities and surface morphology of the carboxylate-alumoxanes are dependent on the identity of the carboxylate substituent. Thus, MEEA-A and MEA-A are white crystalline powders which readily dissolve in water to form clear glassy materials. In addition, they are soluble in methylene chloride, chloroform, and ethanol, which facilitates spectroscopic characterization. MA-A is a white solid which dissolves in water sparingly and is only slightly soluble in methanol. Similarly, A-A is soluble in water and methanol. The surface morphologies of the carboxylate alumoxanes, as determined with SEM, appear as 'fluffy' conglomerates for MA-A and as smooth, flowing 'waves' for MEA-A, MEEA-A, and A-A. Figure 1 shows SEI micrographs of MA-A and MEEA-A.

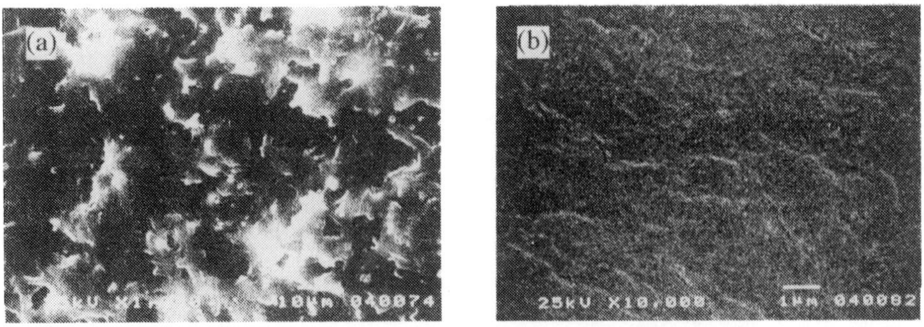

Figure 1. SEM images of (a) methoxyacetate-alumoxane (MA-A) and (b) methoxyethoxyethoxyacetate-alumoxane (MEEA-A).

The similarity in the physical appearance of A-A, MEA-A and MEEA-A and their distinction from MA-A is accompanied by the observation that the size of individual alumoxane particles as measured by PCS was found to depend on the carboxylic acid used. In all cases a range of particle sizes was observed, however, the average particle size for A-A (23 nm), MEEA-A (42 nm), and MEA-A (55 nm), increased with increasing alkyl chain length. Once more the anomaly is MA-A with a broad range of particles between 242 nm and 1000 nm. The reasons for this anomaly are not clear at the present time, however, we are continuing our investigation. The PCS measurements are consistent with GPC molecular weight measurements.

The IR spectra of all the carboxylate-alumoxanes contain bands at 1596 - 1586 and 1473 - 1466 cm^{-1}, consistent with a bridging mode of coordination of the carboxylate to the boehmite core.[13] Solution ^{1}H NMR and ^{13}C NMR spectra of the carboxylate-alumoxanes indicate a single environment for the bridging carboxylates. The ^{27}Al NMR spectra of the alumoxanes consist of a broad resonance at *ca.* 6 ppm indicative of aluminum in an octahedral AlO_6 coordination environment.

Conversion of Carboxylate-Alumoxanes to Alumina

All of the carboxylate-alumoxanes decompose above 180 °C to give Al_2O_3. Sintering at 1000 °C results in the expected formation of α-alumina, as confirmed by XRD. No carbon is detected in the resulting ceramic if an oxidizing atmosphere is employed. In general the thermogravimetric/differential thermal analysis (TG/DTA) of the carboxylate-alumoxanes show two major decomposition regions; the relative mass loss and temperatures at which these regions occur is dependent on the identity of the carboxylic acid. As may be expected, the ceramic yield is dependent on the identity of the carboxylic acid: greatest for A-A (*ca.* 75 %), lowest for MEEA-A (*ca.* 20 %).

Although the polyether-substituted alumoxanes (MEEA-A, MEA-A, and MA-A) give fragile ceramic bodies in the absence of high temperature sintering, pyrolysis of A-A allows for the formation of durable significantly sized bodies.

SEM analysis indicates that the ceramic body has a high density with no macroscopic porosity (see Figure 2). However, TEM studies confirm the presence of 50 nm sized pores.

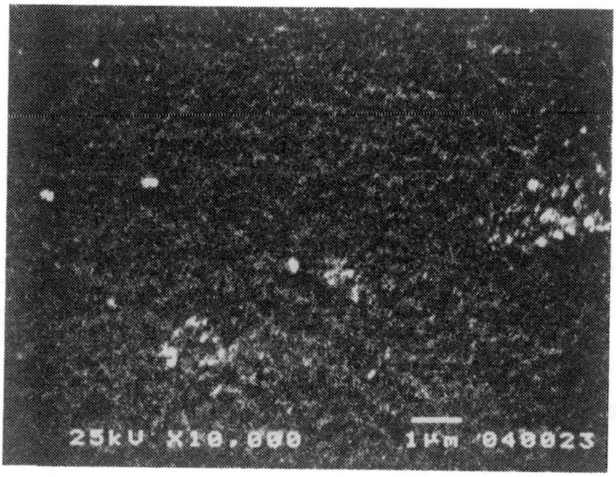

Figure 2. SEM of alumina surface formed by the pyrolysis of acetate-alumoxane (A-A) at 1000 °C for 3 h. This should be compared to the un-fired sample in Figure 1b.

Application of the Alumoxanes as Binders for Green Bodies

As was discussed in the introduction traditional ceramic processing involves three basic steps generally referred to as powder-processing, shape-forming, and densification, often with a final mechanical finishing step. Unfortunately, this process inherently results in significant shrinkage during processing, since a ceramic powder is mixed with various binders, solvents, and other agents to form and stabilize a solid ("green") body. These agents are subsequently removed in gaseous form by direct evaporation or pyrolysis. In many cast-parts, the liquid solvent alone consists of over 50 % of the initial volume of material. All solvents and additives must be removed as gaseous products via evaporation or pyrolysis, resulting in extensive shrinkage. These issues could be mitigated to a great extent if a single species could be employed as a binder that is itself converted to a ceramic. Furthermore, if the extent of shrinkage could be limited to acceptable values near-net shape fabrication of ceramic bodies could also be achieved. Our approach involves the application of processable alumoxanes that are converted into alumina-based ceramics by firing.

Using wax molds we have been able to prepare shaped green bodies with 70 wt% solutions of our water soluble alumoxane and alumina powder in a 2:1 mixture. These mixtures were pourable, thus the alumoxane acts as both the transport media (solvent) and the binder. The material can be cast into a wax mold and dried. After removal of the wax binder by melting, the green bodies were set (120 °C for 5 h.). The setting occurred without any addition of cross-linking agents. Calcination results in the formation of an alumina shaped body with less than 6 % shrinkage in going from green to fired material. An example of a simple hexagonal bolt is shown in Figure 3.

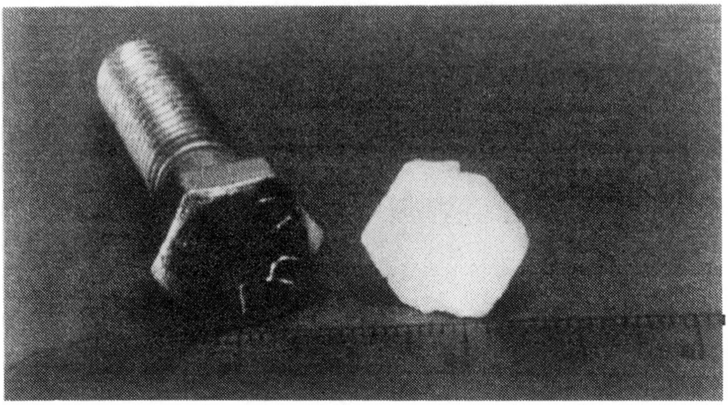

Figure 3. Photograph of a bolt head used to make a wax mold (left) and the alumoxane:alumina green body thus formed (right).

Infiltration of Alumoxane Slips into Porous Substrates

In order to ascertain whether the alumina formed from the thermolysis of the carboxylate alumoxane retains the shape of the alumoxane precursor we have investigated the infiltration of porous objects including: ceramic, glass, and paper filters. In addition, it is of interest to determine whether the alumoxane will infiltrate (and wet) complex shapes such as fiber tows, or repair the porosity that exists in finished ceramic bodies due to the significant shrinkage observed upon sintering a green body.

An automotive fuel filter was impregnated with MEEA-alumoxane and subsequently heated to 1000 °C in air to burn-out all the organic residues. Due to the relatively low sintering temperature the resultant alumina "filter" had very little strength, however, as can be seen from Figure 4 the macroscopic shape of the filter was retained.

SEM analysis was undertaken of filter paper (pretreatment and alumoxane coated) and the ceramic product so as to determine if full infiltration of the papers fibers occurred. Impregnation of filter paper (Whatman #1) with MEEA-alumoxane from aqueous solution, followed by pyrolysis gave a free standing ceramic. The SEM of the untreated-uncoated sample (see Figure 5) shows the paper to consist of interwoven fibers, 5 to 40 μm in diameter.[14] The MEEA-alumoxane treated paper is visually indistinguishable from their untreated equivalents. As can be seen from Figure 5, the alumina ceramic formed upon oxidative thermolysis at 600 °C retains the structure of the paper fibers with little evidence for particulate oxide residues. These results indicate that complex shapes may be readily infiltrated by alumoxane solutions.

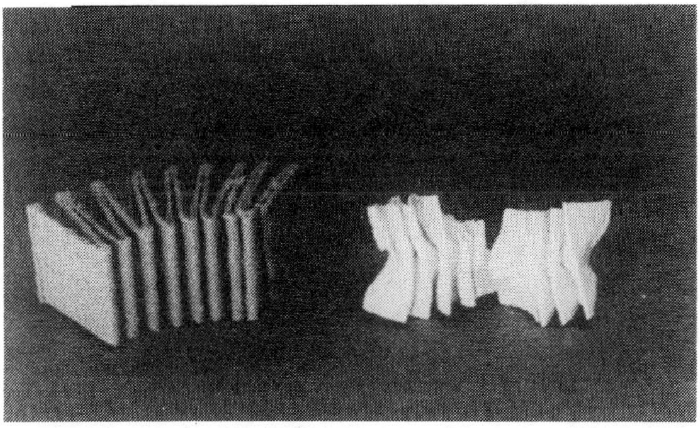

Figure 4. Comparison of paper fuel filter (left) and the ceramic form after calcination (right).

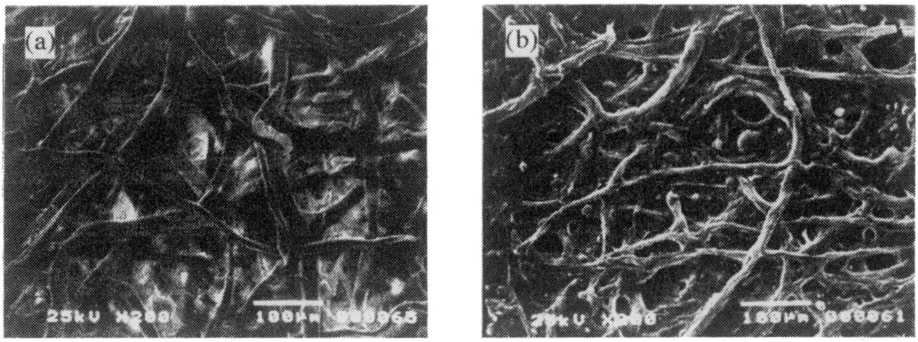

Figure 5. SEM images of Whatman #1 filter paper; (a) uncoated and (b) coated with an alumoxane, and fired in O_2 at 600 °C, demonstrating the complete infiltration of the fiber structure.

Potential Environmental Impact of the Carboxylate-Alumoxanes.

The proposed synthetic methodology would impact the release of contaminants to the environment on at least two distinct levels. First, the new synthetic methodology for producing these ceramic materials will eliminate the use of toxic solvents and reduce energy consumption. With respect to the use of the carboxylate alumoxanes as binders in green body processes byproducts formed from the combustion of plasticizers and traditional binders will be minimized, and with regard to sol-gel synthesis the use of acids may be eliminated. Thus, the process will reduce the potential for environmental release through the use of more environmentally benign feedstocks and an alternative synthetic procedure which will improve energy efficiency.

Second, due to the very versatile nature of the process, the proposed alumoxane process holds the potential for fabricating new ceramic nanofiltration and ultrafiltration membranes with enhanced specificity. Due to their stability under exposure to high temperatures, organic solvents, and oxidants, ceramic membranes are particularly well suited to separations required in less environmentally benign industrial processes in which options for pollution prevention must focus on materials recovery and reuse in the absence of suitable alternative processes.

CONCLUSION

We have shown that the reaction of boehmite with carboxylic acids in water results in the formation of carboxylate-alumoxanes. The structure of the carboxylate alumoxane is best described as being an Al-O nano-particle capped with carboxylate moieties. IR spectroscopy suggests that the carboxylate acts as a bridging ligand to two aluminum centers. The carboxylate alumoxanes are infinitely stable under ambient conditions, and may be readily processed in aqueous media. While polyether-substituted carboxylic acids provide good solubility in water, acetic acid substituted alumoxanes prepared in water are also soluble. The most significant result of this study is the synthesis of water soluble acetate-alumoxanes from acetic acid and boehmite in water. The acetate-alumoxane requires no VOCs in its synthesis, is water soluble, and readily converted to alumina > 500 °C with a ceramic yield greater than 80 %. The low cost (36 $\$.L^{-1}$) of acetic acid and low toxicity of its aqueous solutions offers

Pyrolysis of the carboxylate-alumoxane yields α-alumina with low carbon contamination. The ceramic yield and the resulting physical properties of the alumina thus formed are dependent on the organic substituent. The acetate derivative A-alumoxane gives the highest ceramic yield and allows for the formation of high density alumina bodies. In addition, to being processable pre-ceramic materials, the aqueous processability of the carboxylate-alumoxanes allows for their use as the binder in traditional ceramic "green body" processing. The nano-particle size of the carboxylate alumoxanes allows for full infiltration into micro-porous substrates which upon pyrolysis and sintering offers a method for their use as chemical infiltration and surface repair agents.

EXPERIMENTAL SECTION

Research grade pseudo-boehmite (100%) was kindly provided by Vista Chemical Company. All carboxylic acids were obtained commercially (Aldrich) and were used as received.

Thermogravimetric/differential thermal analyses were obtained on a Seiko 200 TG/DTA instrument using a carrier gas of either dry nitrogen, air or oxygen. FTIR spectra (4000-400 cm^{-1}) were obtained using Nicolet 760 spectrometer, as

Nujol mulls on KBr plates. Solution ^{1}H, ^{13}C, and ^{27}Al NMR spectra were obtained on either Bruker AM-250 and Bruker AM-300 spectrometers. Chemical shifts are reported relative to external $SiMe_4$ (^{1}H, ^{13}C) and $[Al(H_2O)_6]^{3+}$ (^{27}Al). Photon correlation spectra were obtained using a Malvern Automeasure 4700 V4 spectrometer. Additional particle size analyses were obtained using a Coulter LS230 particle size analyzer. Gel Permeation Chromatography data were obtained on a Shodex instrument utilizing a Waters HPLC 150C. Samples were dried for 12 h at 110 °C, then degassed in helium for 2 h at 110 °C for surface measurements using a BET model 4203 gas desorption instrument. XRD data was collected on a Siemens D5000 diffractometer. SEM studies were performed on JEOL JSM-5300 scanning microscope. A small amount of material was attached to an aluminum stub with graphite paint. Due to the insulating nature of the materials, focusing and visual examination was problematical. Therefore, in some cases a thin layer of gold was sputtered onto the sample to provide a conducting surface.

Synthesis of MEEA-A, MEA-A, and MA-A

Pseudoboehmite (10.0 g, 167 mmol) and the appropriate carboxylic acid (334 mmol) were refluxed in water (200 mL) which resulted in a clear solution after 72 h. The solution was centrifuged at 6000 rpm for 1 hour and decanted. The water was removed in vacuo (10^{-2} Torr) at 50 °C resulting in a gel which was washed with diethyl ether (3 x 75 mL) then dissolved in ethanol (50 mL) while stirring. The alumoxane was precipitated as a white powder via the addition of diethyl ether (100 mL). Redissolving the powder in water and drying at 50 °C for 24 h yielded a clear glassy material. The MEEA-A and MEA-A are soluble in water, methanol, chloroform, and methylene chloride. MA-alumoxane is soluble in water and methanol.

Synthesis of A-A

Pseudoboehmite (20.0 g, 333 mmol) was slowly added to a vigorously stirring mixture of acetic acid (51.0 mL, 667 mmol) in water (200 mL). The resulting slurry was decanted after 10 minutes and then centrifuged at 6000 rpm for 1 hour to yield a clear viscous solution. Removal of the volatiles in vacuo (10^{-2} Torr) at 90°C results in clear, white granules. The granules were dissolved in water and dried for 24 hours at 80 °C to yield a clear glassy material. The acetate-alumoxane is soluble in water and slightly soluble in methanol.

ACKNOWLEDGMENTS

Financial support for this work was provided by the Environmental Protection Agency and the National Science Foundation.

REFERENCES

[1] K. Wefers and C. Misra, *Oxides and Hydroxides of Aluminum*, Alcoa Laboratories, 1987.

[2] W.D. Kingery, H. K. Bowen, and D. R. Uhlmann, *Introduction to Ceramics*, 2nd ed. Wiley, New York, 1976, Chapter 1; D. W. Richerson, *Modern*

Ceramic Engineering, Marcel Dekker, New York, 1992.

[3] *Advances in Ceramics*, Vol. 9., ed. J. A. Mangels and G. L. Messing, American Ceramic Society, Westville, OH, 1984.

[4] C. N. Sawyer, P. L. McCarty, and G. F. Parkin, *Chemistry for Environmental Engineering*, McGraw-Hill, New York, 1994.

[5] M. D. LaGrega, P. L. Buckingham, and J. C. Evans, *Hazardous Waste Management*, McGraw-Hill, New York, 1994.

[6] R. V. Thomas, *Systems Analysis and Water Quality Management*, McGraw-Hill, New York, 1972.

[7] K. A. Andrianov and A. A. Zhadanov, "Synthesis of new polymers with inorganic chains of molecules", *J. Polym. Sci*, **30**, 513-524 (1958).

[8] A. W. Apblett, A. C. Warren, and A. R. Barron, "Synthesis and characterization of triethylsiloxy-substituted alumoxanes: their structural relationship to the minerals boehmite and diaspore", *Chem. Mater.*, **4**, 167-182 (1992).

[9] C. C. Landry, J. A. Davis, A. W. Apblett, and A. R. Barron, "Siloxy-substituted alumoxanes: their synthesis from polydialkylsiloxanes and trimethylaluminium, and application as aluminosilicate precursors", *J. Mater. Chem.*, **3**, 597-602 (1993).

[10] A. W. Apblett, C. C. Landry, M. R. Mason, and A. R. Barron, "From Minerals to Materials: A facile synthesis route to preceramic polymers for aluminum oxide", *Mat. Res. Soc., Symp. Proc.*, **249**, 75-80 (1992).

[11] C.C. Landry, N. Pappè, M. R. Mason, A. W. Apblett, A. N. Tyler, A. N. MacInnes, and A. R. Barron, "From minerals to materials: synthesis of alumoxanes from the reaction of boehmite with carboxylic acids", *J. Mater. Chem.*, **5**, 331-341 (1995).

[12] A. Kareiva, C. J. Harlan, D. B. MacQueen, R. Cook, and A. R. Barron, "Carboxylate substituted alumoxanes as processable precursors to transition metal-aluminum and lanthanide-aluminum mixed metal oxides: atomic scale mixing via a new transmetalation reaction", *Chem. Mater.*, **8**, 2331-2340 (1996).

[13] Y. Koide and A. R. Barron, "$[Al_5(^tBu)_5(\mu_3\text{-}O)_2(\mu_3\text{-}OH)_3(\mu\text{-}OH)_2(\mu\text{-}O_2CPh)_2]$: A model for the interaction of carboxylic acids with boehmite. ", *Organometallics*, **14**, 4026-4029 (1995).

[14] A. N. MacInnes and A. R. Barron, "A spectroscopic evaluation of the efficacy of two mass deacidification processes", *J. Mater. Chem.*, **2**, 1049-1056 (1992).

MODELING CORROSION OF CERAMICS IN HALOGEN-CONTAINING ATMOSPHERES*

Dennis W. Readey
Colorado Center for Advanced Ceramics
Colorado School of Mines
Golden, CO 80401 USA

ABSTRACT

Most single component oxide ceramics undergo active corrosion in halogen-containing gases such as HCl to form gaseous corrosion products only. At high temperatures, such corrosion reactions are controlled by gaseous diffusion of the product gases away from the solid surface. As a result, the reaction kinetics and the corrosion rate can be predicted if the thermodynamics of the reaction are known. Similar behavior is expected for nitride ceramics. However, several multicomponent oxides, typified by $NiAl_2O_4$, and carbides may react to form a porous solid product layer through which diffusion takes place and controls the corrosion rate. Nevertheless, even in these cases, if gaseous diffusion still controls, the rate of corrosion can be modeled and predicted.

INTRODUCTION

Active corrosion usually means atmospheric degradation with the formation of gaseous, non-passivating, corrosion products such as SiO and CO which result from oxidizing SiC in water vapor [1]:

$$SiC(s) + 2\,H_2O(g) = SiO(g) + CO(g) + 2\,H_2(g). \qquad (1)$$

* Sponsored by the National Science Foundation, Division of Materials Research, and the DuPont Corporation.

Kinetic data on this system [1] and on the active corrosion of SiO_2 in hydrogen [2]

$$SiO_2(s) + H_2(g) = SiO(g) + H_2O(g) \qquad (2)$$

show that, at high temperatures, diffusion of the product gases away from the corroding surface controls the rate of corrosion. As a result, the kinetics of the reaction can be easily modeled and quantitative comparisons can be made with experimental data to confirm diffusion control [2]. In fact, thermodynamic data for the reaction provide the necessary information to permit quantitative prediction of corrosion rates. Of course, if one of many possible surface reactions controls the rate of corrosion, then prediction of corrosion rates without experimental data is impossible. In any event, for any given material, there should be a temperature regime sufficiently high so that gas diffusion will control the corrosion rate and predictions of corrosion behavior and corrosion rates can be made.

The corrosion of ceramics in halogen-containing gases such as Cl_2, HCl, F_2, and HF is of interest for several reasons, both scientific and technical. First, the number of ceramic compounds that can undergo active gas corrosion in halogens is much larger than those that react to form only gaseous species in either reducing or oxidizing environments. Therefore, the active gas corrosion of a wide variety of ceramics (oxides, carbides, nitrides, borides, etc.) can be studied to determine the systematics, similarities, and differences in their behavior.

In addition, gas corrosion of ceramics by halogens is technologically important and becoming increasingly so. As examples, chemicals such as TiO_2 pigments and fluorocarbons are processed in halogen environments at high temperatures where refractory corrosion is a problem. Ceramics are being used more frequently as supports and furnace fixturing for silicon wafers in integrated circuit manufacturing. Many of the processing steps are carried out in halogen-containing gases either at high temperatures or in plasmas. Not only is corrosion resistance important, but also the mechanism of degradation is critical in these applications. For example, preferential attack of the glass phase in a high alumina ceramic by a reactive atmosphere can lead to release of alumina particles which would pose a serious problem in subsequent wafer processing steps. The behavior of ceramics in incinerators is becoming of increasing importance. The existence of halogens in the combustion gases is assured by the presence of polyvinyl chloride and other halogen-containing polymers in the waste stream. Not only can ceramics corrode under the atmosphere and temperature conditions in an incinerator, but the corrosion of metal components in the same environment depends critically on the resistance of any passive oxide to the halogen gases that are present. Finally, any high temperature system operating near salt water ingests NaCl from the atmosphere producing halogen-containing gases that can cause ceramic and passive oxide degradation.

In another area of technical importance, the presence of reactive atmospheres, such as HCl, during sintering leads to enhanced vapor transport, little shrinkage, and significant grain growth [3]. Reactive vapor phase sintering can be used to fabricate materials with controlled porosity and pore size that can have a variety of applications including filters, sensors, biomaterials, and the backbone for ceramic-metal composites [4]. Knowing whether vapor transport is controlled by gas diffusion or by a surface reaction is important in understanding the kinetics of vapor phase sintering and other applications of reactive vapor transport.

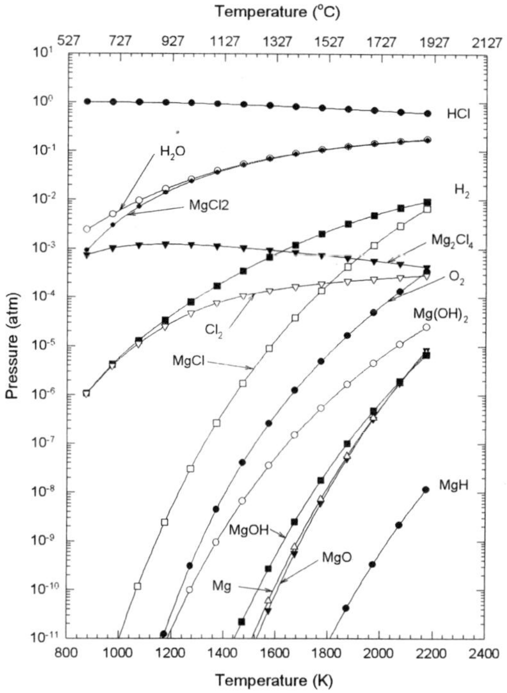

Figure 1 Pressures over MgO in HCl.

Therefore, based on previous experimental data [1,2] that demonstrate the ability to model gas phase corrosion, thermodynamic calculations are used to predict the high temperature, gas diffusion controlled corrosion of several different oxides and nonoxides.

THERMODYNAMICS AND KINETICS

For a reaction such as:

$$MgO(s) + 2\,HCl(g) = MgCl_2(g) + H_2O(g) \qquad (3)$$

available thermodynamic data [5] permit calculation [6] of the gas species in equilibrium. Figure 1 gives the results for the reaction of MgO in HCl and shows that Eqn. (3) is the dominant reaction. Since the reaction does not go to completion, it can be shown that [2] the rate of corrosion given by the corrosion flux density, J(MgO) [moles/cm^2/s], in very dry HCl is:

$$J(MgO) = -[h(MgCl_2)\,h(H_2O)]^{1/2}\,K_e^{1/2}\,p_o(HCl) \tag{4}$$

where $h(MgCl_2)$ and $h(H_2O)$ are the mass transfer coefficients for $MgCl_2$ and H_2O respectively, K_e is the equilibrium constant for the reaction, and $p_o(HCl)$ is the ambient HCl pressure. In a high dew point gas, the corrosion rate is:

$$J(MgO) = -\frac{h(MgCl_2)\,K_e\,p_o^2(HCl)}{p_o(H_2O)} \tag{5}$$

where $p_o(H_2O)$ is the water vapor pressure in the ambient gas. The mass transfer coefficients are given by:

$$h = \frac{D\,Sh}{dRT} \tag{6}$$

where D is the gas diffusion coefficient, d is a sample dimension, and Sh is the Sherwood coefficient that takes into consideration fluid convection. Since the mass transfer coefficients can be accurately calculated from the kinetic theory of gases [7], Eqns.(4) and (5) show that the corrosion rate is predicted by the ambient HCl and H_2O pressures, $p_o(HCl)$ and $p_o(H_2O)$, and the equilibrium constant for the reaction which in this case is:

$$K_e = \frac{p_s(MgCl_2)\,p_s(H_2O)}{p_s^2(HCl)} \tag{7}$$

where the subscript "s" refers to the pressures at the MgO surface. Since K_e can be calculated from thermodynamic data, the rates of corrosion can be calculated and predicted. Furthermore, Eqns.(4) and (5) show that in dry hydrogen, J(MgO) $\propto \exp - \Delta H^o/2RT$ while in wet hydrogen, J(MgO) $\propto \Delta H^o/RT$; that is, the apparent activation energy for corrosion is twice as large in wet HCl than it is in dry and its value can be calculated from tabulated enthalpy data and compared to experimental data.

HALOGEN CORROSION OF CERAMICS

Thermodynamic calculations for several single oxides, multiple oxides, carbides and nitrides in atmospheres containing HCl as the corroding halogen gas have been carried out. In all cases, the predicted results are similar to those described above: the reactions do not go to completion so product gas diffusion controls the

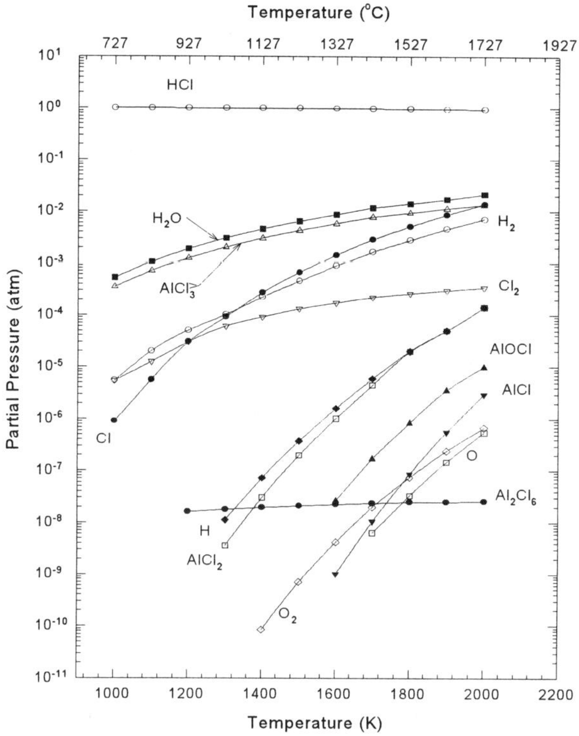

Figure 2. Gas pressures over Al_2O_3 in HCl.

corrosion rate [2]; and that high product gas pressures are predicted down to low temperatures where surface reaction control will undoubtedly dominate the rate of corrosion.

SINGLE OXIDES

They are all very similar and typical of that described above for MgO. The pressures in HCl over Al_2O_3 in Figure 2 is another typical example of a single oxide in HCl.

TERNARY OXIDES AND SOLID SOLUTIONS

In corrosion of multicomponent oxides, two limiting cases are possible. Take for example, $NiFe_2O_4$. In this case, both constituent oxides, NiO and Fe_2O_3, have high product pressures in HCl. In addition, the pressures of $FeCl_2$ and $NiCl_2$ are both

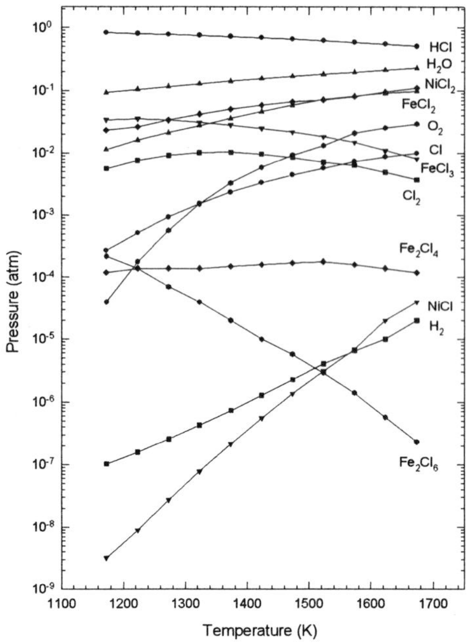

Figure 3. Gas pressures over $NiFe_2O_4$ in HCl.

also high over $NiFe_2O_4$, as Figure 3 shows, and not very different from those over pure NiO and Fe_2O_3 individually. Therefore, $NiFe_2O_4$ will probably corrode congruently in HCl. In contrast, the $AlCl_3$ and $NiCl_2$ pressures over $NiAl_2O_4$ are very different, Figure 4. As a result, NiO will be preferentially corroded from $NiAl_2O_4$ leaving porous Al_2O_3 behind and continued corrosion will depend on gaseous diffusion of $NiCl_2$ through the porous Al_2O_3 as schematically depicted in Figure 5. One of the features of gas transport control is that the gas diffusion coefficient, D_{AB}, can easily be calculated from kinetic theory with the Chapman-Enskog equation [6]:

$$D_{AB} = 1.8583 \times 10^{-3} \left(\frac{1}{M_A} + \frac{1}{M_B} \right)^{1/2} \frac{T^{3/2}}{P \sigma^2 \Omega_{D,AB}} \tag{8}$$

where P = total pressure, σ = collision diameter, $\Omega_{A,B}$ = collision integral, and M_A and M_B are the molecular weights of A and B. For diffusion through a porous region in which the pore diameter is on the order of the mean-free-path of the gas atoms, Knudsen diffusion must be considered in which the Knudsen diffusion coefficient, K_A is given by [8]:

$$D_{KA} = 9.7 \times 10^3 \, \bar{r} \sqrt{\frac{T}{M_A}} \tag{9}$$

where $\bar{r}$ is the mean pore radius. Typically, for porous layers similar to those expected from incongruent corrosion of a compound such as $NiAl_2O_4$ [9], the two diffusion coefficients are similar so an effective diffusion coefficient must be used, $D_{A,eff}$:

$$\frac{1}{D_{A,eff}} = \frac{1}{D_{KA}} + \frac{1}{D_{AB}}. \qquad (10)$$

Depending on the equilibrium product gas pressures and the diffusion coefficients, the thickness of the porous layer may reach a "steady state" and the multicomponent oxide can continue to corrode congruently or the less volatile species may simply be left behind as "ash." In the case of $NiAl_2O_4$, the differences in partial pressures is sufficiently large that in virtually all cases the NiO would be preferentially corroded out leaving behind porous Al_2O_3.

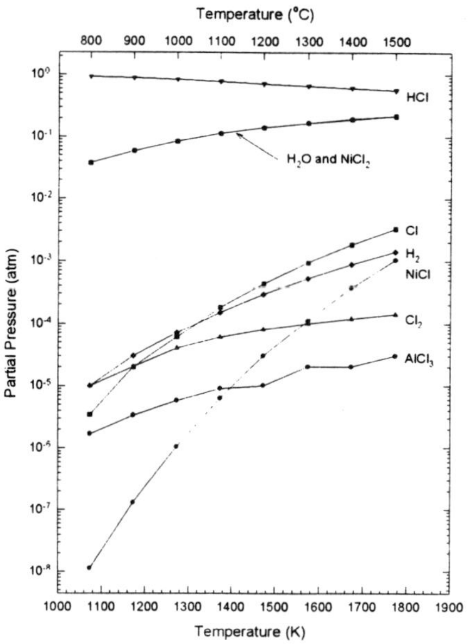

Figure 4. Gas pressures over $NiAl_2O_4$ in HCl.

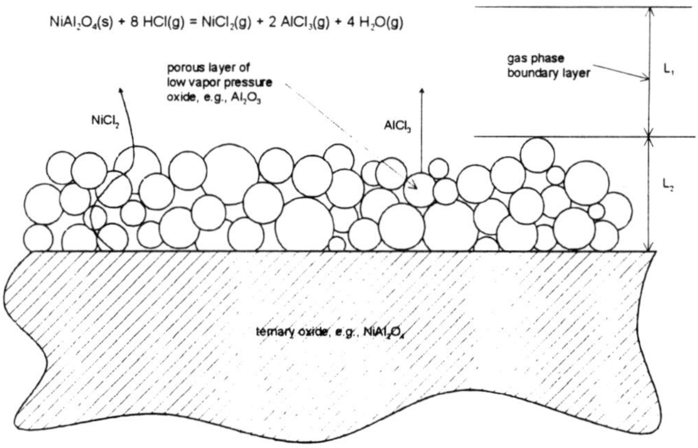

Figure 5. Schematic representation of incongruent corrosion of $NiAl_2O_4$ with the formation of $NiCl_2$ gas and a residual porous layer of Al_2O_3.

CARBIDES

The corrosion of SiC in HCl typifies what is expected for most carbides. Figure 6 shows the equilibrium partial pressures in a starting gas of one atmosphere HCl over what results in a mixture of C and SiC. Even at the lowest temperatures where CH_4 is an important gas species, the primary reaction appears to be:

$$SiC(s) + 4\ HCl(g) = C(s) + SiCl_4(g) + 2\ H_2(g). \qquad (11)$$

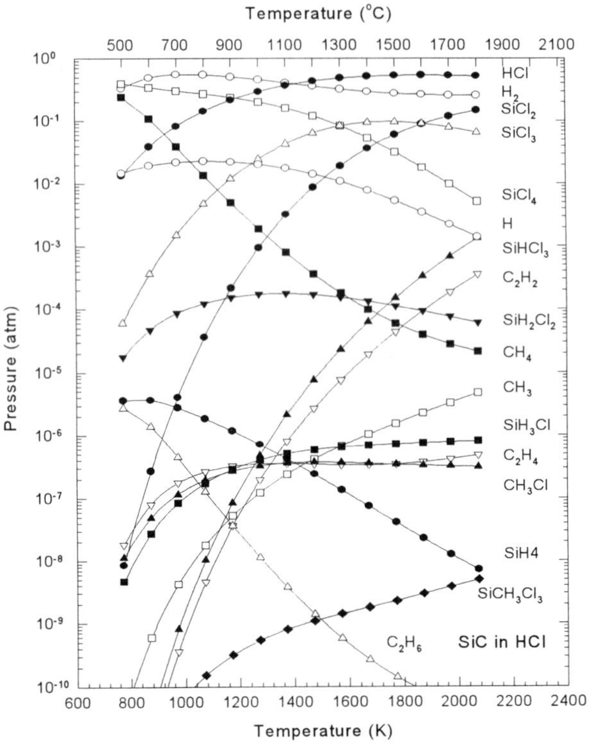

Figure 6. Gas pressures over SiC in HCl.

As the temperature goes higher, the equilibrium CH_4 pressure goes even lower. So in this case, residual porous carbon (or graphite) is left behind and the kinetics will depend on gas diffusion through this po layer as discussed above. It is also interesting to note that the reaction goes essentially to completion at low temperatures so that diffusion of HCl to the solid surface could control the rate of corrosion . However, at low temperatures, surface reaction is far more likely to be the rate controlling process. Other carbides should behave similarly.

NITRIDES

In contrast, Si_3N_4 reacts to form completely gaseous products as shown in Figure 7. In this case the main equilibrium reaction over a large temperature range is:

$$Si_3N_4(s) + 12\ HCl(g) = 3\ SiCl_4(g) + 6\ H_2(g) + 2\ N_2(g) \qquad (12)$$

and the other silicon chloride species become more important at higher temperatures. Interestingly, for this reaction the equilibrium pressures virtually do not vary for almost 1000 °C. As a result, the regions of surface reaction control and gas transport control should be easy to separate experimentally.

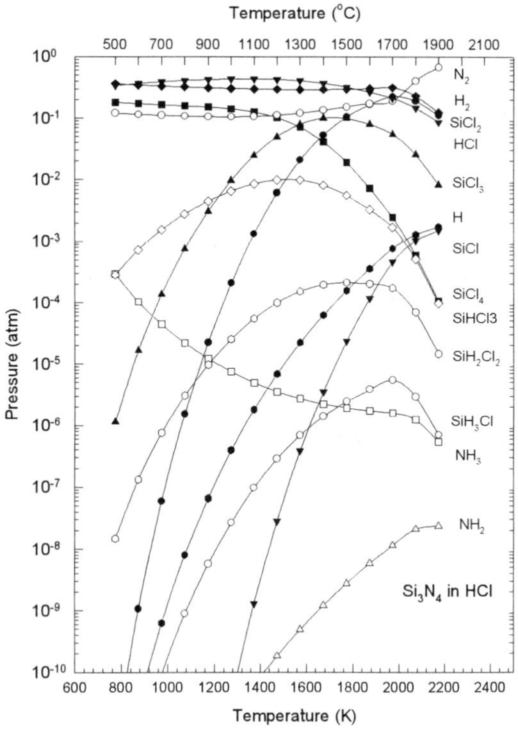

Figure 7. Gas pressures over Si_3N_4 in HCl.

SUMMARY

At high temperatures, ceramics (e.g. SiC in H_2O [1] and SiO_2 in H_2 [2]) have been shown to corrode by gas diffusion control. In halogen-containing atmospheres such as HCl, most ceramics at sufficiently high temperatures will corrode by gas phase diffusion and the rate will be controlled by the ambient gas pressures and the thermodynamics of the reaction. Thus, the general corrosion behavior for specific ceramic materials can be predicted. Even in those cases in which the corrosion is incongruent resulting in a residual porous product layer, the theory of gas diffusion can be used to quantitatively predict corrosion rates. Experiments are necessary to determine when specific ceramic materials corrode by gas phase diffusion and how well the model predicts rates and congruent or incongruent corrosion.

REFERENCES

1. H. E. Kim and D. W. Readey, Corrosion of SiC in Hydrogen Above 1400°C, Ceramic Trans. **2** 301-312 (1989).

2. D. W. Readey, Gaseous Corrosion of Ceramics, Ceramica Acta **3** 11-27 (1991).

3. D. W. Readey, Vapor Phase Sintering of Ceramics, Ceramic Trans. **7** 86-110 (1990).

4. M. A. Ritland and D. W. Readey, Alumina-Copper Composites by Vapor Phase Sintering, Ceramic Engineering and Science Proceedings, **14** [9-10] 896-907 (1993).

5. D. R. Stull and H. Prophet, JANAF Thermochemical Tables. 2nd ed. National Bureau of Standards, Washington, D. C. Report No. NSRDS-NB537, June, 1971.

6. T. Besmann, T. SOLGASMIX-PV, A Computer Program to Calculate Relationships in Complex Chemical Systems. Oak Ridge National Laboratories, Oak Ridge, Tenn. Report No. ORNL/RM-5775, 1977.

7. R. B. Bird, W. E. Stewart, and E. N. Lightfoot, Transport Phenomena , (Wiley, NY), 1960.

8 . C. J. Geankoplis, Mass Transport Phenomena, (C. J. Geankoplis, Columbus), 1972.

9. R. R. Sickafoose, R. R., Jr. and D. W. Readey, Active Gaseous Corrosion of Porous SiC, J. Am. Ceram. Soc. **76** [2] 316-24 (1993).

ELEMENTAL CONCENTRATION MEASUREMENT IN VITRIFIED GLASS BY LASER-INDUCED BREAKDOWN SPECTROSCOPY

Fang-Yu Yueh, Jagdish P. Singh, Chun-Fu Su and Ram Vasudev
Diagnostic Instrumentation and Analysis Laboratory
Mississippi State University, MS 39762
and
James Marra
Westinghouse Savannah River Co.
Aiken, SC 29808

ABSTRACT

Laser-induced Breakdown Spectroscopy (LIBS) spectra of several glass samples have been recorded. The elemental concentrations are inferred by analysis of the experimental lineshapes and these are compared with the concentrations inferred from peak area and peak height analyses. The precision and accuracy with these three data analysis techniques have been estimated. The errors in the inferred concentrations due to errors in plasma temperature and electron density have been investigated. This study shows that inferred concentration for some species are quite sensitive to these two parameters. Detailed analysis of various glass LIBS spectra are described. In addition, we also present preliminary results on vitrified glass which may be eventually used for Pu immobilization.

INTRODUCTION

LIBS is a laser based advanced diagnostic technique for measuring the concentration of elements in a test medium.[1-5] A pulsed laser beam is focussed at the test point and produces spark at the focal point due to high electric-field. The spark generates a high density plasma which excites various atomic elements present in the focal volume. The atomic emission from the plasma is collected with a collimating lens and sent to the detection system. The intensity of the atomic emission lines observed in the LIBS spectrum are then used to infer the concentration of the atomic species.

Recently, a mobile and versatile LIBS system has been developed at the Diagnostic Instrumentation and Analysis Laboratory (DIAL), Mississippi State

University (MSU) to monitor toxic metal concentrations in the off-gas of a plasma hearth process system and also glass which is a product of thermal treatment of hazardous and radioactive waste.[6] It has been used to measure the composition of the glass from an EnVitCo Joule-heated melter at Clemson University[7] and the Westinghouse Savannah River Company's (WSRC) Transportable Vitrification System (TVS).[8] Here, we described the analysis of LIBS spectra of EnVitCo and TVS glass samples. We also report preliminary LIBS spectra of glass that may be eventually used in the DOE complex for Pu immibilization.

EXPERIMENTAL SETUP

The experimental arrangement of the LIBS system is described in reference 6 and is shown in Figure 1. The output (1064 nm) of a Nd:YAG laser (Continuum Surelite I) was frequency doubled to 532 nm using a doubling crystal. A dichroic mirror was used to separate the frequency doubled output from its fundamental frequency. An ultraviolet (UV) grade quartz lens of 300 mm focal length was used to focus the laser beam to a spot size of approximately 0.03 mm. The same lens was used to collect light from the laser-induced spark. Two UV-grade quartz lenses of focal lengths 100 mm and 50 mm were used to couple the LIBS signal to an optical fiber bundle. The core diameter of the single fiber was 0.1 mm and the numerical aperture (NA) of the fiber bundle was 0.16. The other end of the fiber bundle was coupled to an optical spectrograph (Instruments SA Model HR460). The entrance slit of the spectrograph is replaced by the rectangular exit end of the optical fiber. The spectrograph was equipped with 1200 and 2400 lines/mm diffraction gratings of dimensions 75 mm x 75 mm and a 1024-element intensified diode array detector (Model IDAD-1024,

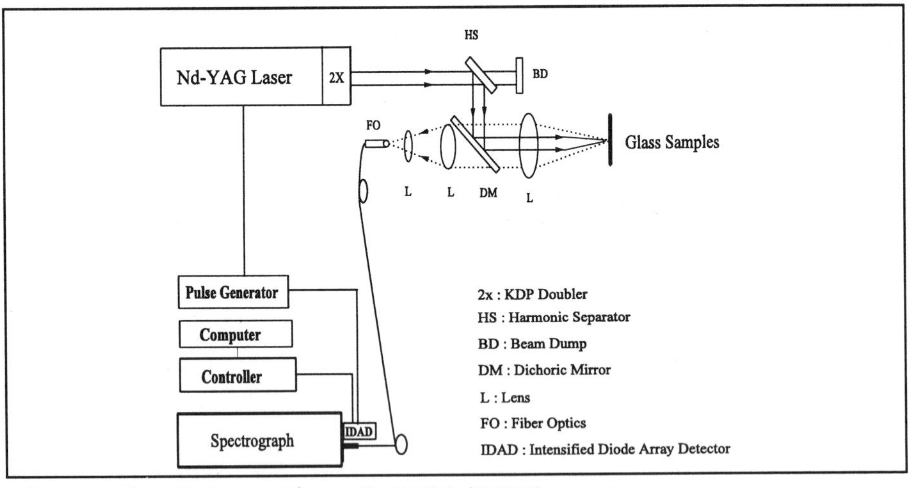

Figure 1. Schematic of DIAL/LIBS system.

Princeton Applied Research Corp.) with a pixel width of 0.022 mm. The detector was used in a gated mode and was synchronized to the laser pulse. To maximize the signal, a gate pulse delay of 5-10 μsec and width of 10 to 20 μsec was used in most of the work. The spectral window monitored by the array was 19 nm wide with a resolution of 0.16 nm with the 2400 l/mm grating. Data acquisition and analysis were performed using a notebook computer (Model T-4700CS, Toshiba).

ANALYSIS

Under local thermal equilibrium (LTE), the intensity of an emission line at ν_{ji} can be expressed as

$$I_{ji} = N \cdot (1 - \alpha) \cdot (h\nu_{ji}) \cdot \frac{A_{ji} g_j}{Q(T)} \cdot \exp^{-hcE_j/kT} \qquad (1)$$

Here, N is the number of atoms, g_j is the statistical weight of the excited state, A_{ji} is the transition probability, α is the degree of the ionization, E_j is the energy of the excited state in units of cm^{-1}, h is Planck's constant, k is the Boltzmann constant, T is the plasma temperature, and Q(T) is the partition function.

The degree of ionization of a species in the plasma depends on the ionization potential V of the neutral atom, the plasma temperature and partition function of that element by[9]:

$$\log\left(\frac{\alpha}{1-\alpha}\right) = -\log n_e + \frac{3}{2}\log T - \frac{5040\,V}{T} + \log\frac{Q_i}{Q_a} + 15.684$$

and

$$\alpha = \frac{n_i}{n_i + n_a}. \qquad (2)$$

Here n_e, n_a, and n_i are electron density, neutral species number density, and ionized species number density, respectively.

The electron density can be calculated from the spectral line intensities and the spectroscopic constants of a pair of neutral and ion lines of a selected element via the following equation[10]

$$n_e = 4.83\times10^{15}\cdot\left(\frac{gA}{I\lambda}\right)_i\cdot\left(\frac{I\lambda}{gA}\right)_a\cdot T^{3/2}\exp\left(\frac{-11605\,V}{T}\right)\cdot \exp\left(\frac{-1.439\,(E_i-E_a)}{T}\right). \qquad (3)$$

The Boltzmann temperature can be determined from Equation (1) using the relative intensity of two lines of a given atomic species. If more than two lines intensities are available, temperature can also be extracted through the Boltzmann plot method.[11] To determine the degree of ionization, a pair of neutral and ion line is needed to determined electron density. The inferred plasma temperature and electron density can then be used to determine the degree of ionization for each species. We note that the electron density and plasma temperature inferred from LIBS spectra of glass are in the range of 10^{15} - 10^{16} cm^{-3} and 5000 - 7000 K, respectively. The degree of ionization for most elements at this temperature and electron density are not negligible. Figure 2 shows the calculated degree of ionization of Fe, Ti, Mn, Cr, Mg and Ni at different values of T and n_e. It is clear that these calculated values are different for different elements. Therefore the degree of the ionization should always be included in the LIBS analysis to obtain accurate concentration ratios.

If the temperature and electron density are inferred from LIBS spectra of some selected elements, then the concentration of other species can be extracted from their line intensities. We use three methods to analyze the LIBS spectra. These are based on the peak areas, peak heights and shapes of the atomic lines. The line area and peak height methods are simple and give reasonable results for well-separated lines. The lineshape fitting method can be used to analyze multiple line spectra, including blended lines.

RESULTS

To determine the most efficient/accurate way to analyze the LIBS spectra of glass, the spectra were analyzed by the threee methods mentioned above. The plasma temperature and electron density were obtained using Fe neutral and ion lines. Figure 3 shows a spectrum from a glass produced in the EnVitco melter recorded in the 260 nm region. Identified lines are marked on the figure. The best fit of this spectrum is also shown. This spectrum was also analyzed by using peak heights and peak areas. The results from these three analysis methods are listed in Table 1. These are compared with the atomic absorption analysis of the same glass sample by the Mississippi State Chemical Laboratory (MSCL) at Mississippi State University. It is clear that the spectrum line-shape fitting gives the best agreement with the MSCL analysis. The degree of interference strongly influences the analysis by either the

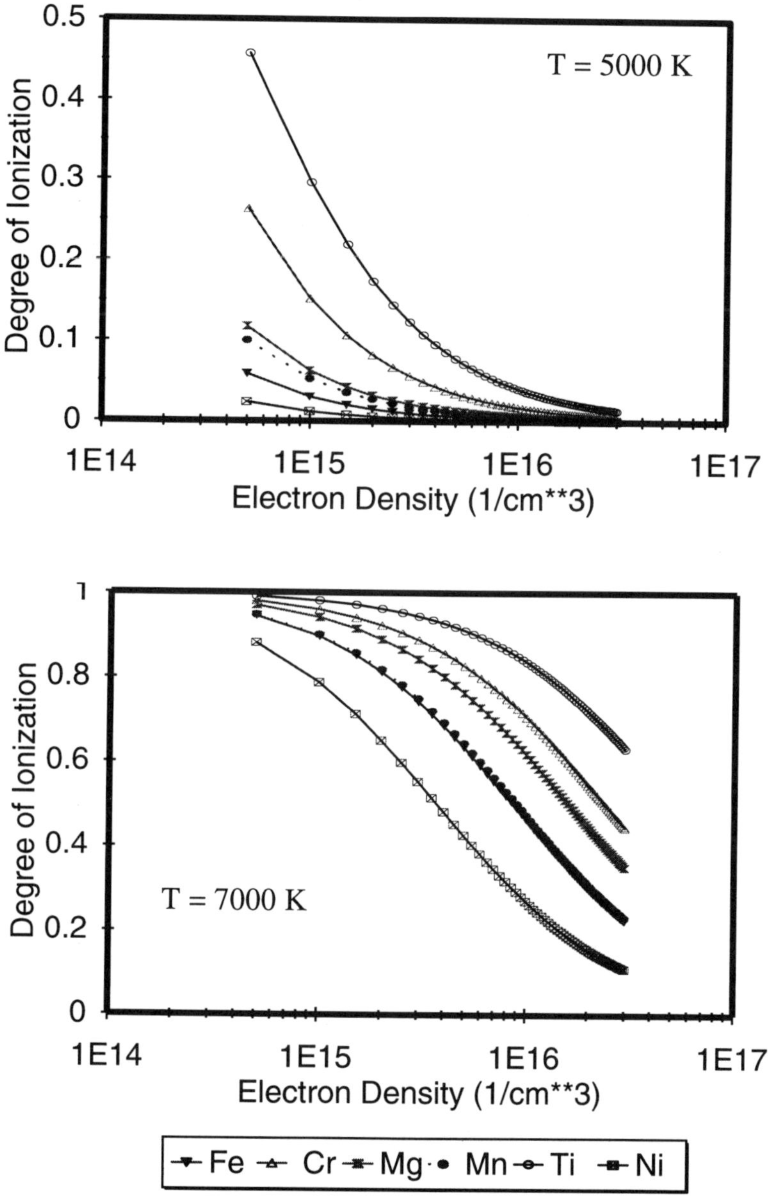

Figure 2. The degree of ionization of several elements at various T and n_e.

peak height or peak area method. If the analyte line is interference-free, the three methods are in very close agreement. Since the peak height and peak area analyses are fast and easy to implement via a real-time data acquisition program than the multiparameter fitting, they are more suitable for real-time monitoring. However only interference-free analyte lines should be used.

The above methods were also applied to the spectra recorded with melt-glass pour during the TVS test. Fe, Cr, Ni, Ca, Ti, Si, Mg, Al, Mn, Sr, Li, Na, K and Cr atomic transitions were identified from the recorded spectra. Spectra were also recorded in one spectral region to monitor the concentration of Ni, Cr, Ti, and Ca. The concentration of these elements were displayed in real time on the data acquisition computer and the center display station. A typical result of real-time

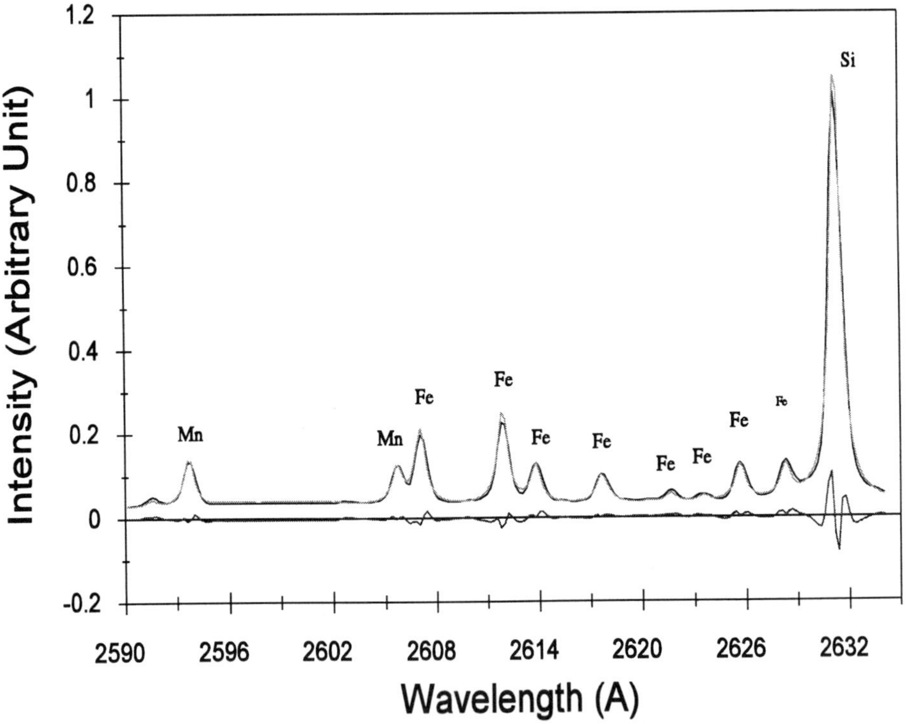

Figure 3. Fitting of a LIBS spectrum for a glass produced in an EnVitCo melter.

Table I. Concentration ratios w.r.t. iron obtained from LIBS spectra of glass produced in the EnVitco melter using line peak intensity, line area, and spectrum fitting.

	Mn		Si
Line Area	0.038[a]	0.023[b]	62.8[c]
Line Peak Height	0.037[a]	0.033[b]	51.94[c]
Spectrum Fitting	0.041		43.8
MSCL	0.042		40.1

[a] using Mn 2593.7 Å line
[b] using Mn 2605.6Å line
[c] using Si 2631.3 Å line

Table II. Results of TVS Glass Stream LIBS Measurements Oxide weight (Wt%)

	NiO	TiO_2	Cr_2O_3	CaO	SiO2	MnO_2	MgO	CuO	SrO
A	0.16 (.016)	0.51 (0.051)	0.059 (0.028)	10.4 (0.77)	64.7 (5.55)	0.40 (0.04)	0.40 (0.021)		
B	0.21 (.019)	0.54 (0.047)	0.056 (0.034)	10.5 (0.89)					
C	0.17 (.01)	0.63 (0.033)	0.059 (0.004)	11.8 (0.54)	68.8 (6.95)	0.33 (0.09)	0.33 (0.037)	0.021 (0.007)	0.014 (.0008)

A: Data taken on 2/18/96
B: Data taken on 2/19/96
C: Data taken on 2/20/96

concentration monitoring is shown in Figure 4. Peak area analysis was used in this real-time monitoring. Since Fe emission lines are found in all the recorded spectral regions, they are used as an internal reference. The emission lines used in the analysis are selected to minimize undesirable effects due to overlap with other lines or due to self-absorption. The inferred oxide weight ratio of the various elements are listed in Table II. Na, Li, K, and Al are not included due to self-absorption problems. The errors in the inferred concentrations are mainly from the uncertainties in the plasma temperature and electron density used in the analysis. Since not all the spectral regions recorded in the test contained suitable lines for plasma temperature and electron density determination, some analyses are based on temperature and

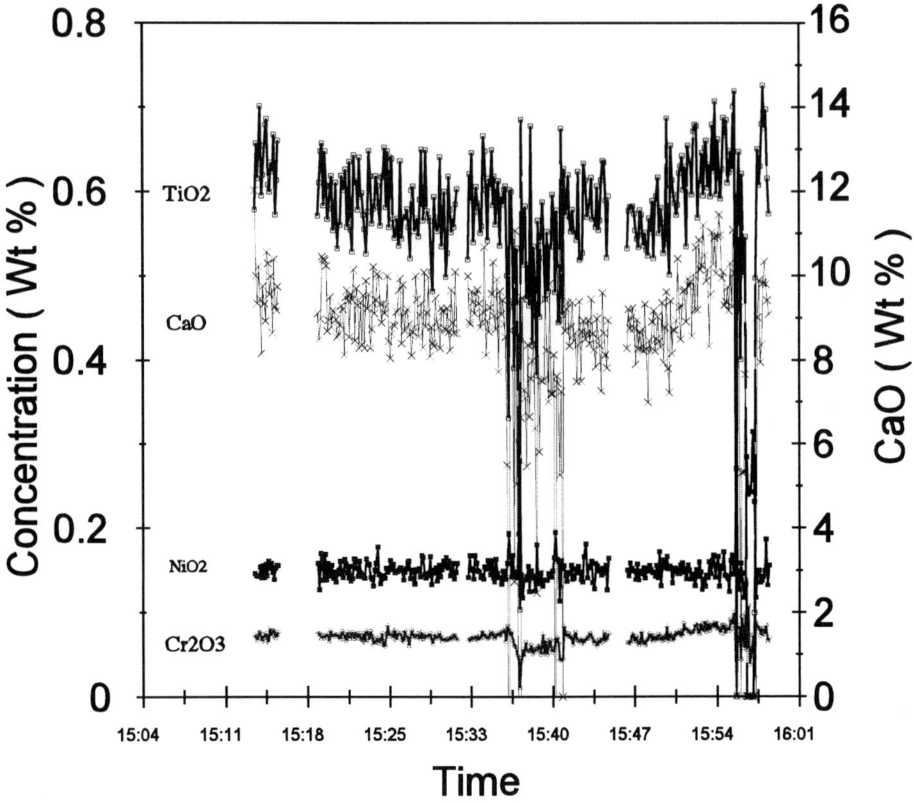

Figure 4. Variation of oxide weight with time during 2/19/96 TVS measurement.

electron density obtained from spectra in other regions.

Finally, we report preliminary exploratory experiments with a vitrified glass sample prepared at WSRC. This glass formulation, or modified versions of it, may be eventually used in the DOE complex for Pu immobilization. Figure 5 shows two typical LIBS spectra. Lines due to Si, Al and La (major consituents) are readily identified. In field experiments at WSRC, planned for the near future, such spectra together with Pu emission near 453 nm will be used for real-time probing of elemental concentrations (a) in melt-glass under preparation (in melter), (b) in melt-glass while being poured into containers for transportation, and (c) in the final solidified glass product.

(A)

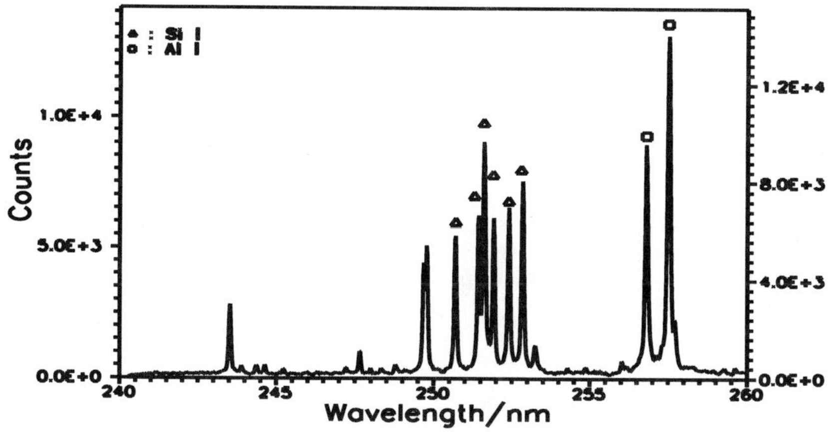

(B)

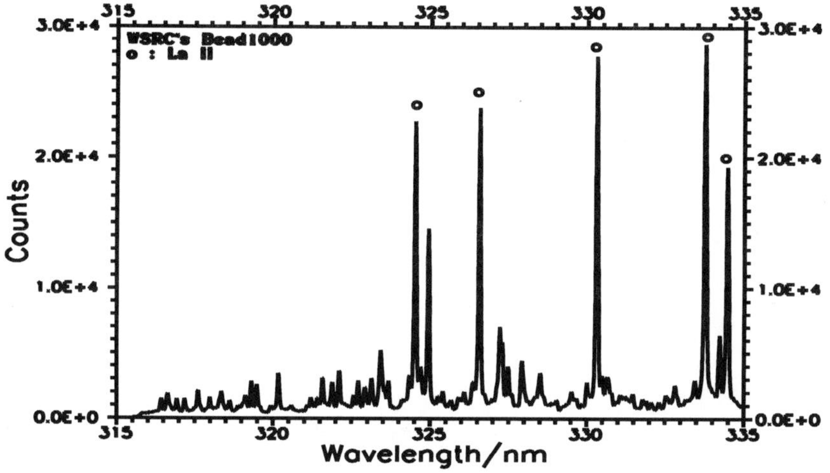

Figure 5. LIBS spectra of WSRC's vitrified glass
(A) near 250 nm (B) near 325 nm.

SUMMARY

Three LIBS analysis methods are compared. The peak height and peak area methods are suitable for real-time analysis for interference-free lines. The real-time monitioring of composition of molten glass with peak area analysis has been demonstrated. LIBS specta of WSRC glass proposed for Pu immobilization are reported.

ACKNOWLEDGEMENT

Major portion of this work is supported by DOE grant DE-FG02-93CH-10575. The project on LIBS of glass proposed for Pu immobilization is funded by WSRC contract AB62512-0.

REFERENCES

[1]Cremers, D. A. and Radziemski,L. J. "Laser Plasmas for Chemical Analysis" in *Laser Spectroscopy and its Application,* L.J. Radziemski, R.W. Solarz, J.A. Paisner, eds. (Marcel Dekker, New York, NY, 1987), Chap. 5, p.351.

[2]Zhang, H., Singh, J.P.,Yueh, F.Y. and Cook, R.L. 1995, Appl. Spectrosc. **49**:1617.

[3]Ottesen,D.K., Baxter, L.L., Radziemski, L. J. and Burrows, L.F. 1991. Energy & Fuels **5:** 304.

[4]Grant,K.J., Paul, G.L. and O'Neill, J.A. 1991. Appl. Spectrosc. **45:** 701.

[5]Ottesen, D.K., Wang, J.C.F. and Radziemski, L.J. 1989. Appl. Spectrosc. **43:**1967.

[6]Singh, J.P., Yueh, F.Y., Zhang, H., and Su, C.F., "Application of Modern Diagnostic Methods to Environmental Improvement: Laser-Induced Breakdown Spectroscopy (LIBS)", in DIAL/MSU Annual Technical Progress Report, DIAL 10575/FY 95 Annual (1995).

[7]Singh, J.P., Yueh, F.Y., and Su, C.F., "LIBS Measurements at Stir Melter", in DIAL 10575 Trip Report 94-1.

[8]Singh, J.P., Zhang, H. and Yueh, F.Y., "Laser Induced Breakdown Spectroscopy Measurements" in "Transportable Vitrification System Shakedown Test," Westinghouse Savannah River Corporation, DIAL Trip Report 96-1.3.

[9]Boumans, P.W.J.M. ,1968, spectrochimica Acta, **23B**, 559.

[10]Boumans, P.W.J.M. and Boer, F.J.De, 1977, Spectrochimica Acta. **32B**, 365.

[11] Faires, L.M., Palmer, B.A., Engleman,Jr. R, Niemczyk, T.M. , 1984 Spectrochimica Acta. **39B**, 819.

AQUEOUS CORROSION OF CRYSTALLINE CERAMICS*

Dennis W. Readey
Colorado Center for Advanced Ceramics
Colorado School of Mines
Golden, CO 80401

ABSTRACT

The aqueous corrosion of crystalline ceramics has received little attention because ceramics are generally thought to be corrosion resistant. However, ceramics can corrode and their corrosion resistance can be a measure of their potential for nuclear waste hosts. The kinetics controlling the corrosion of ceramics varies from transport of diffusing ions to surface reactions. Both the thermodynamics of dissolution and the rates of reaction are sensitive to system variables such as the temperature and the solution composition. These include the pH and the presence and concentration of redox ions. In addition, the composition and the crystal structure of a given ceramic can have a profound effect on dissolution kinetics.

INTRODUCTION

The fundamental mechanisms controlling the aqueous corrosion of crystalline ceramics are not well understood. Ceramics are comparatively inert so aqueous corrosion has received little attention by ceramic engineers and scientists. However, ceramics can corrode in aqueous environments. Recent experience with the environmental degradation of $YBa_2Cu_3O_{7-x}$ high T_c superconductors[1], the destabilization of yttria-stabilized zirconia by water [2], and microstructural degradation during sintering caused by preferential leaching during powder

* Sponsored by the National Science Foundation, Division of Materials Research

processing [3] emphasize the recurring importance of ceramic corrosion [4].

Ceramics have been considered for nuclear waste hosts for many years [5]. The leaching and corrosion of ceramics and glass have been paramount considerations for waste host applications. Glass has been the primary candidate as a potential waste host because of its stability over a wide range of composition. As a result, there have been extensive studies on the leaching of silicate-based glasses [6]. The corrosion of glass is complex because of the extensive ion exchange that occurs between silicate glasses and solutions. Nevertheless, the corrosion kinetics of glasses has been far better studied and are far better understood than those of crystalline ceramics. The question that needs to be answered is, "What are the factors that control the rate of dissolution that can be used to predict the corrosion behavior of a given ceramic or glass." The purpose here is to present a brief review of some of the principles that are involved in the dissolution of crystalline ceramics rather than give an exhaustive review of the literature. In fact much of the literature on ceramic corrosion, particularly that of oxides, is not to be found in the ceramic literature. The information is widely scattered in fields ranging from aquatic biology to geochemistry which means experimental data are evaluated from widely different points of view. As a result, information on ceramic corrosion takes time to obtain and it is even more difficult to resolve differences in proposed mechanisms for the observed behavior.

AQUEOUS CORROSION

Thermodynamics

Corrosion of oxide ceramics is essentially a dissolution process. Two types of dissolution can be distinguished [7]. One is chemical dissolution involving no charge transfer or oxidation/reduction step such as:

$$MgO_{(s)} + 2\ H^+ = Mg^{2+} + H_2O. \tag{1}$$

The other is electrochemical dissolution with electron transfer similar to that which occurs during the dissolution or aqueous corrosion of metals:

$$Fe_2O_{3(s)} + 3\ H^+ + 2\ e^- = 2\ Fe^{+2} + 3\ OH^-. \tag{2}$$

The thermodynamics of aqueous corrosion is concerned with solubility. The solubility is determined by the stability of the constituent ions in solution relative to their stability in the solid or crystal. For purely ionic compounds, the relative values of the crystal lattice energy and the ion hydration energy in solution determine solubility.[8] For example, SiO_2 has a low solubility in acidic solutions[9] because electrically neutral $Si(OH)_4$ is formed which is not strongly solvated by water molecules. The $Si(OH)_4$ "molecule" is uncharged and nonpolar so there is virtually no hydration energy as it enters solution. Its solubility would not be expected to be much different than that of methane. At high pH, $Si(OH)_4$ behaves as a weak acid, it ionizes and hydrates, and the solubility of SiO_2 increases with pH. On the other hand, in an solutions containing fluorine, the stable SiF_6^{2-} ion is formed [9] and silica is quite soluble in acidic solutions.

Most oxides are amphoteric,[9] acting as weak bases in acidic solutions and weak acids in basic solutions. For example, in acids, Al_2O_3 will dissolve as:

$$Al_2O_{3(s)} + 6\ H^+ = 2\ Al^+ + 3\ H_2O \tag{3}$$

while in basic solutions,

$$Al_2O_{3(s)} + 2\ OH^- = 2\ AlO_2^- + H_2O. \tag{4}$$

As a result, there is frequently a minimum in oxide solubility as a function of pH as shown in Figure 1. The solubilities of most oxides are available [9] and the solubilities of both oxide and nonoxide ceramics can, in principle, be calculated[10] from thermodynamic data.[11,12] However, frequently, the ionic species in solution can be considerably more complex.[13] For example, for Al_2O_3, equilibria can exist between several polymeric species, particularly near the solubiliy minimum:[14]

$$Al^{+3} = [Al(OH)]^{+2} = [Al_2(OH)_2]^{+4} = [Al_3(OH)_4]^{+5} = [Al_8(OH)_{20}]^{+4} =$$

$$Al_{13}O_4(OH)_{24}]^{+7}. \tag{5}$$

Kinetics: Diffusion or Surface Reaction

Figure 2 shows the steps involved in dissolving a ceramic oxide, NiO as an example. Solution ions must diffuse to the solid surface where they react with surface anion and cations which then enter the solution. The reaction products must then diffuse away from the interface into the solution. For spherical particles, the diffusion flux, J, is given by:

$$J = \frac{D(C_e - C_o)}{a} \tag{6}$$

which can be integrated to give the volume fraction, "f," of the particles dissolved or reacted with time:

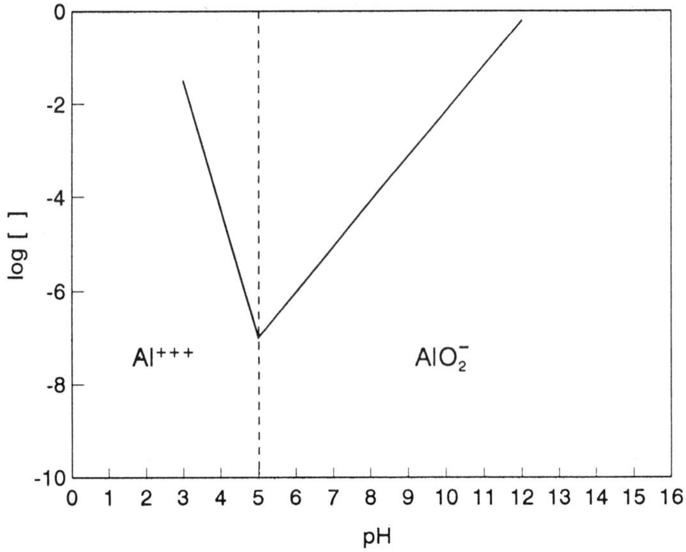

Figure 1. Solubility of Al_2O_3 has a function of pH (based on data from [9]).

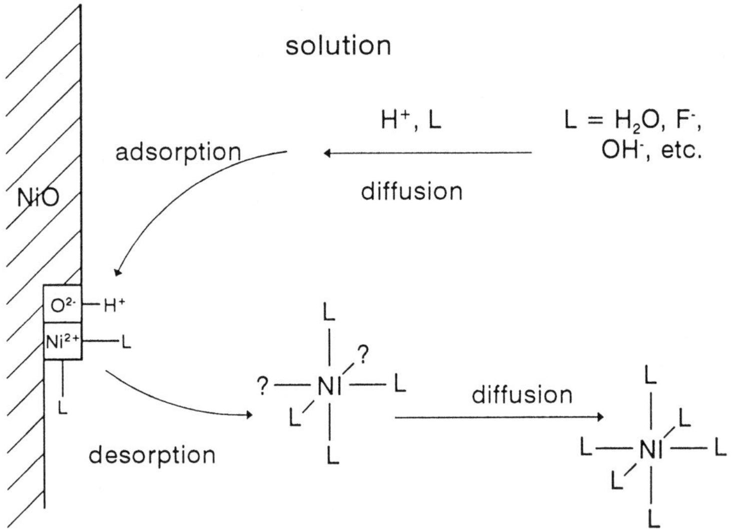

Figure 2. The series steps involved in the dissolution of ceramics.

$$1 - (1 - f)^{2/3} = K_D t \text{ with } K_D = 2\,\Omega D \frac{(C_e - C_o)}{a_o^2}. \qquad (7)$$

where t = time, Ω = molar volume, a_o = sphere radius at t = 0, D = diffusion coefficient in the fluid, C_e = equilibrium concentration in the fluid (solubility), and C_o = initial solution concentration. If diffusion were rate-controlling, the time to dissolve 10 μm Al_2O_3 particles in pure water of pH = 7 would be about 5 hours (based on the solubility data of Fig. 1 and a liquid diffusion coefficient $D \approx 10^{-4}$ cm^2/s) and slip casting alumina at high and low pH values would be virtually impossible. Yet slip casting of Al_2O_3 is routinely performed since the rate of dissolution of alumina, even in strong acids is much slower than predicted by Eqns. 6 and 7. This suggests that some type of surface reaction is rate controlling.

For surface reaction control, there are many possible rate-controlling steps. Some of these are: 1.) adsorption and surface diffusion of the reactants; 2.) decomposition of reactants at specific surface sites such as steps or kinks; 3.) reaction at the surface; 4.) removal of the products from the reaction site; 5.) product surface diffusion; and 6.) desorption. Unfortunately, surface or interface reactions usually do not lend themselves to unambiguous interpretation except in the simplest systems. Therefore, for ceramic corrosion controlled by surface reactions, the kinetic mechanism may only be inferred from agreement between experimental and model functional dependency of the reaction rate constant with experimental variables. For dissolution controlled by a surface reaction, the fraction of spherical particles dissolved can be expressed as:

$$1 - (1 - f)^{1/3} = K_R\, t \text{ with } K_R = \Omega\,(C_e - C_o)\,\frac{k}{a_o} \qquad (8)$$

where k = surface reaction rate.

Chemical Dissolution

The dissolution for some ceramics is controlled by diffusion in the liquid. However, in most cases, a surface reaction controls dissolution with the rate determined by removal of ions from surface sites, probably.[15] Figure 3 shows a typical crystalline surface and Figure 4 schematically depicts the removal of surface cations from surface kinks. The apparent activation energy for chemical dissolution is the energy involved in forming an activated complex consisting of an ion changing its configuration between solution and crystal ligands at reactive sites, Figure 5. Homogeneous ligand exchange reactions in aqueous solutions have been extensively studied[16] and two primary mechanisms have been identified, the associative and dissociative mechanisms. In the dissociative exchange reaction, an ion simply dissociates itself from one of its ligands, reducing

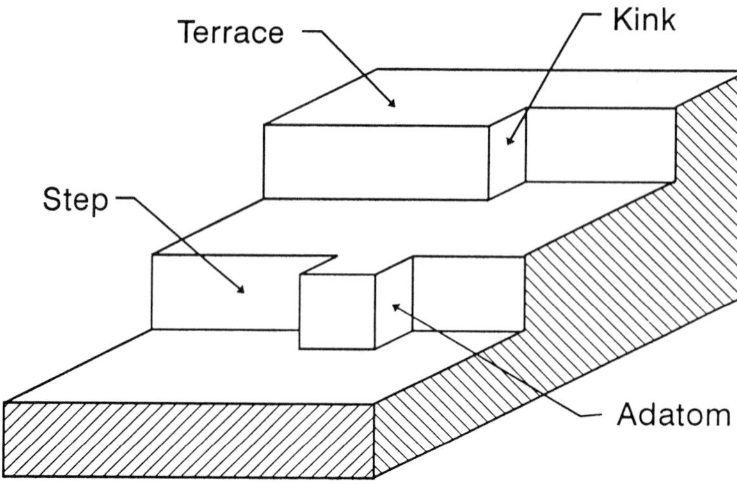

Figure 3. Schematic representation of a crystalline ceramic surface showing kinks, ledges, and terraces.

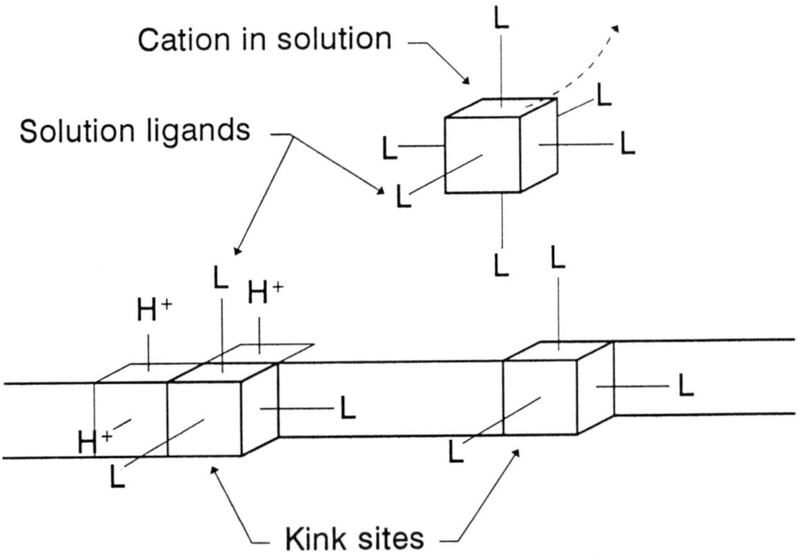

Figure 4. Representation of cations at kink sites exchanging crystal ligands for solution ligands and leaving the crystal surface for the solution.

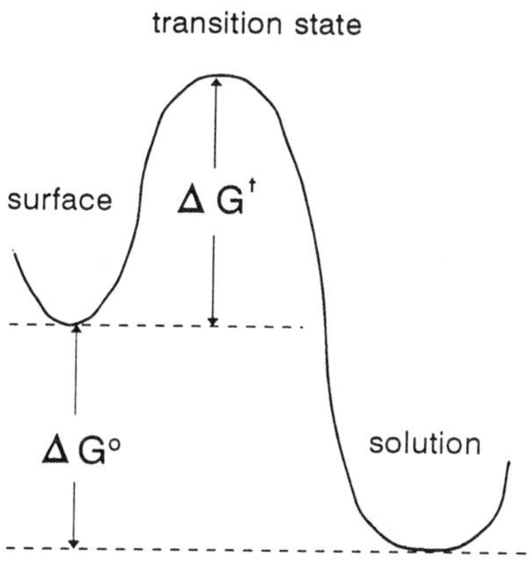

Figure 5. Schematic showing the activation energy for the ligand exchange process and the free energy for dissolution.

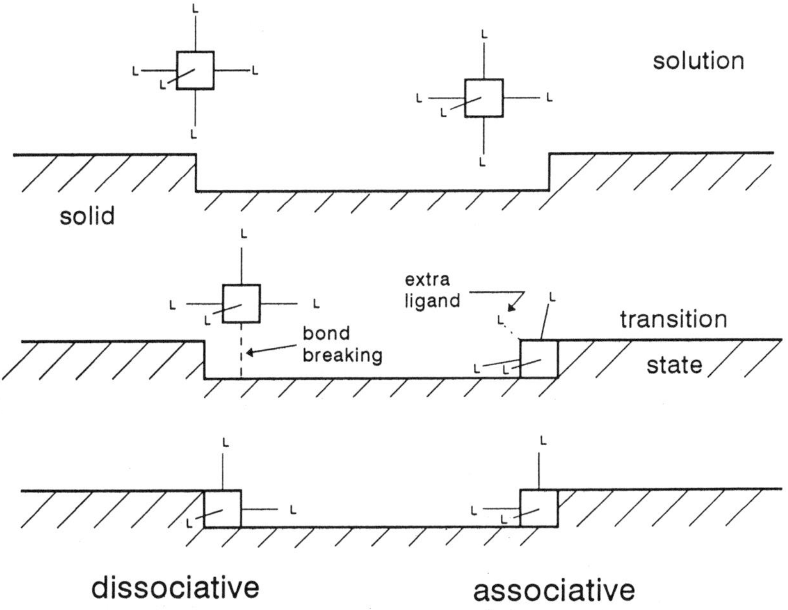

Figure 6. Representation of the two ligand exchange processes (associative and dissociative) at kink sites on a crystal surface.

its coordination number by one, before it recompletes its original coordination with the new ligand; for example,

$$Ni(H_2O)_6^{+2} = Ni(H_2O)_5^{+2} + H_2O$$
$$Ni(H_2O)_5^{+2} + Cl^- = Ni[Cl(H_2O)_5]^{+1}. \quad (9)$$

The activation energy for this process is the energy necessary to remove one of the original ligands. In the associative ligand exchange reaction, an extra ligand enters the inner coordination sphere of the cation increasing the coordination number by one. One of the ligands then leaves the coordination sphere; for example,

$$Ni(H_2O)_6^{+2} + Cl^- = Ni[Cl(H_2O)_6]^{+1}$$
$$Ni[Cl(H_2O)_6]^{+1} = Ni[Cl(H_2O)_5]^{+1} + H_2O. \quad (10)$$

In this case the activation energy is the energy necessary to "squeeze" the extra ligand into the inner coordination sphere. Similar principles and mechanistic models can be applied to the ligand exchange process at the solid-solution interface during the corrosion of ceramics. Figure 6 depicts these two ligand exchange processes taking place at kink sites on the surface of a solid. Unfortunately, these principles have not been applied consistently to the interpretation of much of the data on ceramic corrosion.

It is known from homogeneous solution chemistry that the rate of ligand exchange decreases as the cation charge increases and the cation radius decreases.[17] Ligand exchange kinetics would predict that the rates of dissolution would decrease in the sequence $Li_2O > MgO > Al_2O_3 > TiO_2 > Nb_2O_5$, for example. Thus, the presence of small, highly charged cations on the solid surface as solid impurities or adsorbed from solution can significantly alter the dissolution rate. Deviations from stoichiometry or dopant additions which produce highly charged cations should also decrease the rate of dissolution. A possible example is the presence of Ni^{+3} in NiO produced by deviations from stoichiometry or by the addition of an acceptor dopant such as Li_2O:

$$2\ Ni_{Ni} + 1/2\ O_{2(g)} = 2\ Ni_{Ni}^{\bullet} + V_{Ni}'' + O_O$$
$$2\ Ni_{Ni} + 1/2\ O_{2(g)} + Li_2O \overset{NiO}{=} 2\ Ni_{Ni}^{\bullet} + 2\ Li_{Ni}' + 2\ O_O. \quad (11)$$

In both cases, the rate of dissolution should decrease because of the increase in the concentration of Ni^{+3}. Similarly, doping with a higher valent ion, such as aluminum, would be expected to reduce the rate of dissolution. Therefore, for a ceramic such as NiO which can become metal deficient, deviations from stoichiometry and aliovalent doping might always decrease the rate of dissolution. On the other hand, the Li^+ ion might be preferentially leached producing additional surface kink sites increasing the rate of dissolution. There have been few studies performed to determine which of these possible mechanisms actually occurs.

For oxygen deficient compounds, such as Fe_2O_3 or TiO_2, both deviations from stoichiometry and doping with lower valent cations than those of the host should have little effect on dissolution rates. Doping with a higher valent cation should decrease the rate.

Other factors which change the surface concentration of highly charged cations should also increase or decrease the rate of dissolution. Many ceramics have relatively narrow band gaps. Band bending at the surface of the solid will change the surface concentration of highly charged cations. For example, Figure 7 shows the band bending that can occur in both p-type NiO and n-type Fe_2O_3 in a solution with a cerium redox couple at its standard potential. In such a solution, the concentration of the Ni^{+3} ions at the solid surface of p-type NiO would be reduced. This should increase the dissolution rate over that in a solution without the redox couple present. Thus, adsorbed charge, the presence of redox couples in solution, or the application of a potential can alter band banding and change the dissolution rate without any electron transfer between the solid and solution.[18]The exact behavior of a given semiconducting ceramic in a specific solution will depend on the relative energies of the band edges, Fermi level, redox potential, applied voltage, etc.[19]

The presence of the electrical double layer at the surface of a ceramic will also affect the rate of chemical dissolution.[20] The activation energy for dissolution, ΔG^*, will depend on the potential drop $\Delta\phi$ across the double layer. The double layer forms on the surface of oxides due to adsorption of negative ions (F^-, Cl^-, OH^-, etc.) on cations and H^+ ions on the oxygen ions[21] which behave as Lewis acid and base sites respectively.[22] Specifically,

$$\Delta G^* = \Delta G_o^* + zbF\Delta\phi \qquad (12)$$

where z = rate controlling ion valence, F = Faraday, b = symmetry coefficient, and ΔG_o^* = activation free energy at point of zero charge. This predicts that the rate of dissolution will be affected by the amount of adsorbed surface charge. Typically, double layer potentials are on the order of zero to 100 millivolts. A 100 mV potential drop corresponds to about a 2 kJ change in activation energy. This is to be compared with observed activation energies that are in the neighborhood of 100 kJ. This also corresponds to an adsorbed net charge of about 10^{12} cm^{-2} [23] which can be compared to the total number of ions per unit surface of about 10^{15} cm^{-2}.

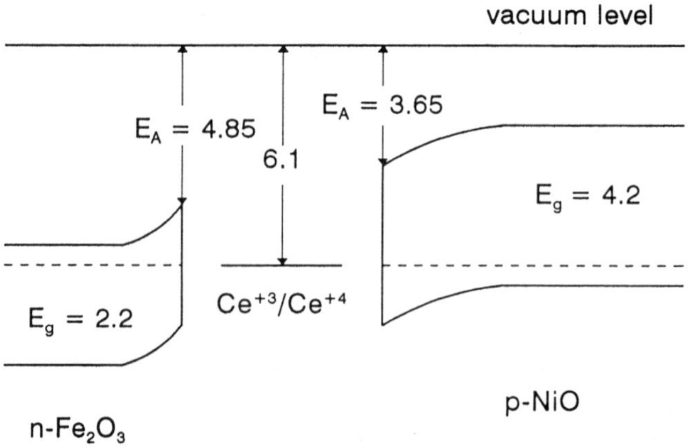

Figure 7. The electronic band bending that is predicted for p-NiO and n-Fe_2O_3 in a solution with a Ce^{+3}/Ce^{+4} redox couple at its standard potential.

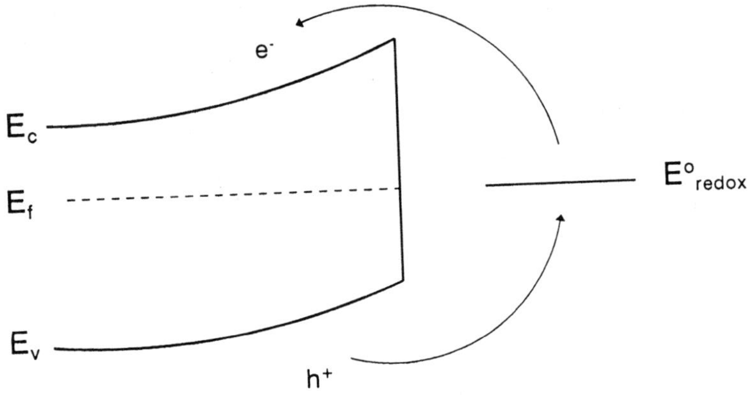

Figure 8. Band bending and electron and hole change between a ceramic and a solution redox couple.

Eqn. (12) also implies that the dissolution rate can also be controlled by an external voltage similar to the potential dependent rate of corrosion of metals as expressed in the Butler-Volmer relation:[24]

$$i = i_o \left\{ e^{[(1-b) F \eta / R T]} - e^{[-b F \eta / R T]} \right\} \quad (13)$$

where i = current density, i_o = exchange current density, and η = overpotential = applied potential - equilibrium potential. This assumes, of course, that the applied voltage drop occurs entirely across the solution double layer. However, for insulating ceramics, the potential drop will occur across the solid and no effect of voltage should be observed.

Electrochemical Dissolution

In contrast, for a ceramic which is a good electronic conductor and which is undergoing electrochemical dissolution with charge transfer, then an applied potential or the presence of a redox couple in solution can play a much more direct role. Application of a voltage or control of the solution Eh with a redox couple, can change the stability of a compound.[9] This is the thermodynamic basis of leaching in many hydrometallurgical systems.[25] For example Fe_2O_3, can be reduced by Ti^{+3}/TiO^{+2} or V^{+3}/V^{+4} couples in acidic solutions:

$$Fe_2O_{3(s)} + 2\,Ti^{+3} + H^+ = 2\,Fe^{+2} + 2\,TiO^{+2} + OH^-, \Delta G^o = -51 \text{ kJ/mole.} \quad (14)$$

Presumably Fe^{+3} is reduced to Fe^{+2} at the solid surface and then dissolution of the lower charged Fe^{+2} cation proceeds rapidly.[26] For electrochemical dissolution, the charge transfer reaction, regardless of the detailed mechanism, is usually thought to be rate controlling. One postulated rate-controlling step in electrochemical dissolution of semiconducting compounds is minority charge carriers reaching the surface by emission over, or tunneling through, the surface Schottky barrier.[27] This is shown schematically in Figure 8.

In homogeneous solution chemistry, the charge transfer between ions surrounded by ligands depends greatly on the character of the ligands. Here again, two distinct mechanisms have been identified.[16] The first is the "inner sphere" mechanism in which electron charge transfer takes place during or after ligand exchange. The second mechanism is the "outer sphere" mechanism in which electron transfer takes place between the ligand shells surrounding the two cations without any ligand exchange. The latter process is facilitated if the ligands surrounding both ions have delocalized electrons.

Choosing a redox couple for leaching of ores in hydrometallurgy requires application of similar principles. To obtain the most rapid leaching by electrochemical dissolution, a redox couple must not only provide the proper Eh for the reaction but also should exchange electrons rapidly with the solid surface to produce a high current so that dissolution proceeds rapidly.[28] This is reflected in a low value of the overpotential in the Butler-Volmer equation above. This is illustrated in Figure 9 for a ceramic undergoing electrochemical dissolution in the presence of a redox couple. The same principles are applied to prevent corrosion of metals in which electrochemical dissolution essentially always occurs.[29] In this case, however, it is desirable to have a low electron exchange rate, or a high overpotential. Unfortunately, in many of the studies on electrochemical dissolution of ceramics, the differences between the thermodynamics of applied voltages and redox potentials have not been carefully separated from the kinetics of electron exchange.

EXPERIMENTAL EXAMPLES

Chemical Dissolution

The rate of sodium chloride dissolution in aqueous solutions is diffusion controlled.[30] Magnesium oxide dissolution shows both diffusion and surface reaction control under differing conditions.[31,32] There have been numerous studies on the dissolution in HF solutions of crystalline and amorphous SiO_2.[33] The general conclusions of these results are that amorphous silica dissolves more rapidly than quartz, that the dissolution rate of quartz is highly anisotropic, and that the rate is determined by some unspecified surface reaction.

Dislocations are sources of surface kink sites. Thus, the rate of chemical dissolution might be expected to increase as the concentration of dislocations increases. However, a study of rutile dissolution showed that dislocations had little effect on the dissolution rate.[34] Studies on the dissolution of silicate mineral, other ternary compounds and solid solutions generally agree that a surface reaction is rate controlling but disagree whether preferential leaching or complete dissolution occurs. However, there is evidence that in some silicate minerals preferential leaching does occur leaving behind a hydrated silica-rich layer that also slowly dissolves.[35] Several studies have also shown that preferential leaching of the faster dissolving constituent in a ternary compound such as $BaTiO_3$ can occur.[3,36] Similarly, a faster dissolving constituent can be preferentially leached from a solid solution.[37] There is a very large volume of literature on the dissolution kinetics of glass because of its potential as a nuclear waste isolation host.[38-41] Although the details of the kinetics still have escaped complete elucidation, the general principles of dissolution are agreed upon. In most glasses

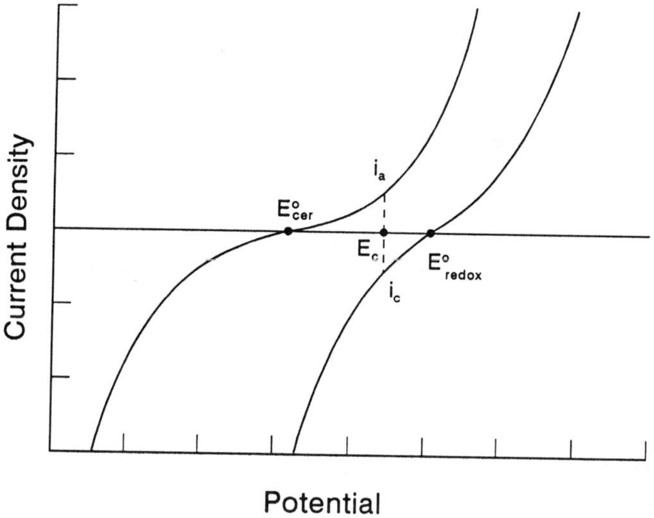

Figure 9. Current density for electron exchange for both a ceramic undergoing electrochemical dissolution and a solution redox couple. The corrosion potential occurs when the exchange currents, i_a and i_c, are equal.

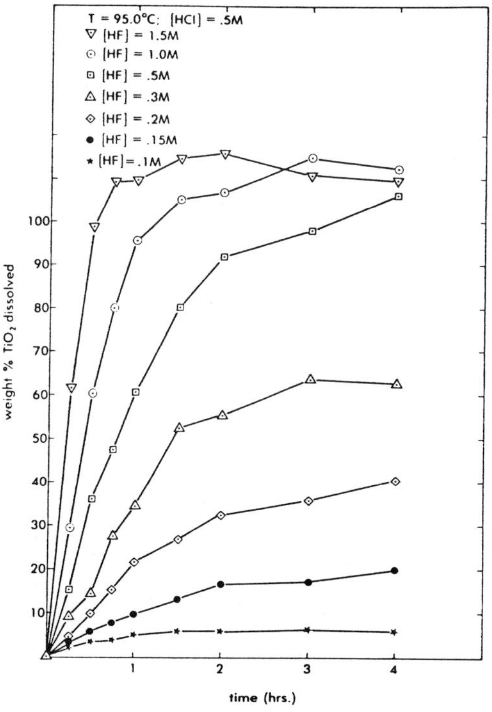

Figure 10. The effect of HF concentration on the dissolution kinetics of TiO_2 powder in 0.5 M HCl at 95°C.

around neutral solution pH, an ion exchanged layer exists near the surface with the alkali ions, Na^{+}, K^{+}, etc. being leached and replaced by hydronium ions, H_3O^{+}. The hydronium ions continue to attack the residual silica rich layer leading to its dissolution. Ionic species which are not soluble form crystalline precipitates at the solid surface or in solution.[42]

Dissolution Kinetics of TiO_2

The dissolution kinetics of TiO_2 in HF-HCl solutions suggest that the rate-controlling step is most likely an associative ligand rearrangement process between the fluorinated surface and adsorbed HF similar to that postulated for silica dissolution.[43] Figure 10 shows the dissolution of TiO_2 in HF-HCl solutions at 95°C. Further insight into the mechanism can be gained by comparing calculated rates with those obtained from experiment. The dissolution rate dependence on temperature and acid concentration can be compared a model of associative surface ligand exchange:

$$J = [HF] \frac{kT}{h} f\, d \exp(\Delta S^{*} / R) \exp(-\Delta H^{*} / RT) \qquad (15)$$

where f = fraction of reaction sites, d = intermolecular distance in the liquid, ΔH^{*} = enthalpy of formation of the activated complex, ΔS^{*} = entropy of formation of the activated complex, and k, T, h, and R have their usual meanings. From the experimental rate of dissolution, the product $f\exp(\Delta S^{*}/R)$ is calculated to be about 3×10^{-6}. The major contribution to the entropy change is arguably that of the HF molecule giving up its translational and rotational degrees of freedom in the solution phase as it enters the activated surface complex ion. A rough estimate of the entropy change can be obtained from the heat of fusion of HF, about 20 J/mole/K.[10] If this is roughly equal $-\Delta S^{*}$, then the fraction of active sites is calculated to be about 3×10^{-5}. If the number of surface atoms is on the order of 10^{14} cm^{-2}, then the number of active sites is on the order of 10^{9} cm^{-2}, a not unreasonable value for the concentration of kink sites. Even with very high specific area powders, this low surface concentration makes it very difficult to investigate directly the activated state with any kind of experimental probe.

Dissolution Kinetics of $SrTiO_3$

$SrTiO_3$ exhibits congruent dissolution followed by TiO_2 precipitation as shown in Figure 11.[44] This is a significant result since it demonstrates that, under conditions of congruent dissolution, the amount of the "insoluble" species in solution, titania in this case, can far exceed its thermodynamic solubility. Also note how much more rapidly the $SrTiO_3$ dissolves than TiO_2 under similar conditions.

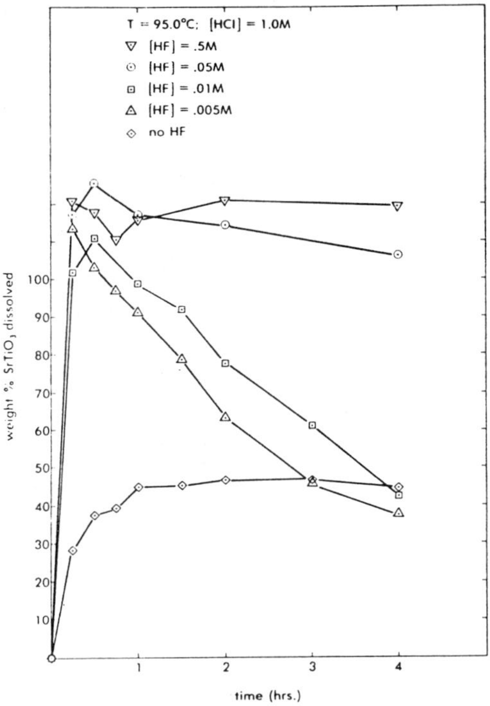

Figure 11. The dissolution kinetics of $SrTiO_3$ powder at 95°C in 1.0 M HCl as a function of the HF concentration. The data show that for low HF concentrations, the $SrTiO_3$ dissolves congruently with subsequent precipitation of TiO_2. At higher HF concentrations, TiO_2 remains in solution well above its calculated solubility.

The data also show that both the dissolution and precipitation rates are strongly catalyzed by only a small concentration of fluoride ion suggesting that the fluoride ion plays a central role in the activated complex on the solid surface. These data, plus the temperature dependence, again suggest a ligand rearrangement rate-controlling process on the crystal surface. Finally, the results show that for a ternary compound or solid solution in which one of the constituents oxides dissolves much more rapidly than the other, then whether there is preferential leaching, recrystallization of the more insoluble constituent on the surface, or dissolution followed by precipitation, will depend on the system and the experimental conditions.

Dissolution Kinetics of Al_2O_3

The dissolution kinetics of alumina in HF-HCl solutions show an important effect of crystal structure on dissolution kinetics.[45] Gamma alumina dissolves much more rapidly than alpha. The dissolution rate of alpha alumina is about two orders of magnitude slower than that of TiO_2 and about five orders of magnitude slower than that of amorphous SiO_2, Table 1. The fact that alpha alumina dissolves much more slowly than either silica or titania is not consistent with the theory of ligand exchange rates since the Al^{+3} ion has a lower charge than either Ti^{+4} and Si^{+4} while they all have about the same ionic radius. On the other hand, a correlation between the local "crystal" structures of these various materials can be made. Both titania and silica have more open crystal structures than do alumina. The SiO_4^{4-} tetrahedra in silica

Table 1

Dissolution Rates of Oxides in HF-HCl Solutions

Oxide	Rate (moles/cm^2/s)
TiO_2, rutile	3.4×10^{-11}
SiO_2, amorphous	2.4×10^{-8}
Al_2O_3, gamma	2.8×10^{-11}
Al_2O_3, alpha	3.0×10^{-13}

share corners, the TiO_6^{8-} octahedra in rutile share edges and the AlO_6^{9-} octahedra in alpha alumina form an interpenetrating lattice. Therefore, these dissolution rates correlate more with the structure surrounding the immediate coordinating ligands than they do with cation charge only. Gamma alumina has a defect spinel structure with some of the aluminum ions in tetrahedral coordination.[46] These tetrahedra share only corners with the remaining octahedrally-coordinated aluminum ions. As a result, gamma has a more open structure than alpha and would be expected to dissolve more rapidly. The proposed effect of local structure on dissolution rate is consistent with anisotropic dissolution of SiO_2 as well. However, in this case it is less clear what the differences in local structure are on different crystal faces that would lead to a difference in dissolution rates.

CONCLUSIONS

Ceramics are inherently corrosion-resistant materials. However, they can and do corrode in water and water solutions. Many principles that should be applicable to the corrosion and leaching of ceramics can be identified. However, because of paucity of experimental data, the most important factors controlling aqueous corrosion remain obscure. As a result, a complete predictive capability for the design of a corrosion resistant ceramic for waste encapsulation does not exist.

ACKNOWLEDGEMENTS

The aqueous corrosion experimental work reported on was carried out by Eric Bright, Da-Tung Liang, and Carole Markhoff, all former graduate students at the Ohio State University, Columbus, Ohio. The research has been supported by the Division of Materials Research of the National Science Foundation.

REFERENCES

1. Fitch, L. D.; Burdick, V. L. Water Corrosion of $YBa_2Cu_3O_{7-x}$ Superconductors. J. Am. Ceram. Soc.; 1989; 72(10): 2020-23.

2. Lilley, E. Review of Low Temperature Degradation in Y-TZPs. Corrosion and Corrosive Degradation of Ceramics: R. E. Tressler and M. McNallan, eds. [Ceramic transactions, Vol. 10]. Columbus, OH; The Am. Ceramic Society: 1990; 387-410.

3. Anderson, D. A. [and others]. Surface Chemistry Effects on Ceramic Processing of $BaTiO_3$ Powder. in: Messing, G. L.; Fuller, E. R., Jr.; Hausner, H., eds. Ceramic Powder Science II. Columbus: Am. Ceramic Soc.; 1988: 485-92. (Ceramic Transactions; v. 1).

4. Clark, David E. and Zoitos, Bruce K., eds. Corrosion of Glass, Ceramics

and Ceramic Superconductors. Noyes Publications, Park Ridge, NJ. 1992.

5. D. W. Readey and C. R. Cooley, "Ceramic and Glass Radioactive Waste Forms," U. S. Energy Research and Development Administration Report No. CONF-770102, 1977.

6. D. E. Clark and B. K. Zoitos, "Corrosion of Glass, Ceramics and Ceramic Superconductors," (Noyes Publications, Park Ridge, NJ), 1992.

7. Diggle, J. W. Dissolution of Oxide Phases, pp. 281-306 in Oxides and Oxide Films, v. 2, J. W. Diggle, ed. (Dekker, New York), 1973.

8. Burgess, J. Ions in Solution:Basic Principles of Chemical Interactions. New York: Wiley; 1988: 15.

9. Pourbaix, M. Atlas of Electrochemical Equilibria in Aqueous Solutions. New York: Pergamon; 1966.

10. Gaskell, D. R. Introduction to Metallurgical Thermodynamics. 2nd ed. New York: Hemisphere Publ.; 1981: 542-55.

11. Latimer, W. M., The Oxidation States of the elements and Their Potentials in Aqueous Solutions, (Prentice Hall, New York), 1952.

12. Bard, A. J.; Parsons, R.; Jordon, J. Standard Potentials in Aqueous Solution. New York: Dekker; 1985.

13. Browne, B. A. and Driscoll, C. T. Soluble Aluminum Silicates: Stoichiometry, Stability, and Implications for Environmental Geochemistry. Science 256 1667-70, 19 June 1992.

14. Baes, C. F., Jr, and Mesmer, R. E., The Hydrolysis of Cations, (Wiley, New York), 1976.

15. Valverde, N.;Wagner, C. Considerations on the Kinetics and the Mechanism of the Dissolution of Metal Oxides in Acidic Solution. Ber. Bunsen. Gesellschaft; 1976; 80: 330-333.

16. Burgess, J. Ions in Solution:Basic Principles of Chemical Interactions. New York: Wiley; 1988: 111-169.

17. Huheey, J. E., Inorganic Chemistry, 2nd ed., (Harper and Row, New York), 489-510 (1978).

18. Morrison, S. R. Electrochemistry at Semiconductor and Oxidized Metal Electrodes. New York: Plenum; 1980.

19. Butler, M. A.; Ginley, D. S. Review: Principles of Photochemical Solar

Energy Conversion. J. Maters. Sci.; 1980; 15: 1.

20. Vermilyea, D. A. The Dissolution of Ionic Compounds. J. Electrochem. Soc.; 1966; 113: 1067.

21. Osseo-Asare, K. Interfacial Phenomena in Leaching Systems. in: Bautista, R. G. Hydrometallurgical Process Fundamentals. New York: Plenum; 1984: 27-268.

22. Adamson, A. W. Physical Chemistry of Surfaces. 4th ed. New York: Wiley; 1982: 626.

23. Adamson, Arthur W. Physical Chemistry of Surfaces, 4th ed. Wiley, NY. p. 392. 1982.

24. Bockris, J. O'M., and Reddy, A. K. N., Modern Electrochemistry, Vol. 2, (Plenum, New York), 1970.

25. Warren, I. H.;Devuyst, E. Leaching of Metal Oxides. in: Evans, D. J. I.;Shoemaker, R. S., eds. International Symposium on Hydrometallurgy; 1973. New York: AIME; 1973: 229-263.

26. Frenier, W. W.; Growcock, F. B. Mechanism of Iron Oxide Dissolution-A Review of Recent Literature. Corrosion; 1984; 40(12): 663-668.

27. Morrison, S. R. The Chemical Physics of Surfaces. New York: Plenum; 1977: 42.

28. Wadsworth, M. E. and Miller, J. D., Hydrometallurgical Processes, pp. 133-241 in Rate Processes in Extractive Metallurgy, Sohn, H. Y. and Wadsworth, M. E., eds., (Plenum, New York), 1979.

29. West, J. M. Basic Corrosion and Oxidation. 2nd ed. Chichester, England: Ellis Horwood Ltd.; 1986: 79-94.

30. Cooper, A. R., Jr.; Kingery, W. D. Kinetics of Solution in High Viscosity Liquids. J. Phys. Chem.; 1962; 66: 665.

31. Vermilyea, D. A., The Dissolution of MgO and $Mg(OH)_2$ in Aqueous Solutions, J. Electrochem. Soc. 116 1180 (1969).

32. Macdonald, D. D.; Owen, D. The Dissolution of Magnesium Oxide in Dilute Sulfuric Acid. Canad. J. Chem.; 1971; 49: 3375.

33. Liang, Da-Tung, and Readey, Dennis W., Dissolution Kinetics of Crystalline and Amorphous Silica in Hydrofluoric-Hydrochloric Acid Mixtures, J. Am. Ceram. Soc. 70 [8] 570-577 (1987).

34. Casey, W. H.; Carr, M. J.; Graham, R. A. Crystal Defects and the Dissolution Kinetics of Rutile. Geochimica et Cosmochimica Acta; 1989; 52: 1545-1556.

35. Casey, W. H. [and others]. Aqueous Surface Chemistry and Corrosion of Minerals. MRS Bulletin. pp. 23-29, May 1992.

36. Lu, Z.; Muir, D. M. Dissolution of metal ferrites and iron oxides by hydrochloric acid under oxidizing and reducing conditions. Hydrometallurgy; 1988; 21(1): 9-21.

37. Valverde, N. Investigations on the Rate of Dissolution of Ternary Oxide Systems in Acidic Solutions. Ber. Bunsen Gesellschaft; 1977; 81: 380-384.

38. Bourcier, W. L., [and others]. A Kinetic Model for Borosilicate Glass Dissolution Based on the Dissolution Affinity of a Surface Alteration Layer. Mat. Res. Soc. Symp. Proc. Vol. 176. 209-216. 1990.

39. Bunker, B. C., Arnold, G. W., Beauchamp, E. K., and Day, D. E. Mechanisms for Alkali Leaching in Mixed-Na-K Silicate Glasses. J. Non-Cryst. Solids 58 295-322 (1983).

40. Bourcier, W. L. Overview of Chemical Modeling of Nuclear Waste Glass Dissolution. Mat. Res. Soc. Symp. Proc. Vol. 212 . 3-18; 1991.

41. Bunker, B. C. Waste Glass Leaching: Chemistry and Kinetics. Mat. Res. Soc. Symp. Proc. Vol. 84. 493-507; 1987.

42. Bates, J. K. [and others]. Colloid Formation During Waste Form Reaction: Implications for Nuclear Waste Disposal. Science 256 649-651, 1 May, 1992.

43. Bright, E.; Readey, D. W. Dissolution Kinetics of TiO_2 in HF-HCl Solutions. J. Am. Ceram. Soc.; 1987; 70(12): 900-906.

44. Bright, E. Dissolution Kinetics of TiO_2 and $SrTiO_3$ [M.S. Thesis]. Columbus, Ohio: The Ohio State University; 1982.

45. Markhoff, C. J., Dissolution of Alumina in HF-HCl Solutions, M. S. Thesis, The Ohio State University, Columbus, Ohio, 1986.

46. Evans, R. C. An Introduction to Crystal Chemistry. Cambridge: Cambridge University Press; 1964: 174.

MERCURY STABILIZATION IN CHEMICALLY BONDED PHOSPHATE CERAMICS

Arun S. Wagh, Seung-Young Jeong, and Dileep Singh

Energy Technology Division
Argonne National Laboratory
9700 S. Cass Avenue
Argonne, IL 60439

Abstract

We have investigated mercury stabilization in chemically bonded phosphate ceramic (CBPC) using four surrogate waste streams that represent U.S. Department of Energy (DOE) ash, soil, and two secondary waste streams resulting from the destruction of DOE's high-organic wastes by the DETOX^{SM} Wet Oxidation Process. Hg content in the waste streams was 0.1 to 0.5 wt.% (added as soluble salts). Sulfidation of Hg and its concurrent stabilization in the CBPC matrix yielded highly nonleachable waste forms. The Toxicity Characteristic Leaching Procedure showed that leaching levels were well below the U.S. Environmental Protection Agency's regulatory limits. The American Nuclear Society's ANS 16.1 immersion test also gave very high leaching indices, indicating excellent retention of the contaminants. In particular, leaching levels of Hg in the ash waste form were below the measurement detection limit in neutral and alkaline water, negligibly low but measurable in the first 72 h of leaching in acid water, and below the detection limit after that. These studies indicate that the waste forms are stable in a wide range of chemical environments during storage.

Introduction

Technology development for nonthermal stabilization and solidification of low-level mixed wastes containing volatile contaminants has been a major effort in the U.S. Department of Energy (DOE) EM-50 program. Among the hazardous contaminants that must be immobilized, mercury (Hg) in DOE mixed wastes has a high priority. This is because many Hg-bearing DOE wastes are nonaqueous and partially stabilized sludges, adsorbed liquids, and contaminated soils. Hg from these wastes is not easily accessible to leaching agents or thermal desorption, but is leachable in excess of the prescribed limits. Therefore, a suitable technology is needed to stabilize Hg from these wastes. In our study, we have explored a room-temperature-setting phosphate bonded ceramic (CBPC) process to address this need.

CBPCs are formed by reaction of MgO and KH_2PO_4 in solution [1-2]; solid or liquid waste streams are added during mixing of the two. Once these components are thoroughly mixed, the slurry is allowed to set. Setting occurs in approximately 2 h and a ceramic waste form of high strength and low open porosity is formed. The contaminants react with the acid solution to form their respective phosphate or hydrophosphate salts that are chemically insoluble in groundwater. They are further encapsulated physically in the dense ceramic matrix. Using this method, we have demonstrated the superior stabilization of several contaminants [2].

Chemical Immobilization of Hg

Our preliminary investigations on the conversion of Hg compounds in the presence of phosphoric acid solution showed that the major product is $Hg_3(PO_4)_2$. Because the reaction in the CBPC is in an aqueous environment, it is likely that hydrophosphates such as $HgHPO_4$ may also form. Table 1 indicates that the solubility of these phosphates is very low [3] and hence Hg from the waste form will not leach out easily into the groundwater. To test this, we fabricated CBPC waste forms of ash and ferric oxide wastes and tested the leaching of Hg by using the U.S. Environmental Protection Agency (EPA) Toxicity Characteristic Leaching Procedure, or TCLP [4]. These initial tests showed that leaching levels had declined from 40 and 138 mg/L in untreated ash and ferric oxide, respectively, to 7.7 and 51 mg/L in the treated wastes. The actual regulatory limits for nonwastewater retort and other are 0.2 and 0.025 mg/L, respectively. This implies that while the CBPC treatment significantly reduces Hg leaching, it may not be sufficient for treating a wide variety of wastes, particularly those characteristic wastes that require retorting. This means that an additional stabilization mechanism is needed to immobilize Hg.

Table 1. Solubilities of mercuric sulfide and mercury phosphates [3]

Species	Solubility Product Constant (K_{sp})	Solubility (mol/L)
HgS	2.0×10^{-49}	4.5×10^{-25}
$Hg_3(PO_4)_2$	7.9×10^{-46}	1.4×10^{-8}
$HgHPO_4$	7.9×10^{-14}	2.8×10^{-7}

This additional stabilization is provided by treating Hg-containing wastes with either hydrogen sulfide or one of the soluble alkali sulfides, the most common being Na_2S and NaHS [5]. This treatment converts Hg compounds into Hg sulfide, which has the lowest solubility shown in Table 1. Although a number of such sulfide stabilizers are available, because CBPC binder is a potassium-based material, we used K_2S for stabilization in our system.

K_2S powder formed an integral part of our process and was mixed with MgO and KH_2PO_4 powders to form one binder powder. It was added in a near-stoichiometric amount so that all of the Hg would be converted to HgS. Thus, no pretreatment step was needed. It has been reported that in the conventional cement stabilization process, addition of excess amount of sulfide affects the setting of cement [6]. This is partly because cement stabilization occurs at the high pH of ≈ 11. Therefore, acidic sulfides will affect stabilization at least locally in the cement matrix. On the other hand, because the CBPC process is based on an acid-base reaction, it is not sensitive to alteration of pH caused by addition of sulfides. The pH of the reaction slurry increases from its initial value of 4 to a more neutral 8 when the slurry sets, and chemical immobilization occurs in acidic to slightly above the neutral pH range. This more neutral environment, therefore, is very favorable for long-term storage of the chemically immobilized sulfide of Hg.

Treatability Study on Surrogate Waste Streams

Table 2 lists the four surrogate waste streams that were part of this treatability study; they represent either actual waste streams in storage at DOE facilities, or secondary waste streams that may be generated during destruction of organics. For example, the DOE complex has stored several ash waste streams that were generated by destruction of combustibles. The surrogate ash waste in Table 2 represents these waste streams [7]. On the other hand, secondary waste streams will be generated during destruction of

Table 2. Surrogate wastes and their formulations

Waste Identification	Composition (wt.%)		Contaminants (wt.%)
DOE ash waste	Activated carbon	5	$HgCl_2$ added to bring Hg level to 0.5.
	Vermiculite	20	
	Class F fly ash	40	
	Coal bottom ash	33	
Delphi DETOX			
Oxide waste	Fe_2O_3	93.6	In both waste streams, $HgCl_2$,Ce_2O_3, and $Pb(NO_3)_2$ were 0.5 each.
	$FeCl_2$	4.9	
Phosphate waste	$FePO_4$	98.5	
Soil	Topsoil from Argonne grounds		$HgCl_2$ added to bring Hg level to 0.1. Original waste had 2.7 ppm of Hg.

organics in certain DOE waste streams by the DETOX^{SM} wet oxidation process developed by Delphi Research [8]. These secondary waste streams contain a mixture of ferric oxide and ferric chloride or ferric phosphate as the major components; the second and third waste streams in the table represent these mixtures. The soil composition, i.e., the last waste stream given in Table 2, represents waste from Argonne's inventory that is destined for treatment according to Argonne's site treatment plan.

To prepare the surrogates, components were mixed thoroughly for 24 h on a vibratory shaker. The resulting mixture was then added to the binder mixture.

To prepare the reaction slurry, the powder mixtures were added to the stoichiometric or slightly higher amount of water and mixed thoroughly in a

Hobart tabletop mixer for 30 min. The resulting slurry was a slightly viscous liquid that poured easily. Once poured in a mold, it set in ≈2 h into a hard and dense ceramic waste form. Using cylindrical polyethylene containers, we made samples of ≈100 g.

The samples were stored for 3 weeks to ensure good curing. Each sample was then crushed into a powder with particle sizes of ≈0.5 mm or less and then subjected to the TCLP test. We also performed TCLP tests on the surrogate wastes. The results of these tests are given in Table 3.

Table 3. TCLP results on wastes and waste forms

Waste	Waste (mg/L)	Waste forms (mg/L)
Delphi DETOX		
Iron oxide	138	<0.00002
Iron phosphate	189	0.01
DOE ash waste	40	0.00085
Soil	2.27	0.00015
EPA limits	0.2	0.2 (noncharacteristic wastes) 0.025 (characteristic wastes)

The TCLP results on the wastes themselves show that the leaching levels are well above the regulatory requirement limits and fail the test. The leaching levels of Hg from the waste forms, on the other hand, are well below the limits and at least one order of magnitude below the Universal Treatment Standard of the Environmental Protection Agency. This implies that the process can be used to treat even characteristic wastes for which the UTS limits are applicable. Considering that the original Hg levels in the surrogate waste were very high and much above 260 ppm that is the limit for waste destined for stabilization, immobilization of Hg in the phosphate ceramic waste forms, coupled with sulfide immobilization, has been very superior.

Long-Term Leaching Behavior

The TCLP test only qualifies the waste form for land disposal. However, to ensure that the waste will form retain the contaminants within the various chemical environments of a landfill (indicated by different pH levels), it is necessary to demonstrate that the contaminants are retained in the waste form in a long-term leaching test. This was done by subjecting the samples to the American Nuclear Society's ANS 16.1 test [9], which allows us to study the leaching behavior in an aqueous environment. Neutral water is used as the leachant, and monolithic samples of the waste forms are immersed in the water for 90 days. Samples of the water are collected and tested for chemical constituents at prescribed periods, and contaminant levels in the leachate water are determined for those periods. The resulting data are then used to calculate the diffusion constant of the contaminant in the waste form which is then expressed as the leaching index that is the negative natural logarithm of the ratio of the diffusion constant.

We used this test to assess the waste forms for retention of the contaminants over 90 days. Samples of ≈20 g were used. The study was done for all four waste streams with neutral water. For ash waste, the leaching levels were consistently below the measuring instrument's detection limit of 0.025 μg, and thus we could not calculate the leaching index. For the rest of the waste forms, the data are given in Table 4.

Table 4. Leaching indices obtained in 90-day immersion tests

Waste form	Iron Oxide	Iron Phosphate	Soil
Leaching index	16.34	16.33	16.42

The leaching indices in Table 4 are much higher than the value of 6 found in cement systems [7]. This implies that Hg diffusion is lower than that in cement systems by 10 orders of magnitude. This extremely superior retention is due to the very superior chemical immobilization of Hg and its microencapsulation in the dense phosphate ceramic matrix.

Because the ash waste form showed the best leach resistance, we studied its leaching performance in both acidic and alkaline environments. The pH of the acidic solution was 3.5 and was obtained by adding acetic acid to the leachate water. The alkaline solution was that of an NaOH solution with a pH of 11. This study at low and high pH simulates extreme conditions in landfills or storage areas. In most cases, leaching levels were below the detection limits and hence the leaching indices could not be calculated. For this reason, we have presented in Table 5 the actual amount leached out during each measurement period.

Table 5. Leaching levels (μg) of Hg from ash waste forms in acidic and alkaline water

	Time (h)					
Hg Leached	2	7	24	48	72	96-2136
In acidic water	0.032	0.025	0.045	0.04	0.045	<0.025
In alkaline water	--- <0.025 at all time intervals ---					

The data in Table 5 show that, just as in the neutral aqueous environment, the leaching levels are undetectable in the alkaline environment. They are also extremely low and very close to the detection limit even in the acidic environment for the first 72 h of the measurement. After that, the leaching levels fall below the detection limit. This suggests that the waste form is very stable and can withstand a range of chemical environments from acidic to alkaline.

Discussion

Treatment of Hg-bearing mixed wastes is a difficult problem. This is because Hg in these wastes is not easily accessible to leaching agents or thermal desorption, and its successful removal has not been demonstrated. Therefore, stabilization of Hg becomes an important issue for DOE wastes. Coupled with this, many Hg-containing wastes are characteristic wastes for which the treatment standard requires that the TCLP limit be as low as 0.025 mg/L; this demands a superior stabilization technology. The data presented here indicate that CBPC technology may fulfill this need.

Because CBPC and Portland cement are both nonthermal technologies and use similar process steps, a comparison of the two is inevitable. The CBPC technology is based on an acid-base reaction, in which acidic to alkaline wastes can be incorporated. On the other hand, cement stabilization occurs only in an alkaline environment and the pH of the cement is ≈11 at the onset of setting. This limits its applications in waste management. Hg stabilization as HgS, although superior, may affect setting of the cement at least locally [6], but due to the acid-base nature of the CBPC binders, adding a component such as HgS does not affect the setting of

CBPC binders. In addition, waste loading in cement is comparatively low. In the CBPC matrix waste loading, can range from 50 wt.% for soil to 70 wt.% for ash, while a similar range for cement is 25 to 40 wt.%. The cost of the CBPC powders is slightly higher than that of cement, but the higher waste loading and shorter curing time offset these costs.

Acknowledgments

Work was supported by the U.S. Department of Energy, Office of Technology Development, as a part of the Mixed Waste Focus Area, under Contract W-31-109-Eng-38, and Delphi Research, Inc., of Albuquerque, NM.

References

1. A. S. Wagh, S. Y. Jeong, D. Singh, R. Strain, H. No, and J. Wescott, "Stabilization of Contaminated Soil and Wastewater with Chemically Bonded Phosphate Ceramics," Proc. Waste Management Annual Meeting, WM '97, Tucson, AZ (1997)

2. S. Jeong, A. S. Wagh, and D. Singh, "Chemically Bonded Phosphate Ceramics for Stabilizing Low-Level Radioactive Wastes," Amer. Ceram. Soc. Annual Meeting, Indianapolis (1996).

3. J. C. Conner, "Chemical Fixation and Solidification of Hazardous Wastes," Van Nostrand Reinhold (1990).

4. Environmental Protection Agency Method 1311, Toxicity Characteristic Leaching Procedure (TCLP), March 15, 1992, Revision II, pp. 138-139.

5. K. F. Cherry, "Plating Waste Treatment," Ann Arbor Science, Ann Arbor, MI (1982).

6. C. H. Mattus, and T. M. Gilliam, "A Literature Review of Mixed Waste Components: Sensitivites and Effects Upon Solidification/Stabilization in Cement-Based Matrices," Report no. ORNL/TM-12656 (1994) 114.

7. J. Mayberry et al., "Technical Area Status Report for Low-Level Mixed Waste Final Waste Forms," Vol. I, DOE/MWIP-3 (1992).

8. P. M. Dhooge, "Bench-Scale Operation of the DETOX Wet Oxidation Process for Mixed Wastes," Proc. Waste Management 93, ed. R. G. Post, Vol. 2 (1993) 1399-1402.

9. American Nuclear Society, "American National Standard Measurement of the Leachability in Solidified Low-Level Radioactive Wastes by a Short Term Test Procedure," Method ANSI/ANS 16.1-1986, American Nuclear Society, La Grange Park, IL 1986.

ENVIRONMENTAL ISSUES IN CERAMIC MANUFACTURING
1997 Annual Meeting
Cincinnati, OH

Environmental Regulations and the Ceramic Industry

THE NORM FACTOR: RADIONUCLIDE REGULATIONS AND THE CERAMICS INDUSTRY

Charles T. Simmons
Kilpatrick Stockton LLP
700 13th Street, N.W.
Suite 800
Washington, DC 20005

INTRODUCTION

Historically, the legal and regulatory implications of trace concentrations of naturally radioactive elements in minerals and mineral products used in the ceramics industry (e.g., zircon, zirconia, bauxite, and alumina) have not been viewed as a significant concern by producers and industrial consumers. While it is generally recognized that most minerals and many common substances contain low -- but detectable -- concentrations of natural radionuclides (i.e., uranium-238, thorium-232, and their respective decay elements, radium-226 and radium-228, respectively), the primary focus has been on the U.S. Nuclear Regulatory Commission's ("NRC's") licensing threshold for "source material." Under NRC regulations, materials are exempt from NRC classification as licensable "source material" provided the uranium and thorium content does not exceed 0.05 percent by weight.* Because typical mineral products used in the ceramics industry contain less than the source material threshold of uranium and thorium, NRC regulations have not been a concern for most suppliers and consumers, unless the uranium and thorium content becomes elevated above source material levels by processing. Maintaining regulatory compliance with NRC's concentration-based standard depends on accurate analytical data regarding the uranium and thorium assay, and is therefore relatively straightforward.

* 10 C.F.R. Part 40.

STATEMENT OF THE PROBLEM

Regulation of "hazardous substances" is based on an actual or perceived threat to human health or the environment. Regulatory programs in the 1970s and 1980s focused initially on obvious sources of environmental contamination such as smokestacks, wastewater discharge pipes, and landfills. Now, environmental regulation is predicated increasingly on "exposure," "dose," and "risk." Thus, the regulatory landscape for materials containing very low levels of radioactivity is changing.

The list of regulatory targets has been expanded to include substances that are deemed "hazardous" at very low concentrations -- sometimes below instrument detection limits -- because sophisticated mathematical models and conservative "exposure" and "dose" assumptions lead regulators to conclude that such substances may pose an unacceptable "risk." Federal and State agencies recently targeted diffuse, low level naturally occurring radioactive materials ("NORM"), without clear evidence that these materials pose an incremental health risk.

Stringent radiation regulations are predicated on the theory that the risk of cancer from radiation exposure increases linearly with no threshold dose. This theory employs extrapolation to conclude that very low doses of radiation are harmful based on the effects of very high instantaneous doses of radiation received by the atom bomb survivors. The "linear no-threshold" ("LNT") theory is controversial and its scientific validity has been questioned by some national and international authorities on health physics.

During the last several years, regulation of radioactive substances has been the subject of increased attention by federal authorities in the United States. In May 1994, the U.S. Environmental Protection Agency ("EPA") published draft regulations for cleanup of radioactive contamination that would establish an exposure-based standard and require cleanup to very low levels (15 millirem per year above background).* In August, 1994 the NRC proposed decommissioning standards that would apply an identical 15 millirem per year cleanup standard to all NRC-licensed facilities. Late in 1994, legislative efforts were initiated by Congress to revise the Comprehensive Environmental Response, Compensation and Liability Act ("CERCLA" or "Superfund") to expressly classify all radionuclides as "chemical carcinogens." The one-in-a-million-risk standard for carcinogens would have applied to all radionuclides under that approach, including radionuclides in

---

* A millirem ("mrem") is a measurement of radiation dose and equals 1/1000 rem.

common raw materials (e.g., zircon, zirconia, baddeleyite). For cleanup intended to achieve a one-in-a-million cancer risk level, substances containing radioactivity would have to be removed or diluted to *less* than background radiation in many locations. Such levels are well below the limits of accurate quantitation of field survey instrumentation. Finally, in December 1994, EPA proposed revised federal guidelines for protection of the general public against ionizing radiation that, while nominally adopting the internationally accepted public protection standard (100 millirems per year above background), also recommended limiting public exposures from specific sources of radiation to *only a fraction* of that standard.

The trend towards increasingly stringent control over radioactive substances is also apparent at the State level. Public health and environmental authorities in many States have either adopted or proposed regulation of substances containing NORM. The Conference of Radiation Control Program Directors ("CRCPD") has proposed "model NORM rules" that are intended to serve as the basis for State regulation of "technologically enhanced NORM" or "TENORM." For jurisdictional reasons arising from delegation of authority over source material under the Atomic Energy Act of 1954*, only NRC or NRC "Agreement States" have authority to regulate uranium and thorium. Accordingly, most State NORM regulations are based on the radium (e.g., Ra-226 and Ra-228) isotopic content of a material.

In sum, the trend in regulation of radioactive substances in the United States is shifting towards targeting materials at lower radioactivity levels and seeking to lower permissible exposures to workers and the general public. Federal authorities have asserted that stringent radiation and exposure standards are justified for adequate remediation of numerous high level radiation sites left over from Cold War government defense activities. The reality of the situation is that low numerical cleanup standards proposed by EPA for defense facility cleanup, once established by law, are likely to be applied to <u>any</u> location containing low specific activity materials, even disposal sites for industrial materials containing detectable levels of radionuclides.

The essence of the problem is that U.S. government agencies have not established an adequate technical basis for applying stringent cleanup or exposure standards to <u>very low level natural radioactive materials</u>. Very stringent regulatory standards are the result of extremely conservative radiation exposure pathway models, unrealistic exposure scenarios (<u>e.g.</u>, continuous exposure, influence of multiple radiation sites, no landfill cover), and outdated assumptions regarding the relationship between radiation and adverse health effects. Moreover, the economic

---

* 42 U.S.C.A. Section 2011, *et seq.*

consequences of restricting useful materials containing NORM or mandating cleanup of all facilities containing high volumes of low level NORM are staggering.

While stringent regulation of low level radioactivity could adversely affect those segments of the ceramics industry that use NORM-containing materials, federal proposals and many State regulatory proposals have not yet been finalized. Federal rules remain in the conceptual stage and are subject to modification. Given the current Congressional emphasis on risk-assessment and economic ("cost-benefit") justification of new regulations, the opportunity presently exists to achieve legislative solutions to NORM disposal and cleanup problems without imposing undue burdens on the regulated community. Regulators in many States are aware that significant adverse economic consequences for industry will follow over-regulating beneficial materials that do not pose significant health risk. Absent accurate data and appropriate education of regulators on the actual radiation exposure risks presented by minerals and mineral products, however, regulatory decisions will likely be based on arbitrary "worst case" assumptions.

THE SOLUTION TO THE PROBLEM

Because of the uncertain status of radiation regulation in the United States and the potential for adverse impacts on many industries, members of the zircon/zirconia community have joined together in a collective effort to address regulatory issues at federal and state levels. Following its first meeting in July 1994, the Zirconium Environmental Committee ("ZEC") has expanded to include over seventeen Members and Associate Members, and represents primary stakeholders in the zircon/zirconia industry as well as significant industrial consumers. The Committee has established the following objectives:

- Maintain awareness of proposed regulations and legislation that could adversely affect the interests of the zircon industry. Provide a communication network to alert Committee Members of regulatory developments.
- Provide appropriate comments on regulatory proposals during the administrative "notice and comment" period required for rulemaking. Educate regulators through comments on the adverse implications of regulatory proposals; seek more reasonable regulations based on science, not speculation.
- Implement a legislative strategy at the Congressional level to exempt NORM-containing materials from cleanup liability under Superfund.
- Develop a scientifically defensible generic risk assessment of the disposal of NORM-containing wastes (e.g., foundry and refractory

materials) in landfills that supports the exemption of these materials from State NORM regulations.

- Proactively pursue reasonable regulatory decisions that are risk- or dose-based, rather than arbitrary numerical limits on specific activity.

The purpose of the Zirconium Environmental Committee is not to seek to avoid appropriate regulation, but to ensure that regulatory controls imposed on diffuse NORM materials are reasonable and based on sound scientific principles. In sum, all industries that utilize NORM-containing materials must act proactively and collectively to educate federal and state regulators that stringent regulation of low level diffuse NORM is neither appropriate nor warranted.

OVERVIEW OF REGULATORY CONTROL OF NORM

The following discussion summarizes the development of legislative and regulatory initiatives at the federal and state levels that require close monitoring by the zircon industry.

Federal Legislative and Regulatory Initiatives

Superfund: The Comprehensive Environmental Response, Compensation and Liability Act ("CERCLA" or "Superfund") is a federal program that imposes strict, joint and several liability retroactively on persons responsible for the release or threatened release of "hazardous substances," as that term is defined under the Act. EPA's Superfund cleanup standards seek to achieve a human health risk level of 10^{-4} to10^{-6} (e.g., a risk probability ranging from one adverse effect in ten thousand to one chance in a million).

Legislation for Superfund reauthorization is virtually assured in this Congressional term. In 1994, both House and Senate versions of Superfund legislation proposed a single numerical risk-based cleanup standard for chemical carcinogens. Early discussion in the House made it clear that Congress did not intend to distinguish radionuclides from other "chemical carcinogens" subject to numerical risk-based cleanup standards. Thus, all radionuclides, including low level diffuse NORM, would have been subject to Superfund liability and cleanup standards.*

---

* There is judicial precedent for remediation of radionuclide contaminated sites under CERCLA. See: Amoco Oil Company v. Borden, Inc., 889 F. 2d 664 (5th Cir. 1989) (In a private Superfund action seeking a judgment of liability and recovery of cleanup costs, the Court held that radionuclides are "hazardous substances" under

The application of federal Superfund cleanup standards to NORM-containing materials could require cleanup to the 10^{-4} to 10^{-6} Superfund risk level. Using an annual risk coefficient of 5 x 10^{-4} health effects per rem, the 10^{-6} lifetime risk value theoretically could yield a continuous lifetime radiation dose rate in the range of 0.01 to 0.1 mrem per year.* The practical consequences of this cleanup requirement would require NORM contaminated sites to be remediated to very low levels of radiation above background, which in reality are indistinguishable from background.** Legislative staff and regulatory proponents were not aware that such a cleanup requirement is neither achievable nor necessary.

The strategic problem in addressing Superfund legislation is that proposals for Superfund reform are often the product of very delicate, last minute compromises worked out between industrial, environmental and Administration interests to address the major problems with existing Superfund law (e.g., disputes over insurance for costs of cleanup, municipal liability, and community participation). The potential adverse effects to industrial users of NORM-containing materials are, in our opinion, the unintended result of targeting Department of Energy and Defense Department weapons facilities for Superfund cleanup. It does not appear that low-level NORM characteristic of zircon and other minerals is an intentional target. However, absent the appropriate language to clarify that Congress does not intend very low levels of NORM characteristic of useful materials to be addressed by Superfund, legislative changes could extend Superfund cleanup liability to all sites containing diffuse NORM.

EPA Draft Regulations for Radiation Site Cleanup: On May 18, 1994 EPA released a proposed regulation establishing radiation cleanup standards to the National Advisory Council for Environmental Policy and Technology (NACEPT).

CERCLA and a release of radionuclides, in any amount, triggers CERCLA cleanup liability.)

* The value of 5X10^{-4} effects per rem was developed by the Nuclear Regulatory Commission on the basis of findings made by the International Commission on Radiological Protection (ICRP). See 55 Fed. Reg. 27522 (July 3, 1990).

** Background radiation exposure in the United States is estimated to be 360 mrem per year. See: NRCP Report No. 93. See also: U.S. Department of Energy, Office of Environmental Restoration and Management Fact Sheet (November 1991).

EPA is proposing to establish a site cleanup standard of 15 mrem/yr above background for unrestricted use, and a groundwater cleanup standard of 4 mrem/yr. If remediation of a site to meet the 15 mrem standard for unrestricted use is not feasible, an alternative cleanup standard of 75 mrem/yr over background could be permitted for "commercial use." EPA has tentatively selected the Atomic Energy Act as statutory authority for implementing the draft regulations.

Significantly, EPA's position on radionuclide cleanup is more stringent than current nationally and internationally accepted standards for radiation protection. The National Council on Radiological Protection and Measurements ("NCRP") and International Commission on Radiological Protection ("ICRP") recommend a 100 mrem annual dose limit above background for the protection of the general public. In contrast to EPA's proposed cleanup standard, the NCRP expressly recommends that remedial action be undertaken when continuous exposures from natural sources are expected to exceed 500 mrem above background annually.* EPA, however, has not yet provided any scientifically defensible basis or realistic economic analysis for its recommended cleanup standard. Moreover, remediation of NORM containing soils to achieve EPA's recommended unrestricted standard (assuming that 15 mrem/yr over background can be reliably quantified) may cost trillions of dollars because very large volumes of NORM-containing material are involved nationwide.

On December 19, 1996, EPA informed the Office of Management and Budget ("OMB") that it was withdrawing the radiation site cleanup rule indefinitely. Prior review and approval by OMB is required before a proposed rulemaking is published in the *Federal Register*. EPA has assumed responsibility for developing residual radioactivity cleanup levels that will apply to all sites contaminated with radioactive materials, including facilities regulated by the NRC and the states. NRC, on the other hand, is developing its own site cleanup rule that would involve less restrictive cleanup standards than sought by EPA. EPA has expressly opposed NRC's less stringent cleanup standard of 25 to 30 mrem/yr, not including groundwater.

The broader implication of EPA's position on site cleanup rule is to establish a radiation cleanup standard that (1) may be used to assess Superfund site cleanups and (2) may serve as precedent for State regulatory programs. While it is possible that EPA could revive its site cleanup rulemaking, recent correspondence from EPA suggests that the Agency will attempt to persuade NRC to adopt stringent site

---

* See NCRP Report No. 116, Limitation of Exposure to Ionizing Radiation (March 1993) at page 50.

cleanup standards comparable to EPA's. NRC's Commissioners are scheduled to vote on the appropriate cleanup standard this month. Importantly, unlike other environmental rulemakings under the Clean Air or Clean Water Acts, EPA is not under any statutory mandate to engage in radiation rulemaking. Thus, EPA's impetus for rulemaking is entirely its own.

EPA Proposed Federal Guidance on Protection of the General Public from Radiation: On December 23, 1994 EPA published proposed Federal Radiation Protection Guidance for Exposure of the General Public (59 Fed. Reg. 66414). EPA asserted that its proposed guidance would result in a "five fold reduction in the maximum allowable risk of cancer" by reducing public exposure to sources of ionizing radiation to a fraction of 100 mrem/yr. EPA's proposal was criticized by national and international experts on health physics at a public hearing on technical and legal grounds. The Agency is considering the comments received and has not yet finalized this policy.

EPA NORM Study: In April, 1993, EPA published *Diffuse NORM Wastes - Waste Characterization and Preliminary Risk Assessment (Draft)* which attempted to define the extent of public exposure and risk from NORM-related industrial activities in the U.S. The EPA Science Advisory Board ("SAB") determined that, because the draft study contained many assumptions and insufficient data on activity levels and exposures, the study was not an appropriate basis for NORM regulation. EPA has recently undertaken a followup study to fill the data gaps in the 1993 report, with particular emphasis on NORM in the mineral and petroleum industries. EPA anticipates completing its followup study by late 1997.

Regulation of NORM by the States

Many States have already adopted, or are in the process of adopting, regulations governing the use and disposal of "NORM" materials based on the concentration of radium-226 and radium-228 (collectively, "radium"), which are the radioactive decay products of natural uranium (uranium-238) and thorium (thorium-232), respectively. Because radium is naturally associated with uranium found in materials used in the ceramics industry (e.g., zircon, zirconia), State NORM regulations restricting the use, transportation, and disposal of radium may adversely affect the ceramics community. Problems at the State level most often arise in the context of site inspections or waste disposal. A typical scenario triggering State restrictions on NORM disposal is encountered when a load of NORM-containing waste materials (e.g., foundry sand or ceramic shell from investment casting) is taken to a landfill for disposal. Many non-hazardous solid waste disposal facilities have recently installed very sensitive scintillation detectors because of concerns about unauthorized disposal of radioactive medical wastes or contaminated scrap metal. Detection of very low levels of radiation emanating from NORM-containing wastes has triggered rejection of waste shipments, without any consideration of the actual risk presented. For some zircon users, higher cost waste disposal at a specially licensed NORM disposal facility may be the only alternative.

At present, not all States regulate NORM, and existing regulations are inconsistent: some states (e.g., Arkansas) impose a stringent 5 picocurie per gram (5 pCi/g) radium restriction on the use and disposal of NORM-containing materials, while other States are less restrictive (e.g., Mississippi exempts materials containing up to 30 pCi/g radium-226, provided low radon emanation occurs; landfill disposal of wastes containing up to 50 pCi/g radium is permitted in Michigan; and zircon sands and products made from zircon sands are exempt from regulation as NORM in Ohio, provided that the NORM content is stated on the Material Safety Data Sheet) .

Conference of State Radiation Control Program Directors

Efforts are currently underway by the Conference of State Radiation Control Program Directors ("CRCPD") to harmonize State regulations and establish uniform control over NORM. The CRCPD, an affiliation of State radiation regulators, recently announced revisions to the Part N Suggested State Regulations ("SSRs") which address NORM. The SSRs are not legally binding regulations, but are "model rules" that will substantially influence a State's adoption of NORM regulations.

The CRCPD has proposed the following revisions to the NORM SSRs:

1. Uranium and thorium will be exempt because these radionuclides are regulated under the Atomic Energy Act;
2. Materials containing radium exceeding 5 pCi/g should be licensed;
3. Public exposure from technologically enhanced NORM ("TENORM") should be limited to a fraction of 100 mrem/yr.

Importantly, CRCPD is willing to consider state flexibility in exempting certain NORM uses, raw materials and disposal (or recycling) practices from regulation, provided such exemptions are technically supported

Hazard Communication under the Occupational Safety and Health Act (OSHA)

Regulations promulgated under the Occupational Safety and Health Act ("OSHA") require chemical manufacturers or importers to assess the hazards of chemicals which they produce or import. Under the Act, all employers must provide information to their employees about the hazardous chemicals to which they are exposed, by means of a hazard communication program, labels and other forms of warning, Material Safety Data Sheets ("MSDS"), and training. (See 29 C.F.R. §1910.1200.)

A survey of MSDS for NORM containing minerals and mineral products indicated that many, but not all, manufacturers have included appropriate information on potential exposure hazards for all hazardous substances present, including NORM and crystalline silica (silicon dioxide). In addition, the MSDS can alert the user that some states have adopted restrictive NORM disposal regulations. The MSDS may also contain a Proposition 65 statement, applicable in the State of California, that the product contains radionuclides and crystalline silica, which are substances known to the State of California to cause cancer.*

Because substantial fines and penalties can be imposed under OSHA for failure to comply with the Hazard Communication Standard, producers and distributors should review their product's MSDS to confirm that all OSHA communication requirements are met.

Regulation of Radioactive Substances in the European Union

The international nature of the ceramics industry mandates that producers and users of NORM-containing materials maintain an awareness of the current status of radiation regulations in a multitude of producer and user nations. For example, the trend towards increasingly stringent regulation of radioactive substances is not limited to the United States. In May, 1996 the European Commission published Council Directive No. 96/29/Euratom amending 80/836/Euratom and substantially reduces the current reporting exemption levels for materials containing natural uranium, radium, or thorium.** At present, Directive 80/836/Euratom exempts solid phase materials containing natural uranium and thorium from reporting requirements provided the radionuclide content does not

---

* California Proposition 65 requires a warning label for products that contain "substances known to the State of California to cause cancer or reproductive harm."

** COM94/298 dated 8.7.1994.

exceed 500 Bq/g; all other materials less than 100 Bq/g are exempt from reporting. The May 1996 Directive lowers the reporting exemption level to 10 Bq/g for radium-226 and 1 Bq/g for thorium-232(sec) and uranium-238(sec).* Such lower exemption levels will serve as a basis for regulatory control of workplace exposureto natural sources of radiation by Member States under Title VII of the Directive. Member States have until May, 2000 to adopt national legislation consistent with the terms of the Directive.

INTERNATIONAL ATOMIC ENERGY AGENCY (IAEA)

The International Atomic Energy Agency ("IAEA") will meet on May 6 through 9, 1997 in Vienna, Austria to discuss international regulation of naturally occurring radioactive materials. The IAEA will discuss whether NORM-containing materials that are currently exempt should be re-evaluated for possible regulation, including materials used in industrial processes and wastes from industrial and mining operations. The following topics are on IAEA's agenda:

- Definition and clarification of the terms: exclusion, exemption, clearance and authorized release;
- Problems concerning regulation of practices involving very low exposures or very low risk of exposure to radiation;
- Problems concerned with the regulation of wastes containing naturally occurring radionuclides;
- Agency review of the management of very low level wastes in its member countries
- Regulation of the metal recycling industry;
- Management of very low level radioactive wastes, including large volume long-lived wastes.

It is our understanding that a representative of the U.S. Environmental Protection Agency ("EPA") will be attending the meeting to present EPA's view of how NORM should be regulated. It would be reasonable to expect EPA's position to be very conservative. While the recommendations of the IAEA are not legally enforceable in themselves, they serve as the basis for national and international regulation of radioactive materials. In addition, many contracts for the sale of goods specify that shipment shall be consistent with IAEA recommended practices.

---

* The notation "(sec)" designates a radionuclide in "secular equilibrium" with all of the radioactive progeny in its decay chain.

CONCLUSION

The foregoing discussion is not intended to present an alarmist perspective of impending regulation of NORM. On the contrary, because federal and State regulatory initiatives are currently in a state of flux and amenable to change, the stakeholders in the ceramics industry have opportunities to ensure that regulation of NORM is appropriate and grounded on sound scientific principles reflecting actual risk. The key to successful development of reasonable regulations is involvement of affected parties before regulations become final. Because no individual company can manage to monitor all federal, State and international regulatory initiatives directed towards NORM, the most economically efficient approach for addressing such issues is through collective participation in a group. This is the approach taken by the Zirconium Environmental Committee.

ACCIDENTAL RELEASE PREVENTION

Peter R. Charrington, P. E., DEE
Sr. Program Director
Environmental Resources Management
855 Springdale Drive
Exton, PA 19341

ABSTRACT

Under Title III of the Clean Air Act Amendments of 1990 (CAAA), both the Occupational Health and Safety Administration (OSHA) and the Environmental Protection Agency (EPA) had to promulgate rules for facilities that handle regulated substances requiring them to prepare hazard assessments of facilities. OSHA is accomplishing this activity through its Process Safety Management (PSM) program, while EPA is addressing its requirements with the Risk Management Program Rule under Section 112 (r) of the CAAA. The final rule was signed on 24 May 1996 and published in the Federal Register on 20 June 1996. It is estimated that approximately 66,000 facilities will be effected by the EPA's Rule, some of which are already covered under the OSHA PSM regulations. Even though compliance with the 112 (r) requirements is 21 June 1999, the time to determine applicability and develop cost-effective compliance strategies is now. This paper will assist you in understanding the regulations and in developing an effective strategy to reduce the risk of accidental releases.

BACKGROUND

By the end of 1990, Congress had passed two significant pieces of legislation dealing with the prevention of accidents involving hazardous chemical substances. The OSHA and EPA regulations addressed many of the concerns raised by the environmental community and organized labor, concerns which were heightened by a series of catastrophic incidents in the chemical and petroleum manufacturing sectors in the preceding years. Following their passage into law, however, the twin programs evolved under quite different conditions. OSHA rulemaking proceeded relatively swiftly, and by February, 1992 a final rule 29CFR1910.19 was issued with a five-year compliance phase-in. Early audit results indicate widespread compliance, a fairly uniform approach to compliance programs in affected industries, and a broad consensus among affected industries, consultants and regulators regarding even the smaller details of program requirements for compliance.

EPA's Risk Management Program Rule 40CFR68, on the other hand, experienced a rocky start and seemed to be impeded by both external (legal challenges) and internal (resource) difficulties. The initial proposed list of regulated substances was initiated by Congress who mandated both its size (a minimum of 100 chemicals were to be listed for regulation) and, in part, content (EPA was explicitly required to "include chlorine, anhydrous ammonia, methyl chloride, ethylene oxide, vinyl chloride, methyl isocyanate, hydrogen cyanide, ammonia, hydrogen sulfide, toluene diisocyanate, phosgene, bromine, anhydrous hydrogen chloride, hydrogen fluoride, anhydrous sulfur dioxide, and sulfur trixoide" by the language of the law itself).

The relative success of OSHA's program in its early stages led some industry groups such as The Chemical Manufacturers Association (CMA) to press for a regulatory 'rationalization' of EPA's rule to prevent wasteful duplication of effort. EPA settled on a list comprised of 77 toxics (Table I) and 63 flammables (Table II), most with threshold quantities (TQs) above the OSHA Process Safety Management (PSM) TQ. Additionally, EPA clarified that although certain materials were not listed as a result of industry petition, owners and operators of processes containing hazardous materials are subject to the general duty clause of section 112(r)(l) "to identify hazards which may result from such releases using appropriate hazard assessment techniques ."

COMPARISON TO PSM

Following publication of EPA's list in early 1994, industry submitted comments critical of the Agency's failure to replicate the OSHA PSM regulations as requested by many affected industries.·The result of a combination of challenging, well-founded commentary with an unprecedented level of industry-EPA collaboration is seen in Table III, which compares the requirements of the final rule with OSHA's PSM rule.

In response to comments, EPA divided compliance requirements under the final rule into three categories called Program Level 1, 2 and 3. Figure I illustrates the process for determining the Program Level of a "covered process", and Table IV lists the principle requirements of each Program Level.

A major difference between the EPA and OSHA rules arises from the EPA position on exemptions. Basically, there are none under EPA's rule: that is, the nature of the business is not considered in determining compliance requirements. Only the presence of a process containing a regulated substance above its threshold quantity determines applicability. There are no exemptions under the EPA rule for government facilities, retail establishments, or for fuel-storage of dispensing processes for on-site use. To date, EPA has responded to criticisms of burdening small businesses by emphasizing the minimal requirements of Program Level 1 (for which they believe most PSM-exempted processes will qualify), and by publishing model risk management plans (i.e., ammonia refrigeration facilities) and soliciting cooperative support from affected industry associations and trade groups in developing additional standardized models for appropriate covered process (e.g., propane vehicle refueling).

STATUS OF THE RULE

The Risk Management Program Rule included a "stay of effectiveness" on the rule's applicability to gasoline for internal combustion engines, naturally occurring hydrocarbon mixtures at oil/gas production facilities, and explosives. Although final guidance on the required reporting format for Risk Management Plans (RMP) is still pending, no substantive change in the Rule's requirements is anticipated. In other words, now is the time to roll up your sleeves and begin working on compliance with EPA's Risk Management Program Rule.

WHAT SHOULD YOU DO NOW?

First, if your process contains a significant quantity of one or more regulated substances, and you were not required to comply with the OSHA PSM rule because of OSHA exemptions, you should review the EPA rule requirements. Review Tables I and II carefully, as there are differences between the OSHA and EPA lists and some EPA TQs are higher than OSHA's. If you determine that you have a "covered process", you will have a lot of work ahead of you — particularly if you fall into Program Level 2 or 3 for the EPA program.

Next, for each "covered process", compare your information to the requirements presented in Figure 1. If you operate a covered process which satisfies most but not all of the requirements for Program Level 1, you will have more time to address these areas if you identify them soon. Looking at the increase in requirements for compliance with Program Level 2 or 3 versus Program Level 1 (as shown in Table IV), facilities would be better served to investigate mitigative measures (passive or active) to reduce the program level requirements. While a wholesale reengineering of a process may not be practical in the time-frame available, it may be possible to effect changes which allow a "covered process" to qualify for Program Level 1. Note that administrative controls may be invoked to limit a process' inventory of a listed compound to an amount below the TQ. In such a case, the subject process could be promoted out of the Risk Management Program Rule altogether!

Finally, EPA anticipates that very few of the covered processes affected by the rule will be in Program Level 2. This reasoning is based on the similarities between the EPA and OSHA lists and the regulatory provision that PSM-covered processes which do not qualify for Program Level 1 are placed in Program Level 3. Therefore, it will be worthwhile to examine all covered processes to see if any of them belong in Program Level 2. Again, refer to Table IV and note how many of the Program elements in Level 2 and 3 are substantially identical to elements of OSHA PSM.

SUMMARY

In summary, we have seen that the differences between EPA's Risk Management Program and OSHA's PSM rules mean additional effort, even for facilities already in full compliance with PSM; and that the distinct requirements of the EPA rule place a premium on specialized risk assessment skills as well as detailed process and facility knowledge. The rule-making process is complete and no substantial changes are anticipated for the EPA rule under section 112(r).

Resources are available for cooperative development of RMP elements dealing with communication, emergency response and mitigation through local emergency planning committees (LEPCs) and trade associations. Recent experience with Title V operating permits indicates that it can expected that a shortage of quality services as consultants with relevant experience 'load up' with RMP-related assignments, particularly in the area of computer dispersion modeling for refined estimates of offsite consequences.

This all suggests that facilities which know they will be subject to this rule should be in the planning process today to assure compliance in less than three years. For facilities which are still unsure of their status under the rule, start today with the guidance above.

Table I ***Toxics***

Chemical Name	CAS No.	TQ (Lbs.)
Acrolein [2-Propenal]	107-02-8	5,000
Acrylonitrile [2- Propenenitrile]	107-13-1	20,000
Acrylyl chloride [2-Propenoyl chloride]	814-68-6	5,000
Allyl alcohol [2-Propen-l-ol]	107-18-61	15,000
Allylamine [2-Propen-l-amine]	107-11-9	10,000
Ammonia (anhydrous)	7664-41-7	10,000
Ammonia (conc. 20% or greater)	7664-41-7	20,000
Arsenous trichloride	7784-34-1	15,000
Arsine	7784-42-1	1,000
Boron trichloride [Borane, trichloro]	10294-34-5	5,000
Boron trifluroide [Borane, trifluoro]	7637-07-2	5,000
Boron trifluoride compound with methyl ether (1:1)[Boron, trifluoro[oxybis[metane]]-, T-4-	353-42-4	15,000
Bromine	7726-95-6	10,000
Carbon disulfide	75-15-0	20,000
Chlorine	7782-50-5	2,500
Chlorine dioxide [Chlorine oxide ($Cl0^2$)]	10049-04-4	1,000

Table I Toxics (con't)

Chemical Name	CAS No.	TQ(Lbs.)
Chloroform [Methane, trichloro-]	67-66-3	20,000
Chloromethyl ether [Methane, oxybis[chloro-]	542-88-1	1,000
Chloromethyl methyl ether	107-30-2	5,000
Crotonaldehyde [2-Butenal]	4170-30-3	20,000
Crotonaldehyde, (E)-[2-Butenal, (E)-]	123-73-9	20,000
Cyanogen chloride	506-77-4	10,000
Cyclohexylamine [Cyclohexanamine]	108-91-8	15,000
Diborane	19287-45-7	2,500
Dimethyldichlorosilane [Silane, dichlorodimethyl-]	75-78-5	5,000
1,1-Dimethylhydrazine [Hydrazine, 1,1-dimethyl-]	57-14-7	15,000
Epichlorohydrin [Oxirane, (chloromethyl)-]	106-89-8	20,000
Ethylenediamine [1,2-Ethanediamine]	107-15-3	20,000
Ethyleneimine [Aziridine]	151-56-4	10,000
Ethylene oxide [Oxirane]	75-21-8	10,000
Fluorine	7782-41-4	1,000
Formaldehyde (formalin)	50-00-0	15,000
Furan	110-00-9	5,000
Hydrazine	302-01-2	15,000
Hydrochloric acid (conc 30% or greater)	7647-01-0	15,000
Hydrocyanic acid	74-90-8	2,500
Hydrogen chloride (anhydrous) [Hydrochloric acid]	7647-01-0	5,000
Hydrogen fluoride/Hydrofluoric acid (conc 50% or greater) [Hydrofluoric acid]	7664-39-3	1,000
Hydrogen selenide	7783-07-5	500
Hydrogen sulfide	7783-06-4	10,000
Iron, pentacarbonyl- [Iron carbonyl (Fe(CO)5)]	13463-40-6	2,500
Isobutyronitrile [Propanenitrile, 2-methyl-]	78-82-0	20,000
Isopropyl chloroformate [Carbonochloridic acid, 1-methylethyl ester]	108-23-6	15,000
Methacrylonitrile [2-Propenenitrile, 2-methyl-]	126-98-7	10,000
Methyl chloride	74-87-3	10,000
Methyl chloroformate	79-22-1	5,000
Methyl hydrazine	60-34-4	15,000
Methyl isocyanate	624-83-9	10,000
Methyl mercaptan	74-93-1	10,000
Methyl thiocyanate [Thiocyanic acid, methyl ester]	556-64-9	20,000
Methyltrichlorosilane	75-79-6	5,000
Nickel carbonyl (nickle tetracarbonyl)	13463-39-3	1,000
Nitric acid (conc. 80% or greater)	7697-37-2	15,000
Nitric oxide [Nitrogen oxide (NO)]	10102-43-9	10,000

Table I Toxics (con't)

Chemical Name	CAS No.	TQ (Lbs.)
Oleum (Fuming Sulfuric acid) [Sulfuric acid, mixture with sulfur trioxide]	8014-95-7	10,000
Peracetic acid [Ethaneperoxoic acid]	79-21-0	10,000
Perchloromethylmercaptan [Methanesulfenyl chloride, trichloro-]	594-42-3	10,000
Phosgene [Carbonic dichloride]	75-44-5	500
Phosphine	7803-51-2	5,000
Phosphorus ocxychloride [Phosphoryl chloride]	10025-87-3	5,000
Phosphorus trichloride [Phosphorous trichloride]	7719-12-2	15,000
Piperidine	110-89-4	15,000
Propionitrile [Propanenitrile]	107-12-0	10,000
Propyl chloroformate [Carbonochloridic acid,propylester	109-61-5	15,000
Propyleneimine [Aziridine, 2-methyl-]	75-55-8	10,000
Propylene oxide [Oxirane, methyl-]	75-56-9	10,000
Sulfur dioxide (anhydrous)	7446-09-5	5,000
Sulfur tetrafluoride [Sulfur fluoride (SF^4), (T-4-)-]	7783-60-0	2,500
Sulfur trioxide	7446-11-9	10,000
Tetramethyllead [Plumbane, tetramethyl-]	75-74-1	10,000
Tetranitromethane [Methane, tetranitro-]	509-14-8	10,000
Titanium tetrachloride	7550-45-0	2,500
Toluene 2,4-diisocyanate	584-84-9	10,000
Toluene 2,6-diisocyanate [Benzene, 1,3-diisocyanato-2-methyl-]	91-08-7	10,000
Toluene diisocyanate (unspecified isomer) [Benzene, 1,3-diisocyanatomethyl-]	26471-62-5	10,000
Trimethylchlorosilane [Silane, chlorotrimethyl-]	75-77-4	10,000
Vinyl acetate monomer [Acetic acid ethenyl ester]	108-05-4	15,000

Table II Flammables

Chemical Name	CAS No.	TQ (Lbs.)
Acetaldehyde	75-07-0	10,000
Acetylene [Ethene, bromotrifluoro-]	74-86-2	10,000
1,3-Butadiene	106-99-0	10,000
Butane	106-97-8	10,000
1-Butene	106-98-9	10,000
2-Butene	107-01-7	10,000
Butene	25167-67-3	10,000
2-Butene-cis	590-18-1	10,000
2-Butene-trans [2-Butene, (E)]	624-64-6	10,000

Table II Flammables (con't)

Chemical Name	CAS No.	TQ (Lbs.)
Carbon oxysulfide [Carbon oxide sulfide (COS)]	463-58-1	10,000
Carbon monoxide [Chlorine oxide]	7791-21-1	10,000
2-Chloropropylene [1-Propene, 2-chloro-]	557-98-2	10,000
1-Chloropropylene [1-Propene, 1-chloro-]	590-21-6	10,000
Cyanogen [Ethanedinitrile]	460-19-5	10,000
Cyclopropane	75-19-4	10,000
Dichlorosilane [Silane, dichloro-]	4109-96-0	10,000
Difluoroethane [Ethane, 1,1-difluoro-]	75-37-6	10,000
Dimethylamine [Methanamine, N-methyl-]	124-40-3	10,000
2,2-Dimethylpropane [Propane, 2,2-dimethyl-]	463-82-1	10,000
Ethane	74-84-0	10,000
Ethyl acetylene [1-Butyne]	107-00-6	10,000
Ethylamine [Ethanamine]	75-04-7	10,000
Ethyl chloride [Ethane, chloro-]	75-00-3	10,000
Ethylene [Ethene]	74-85-1	10,000
Ethyl ether [Ethane, 1-1'-oxybis-]	60-29-7	10,000
Ethyl mercaptan [Ethanethiol]	75-08-1	10,000
Ethyl nitrite [Nitrous acid, ethyl ester]	109-95-5	10,000
Hydrogen	1333-74-0	10,000
Isobutane [Propane, 2-methyl]	75-28-5	10,000
Isopentane [Butane, 2-methyl-]	78-78-4	10,000
Isoprene [1,3-Butadinene, 2-methyl-]	78-79-5	10,000
Isopropylamine [2-Propanamine]	75-31-0	10,000
Isopropyl chloride [Propane, 2-chloro-]	75-29-6	10,000
Methane	74-82-8	10,000
Methylamine [Methanamine]	74-89-5	10,000
3-Methyl-1-butene	563-45-1	10,000
2-Methyl-1-butene	563-46-2	10,000
Methyl ether [Methane, oxybis-]	115-10-6	10,000
Methyl formate [Formic acid, methyl ester]	107-31-3	10,000
2-Methylpropene [1-Propene, 2-methyl-]	115-11-7	10,000
1,3-Pentadinene	504-60-9	10,000
Pentane	109-66-0	10,000
Pentene (1-)	109-67-1	10,000
Pentene, (E)-(2-)	646-04-8	10,000
Pentene, (Z)-(2-)	627-20-3	10,000
Propadiene [1,2-Propadiene]	463-49-0	10,000
Propane	74-98-6	10,000
Propylene [1-Propene]	115-07-1	10,000
Propyne [1-Propyne]	74-99-7	10,000
Silane	7803-62-5	10,000

Table II ***Flammables (con't)***

Chemical Name	CAS No.	TQ (Lbs.)
Tetrafluoroethylene [Ethene, tetrafluoro-]	116-14-3	10,000
Tetramethylsilane [Silane, tetramethyl-]	75-76-3	10,000
Trichlorosilane [Silane, trichloro-]	10025-78-2	10,000
Trifluorochloroethylene [Ethene, chlorotrifluoro-]	79-38-9	10,000
Trimethylamine [Methanamine, N,N-dimethyl-]	75-50-3	10,000
Vinyl acetylene [1-Buten-3-yne]	689-97-4	10,000
Vinyl chloride [Ethene, chloro-]	75-01-4	10,000
Vinyl ethyl ether [Ethene, ethoxy-]	109-92-2	10,000
Vinyl fluoride [Ethene, fluoro-]	75-02-5	10,000
Vinyl methyl ether [Ethene, methoxy-]	107-25-5	10,000
Vinylidene chloride [Ethene, 1,1-dichloro-]	75-35-4	10,000
Vinylidene fluoride [Ethene, 1,1-difluoro-]	75-38-7	10,000

Table III EPA Rule Requirements vs. OSHA PSM Elements

Program Requirement	OSHA PSM 29CFR1910.119	EPA RMP 40CFR68
Employee participation	required	Req'd for Prog. 3
Process safety information	21 items addressed	Req'd for Prog. 3;
Process hazard analysis	Required; repeat every 5 years or as needed for changes. Five years to complete first round	Req'd for Prog. 3; may use exist. PSM analyses for initial compliance.
Operating procedures	Req'd to address seven different phases of operation	Req'd for Prog. 3; Partial for Prog. 2
Training	initial & refresher	Req'd for Prog. 3; Partial for Prog. 2
Contractors	safe work practice; periodic evaluation injury/illness log	Req'd for Prog. 3; Partial for Prog. 2; no inj/illness log
Pre-startup safety review	four areas of emphasis	Req'd for Prog 3; Partial for Prog. 2

Table III (con't) EPA Rule Requirements vs. OSHA PSM Elements

Program Requirement	**OSHA PSM 29CFR1910.119**	**EPA RMP 40CFR68**
Mechanical integrity	pumps, vessels & tanks, piping, controls, relief/vent systems, emergency shutdown systems	Req'd for Prog. 3; Partial for Prog. 2; to inc. maintenance procedures & trang
Hot work permitting	required "for work on or near a covered process"	Req'd for Prog. 3
Management of change	written procedures	Req'd for Prog. 3
Incident investigation	for each incident which "could have reasonably resulted in a catastrophic release in the workplace"	Req'd for Prog. 3 for incidents presenting imminent and sig. endangerment to public & environmt
Emergency planning contingency and response	req'd an emergency action plan per 29CFR1910.38(a)	integrated plan ("one-plan")
Compliance audits	every 3 years need not be third-party last 2 audits kept on file	every 3 yrs. by person w/knowledge last 2 audits kept not longer than 5 yr.
Trade secrets	must not be withheld from employees if part of process safety information	cannot be withheld if relevant to offsite conseq. analysis
Plan submittal	not required	Req'd - will be made available to public

Table IV Comparison of Program Level Requirements

Program 1	**Program 2**	**Program 3**
Hazard Assessment		
Worst-case analysis	Worst-case analysis	Worst-case analysis
	Alternative releases	Alternative releases
5 yr. accident history	5 yr. accident history	5 yr. accident history

Table IV Comparison of Program Level Requirements

Program 1	Program 2	Program 3
Prevention Program		
Certify no additional steps needed	Safety Info. Hazard Review	Process Safety Info. Process Hazard Analy
	Operating Procedures	Operating Procedures
	Training	Training
	Maintenance	Mechanical Integrity
	Incident Investigation	Incident Investigation
	Compliance Audit	Compliance Audit
		Mgmt of chg., Pre-startup rev., employee part. & hot work pmts
Management Program		
None required	Doc. mgmt system	Doc. mgmt system
Emergency Response Program		
Coordinate with LEPC	Develop plan & prog.	Develop plan & prog.
Risk Management Plan Contents		
Exec. Sum & Regist.	Exec. Sum. & Regist.	Exec. Sum. & Regist.
Worst-case data	Worst-case data	Worst-case data
	Alt. release data	Alt. release data
5 yr. accident history	5 yr. accident history	5 yr. accident history
	Prevention prog. data	Prevention prog. data
	Emergency response	Emergency response
Certification	Certification	Certification

Figure I
Accidental Release Prevention Applicability

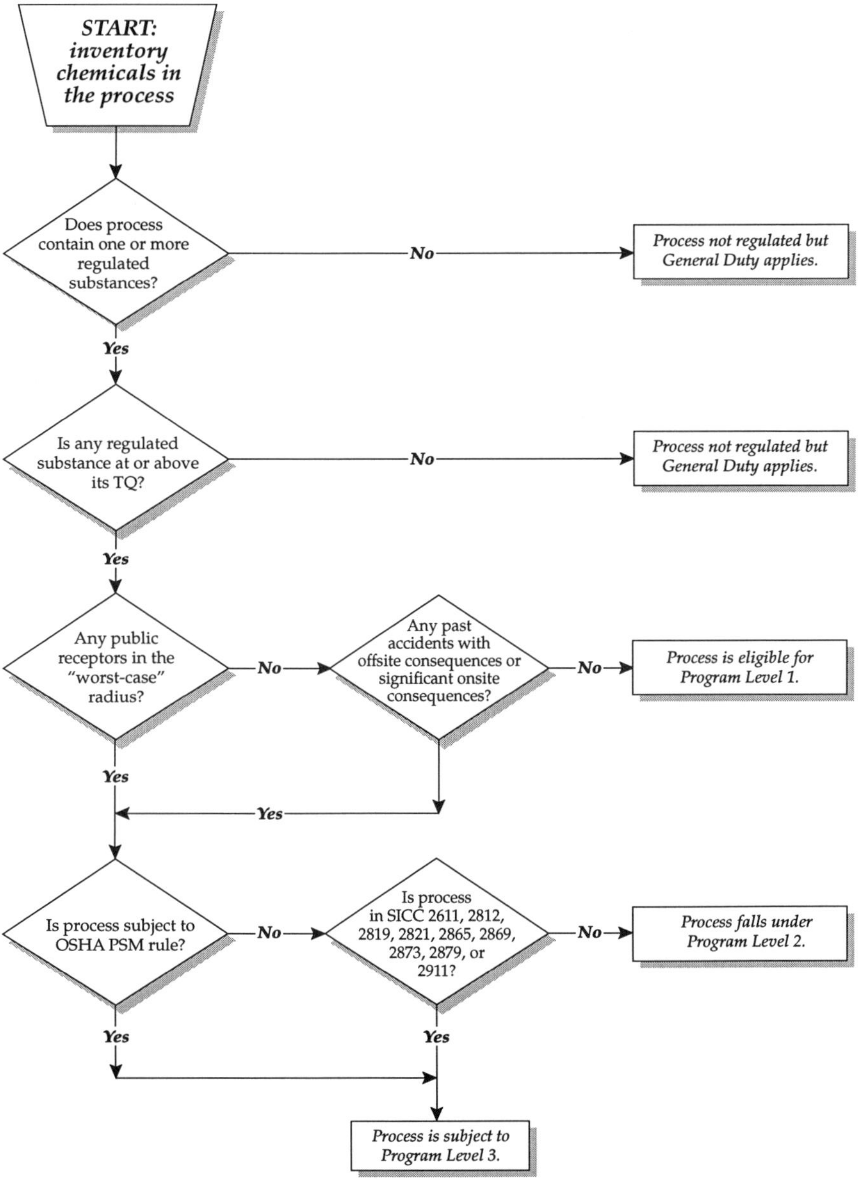

REVISED AIR QUALITY STANDARDS FOR PARTICULATE MATTER

George J. Schewe
Environmental Quality Management, Inc.
1310 Kemper Meadow Drive, Suite 100
Cincinnati, Ohio 45240

D. Kent Berry
Environmental Quality Management, Inc.
3325 Durham-Chapel Hill Blvd., Suite 250
Durham, North Carolina 27707-2646

ABSTRACT

This paper summarizes the U.S. Environmental Protection Agency's (EPA's) proposed changes to the National Ambient Air Quality Standards (NAAQS) for particulate matter (PM) and reviews the basis for the proposed changes. The paper discusses the impacts of the proposed standards in terms of the number and location of expected nonattainment areas, sources affected, and costs and benefits achieved. A number of implementation issues and control approaches EPA is considering are discussed, including the development of integrated control programs addressing revised NAAQS for PM and ozone and visibility impairment problems in national parks and wilderness areas.

PROPOSED STANDARDS

On December 13, 1996, the EPA proposed major revisions to the NAAQS for PM.[1] The current standards are for particles with an aerodynamic diameter less the 10 micrometers (μm), known as PM_{10}. EPA proposed to revise the current primary (health based) PM standards by adding a new annual $PM_{2.5}$ standard set at 15 micrograms per cubic meter of air (μg/m^3) and a new 24-hour $PM_{2.5}$ standard set at 50 μg/m^3. The current PM_{10} standards are 50 μg/m^3, annual average, and 150 μg/m^3, 24-h average (expected exceedances less than one per year averaged over 3 years). EPA proposed to retain the current PM_{10} standards, but to revise the form of the 24-hour standard to be based on the 98th percentile value rather then the maximum concentration in a year.

The EPA proposal contained some very innovative provisions concerning the form of the new $PM_{2.5}$ standards not found in any of the NAAQS for other pollutants. EPA proposed that the annual $PM_{2.5}$ standard would be based on a 3-year average concentration, spatially averaged across designated air quality monitors in an area. The 24-hour $PM_{2.5}$ standard would be based on the 98th percentile value, averaged over three years, at the single population oriented monitoring site with the highest values in an area. These provisions have the effect of "dampening" the impact of unusually poor meteorology such that stringent controls are not required in a generally clean area that experienced an unusually bad year or few days.

EPA's proposal was based largely on a series of community epidemiology studies showing that "fine" particles (less than 2.5 μm) penetrate more deeply into the lung and are more likely to contribute to health effects (e.g., premature mortality and hospital admissions) than the "coarse mode" particles (from 2.5 to 10 μm in diameter). Coarse particles are generally "crustal" in nature, and come from windblown dust from disturbed land (e.g., agricultural fields), vehicle traffic on unpaved or even paved roads, and transportation and processing of metallic or nonmetallic minerals. Fine particles are generally emitted from activities such as industrial and residential combustion and from vehicle exhaust. Fine particles are also formed in the atmosphere from gases such as sulfur dioxide, nitrogen oxides, and volatile organic compounds (VOCs) that are emitted from combustion activities and then become particles as a result of chemical transformations in the air.

Also on December 13, 1996, EPA published two other major proposals which will affect many of the same sources as the new PM NAAQS. EPA proposed to tighten the NAAQS for ozone (smog)[2] and also announced that it is developing regional haze regulations [required by Section 169A of the Clean Air Act (CAA)] to address visibility impairment in so-called Federal Class I areas (certain national parks and wilderness areas).[3] Because of the great overlap of the sources, atmospheric transport and chemistry, and receptor areas related to these programs, EPA announced it was attempting to develop an integrated implementation program that would encourage efficient, least-cost control programs.

EPA is under a court order to issue a final rulemaking on the PM standard by July 19, 1997 and intends to issue a final rule on the ozone standard at the same time. EPA also intends to propose regional haze program rules in mid-1997. Because of this tight schedule, the comment period on the proposed PM and ozone standards was held to 82 days, closing on March 12, 1997 (despite numerous requests to extend the comment period by as much as a year). In addition, EPA held four public hearings in January 1997 in various areas of the country to take comment on the proposals.

IMPACTS OF PROPOSED STANDARDS

EPA has designated approximately 70 areas as nonattainment with respect to the current PM_{10} NAAQS. Most of these areas are narrowly drawn and are limited to county or subcounty areas (as opposed to ozone nonattainment areas which frequently correspond to multicounty metropolitan statistical areas). Most of these current PM_{10} nonattainment areas are located in the Western U.S., where fugitive dust, residential wood combustion, and prescribed fires for forestry management are frequently a significant part of the problem. As many as 30 of these areas would come into attainment with the revised PM_{10} standards as a result of the proposed 98th percentile form of the standards.

Based on very limited $PM_{2.5}$ monitoring and standard ratios of $PM_{2.5}$ to PM_{10}, EPA has estimated that 167 counties would become nonattainment with respect to the proposed $PM_{2.5}$ standards. Thus, the net impact of the proposed standards is to add about 130 nonattainment counties that are currently in attainment; many of these new counties are in the East.

EPA conducted a cost-benefit analysis of the proposed PM standards, although by law EPA is not allowed to consider the relative costs and benefits in deciding on the level of the standards. EPA estimated that the annual cost of this program in 2007 would be $6 billion, while the benefits (monetized effects of reduced mortality and morbidity, etc.) were estimated at $58-119 billion (in 1990 dollars). The control measures that EPA used in this analysis did not show attainment in 57 areas, representing 29 million people. Thus these costs and benefits represent only a partial attainment scenario. They represent the incremental costs and benefits relative to attainment of the current PM_{10} standards.

INTERIM IMPLEMENTATION POLICY

Also on December 13, 1996, EPA proposed an interim implementation policy[4] regarding both the ozone and PM control programs that would apply during the period between promulgation of revised standards in July 1997 and the date revised State Implementation Plans (SIPs) would be due in July 2002. The policy is generally intended to ensure that current nonattainment areas continue to make progress in reducing emissions under the current regulatory framework, where it makes sense to do so (an "antibacksliding" policy).

In the spring of 1995, EPA analyzed available PM air quality data and determined which areas were attaining the PM_{10} standard as of December 31, 1994 (the date all moderate nonattainment areas were to have reached attainment). At that time EPA determined:

- Out of the 43 areas with approved SIP's in place, 22 of the areas classified as moderate nonattainment had 3 years of clean air quality data making them eligible to redesignate to attainment, which EPA encouraged.
- Thirteen of the 43 areas with approved SIP's in place qualified for a 1-year attainment date extension. These areas should be able to demonstrate attainment of the current PM_{10} standard with 3 years of clean air quality data prior to the July 1997 NAAQS promulgation date and were urged to proceed with redesignation requests to attainment.
- Eight of the 43 areas with approved SIP's in place had clean, but incomplete, air quality data. Additional data needed to be collected before EPA could determine whether the areas were attaining the standard.
- Five of the initial moderate areas failed to attain by the December 31, 1994 attainment date and did not qualify for attainment date extensions. Under the CAA these areas should be reclassified to "serious" nonattainment areas for PM_{10}. Consistent with the no backsliding principle, these areas were instructed to proceed with adoption of best available control measures for PM as required by the CAA.

With respect to the new source review (NSR) program for major new sources and modifications, EPA proposed that the current requirements remain in place based on the current nonattainment designations and classifications (as they may be modified) with respect to PM_{10}. The NSR procedures during the interim period before $PM_{2.5}$ designations are made would be based only on PM_{10}, not on fine particle emissions and their ambient impacts. Under the prevention of significant deterioration (PSD) programs of the CAA, EPA has until June of 1999 to set new $PM_{2.5}$ air quality increments or otherwise address air quality deterioration in "clean" $PM_{2.5}$ areas.

IMPLEMENTATION PROGRAM

In conjunction with the proposal of revised PM NAAQS, EPA also published an Advance Notice of Proposed Rulemaking[3] discussing and soliciting comment on a number of policy and technical issues associated with the development of an integrated implementation program addressing ozone, $PM_{2.5}$, and regional haze. EPA plans to publish implementation guidance rules in two phases. Phase I will give guidance to States concerning actions prior to and including designation of nonattainment areas with respect to the new standards and will be proposed in mid-1997. In mid-1998, EPA plans to finalize the Phase I rules and to propose Phase II rules, which will focus on development of control strategies and SIP revisions to attain the revised PM NAAQS. Phase II will be promulgated in mid-1999. EPA also plans to propose and promulgate regional haze rules on the same schedule as Phase I. These rules will not set a specific ambient pollutant standard, but will focus on criteria for making reasonable progress in improving visibility in Federal Class I areas and will establish methods to measure such progress.

To assist in the development of these implementation programs, EPA has established an advisory committee called the Subcommittee for Ozone, PM, and Regional Haze Implementation Programs composed of 60 representatives of State and local organizations; environmental groups; industry and trade groups (including small business representatives); academics; and other Federal agencies. The Subcommittee was formed under the Federal Advisory Committee Act (FACA) and thus all meetings must have prior public notice and be open to public participation. The Subcommittee and its four Work Groups have held a number of meetings in various areas across the country and have developed a number of issue/option papers designed to lead to recommendations to EPA on a wide variety of implementation issues, several of which are discussed below.

EPA has released a preliminary schedule for implementing the revised PM and ozone standards based on an integrated implementation approach. This schedule calls for designation of nonattainment areas for both pollutants by July 1999, submittal of SIP's by July 2002, and attainment of the standards for all areas that can do so by July 2004, which may be extended to July 2009 for severe nonattainment areas.

Because of the lack of monitoring data for $PM_{2.5}$, this schedule and the integration with the ozone control program schedule may not be possible for many areas. One of the draft issue papers being prepared by the FACA Subcommittee Work Group dealing with designation of $PM_{2.5}$ attainment/nonattainment areas expresses a strong preference for using 3 years of measured air quality data for such designations (as opposed to a statistical approach using $PM_{2.5}$ to PM_{10} ratios or fewer years of data). The paper estimates, however, that it will be December 2000 before a fully deployed nationwide $PM_{2.5}$ network is in place. Since the CAA requires designations to be made no later than 3 years after a new or revised NAAQS is promulgated, the paper examines other options besides 3 years of measured data. These options involve a phased designation schedule and either establishing interim designations based on fewer years of data and/or using a statistical probability approach. The SIP development and submittal process and schedule would be based on the interim designations, with adjustment and mid-course corrections made to account for final designation using three full years of data.

Important new concepts of designating "Areas of Violation" (AOV) and "Areas of Influence" (AOI) have been developed by the FACA Subcommittee to frame the implementation programs for $PM_{2.5}$, ozone, and regional haze. The AOV/AOI concept was developed because the traditional "nonattainment area" approach to dealing with violations of the ambient standards has not proven to be workable when the violations are the result of transported as well as locally generated pollutants. Nonattainment areas previously were designated to identify both the area in which a violation is occurring and the area which is causing or influencing the violation and therefore is where controls are required to bring the area into attainment with the standard. The designation of an area results in a number of automatic regulatory requirements under the CAA which creates a number of problems.

In order to overcome the problems which were identified with the current approach, the FACA work groups are recommending the AOV/AOI approach. This approach separates the two concepts of violation and influence and establishes a process to develop attainment plans which is fundamentally different from the current "nonattainment" process.

An AOV is defined in the draft issue paper as a region which actually exceeds the ambient air quality standard. It thus provides the information on where the population is being exposed to unhealthy air. The boundary of the region would be based on monitored data or, where available on a combination of both monitored and modeled data. Modeling alone would not be the basis for determining AOV's, except that for the purpose of encouraging monitoring, modeling could be used as a basis for interim AOV determinations. Where AOV's for ozone and $PM_{2.5}$ overlap, co-AOV's may be defined.

An AOI is a specified domain containing the natural and man-made sources potentially contributing to a downwind AOV. States and tribes which are a part of an AOI would be required to participate in the development of a regional air quality management plan. An AOI designation does not mean, however, that all portions of that area will be required to implement regulation, nor does it indicate that sources in all portions of the area will be subject to identical regulations. This assumption is extremely important to understand and accept, because it allows for a more inclusive approach to identifying the AOI. It allows designation of large enough AOI's to encompass all of the potentially significant sources and groups of sources. AOV's may or may not be a part of the AOI. The AOI determination may be made early in the designation process or may be determined during the planning process.

The regional plan would address issues such as the definition of the portion of the AOI subject to regulation, what the levels of control will need to be in each of the states covered by the AOI, and what type of new source controls should be applied in the various portions of the AOI. It might also make recommendations concerning the structure and type of market-based incentive programs and Federal regulatory action to assist in the effort to meet the standards.

Following development of the regional air quality management plan for the AOI, States would need to adopt and submit to EPA a SIP that contains rules to implement the regional plan. The draft FACA issue paper goes on to suggest that AOI's might be as broad as two regions (east and west of the Mississippi) that would cover the entire country.

If this AOI/AOV approach is adopted, it is likely that the current NSR system for major new or modified sources would need to be changed to be consistent. A variety of options are under consideration by the FACA NSR Workgroup including the establishment of an emission budget for AOI's which would be implemented through a market-based cap and trade program. This could allow the control technology drivers in the current NSR programs [Best Available Control Technology (BACT) in attainment areas and Lowest Achievable Emissions Rate (LAER) in nonattainment areas] to be reduced or eliminated. It could also allow the PSD air quality increment analysis to be eliminated within the AOI. The cap and trade approach could be extended to smaller (non-major) new sources. While simplifying the NSR program, the effect of a cap and trade program within the AOI's would be to extend the requirement to obtain emission offsets to many more sources than it currently does. At least there would be an established market mechanism for new sources to obtain the needed emission allowances, but such allowances could become quite expensive. Nonetheless, it should reduce the delays and economic inefficiencies of the current BACT/LAER and air quality modeling and monitoring approach of the NSR program.

Besides the lack of $PM_{2.5}$ air quality data discussed above, there are other technical and informational deficiencies that will inhibit the development of effective and efficient $PM_{2.5}$ control programs. One major deficiency involves the lack of emission factors for sources of $PM_{2.5}$. While EPA's emission factor data base contains particle size distribution information for some types of sources, it is lacking for many others. In addition, many of EPA's emission factors only deal with the "filterable" PM (particles measured at stack temperature during a source test). Much less information is available on the so-called "condensible" PM, which is material that passes through the filter as a vapor but ultimately condenses into solid or liquid particles in the ambient air. This condensible PM is more important under a $PM_{2.5}$ standard than the PM_{10} standards. Finally, there is the role of gaseous precursors such as sulfur and nitrogen oxides, volatile organic compounds, and ammonia which are converted in the ambient air to fine particles. While emissions and emission factors are fairly well defined for most of these precursors (except for ammonia), the atmospheric transformation processes and modeling techniques are not. Better emissions and emission factors for ammonia will be needed in areas where ammonium nitrate and sulfate are important contributors to $PM_{2.5}$ concentrations. Unfortunately, obtaining better information in these areas is very expensive, and EPA does not have adequate resources to develop this information.

CONCLUSION

The decision on the revised PM NAAQS is one of the most significant actions EPA has made since it was formed back in 1970, both in terms of control program costs and the health benefits that would be produced. The proposed changes would result in more stringent controls for many sources and redirect controls from sources of coarse mode particles to sources of fine particles and gaseous precursors. Nonattainment problems will be shifted from many areas in the Western U.S. to broad areas in the East. The NAAQS changes provide the opportunity to develop new and more efficient control programs, some of which may require changes to the CAA. In any case, revision of the PM NAAQS as proposed would be a major factor in shaping the Nation's air pollution control program for the next 20 years.

REFERENCES

[1]Federal Register, 61 FR 65638, Dec. 13, 1996.
[2]Federal Register, 61 FR 65716, Dec. 13, 1996.
[3]Federal Register, 61 FR 65764, Dec. 13, 1996.
[4]Federal Register, 61 FR 65752, Dec. 13, 1996.

NATURALLY OCCURRING RADIOACTIVE MATERIALS (NORM)

Charles G. Marvin
The Refractories Institute
Pittsburgh, PA

ABSTRACT

All soils and rocks contain radioactive materials. The three most common natural radioactive materials (NORM) are uranium - 238, uranium 235, and thorium - 232. Two ceramic raw materials which contain large quantities of NORM materials include zirconia and bauxite. NORM materials are not usually covered by the Atomic Energy Act of 1954; and in general, are not specifically covered by Environmental Protection Agency (EPA) regulations. Issues surrounding these naturally occurring redioactive materials have recently become more intense as individual states, such as Ohio, initiate actions to propose regulations designed to control the handling, transfer, storage and disposal of NORM containing materials. Some of these issues shall be reviewed to increase the ceramic community's awareness of an issue that deserves the attention of those utilizing NORM containing materials.

INTRODUCTION

The Nuclear Regulatory Commission held a meeting to discuss "Below Regulatory Concerns" related to low level radioactive materials in the middle 1980's. Eigthty seven registered for the meeting, and over 200 attended. The extra attendees wore green t-shirts, carried placards and complained that the meeting site, Holiday Inn-Ohara Airport, was such a remote location for the meeting. They wanted it held in downtown Chicago. The NRC had proposed exempting low level radioactive materials from regulation. One State Senator from the state of Illinois pounded the table and demanded that laws and regulations requiring zero radiation be implemented. The NRC representatives explained that, such action would require digging up the Senator's back yard, and there would be no material which could be substituted. The Senator still pounded the table demanding zero radiation limits. This demonstrates that the government, which developed the atomic bomb, under the supervision of General Leslie Groves, has not educated the public about radioactive materials, their occurrance in nature. We live with low level radiation every day, expecially if you fly to the American Ceramic Society meetings, visit the Washington Monument and our Nation's Capitol building in Washington DC. which emite more radiation than the majority of naturally occurring radioactive materials used in the ceramics industry.

Peter Gray's quarterly NORM Report states that Naturally Occurring Radioactive material (NORM) means any radionuclides or radioactivity in its natural physical state, disturbed by man-made activities, or technologically enhanced state, but does not include source, byproduct or special nuclear materials as defined by the Atomic Energy Act, as amended.[1]

Some ceramic materials and souces associated with elevated norm levels include: bauxite, lignite, monazite, pumice and scoria, titanium, zircon. There are also many common household items that contain low levels of NORM. This includes antiacids, lantern mantles (throium), brazil nuts (potassium - 40), smoke detectors, luminous clocks and watch dials (tritium or radium-226) natural gas (radon-222) porcelain dentures (potassium - 40 and uranium) and tobacco products (lead-210 and polonium -210).[2]

One of the reasons that interest in NORM has been on the increase is that during the past ten years, more than thirty incidents were reported, where radioactive material found its way into scrap metal. One-third of these cases involved NORM. Eleven cases resulted in widespread contamination of smelters. The NRC estimates that there are another 200 cases concerning contaminated scrap metal which go undetected every year.[3] That is the main reason steel mills have installed radiation detectors on their gates where scrap metals are delivered. There have been instances where trucks delivering refractory products have gone thru these gates and set off the detectors.

The result is the development of NORM regulations, at the state level which address the handling and release of NORM materials. Seventeen states have existing regulations. Regulations are being developed in twelve states and 21 states have no NORM regulations at this time.[4]

In June 1996, Ohio announced its program to promulgate new regulations for materials, products and waste containing NORM.[5] Facilities that process, store, distribute or otherwise handle NOMR materials within the state will be subject to these new regulations. the announcement stated that NORM materials used by the ceramic and refractories industries would be included in these regulations. The provisions in the initial draft were so stringent that it would have put Zircoa, in Solon, Ohio out of business. The state health authorities have very little knowedge of the extent of materials containing NORM and their importance in everyday commerce. Active participation by those companies, which would have been impacted has corrected some of the situation, however there is much work yet to be done.

Materials which contain more than 0.05% NORM material are classified as source materials. Persons who use more that 150 pounds of source materials in one calendar year are required to obtain a license from the Nuclear Regulatory Commission. Some individual states are requiring those individuals handling materials with uranium and thorium levels below 0.05% to obtain a license.[6] The majority of NORM containing materials are listed in the Threshold Limit Value (TLV) tables published by the National Institute for Occupational Safety and Health (NIOSH)[7] and American Conference of Governmental Industrial Hygienists (ACGIH)[8]. OSHA (29CFR 1910.96 requires that all individuals working in or frequenting an area where radioactive materials are located be informed of the occurance of radioactive materials, and shall be instructed in the safety problems associated with exposure to materials with radiation[9]. Employers are required to post the OSHA provisions and a copy of the company's operating procedures applicable to the work. Therfore it is prudent for employers to know what NORM materials they have, the levels of radioactivity and keep their employees informed of the safety aspects of these materials.

The NORM Report, published quarterly by Peter Gray and Associates,, in Tulsa Ok, does an excellent job providing updates on the status of regulations for the contorl of norm. Peter Gray also teaches a course: NORM Contamination - An Emerging Environmetal Problem, and provides a text on the subject.

Charles Simmons, Kilpatrick and Cody, Washington, DC(202-508-5806) manages the Zirconia Coalition and keeps its members up to date on both federal and state activities which impact both suppliers and users of zircon and zirconia compounds[10]. This service is invaluable, since in is almost impossible for individuals and individual companies to keep up to date with all the current activities concerning NORM both at the federal and state levels. The NORM issues, which have the potential of impacting the ceramic industry in may ways, shall be active for a number of years and require close monitoring to ensure that these proposals are consistent with the safe and useful handling of raw materials and finished ceramic products.

REFERENCES;

[1,2] P. Gray, "Living Habits and Radiation", NORM Report, Fall 1995, P. Gray & Associates, Tulsa, OK.

[3]R. Turner, "Federal and State Requirements for Handling Radioactive Scrap Problems",Iron and Steelmaker, pp35-36, May 1995.

[4,6,10]P. Gray, "Regulations for the Control of NORM-Update", Winter 1997, P. Gray & Associates, Tulsa, OK.

[5]"Licensing of naturally Occurring and Accelerator Produced Radioactive Material", Draft Ohio Administrative Code 3701-39, Ohio Dept. of Health, Columbus, OH 1996.

[7]"Pocket Guide to Chemical Hazards", National Institute for Occupational Safety & Health (NIOSH), Washington, DC, 1990.

[8]"1995-1996 Threshold Values for Chemical Substances and Physical Agents", American Conference of Government Hygienists, Cincinnati, OH, 1995.

[9]"Ionizing Radiation",29CFR 1910.96, P218,OSHA, USGPO,1993.

[10]C. Simmons, "Current Issues Concerning Naturally Occurring Radioactive Materials", Presented to The Refractories Institute, March 28, 1996.

Air Emissions and Treatment of Secondary Wastes

NITROGEN EMISSIONS FROM JOULE HEATED GLASS MELTER

Marek Brandys, Keith S. Matlack, Hamid Hojaji, and Pedro B. Macedo, Vitreous State Laboratory, The Catholic University of America, Washington, D.C. 20064

ABSTRACT

Controlling nitrogen oxides emissions (primarily NO and NO_2) is of great concern for many industries, including the ceramic. Over the past four years VSL researchers have examined several methods for limiting NO_x during the vitrification of nitrate and nitrite containing feeds. Tests were conducted on glass melters with nominal production rates ranging from 0.4 to 40 kg of glass per hour. The feed consisted of surrogate or actual nuclear and hazardous wastes, sometimes containing organic materials. Strategies, such as thermal inversion in the melter plenum, addition of urea to the feed and peroxide spraying into the exhaust gases, as well as various liquid scrubber arrangements were evaluated. The success of various technologies was measured with relation to chemical composition of the feed and the melter conditions, such as cold-cap formation, plenum temperature and glass pool bubbling rate. Acquired process material balance data suggests involvement of fairly complex mechanisms.

INTRODUCTION

To date, in all DOE remediation efforts involving high nitrogen content wastes, vitrification has been selected as a principal thermal treatment technology. These include West Valley Demonstration Project (WVDP), Savannah River Site (SRS) M-Area Vendor Treatment Facility (VTF) and Hanford Tank Waste Remediation System - Privatization (TWRSP). Since nitrogen solubility in glass in any form is negligible, the secondary effluents from operations cannot be recycled, making emissions control of primary importance. Nitrogen oxides (NO_x) are EPA criteria air pollutants, while nitrate and nitrite solutions releases are regulated under NPDES permits.

Nitrogen emissions from thermal processes have been investigated extensively in the past[1], mostly in conjunction with combustion and power generation systems[2]. Nitrate waste processing in a glass furnace was the subject of the 1985 Mound Laboratory report[3]. A variety of catalytic and non-catalytic, selective and nonselective control processes were proposed. However, the high temperature chemistry of nitrogen is complex, involving hundreds of reactions[4] and making the whole process inherently condition-sensitive. Experimental evaluation on a case by case basis is, therefore, necessary.

The following paper describes testing of several nitrogen emission control strategies developed at the VSL in support of M-Area and Hanford projects using Bench-scale experiments with standard gases and the DuraMelter™ runs on actual and surrogate feed. The design of the DuraMelter™ family of furnaces has been described elsewhere[5].

GENERAL PROCESS CHEMISTRY OVERVIEW

Alkali metal nitrates decompose at ~350°C[6]:

(1) $2NaNO_3 \rightarrow 2NaNO_2 + O_2$
(2) $2NaNO_2 \rightarrow Na_2O + N_2 + {}^3/_2O_2$
(3) $NaNO_2 \rightarrow Na_2O + NO + NO_2$
(4) $2NO_2 \rightleftharpoons 2NO + O_2$

The equilibrium reaction (4) shifts to the right as temperature increases, with the ratio $p_{NO2}/p_{NO} = 1$ (in air) at about 430°C[6]. Only NO_2 exhibits appreciable solubility in water (H_{NO2}=113 atm/mol vs. H_{NO}=28700 atm/mol at 25°C) reacting according to the overall process:

(5) $3NO_2 + H_2O \rightarrow 2HNO_3 + NO$

Consequently, an effective, oxidative abatement of the nitrogen oxides must involve processes such as:

(6) $NO + H_2O_2 \rightarrow H_2O + NO_2$

which occur, specifically, in the gaseous phase through a series of reactions involving OH· radicals[7]. From an emissions control point of view, however, a reductive system is much more attractive, since no secondary liquid wastes are involved. For example:

(7) $4NH_3 + 4NO_2 \rightarrow 4N_2 + 6H_2O + O_2$
(8) $4NH_3 + 4NO + O_2 \rightarrow 4N_2 + 6H_2O$

The ability of ammonia to selectively reduce NO_x to nitrogen has been recognized for some time[8]. In the case of a liquid fed melter, amine groups can be conveniently delivered *in situ* using urea, which decomposes at ~120°C:

(9) $(NH_2)_2CO + H_2O \rightarrow 2NH_3 + CO_2$.

SUMMARY OF TESTS RESULTS

The focus of this report is on the results from a series of DuraMelter™ 1000 runs. We report, however, a summary of the major conclusions from each of the other tests as well as data collected during the actual M-Area VTF operations.

1. Bench Scale Experiments with Peroxide

Bench scale experiments were conducted with standard gases to identify the critical parameters associated with liquid scrubbing of NO_x laden gases. NO_x removal of over 50% was demonstrated with liquid scrubbing alone. This process is highly dependent upon the residence time of the gases in the scrubber column. Solution pH was observed not to be a factor between 3 and 13, with only a slight increase in efficiency above 13. Scrubber solution temperature was observed to affect NO_x removal in that the lower the temperature, the greater the amount of NO_x removal. Spraying peroxide at vaporizing temperatures was observed to reduce NO_x emissions by an order of magnitude.

2. DuraMelter™ 10 Runs with Actual M-Area Waste

During this run over 40 kg of glass were produced from actual Savannah River M-Area waste. Various amounts of urea were added to the feed to determine the effect on NO_x emissions. Overall NO_x emissions were reduced by a factor of 5 to 10 depending on melter conditions, such as cold cap size and melter plenum temperature. This reduction was achieved at a urea-to-feed-nitrate ratio of approximately 0.5 (stoichiometric). Higher concentrations of urea in the feed did not result in significantly further reductions of NO_x emissions. Based on these results, the urea-to-feed-nitrate ratio of about 0.5 was selected for subsequent DuraMelter™ 1000 runs. The analytical instrumentation (ENERAC 2000) was also checked for potential interferences with ammonia; none were observed.

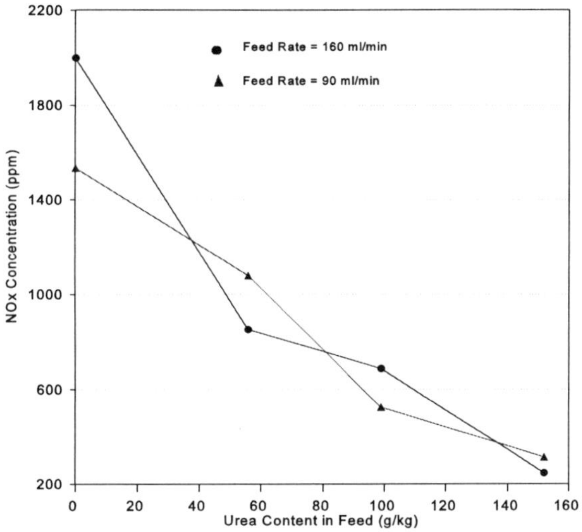

Figure 1. The effect of urea concentration in feed on NO_x emissions during the 100 kg/day Hanford run.

3. 100 kg glass/day Hanford Runs

Over 600 kg of glass were produced using the VSL DuraMelter™ 100. The nitrate-rich Hanford simulant was approximately 50% of the feed, resulting in a nitrogen content of 2.2% by weight. The primary goal of the work was to demonstrate that Hanford LLW simulant could be vitrified in such a system - not to determine operating parameters most conducive to NO_x emission reduction. However, experiments were conducted to determine the effect of urea as a feed additive on NO_x reduction. The amount and nature of particulate material leaving the melter was also monitored.

Urea was added in various amounts to the feed. Its effect was not correlated to cold cap size or to bubbling rate through a glass pool, although cold caps between 80 to 100 percent were generally maintained. Figure 1 shows the effect of urea additions on NO_x emissions. Nitrogen oxides emissions were reduced by an order of magnitude at the highest feed rate.

4. DuraMelter™ 1000 Test Runs

A. Methods and overview

Several experiments were conducted to determine optimum conditions for the reduction of NO_x emissions. M-Area feed was simulated by a three-fold volume dilution of the Hanford simulant waste. The composition of the Hanford feed can be summarized as 1:1 (by weight) mixture of a solid glass forming additives and the DSSF simulant, an alkaline, 10M sodium solution incorporating also 1M aluminum, 0.5M potassium, 3.1M nitrate and 1.7M nitrite. The dry chemicals, the DSSF simulant and water (purified by reverse osmosis) were combined in 500-gallon tanks and fed into the melter. The dilution resulted in an elemental nitrogen concentration of 13 g/liter or 1.06 weight percent (subsequent M-area field measurements indicated a dilution of 6 should have been used).

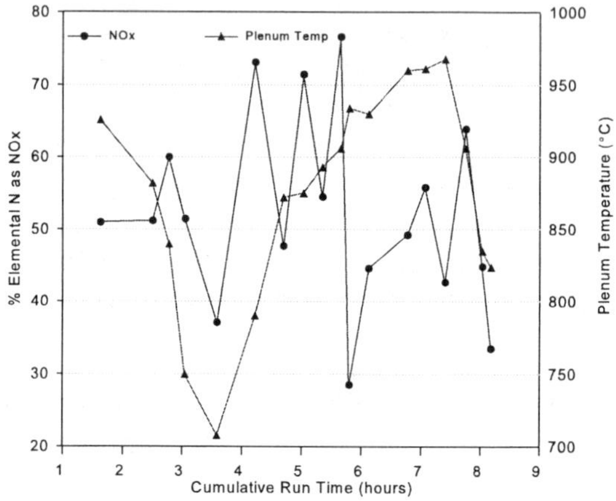

Figure 2. Effect of plenum temperature on nitrogen emissions with diluted Hanford feed.

Data was collected under a variety of melter conditions including various plenum temperatures, cold cap sizes, feeding conditions (such as times and rates), scrubber conditions (such as liquid types and volumes) and off-gas conditions (such as air flow rates). The glass pool temperature was maintained at 1100 (±20)°C. Feed rates were monitored by calibrating the feed pump and observing the drop in feed level in the feed tank.

EPA Test Methods 1A and 2C were utilized to locate off-gas sampling points and to measure gas velocity with a standard pitot tube, respectively. Each velocity determination was the average of two perpendicular 12 point traverses. The density of air (as a function of temperature and humidity) at the sampling locality was calculated to be 1.14 kg/m^3. This value was then used to calculate velocity. Measurements of NO, NO_2, SO_2, NO_x, CO, O_2 and stack temperature were performed using an ENERAC 2000 system.

B. Effects of Plenum Temperature and Cold Cap Size

The first set of experiments was performed to determine the effect of plenum temperature and cold cap on NO_x emissions. These experiments were conducted during the first nine hours of run time and no urea or peroxide was used. The data collected is shown in Figure 2. The fraction of elemental N emitted as NO_x is a convenient measure of process effectiveness in that emissions are normalized to feeding rates and the weight of oxygen is factored out. It is calculated by dividing the mass of elemental nitrogen leaving the system as NO_x (air volume times NO_x concentration that is corrected to a nitrogen basis) by the amount being fed into the melter (feed nitrogen concentration times feed rate) at a given point in time. There was no clear connection between plenum temperature and NO_x emissions. The amount of elemental nitrogen leaving as NO_x averaged approximately 50% and ranged between 25 and 80 percent with no clear correlation to cold cap size or plenum temperature.

C. Effects of Urea and Hydrogen Peroxide

The second set of experiments were to determine the effect of urea and hydrogen peroxide in reducing NO_x emissions. Both technologies were effective. The data from these experiments is depicted in Figures 3a and 3b. A concentrated (50%) hydrogen peroxide solution was pumped at 175 ml/min and sprayed (atomized) against the exhaust gases in the quencher to cause flash evaporation of the droplets. The melter was fed at a rate of 2 liters/min resulting in a 1:1 molar ratio of peroxide to feed nitrate. Urea was added to the feed to achieve a 5:12 weight ratio (0.5 stoichiometry) of urea to feed nitrate.

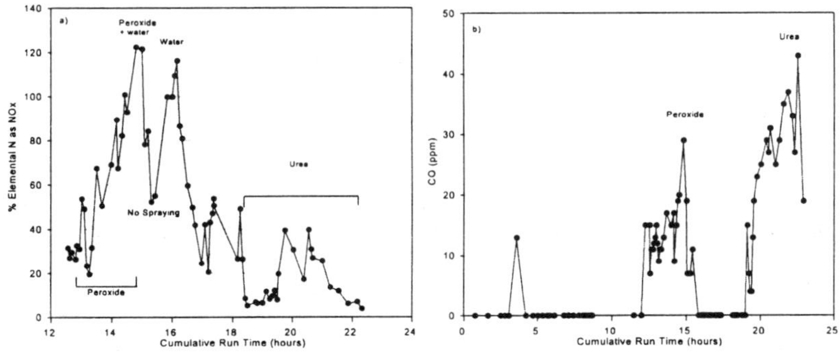

Figure 3. Effects of hydrogen peroxide and urea: a) nitrogen oxides, b) carbon monoxide.

Figure 3a illustrates the vast difference between spraying peroxide on the exhaust gases as opposed to water. The values above 100% relate to the destruction of the cold cap; for example: NO_x is released from unmelted feed on top of the melt pool even if feeding is stopped. The conversion of NO_x to N_2 is hampered if the temperature of the quenched gases is reduced by the spraying of water in addition to peroxide. The spraying of hydrogen peroxide on exhaust gases is, therefore, effective at reducing NO_x emissions only at high temperatures due, most probably, to the competition between the rate of peroxide evaporation and its decomposition to oxygen and water. Urea additions to the feed reduced emissions to as low as 5% of the total nitrogen fed into the melter. Small concentrations of CO are released when urea and peroxide are used, as shown in Figure 3b. It is likely, that small amounts of carbon monoxide are produced upon the breakdown of urea ($CO(NH_2)_2$). The mechanism for the CO production when only the peroxide was used is unclear, however, it could be a result of the decomposition of a small amount of EDTA in the DSSF solution.

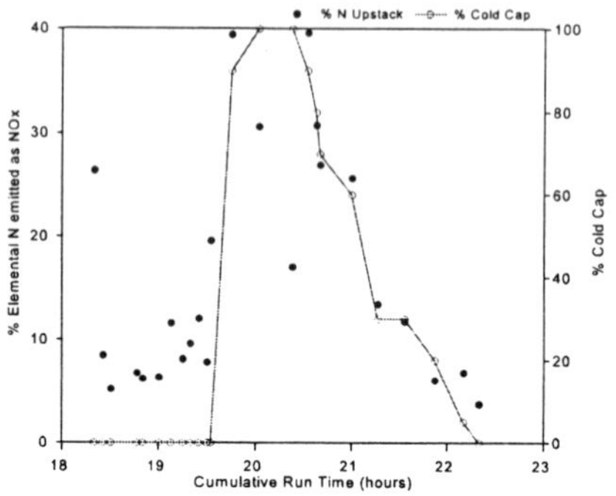

Figure 4. The correlation between cold cap size and NO_x emissions

The effectiveness of the urea in reducing NO_x emissions is affected by the extent of the cold cap (Figure 4). As the cold cap coverage increases to 100%, the NO_x emissions increase by a factor of four. This trend is reversed as the cold cap is destroyed, suggesting strong dependance on the plenum temperature.

D. Additional Experiments with Hanford Feed

Two additional experiment sets using an undiluted Hanford feed were conducted prior to the Savannah River runs. The first was to test the feed system and establish an emission baseline for the nitrogen-rich Hanford feed. The second was conducted to provide more data on hydrogen peroxide quenching particularly, when the feed contained urea.

The first set of experiments demonstrated that, either due to the higher concentrations of nitrate in the feed or due to the decreased water-to-nitrate ratio, the overall NO_x emissions decreased. These experiments were conducted with cold caps of between 50 and 65% and moderate bubbling rates. No peroxide or urea was used, yet 50 to 70% of the nitrogen was decomposed to N_2 in the plenum area. It is not possible to separate the effects of nitrate concentration and the water-to-nitrate ratio because water was used to dilute the Hanford feed.

The second set of experiments further demonstrated the effectiveness of hydrogen peroxide and urea at reducing NO_x emissions. Figure 5 illustrates the dramatic effect of both hydrogen peroxide and urea on emissions. NO and SO_2 are shown, because both remained in-scale on the ENERAC analyzer for the entire experiment. NO_x emissions from melting a urea-containing feed can further be reduced by using hydrogen peroxide to quench the exhaust gases. This reduction, however, is only an additional 20% of the baseline NO concentration.

In these experiments, an 80% cold cap was maintained while the glass pool was bubbled vigorously. The result was reduced NO_x emissions. This is in contrast to the prior urea run in which large cold caps contributed to an increase in NO_x emissions. Evidently, vigorous bubbling and the lack of a cold cap resulted in higher reaction temperatures which facilitate nitrate decomposition to N_2 rather than NO_x.

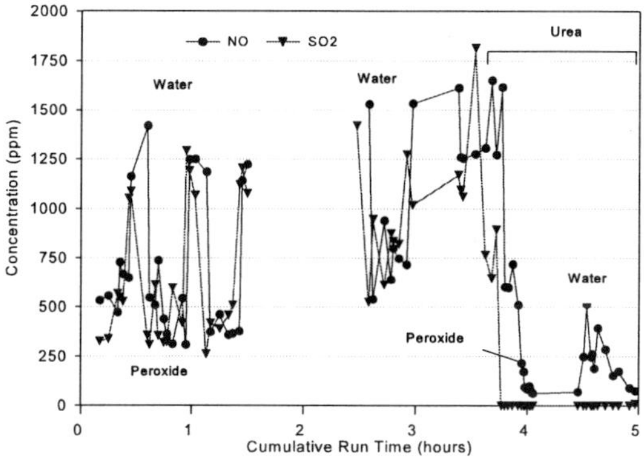

Figure 5. Effect of peroxide and urea on NO nad SO_2 emissions with Hanford feed

E. Nitrogen Mass Balances

During experiments with diluted Hanford feed, data was gathered to determine the distribution of nitrogen between scrubber liquids and the various kinds of emissions. Scrubber solution volumes were combined with the concentration data to calculate the nitrogen accumulation in scrubber liquids. Elemental nitrogen emitted as NO_x values were time weighted and averaged to give the total mass of nitrogen emitted as NO_x over given periods of time. These values were subtracted from the amount of nitrogen fed into the melter to determine the amount of nitrogen decomposed to N_2. The results are given in Table 1 along with results obtained from feeding a more concentrated material.

Table 1. Mass Distribution of Nitrogen

Run	Total Run Time (hr)	Feed		Elemental Nitrogen (%)			Total Conditions
		Conc. (g/l)	Total Mass (kg)	In Scrubber Solutions	Emitted as NO_x	Emitted as N_2	
A	3.5	39	9.6	9.4	34.4	56.2	No urea
B	11	39	51	8.0	20.0	72.0	No urea
C	6.6	13	5.2	10.8	51.3	37.9	No urea
D	4	13	4.3	6.7	20.9	72.4	Basic scrubber, urea in feed

*NO_x values not time averaged, differential pressure readings not adjusted for humidity and temperature, hence, 10% error possible.

A large portion of the nitrates fed into the melter were thermally decomposed to N_2 in all of these time periods. The major factors affecting this process appear to be concentrations of nitrate and urea in feed. The amount of nitrogen trapped in scrubber liquids was relatively constant and the efficiency of the scrubber was also relatively constant over the range of conditions tested.

The ratio of nitrate to nitrite in the scrubber liquid was drastically affected by the pH. Nitrite is kinetically stable in basic solutions and can only be converted to nitrate upon acidification. Unfortunately, acidification alone results in significant release of NO, most likely due to catalytic disproportionation[9]:

$$(10) \quad 3NO_2^- + H^+ \rightleftharpoons NO_3^- + 2NO + OH^-$$

Samples from the bag house were analyzed to determine the potential of ammonium nitrate formation. Analysis showed ammonia values in all cases below 0.5% and significantly lower than nitrate. In any event, material accumulations in the bag house can be reduced by an efficient mist elimination and multiple stage liquid scrubbing.

5. Savannah River M-Area VTF Operations

The DuraMelter™ 5000 at the M-Area VTF utilizes a peroxide oxidation followed by an acidic and basic scrubbing to control nitrogen emissions. Data was collected over a three hour period to determine an operational efficiency of the off-gas system. Overall, 26% (by mass as NO_2) of the nitrogen fed into the system was destroyed (i.e. converted into a non-NO_x form) in the melter with an additional 20% removed in the off-gas system. One reason for the diminished thermal NO_x reduction was probably due to the decreased concentration in the feed (half that of the diluted Hanford feed). Currently, it is unclear how sensitive NO_x destruction is in this melter to the effects of temperature and nitrite/nitrate ratio in the scrubbers. Additional tests are being performed to better characterize parametric sensitivity of NO_x destruction in large scale operations. Also, investigation into the extent of nitrogen species being stored in the cold cap and the glass is being conducted.

CONCLUSIONS

In pilot scale experiments, fifty to seventy percent of the nitrogen entering the melter as diluted Hanford feed was decomposed to N_2 in the plenum area with no additional engineering controls. This was demonstrated very clearly with the DuraMelter™ 1000 (20% the size of the VTF unit at SRS M-Area). Once the cold cap was stabilized at 50-65% coverage, factors such as plenum temperature and feed rate show little effect on this process. The nitrogen content of the feed (or the ratio of water to nitrogen) did have an effect in that, the greater the percentage of nitrogen in the feed, the greater the percentage being decomposed to N_2.

The addition of urea to the feed increased thermal decomposition of nitrates to more than 70 percent in every case. The effectiveness of this was reduced by the formation of an extensive cold cap. However, if a cold cap of approximately 80% was maintained with vigorous bubbling, maximum decomposition could still be achieved. Effective decomposition of nitrates was achieved at urea/nitrate weight ratio as low as 5/12. Potential problems with urea, such as ammonia slippage and ammonium nitrate accumulation in the filters, can be eliminated by an efficient de-mister.

At high temperatures, 400 to 600°C, hydrogen peroxide is effective in reducing NO_x emissions. The vapor-phase reaction involves oxidation of NO to NO_2 and NO_2 to nitric acid. NO cannot be removed by liquid scrubbing and therefore even conversion of NO to NO_2 alone will result in the overall lower NO_x emissions. This was demonstrated with extensive bench scale experiments and DuraMelter™ 1000 runs. However, the field scale results from the M-Area operations were less spectacular. At the time, the VTF DuraMelter™ 5000 was operated with a cold cap approaching 90% and with a minimal bubbling. The field data is, therefore, qualitatively consistent with the pilot study.

Bench scale experiments demonstrated, that significant removal of NO_x (primarily NO_2) by spraying exhaust gases in a packed bed column is possible, but highly dependent, as expected, on residence time[10]. Using residence times feasible in the scrubber tower of the DuraMelter™ 1000, 6 to 11% of nitrogen fed into the melter was removed which corresponds to a capture of about 25% of the NO_x reaching the scrubber. A greater proportion of the NO_x entering the scrubber was removed because 40 to 70% has already been converted to N_2 in the plenum. The pH of the scrubbing liquids did not exert a marked effect on the scrubbing efficiency. It did, however, determine the final nitrite/ nitrate ratio in the scrubbing liquids. Nitrite/ nitrate ratios were found to be 0.01 and 6 in an acidic and a basic scrubbers, respectively. Nitrite was not converted to nitrate in basic solutions by the addition of hydrogen peroxide, at least, at any useful rate.

The results from these tests can be summarized as the following: 1) unaided thermal decomposition in the melter plenum can result in significant NO_x destruction provided high enough plenum temperature and bubbling; 2) the amount of NO_x destruction is highly dependent on operating conditions such as temperatures, cold cap and feed NO_x concentration; 3) urea addition to the feed at, or below, 5/12 weight ratio of urea to feed nitrate can significantly reduce NO_x emissions even further; 4) hydrogen peroxide is effective in reducing NO_x emissions by increasing the amount captured by liquid scrubbing. Further research is required to determine the causes of different pilot and field scale results and to determine the extent in which nitrogen species are temporally contained in areas such as the cold cap.

ACKNOWLEDGMENTS

The authors would like to thank Dr. John B. Pickett of the Westinghouse Science and Technology Center at SRS for providing valuable information and for insightful discussions.

REFERENCES

1. a) Y.N. Lee and S. Schwartz, *Reaction Kinetics of Nitrogen Dioxide with Liquid Water at Low Partial Pressure*, J. Phys. Chem., 840, **85**, 1981; b) B.E Hurst, *The Noncatalytic Denitrification Process for Glass Melting Furnaces*, Glass Technology, 97, **24**(2), 1983; c) R.M. Counce, D.B. Crawford and A.J. Lucero, *Performance Models for NO_x Absorbers/Strippers*, Environmental Progress, 87, **9**(2), 1990.
2. Interagency Energy-Environment Research and Development Program Report, *Technical Assessment of NO_x Removal Processes for Utility Application*, TVA Y-120, **EPA-600/7-77-127**, EPRI AF-568, Nov. 1977.
3. K.M. Armstrong and L.M. Klinger, *Nitrate Waste Processing by Means of the Joule-Heated Glass Furnace*, Mound Laboratory Report **MLM-3304**, Oct. 1985.
4. J.A. Miller and C.T. Bowman, Mechanism and Modeling of Nitrogen Chemistry in Combustion, Prog. Energy Combust. Sci., 287, **15**, 1989.
5. B.W. Bowen, R.K. Mohr, M. Brandys and P.B. Macedo, *Design of a Vitrification Process for the Destruction of Asbestos Wastes*, Ceramic Trans., G.B. Mellinger Ed., 53, **39**, 1994.
6. Adapted from: J.M. Smith and H.C. Van Ness, *Introduction to Chemical Engineering Thermodynamics*, p. 423, 2nd Ed., McGraw-Hill Book Company, New York, 1959.
7. J.M. Casper, C.A. Clausen III and C.D. Cooper, *Control of Nitrogen Oxide Emissions by Hydrogen Peroxide-Enhanced Gas-Phase Oxidation of Nitric Oxide*, J. Air & Waste Manage. Assoc., 127, **46**, 1996.
8. a) R.K. Lyon (Exxon Research and Engineering Company), *Method for the Reduction of the Concentration of NO in Combustion Effluents Using Ammonia*, **U.S. Patent 3,900,554**, Aug. 19, 1975; b) R.K. Lyon, *The NH_3-NO-O_2 Reaction*, Int. J. Chem. Kinet., 315, **8**, 1976.
9. R.M. Counce and G.W. Selby, *Aqueous Scrubbing of Dilute Nitrogen Oxide Gas Mixtures*, Ind. Eng. Chem. Res., 1917, **27**, 1988.
10. R.M. Counce and J.J. Perona, *Scrubbing of Gaseous Nitrogen Oxides in Packed Towers*, AIChE. J., 26, **29**(1), 1983.

A METHOD FOR PERMANENT DISPOSAL OF CO_2 IN SOLID FORM

Darryl P. Butt, Klaus S. Lackner, Christopher H. Wendt, Alana S. Benjamin, David M. Harradine, Terry G. Holesinger, Youngsoo S. Park, and Meridith Rising
P. O. Box 1663, M. S. G755
Los Alamos National Laboratory, Los Alamos, NM 87545

ABSTRACT

We describe a potential method for binding the greenhouse gas carbon dioxide as magnesium carbonate, a thermodynamically stable solid, for safe and permanent disposal, and with minimal environmental impact. The technique is based on extracting magnesium hydroxide from common ultramafic rock for thermal carbonation and subsequent disposition. The economics of the method appear to be promising, however, many details of the proposed process have yet to be optimized. Initial estimates indicate that binding and disposal would impose a burden of approximately 3¢/kWH onto the cost of electricity. This cost could be reduced significantly in the short term through niche markets for various materials inherent in the process. Realization of a cost effective method requires development of optimal technologies for efficient extraction and thermal carbonation. In this paper, we describe some of the kinetic limitations and opportunities. The proposed disposal technique may be viewed as a sort of insurance policy in case global warming, or the perception of global warming, causes severe restrictions on CO_2 emissions.

INTRODUCTION

There is a great deal of evidence that CO_2 levels in the atmosphere affect the environment and are increasing at a potentially dangerous rate. Today, the level of carbon dioxide in the atmosphere is 30% higher than at the beginning of the 19th century, with half of this increase occurring in the last 30 years.[1,2] Consequently, there is little doubt that this increase is due to anthropogenic causes. However, the potential consequence of this rise is and will continue to be argued for years to come. There are a significant number of researchers who predict highly deleterious effects of carbon dioxide, in particular, on the global climate.[3-7]

Regardless of whether or not the more ominous predictions will turn out to be correct, they are sufficiently alarming to motivate the search for alternative means for CO_2 disposal. If society could develop ways to safely and economically dispose of CO_2 in thermodynamically stable forms, we could eliminate the potential danger of climatic changes due to greenhouse gas emissions from burning of fossil fuels. Toward this end, we have recently outlined and developed a methodology for binding carbon dioxide in a solid form that is thermodynamically stable at the Earth's surface.[8-10] The carbonation of $Mg(OH)_2$ is of particular interest because of the abundance of magnesium on earth and the relative ease with which it can be extracted from Mg-bearing minerals. As will be discussed in greater detail below, the major focus of our research has been on developing ways to extract $Mg(OH)_2$ from common ultramafic rock, such as serpentinite and peridotite, and then rapidly and efficiently carbonating the powder to $MgCO_3$. The quantities of accessible deposits of these minerals vastly exceed the quantities of available fossil fuel in the world. Because this active approach would result in a waste that is thermodynamically stable, it would be possible to permanently dispose of great quantities of CO_2 with minimal environmental impact and without the danger of a sudden accidental release of gaseous CO_2, which, as will be discussed in more detail below, has proven fatal even in comparatively small releases.

The 10,000 Gigatons of coal reserves, in particular, could provide the world with energy for tens of generations.[11,12] By comparison, we currently consume about 6

Gigatons of coal annually.[7,11] However, coal's long-term use may be severely curtailed if we continue to dispose of CO_2 into the atmosphere. The availability of a CO_2 fixation technology would serve as insurance in case global warming, or the perception of global warming, causes severe restrictions on CO_2 emissions. If the increased energy demands of a growing world population are to be satisfied from coal, the implementation of such a technology would quite likely be unavoidable. There are now a number of proposed schemes other than our own for collecting and/or disposing of CO_2, including ocean disposal as liquid CO_2 or clathrates, the use of bacteria in the ocean to collect CO_2, planting of trees, and underground disposal. As with our technology, there is often a cost penalty for any technology which can reduce global emissions by a large fraction. We emphasize that there are other techniques that can provide small reductions in CO_2 emissions at little cost.[13] Thus, we view our CO_2 fixation concept as a viable option in case greater measures are required due to political or environmental pressures.

The economics of binding CO_2 in a solid form is strongly dependent on the kinetics of reactions between CO_2 and the host substrate. The kinetics of simultaneous thermal dehydroxylation and carbonation of precipitated $Mg(OH)_2$ has been studied in some detail.[10] We demonstrated that during carbonation, $MgCO_3$ precipitates on the surface of disrupted $Mg(OH)_2$ crystals acting as a kinetic barrier to both the outward diffusion of H_2O and the inward diffusion of CO_2. In this paper we describe our experimental work on extracting and carbonating $Mg(OH)_2$. The carbonation kinetics appear to be rapid enough for an economical, industrial process using fine particle size distributions and high pressures.

THE IMPORTANCE OF THE SAFE AND PERMANENT DISPOSAL

In many respects, the disposal or storage of greenhouse gases such as CO_2 can be viewed as having safety issues analogous to those associated with storage of other hazardous materials. Nature has given us a tragic, but valuable example of the potential danger of stored CO_2. In 1986, there was a massive release of CO_2 from Lake Nyos, which sits above a small village in Cameroon, Africa.[14-17] Being heavier than air, the cloud of gas flowed down the mountainside over the village and for distances of approximately 10 km into the surrounding countryside, asphyxiating all but plant life, including more than 1700 people and 3000 cattle.

All evidence suggests that the release of CO_2 was caused by a hydrodynamic instability. The most plausible explanation for the event seems to be that described by Zhang.[17] He proposed and demonstrated the feasibility that, due to a natural perturbation in the lake, the dense, CO_2 rich water at the bottom rose to a point where $p_{CO_2} \geq p_{total}$. Under this condition bubbles nucleated and grew and, subsequently, rose quickly to the lake's surface. The force associated with the rising gas lead to an overturning of the lake, causing increasingly vigorous outgassing, and ultimately a kind of limnic eruption of CO_2.

The 0.1 km^3 of CO_2, reported to be released in the disaster, was the equivalent of a week of emissions from a 1 GW, coal-fired powerplant. Thus, a large accidental release of CO_2 from an unstable storage deposit could be catastrophic. Furthermore, consider the safety issues that would be associated with storing the output of hundreds or thousands of power plants over many decades. Additionally, even slow leaks of CO_2 could be problematic since it only postpones, and could, therefore, compound the problem for future generations. Thus, it has been our objective in this research to offer a very safe method for binding CO_2 in a form that is stable virtually forever.

THE MAGNESIUM HYDROXIDE EXTRACTION PROCESS

A significant fraction of the world is covered with carbonated rock, mostly rich in Ca and to a lesser extent Mg. Essentially, what we are proposing is to make use of the huge quantity of rock in the world that has not been carbonated by natural events, but that could be carbonated. Magnesium bearing minerals constitute the most abundant and technologically available source of such material. Uncarbonated, Ca bearing materials are comparatively less common and we believe would be more costly to process. In particular, peridotites and serpentinite, which are approximately 30-50% MgO by weight, are available in quantities far exceeding what is required to consume the world's anthropogenic CO_2. Most of our research to date has focused on developing processes relying on serpentinite, which in its purest form has the composition $3MgO{\bullet}2SiO_2{\bullet}2H_2O$, and, therefore, contains 43.6 wt% MgO.

The direct carbonation of serpentinite, for example, is relatively slow. In order to economically bind the CO_2 in the host substrate, the magnesium must be extracted from the mineral. Fortunately, methods for extracting magnesium from minerals were worked out in some detail around the time of World War II and the advent of sea water extraction processes.[18,19] The economics of the processes were never optimized but appeared to be quite viable. Making use of this early research, we have set out to develop an economical process whereby the magnesium is extracted by first dissolving the mineral in hydrochloric acid. Through a multi-step extraction process, $Mg(OH)_2$ is precipitated out of solution and then thermally carbonated. Most of the HCl is recovered and reused in the extraction process. A simple schematic of the basic material flow is shown in Fig. 1 for a system using a peridotite with 45.8 wt% MgO. Technologies for collecting and transporting CO_2 are very well understood.[13] Thus, the focus of our work has been on the bottom half of Fig. 1 including the direct carbonation step. Many of the details of each step and the economics have been discussed preliminarily elsewhere[8,9], and are being described in greater detail in a forthcoming paper.[20,21]

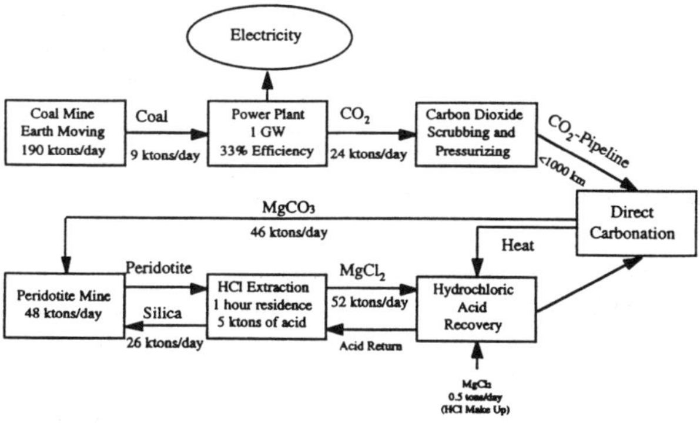

Figure 1. A simple diagram showing the flow of material in a CO_2 disposal scheme for a process using peridotite as the source for magnesium. The processing rates are geared to eliminate the emissions of a coal fired power plant producing 1 GW of electric power at 33% conversion efficiency. The focus of our research has been primarily the bottom half of the diagram including the direct carbonation step.

Table 1 summarizes the thermodynamics and mass balance equations relevant to the process outlined in Fig. 1. The overall process is exothermic and is defined by the following net reaction:

$$Mg_3Si_2O_5(OH)_4 + 3CO_2(g) \rightarrow 3MgCO_3 + 2SiO_2 + 2H_2O \qquad (1)$$

As written, equation 1 occurs very slowly except at very high pressures and temperatures. For example, in relatively unoptimized experiments, we have achieved near 25% completion in this reaction by exposing 100 μm serpentine powder to 340 atm of CO_2 at 500°C for two hours. Thus, in order to make the reaction go to completion under less costly conditions, we extract a more reactive form of magnesium by first dissolving the serpentine in hydrochloric acid as follows:

$$Mg_3Si_2O_5(OH)_4 + 6HCl(aq) + 13H_2O \rightarrow 3(MgCl_2{\cdot}6H_2O) + 2SiO_2 \qquad (2)$$

By heating the solution and distilling off water and HCl according to the series of reactions defined in table 1, solid Mg(OH)Cl is produced. Adding water to react with this solid, we then precipitate out $Mg(OH)_2$ in $MgCl_2{\bullet}6H_2O$ solution. In the laboratory, we can do this in essentially a two step process as follows:

$$MgCl_2{\bullet}6H_2O \xrightarrow{\sim 220°C} MgCl(OH) + HCl + 5H_2O \qquad (3)$$

followed by:

$$2MgCl(OH) \xrightarrow{H_2O} Mg(OH)_2 + MgCl_2 \qquad (4)$$

The $Mg(OH)_2$ is easily filtered from the solution for carbonation, whereas the $MgCl_2$ is recycled to the first step. In the following section, we describe the end process; that is, the essential and perhaps critical step of converting $Mg(OH)_2$ to a stable carbonate.

Table I. Heat balances for reactions important to the extraction of $Mg(OH)_2$ from serpentine and its subsequent carbonation. Note equation 1 is the net reaction of equations 2 and 12, 2 is the net of 3 an 4, 4 is the net of 5 and 11, 5 is the net of 6 and 7, and 8 is the net of 9 and 10. Thus, the molar heat of reaction 1 equals the sum of 2 and 12, and so on. The data were obtained from several sources.[29-31]

Eqn. No.	Reaction	Heat/ Equation, kJ	Corr. Factor	Heat/mole CO_2 kJ/mol
1	$Mg_3Si_2O_5(OH)_4 + 3CO_2 \rightarrow 3MgCO_3 + 2SiO_2 + 2H_2O(l)$	-190.8	1/3	-63.6
2	$Mg_3Si_2O_5(OH)_4 + H_2O(l) \rightarrow 3Mg(OH)_2 + 2SiO_2$	52.5	1/3	17.5
3	$Mg_3Si_2O_5(OH)_4 + 6HCl(aq) + 13H_2O(l) \rightarrow 3(MgCl_2{\bullet}6H_2O) + 2SiO_2$	-234.3	1/3	-78.1
4	$MgCl_2{\bullet}6H_2O \rightarrow Mg(OH)_2 + 2HCl(aq) + 4H_2O(l)$	96.7	1/1	96.7
5	$MgCl_2{\bullet}6H_2O \rightarrow MgCl_2{\bullet}H_2O + 5H_2O(l)$	103.3	2/1	206.5
6	$MgCl_2{\bullet}6H_2O \rightarrow MgCl_2{\bullet}H_2O + 5H_2O(g)$	323.3	2/1	646.6
7	$H_2O(g) \rightarrow H_2O(l)$	-44.0	10/1	-440.1
8	$MgCl_2{\bullet}H_2O \rightarrow Mg(OH)Cl + HCl(aq)$	-0.15	2/1	-0.3
9	$MgCl_2{\bullet}H_2O \rightarrow Mg(OH)Cl + HCl(g)$	74.7	2/1	149.4
10	$HCl(g) \rightarrow HCl(aq)$	-74.9	2/1	-149.7
11	$2Mg(OH)Cl + 6H_2O \rightarrow Mg(OH)_2 + MgCl_2{\bullet}6H_2O$	-109.5	1/1	-109.5
12	$Mg(OH)_2 + CO_2 \rightarrow MgCO_3 + H_2O(l)$	-81.1	1/1	-81.1

THE DIRECT CARBONATION STEP

Preliminary studies of the carbonation kinetics of $Mg(OH)_2$ have been completed under atmospheric conditions[10] and more recently at 30 to 440 atmospheres. The results generally follow trends predicted from thermodynamics. However, the processes can be quite complicated and the mechanisms are not fully understood at this point. Understanding the mechanisms could lead to significant improvements in the efficiency of the process. Figure 2, which was calculated from thermodynamic data, shows the effect of pressure on the dissociation temperatures of $MgCO_3$ and $Mg(OH)_2$. It is apparent that the dissociation temperatures of both the carbonate and hydroxide increase significantly with pressure. Thus, the carbonation of $Mg(OH)_2$ is inherently limited by the dissociation temperature, which is a function of the pressure of carbon dioxide. Above the dissociation temperature, which is near 400°C at atmospheric pressure, the carbonate will decompose according to the reaction:

$$MgCO_3 \rightarrow MgO + CO_2(g) \qquad (5)$$

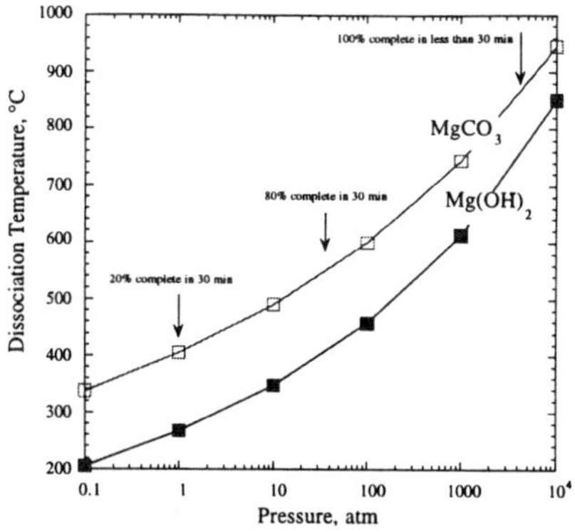

Figure 2. Calculated thermodynamic dissociation temperatures for $MgCO_3$ and $Mg(OH)_2$ to MgO and CO_2 or H_2O, respectively. The arrows show how the efficiency of carbonation improves with pressure and temperature for a precipitated $Mg(OH)_2$ powder heat treated 30 minutes. Note the x-axis refers to the total CO_2 or H_2O pressures.

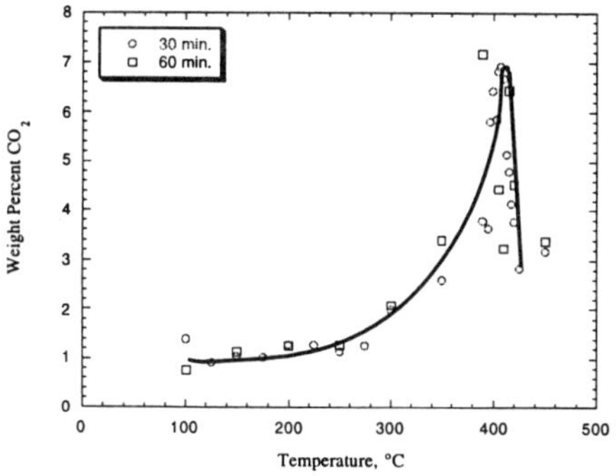

Figure 3. Measured CO_2 content of precipitated $Mg(OH)_2$ carbonated for 30 and 60 minutes in 0.765 atm of CO_2. The powder had a 20 μm average particle or agglomerate size, where the agglomerates were made up of approximately 100 nm $Mg(OH)_2$ crystallites. Note these data were obtained by measuring the volume of CO_2 liberated during reaction with 6M HCl as described in detail elsewhere.[10,22]

Our studies have shown that there is a strong interconnectivity between the dehydroxylation and carbonation reactions. As illustrated in Fig. 3, the rate of carbonation is most rapid very near the dissociation temperature of $MgCO_3$ (near 390°C at the altitude of our laboratory). Note that the carbonate fractions in Fig. 3 were determined by measuring the CO_2 evolved during reaction with 6M HCl.[10,22] During carbonation of $Mg(OH)_2$, there is simultaneous dehydroxylation and carbonation. Thus, it is important to fully understand both the kinetics and mechanisms of each process. The kinetics and mechanisms of dehydroxylation of single crystal $Mg(OH)_2$ are relatively well understood.[23,24] In the case of precipitated $Mg(OH)_2$ (which is polycrystalline) the kinetics of dehydroxylation can vary significantly from one material to another.[10,23-28] Also, the mechanism of dehydroxylation of precipitated $Mg(OH)_2$ has not been fully elucidated. The variations between the different studies appear to be real; that is, the kinetics are strongly dependent on purity, sample size, crystallite and agglomerate size, thermomechanical history, and environment. Despite these complications, we can still make some general statements about the process of simultaneous dehydroxylation and carbonation.

As shown in Fig. 3, using a 20 μm precipitated $Mg(OH)_2$ powder, the carbonation reaction reached approximately 20% completion in 30-60 minutes, assuming that the carbonate that forms is $MgCO_3$. We have been able to get substantially more carbonation (approximately 50% completion) under the same conditions by grinding our powders to a finer particle size. However, the efficiency of carbonation at atmospheric pressure is relatively slow. This is due to the fact that the carbonate that forms is a barrier to both the outward diffusion of water, thus inhibiting dehydroxylation, and the inward diffusion of CO_2, thus inhibiting carbonation.[10] This is illustrated by the X-ray diffraction data (XRD) shown in Fig. 4. The sample that was heated to 350°C in helium was converted almost entirely to MgO, while samples heated in CO_2 retained a significant amount of $Mg(OH)_2$ up to the $MgCO_3$ dissociation temperature.

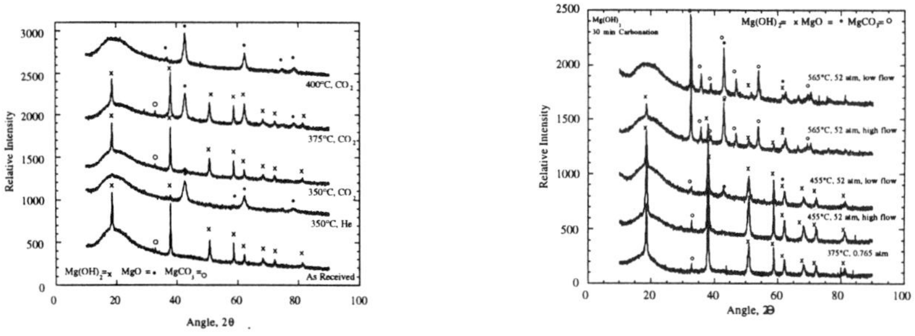

Figure 4. X-ray diffraction patterns of as-received $Mg(OH)_2$ powder and those after 12 hour heat treatments in 0.765 atm of carbon dioxide and helium (left) and after carbonation in 0.765 to 52 atm of CO_2 (right).

Typically, the rate of a thermally activated reaction increases with temperature. As indicated in Fig. 2, by increasing the pressure of CO_2, we can substantially increase the temperature at which $MgCO_3$ dissociates. Therefore, by increasing the pressure of CO_2, we can increase the temperature at which we perform the carbonation step and presumably accelerate the carbonation rate. We have done this and found that relatively modest increases in pressure give substantially more rapid rates of carbonation. For example, at 565°C and 52 atm, we have achieved 90% carbonation in 30 minutes using the same 20 μm powder described above. The final product is significantly different from that obtained at lower pressures and temperatures. Figure 5 shows the XRD spectra, which can be contrasted to those in Fig. 4. By pushing the dissociation temperature to higher values through increase pressure, we get substantial carbonation. As Fig. 4 illustrates, the carbonation kinetics increase substantially between approximately 450 and 550°C. This enhance kinetics is also apparent in the microstructures. Figures 5 and 6 contrast the microstructures of the as received $Mg(OH)_2$, and those after heat treatments in 0.765 atm of CO_2 at 375°C, and 52 atm of CO_2 at 565°C, respectively. As shown in Fig. 6, the crystallites within the particles heat treated at the higher pressure and temperature are transformed to a mixture of $MgCO_3$ with a minor amount of MgO with a cubic morphology. Also, the final product is more chemically stable compared with material heat treated at lower temperatures. This is illustrated in Fig. 7 where we show the rate at which CO_2 is liberated upon reaction with 6M HCl. The powders heat treated at 540 to 565°C react relatively slowly with the acid compared with samples treated at 430 to 455°C.

Based on these results it appears that the final process will require the use of moderately high pressures. We do not view this as a problem because CO_2 is typically transported under similar pressures. Thus, there may be no need to add extra energy to the system since the pipeline will supply at least part of the required pressure.

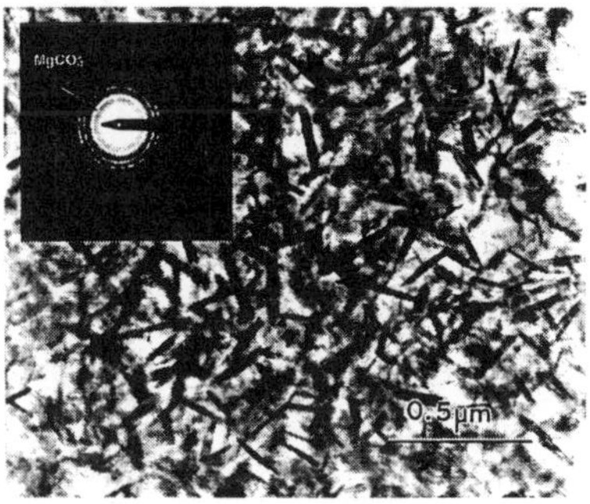

Figure 5. Transmission electron micrograph showing the crystalline morphology inside an agglomerate of precipitated $Mg(OH)_2$. The platelet shaped crystallites are $Mg(OH)_2$. There is a slight diffraction ring associated with $MgCO_3$ which is present as an impurity (≈2 wt. %) in the powder.

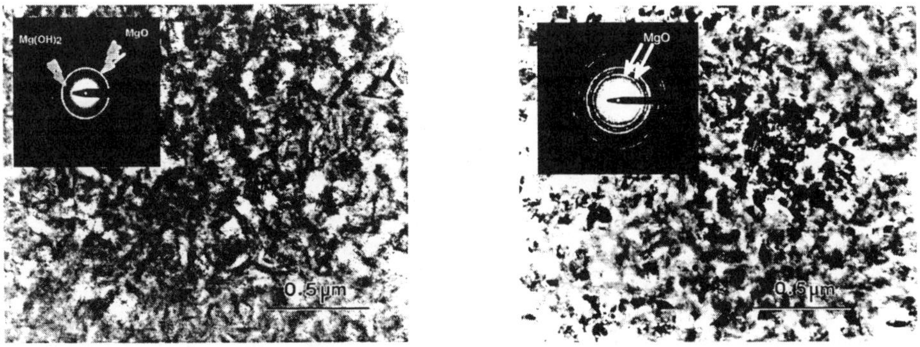

Figure 6. Transmission electron micrographs showing (at left) the crystalline morphology inside an agglomerate of precipitated $Mg(OH)_2$ after carbonation for 12 hours in 0.765 atm of CO_2 at 375°C. What's left is a collection of disrupted $Mg(OH)_2$ crystallites in which MgO has precipitated. Very little $MgCO_3$ can be detected by TEM or XRD, although from gas analysis it was determined that the sample contained approximately 6% CO_2 by weight. And (at right) crystalline morphology inside an agglomerate of precipitated $Mg(OH)_2$ after carbonation for 30 minutes in 52 atm of CO_2 at 565°C. What's left is a collection of nanometer sized $MgCO_3$ crystallites with a minor amount of unconverted MgO. From gas analyses it was determined that the sample was 42% CO_2 by weight, that is, the carbonation reaction was nearly 90% complete.

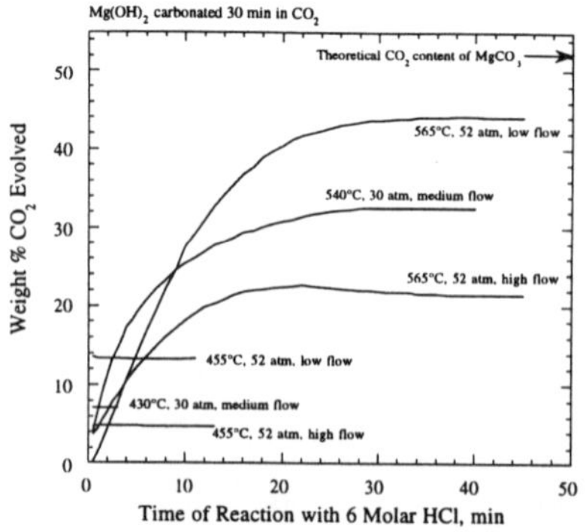

Figure 7. Plot of weight percent carbon dioxide evolved during reaction with 6M HCl from $Mg(OH)_2$ samples carbonated between 30 and 52 atm at various temperatures.

ECONOMICS AND NICHE MARKETS

Binding our anthropogenic CO_2 may sound like a colossal undertaking and difficult philosophical change for society; however, it may be possible to bootstrap our way to the final solution by first taking advantage of niche markets. For example, in our operation we can accept and process waste asbestos, a hazardous material that contains approximately 40 wt. % MgO. Thus, it would be possible to charge a fee to dispose of this waste making the operation more profitable. Or, in the short term, some fraction of the final, carbonated products, as well as the silica and iron impurities, could have some value. For example, the raw materials could be sold for use in glass, concrete, or other building products. We estimate that in the short term, sales of iron could effectively reduce the cost of CO_2 collection by approximately $10/ton.[20,21] Other niche markets could reduce the process costs to as little as 1¢ versus our more conservative estimate of 3¢/kWh.[8]

However, in order to consume a sufficient quantity of CO_2, we will in all likelihood need to ignore the niche markets in our final economic analysis because of the great quantities of material that will necessarily be generated. That is, the value added to the final, carbonated product will be negligible. Thus, disposal ultimately becomes our only real option. As noted above, we are in the process of doing a rigorous economic assessment of our process, using the data that we have available, and will report those results in at least one forthcoming publication.[20,21] However, in a previous paper we assessed the probable cost of our process by comparing it to other, similar large-scale mining and chemical operations.[8] (For the purpose of comparison, in these estimates we ignored the additional cost associated with extracting and shipping the CO_2 because these are upfront costs that must also be incurred by other competitive technologies that aim at removing CO_2 by an engineered route. The cost of extracting and shipping CO_2 is likely to be on the order of 3¢ per kWh and would be incurred by all CO_2 collection technologies.) Using relatively simple but logical assumptions, it was shown that the cost of binding CO_2 by our process would be on the order of $30 per ton of carbon dioxide. This translates into about 3¢ per kWh based on the U. S. emissions of CO_2 from coal, which amount to approximately 90 gms of CO_2 per MJ of produced energy.[32] Thus, if implemented immediately, our process would roughly triple the cost of electricity. However, the efficiency of coal fired processes, in particular, has steadily improved during the last 30 years. Thus, through relatively gradual implementation of our technology and continued improvements in efficiency, the economic impact of our proposed technique would be less difficult to absorb.

SUMMARY REMARKS

Regardless of one's view of the consequences of global warming, it would prudent for the U. S. and other countries to invest in research aimed at understanding both the effects of and how to better mitigate CO_2 emissions. The Japanese, for example, have recognized the importance of CO_2 mitigation technologies and recently established the Research Institute of Innovative Technologies for the Earth (RITE) which is an organization largely dedicated to studying ways to collect, bind, and reuse CO_2. The political pressure to curb CO_2 emissions into the atmosphere is gradually rising. At the 1996 Earth Summit in Geneva, the United States agreed for the first time to the principle of setting legally binding limits on total carbon dioxide emissions. And in his 4 February 1997 State of the Union Address, President Clinton stated his commitment to reducing greenhouse gas emissions. This trend, regardless of its scientific merits, poses a risk to fossil fuel power and mining industries. However, opportunities may also be created as the market for mitigation technologies could become quite large.

Our research has been aimed at binding CO_2 in a thermodynamically stable solid, namely magnesium carbonate. In our laboratory we have demonstrated the feasibility of extracting $Mg(OH)_2$ from common minerals, and that the $Mg(OH)_2$ can be converted by a direct thermal treatment to a stable carbonate for subsequent disposition using CO_2 pressures comparable to those used in its normal transportation. The processes that we have developed can be carried out on a massive scale, sufficient enough to bind in theory all anthropogenic CO_2. The quantity of useful minerals in the world, namely serpentinite and peridotites, are far in excess of what is needed to collect the CO_2 emissions from the remaining world's supply of coal. The kinetics of extraction and carbonation have yet to be optimized. Likewise, thermal management in the $Mg(OH)_2$ extraction process has yet to be optimized. Further research at both our laboratories (Los Alamos and Chichibu Onada Cement Co.) is aimed at enhancing the reaction kinetics. However, we are nearing the stage where a pilot demonstration of the entire process can be justified. It is hoped that the first such system will be developed within the next few years.

REFERENCES

1. U. Siengenthaler and H. Oeschger, "Biospheric CO_2 Emissions During the Past 200 Years Reconstructed by Deconvolution of Ice Core Data," *Tellus*, **39B**, 140-154 (1987).
2. C. D. Keeling, T. P. Whorf, M. Wahlen and J. van Der Plicht, "Interannual Extremes in the Rate of Rise of Atmospheric Carbon Dioxide Since 1980," *Nature*, **375**, 666-670 (1995).
3. S. Manabe and R. T. Wetheral, "Thermal Equilibrium of the Atmosphere with a Given Distribution of Relative Humidity," *J. Atmos. Sci.*, **24**, 241-259 (1967).
4. V. Ramanathan, "The Greenhouse Theory of Climate Change: A Test by an Inadvertent Global Experiment," *Science*, **240**, 293-299 (1988).
5. J. Darmstadter and J. Edmonds, "Human Development and Carbon Dioxide Emissions: The Current Picture and the Long-Term Prospects," in Greenhouse Warming: Abatement and Adaptation, Eds. N. J. Rosenberg, W. E. Easterling III, P. R. Crosson, and J. Darmstadter, Resources for the Future, Washington, D. C., 1989.
6. S. H. Schneider and N. J. Rosenberg, "The Greenhouse Effect: Its Causes, Possible Impacts, and Associated Uncertainties," in Greenhouse Warming: Abatement and Adaptation, Eds. N. J. Rosenberg, W. E. Easterling III, P. R. Crosson, and J. Darmstadter, Resources for the Future, Washington, D. C., 1989.
7. G. M. Hidy and D. F. Spencer, "Climate Alteration," *Energy Policy*, **22**, 1005-1027 (1994).
8. K. S. Lackner, C. H. Wendt, D. P. Butt, D. H. Sharp, and E. L. Joyce, "Carbon Dioxide Disposal in Carbonate Minerals," *Energy*, **20** [11] 1153-1170 (1995).
9. K. S. Lackner, C. H. Wendt, D. P. Butt, and D. H. Sharp, "Carbon Dioxide Disposal in Solid Form," in the proceedings of *21st International Technical Conference on Coal Utilization and Fuel Systems*, B. A. Sakkestad Editor, Coal and Slurry Technology Association, Washington, D. C., pp. 133-144, 1996.

10. D. P. Butt, K. S. Lackner, C. H. Wendt, S. Conzone, H. Kung, Y.-C. Lu, and J. K. Bremser, "Kinetics of Thermal Dehydroxylation and Carbonation of Magnesium Hydroxide," *J. Am. Ceram. Soc.*, **79** [7] 1892-1898 (1996).
11. United Nations, *1991 Energy Statistics Yearbook*, New York, 1993.
12. M. G. Schomberg, Ed., *1992 Survey of Energy Resources*, World Energy Council, 16th Edition, London, United Kingdom, 1993.
13. H. Audus, P. W. F. Riemer, and W. G. Ormerod, "Greenhouse Gas Mitigation Technology Results of CO_2 Capture and Disposal Studies," in the proceedings of *20th International Technical Conference on Coal Utilization and Fuel Systems*, Coal and Slurry Technology Association, Washington, D. C., pp. 349-358, 1995.
14. G. W. Kling, M. A. Clark, H.R. Compton, J. D. Devine, W. C. Evans, A. M. Humphrey, E. J. Koeningsberg, J. P. Lockwood, M. L. Tuttle, and G. N. Wagner, "The 1986 Lake Nyos Gas Disaster in Cameroon, West Africa," *Science*, **236**, 169-175 (1987).
15. R. Ladbury, "Model Sheds Light on a Tragedy and a New Type of Eruption," *Physics Today*, 20-22, May 1996.
16. S. J. Freeth, "Lake Nyos: Can Another Disaster be Avoided?," *Geochemical Journal*, **28** [3] 163-172 (1994).
17. Y. Zhang, "Dynamics of CO_2-Driven Lake Eruptions," *Nature*, **379**, 57-59 (1996).
18. E. C. Houston, "Magnesium from Olivine," *A.I.M.E. Tech. Pub.*, No. 1828, **85**, 1-14 (1945).
19. E. A. Gee, C. E. McCarthy, F. S. Riordan, Jr., and M. T. Pawel, "Magnesia from Olivine," U. S. Bureau of Mines Report 3938, 1946.
20. K. S. Lackner, D. P. Butt, and C. H. Wendt, "Progress on Binding CO_2 in Mineral Substrates," *Energy Convers. Mgmt.*, **38**, S259-S264 (1997).
21. K. S. Lackner, D. P. Butt, and C. H. Wendt, "Carbon Dioxide Disposal in Mineral Form: Keeping Coal Competitive," Los Alamos National Laboratory Technical Report, in preparation, 1997.
22. D. L. Pile, K. S. Lackner, D. P. Butt, A. S. Benjamin, and C. H. Wendt, "A Simple Method for Determining the CO_2 Content of Materials," submitted to *J. Chem. Ed.*, 1997.
23. R. S. Gordon and W. D. Kingery, "Thermal Decomposition of Brucite: I, Electron and Optical Microscope Studies," *J. Am. Ceram. Soc.*, **49** [12] 654-660 (1966).
24. R. S. Gordon and W. D. Kingery, "Thermal Decomposition of Brucite: II, Kinetics of Decomposition in Vacuum," *J. Am. Ceram. Soc.*, **50** [1] 8-14 (1966).
25. S. J. Gregg and R. I. Razouk, "The Kinetics of the Thermal Decomposition of Magnesium Hydroxide," *J. Chem. Soc. London*, S36-S44 (1949).
26. P. J. Anderson and R. F. Horlock, "Thermal Decomposition of Magnesium Hydroxide," *Faraday Soc. Trans.*, **58**, 1993-2004 (1962).
27. Y. Laureiro, A. Jerex, C. Pico, and M. L. Veiga, "Controlled Decomposition Rate Thermal Analysis of $Mg(OH)_2$ and $Cd(OH)_2$. Kinetic Study," *Thermochimica Acta*, **182**, 47-56 (1991).
28. I. Halikia and A. Economacou, "Application of Various Methods of Nonisothermal Kinetic Analysis to Magnesium Hydroxide Decomposition," *Int. J. Chem. Kinetics*, **25**, 609-631 (1993).
29. R. A. Robie, B. S. Hemingway, and J. R. Fisher, "Thermodynamic Properties of Minerals and Related Substances at 298.15 K and 1 Bar (105 Pascals) Pressure and Higher Temperatures," United States Government Printing Office, U. S. Geological Survey Bulletin, 1452, Washington, D. C. (1978), reprinted with corrections, 1979.
30. I. Barin, *Thermochemical Data of Pure Substances*, 2nd Edition, Verlagsgesellschaft, Germany, 1993.
31. M. W. Chase, Jr., C. A. Davies, J. R. Downey, Jr., D. J. Frurip, R. A. McDonald, and A. N. Syverud, *JANAF Thermochemical Tables*, 3rd Ed., American Institute of Physics, New York, 1985.
32. Energy Information Administration, "Carbon Dioxide Emission Factors for Coal: A Summary," Monthly Energy Review, Sept., 1994.

THE RECYCLING OF SPENT REFRACTORIES FROM A SECONDARY BRASS PROCESSOR

K.S. Kwong, J. P. Bennett
Albany Research Center, DOE, Albany, OR 97321

A. E. Wynne III
National Metals, Inc., Leeds, AL 35094

ABSTRACT

The Albany Research Center is conducting research on recycling/alternate uses for spent refractories. Refractory bricks were sampled in a rotary reverberatory furnace from a secondary brass processor. These spent bricks consisted primarily of 70% alumina shaped refractory. Physical, chemical, and microstructural properties of the spent refractory material have been studied. The results showed that Pb existed in some parts of the furnace. The location of Pb components and how these components are present will be discussed. Chemical reactions between metal, slag, and refractory in the furnace environment will be addressed. Some suggestions for beneficiation processes and refractory improvements will be proposed.

INTRODUCTION

Refractories are essential materials in the manufacture of a wide variety of products, ranging from iron and steel, nonferrous metals, glass, chemicals, petroleum base products to metal casting. Refractories have been used as insulation, molten metal containment, and erosion resistant materials. According to statistical survey information from Bureau of Census, approximately 3.4 million metric tons of refractory materials are used annually with a value of two billion dollars (1). After removal from service, landfill has been the primary disposal means for spent refractories (2). Depending on the user location and local regulations, the cost of landfill of nonhazardous spent refractories can range from \$40 to \$120 per ton or even higher (3). Concerns for the environment, landfill costs, environmental legislation, product liability associated with disposed materials, and product stewardship programs are the driving forces for recycling spent refractories or finding alternate uses of spent materials (4). Landfill disposal of industrial wastes has become or will be restricted or eliminated in many developed countries. The major benefits for recycling spent refractories will be to reduce operating costs, waste disposal costs, future liability for waste materials, or even generating income from wastes that can be sold (5).

Refractories range from acid materials (such as silica) to neutral (such as alumina) to basic materials (such as MgO and dolomite). Depending on service environments and mechanical requirements, different types of refractories are used in different industries. For example, the refractory used by some secondary brass producers in their rotary reverberatory furnaces is a 70% Al_2O_3 material.

Brass and bronze may contain from 55% to 95 wt.% copper as the main ingredient with zinc and a small amount of lead, tin, iron, aluminum, silicon or other elements (6). Based on the standard free energy of formation for metal oxides, Pb can be reduced to the metal easier than elements such as Cu and Zn. Zn can be oxidized easier than Pb and Cu. Both zinc and lead have high vapor pressures. At the furnace operating temperature of 1650 °C reached in brass production, zinc (Zn) and lead oxide (PbO) have a vapor pressure at 760 mm and lead has about 400mm (7). Lead can form several oxides and is denser than zinc. It is expected that the 70% Al_2O_3 refractory materials present in the secondary brass furnace may react with compounds of these elements by solid-solid, solid-liquid, or solid-vapor reactions at different locations within the furnace.

Recycling of spent refractories from a rotary reverberatory furnace used in brass and bronze production is being evaluated. The rotary reverberatory furnace has a capacity of 11,000 Kg and is heated from room temperature to 1650° C with a borax flux in a batch mode. The gas burner system is oxygen enriched and the furnace can pivot on its axis up to 100°. In general, the furnace can last about 170 heats (5 months) before installing a new refractory lining. It is estimated that approximately 7 tons of spent refractory bricks are removed each time the furnace is repaired/relined..

The characterization of spent refractories from this furnace is a critical step in developing an optimum method or strategy to recycle refractories and to find alternate uses for them. Because of refractory contaminations, desirable materials and undesirable are separated or concentrated using beneficiation. Beneficiation processes include physical methods such as hand sorting, screening, magnetic and gravitation separation, or chemical methods such as leaching and flotation. The selection of a beneficiation process is based on the characteristic of the "wanted" or "unwanted" materials, process costs, the degree of beneficiation desired and other factors such as plant space. The characterization of spent refractory materials may also contribute to the understanding of how refractories change during service, which may lead to improvements in their performance and/or service life.

The goals of this study are to characterize spent refractories from a secondary brass producer, to suggest suitable beneficiation methods, to understand the reactions of refractory and molten slags and metals in the secondary brass manufacturing processes and to explore the evolution of Pb compounds in the refractory.

EXPERIMENTAL PROCEDURE

Spent 70% Al_2O_3 refractory bricks were sampled from eight locations in a rotary reverberatory furnace. These areas represent locations in the roof (A), above (B) and below (C) the metal-slag interface, the furnace base (D), and exhaust (E) and the burner port (F). Figure 1 shows the schematic of the furnace and the sampling locations. After sampling, bricks were photographed, visual observations noted and the sample size and weight recorded. Every spent refractory sample has a distinct transition zone when moving from the hot face to the cold face. This line indicated two different layers. Usually, the layer closest to the hot face is more contaminated than the one closest to the cold face. Several measurements of the line depth were taken for each sample to determine the average of penetration at the metal/slag line.

The eight samples of spent refractories and one unused brick were analyzed for chemistry, X-ray diffraction crystalline phases and SEM microstructure. Two samples were taken from each brick except sample D; one on the brick surface facing molten metals, slag, or vapors (the hot face); the other taken from in the interior of the brick. One extra sample was taken on the cold face of sample D (bottom) because of its interesting surface appearance. Chemical analysis was done by atomic absorption spectrophotometry and wet chemistry methods. SEM microstructural analysis revealed Cu, Zn and Pb compounds, the location and the size of these compounds, and other characteristics of interest. This information will provide the basis for selecting or developing beneficiation processes.

Figure 1. Schematic of where samples were removed from the rotary reverberatory furnace used in secondary brass and bronze processing.

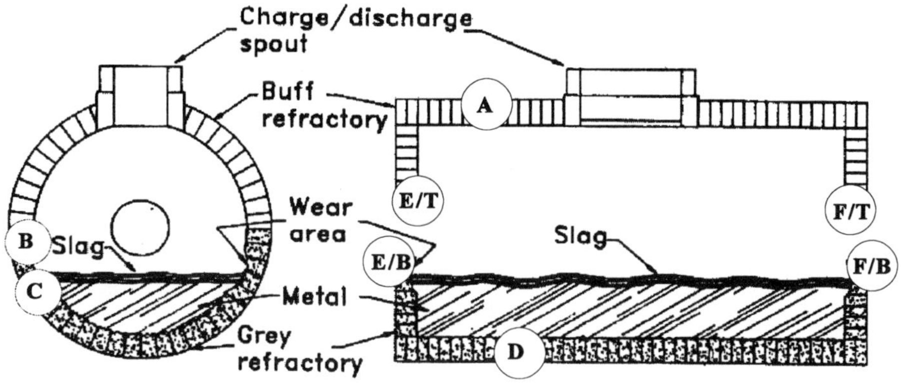

RESULTS AND DISCUSSION

Observation:

Table I shows the penetration depth, the brick surface and interior color, size, and weight of spent refractory samples. Brick A has a thin metal/slag reaction layer on the hot face. Once the surface layer is removed, brick A looked like the original brick. Brick A represented material from the roof and comprised about 40% of the refractory material removed from the furnace. Brick B was taken from the side of the furnace above the molten metal/slag line and was much thinner (more eroded/corroded) than the other bricks. This indicated severe corrosion and erosion occurred in this area due to the rocking of the furnace and constant "washing" and exposure of the refractory to metal, slag, and air. On the other hand, brick C did not experience the severe corrosion or erosion of brick B. It was on the side of the furnace below the molten metal/slag line. Brick C's physical appearance was similar to that of brick A. Brick D's surfaces were completely covered with glass and slag of brown and green colors. The interior of brick D shows black-gray colors. It was not severely worn as brick B. Brick D was on the furnace bottom and represented about 40% of the brick in the furnace. Brick E is located at the exhaust side. Brick F located at the burner side of the furnace. Brick F was thinner than others bricks with the exception of brick B. The flame temperature of the rotary reverberatory was not monitored but was thought to be as hot as 1800°C.

Chemical and X-ray Diffraction Analysis:

Table II lists the chemical analysis of six elements found in the different samples. Copper

Table I. The location and physical properties of spent refractory samples.

Sample	location	Penetration Visible Depth (mm)	Color		Length (mm)	Width (mm)	Thickness (mm)	Weight (g)
			exterior	interior				
A	roof	30	black	light yellow	218	230	68	7800
B	above metal/ slag line	66 - 80	black	gray	108	190	70	4330
C	below metal/ slag line	40 - 90	black	gray/yellow	180	227	73	7260
D	bottom	45 - 70	green/brown	gray/black	180	225	70	7700
E/B	exhaust bottom	45 - 80	black	light gray	181	114	77	4660
E/T	exhaust top	25 - 125	black	white yellow	188	115	77	4830
F/B	burner bottom	30 - 60	black	gray	169	113	76	4070
F/T	burner top	45 - 60	black	light yellow	139	112	77.5	3240
Original			yellow	yellow	226	114	76	4860

Table II The chemical composition and crystallography of spent refractory samples

Samples		Si (%)	Al (%)	Pb (ppm)	Cu (ppm)	Zn (ppm)	Sn (ppm)	Crystallography
A	surface	4.06	12.70	521	205000	203000	7640	$ZnAl_2O_4$, Fe_3O_4
	interior	10.50	36.10	79	45	44	<100	Al_2O_3, $Al_6Si_2O_{13}$, SiO_2
B	surface	8.52	27.60	10900	45800	62400	3220	Al_2O_3, $ZnAl_2O_4$, $Al_6Si_2O_{13}$, SiO_2, TiO_2
	interior	10.20	35.50	8550	332	236	621	Al_2O_3, $Al_6Si_2O_{13}$, SiO_2, TiO_2
C	surface	9.18	31.40	1400	12200	9480	623	Al_2O_3,$ZnAl_2O_4$, $Al_6Si_2O_{13}$, TiO_2, Fe_3O_4
	interior	10.40	35.30	340	41	26	<100	Al_2O_3, $Al_6Si_2O_{13}$, Fe_2O_3, SiO_2, TiO_2
D	surface	9.34	30.00	13600	13700	14400	1790	$ZnAl_2O_4$, Al_2O_3, $Al_6Si_2O_{13}$, TiO_2, SiO_2
	interior	10.40	35.30	10600	589	1570	1400	Al_2O_3, $Al_6Si_2O_{13}$, TiO_2, $ZnAl_2O_4$, SiO_2
E/T	surface	2.66	17.40	2970	244000	163000	3030	Al_2O_3, $ZnAl_2O_4$, $Al_6Si_2O_{13}$, TiO_2
	interior	3.58	43.40	<10	22	20	240	Al_2O_3, $Al_6Si_2O_{13}$, SiO_2, TiO_2
E/B	surface	3.91	37.10	5950	8240	9950	403	$ZnAl_2O_4$, Al_2O_3
	interior	3.44	43.20	41	40	82	<100	Al_2O_3, $Al_6Si_2O_{13}$
F/T	surface	10.20	34.00	1320	4140	6840	<100	Al_2O_3, $Al_6Si_2O_{13}$, $ZnAl_2O_4$, TiO_2, SiO_2
	interior	10.60	36.80	44	425	806	<100	Al_2O_3, $Al_6Si_2O_{13}$, TiO_2, $ZnAl_2O_4$, SiO_2
F/B	surface	9.77	32.90	2690	113	8510	494	Al_2O_3, $Al_6Si_2O_{13}$, $ZnAl_2O_4$, TiO_2, SiO_2
	interior	10.40	32.80	941	265	1610	<100	Al_2O_3, $Al_6Si_2O_{13}$, SiO_2, TiO_2, SiO_2

and zinc each comprised about 20% of the surface material of brick A (roof area), dropping to very low levels in the brick interior. Lead was not present in significant concentration on the roof (brick A) but was present in high concentrations on both the surface and interior of brick D (bottom of the furnace) and in location B. From the ratio of Al to Si, it seems that brick E was made from higher Al_2O_3 starting material than other bricks. Brick E was located at the exhaust side of the furnace. X-ray analysis indicated that Al_2O_3, $ZnAl_2O_4$, and $Al_6Si_2O_{13}$ were major phases in most samples. No crystalline phases were identified containing lead by X-ray diffraction. Based on chemical analysis of Table II and past experience, it was estimated that less than 50%of the refractory removed after usage required beneficiation for removal of hazardous lead. High lead levels are confined to material below the molten metal area.

SEM Micrographs:

As previously discussed, the selection of beneficiation processes requires information such as the beneficiated particle size, the physical and chemical characteristics of beneficiated particles, the location and the abundance of beneficiated particles, etc. SEM analysis can supply much of this information while X-ray diffraction and chemical analysis can't. This information can also be used to infer service environment and to recommend possible refractory improvements. Because this paper focused on the characterization of hazardous Pb element and compounds, only microstructure information related to goals of this study will be discussed in this section.

Table III summarizes the Pb-containing phases found in the refractories, their forms, and their location (occurrence). Nine different formations of Pb compounds were identified by SEM while none were found by X-ray crystalline phase analysis. None of the lead phases were predominant in the matrixes. Each type formation will be discussed separately:

Figure 2. shows bright acicular particles about 40-100 μm long and 10-20 μm wide on sample A's hot surface. EDAX indicated these particles consisted of about 9 wt.% Pb, 4 wt.% Cu, 8 wt.% Ti, 6 wt.% Fe, and 16 wt.% Al. This type formation was only observed on the surface of sample A and was not identified as any known crystalline species. It seems that the vapor or compounds of Pb, Cu, Ti, and Fe deposited on the refractory surface and reacted with Al_2O_3 to form these acicular particles. These phases are in a glassy matrix.

Figure 3 shows tiny bright spherical particles in the matrix of the sample A hot surface. These particles contained 17 wt.% Cu, 9 wt.% Pb and some Na, Al, and Ti. Because these particles are so small (1-3 μm), the surrounding material may affect the accuracy of EDAX analysis. The same spherical particles were found on the hot surface of samples D.

Figure 3 also shows a glassy matrix containing Pb compounds with a smooth surface appearance. EDAX indicated it usually consists of 1-5 wt.% Pb and coexists with Si, Fe, Al, Ca, and O. It seems that lead oxide compounds prefer to interact with SiO_2 rather than Al_2O_3. This type of lead compound formation was also observed on the surface of samples A, B, D, E/B, E/T, F/B and F/T and in the interior of brick D, but not at D's cold face (bottom).

Figure 4 shows a lead-copper alloy particle in a sample from C's surface. Some lead metal is phase separated in a copper particle. Many of these alloy particles are spheres with a size about 1000 μm (1 mm). This formation was also observed on the surface of samples taken from D.

Figure 5 shows lead oxide particles in a crack on the hot face of the sample C surface. These particles were about 3-5 μm in size and contained about 50-70 wt.% Pb. This indicated lead/ lead oxide vapor could diffuse through and condense in fine cracks. Lead oxides may have five different oxide states -- PbO_2, PbO, Pb_3O_4, Pb_2O_3, and Pb_2O. PbO melts at approximately 888 °C; all others decompose below 500 °C. Based on EDAX analyses, tiny particles observed in figure

Table III. The Pb phases, forms, and locations of spent refractory samples

	forms	acicular	Tiny Pb compound	Glassy	Pb-Cu Alloy	PbO particle	Pb-Si-O Sphere	Pb compound band	Pb compound	PbO chunk	pb metal
	location	matrix	matrix	matrix	matrix	crack	matrix	crack	matrix	crack	crack
	Figure No.	2	3	3	4	5	6	7	8	9	10
A	surface	X	X	X							
	interior										
B	surface			X							
	interior										
C	surface				X	X					
	interior										
D	surface		X	X	X	X	X	X	X		
	interior			X							
	bottom									X	X
E/T	surface			X							
	interior										
E/B	surface			X							
	interior										
F/T	surface			X							
	interior										
F/B	surface			X							
	interior										

5 may be PbO_2. If so, the formation of PbO_2 indicates that the temperatures at the deposition sites were low. This type of phase formation (PbO_2) was observed only in cracks on the surface of sample C and D.

Figure 6 shows spherical particles containing 17 wt.% Pb with 16 wt.% Si (lead silicate) that were located in cracks surrounding a matrix rich in Si, Fe, Zn, Ca, and containing 12 wt. % Pb. These lead silicate particles are about 3μm in size. This type formation was observed in the sample D surface.

Figure 7 shows lead phases that filled a crack and form a band. EDAX indicated these phases contained about 9 wt.% Pb, 16 wt.% Si and 8 wt.% Al. This type of formation was observed in the sample D surface.

Figure 8 shows compounds containing Pb as particles precipitated in a glassy matrix formed in the hot face of sample D. These particles appear to have crystal-like faces. It indicates that the Pb compounds were saturated and crystallized in this local glassy matrix. The Pb particles are about 1-3 μm in size.

Figure 9 shows lead compounds about 20-30 μm in size deposited in a large crack on the cold face of sample D. This type of formation is different from the formation of tiny PbO_2 deposition in the crack (figure 5). The deposit shown in figure 5 probably forms through gas transport and deposition. In figure 9, however, the molten lead compounds appear to have flowed through a hole and the lead oxide compound was pushed outward and deposited in the crack.

Figure 10a shows a large area of lead-containing material in a crack; 10b shows the X-ray map for the element of lead, and 10c for the element of oxygen. As shown in figure 10b, Pb metal is present in the crack. Formations of this type were only observed on the cold face of sample D. As previously mentioned, Pb oxide can easily be reduced to metallic lead. Because CO_2 is more stable thermodynamically than CO below 700 °C. The following reaction may be possible for the metallic Pb formation

$PbO + CO \Rightarrow Pb + CO_2$

For PbO to be reduced, the environment occurring in the cold face of sample D would have to be reducing. In support of this, free carbon was noted visually in this area of the furnace.

The following mechanism is proposed to describe the evolutions of Pb compounds in the refractory liner during melting and processing of copper alloys. It is assumed that one side of the brick is hot and in contact with the molten copper-zinc-lead metal, slag, and/or hot gases. The temperature gradually decreases when moving from the hot face to the cold face. In the beginning, Pb molten metal may form an alloy with Cu (figure 4) or Pb vapor may react with Cu oxide to form an acicular crystal (figure 2) on the hot surface. Pb may also be oxidized and react with free quartz, aluminosilicates and some alkaline or alkali oxide to form glassy matrixes (figure 3). Because, Pb has a high vapor pressure, some Pb or lead oxide may volatilize from a glassy matrix in the refractory liner and deposit cracks to form a particle or a band (figure 5 and 7). If Pb was saturated in the molten glass, then Pb compounds may be precipitated during cooling (figure 8). Because of gravity, the glassy molten fluid continues to penetrate into a lower temperature zone of the refractory liner. Lead compounds then start to crystallize and are pushed outward into cracks (figure 9). If the environment is in slightly oxygen deficient, then metal lead may be formed (figure 10).

RECOMMENDATIONS

Suggestions given below describe 1) ways to beneficiate spent refractories from a secondary

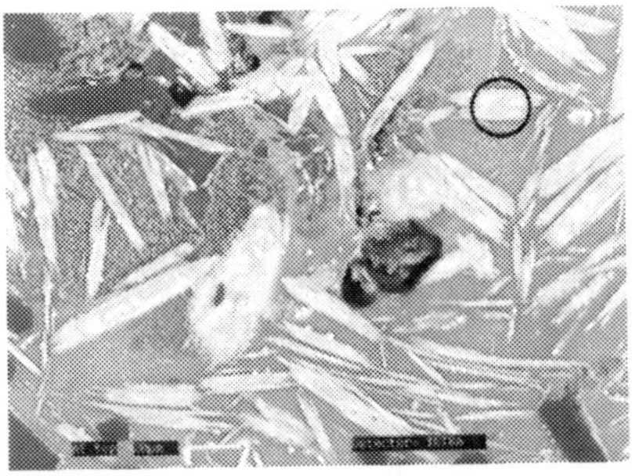

Figure 2. SEM micrograph of a lead containing phase identified on the hot zone surface of sample A (roof area).

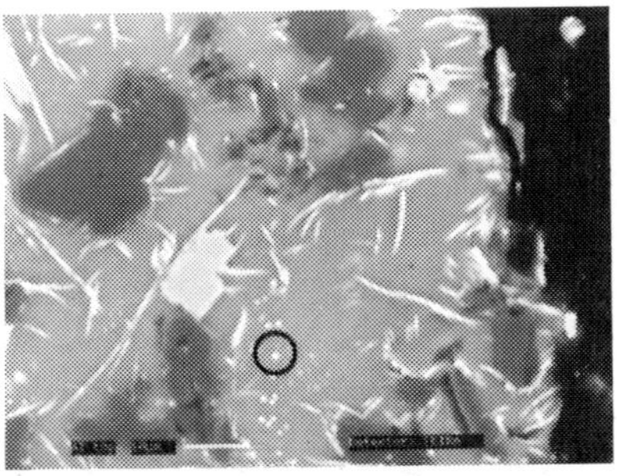

Figure 3. SEM micrograph of a lead containing spherical phase identified in the hot zone surface of sample A (roof)

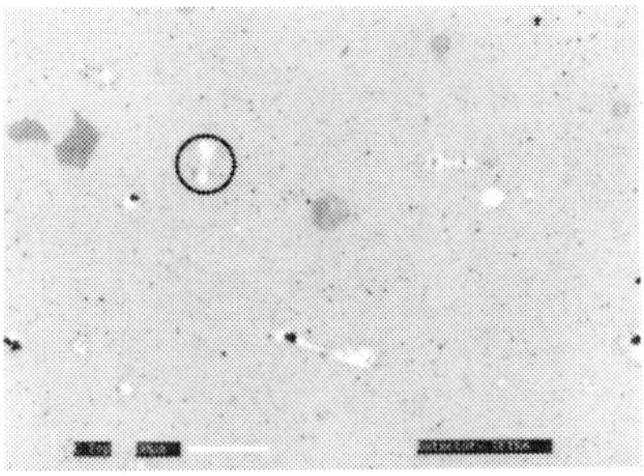

Figure 4. SEM micrograph of a Pb-Cu alloy particle taken from the surface of the hot zone surface of sample C.

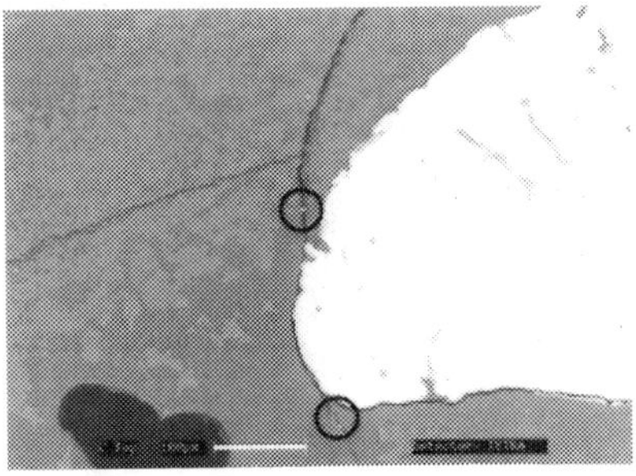

Figure 5. SEM micrograph of lead oxides particles present in a crack taken on the hot face of sample C.

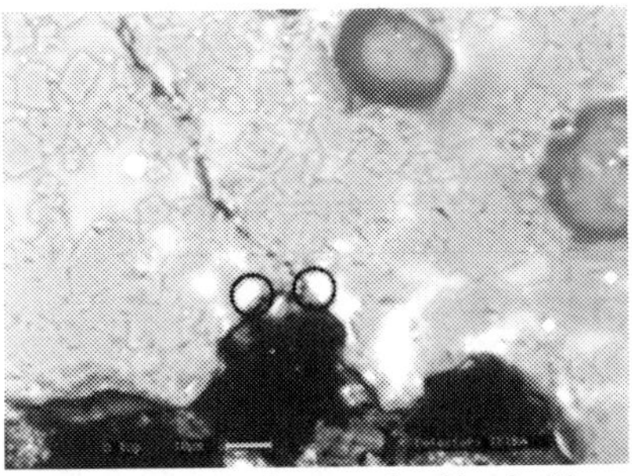

Figure 6. SEM micrograph of lead silicate spherical particles on the hot face surface of samples.

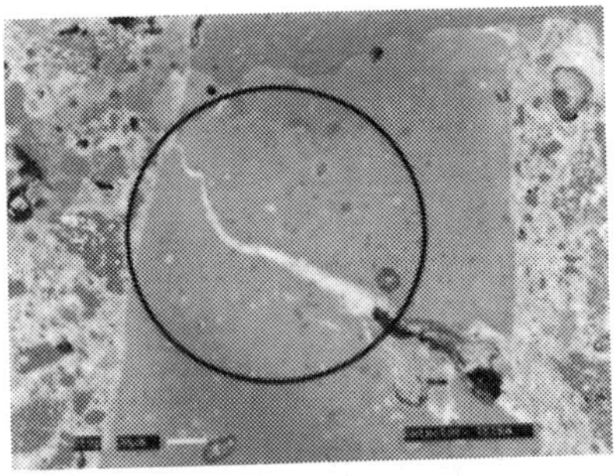

Figure 7. SEM micrograph of a lead phase filling the crack of a sample removed from the hot face of sample D.

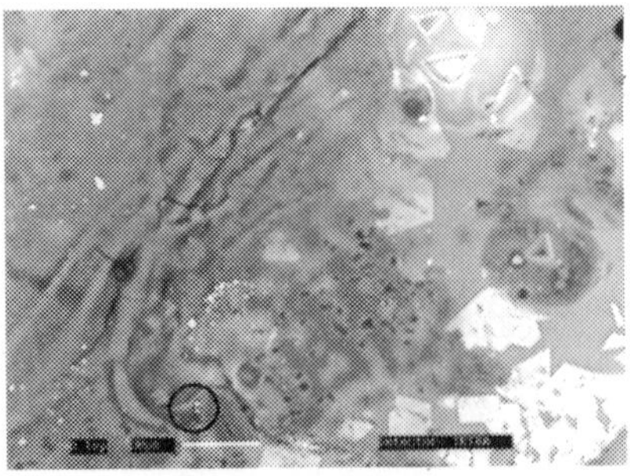

Figure 8. SEM micrograph of lead compounds precipitated near the hot face of sample D's surface.

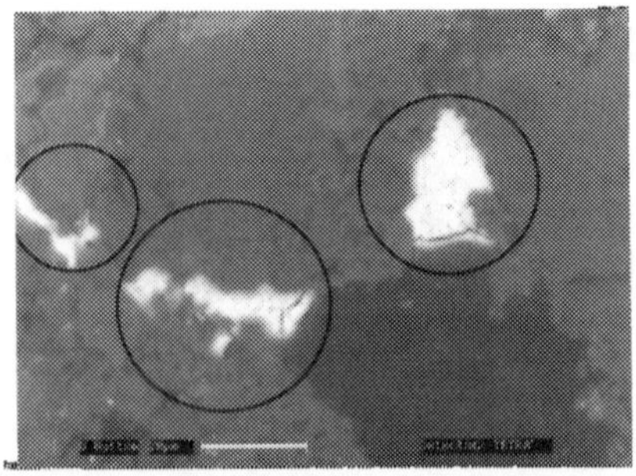

Figure 9. SEM micrograph of lead oxide compounds deposited in a crack near the cold face of sample D..

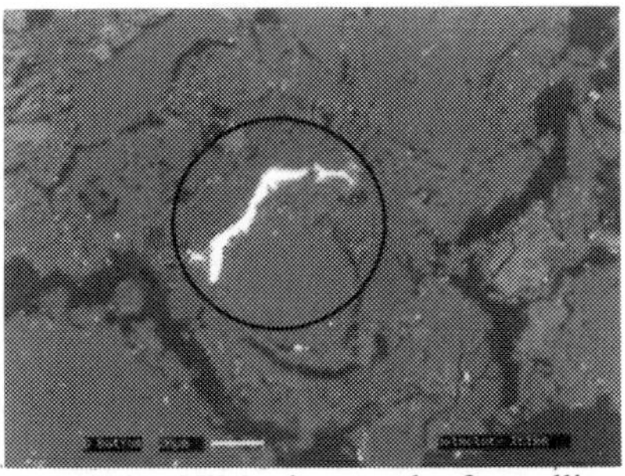

Figure 10a. SEM micrograph of metallic lead and lead oxides present in the cold face of sample D.

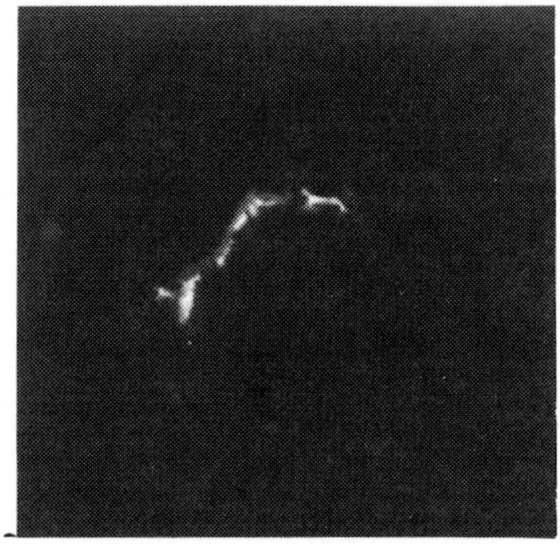

Figure 10b. Elemental map of lead in the cold face of sample D.

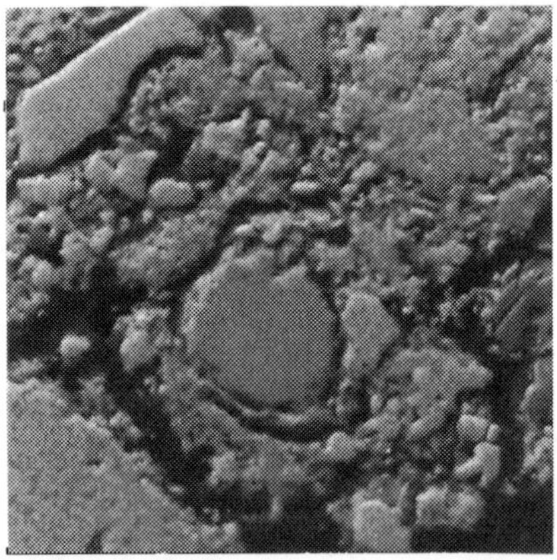

Figure 10c. Element map of oxygen in the cold face of sample D.

brass producer's rotary reverberatory furnace, 2) possible refractory improvements to reduce recycling and to concentrate unwanted materials, and 3) possible changes in furnace operating conditions to reduce refractory waste.

A) Beneficiation:

Most of the roof refractory bricks are clean, containing only a thin surface layer of hazardous lead compounds. Therefore, putting a thin layer of a gunning mixture, castable or "mold wash" over the brick's surface on the interior of the furnace may help in the removal of the hazardous surface layer or reduce its adherence or may reduce material penetration into the surface of the refractory. Mechanical abrasion of the surface before the furnace is torn apart may also be possible to remove the "slag" layer.

Because the glassy matrix contains the majority of hazardous lead components and tiny lead oxide particles, a leaching process is suggested to remove these materials from the spent refractories or from the slag/metal surface. After leaching, electrolytic or chemical deposition processes could be used to remove lead from the leaching solution.

After concentrating Pb containing materials, alternate uses should be found to reuse clean spent refractory and Pb materials.

B) Refractory Improvements:

For ease of beneficiation, the use of higher or improved refractories should be considered to avoid the formation of the glassy matrix where hazardous lead concentrates in a thin surface layer. Possible approaches for improving refractories are as follow:

1) Decrease the impurity content in the refractories

Impurities such as alkali, and alkaline earths, or iron will decrease the viscosity of a glassy matrix, allowing the molten matrix to penetrate deep into a brick. This makes lead components harder to concentrate. A reduction in brick impurities would help minimize Pb contamination into the refractory.

2) Decrease the fine free quartz content and increase particle size of raw material

It was observed that the glassy matrix formed by lead compounds prefers to react with fine free quartz and other fine particle materials. By increasing the particle size of these materials, smaller quantities of glassy material may form.

3) Use low porosity refractories in metal contact areas of the furnace through zoning

Low porosity refractories may reduce the penetration of glassy matrixes into bricks.

4) Use carbon containing refractories

Optimum carbon content in a refractory material may reduce the lead oxide to metalic lead before it has a chance to react with quartz or aluminosilicates forming glassy materials. This may concentrate Pb metal in the copper alloys.

5) Use monolithic refractories

A concept similar to the endless lining concept utilized in the steel industry may be possible. Only thick layers of metal and slag need be removed from the furnace hot face, then a new lining is cast over the old for use as metal lining (8).

6) Apply a mold wash or a thin vanier layer over bricks to reduce slag and metal penetration during use.

7) Wash the furnace with slag before first use to seal the furnace

8) Develop and improve performance refractories.

C) Furnace Operation:

Furnace operating conditions are another critical factors that influence hazardous lead pickup in refractories. The following suggestions would reduce hazardous lead pickup but may not be practical because of their adverse effect on productivity. The recommendations are as follows:

1) Reduce/eliminate furnace cycling

The reduction/elimination of furnace cycling may prevent crack growth and surface spalling which allows Pb penetration into refractory.

2) Modify the flux formula to reduce the Pb compounds' penetration into the refractory lining.

3) Lower furnace operating temperature

Lowing the furnace operating temperature may decrease Pb volatilization in the glass matrix and increase the viscosity of glass. and therefore decrease Pb penetration into and interaction with refractories.

CONCLUSION

Spent refractory bricks were sampled in a rotary reverberatory furnace from a secondary brass-bronze processor. These spent bricks consisted primarily of 70% Al_2O_3 shaped refractory. Physical and chemical analyses indicated that more than 50% of the spent refractory in the furnace may be easily recycled. Microstructural analysis of refractories in different areas of the furnace indicated at least nine different formations of Pb compounds and their sizes and locations. A mechanism was proposed to describe the evolution and formation of Pb compounds in the refractory. Recommendations were also made to beneficiate spent refractories from the furnace, to improve refractories to avoid contamination, and to modify the furnace firing schedule with the goal of reducing furnace contamination.

REFERENCES

[1]Bureau of Census, U.S. Department of Commerce News, Current Industrial Report, 1995 Annual report on Refractories, Report MA32C(95)-1.

[2]J. P. Bennett and J. K. Kwong, "Recycling/Alternative Use of Spent Refractories" Iron and Steel [1] 23 (1997).

[3]Conversation between numerous people across country

[4]J. P. Bennett, J. K. Kwong, and S. W. Sikich, "Recycling and Disposal of Refractories" American Ceramic Society Bulletin. 74[12] 71 (1995).

[5]Center for hazardous materials research "Pollution prevention: The Environmental Strategy of the Nineties" Fact sheet downloaded from Internet.

[6]R. E. Kirk and D. F. Othmer, Encyclopedia of Chemical Technology, The Interscience Encyclopedia Inc. New York. NY 1960.

[7]R. C. Weast, Handbook of Chemistry and Physics, 45th edition, The Chemical Rubber Co. Cleveland, Ohio 1964.

[8]E. Alasaarela and W. Eitel, "How Infinite is Endless Lining of Ladles" Paper presented at the Unififed International Technical Conference on Refractories. Sao Paulo, Brazil, October 31-November 3, 1993. pp.1267-1278.

WASTE MANAGEMENT TECHNOLOGIES IN THE CERAMIC AND NUCLEAR INDUSTRY
1997 Annual Meeting
Cincinnati, OH

Historical Aspects of Radioactive Materials

Rutherford and Soddy: The Cause and Nature of Radioactivity

Montague Cohen, PhD., F.Inst.P., FCCPM
Physics Dept., McGill University, Montreal, Qc, Canada

Radioactivity is a physical phenomenon in which the atom disintegrates, with the emission of particles and/or radiation. As a result, the atom converts into an atom of another element, with different chemical properties. The study of radioactivity therefore involves both physics and chemistry and, at least in the early days, required the collaboration of a physicist and a chemist. Two such partnerships immediately come to mind: Pierre and Marie Curie, and Ernest Rutherford and Frederick Soddy.

You will, of course, protest that both Pierre and Marie Curie were physicists and neither had a formal qualification in chemistry. That is true, but when we examine their joint work, we find that Marie Curie effectively turned herself into a chemist. There was no such ambiguity in the case of Rutherford and Soddy - Rutherford was a physicist by training and in practice, while Soddy was a chemist by the same criteria. My paper today is concerned solely with the Rutherford-Soddy collaboration.

Before discussing the partnership a few words are needed on the background to their collaboration. In the early 1890s it was widely assumed that physical science had reached an end-point: everything about the physical world was known and it remained only to measure physical constants with ever-greater precision. This cosy picture was rudely shattered by Roentgen's discovery of X-rays in November 1895, followed by Becquerel's discovery of radioactivity in March 1896. The two discoveries were closely linked, since Becquerel was intrigued by the fact that the emission of penetrating radiation in a vacuum discharge tube was accompanied by visible fluorescence in the glass walls of the tube. Becquerel reasoned that the phosphorescence of naturally occuring compounds of uranium might also be accompanied by an invisible penetrating radiation.

The new phenomenon of radioactivity was chosen by Marie Curie in 1897 as the subject of her doctoral thesis. Almost immediately she reported that radioactivity is an *atomic* phenomenon and by December 1898 she and her husband Pierre had discovered two new radioactive elements, polonium and radium. Meanwhile Gerhard Schmidt, in Berlin, had discovered that thorium and its compounds were also radioactive. By 1902 the Curies had succeeded in isolating a sample of one-tenth of a gram of pure radium.

Ernest Rutherford

Rutherford was born in 1871 in the South Island of New Zealand. In 1892 he obtained his B.A. degree at Canterbury College in Christchurch, and his M.A. the following year. He remained at Canterbury to undertake research for the B.Sc., which at that time was a research degree. The subject of his research was the magnetization of iron by high-frequency electrical discharges. He did not investigate radioactivity, for the simple reason that this phenomenon had not yet been discovered.

In 1895 Rutherford was awarded a scholarship to carry out research at Cambridge University in England. Only one award was available to New Zealanders, and Rutherford was the runner-up, but the winner was unable to take up the scholarship for personal reasons and the opportunity fell to Rutherford. This was the only time in his career when luck played a significant role. He arrived in Cambridge in the autumn of 1895, bringing his New Zealand apparatus, and immediately resumed his research under the direction of Professor J.J. Thomson.

During 1896 - it is not clear exactly when - Rutherford gave up his magnetization research and joined Thomson in investigating the phenomena of ionization in gases [1,2]. At first Rutherford regarded the newly-discovered X- and uranium-radiations merely as tools to induce ionization, but by 1898 these priorities had been reversed and the measurement of ionization became primarily a tool for investigating the new radiations. Rutherford's final submission from Cambridge, published in January 1899, was a 55-page paper entitled "Uranium Radiation and the Electrical Conduction Produced by it." [3]

In the autumn of 1898 Rutherford moved to

Montreal, where he was appointed Professor of Physics at McGill University. During his first three years in this post, before the onset of his partnership with Soddy, Rutherford published 10 papers on various aspects of radioactivity. In particular, in January 1900, Rutherford reported that thorium compounds "continuously emit radioactive particles of some kind, which retain their radioactive power for several minutes." [4] Furthermore, this "emanation" (also emitted by radium) is able to induce radioactivity on all surfaces on which it falls. This was the beginning of Rutherford's investigations of radioactive series and of "excited" or "induced" radioactivity. In particular, he attempted to determine the nature of the radium emanation by measuring its rate of diffusion and hence its atomic weight.(Fig. 1) [5] This experiment yielded an answer of between 40 and 100, which was much too low - the actual value is 222 - but the result convinced Rutherford that emanation was not simply radium vapor, since radium was known to have an atomic weight of at least 140. This experiment was, in fact, an important step in the process whereby Rutherford came to realize that radioactivity involves the conversion of an atom of element "A" into an atom of element "B".

<u>Frederick Soddy</u>

Soddy was born in England in 1877 and educated at Oxford University, where he graduated (with first-class honors) in 1898. He remained at Oxford for two more years and undertook some independent research. In the spring of 1900, after an unsuccessful application for the post of Professor of Chemistry at the University of Toronto, he was appointed a Demonstrator in Chemistry at McGill University.

Soddy's first encounter with Rutherford was in March 1901, when he defended the classical concept of the indivisible atom in a debate at the McGill Physical Society on "The Existence of Particles Smaller than the Atom." In spite of his initial opposition to Rutherford's ideas, in the autumn of 1901 Soddy joined Rutherford in research that led, in 1902-03, to the disintegration theory of radioactivity.

<u>The Rutherford-Soddy Collaboration</u>

The Rutherford-Soddy partnership lasted only 18 months, from October 1901 to March 1903, but resulted in nine papers totalling almost 150 pages of print. As far as we know, they never worked together in the same laboratory, but the Physics and Chemistry Buildings were adjacent on the McGill downtown campus.

The first joint paper, titled "The Radioactivity of Thorium Compounds," was published in 1902 in the Transactions of the Chemical Society. Part I of this paper was subtitled "An Investigation of the Radioactive Emanation."[6] The authors showed that the production of thorium emanation is a specific property of thorium, and the amount of emanation is directly proportional to the weight of thoria (thorium oxide) present. Furthermore, the process of "de-emanating" thoria by heating does not permanently destroy its emanating power, which fully recovers with time. Finally, the experiments indicated that emanation was a chemically inert gas, like argon, but its origin remained in doubt. One possibility was that an inert constituent of the atmosphere was rendered radioactive in the presence of thoria. Alternatively, the emanation might be an inert, radioactive gas continuously emitted by thoria, but "in the present state of knowledge, it would be premature to choose between these two alternatives."

Part II of "The Radioactivity of Thorium Compounds" was subtitled "The Cause and Nature of Radioactivity." The authors showed that about 54% of the radioactivity of thorium is due to a substance which they named "thorium-X." This has distinct chemical properties and is temporarily radioactive with a half-life of about four days. The constant radioactivity of thorium is maintained by the production of ThX at a constant rate independent of the physical and chemical condition of the system. One of the products of ThX is the radioactive gaseous emanation produced by exciting radioactivity on surrounding inactive matter. Finally, Rutherford and Soddy concluded that "radioactivity is at once an atomic phenomenon and the accompaniement of a chemical change in which new kinds of matter are produced. The two considerations force us to the conclusion that radioactivity is a manifestation of sub-atomic change."[7]

The Cause and Nature of Radioactivity

The next Rutherford-Soddy joint paper, published in two parts in the September and November 1902 issues of the *Philosophical Magazine*, was a landmark in the history of radioactivity.[8] The title was "The Cause and Nature of Radioactivity" and the paper presented the first concise and logical explanation of the phenomena. Part I of the paper was concerned with the properties of thorium-X and the relationship between thorium and thorium-X. The authors showed that the radioactivity of thorium at any given time is the resultant of two opposing processes, i.e. the production of fresh radioactive material at a constant rate by the thorium compound, and the decay of the active material with time. They concluded that throium's constant radioactivity is an equilibrium value depending on a balance between the rate of production of fresh radioactive material and the rate of decay of radioactive material already formed.[9] (The half-life of thorium is over 10^{10} years, virtually infinite, whereas that of thorium-X is 3.64 days.)

The experimental curves for the recovery of the activity of thorium after the chemical separation of thorium-X, and for the decay of thorium-X, showed an anomalous behavior in the first two days. Rutherford and Soddy investigated this problem by separating the thorium-X from the thorium by repeated precipation, and concluded that the initial irregularities in the curves were due to secondary radiation excited by ThX in the surrounding matter. (It must be borne in mind that the thorium decay series was unknown at the time of this investigation by Rutherford and Soddy. There are actually three intermediate nuclides between thorium and thorium-X or, more precisely, between thorium-232 and radium-224.)

In Part II of the "Cause and Nature" paper Rutherford and Soddy investigated varius aspects of the gaseous emanation produced by thorium. They concluded that thorium emanation, like radium emanation, behaves in all respects like a temporarily radioactive gas, including its ability to diffuse through porous substances such as thick cardboard. Fig. 2 shows the apparatus they used for comparing the emanating power of different substances. The half-life of thorium emanation was found to be about one minute. The paper included a discussion of the chemical nature of the emanation. Rutherford and Soddy rejected the possibility that emanation was the result of excited radioactivity in the atmosphere and concluded that "... the emanation is a chemically inert gas analogous in nature to members of the argon family.[10]

The Condensation of the Radioactive Emanations

Towards the end of 1902 Rutherford and Soddy published a short note announcing that they had succeeded in condensing both thorium and radium emanations.[11] A full account of the investigation appeared in the May 1903 issue of the *Philosophical Magazine*.[12] The apparatus used (Fig. 3) comprised a spiral copper tube immersed in a Dewar (thermos) flask containing liquefied ethylene. The copper spiral also served as a resistance thermometer. The result indicated that thorium emanation condenses at about -150°C. The authors concluded that this result, as well as the properties of the emanation previously described, leave no doubt that the emanation must consist of matter in the gaseous state.[13]

The Radioactivity of Uranium

The April 1903 issue of the *Philosophical Magazine* included two papers by Rutherford and Soddy. The first was a short paper on the radioactivity of uranium.[14] In 1900 Crookes showed that the photographic activity of uranium is due to the presence of a minute amount of a substance which he called uranium-X.[15] The latter emits ß-rays, which are readily deviated in a magnetic field, but uranium freed from uranium-X emits only α-rays, which are mainly responsible for the ionization effect but, at least at the time, could not be deviated in a electric or magnetic field. (Three years later Rutherford showed that the α-rays are indeed deviated in a powerful field.[16]) Rutherford and Soddy showed that uranium-X loses its activity with time according to an exponential law, with the exponential constant having an approximate value of 0.031/day. They also showed that uranium, free from uranium-X, recovers its activity by the same law and with the same value of the constant.

The Radioactivity of Radium and Thorium

In the second of the two joint papers published in April 1903, Rutherford and Soddy compared the radioactivity of radium and thorium.[17] They questioned the existence of a "radium-X" analogous to "thorium-X," but the amount of radium at their disposal was so small that they were unable to obtain a definite answer to this problem. They also measured the half-life of radium emanation (radon) and obtained a value of 3.71 days, which is close to the present value of 3.82 days. Rutherford and Soddy also found that the emanating power of solid radium chloride is only a tiny fraction of that of a solution of radium chloride, and the same applied to solid thorium mitrate. In other words, most of the emanation is occluded in the solid compound.

Finally, in this paper, Rutherford and Soddy noted that radium itself, emanation and the activity excited by emanation all emit only α-rays. Thus the evidence indicated that the ß-rays are emitted only in the second or third change of the excited activity.[18]

Radioactive Change

Rutherford and Soddy published their final joint paper, titled "Radioactive Change," in the May, 1903 issue of the *Philosophical Magazine*.[19] In many ways this was their most important joint publication, since it discussed in detail the nature of the radioactive process. The paper points out that the radioactivity of radium, thorium and uranium involves the continuous production of new kinds of matter which possess temporary activity. Some of these new bodies are gaseous in character. Unfortunately the quantity of these new materials is far too small to enable chemical and spectroscopic studies to be made. When several changes occur together they are successive, not simultaneous. The radioactivity of each substance is not connected with the change in which it was itself produced, but with the change in which it produces the next new substance. Rutherford and Soddy expressed this rule as follows: "Hence it is not possible to regard radioactivity as a consequence of changes that have already taken place. The rays emitted must be an accompaniment of the change of the radiating system into the one next produced."[20]

The paper continues with a discussion of the exponential law of radioactive decay, and the introduction and definition of the "radioactive constant, λ. Rutherford and Soddy pointed out that "... the complexity of the phenomena of radioactivity is due to the existence ... of several types of matter changing at the same time into one another, each type possessing a different radioactive constant."[21]

The next section of the paper discusses the relation of radioactive change to chemical change. In this section, Rutherford and Soddy came close to defining the existence of isotopes, for which they suggested the name *metabolon*, but fortunately this term was not generally accepted.[22]

The paper concludes with a detailed analysis of the energy involved in a radioactive change and the internal energy of the chemical atom. In February 1903 Rutherford had measured the velocity and charge/mass ratio of the α-particle emitted by a radium source.[23] From these values Rutherford and Soddy calculated the kinetic energy of each projected α-particle as 10^{-5} erg. Taking 10^{20} as the probable number of atoms in one gram of radium (the actual number is somewhat larger, about 3×10^{21}), they calculated the total energy of the rays from one gram of radium as 2.4×10^{7} gram-calories. Since five successive disintegrations were known, each corresponding to the projection of at least one alpha-ray, the total energy emitted during the disintegration of 1 gram of radium could not be less than 10^{8} gram-calories. Furthermore, the energy radiated may be only a part of the total energy of disintegration. In comparison, the union of hydrogen and oxygen liberates approximately 4×10^{3} gram-calories per gram of water produced, and this is the maximum for any known chemical change. Rutherford and Soddy concluded that "... the energy of radioactive change must therefore be at least twenty thousand times, and may be a million times, as great as the energy of any molecular change."[24] In conclusion, Rutherford and Soddy pointed out that "... the energy of radioactive change ... can only be a portion of the internal energy of the atom, for the internal energy of the resulting products remains unknown." Hence "... the energy latent in the atom must be enormous compared with that rendered free in ordinary change ... there is no reason to assume that this enormous store of energy is possessed by the radio-elements alone. It seems probable that atomic energy in general is of a similar, high order of magnitude ..."[25] This amazing conclusion was reached 40 years before the detonation of the first atomic bomb.

Conclusion

Frederick Soddy left McGill in March 1903 to take up a post in University College, London. Thus four of the joint Rutherford-Soddy papers were actually published after Soddy had left Canada. Rutherford remained at McGill another four years, until 1907, when he moved to Manchester University in England. During these four years, Rutherford published several papers with co-authors, but none of his later partnerships were as significant, or as fruitful, as that with Frederick Soddy.

References

1. J.J. Thomson and E. Rutherford, "On the Passage of Electricity through Gases exposed to Roentgen Rays," Philosophical Magazine, ser. 5, **42** 392-407 (1896).
2. E. Rutherford, "The Velocity and Rate of Recombination of the Ions of Gases exposed to Roentgen Radiation," Philosophical Magazine, ser.5, **44**, 422-440 (1897).
3. E. Rutherford, "Uranium Radiation and the Electrical Conduction Produced by it," Philosophical Magazine, ser.5, **47**, 109-163 (1899).
4. E. Rutherford, "A Radioactive Substance emitted from Thorium Compounds," Philosophical Magazine, ser.5, **49**, 1-14 (1900).
5. E. Rutherford and H.T. Brooks, "The new Gas from Radium," Transactions of the Royal Society of Canada, section iii, ser.2, **7**, 21-25(1901).
6. E. Rutherford and F. Soddy, "The Radioactivity of Thorium Compounds," Part I, "An Investigation of the Radioactive Emanation." Transactions of the Chemical Society, **81**, 321-350(1902).
7. E. Rutherford and F. Soddy, "The Radioactivity of Thorium Compounds," Part II, "The Cause and Nature of Radioactivity," Transactions of the Chemical Society, **81** 837-860 (1902).
8. E. Rutherford and F. Soddy, "The Cause and Nature of Radioactivity," Philosophical Magazine, ser.6, **4**, 370-396 and 569-585 (1902).
9. Ibid., p. 378.
10. Ibid., p. 583.
11. E. Rutherford and F. Soddy, "Note on the Condensation Points of the Thorium and Radium Emanations," Proceedings of the Chemical Society, 1902, pp. 219-220.
12. E. Rutherford and F. Soddy, "Condensation of the Radioactive Emanations," Philosophical Magazine, ser.6, **5**, 561-576 (1903).
13. Ibid., p. 576.
14. E. Rutherford and F. Soddy, "The Radioactivity of Uranium," Philosophical Magazine, ser.6, **5**, 441-445 (1903).
15. W. Crookes, "Radio-activity of Uranium" Proc. Royal Society, **66**, 409-423 (1900).
16. E. Rutherford, "Magnetic and Electric Deflection of the Alpha-Rays from Radium," Physical Review, **22**, 122-123 (1906).
17. E. Rutherford and F. Soddy, "A Comparative Study of the Radioactivity of Radium and Thorium," Philosophical Magazine, ser.6, **5**, 445-459 (1903).
18. Ibid, p. 457.
19. E. Rutherford and F. Soddy, "Radioactive Change," Philosophical Magazine, ser.6, **5**, 576-591 (1903).
20. Ibid, p. 578.
21. Ibid, p. 581.

22. Ibid, p. 586.
23. E. Rutherford, "The Magnetic and Electric Deviation of the Easily Absorbed Rays from Radium," Philosophical Magazine, ser.6, 5, 177-187 (1903).
24. Ref. 19, p. 589.
25. Ref. 19, p. 591.

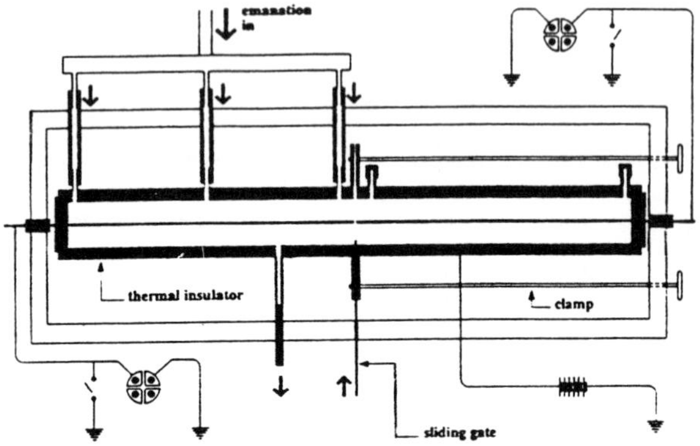

Fig. 1
(Ref. 5)

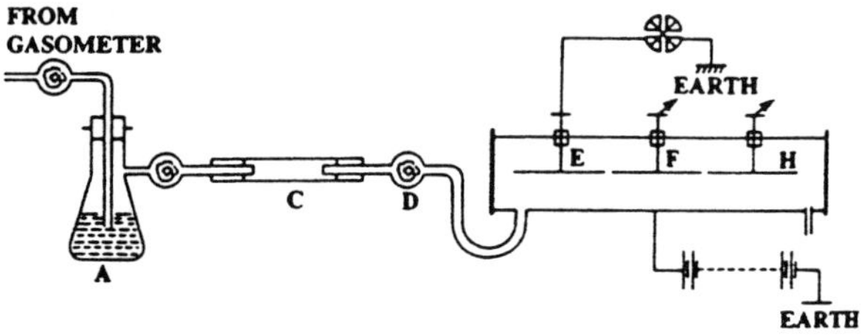

Fig. 2
(Ref.8, Part II)

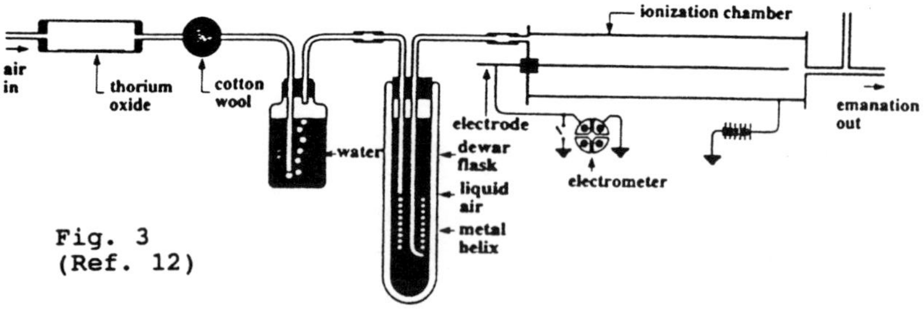

Fig. 3
(Ref. 12)

ATOMIC ENERGY COMMISSION (AEC) NUCLEAR EDUCATIONAL PROGRAMS 1947-1952 *

Kimberlee J. Kearfott
Department of Nuclear Engineering and Radiological Sciences
University of Michigan
Ann Arbor, MI 48109-2104

Laura M. Calkins
Department of Nuclear Engineering and Radiological Sciences
University of Michigan
Ann Arbor, MI 48109-2104

ABSTRACT

Immediate post-World War II Atomic Energy Commission (AEC) activities were not restricted to weapons production and research, but also included significant educational activities. Within two years of the Hiroshima and Nagasaki atomic bombings, the AEC embarked upon widespread efforts to enhance the education of the general public about radiation, develop K-12 nuclear science pipelines, and establish graduate training programs in nuclear engineering and radiation protection. The shortage of nuclear specialists and increasingly negative public views of nuclear projects and their funding were both recognized as serious problems of vital importance to the optimal expansion of US nuclear activities. In 1949, the Chair of the Congressional Joint Committee on Atomic Energy called for hearings on the issue of public education, noting that the US public's knowledge of nuclear issues appeared to be limited by an "iron curtain of apathy", compared with an "iron curtain of calculated ignorance" being intentionally constructed around nuclear issues in the Soviet Union. A review of early AEC and related educational activities designed to address this problem is made.

---

* This project was supported in part by the University of Michigan, Rackham School of Graduate Studies, Humanities Division, and by National Science Foundation Grant No. SBR-9411952.

INTRODUCTION

Much of the fundamental knowledge of nuclear reactor engineering and the radiation protection practice was developed under secret wartime conditions during World War II. These advances were dramatically revealed to the general public through the deployment of atomic weapons against Japan in 1945 and the aftermath of that bombing. A transitional period was required to transfer nuclear technology from secret military to peaceful public application: the transfer was complicated both by the wartime origins of the scientific and engineering advances and the fact that the initial demonstration of nuclear capability was destructive. Important components of the transition period included the declassification of fundamental scientific information, the necessity of development of formalized, formal educational efforts in a field which was previously nonexistent, and a growing general mistrust by the public of radiation and nuclear scientists. Soon after its establishment in 1947, the Atomic Energy Commission (AEC) recognized the challenges it faced in the area of educating the public and the development of a trained nuclear work force. This paper will explore the initial need for public educational programs and the developing shortages of technically trained personnel, the relationship between these needs and the political rhetoric of the day, the spontaneous attempts by the US educational system and civic groups to promote knowledge about nuclear science, and the earliest AEC investments in the educational area.

ATOMIC ENERGY COMMISSION (AEC) AND RELATED NUCLEAR EDUCATIONAL PROGRAMS

The Need for General Public Nuclear Educational Programs

By early 1950, more than 60,000 individuals were employed in the atomic energy field in the United States, with the vast majority working for the Atomic Energy Commission. Of nuclear workers at Knolls Atomic Power Laboratory, a national laboratory devoted specifically to the development of nuclear power and breeder reactor technology, only 15% had expert knowledge of neutron physics. Less than 3% of the technical personnel employed at that facility were concerned directly with the practice of health physics, or radiation protection[1]. If Knolls Atomic Power Laboratory is assumed to be representative of the US nuclear laboratory infrastructure as a whole, this meant that less than 1800 individuals, or 0.001% of a total US population of 151 million [2], were, by virtue of their employment, familiar with radiation health effects.

According to the Industrial Advisory Committee to the Atomic Energy Commission, prior to 1950 the field of commercial uses of atomic energy was "so shadowy that business men neither know where to look nor what to look for" [1]. Information was not even readily available concerning the legalities of mining of strategic materials or the prices which would be paid for such materials [3]. Although President Eisenhower's 'Atoms for Peace' speech delivered on December 8, 1953 [4] was to greatly promote awareness among US business leaders of the potential peaceful applications of nuclear technology, it is clear that during the critical years following the end of World War II nuclear technology remained largely unknown even to those likely to profit from its industrial applications. The absence of tangible, everyday non-military products and processes arising from nuclear science, which had been developed under conditions of wartime secrecy, promised to make ready public acceptance of the risks and benefits of development of nuclear applications even more difficult.

Five years after the bombing of Hiroshima, the teaching of fundamental modern physics and applied radiation physics remained absent from public school curricula, and there was no nation-wide scientific educational campaign. In the absence of an effective widespread public information campaign, the details of reactor and nuclear fission technology remained mysterious to the majority of US citizens except for an elite group which was isolated, primarily for security reasons, within the atomic energy government labs. The importance and magnitude of the problem of bringing knowledge of nuclear technology and radiation health effects to the public during the 1947-1952 time-period, however, had not been entirely overlooked by the Atomic Energy Commission. The AEC recognized that in order for informed public participation in policy decisions on nuclear matters to be possible, public education on nuclear issues was imperative.

Projected Human Resources Deficiencies in the Nuclear Field

After the war, many members of the expert technical staff assembled for the Manhattan Project returned to their university positions, often abandoning their wartime scientific specialties. Scientific development and production activities at the government nuclear laboratories, however, continued at wartime levels, or, in some instances, intensified. US defense policy dictated continued utilization of the physical and human resources infrastructures created by $2 billion in investments for the Manhattan Project, especially in light of growing threats that the Soviet Union would develop atomic weapons: indeed, work had already begun on the next generation of atomic weapons. In addition, peaceful nuclear applications were targeted for research and development, partially in hopes of softening the public reaction to growing expenditures on mysterious and terrifying nuclear weapons. The competitive engineering of these non-military applications would also require an elite scientific work force. It was further expected that commercial interest, and thus private sector demands for nuclear scientists, would grow as information spread about potentially profitable nuclear applications. The growing shortages of scientists, especially those competent in the nuclear and radiation fields, were further amplified by the fact that women, a sizable fraction of the scientific and technical work force of World War II, were increasingly relegated to unpaid and unskilled domestic employment. This had the effect of immediately removing trained individuals from the skilled technical labor pool, while radically diminishing the numbers of scientists and engineers who could be recruited and trained for the continuing nuclear efforts.

For the academic year ending in 1948, US engineering colleges had produced 5,400 master's degrees and 198 doctoral degrees in engineering [1], a modest number relative to escalating national demands. Human resources shortages in most technical areas were quickly becoming evident; the newly created demand for nuclear experts and trained individuals in the supporting areas of physics, engineering, and mathematics promised to further enhance personnel shortfalls. It was clear that national support for the development of a trained nuclear work force was imperative.

Political Justification for a National Nuclear Education Policy

In early 1946, debate arose as to whether military control of the atomic sciences should be continued, or whether civilian control should be instituted. The debate was resolved with the creation of the Atomic Energy Commission (AEC), which was responsible for both the nuclear weapons programs and civilian nuclear applications. The AEC was overseen by the Congressional Joint Committee on Atomic Energy (JCAE). Senator Brien McMahon, a popular Democrat from Connecticut whose presidential aspirations were later thwarted by a premature death from cancer, became Chair of the JCAE in 1949, a position which he skillfully exploited for political gain, and through which he addressed both national and international nuclear policy issues. In his first speech as JCAE Chair, delivered at the end of January 1949 to the Colorado Press Association, Senator McMahon claimed that "public fear, ignorance and apathy" already surrounded the entire US Atomic Energy program. He implored the press to assume responsibility for public dissemination of information about the nuclear field, now that civilian control had been achieved. Ten months previously, the American Society of Newspaper Editors' Special Committee on Atomic Information had urged the government to disclose the maximum possible information about atomic science, except in cases where such disclosures would be detrimental to national security. [3] In 1949, then, both the press and the Chair of the JCAE agreed that greater public awareness of nuclear issues was essential.

Senator McMahon argued for greater public information by linking education on nuclear and radiation issues to the growing public apprehension about the Cold War with the Soviet Union. He proposed that the people of Russia and its satellites were being deliberately kept in the dark about atomic energy, with distortions about US weapons capabilities intentionally propagated by the controlled Soviet press to trivialize the military importance of atomic energy, while at the same time the positive social and economic implications of atomic energy were under-reported. Senator McMahon, placing partial blame for lack of knowledge upon the complacency of the US population itself, contrasted an "iron curtain of apathy" on nuclear issues in the US with an "iron curtain of calculated ignorance abroad" in the Soviet Union and Eastern Europe.[3] This argument, equating freedom with nuclear knowledge, was strengthened when Soviet success in the production of nuclear weapons was revealed later in 1949. Public dissemination of nuclear information was seen as an "unequaled opportunity for patriotic service of the highest order", as long as the many small secrets essential for the assembling of nuclear weapons were preserved. McMahon believed that the issue of public education on nuclear issues was so significant that he arranged for the JCAE to hold public hearings on the topic within the next year.

Early Nuclear Educational Programs

A number of educational programs about nuclear technology involving teacher training, high school programs, elementary school programs, and adult education programs were initiated in the US in 1948-1949. These programs were conducted and sponsored by a variety of public and private groups. Identifiable sponsors of the earliest teacher training programs included state departments of education (Arkansas, Colorado, Connecticut, Florida, Iowa, Nebraska, Oregon, Rhode Island, Virginia), city governments (Baltimore, Battle Creek, Buffalo, Chicago, Cincinnati, District of Columbia, Minneapolis, New York City, Philadelphia, Reno), a small selection of colleges (Rhode Island College of Education, Keene Teachers College), public universities (New York University, University of Colorado, University of Denver, University of Illinois, University of Maine, University of Nebraska, University of Scranton, University of Virginia), private universities (Columbia University, Harvard University, Massachusetts Institute of Technology, Rensselaer Polytechnic Institute), and religious groups responsible for parochial schools (The Diocese of Boston). Before the end of 1949, few high schools had actually implemented substantive instructional programs in atomic energy, although there were rudimentary programs in select schools in 15 states. Only five states could boast public elementary school programs (California, Illinois, New Hampshire, New Jersey, and Rhode Island). Several universities and colleges sponsored adult education programs ranging in duration from individual lectures to semester-long courses. [5]

Local civic organizations and professional groups also provided a variety of programs such as exhibits, shows and evening lectures. An extensive program of atomic energy education was, for example, the top priority of the National League of Women Voters in 1947-48. Efforts were also undertaken to draft policies on atomic energy education by the National Education Association, the North Central Association of Colleges and Secondary Schools, the National University Extension Association, and the American Council on Education. The American Council on Education was particularly concerned with whether or not the AEC would ensure that sufficient unclassified material would be available for substantive public and school educational programs. [5]

In June 1948, an Interdivisional Committee on Educational Implications of Atomic Energy, which included representatives of the Atomic Energy Commission's Division of Public and Technical Information and the US Office of Education's Division of Higher Education, was formed. This committee was tasked with coordinating data concerning high school atomic energy courses, serving as a clearinghouse for materials concerned with the educational implications of atomic energy, assisting public and private groups involved in the planning of atomic energy-related conferences, workshops, and exhibits, and cooperating with other government agencies in developing general public awareness and understanding of the implications of nuclear and radiological sciences. This committee quickly developed a 13-page report, designed to assist elementary and secondary school teachers in integrating atomic energy into the school curriculum, which was distributed to 45,000 educators. Additional informational reports and bibliographies were compiled and distributed by the Interdivisional Committee. [5]

In 1948 the AEC/Office of Education Interdivisional Committee conducted an extensive survey of 2,000 high schools and 1,700 colleges and universities to assess atomic energy education in the US. This survey revealed that the nuclear educational effort of colleges and universities was focused on scientific, rather than social, aspects of atomic energy: students in the sciences rather than those in the humanities were receiving nuclear information. Institutions of higher education requested additional government support and funding for nuclear education, arguing that "the Federal Government, which spent billions releasing the genie, ought to bear some responsibility for paying the costs of getting him back into the bottle". The Interdivisional Committee's survey of public schools received a less than 5% response rate, itself indicative of general apathy within the school system. The scarcity of suitable instructional materials and visual aids at appropriate levels of complexity for elementary and high schools was cited as the reason for lack of involvement by many respondents.[5] The results of this survey by the Interdivisional Committee provided support for a national initiative aimed at educating students and the public on nuclear developments.

Atomic Energy Commission Nuclear Educational Programs

By the early 1950s, the Atomic Energy Commission supplemented its study of the problems of public education on nuclear technology with a series of outreach and educational programs and activities designed both to address the problem of inadequate public information and education and to fill projected human resources requirements for skilled technical workers. Such efforts included the granting of research funds to a variety of institutions of higher learning and the provision of experts from the national nuclear laboratory system to serve as emissaries and resources for the public educational system. For example, outreach activities were in evidence by 1950 at the Oak Ridge National Laboratory (ORNL) Health Physics Division. The group, working in collaboration with the newly formed Oak Ridge Institute of Nuclear Science (ORINS), spent considerable time planning training courses in radiological defense for high school teachers. [6] Personnel within the Education and Training Division of the ORNL Health Physics Division traveled extensively to promote graduate education in nuclear-related at a variety of schools, including King's College, University of West Virginia, and Georgia Institute of Technology.[7] Members of the ORNL Health Physics Division later organized public speaking and demonstration activities through the newly established Oak Ridge National Laboratory Traveling Lecture Program. [8]

In October 1946, Oak Ridge Institute of Nuclear Studies (ORINS) had been incorporated for the purpose of developing and managing educational and training activities for the AEC. Some of ORINS' early activities, funded by the AEC, promoted the development of nuclear activities at US universities through summer and thesis research appointments for university faculty and graduate students at the national nuclear laboratories, with the first appointments arranged at Oak Ridge National Laboratory. ORINS, relying upon staff from the AEC laboratories, developed a continuing series of intensive month-long training courses in nuclear topics. One particularly successful example of these activities was a course in basic radioisotope techniques, first offered in summer of 1948, which eventually resulted in the training of several thousand US and international scientists in the use of radioactive tracers in a variety of fields of basic and applied science. [9]

The Atomic Energy Commission recognized, however, that the shortages of individuals trained in nuclear reactor technology were such that there were too few individuals with sufficient training to conduct nuclear educational programs within universities, let alone to supply the demands of the national nuclear laboratories and a growing nuclear industry. In direct response to this situation, the AEC established its own training school, known as the Oak Ridge School of Reactor Technology (ORSORT). The limited number of US nuclear experts available at the time served as faculty for ORSORT. Students were primarily university faculty and professional engineers from both the public and commercial sectors, who later became advanced instructors in the nuclear field. ORSORT graduates included Admiral (then Captain) Hyman G. Rickover, who later guided the Navy's nuclear submarine and ship program. ORSORT was discontinued in the late 1950s when formal programs of study in nuclear engineeringwere initiated at a number of US universities. [9]

Atomic Energy Commission Post-Graduate Fellowship Programs

On May 1, 1948, a major new federal fellowship program in science was established. This fellowship program focused on the physical and biological sciences and was funded by the AEC, which contracted with the National Research Council of the National Academy of Science to administer the program. ORINS was granted partial responsibility for the administration of these fellowships for 1950, and full responsibility for program administration beginning in 1951. Substantial investments were made by the AEC in these programs, which continued until 1973 and remained independent of the similar fellowship programs initiated later by the National Science Foundation. Prior to 1952, the AEC fellowship programs included predoctoral and postdoctoral fellowships in physical and biological sciences, and health physics (radiation protection) graduate fellowships. Between 1948 and 1952, a total of 920 individuals received predoctoral and postdoctoral fellowships in the physical and biological sciences through the AEC fellowship program. This program proved quite successful, since of these students 84 became employees of the AEC, 300 worked for universities or colleges, and 200 continued their education eventually to contribute directly or indirectly to the nuclear field. The health physics graduate fellowships had supported the training of more than 100 individuals by 1952. Over the immediate and long term, these fellowships not only had a greater than 80% retention of individuals within the health physics profession, but accounted for a sizable fraction of the professional health physics community. The physical and biological sciences and health physics fellowship programs established the basic structure followed by the AEC nuclear science and engineering fellowship programs, which were begun in 1957, when nuclear engineering and sciences programs had been established in universities with the assistance of ORSORT graduates.

CONCLUSIONS

Following World War II the difficulties of dissemination of nuclear technology and training of future nuclear scientists and engineers were quickly recognized by the AEC, by influential political figures, by a variety of professionals in the media and the educational system, and by many civic and scientific groups. Significant achievements resulted from the AEC's specialized and advanced training initiatives for scientists and educators. Substantial efforts were made prior to 1952 to disseminate knowledge and inspire public interest and confidence in nuclear technologies but these efforts appear to have been largely ineffective because of the sheer magnitude of educational effort required at the local level, and growing widespread mistrust of nuclear scientists and of radiation. An analysis of the failure of early public educational efforts is needed in order to assess present-day educational efforts concerned with radiation, many of which appear duplicate earlier efforts. More positively, the models of graduate education and training by the AEC appear to have been phenomenally successful in providing specialized education and in meeting critical human resources shortages, while providing promise as approaches to new scientific challenges of the future.

REFERENCES

[1] Kingdon, K. H., "Engineering Education for Atomic-Energy Development," Report Number KAPL-P-235, Knolls Atomic Power Laboratory, Schenectady, New York, February, 1950.

[2] Famighetti, Robert, Editor, *The World Almanac and Book of Facts, Funk and Wagnalls Corporation*, Mahwah, New Jersey, 1995, p.377.

[3] Address by Senator Brien McMahon (D. Conn.), Chairman, Joint Committee on Atomic Energy, Before the Colorado Press Association, Denver, Colorado, January 29, 1949. NARA Archives I, RG 128: Records, Joint Committees of Congress, Joint Committee on Atomic Energy General Correspondence Files, Box 277.

[4] Holl, Jack M., Anders, Roger M. Atoms for Peace: Dwight D. Eisenhower's Address to the United Nations. Milestone Documents in the National Archives, National Archives and Records Administration, Washington, DC, 1990.

[5] Atomic Energy Educational Programs 1947-49: As reported to the Atomic Energy Commission. NARA Archives I, RG 128: Records, Joint Committees of Congress, Joint Committee on Atomic Energy General Correspondence Files, Box 277.

[6] Health Physics Division Quarterly Progress Report: Period ending 15 April 1950. ORNL-695

[7] Health Physics Division Quarterly Progress Report: Period ending 20 July 1951. ORNL-1086

[8] Health Physics Division Quarterly Progress Report: Period ending 20 January 1953. ORNL-1277

[9] Scientists and Engineers for the Nuclear Age: Final Report on the Atomic Energy Commission Fellowship Programs 1948-1973. Oak Ridge Associated Universities.

RADIOACTIVE WASTE MANAGEMENT IN THE EARLY NUCLEAR ERA: A SURVEY OF PRACTICES AND POLICY TRANSITIONS AT US NUCLEAR SITES 1946-54 *

Laura M. Calkins
Department of Nuclear Engineering and Radiological Sciences
University of Michigan
Ann Arbor, MI 48109-2104

Kimberlee J. Kearfott
Department of Nuclear Engineering and Radiological Sciences
University of Michigan
Ann Arbor, MI 48109-2104

ABSTRACT

Under the auspices of the Manhattan Engineering District (MED), early postwar radioactive waste management was designed independently by each MED research and production site. Disposal methods varied widely, including land burial, open pit storage, direct discharge to water systems, and ocean dumping. Waste volumes and characteristics differed, and only basic waste handling protocols were developed. Under the Atomic Energy Commission (AEC), more cohesive policies on waste management were introduced, and R & D efforts were initiated on volume reduction and improved processing, packaging, and isolation. Waste reduction research focused on reprocessing, decontamination of equipment, and filtering effluent streams. Private contractors participated, but the AEC retained responsibility for waste disposition. By the early 1950s, a Commission-wide framework of regulations, including permissible radiation levels and a goal of zero-releases, was in place, but waste volumes continued to grow

* This project was supported in part by the University of Michigan, Rackham School of Graduate Studies, Humanities Division, and by National Science Foundation Grant No. SBR-9411952.

at many sites. Cross-country waste shipments increased, permitting the AEC to concentrate land burials, and its future waste disposition difficulties, at a few sites.

RADIOACTIVE WASTE MANAGEMENT PRACTICES AND POLICIES

In late 1946, as the US Army's Manhattan EngineeringDistrict (MED) was preparing for the transfer of US nuclear facilities and programs to civilian control, the members of the new civilian agency, the Atomic Energy Commission (AEC), were "inspect[ing] the scattered empire of plants and laboratories to be inherited from the Army."[1] The paramount point at issue was the transfer of control over atomic weapons; the management of radioactive wastes and contaminated equipment produced at MED facilities was a relatively unimportant issue at this time.

In fact, however, the Manhattan Project had produced huge amounts of radioactive wastes. During World War II each Manhattan Project facility had developed waste management procedures locally, without central MED coordination or standards for environmental releases or worker exposure tolerances. Only in July 1946, with the war long over and the transition to civilian authority in view, did MED scientists at Oak Ridge create a "Committee on Waste Disposal." This Committee had as its goal the creation of radiological criteria for the handling and disposition of radioactive wastes and contaminated materials.[2]

The Disposal Committee established general operating principles which, while not radically altering the MED's decentralized waste management arrangements, did for the first time provide system-wide rules, including a recognition of the utility of sea disposal and a determination that land burial of wastes was to be concentrated at four sites: Santa Fe/Los Alamos, Chicago/Argonne,Hanford, and Oak Ridge. Individual facilities were allowed to decide on waste disposal methods. The District's Berkeley Area Office, for example, disposed of its wastes by ocean dumping.[3]

The Disposal Committee was not authorized to create radiation release and exposure standards. It noted that no District-wide standards were in place in mid-1946, and that a standards-setting exercise was to be undertaken later that year by an expert medical advisory group. Meanwhile, the Committee decided, levels of activity in wastes were to continue to be set "according to the standards of exposure pertaining at the place of use."[4]

The Disposal Committee was eliminated with the creation of the AEC in January 1947. In its final meeting Committee members made an urgent appeal to the AEC: waste management required

> "an overall policy... applying to every unit and broad enough to apply to the whole US. The policy should cover all phases of disposal of contaminated materials, whether production wastes, solutions, gases, scrap, precious metals, machines, equipment, dead animals or the remains of individuals." [5]

The AEC initially followed the MED's practice of decentralized decision-making on waste management. Its highest priorities were weapons development, fissionable materials production, and reactor design research. [6] Contractor and management changes at key sites, including Oak Ridge and Hanford, diverted attention from the Disposal Committee's advice. Only in December 1947 did a major policy-setting body, the AEC's Advisory Committee on Biology and Medicine, recommend that the AEC underwrite a major research and regularization initiative on waste disposal methods. [7]

In April 1948 at a meeting of senior site managers, fundamental decisions were reached on AEC-contractor relations; these decisions affected both contractors' and the AEC's waste management responsibilities. Private contractors were charged with conducting research on waste concentration, temporary storage methods, and the reprocessing and recovery of uranium from spent fuels and other wastes. The AEC agreed to develop methods for "collection, transportation and permanent storage and for disposition of radioactive wastes." [8]

This division of labor became the basic framework for the AEC's system-wide waste management strategy in the late 1940s. In June 1948, for example, AEC informed the Dow Chemical Company that the Commission wished to establish waste minimization and research on processing, packaging and interim isolation as the responsibility of private industry, in this case AEC site contractors.[9] Meanwhile, cross-country waste shipments to the four key land burial sites continued. Oak Ridge received the largest number of shipments, including large quantities of wastes from Iowa, the Chicago area, and AEC facilities in Ohio and New York.

By late 1948 the AEC had announced its intention "to establish an integrated and coordinated program to attack liquid waste disposal problems."[10] A nationwide survey of waste production and management conducted in late 1948 produced the first full picture of the wastes being produced, stored, buried, burned, released or shipped from AEC facilities.[11] At the same time the AEC developed a program of research and development on wastes to be carried out at each site, and sought budget increases to support this new initiative. [12]

Gaps in the waste disposal program persisted. One AEC scientist pointed out in mid-1948 that the management of gaseous wastes had not yet been organized, and that control of gases posed problems which differed technically from liquid waste management, in that it must be based on a "system of maximum dispersion, as contrasted to a system of concentration and control."[13] The quantities of liquid wastes in storage were mounting, and accumulations outpaced storage capacity at many sites. With the characteristics of liquid wastes varying greatly, and with the accelerating waste disposal research program and inter-site transportation of wastes, the costs of managing wastes escalated rapidly in 1949-50.[14]

The AEC adopted a waste management target of zero radioactivity releases from waste treatment and storage facilities in late 1949, but the growing pressures on the waste management program prevented realization of this goal.[15] More expedient methods were sought for reducing waste volumes; these included "controlled open field burning."[16] As Oak Ridge declined to accept further shipments because it lacked storage capacity, the Knolls Atomic Power Laboratory in New York began discharging radioactive wastes to the Mohawk River, and the Mound Laboratory in Ohio increased its effluent releases to the Miami River.[17] Some significant volume reductions were realized in certain categories of liquid wastes, notably at Oak Ridge where research on evaporation techniques had produced improved processes.[18] As new facilities were planned to support the growing nuclear weapons program, however, the AEC faced new challenges in its waste management program.

The passage into law of the Atomic Energy Act of 1954 marked a turning point in the AEC's relationships with private industry, which again affected waste management policies. The Act empowered the AEC to license private firms to use radioactive materials in a range of new activities, including the production of electricity at commercial nuclear reactors. The AEC took this opportunity to reintroduce some elements of earlier waste management programs. Increasing emphasis was placed upon reprocessing uranium from spent reactor fuel, both to meet fuel requirements in a period of uranium shortages and as a means of reducing waste volumes. [19] The shipment of wastes to selected land burial sites was also renewed. These sites again included Oak Ridge, where research into calcination had produced a process for turning high-level liquid wastes into a ceramic material "for easier storage and handling." [20]

Furthermore, as the MED Disposal Committee had urged in July 1946, a new set of radiological criteria for waste management was finally developed by the AEC in 1954. These regulations prescribed maximum permissible concentrations (MPCs) of radioisotopes in air and in consumables, established maximum rates of human exposures to radioisotopes released from AEC facilities, and created guidelines for concentrations of radioactive materials in natural water systems. Most important, the AEC reaffirmed the responsibility of private industry (and AEC contractors) for "design, construction and operation of waste treatment and disposal facilities and methods of release of wastes to the environment," while also confirming the AEC's "responsibility... for evaluation of methods of treatment and release of wastes and for control over these practices and their effect on the environment." [21]

As it had in 1948, the AEC again acknowledged in 1954 that, even in the context of increasing utilization of nuclear materials for peaceful purposes by private industry, the US government through the AEC retained "basic and continuing responsibility for public health and safety in connection with these waste disposal practices and their effects on man, property and natural resources." [22]

REFERENCES

[1]RG Hewlett and OE Anderson, *The New World: A History of the United States Atomic Energy Commission, Vol I: 1939-1946.* Berkeley: University of California Press, 1990, 641.

[2]Memo To EE Kirkpatrick Fr WD Fleming, EIDMT MD-400.7, 11 July 1946 in US National Archives Record Group (RG) 326, Records of the Atomic Energy Commission, Records of the Oak Ridge Operations Office (OROO), Research and Development Divison Correspondence Files 1947-63, Box 23.

[3]Memo To District Engineer, Oak Ridge Fr EB Kelly, EIDM A-4, 4 Nov 1946 in RG 326, OROO Research and Development Division Correspondence Files 1947-63, Box 34.

[4]EIDMT MD-400.7, 11 July 1946, op cit.

[5]Memo To H Speed Fr WR Smith, Jr., et al., EIDMZ, 2 Dec 1946 in RG 326, OROO, Research and Development Division Correspondence Files 1947-63, Box 23.

[6]RG Hewlett and F Duncan, Atomic Shield: *A History of the United States Atomic Energy Commission, Vol II: 1947-1952*. Berkeley: University of California Press, 1990, 35.

[7]Draft Minutes, Fourth Meeting, Advisory Committes for Biology and Medicine, 12-13 Dec 1947 in RG 326, OROO Research and Development Division Correspondence Files 1947-63, Box 30.

[8]Memo To RS Warner Fr ER Jette, 15 Apr 1948 in RG 326, OROO Research and Development Division Correspondence Files 1947-63, Box 73.

[9]Letter To WH Dow Fr RS Warner, 17 June 1948 in RG 326, OROO Research and Development Division Correspondence Files 1947-63, Box 73.

[10]Memo To Area Managers Fr RS Warner, 20 Oct 1948 in RG 326, OROO Research and Development Division Correspondence Files 1947-63, Box 73.

[11]Memo To JC Franklin Fr RS Warner, 3 Jan 1949 in RG 326, OROO Research and Development Division Correspondence Files 1947-63, Box 73.

[12]"Decontamination and Disposal Program," 3 Dec 1948 in RG 326, OROO Research and Development Division Correspondence Files 1947-63, Box 73.

[13]K.Z. Morgan, "Effectiveness of Our Present Waste Disposal Program," in *Minutes of a Conference on Liquid Waste Disposal*, 23-25 August 1948, ORNL-163, 2 November 1948.

[14]"Notes on Liquid Waste Disposal Picture," 11 Jan 1949 in RG 326, OROO Research and Development Division Correspondence Files 1947-63, Box 73.

[15]Memo To JH Robertson Fr WS Allred, 18 Oct 1949 in RG 326, OROO Research and Development Division Correspondence Files 1947-63, Box 73.

[16]Memo To B Sparks Fr WB Harris, 20 Apr 1954 in RG 326, OROO Records of the Feed Materials Division, Correspondence Files 1946-66, Box 16.

[17]"Proposed Policy Governing Radioactivity Allowable in AEC-Controlled Liquid Waste: Appendix 3," 8 Feb 1951, in RG 326, OROO Research and Development Division Correspondence Files 1947-63, Box 102.

[18]Memo To SR Sapirie Fr CE Larson, 10 Aug 1954 in RG 326, OROO Research and Development Division Correspondence Files 1947-63, Box 100.

[19]RG Hewlett and JM Holl, *Atoms for Peace and War 1953-61: Eisenhower and the Atomic Energy Commission: A History of the United States Atomic Energy Commission, Vol III.* Berkeley: University of California Press, 1989, 125-27.

[20]L Johnson and D Schaffer, *Oak Ridge National Laboratory: The First Fifty Years.* Knoxville: University of Tennessee Press, 1994, 101.

[21]US AEC, "The Release of Radioactive Materials," [draft], 13 May 1954 in US National Archives RG 90 Records of the US Public Health Service, Records of the Division of Sanitary Engineering Services, Correspondence Files 1953-58, Box 68.

[22]ibid.

HISTORICAL REPROCESSING OF SPENT FUEL AND HLW MANAGEMENT AT THE IDAHO CHEMICAL PROCESSING PLANT

Dieter A. Knecht, M. Dan Staiger, Jerry D. Christian, and C. Lee Bendixsen
Lockheed Martin Idaho Technologies
P. O. Box 1625, MS 5218
Idaho Falls, ID 83415-5218
G. W. (Bill) Hogg and Julius R. Berreth
INEEL Retirees

ABSTRACT

The Idaho Chemical Processing Plant (ICPP), which is a part of the Idaho National Engineering and Environmental Laboratory (INEEL), has stored and reprocessed irradiated nuclear fuel since 1953 to recover uranium-235, neptunium-237 and krypton-85 for the U.S. Department of Energy (DOE) and its predecessor agencies. The primary focus of the reprocessing at ICPP was on fuels that contained highly enriched uranium for recovery of the fissile uranium with no plutonium recovery. A total of 31,432 kg uranium containing highly enriched U-235 was recovered from 1953 through 1991. The acidic high-level liquid radioactive waste (HLLW) that resulted from reprocessing was stored in stainless-steel 1100 m^3 single-shell tanks in underground concrete vaults. A solidification process was developed to form a granular calcine solid from the acidic HLLW with a seven-fold volume reduction, processing over 27,000 m^3 of HLLW to over 3800 m^3 calcine from 1963 to 1996. The calcined waste is stored in near-surface, stainless-steel bins, approximately 4-m diameter by 12.5 to 18.5-m high, within concrete vaults. Currently there is and inventory of 6800 m^3 liquid and 3800 m^3 calcine. If the practice of neutralization such as in place at other sites had been used to manage ICPP HLW, increased waste storage volume equal to 195 additional HLLW tanks would have been required. Glass formulations were developed for ICPP calcines and included borophosphate and borosilicate formulations for alumina and zirconia calcines, with waste loadings of 24 to 29 wt% for alumina calcines and 33 wt% for zirconia calcines. Glass-ceramic waste forms were developed with 70 wt% calcine loading and high durability, comparable to glass. Alternative waste processes are being evaluated in the NEPA process, including dissolution and separation of high and low activity wastes for vitrification and grouting, respectively. This paper reviews some of the key actions in the historical development of spent fuel reprocessing and waste management practices which has helped the ICPP successfully accomplish its mission safely and with minimal impact on the environment.

INTRODUCTION AND PURPOSE

The Idaho Chemical Processing Plant (ICPP) was constructed on the site of the National Reactor Testing Station, that is now the Idaho National Engineering and Environmental Laboratory (INEEL), and has stored and reprocessed spent nuclear fuel since 1953 to recover uranium-235, neptunium-237 and krypton-85 for the U.S. Department of Energy (DOE) and its predecessor agencies.[1,2,3] The primary focus of the reprocessing at ICPP was on fuels that contained highly enriched uranium-235 for recovery of the fissile uranium with no plutonium recovery. In 1991, DOE curtailed reprocessing for the recovery of fissile material and plans to dispose of the remaining fuels. An EIS is being developed at the INEEL for the path forward for processing of the remaining wastes.

This paper reviews some of the key actions in the historical development of spent fuel reprocessing and waste management practices which has helped the ICPP successfully accomplish its mission safely and with minimal impact on the environment.[4]

ICPP REPROCESSING DEVELOPMENT

The ICPP was constructed on an arid tract of withdrawn public land previously used by the U. S. Navy for testing guns and munitions. The development of the initial plant processes and the initial design scoping of the plant were carried out by Oak Ridge National Laboratory (ORNL). After completion of the design studies the AEC Processing Advisory Committee recommended in April 1950 that the plant be built at the National Reactor Test Site (NRTS), a precursor of the INEL, in Idaho, and in June 1950, the Foster Wheeler Corporation was selected as the architect-engineer. The American Cyanamid Company was selected as the operating contractor in late 1950, and the Bechtel Corporation was selected as the construction contractor, with excavation starting in 1950 and construction completed in mid-1952. After a preliminary test, cold run, and low radiation level run period, the plant was placed in routine production in March 1953. Late in 1953 Phillips Petroleum Company assumed responsibility for plant operation.[1]

The original main processing building at ICPP consisted of 29 cells each typically 20 feet square and 28 feet deep. The bottom of the cells were lined with stainless steel and maintenance was accomplished by direct contact after decontamination. With the exception of four cells, which had viewing windows and manipulators, the in-cell equipment was controlled remotely from an operating corridor that ran the length of the building between two rows of cells. Figure 1 is a 1956 photograph of the completed reprocessing and fuel storage facilities and location of the underground HLLW storage tanks.

The ICPP was designed for processing of highly enriched uranium fuels[3] from over 100 reactors, including aluminum, zirconium, stainless steel, and graphite fuel types. Aluminum-based fuel was dissolved[5] in nitric acid using a mercuric nitrate catalyst; recently a new catalytic process that eliminated the use of mercury was patented.[6] Zirconium-based fuel was dissolved in hydrofluoric acid and subsequently complexed with aluminum nitrate while using soluble neutron poisons to control criticality; stainless steel-based fuel was initially dissolved in sulfuric/nitric acid dissolution and later by electrolytic dissolution in nitric acid containing gadolinium as a neutron poison. Graphite-based fuel was processed by combustion of graphite followed by nitric and hydrofluoric acid dissolution of the resulting ash. Low-level wastes from throughout the ICPP were collected, concentrated through evaporation, and transferred to the liquid waste tanks.

The ICPP began reprocessing Hanford aluminum fuel in the batch aluminum dissolution facility in 1953,[7] and in 1955 the facility was modified for continuous aluminum dissolution.[4,7,8] In 1956, a batch zirconium dissolver was operational,[9] and stainless steel dissolution using sulfuric acid was developed.[4,10] Semi-continuous zirconium dissolution using the soluble poison (boron) was developed in 1965.[11] A co-processing flowsheet was developed to use the aluminum nitrate from the aluminum fuel dissolver solution to complex the fluoride from the zirconium dissolver solution in order to minimize addition of process reagents.[4] Electrolytic dissolution of stainless steel fuel in nitric acid using soluble gadolinium poison replaced the Carpenter-20 stainless steel dissolver in 1971.[4,12] Processing of graphite based Rover fuel was initiated in 1983.[4,13] A new zirconium based dissolution process, the Fluorinel Dissolution Process (FDP), was constructed in 1986 as part of the Fluorinel and Storage Facility (FAST) that used viewing windows, manipulators and a crane for remote operation and remote maintenance.[14] Cadmium and /or boron soluble neutron poisons were used in reagents that included hydrofluoric acid, nitric acid, and aluminum nitrate in three batch dissolver and complexer trains operating in parallel. A Custom Processing Facility was operated to process small-to-moderate quantities of fissile material from a variety of fuels that could not be processed by conventional facilities and equipment.[4] A total of 31,432 kg uranium containing highly enriched U-235 was recovered at the ICPP from 1953 through 1991. The amount of recovered uranium in kg for each fuel category is aluminum-15,636, zirconium-5,403, stainless-steel-6,036, graphite-3,028, and Custom Processing-1,329.

Figure 1. 1956 Photograph of Fuel Storage Building CPP 603, Fuel Reprocessing Building CPP 601, and Location of Underground HLW Tank Storage.

Neptunium-237 was first recovered from the fuel dissolver product in 1965, and the solutions were stored until it was purified and shipped in 1972.[4] Barium-140 was recovered at the ICPP as a special isotope and source of a short-lived high-gamma producing lanthanum-140 by rapid turn-around processing of MTR fuel.[4]

A Rare Gas Plant with capability of recovering fission product xenon and radioactive krypton-85 from fuel dissolver off-gases was constructed. The first process used liquid nitrogen cooled carbon beds to capture the noble gases from the aluminum dissolver off-gases.[4,15] The carbon beds were replaced in 1958 with a more efficient cryogenic distillation process.[4,16] An ambient-temperature hydrogen recombiner was added in 1978 to permit noble gas recovery from the zirconium fuel dissolver off-gas, containing hydrogen. Over 100,000 Curies of krypton-85 and 8,000 L of fission product xenon were recovered after restart and upgrade of the Rare Gas Plant in 1968.

The dissolver effluent from the dissolution facilities contacted organic solvent counter-currently in three different extraction cycles.[2,31] The basic process used tributyl phosphate in normal paraffin hydrocarbon to extract uranium from the dissolver effluent. A second and third cycle used hexone for improved product purity. Before 1969, the uranyl nitrate was packaged and shipped to fuel fabrication facilities. After 1969, the concentrated uranyl nitrate was converted to uranium trioxide, prior to shipment, using a fluidized bed process developed and implemented at ICPP.[32] The raffinate from the solvent extraction process (the high-level waste) was transferred to liquid waste storage tanks.

CALCINATION TECHNOLOGY DEVELOPMENT

The concept of calcining radioactive liquid waste by atomizing the liquid in a heated fluidized bed of inert particles[18] was originated by ANL. Early development by ANL consisted of brief feasibility studies using a 7.5-cm (3-inch) diameter calciner and more extensive studies using both unshielded and shielded 15-cm (6-inch) diameter calciners.[4, 19] The 7.5-cm diameter calciner tests were all run with cold feed. The 15-cm diameter calciners were run primarily to test the behavior and treatment of fission products in the calcination process, and tests included up to 12% spiked, hot, as-produced feed from the ICPP. Initial development was based on calcining acidic aluminum nitrate solution and was later extended to include Hanford type PUREX waste. ANL pilot plant experience indicated the fluidized bed process was feasible and that ruthenium was the only volatile fission product. The tests showed that virtually all activity except ruthenium remained in the solid product over the temperature range of 350 to 550 °C. At 350 degrees operating temperature about 88% of the ruthenium was volatilized, whereas at 550 °C less than 1% was volatilized. Both sintered metal filters and venturi scrubbers, in which the scrub solution was recycled to the feed to minimize additional waste streams, were tested for removing particulates from the off-gas; silica gel adsorbers were tested for removal of volatile ruthenium. Tests showed that the off-gas system was very effective in removing ruthenium and particulates, and the silica gel could be readily regenerated.

Pilot-plant testing at ICPP used a 15-cm (6-inch) diameter pilot plant calciner patterned after the one used by ANL but with a number of changes to provide equipment and process scale up data and to define process parameters for designing the Demonstration Waste Calcining Facility (DWCF).[4, 20] The 15-cm diameter calciner vessel was 1.7 m (5.5 feet) long and contained a charging tube, a thermowell and off-gas tube at the top of the vessel. The calcined product was removed continuously by means of a one-inch draw-off line extending 61 cm (2 feet) above the distributor plate. Heat was supplied by 18 internal electric heaters with 15-cm (6-inch) heated lengths located horizontally inside the calciner bed. Liquid feed was introduced under pressure through commercially available pneumatic atomizing spray nozzles. To provide operating data, pressure taps and thermocouples were located throughout the bed. Auxiliary equipment included a cyclone, a spray scrubber, condenser, and fluidizing gas pre-heated. Operating variables studied were feed rate, bed temperature, fluidizing velocity, feed nozzle air-to-feed ratio, product draw off, and liquid feed composition. It was difficult to determine the various effects of the operating variables on the product particle size due to the inability to achieve steady state.

The results of the pilot plant tests verified the feasibility of calcining aluminum fuel raffinate waste and completed exploratory runs with simulated stainless steel and zirconia fuel raffinates. The advantages included independence from complicated moving parts, good control of bed temperature with uniform high heat transfer rates, reduced corrosion potential of stored product, remote operation capability, and excellent control of particle size and properties. Low feed rates generally produced smaller particles. The bed temperature measurements showed decreasing bulk density with increasing bed temperature throughout the temperature range studied, and low bed temperatures (300 °C) produced large mass mean particle diameters, whereas, 500 °C tended to produce fines. Changing gas fluidizing velocity appeared to change particle size distribution. A relatively small amount of air moving countercurrent to product overflow was found to be an effective means of controlling particle size withdrawal. The concentration and composition of feed to the calciner was not studied but was expected to have a strong effect on calcine structure and size distribution.

ICPP CALCINATION DEMONSTRATION AND HOT OPERATIONS

Pilot plant tests established target ranges of principal operating values for a full scale demonstration, which included the following: 1) calciner bed temperatures of 400 to 500 °C, 2) fluidizing air velocities of 23 to 40 cm/sec (0.75 to 1.3 feet/second), and 3) air to liquid volume ratios of 500 to 750 through the feed nozzle. The Demonstration Waste Calcining Facility (DWCF) was constructed from 1958 to 1961 at a cost of $6 million. About 70% of the facility was used for off-gas cleanup equipment, including a cyclone, a quench tower, a venturi scrubber, entrainment separators, silica gel adsorbers, and HEPA filters. Calcium nitrate was added to fluoride bearing zirconium type waste to react with fluoride and to reduce corrosion on downstream process equipment and piping by controlling fluoride volatility. The granular calcined solids product was pneumatically transferred to and stored in vented stainless steel bins located in concrete vaults.

The calciner vessel was 1.2 m (4 ft) in diameter. Radioactive feed was sprayed into the bed through three pneumatic atomizing nozzles, at 320 L/hr (85 gallons/hour) total net processing rate including waste, additives, and scrub recycle. During the first three campaigns and until June 1969, heat was supplied by an in-bed heat exchanger bundle using sodium-potassium (NaK) alloy which was heated in an oil-fired furnace at 400 °C. Preheated fluidizing air was introduced through 14 capped orifices equally spaced on a distributor plate. The design operating conditions resulted in average bed particle sizes ranging from 0.6 to 0.75 mm.

One and one half years of cold testing began in 1961. In November 1962, the DWCF was renamed as the WCF. Radioactive feed was introduced on December 8, 1963, and the first run lasted 11 months, until October 15, 1964. The first Calcined Solids Storage Facility (CSSF) was filled by calcining over 1,800 m^3 (500,000 gallons) of radioactive liquid waste to 210 m^3 (7,500 ft^3) solids. The net output exceeded the design rate by 15%, and the process operated at 99% efficiency. Additional CSSF's were built and the WCF was modified for higher feed rates. To allow increased WCF capacity and reliability, an in-bed combustion system using oxygen-atomized kerosene was installed in 1970.[21] The use of in-bed combustion increased the operating temperature to 500 °C and resulted in increased heat transfer rates, a decrease in the volatility of fission product ruthenium, and lower metal wall temperatures. The reduced ruthenium concentrations in the off-gas by a factor of 100 to 200 were offset by increased difficulty in decontamination of WCF internal surfaces. Boric acid was added to the aluminum nitrate feed solution to suppress the formation of alpha alumina and insure the product remains in the amorphous rather than crystalline form. Boric acid also helped to control particle size, minimize decontamination problems, and reduce erosion of pumps, valves, and piping by making the aluminum species soluble in the nitric acid scrub solution. The WCF operated until March 1981, calcining about 15,000 m^3 (4 million gallons) of liquid radioactive waste.[4]

NWCF DESIGN, CONSTRUCTION, AND OPERATION

The New Waste Calcining Facility (NWCF) was designed to replace the WCF because of increased maintenance time of the WCF without remote capabilities and a need to increase throughput to accommodate increased fuel processing needs.[22] The calciner vessel active bed region was increased to a 1.5-m (5-ft) diameter with a 2.1-m (7-ft) diameter expanded upper section fabricated from 347 stainless steel. Sizes of downstream vessels were also increased, and the design net processing rate was increased to 11,300 L/day (3,000 gallons/day). A more corrosion resistant alloy, Nitronic 50, was used because of the fluoride and trace chloride chemical environments and high operating temperatures. Because of the WCF experience, high maintenance equipment (pumps, valves, flanges, electrical connections, and off-gas filters) was placed in corridors readily accessible for remote replacement using master-slave and electro-mechanical manipulators and cranes. The remote capabilities installed at NWCF were first tested in a full-scale mock up facility at the INEEL and resulted in easy removal and replacement of equipment without lengthy shut downs or high radiation exposures to personnel. Failed equipment was transferred directly to integral shielded facilities for remote decontamination and subsequent repair or disposal. Figure 4 shows a 1985 photograph of the WCF, NWCF, Calcined Storage Bin Sets No. 1-6, and location of Subsurface HLW Liquid Storage Tanks.

Figure 4. 1985 View of ICPP Waste Storage and Waste Processing Facilities

The NWCF began hot operations in September 1982, after extensive cold testing, and calcined 6,000 m^3 (1.6 million gallon) of waste before shutting down in March 1984 when all available liquid waste feed was processed. During the down period a new Distributive Control System was installed. The NWCF was restarted on September 30, 1987 and has operated successfully on various feeds and combinations of feed such as Fluorinel, zirconium, and blended Fluorinel-Na waste feeds. To date the NWCF has calcined over 12,100 m^3 (3.2 million gallon) of radioactive liquid waste to 1,670 m^3 (59,000 ft^3) of calcined solids.

HLW IMMOBILIZATION TECHNOLOGY DEVELOPMENT

Currently, there is an inventory of 3,800 m^3 HLW calcine at ICPP consisting primarily of zirconia-based calcines (3,000 m^3), alumina calcine (600 m^3), and other (200 m^3). The major component (wt%) in ICPP HLW alumina calcine includes alumina (82-95), and the major components in zirconia-based calcines include fluorite (41-44), zirconia (17-19), calcia (12-13), alumina (9-14), alkali oxides (5-7), borate (2-4), and cadmium oxide (0-7). Fission product content is typically less than 1 wt%.[23]

Although the calcine is physically and thermally stable, it is not suitable for long-term disposal because it is potentially dispersible and is leachable for several of its radioactive constituents, mainly cesium. Preliminary scoping tests were run in the 1960s, and laboratory testing was started in the 1970s to develop glass formulations for ICPP calcines. A phosphate glass was prepared and leach tested in the mid-1960s with the following components in wt%: 25 alumina calcine, 40.5 P_2O_5, 18.6 Na_2O, and 15.9 PbO using fully radioactive calcine.[24] Other early waste forms which were studied in the late 1960s included calcine particle coating with molten aluminum and steel spray and matrix encapsulation in metal, glass, plastic or grout.[25]

In the 1970s, glass formulations were tested further. For alumina calcine, waste loadings of up to 29 wt% and 24 wt% could be obtained in a borophosphate[25] and borosilicate[26] glass, respectively. For zirconia calcines, waste loadings of 33 wt% were observed using a borosilicate frit. Laboratory and pilot-scale melter experiments showed that the high zirconia-calcium fluoride glass can be produced at practical melter temperatures, and very durable melter refractories are required because of fluoride corrosiveness.[25, 27] Nonradioactive laboratory- and pilot-scale and radioactive laboratory-scale tests were run using the borosilicate Frit 127 (composition in wt%: SiO_2-70.3, Na_2O-12.8, B_2O_3-8.5, Li_2O-6.2, and CuO-2.1). MCC-1 and MCC-2 leach tests indicated that there did not appear to be significant differences in the glasses formed using simulated zirconia calcine at laboratory or pilot scale and using radioactive zirconia calcine at laboratory scale.[26]

Glass-ceramic waste forms were prepared for various calcine compositions with the goal of effectively utilizing the refractory components, alumina, zirconia and fluorite, to form durable glass and crystalline phases.[28] Early testing was run with sintered glass-ceramics, cold-press sintered ceramics, aluminum phosphate ceramics, and fluorapatite phosphate ceramics.[25] Polyphase glass-ceramic experiments were carried out for ICPP zirconia calcine waste compositions using a hot isostatic press (HIP) for high temperature sintering. Durable glass-ceramic forms could be made for zirconia calcine with additives including silica, alkali, and yttria.[25, 29] Detailed microstructure analysis of these waste forms revealed the presence of cubic fluorite, monoclinic zirconia, stabilized cubic zirconia, tetragonal zircon, zirconolite, perovskite and amorphous aluminosilicate phases. Normalized MCC-1 release rates at 90 °C of all major elements were less than 1 g/m^2day.[30] At a calcine loading of 70 wt% and higher density, the glass-ceramics provided a potential 60 % volume reduction compared to glass.

Glass-ceramic formulations were also developed for zirconium-sodium blended calcines using added borosilicate frit and other additives such as titania and metallic Ti and Al powders.[25, 31] Major crystalline phases included fluorite, zirconia, and zircon. Sphene, $Cr_2Ti_7O_{18}$ and other titanates were formed with added titanium, and nepheline and calcium-aluminum silicates were formed with the added aluminum. The higher amounts of metallic powder increased the amount of zirconia at the expense of zircon. MCC-1 leach rates decreased with decreased borate, and the lowest normalized elemental leach rates of <1 g/m^2d were found in the formulations containing 5-6 wt% Ti and 2-3 wt% Al. The amounts (in wt%) of crystalline phases were identified by X-ray diffraction as fluorite (37), zirconia (17), zircon (14), and sphene (16). The major components in wt% of the glass phase were identified by SEM as silica (62), alumina (19), and calcia (19). The glass phase components for the zirconia and zirconia-Na calcines were found to lie in the phase stability triangle for silica, anorthite, and alpha-wollastonite.[30] Additional studies with similar calcine and frit composition were run to determine the effect of process soak time and added metal Si and Al powders on the microstructure and durability.[25, 32] The lowest leach rates were found for the Al concentration at 2 wt% and were found to increased with increasing Si concentration. Constant leach rates were found for soak times of 4, 8, 16 and 24 hr. The results of a statistical analysis study indicate that higher HIPing temperatures and longer HIPing times improve leaching characteristics for many elements in experimental ICPP glass-ceramics.[33]

Recent evaluations have proposed dissolution of calcine and separation of a small high-activity waste fraction, followed by a significantly smaller-scale vitrification process. The low activity waste would be disposed on site as a grout. Evaluations are continuing with a full-scale facility projected to process all of the calcine by 2035 in order to meet a legal agreement with the State of Idaho.[34]

ICPP WASTE MANAGEMENT CONSIDERATIONS AND DESIGN DECISIONS

Basic decisions were made that resulted in a different approach than at the other DOE reprocessing sites, including[4] maintaining acidic liquid wastes, constructing a limited number of stainless-steel liquid waste storage tanks for the acidic wastes, developing calcination to convert the acidic liquid wastes to a solid with a seven-fold volume reduction, and construction of stainless steel storage facilities for the solid calcine. If the acidic liquid waste had been neutralized, there would have been a volume increase from the amount of caustic required, and large quantities of sludge would also have been generated. Thus, by calcining and not neutralizing HLLW at ICPP, construction of 195 new liquid waste storage tanks was avoided.

CONCLUSIONS

The ICPP fuel reprocessing operation that began in 1953 has resulted in the recovery of 31,432 kg of highly enriched uranium and other byproducts. The resulting high-level liquid waste was converted with a seven-fold volume reduction to approximately 3800 m^3 of solid calcined HLW which is stored safely in stainless steel bins with a long design life. The remaining liquid waste and calcine are scheduled for immobilization to be road-ready for shipment off-site by 2035. A set of decisions and focused development activities, including innovative fuel reprocessing and waste management technologies, were developed during the history of the ICPP and have resulted in these current relatively safe conditions, with no major waste leaks to the environment compared to some of the other DOE sites. By calcining and not neutralizing the HLLW, the ICPP avoided constructing up to 195 ICPP tanks or 65 Hanford-sized tanks and all of the potential problems associated with long-term HLLW storage. As a result, the HLW storage at the ICPP is currently under a relatively minimal environmental risk compared to other related DOE sites.

Some of the ICPP pioneering contributions include the following:

1. First to reprocess highly enriched uranium on a production basis.
2. First to process breeder reactor fuels, the EBR-I Core I.
3. First to dissolve spent nuclear fuel continuously on a routine schedule.
4. First to use fixed and soluble neutron poisons for criticality control.
5. First to operate a radioactive rare gas recovery plant in conjunction with a fuel recovery process.
6. First to process stainless steel alloyed fuel by sulfuric acid dissolution.
7. First to recover uranium from high zirconium-uranium alloy fuels by dissolution in hydrofluoric acid and to use an innovative co-processing flow sheet for minimizing waste volumes.
8. First to show that multiple head-end processing of various fuels is compatible with common second- and third-cycle purification.
9. First to dissolve stainless-steel-clad EBR-II reactor fuels using electrolytic dissolution.
10. First to develop and implement the production-scale fluidized-bed denitration of the high enriched uranium product.
11. First to develop and implement a production-scale calciner to solidify the high-level liquid wastes, while significantly reducing the waste volume and avoiding the potential problems associated with long-term liquid waste storage.

ACKNOWLEDGMENTS

This work is sponsored by the U. S. Department of Energy, Idaho Operations Office, under contract DE-AC07-94ID13223. The authors would like to acknowledge all of the pioneers who developed the processes and facilities and completed the successful reprocessing and waste management operations at the INEEL.

REFERENCES

[1]C. E. Stevenson and J. P. Lyon, *Idaho Chemical Processing Plant Description, Operation Experience, and Costs*, PPC-226, Phillips Petroleum Co. (July 1955).

[2]L. A. Decker, "Significant Accomplishments in the History of the Idaho Chemical Processing Plant," presented at the 172nd National Meeting of the American Chemical Society, San Francisco, CA (September 1, 1976).

[3]C. E. Stevenson, "Solvent Extraction Processes for Enriched Uranium," *Proceedings of the Symposium on the Reprocessing of Irradiated Fuel*, TID-7534, U. S. AEC, May 20-25, 1957, pp. 152-179 (1957).

[4]D. A. Knecht, M. D. Staiger, J. D. Christian, G. W. Hogg, and J. R. Berreth, "Historical Fuel Reprocessing and HLW Management," *Radwaste*, **4**, 35-47 (May 1997).

[5]J. A. Buckham and C. E. Stevenson, "Dissolution Equipment," *Proceedings of the Symposium on the Reprocessing of Irradiated Fuel*, TID-7534, U. S. AEC, May 20-25, 1957, pp. 831-847 (1957).

[6]J. D. Christian and P. A. Anderson, "Mercury-Free Dissolution of Aluminum-Clad Fuel in Nitric Acid," U. S. Patent No. 5,364,603 (November 15, 1994).

[7]A. E. Erhard, C. L. Murray, R. F. Murray, B. J. Newby, K. L. Rohde, F. P. Schilling, and F. M. Warzel, *Processing of the Core Fuel Used Initially in the Experimental Breeder Reactor (EBR-I)*, IDO 14331 (December 15, 1955).

[8]A. F. Boeglin, J. A. Buckham, L. Chajson, R. B. Lemon, D. M. Paige, and C. E. Stoops, "Continuous Dissolution of Uranium-Aluminum Fuels," *A.I. Ch. E. Journal*, 2, 190-4 (June 1956).

[9]D. G. Reid, C. E. Stevenson, R. B. Lemon and F. K. Wrigley, "Reprocessing of Fuel Containing Zirconium," *Proceedings of 2nd UN Int. Conf. On Peaceful Uses of Atomic Energy*, UN, Geneva, vol. **17**, pp. 1145-153 (1958).

[10]R. N. Legan, *ICPP Processing of VBWR Stainless Steel Fuel (April 14 to June 2, 1965)*, IDO-14658 (December 1965).

[11]G. F. Offutt and B. R. Wheeler, *First Zirconium Alloyed Fuel Reprocessing Campaign Using Soluble Nuclear Poison*, IN-1021 (March 1968).

[12]L. T. Lakey and W. B. Kerr, "Pilot Plant Development of an Electrolytic Dissolver for Stainless Steel Alloy Nuclear Fuels," *I & EC Process Des. & Dev.*, **6**, 174-179 (April 1967).

[13]A. P. Roeh, J. A. Rindfleisch, P. E. LaMont, L. C. Borduin, G. W. Lohman and M. R. Fox, "Rover Fuel Processing - a Success with Lessons Learned," ASME, 84-WA/PVP-7 (1984).

[14]L. A. Decker, G. E. McDannel, and W. B. Palmer, "Fluorinel Dissolution Process Development and Operational Experience," *Proc. 36th Conf. Remote Systs. Tech.*, Vol. **1**, pp. 29-34 (1988).

[15]A. L. Ayers, W. B. Lewis and C. E. Stevenson, "Production of High Specific Activity Radioisotopes," *Proc. Int. UN Conf. On Peaceful Uses of Atomic Energy*, UN, Geneva, vol. **20**, pp. 27-37 (1958).

[16]C. L. Bendixsen and G. F. Offutt, *Idaho Chemical Processing Plant Rare Gas Recovery Facility*, IN-1221 (1969).

[17]W. J. Bjorklund and G. F. Offutt, *First Product Denitration Campaign with Enriched Uranium at ICPP*, IN-1475 (June 1971).

[18]P. F. Brown, *Minutes of ICPP Waste Disposal Conference*, IDO-10030, Idaho Operations Office (August 1955). Declassified.

[19]J. W. Loeding, E. L. Carls, L. G. Anastasio, and A. A. Jonke, *The Fluid-Bed Calcination of Radioactive Waste*, ANL-6322, Argonne National Laboratory (May 1961).

[20]E. S. Grimmett, *Calcination of Aluminum-Type Reactor Fuel Wastes in a Fluidized Bed*, IDO-14416, Idaho Operations Office (August 1957).

[21]T. K. Thompson, G. E. Lohse, and B. R. Wheeler, "Fluidized Bed Calcination of Radioactive Waste Using In-Bed Combustion Heating," *Nuclear Technology*, **16**, 396-405, (November 1972).

[22]B. R. Dickey, B. R. Wheeler, J. A. Buckham, "High-Level Waste Solidification," *Nuclear Technology*, **24**, 371-382 (December 1974).

[23]J. R. Berreth, *Inventories and Properties of ICPP Calcined High-Level Waste*, WINCO-1050 (1988).

[24]B. E. Paige, *Leachability of Glass Prepared from Highly Radioactive Calcined Alumina Waste*, IDO-14672 (1966).

[25]D. A. Knecht, P. C. Kong and T. P. O'Holleran, "Proposed Glass-Ceramic Waste Forms for Immobilizing Excess Plutonium," *Proceedings of the Embedded Topical Meeting on DOE Spent Nuclear Fuel and Fissile Material Management*, ANS, La Grange Park, IL, pp. 183-191 (1996).

[26]B. A. Staples, H. S. Cole, and D. A. Pavlica, "Properties of Formula 127 Glass Prepared with Radioactive Zirconia," *Scientific Basis for Nuclear Waste Management II*, D. G. Brookins, ed., pp. 125-134 (1979).

[27]D. Gombert, D. B. Chamberlain and E. L. Colton, *Design and Installation of a Pilot Plant to Vitrify ICPP Simulated Calcined High-Level Waste*, ENICO-1087, Exxon Nuclear Idaho Co (May 1981).

[28]D. A. Knecht and J. R. Berreth, *The Management of Defense High-Level Radioactive Waste at the Idaho Chemical Processing Plant*, WINCO-1063 (1989).

[29]A. B. Harker and J. F. Flintoff, "Polyphase Ceramic for Consolidating Nuclear Waste Compositions with High Zr-Cd-Na Content," *J. Am. Ceram. Soc.*, **73**, 1901-1906 (1990).

[30]K. Vinjamuri, "Glass-Ceramic Waste Forms for Immobilization of the Fluorinel-Sodium, Alumina, and Zirconia Calcines Stored at the Idaho Chemical Processing Plant," *Environmental and Waste Management Issues in the Ceramic Industry II*, D. Bickford, S. Bates, V. Jain, and G. Smith, eds., *Ceramic Transactions*, Vol. **45**, pp. 3-13 (1994).

[31]R. S. Baker and J. R. Berreth, "A Glass-Ceramic Composition to Immobilize ICPP HLW," *4th Int. Symp. Ceramics in Nuclear Waste Management, Proc. 91st Amer. Ceram. Soc. Mtg*, April 23-27, 1989, Indianapolis, IN.

[32]K. Vinjamuri, "Candidate Glass-Ceramic Waste Forms for Immobilization of the Calcines Stored at the Idaho Chemical Processing Plant," *Environmental Issues and Waste Management Technologies in the Ceramic and Nuclear Industries*," V. Jain and R. Palmer, eds., *Ceramic Transactions*, **61**, p.439 (1995.)

[33]B. A. Staples and D. V. Miley, "Effects of HIPing Time, Temperature and Pressure on Experimental ICPP Waste Form Properties," *Nuclear Waste Management IV*, G. G. Wicks, D. F. Bickford, and L. R. Bunnel, eds., *Ceramic Transactions*, Vol. **23**, pp. 27-35 (1991).

[34]State of Idaho, *Idaho National Engineering Laboratory Site Treatment Plan*, DOE-ID (1996).

Actinide Stabilization I

Component Solubility in Lanthanum Borosilicate Glasses for the Vitrification of Plutonium Oxide and Plutonium-Bearing Materials

Hong Li, John D. Vienna, Michael J. Schweiger, and Jarrod V. Crum
Pacific Northwest National Laboratory[1], P.O. Box 999, MSIN: P8-37
Richland, WA 99352, (509)373-0256, e-mail: hong.li@pnl.gov

ABSTRACT

Component solubilities in lanthanum borosilicate (LaBS) glasses were studied. The results suggest that LaBS glass may be a candidate frit for vitrification of PuO_2 and Pu-bearing materials that contain CaO, Fe_2O_3, Ga_2O_3, K_2O, MgO, MnO, MoO_3, Na_2O, NiO, P_2O_5, Ta_2O_5, and WO_3. This study also found that LaBS glasses with Na_2O or K_2O, or a mixture of the two were prone to liquid-liquid phase separation with higher than expected concentrations of alkalis from Pu waste streams. The effect of phase separation on durability of the LaBS system is not available in the literature to our knowledge. Further investigation is recommended.

INTRODUCTION

Vitrification technology is one of the options being investigated for the immobilization of PuO_2 and Pu-bearing materials by the U.S. Department of Energy (DOE). Recently, a study of Pu solubility in lanthanum borosilicate glass (LaBS) showed that 11.4 wt% PuO_2 was soluble [1], suggesting that LaBS glass is a candidate glass system for Pu immobilization. The lanthanide silicate glass family contains no alkali metal ions [2], and it has not been well studied for the immobilization of radioactive wastes as compared to alkali borosilicate glasses. Pu waste streams contain Ca, Cl, Cr, F, Fe, Ga, K, Mg, Mn, Mo, Ni, Ta, U, and W, the most of which are in elemental, oxide, and salt form. Their concentrations vary widely with the waste streams. This single component solubility study provides basic information for the assessment of the robustness of the LaBS frit for the vitrification of various Pu streams. This study is a part of efforts to assist DOE in the down selection of waste forms for Pu immobilization. This paper reports the results of the single component solubility study of the selected LaBS glass compositions. The effect of component interaction on single component solubility was selectively investigated.

---

[1] Pacific Northwest National Laboratory is operated for the U.S. Department of Energy by Battelle Memorial Institute under Contract DE-AC06-76RLO 1830.

EXPERIMENTAL

In the solubility study, modified LaBS glasses, LaBS-1 and LaBS-2, were used, which were made by adding 1 and 8 wt% Ga_2O_3 in the LaBS glass, respectively. The Ga_2O_3 concentrations covered a low and high bound as estimated from Pu waste streams. The components under investigation were based on the estimates of Pu waste streams, which were Cl, F, P_2O_5, Na_2O, K_2O, MgO, CaO, Cr_2O_3, Fe_2O_3, MnO, NiO, MoO_3, WO_3, and Ta_2O_5. The LaBS-1 glass was mainly used, whereas the LaBS-2 glass was used only for the solubility measurements of Na_2O, K_2O, MgO, and CaO. Uranium was separately investigated in another study, which will not be reported here.

Several batches of glass, 500 g each, were made for the LaBS-1 and LaBS-2 composition using reagent grade chemicals (cf. Table 1). The batched materials were throughly mixed in an agate disk mill for 5 min and then melted at 1450°C for 1 hour. After 1 hour melting, the melt was poured on a stainless steel plate. The glass was ground into fine powder in a tungsten carbide disk mill. The glass powder was used for component solubility study.

For each test run, 50 g of glass powder was mixed with a particular component of interest; the concentration of that component was varied to measure its solubility limit[2]. The mixture was thoroughly mixed in an agate disk mill for 2 min and then melted at 1450°C for 2 hours in a Pt-10%Rh crucible covered with a Pt lid. The molten glass was stirred with a Pt rod after 1 hour melting. After the second hour melting, the melt was quenched by pouring on a stainless steel plate. Both quenched and annealed samples were used to determine the component solubility. The annealing temperature was 840°C, which was slightly higher than glass transition temperature, T_g , (using a beam bending viscometer, T_g, at which glass viscosity is $10^{11.3}$ Pa.s, was determined to be 787°C for LaBS glass with addition of 7.35 wt% CeO_2). The annealing time was 24 hours. This procedure ensured that the component under investigation was completely soluble in the glass.

Crystalline phases were characterized using powder x-ray diffraction (XRD) analysis or scanning electron microscopy with energy dispersive x-ray spectrometer (SEM/EDXS). Liquid phase separation in glass was mainly evaluated by visual examination for glass opalescence, except for one glass with 10 wt% Na_2O, which exhibited phase separation and was characterized by transmission electron microscopy (TEM).

---

[2] Some components appeared to have higher solubility limits than the maximum concentrations that are found in the Pu waste streams. In these cases, the investigations were terminated at those concentration levels.

Table 1. Compositions (wt%) of LaBS Based Glasses

Glass ID	LaBS (a)		LaBS-1	LaBS-2	Chemical
Oxide	nominal	measured	nominal	nominal	Source
SiO_2	29.12	29.13	26.70	24.82	SiO_2
Al_2O_3	21.49	21.54	19.65	18.26	Al_2O_3
B_2O_3	11.74	11.71	10.76	10.00	H_3BO_3
CeO_2	0.00	N/A	7.35	6.83	CeO_2
Ga_2O_3	0.00	N/A	1.00	8.00	Ga_2O_3
Gd_2O_3	8.59	8.61	7.88	7.32	Gd_2O_3
La_2O_3	12.43	12.35	11.40	10.59	La_2O_3
Nd_2O_3	12.83	12.80	11.77	10.94	Nd_2O_3
SrO	2.51	2.51	2.30	2.13	$SrCO_3$
ZrO_2	1.30	1.29	1.19	1.11	ZrO_2
Total	100.00	99.94	100.00	100.00	

(a) Quantitative x-ray fluorescence - wavelength dispersive spectroscopy (XRF-WDS) was used to check the composition of the LaBS glass.

Because of high volatility of Cl and F in borosilicate glasses at high temperatures [3], the concentrations of Cl and F in the glass were measured using ion chromatography (IC). For the rest of other components, nominal (as batched) concentrations were used.

RESULTS AND DISCUSSION

The LaBS-1 and LaBS-2 glasses with additions of waste components below the solubility limits were homogeneous and free from crystals and liquid phase separation, after being quenched or annealed at 840°C for 24 hours. Table 2 summarizes the component solubility results for the LaBS-1 and LaBS-2 glass. The target concentrations of waste the components that are expected to be incorporated to the glass are also included in Table 2 to compare with the solubility data. The target concentrations were estimated based on 20 wt% of waste loading. Table 3 summarizes crystallinity data for the LaBS-1 glass.

The solubility limit of Cl in the LaBS-1 glass was found to be much lower than the target concentration, 1.6 wt%, which might be expected in the glass with a target maximum waste loading at 20 wt% (cf. Table 2). In general, Cl solubility is low in borosilicate waste glasses, typically less than 1 wt% [4,5]. The incorporation of Cl^- into the glass network by the substitution for bridging oxygen was suggested by a recent x-ray absorption near-edge structure (XANES) study of simulated nuclear waste glasses [6]. However, the ionic radius of Cl^- is 1.81 Å, much larger than that of O^{2-}, 1.38 Å, and hence, the limited Cl solubility in borosilicate based glass should be expected since the network can only accommodate a limited amount of large Cl^- anions. The solubility of F in the LaBS-1 glass was 0.7 wt%. It is about the same as the target concentration, 0.8 wt%, and lower than those found in

Table 2. Component Solubility (wt%) in LaBS-1 and LaBS-2 Glasses

Component	**LaBS-1** [(a)]	**LaBS-2**	Target Value [(b)]
Cl	0.6 (IC)	N/A	1.6
F	0.7 (IC)	N/A	0.8
P_2O_5	2.5	N/A	0.8
Na_2O	3.5	4.5	0.5
K_2O	3.5	3.5	1.2
Na_2O: K_2O	(2 : 2) < S < (3 : 3)	N/A	1.7
MgO	> 20	S > 8	1.7
CaO	6.5	S > 8	3.4
MgO : CaO	(4 : 4) < S < (5 : 5)	N/A	5.1
Na_2O : K_2O : MgO : CaO	S < (2 : 2 : 5 : 5)	N/A	6.8
Fe_2O_3	5.5	N/A	0.6
NiO	8.5	N/A	0.5
MnO	> 7	N/A	N/A
Cr_2O_3	0.6	N/A	0.3
Fe_2O_3 : NiO : MnO : Cr_2O_3	S < (2 : 4 : 1 : 0.5)	N/A	N/A
MoO_3	> 5	N/A	1.2
WO_3	> 5	N/A	2.0
Ta_2O_5	> 8	N/A	0.7

(a) The values in parentheses are for concentrations of multi-components that were added to the glass simultaneously. (b) The target concentrations of the waste components in the glass, which come from Pu waste streams, were obtained from D.K. Peeler at Savannah River Technology Center, Aiken, SC.

alkali borosilicate waste glasses, which range from 1 to 3.4 wt% F [4,5]. Figure 1 represents the SEM analysis of the annealed LaBS-1 glass with 3 wt% F, showing both liquid phase separation and crystallization existed in the glass. The crystalline phase was rich in Ce, La, Nd, and Gd, suggesting the formation of solid solutions among the rare earth oxides [7]. The XRD results of the same sample, however, could not provide phase identification.

The solubility of P_2O_5 in the LaBS-1 glass was about 2.5 wt% which is much greater than the target concentration, 0.8 wt%, and is comparable to the solubility range from 2.1 to 4.9 wt% in alkali borosilicate waste glasses [4,5]. At higher P_2O_5 concentrations in excess of the solubility limit, $CePO_4$ and $(Ce,La,Nd)PO_4$ precipitated in the glass. A possibility of $PuPO_4$ formation during glass melting was examined; the LaBS frit with 11.4 wt% PuO_2 and 2 wt% P_2O_5 was melted at 1500°C for 4 hours (higher temperature and longer melting time were used to ensure PuO_2 completely dissolving in the glass [1]). The quenched glass was free of any crystalline phases according to XRD analysis.

The solubility limit of Na_2O in the LaBS-1 was about 3.5 wt% which is signicantly higher than the target concentration. Above the solubility limit, liquid-liquid immiscibility

Table 3. Secondary Phases Found in the LaBS-1 Glass with Higher Concentrations of Waste Components in Excess of the Solubility Limits

Component	Secondary Phases [a]
Cl	Not observed in the glass
F	ph. sep. or La, Nd, Ce, and Gd cryst. ph. (SEM)
P_2O_5	$CePO_4$, (Ce, La, Nd)PO_4
Cr_2O_3	Cr_2O_3
K_2O	ph. sep. or $Sr_2Nd_8(SiO_4)_6O_2$ and $Sr_2Gd_8(SiO_4)_6O_2$
Na_2O	ph. sep. (TEM), $Ce_{0.75}Nd_{0.25}O_{1.875}$, CeO_2 (XRD)
MgO	Not observed in the glass up to 20 wt%
CaO	Al, Ca, La, Ce, Gd, and Zr cryst. ph. (SEM)
MgO and CaO	Al, Ce, La, Nd, and Zr cryst. ph. (SEM)
Na_2O, K_2O, MgO, and CaO	$Sr_2Gd_8(SiO_4)_6O_2$ and $Sr_2Nd_8(SiO_4)_6O_2$ $Ca_2Gd_8(SiO_4)_6O_2$, $Ca_2Nd_8(SiO_4)_6O_2$ $NaNd_9(SiO_4)_6O_2$, K(Na, K)$_3$A $l_4Si_4O_{16}$
Fe_2O_3	Fe_2O_3
NiO	Ni cryst. ph. (SEM)
MnO	Not observed in the glass up to 7 wt%
Fe_2O_3, NiO, MnO, and Cr_2O_3	spinel
MoO_3	Not found in the glass up to 5 wt%
WO_3	Not observed in the glass up to 5 wt%
Ta_2O_5	Not observed in the glass up to 8 wt%

(a) ph. sep. - phase separation and cryst. ph. - crystalline phase

was first observed; the glasses became opalescent during melt cooling. The phase separation was verified by TEM. Figure 2(a) shows a TEM micrograph of a quenched LaBS-1 sample with 10 wt% Na_2O. EDXS analysis showed that the dispersed phase was rich in Ce, La, Nd, Gd, and Zr as compared to the matrix [cf. Figure 2(b, c)]. By further increasing Na_2O in the LaBS-1 glass, for instance, 20 wt%, XRD analysis shows the formation of a solid solution, $Ce_{0.75}Nd_{0.25}O_{1.875}$, in the quenched sample. No TEM was performed on the partially crystallized samples. However, the quenched sample with excess Na_2O exhibited opalescence. It is likely that liquid-liquid phase separation still occurred in the glass in additional to crystallization. The solubility limit of K_2O was about 3.5 wt% in the LaBS-1 glass. For glasses with addition of K_2O that were well above the solubility limit, the precipitation of $Sr_2Nd_8(SiO_4)_6O_2$ and $Sr_2Gd_8(SiO_4)_6O_2$ in the glass was found by XRD analysis on the quenched samples. These glasses became opalescent during cooling, suggesting the occurrence of liquid phase separation.

The effect of Ga_2O_3 in the glass on the solubility limits of Na_2O and K_2O can be found by comparing the data between the LaBS-1 and LaBS-2 glass. Namely, Na_2O was slightly more soluble in the glass with higher Ga_2O_3, whereas Ga_2O_3 has little effect on K_2O solubility for the composition investigated here (cf. Table 2).

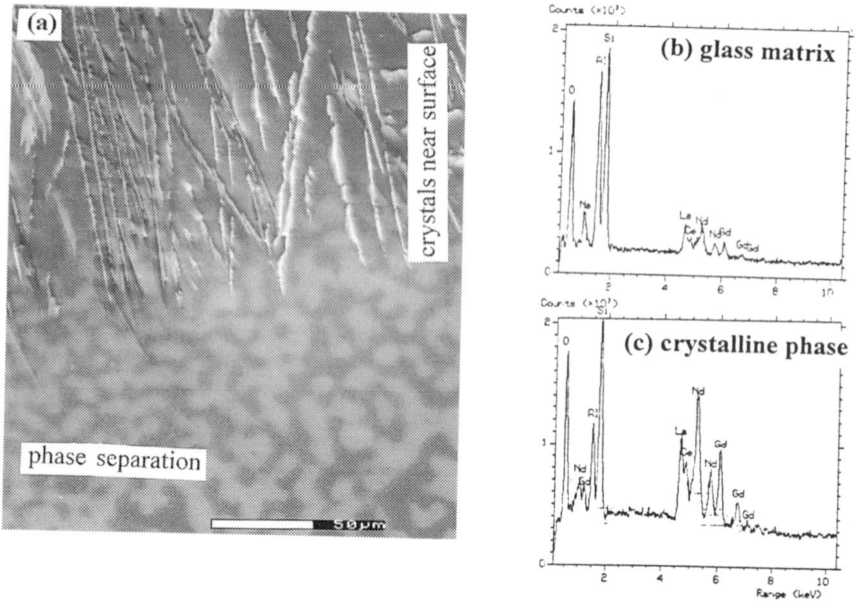

Figure 1. (a) SEM micrograph of the annealed LaBS-1 glass with 3 wt% F and (b, c) EDXS of the matrix and the crystals.

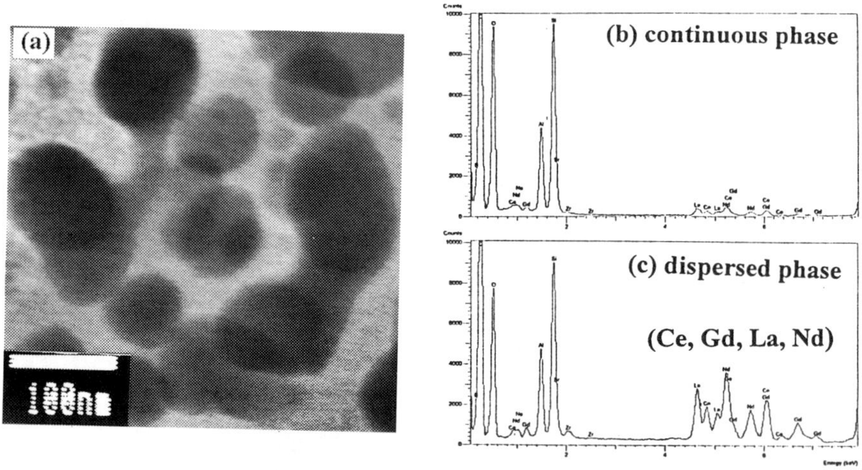

Figure 2. Phase Separation of LaBS-1 Glass with 10 wt% Na_2O: (a) TEM Micrograph and (b,c) EDXS of the Contiguous Phase, and the Dispersed Phase.

The solubility limit of CaO was about 6.5 wt% which is higher than the target concentration, 3.4 wt%. Figure 3 shows SEM/EDXS of the anealed LaBS-1 glass with 15 wt% CaO. The results suggested that high CaO decreased the solubility limits for Al_2O_3, La_2O_3 and Nd_2O_3. No phase separation, nor crystallization was found in the quenched and

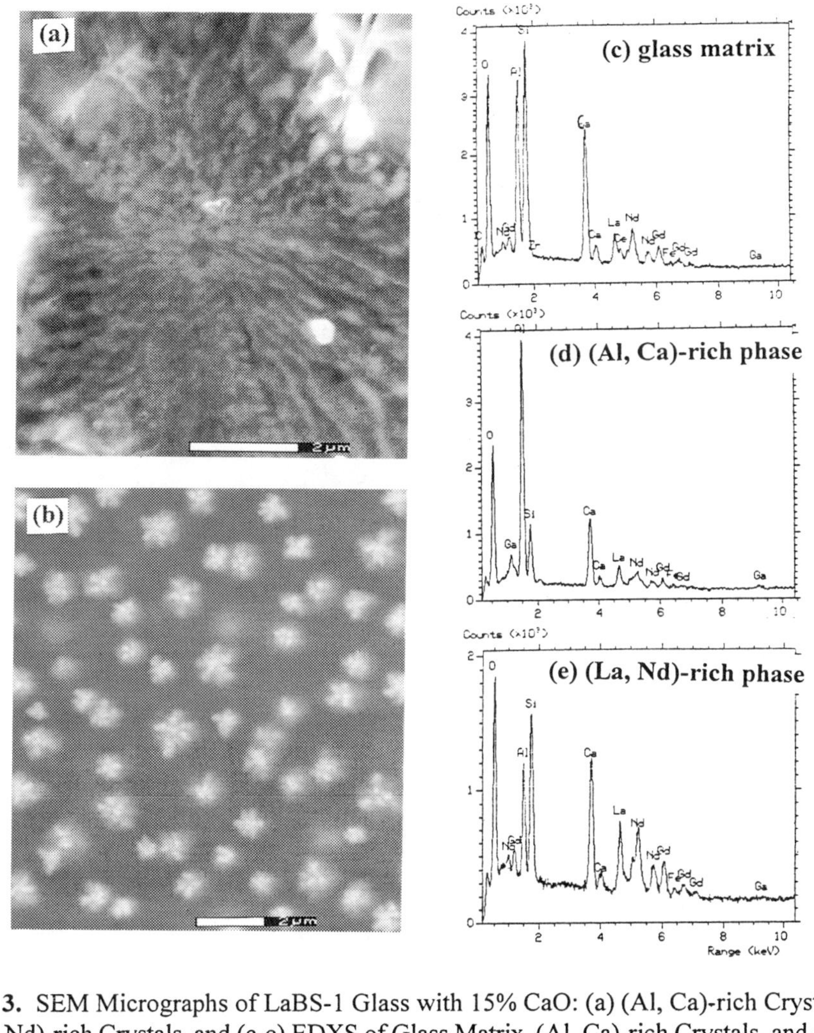

Figure 3. SEM Micrographs of LaBS-1 Glass with 15% CaO: (a) (Al, Ca)-rich Crystals, (b) (La, Nd)-rich Crystals, and (c-e) EDXS of Glass Matrix, (Al, Ca)-rich Crystals, and (La, Nd)-rich Crystals.

annealed LaBS-1 glasses with 20 wt% MgO which is more than one order of magnitude higher than the target concentration, 1.7 wt%. However, the presence of both CaO and MgO in the glass promoted crystallization of other phases. Figure 4 shows SEM/EDXS results of the annealed glass with 10 wt% CaO and 10wt% MgO. Crystalline phases rich

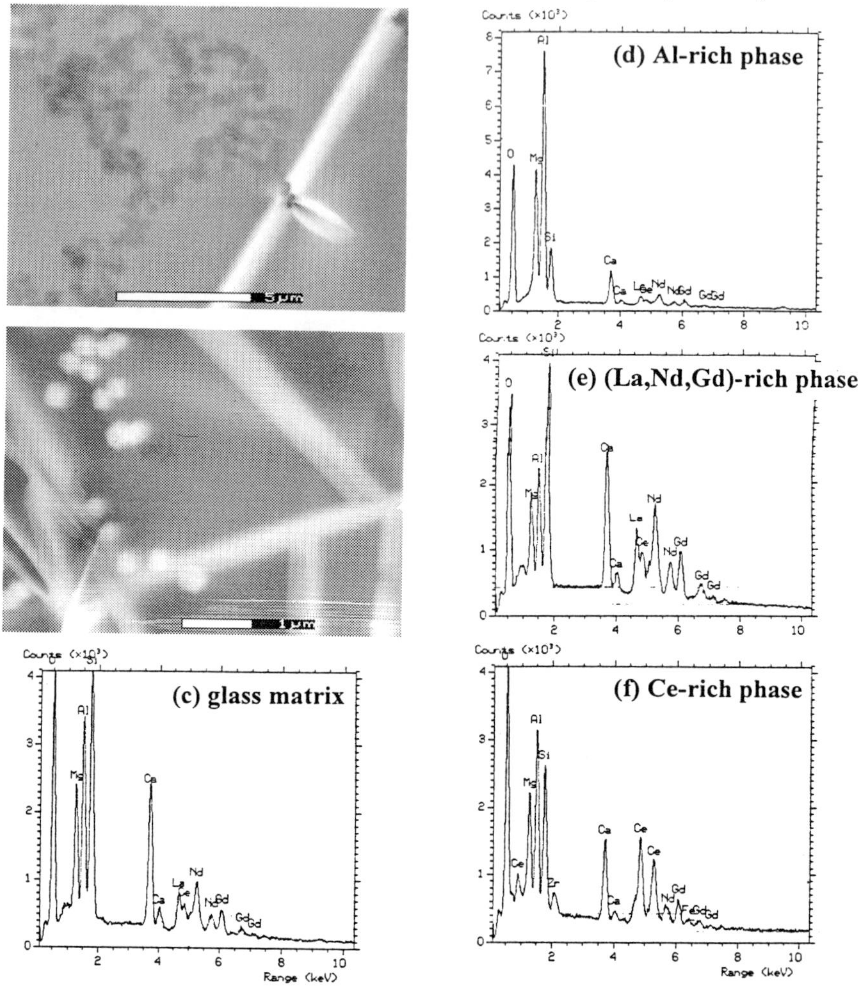

Figure 4. SEM Micrographs of LaBS-1 Glass with 10 wt% CaO and 10 wt% MgO: (a) Al-rich and (La, Nd)-rich Crystals, (b) (La, Nd)-rich and Ce-rich Crystals, and (c-f) EDXS of Glass Matrix, Al-rich Crystals, (La, Nd, Gd)-rich Crystals, and Ce-rich Crystals.

in (Al, Mg, Ca), (La, Nd, Gd), and Ce were found in the annealed glass. Furthermore, crystallization was found in the quenched LaBS-1 glass with 2 wt% Na_2O, 2 wt% K_2O, 5 wt% CaO, and 5 wt% MgO (see Table 3 for phase identifications). The results suggested that the glass was prone to crystallization with mixed alkali and alkaline earth oxides at much lower concentrations than their solubility limits.

The solubility of Cr_2O_3 was 0.6 wt% in the LaBS-1 glass, which is slightly higher than the target concentration. Cr^{3+} was believed to be predominant in the LaBS-1 glass according to previous studies of Cr^{3+}/Cr^{6+} in waste borosilicate glasses [6,8]. Cr^{3+} ions were suggested by extended x-ray absorption fine structure (EXAFS) analysis [6] to occupy interstitial sites of the glass network, forming discrete groups that have a very similar chemical environment as in crystalline Cr_2O_3. In the glass, each Cr^{3+} ion bounds with 6 O^{2-} anions, and the ionic radius of Cr^{3+} is 0.66 Å [6] as compared with 0.62 Å when it is in crystalline Cr_2O_3. Hence, Cr^{3+} is larger than Al^{3+} (0.39 Å) that coordinates with 4 O^{2-} in the glass with sufficient Na_2O or other alkalis for charge balancing Al^{3+}, and it is also larger than Fe^{3+} (0.49 Å) that coordinates with 4 O^{2-} in the glass [9]. Both Al^{3+} and Fe^{3+} participate in the network forming; whereas, the soluble Cr^{3+} is limited by the availability of interstitial sites that are large enough to accommodate Cr^{3+} cations.

The solubility of Fe_2O_3 was about 5.5 wt% in the LaBS glass. The LaBS-1 glass with excessive Fe_2O_3 precipitated hematite (Fe_2O_3) according to XRD analysis. The solubility limit of NiO was 8.5 wt%. Ni-rich crystalline phase precipitated in the LaBS-1 glass with NiO in excess of the solubility limit. The size and the amount of Ni-rich phase were too small to be positively identified by either XRD or SEM/EDXS. The solubility limit of MnO was greater than 7 wt%. Since 7 wt% MnO is the maximum concentration found in Pu bearing waste streams, the exact solubility limit of MnO was not further studied. Similarly, 5 wt% MoO_3, 5 wt% WO_3, and 7 wt% Ta_2O_5 were all soluble in the LaBS-1 glass, the concentrations of which were close to their limits in the Pu waste streams. For all of the above waste components, their solubility limits were significantly higher than the target concentrations (cf. Table 2). However, this study also showed the LaBS-1 glass precipitating spinel crystals when 2 wt% Fe_2O_3, 4 wt% NiO, 1 wt% MnO, and 0.5 wt% Cr_2O_3 were in the quenched glass, demonstrating the component interactive effect on their solubility limits. In this case, Cr_2O_3 is believed to have the strongest effect on spinel formation according to previously published data for high-level waste glasses [10].

SUMMARY

The component solubility data suggest that LaBS glass may be a candidate glass frit for vitrification of PuO_2 and Pu-bearing materials. LaBS based glasses were also found to be susceptible to liquid-liquid phase separation with higher than expected concentrations of alkalis. TEM characterization of a phase separated glass with excessive Na_2O showed that the dispersed phase had slightly lower Si concentration and much higher concentrations of La, Nd, Gd, and possibly Ce. Investigations of the effect of phase separation on glass durability and releases of Gd, Ce, and Pu in aqueous environments are recommended.

ACKNOWLEDGMENT

This work is supported by the U.S. Department of Energy under Contract DE-AC06-76RLO 1830.

REFERENCES

1. J.D. Vienna, D.L. Alexander, H. Li, M.J. Schweiger, D.K. Peeler, and T.F. Meaker, PNNL-11346, Pacific Northwest National Laboratory, Richland, Washington, 1996.
2. M.B. Volf, Chemical Approach to Glass, *Glass Science and Technology*, Vol. 7, (Elsevier, Amsterdam, 1984) p.397-398.
3. H. Li, J.D. Vienna, D.K. Peeler, P. Hrma, and M.J. Schweiger, Plutonium Stabilization and Immobilization Workshop, CONF-951259, U.S. Department of Energy (Washington D.C. 1995) p.241.
4. H. Li, J.G. Darab, P.A. Smith, M.J. Schweiger, D.E. Smith, P. Hrma, *Nucl Mat. Manag.*, Vol. XXIV, (Institute Nucl. Mater. Manag., Northbrook, IL, 1995) p.466.
5. H. Li, J.D. Vienna, M.J. Schweiger, *Ceramic Transactions*, Vol. 72, ed. by V. Jain and D.K. Peeler (Am. Ceram. Soc., Westerville, OH, 1997) p.399.
6. J.G. Darab, H. Li, D.W. Matson, P.A. Smith, *Synchrotron Radiation Techniques in Industrial, Chemical, and Materials Science*, ed. by K.L. D'Amico and D.K. Shuh (Plenum, New York, 1996) p.999.
7. P. Li, I.W. Chen, J.E. Penner-Hahn, and T.Y. Tien, J. Am. Ceram. Soc., 74, 958 (1991).
8. H. Li, M.H. Langowski, P. Hrma, J. Hlavac, O. Vojtech, V. Krestan, P. Exnar, *Ceramic Transactions*, Vol. 72, ed. by V. Jain and D.K. Peeler (Am. Ceram. Soc., Westerville, OH, 1997) p.299.
9. M. Magini, M. de Moraes, A.F. Sedda, G. Paschina and G. Piccaluga, Phys. Chem. Glasses, 27, 95 (1986).
10. M. Mika, M.J. Schweiger, P. Hrma, and J.D. Vienna, Symp. Mat. Res. Sco. Proc., Vol. 465, ed. by W. Gray and I. Triay (Mater. Res. Soc., Pittsburgh, PA, 1997, in press).

DEVELOPMENT OF LOW-TEMPERATURE HIGH-RARE EARTH ELEMENT GLASSES

MICHAEL S. GARDNER, JOHN D. VIENNA, GREGORY F. PIEPEL, and HONG LI
Pacific Northwest National Laboratory, Richland, Washington 99352

ABSTRACT

Vitrification is currently being explored as an option for excess Pu because of its ability to convert Pu into a safe, durable, and accountable immobilization form that meets the spent nuclear fuel standard. To help guide the development of a low-temperature glass for Pu immobilization, the solubility of Ce (a Pu surrogate) and Gd (a neutron absorber) oxides were studied. A test matrix was statistically designed to assess the effects of varying the concentrations of SiO_2, B_2O_3, Na_2O, K_2O, Li_2O, Fe_2O_3, Al_2O_3, and Bi_2O_3 on the solubility of a 1:1 Ce:Gd oxide mixture in glass. The temperature at a viscosity of 5 Pa·s (T_5) varied between 1100°C and 1300°C for these glasses. The solubility of the Ce and Gd oxide mixture varied between 1 and 27 wt% at T_5 for the 37 test matrix glasses. Based on these results, an optimum glass with a maximum concentration of Ce and Gd oxide mixture (29 wt%) was designed with a T_5 of 1299°C.

INTRODUCTION

In the aftermath of the Cold War, the U.S. Department of Energy's (DOE) Office of Fissile Materials Disposition (OFMD) is charged with providing technical support for evaluation of disposition options for excess fissile materials manufactured for the nation's defense. One option being considered for the disposition of excess Pu is immobilization by vitrification. The vitrification option entails immobilizing Pu in a host glass and waste package that are criticality-safe (immune to nuclear criticality), proliferation-resistant, and environmentally acceptable for long-term storage or disposal [1].

To prove the technical and economic feasibility of candidate vitrification options it is necessary to demonstrate that PuO_2 feedstock can be dissolved in glass in sufficient quantity. The dissolution of over 10 wt% Pu has been demonstrated in high-temperature lanthanide borosilicate (LaBS) glasses [2]. However, for processing ease it would be advantageous to incorporate high concentrations of Pu in low-temperature glasses.

The objective of this study was to determine the effect of composition on Ce solubility in glasses that melt below 1300°C. Ce was used as a non-radioactive surrogate for Pu in this study. Both Ce^{4+} and Pu^{4+} exhibit similar coordination chemistries and ionic radii and are believed to enter the glass in the 8-fold coordination [3]. Although Ce^{4+} is more readily reduced to Ce^{3+} in glasses, at low-temperature Ce is assumed to be a reasonable Pu surrogate. Gd is typically added to the radioactive glasses in a 1:1 molar ratio with Pu as a neutron absorber to maintain critically safety, therefore Gd was added to glasses in this study in a 1:1 molar ratio with Ce.

TEST MATRIX DESIGN

The frit components with their minimum and maximum limits for the test matrix design are

---

shown in Table I. All glasses are composed of varied concentrations of eight oxides — SiO_2, B_2O_3, Na_2O, K_2O, Li_2O, Fe_2O_3, Al_2O_3, and Bi_2O_3. Cs_2O was added to each glass in a constant mass fraction of 0.01.

Table I. Glass composition region in mass percentage oxides

Component	Lower limit	Upper limit
SiO_2	46.00	70.00
B_2O_3	0.00	20.00
Na_2O	0.00	10.00
K_2O	0.00	10.00
Li_2O	0.00	10.00
Fe_2O_3	0.00	10.00
Al_2O_3	0.00	17.00
Bi_2O_3	0.00	10.00
Cs_2O	1.00	1.00

The test matrix in Table II is referred to as a *layered design* [4]. It consists of glass compositions on three layers of the polyhedral glass composition region (GCR) defined by the component ranges in Table I and constraints on T_5 and glass durability. The three layers of the GCR are: the boundary, an interior layer, and the center point. Piepel et al. [4] found that layered designs provide for covering the boundary as well as the interior of a variable space of interest, and have good bias as well as variance properties (so that models fit to the resulting data are more accurate as well as precise in their predictions).

Glass #1 in Table II covers the first layer, namely the center point of the GCR. The center composition was obtained by averaging the 966 extreme vertices of the GCR, which were generated using the MIXSOFT [5] software. Glasses # 2 to 13 cover the second layer, namely the boundary of the GCR. Glasses # 2 to 13 are 12 of the 966 extreme vertices selected using the ACED [6] optimal experimental design software. ACED selected the 12 glasses so as to minimize the variances of coefficients and predictions associated with a first-order mixture (FOM) model (see Eq. 1 later) that could be fitted to the 12 glasses plus the center glass. Glasses # 14 to 37 cover the third layer, namely the 0.66 shrunken version of the GCR boundary. This layer can be envisioned as resulting from shrinking each point on the GCR boundary toward the center glass (#1) composition, such that the distance from the center glass to each shrunken boundary glass is 0.66 of the distance to each corresponding boundary glass. Glasses # 14 to 37 were selected from the set of 0.66 shrunken extreme vertices using ACED to augment the glasses already in the test matrix (# 1 to 13). These 24 glasses were selected so that, combined with the 13 glasses already in the test matrix, the variances of coefficients and predictions associated with a full second-order mixture (SOM) model (see Eq. 2 later) would be minimized. Glasses # 1 to 37 barely support fitting the full SOM model for 8 components, which contains 36 coefficients. However, it is expected that a reduced form of the SOM model will be sufficient to adequately approximate the relationship between glass composition and solubility.

Readers wanting to know more about layered designs and the optimal experimental design methods used to construct them should refer to Piepel et al. [4]. While it is important to document the statistical methods used to design the test matrix, it is more important from a practical standpoint that the two-stage application of optimal experimental design software provides even coverages of the GCR boundary and interior layer. This provides a very good basis for understanding and modeling the single and interactive effects of glass components on solubility.

EXPERIMENTAL PROCEDURE

The frit glasses were prepared with the compositions given in Table II. Melting temperatures were predicted by a model [7] developed for Hanford High-Level waste glass [2], based on 0.15 mass fraction $CeO_2 \cdot Gd_2O_3$. Data from a study of composition effects on Pu residue glass properties were used to add Gd and Ce terms to this model [8]. Since these oxides have been shown to have little effect on viscosity in an equimolar ratio [8], the predicted temperatures were used to melt the frit glasses unless the temperature was found to be inappropriate during melting. Temperatures used to prepare the frits are shown in Table III. The viscosities of glasses #1 and #25 were measured using procedures reported by Hrma et al. [7], #1 being the centroid of the composition region, and #25 a representative of the shrunken boundary glasses. Glass #1 had a viscosity of 3.9 Pa·s at 1245°C, compared to the predicted 5 Pa·s. Glass #25 had a viscosity of 5.9 Pa·s at 1129°C, compared to the predicted 5 Pa·s. These two glasses were melted at their predicted T_5 values.

The frits were prepared in the following manner. The necessary reagent grade chemicals, calculated for a 200 g batch, were weighed on a Mettler PM4600 balance. Oxides were added directly for SiO_2, Fe_2O_3, Al_2O_3, and Bi_2O_3. Carbonates were added for Na_2O, K_2O, Li_2O, and Cs_2O. Boric acid was added for B_2O_3. The chemicals were milled in an agate disc mill for six minutes. The milled chemicals were placed in a cylindrical platinum crucible of depth 7.5 cm and diameter 10 cm. This was covered with a platinum lid and placed in a furnace at the predicted T_5 for 1 h. The melt was quenched by pouring onto a steel plate, with the remainder in the crucible quenched by submerging in a water bath. The resulting glass was crushed to a powder by milling in a tungsten carbide disc mill for 4 min.

The solubilities were determined in the following manner. CeO_2 and Gd_2O_3 in an equimolar ratio were added to the pre-melted frit in several percentages, to give a 2% (weight of total Ce and Gd oxides) difference between each sample. Each 10 g batch sample was mixed by milling in an agate disc mill for 5 minutes. For each sample, 1.50 g was added to a cubical platinum crucible measuring 1.9 cm on a side. After melting these initially for six minutes, another 1.50 g of sample was added to the crucibles. Lids were placed on the crucibles, and the samples were heat treated for 24 hours, at the temperatures denoted T_5 in Table III. At these temperatures, and at a depth in the crucible of 5 mm, the glass is predicted to reach redox equilibrium with the atmosphere in 24 hours or less according to redox kinetics reported by Schreiber et al. [9]. The samples were then removed from the furnace and allowed to cool in the air. The treated glass typically came to a depth of 3 mm at the crucible wall, and 2 mm in the center. The samples were removed from the crucibles and examined visually. Each translucent sample was examined for the presence of crystals under a microscope at 200 to 500X magnification. Solubilities are reported as the average of the highest percentage sample to be free of crystals or phase separation, and the lowest percentage sample to contain crystals or amorphous phase separation.

Table II -- Test matrix, frit mass percentage composition, and predicted T_5

Glass	SiO_2	B_2O_3	Na_2O	K_2O	Li_2O	Fe_2O_3	Al_2O_3	Bi_2O_3	Cs_2O	T_5 (°C)
1	57.86	7.74	5.53	6.42	6.13	4.88	6.00	4.44	1.00	1245
2	68.61	0.00	10.00	0.00	10.00	10.00	0.39	0.00	1.00	1115
3	46.00	20.00	9.81	0.00	6.19	0.00	17.00	0.00	1.00	1107
4	46.38	5.86	0.00	10.00	10.00	0.00	17.00	9.76	1.00	1102
5	46.00	19.18	0.00	10.00	5.48	10.00	8.34	0.00	1.00	1107
6	68.17	0.00	10.00	0.00	9.71	0.00	1.12	10.00	1.00	1104
7	53.02	20.00	0.00	0.00	5.12	10.00	0.86	10.00	1.00	1106
8	70.00	1.51	0.00	10.00	10.00	0.00	7.49	0.00	1.00	1257
9	46.00	20.00	10.00	0.00	0.00	10.00	12.91	0.09	1.00	1258
10	47.44	10.64	10.00	10.00	0.00	0.00	10.92	10.00	1.00	1253
11	70.00	20.00	0.00	0.30	4.70	0.00	0.00	4.00	1.00	1263
12	52.26	0.00	0.00	0.00	9.74	10.00	17.00	10.00	1.00	1241
13	54.62	3.28	10.00	10.00	0.00	10.00	1.09	10.00	1.00	1254
14	55.31	15.83	1.88	5.48	2.08	8.26	2.04	8.11	1.00	1227
15	50.69	15.83	1.88	2.18	5.38	1.66	13.26	8.11	1.00	1211
16	50.03	12.34	8.48	8.78	2.08	8.26	7.51	1.51	1.00	1197
17	61.43	15.83	1.88	2.18	5.86	8.26	2.04	1.51	1.00	1183
18	65.87	2.63	5.12	4.89	8.68	1.66	2.04	8.11	1.00	1150
19	62.30	2.63	1.88	8.78	8.68	8.26	2.04	4.43	1.00	1146
20	50.03	12.03	1.88	2.18	8.68	8.26	13.26	2.67	1.00	1129
21	58.16	10.76	1.88	8.78	8.68	1.66	7.56	1.51	1.00	1130
22	50.03	15.83	8.48	2.18	3.61	3.88	6.87	8.11	1.00	1130
23	50.03	15.34	1.88	8.78	5.34	1.66	7.85	8.11	1.00	1130
24	63.69	10.20	1.88	2.18	8.68	2.22	2.04	8.11	1.00	1130
25	58.78	4.88	8.48	2.18	6.27	8.26	2.04	8.11	1.00	1129
26	56.81	2.63	1.88	8.78	8.68	6.28	5.82	8.11	1.00	1129
27	60.87	2.63	8.48	2.18	8.41	1.66	13.26	1.51	1.00	1226
28	50.03	15.83	1.88	8.78	4.32	3.38	13.26	1.51	1.00	1230
29	55.88	2.63	1.88	6.90	8.68	8.26	13.26	1.51	1.00	1223
30	65.87	2.63	2.35	2.18	7.56	8.26	2.04	8.11	1.00	1225
31	54.77	2.63	1.88	8.78	7.90	1.66	13.26	8.11	1.00	1220
32	50.03	2.63	8.48	2.94	5.29	8.26	13.26	8.11	1.00	1218
33	65.87	11.24	8.48	2.18	4.63	1.66	3.43	1.51	1.00	1235
34	56.11	15.83	8.48	6.02	2.08	1.66	7.30	1.51	1.00	1235
35	63.53	4.32	5.39	8.78	5.18	8.26	2.04	1.51	1.00	1231
36	61.50	15.83	4.61	2.18	3.07	1.66	2.04	8.11	1.00	1232
37	64.74	3.11	8.48	6.09	4.78	1.66	2.04	8.11	1.00	1228

RESULTS AND DISCUSSION

The concentration range of Ce and Gd oxides tested, melting temperatures, and measured lanthanide oxide solubility (S_{Ln}) for each glass are listed in Table III. The S_{Ln} values ranged from 1 to 29 wt% in this study. The composition effects on S_{Ln} can be seen in Figure 1, which shows the mole fraction of each component plotted as a function of every other component, where different S_{Ln} values are plotted using different symbols. It can be readily seen that high S_{Ln} values predominantly occur at higher concentrations of SiO_2 and lower concentrations of Al_2O_3 and Fe_2O_3. The formation of an immiscible phase was observed in four of the test matrix glasses (14,

15, 28, and 36). Figure 1 shows that all such glasses had a high B_2O_3 content.

Mixture models were fitted to the S_{Ln} data to evaluate the effect of composition. The first-order mixture (FOM) model is of the form

$$S_{Ln} = \sum_{i=1}^{q} b_i x_i \qquad (1)$$

and the full second-order mixture (SOM) model is of the form

$$S_{Ln} = \sum_{i=1}^{q} b_i x_i + \sum_{i=1}^{q-1} \sum_{j=i+1}^{q} b_{ij} x_i x_j \qquad (2)$$

where x_i is the mole fraction of the i-th component, b_i is the coefficient of the first-order blending term involving x_i, and b_{ij} is the coefficient of the second-order blending term involving x_i and x_j. In this study, reduced SOM models containing the first-order terms and significant squared ($b_{ii}x_i^2$) and crossproduct ($b_{ij}x_ix_j$) terms were developed. The coefficients b_i and b_{ij} were estimated from the data using least squares regression.

Table III -- Lower & upper test limits and S_{Ln} results of 37 test matrix glasses, wt.% $CeO_2 \cdot Gd_2O_3$ Temperatures shown are actual heat treatment temperatures.

Glass	Lower	Upper	T (°C)	SLn
1	8	28	1239	19
2	8	18	1107	11
3	2	22	1093	3
4	2	20	1103	3
5	2	18	1106	5
6	8	18	1102	11
7	2	16	1136[b]	1
8	6	26	1252	16[c]
9	2	16	1301[b]	1
10	6	30	1254	7
11	2	20	1300[b]	1
12	19	29	1332	<19
13	4	24	1254	17
14	5	25	1227	14[a]
15	1	19	1211	14[a]
16	10	20	1194	11
17	3	23	1183	20
18	9	19	1176[b]	18
19	9	19	1145	12
20	3	13	1130	8
21	7	17	1130	12
22	4	14	1129	11
23	6	16	1130	11
24	6	26	1130	17
25	7	17	1129	12
26	1	19	1129	8
27	11	21	1226	12
28	8	18	1229	13[a]
29	3	13	1223	8
30	15	25	1225	22
31	2	22	1220	9
32	3	13	1216	6
33	20	30	1235	27
34	14	24	1235	17
35	13	23	1230	20
36	20	30	1233	27[a]
37	17	27	1229	20

a) Amorphous phase separation is observed in some or all samples.
b) Temperature is increased by approximately 30°C from predicted.
c) 16% was checked twice. In one experiment, 16% contained no crystals. In another experiment, 16% contained crystals. 14% did not contain crystals, and 18% contained crystals.

Table IV lists the FOM and SOM model coefficients. Glasses #1-37 were used to fit the FOM, and glasses #1-37 and 39 were used to fit the SOM. These coefficients confirm the positive effect of SiO_2 and negative effects of Al_2O_3 and Fe_2O_3 seen in Figure 1. The SOM shows large

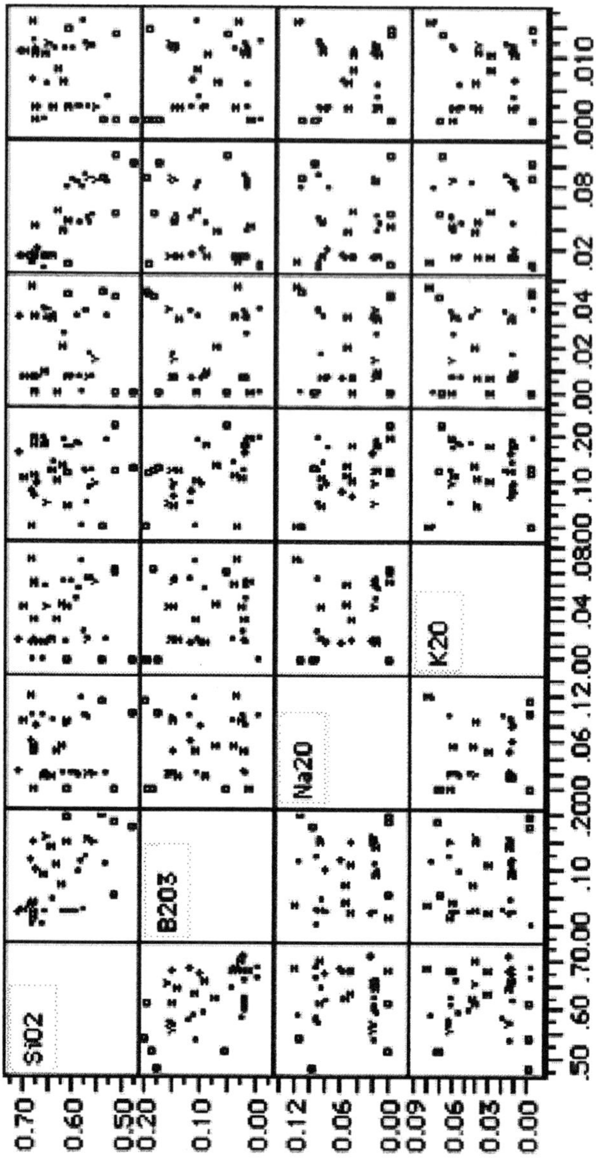

Symbols:

□ $1 \leq SLn \leq 5$ Z $16 \leq SLn \leq 20$

H $6 \leq SLn \leq 15$ + $SLn \leq 21$

Y Phase separated

Figure 1 Scatter plot matrix of SLn vs. mole fraction components.

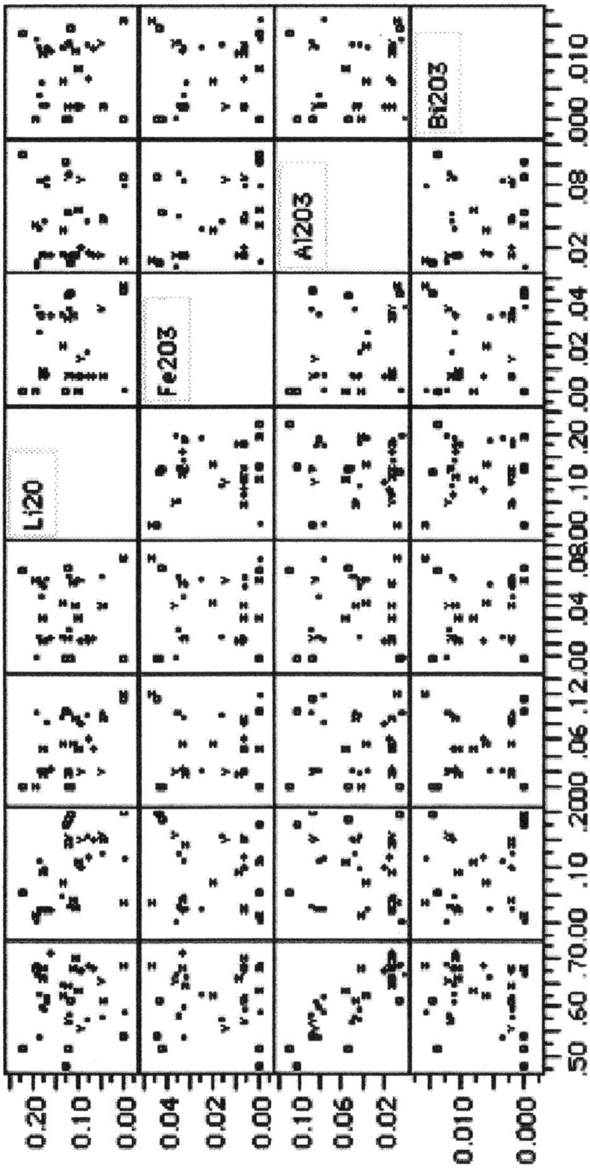

Symbols:

□ $1 \leq SLn \leq 5$ Z $16 \leq SLn \leq 20$

⌘ $6 \leq SLn \leq 15$ + $SLn \leq 21$

Ψ Phase separated

Figure 1 Scatter plot matrix of SLn vs. mole fraction components.

Table IV -- Model coefficients, wt.% $CeO_2 \cdot Gd_2O_3$ per mole fraction of components.

Term	FOM	SOM
SiO_2	61.490	70.091
B_2O_3	-29.163	63.917
Na_2O	-73.962	58.721
K_2O	-40.821	-49.905
Li_2O	-75.678	-92.566
Fe_2O_3	-163.212	-51.741
Al_2O_3	-36.780	-98.823
Bi_2O_3	-320.940	-124.334
Cs_2O	-61.304	-6193.700
$(B_2O_3)^2$	N/A	-324.304
$(Na_2O)^2$	N/A	-1036.745
$(Al_2O_3)X(Li_2O)$	N/A	569.064
R^2	0.770	0.962

negative second-order effects from B_2O_3 and Na_2O, as well as a positive interactive effect between Al_2O_3 and Li_2O. Inclusion of these second order effects increases the models' R^2 from 0.770 to 0.962. Any effects of redox and temperature may be included in the component coefficients. The effect of temperature on solubility is shown in Figure 2 which plots the natural logrithm of S_{Ln} as a function of the inverse melting temperature. Li et. al. [10] showed that Ce solubility in glass strongly depends on the melt redox, which is temperature dependent, as well as on glass composition. The lack of correlation between temperature and S_{Ln} suggests that composition effects on S_{Ln} are dominant over temperature effects for these glasses.

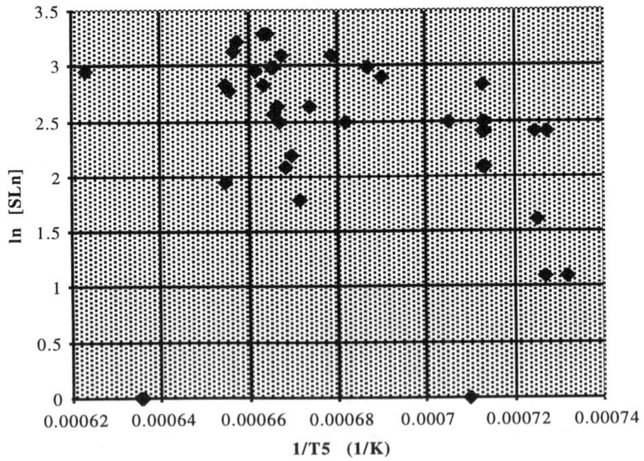

Figure 2. Correlation of S_{Ln} with T_5: ln $[S_{Ln}]$ vs $1/T_5$

The mixture models given in Table IV were used along with previous model versions developed from preliminary data to formulate glasses with maximized S_{Ln} at temperatures below 1300°C. The glass compositions shown in Table V were fabricated and measured. Table VI summarizes the S_{Ln} predictions, 95% confidence intervals for SOM predictions, and measured S_{Ln}. The FOM was used to formulate glass #38 with a predicted T_5 of 1201°C. It had a measured S_{Ln} of less than 22 compared to the FOM and SOM predicted values of 32 and 27. Glass #39 was formulated as the average of glasses #33 and #36, with added SiO_2 and other components decreased proportionally. It had a measured S_{Ln} of 25 compared to the FOM and SOM predictions of 24 and 27. Glass #40 was formulated by evaluating the SOM with restrictions of at least five components greater than 0.01 mass fraction, at least two nonzero alkali components, a maximum B_2O_3 mass fraction of 0.12, and a maximum predicted T_5 of 1250°C. It contained amorphous phase separation, and had a measured S_{Ln} of 23 compared to FOM and SOM predicted values of 27 and 31. Glass #41 was formulated by evaluating the SOM with a maximum B_2O_3 mass fraction of 0.088, a maximum T_5 of 1300°C, and all other restrictions the same. The measured SLn value of 29 closely matches the FOM and SOM predicted values of 30 and 31. The predictions from the two models correspond closely with each other and measured values for two of the formulated glasses (#39, 41), which are within or slightly outside the composition region. Measured S_{Ln} for these glasses was within the 95% confidence interval for the SOM. Glass 40 exhibited amorphous phase separation which cannot be easily predicted by these models, and its measured S_{Ln} was outside of the 95% confidence interval. Glass 38 containing 29.56 wt% B_2O_3 is far outside of the valid composition region for the models. An exact S_{Ln} value was not measured for this glass.

Table V -- High solubility test glasses, mass percentage composition.

Glass	SiO_2	B_2O_3	Na_2O	K_2O	Li_2O	Fe_2O_3	Al_2O_3	Bi_2O_3	Cs_2O
38	67.67	29.56	0.00	0.00	1.78	0.00	0.00	0.00	1.00
39	64.69	13.20	6.27	2.12	3.70	1.61	2.62	4.78	1.00
40	72.45	12.00	6.08	0.00	5.47	0.00	0.00	3.00	1.00
41	68.11	8.77	5.17	14.83	2.12	0.00	0.00	0.00	1.00

Table VI -- Comparison of S_{Ln} predictions and results for formulated glasses.

Glass	Predicted S_{Ln}, FOM	Predicted S_{Ln}, SOM	Lower 95% conf. limit S_{Ln}, SOM	Upper 95% conf. limit S_{Ln}, SOM	Measured S_{Ln}	Predicted T_5
38	32	27	18	36	< 22	1201
39	24	27	25	28	25	1248
40	27	31	29	33	23[a]	1250
41	30	31	28	35	29	1299

a) Amorphous phase separation is observed in some or all samples.

SUMMARY AND CONCLUSIONS

The solubility of the mixture of CeO_2 and Gd_2O_3 in a 1:1 mole ratio were measured with varying frit composition and at constant viscosity. This allowed the solubility to be expressed as a first or second order dependence solely on frit composition. Models derived from experimental data were solved for maximum solubility within a valid composition and temperature region, and the solutions were used to formulate high solubility test glasses. These models agreed closely with experimental data within and slightly outside the composition region studied. Amorphous phase separation, which occurs through a different mechanism than crystallization and is also undesirable in a nuclear waste glass, was observed in some of the test glasses. This study demonstrates that it is possible to model Ce and Gd solubility as a function of composition, and use experimental data to maximize the solubilities of these oxides in glasses.

REFERENCES

[1] Record of Decision for the Storage and Disposition of Weapons-Usable Fissile Materials, Office of Fissile Materials Disposition, U.S. Department of Energy, Washington, D.C. (1997).

[2] J.D. Vienna, D.L. Alexander, H. Li, M.J. Schweiger, D.K. Peeler, T.F. Meaker, Plutonium Dioxide Dissolution in Glass, PNNL-11346, Pacific Northwest National Laboratory, Richland, Washington (1996).

[3] J.G. Darab, H. Li, M.J. Schweiger, J.D. Vienna, P.G. Allen, J.J. Bucher, N.M. Edelstein, D.K. Shuh, "Chemistry of Plutonium and Plutonium-Surrogates In Vitrified Nuclear Wastes", submitted to Chemistry of Materials (1997).

[4] G.F. Piepel, C.M. Anderson, P.E. Redgate, "Response Surface Designs for Irregularly-Shaped Regions" (Parts 1, 2, and 3), 1993 Proceedings of the Section on Physical and Engineering Sciences, 205-227, American Statistical Association, Alexandria, Virginia (1993).

[5] G.F. Piepel, MIXSOFT™ Software for the Design and Analysis of Mixture Experiments, User's Guide, Version 2.08, MIXSOFT--Mixture Experiment Software, Richland, Washington (1993).

[6] W.J. Welch, ACED--Algorithms for the Construction of Experimental Designs, User's Guide Version 1.6.1, Department of Statistics, University of Waterloo, Waterloo, Ontario, Canada (1987).

[7] P.R. Hrma, G.F. Piepel, M.J. Schweiger, D.E. Smith, D.S. Kim, P.E. Redgate, J.D. Vienna, C.A. LoPresti, D.B. Simpson, D.K. Peeler, M.H. Langowski, Property / Composition Relationships for Hanford High-Level Waste Glasses Melthing at 1150°C, PNNL-10359, Vol.1, Pacific Northwest National Laboratory, Richland, Washington (1994).

[8] S.A. Bulkley and J.D. Vienna, "Composition Effects on Viscosity and Chemical Durability of Simulated Plutonium Residue Glasses", Scientific Basis For Nuclear Waste Management, **465**, 65-70 (1997).

[9] H.D. Schreiber, S.J. Kozak, A.L. Fritchman, D.S. Goldman, H.A., Schaeffer, "Redox Kinetics and Oxygen Diffusion in a Borosilicate Melt", Physics and Chemistry of Glasses, **27**, 152-177 (1985).

[10] H. Li, J.D. Vienna, P. Hrma, M.J. Schweiger, D.E. Smith, M Gong, "Borosilicate Based Glasses for Immobilization of Plutonium-Bearing Materials", Environmental Issues and Waste Management Technologies in the Ceramic and Nuclear Industries II, **72**, 399-408, American Ceramic Society, Westerville, Ohio (1996).

COMPARISON OF THE CORROSION BEHAVIOR OF PLUTONIUM GLASSES

C. J. Mertz, A. J. Bakel, J. K. Bates, D. B. Chamberlain, J. A. Fortner, J. M. Hanchar, and S. F. Wolf
Argonne National Laboratory
9700 South Cass Avenue
Argonne, IL 60439

ABSTRACT

The corrosion behavior of two glasses (alkali-tin-silicate and lanthanide borosilicate) developed for the immobilization of plutonium are examined. The durability of these glasses and the disposition of Pu during corrosion are discussed and compared. Test results are presented from various static test methods, including product consistency tests and ANL vapor hydration tests. The extent of corrosion of the Pu-bearing glasses was characterized by both solution (for glass component dissolution) and solid phase analyses (for glass reaction and the formation of alteration phases). Insight gained from testing the Pu glasses is utilized to evaluate the corrosion behavior under potential repository disposal conditions.

INTRODUCTION

Because of the dismantlement of nuclear weapons and the cleanup of weapons production sites, the United States is committed to evaluating options for plutonium disposition. One such option is the immobilization of Pu in a glass matrix and storage in a high-level waste repository. This option involves the development of a glass that can incorporate high Pu loadings while maintaining proliferation resistance and long-term durability. While typical loadings of Pu in high-level waste glasses are about 0.01% [1], glasses for Pu immobilization are required to contain at least 5% Pu [2]. In addition to high Pu loadings, the corrosion resistance and the potential for criticality of the glasses must be examined. Reaction processes that lead to the separation of actinide elements from neutron absorbers during the long-term storage of high-Pu-loaded glasses may lead to criticality concerns. Therefore, the disposition of the neutron absorber, Gd, as well as Pu, is of concern as the glass corrodes. The work discussed in this paper compares the corrosion behavior of a lanthanide borosilicate (LaBS) glass and the alkali-tin-silicate (ATS) glasses.

The ATS glasses were formulated at ANL to have a processing temperature of about 1100-1150°C, to accommodate high Pu loadings, and to be durable under conditions likely to exist in the geological environment during long-term storage [1]. The compositions of the two ATS glasses (ATS-P10 and ATS-P5) are listed in Table I. The ATS glass compositions are high in alkali metals, boron, zirconium, and tin, are low in aluminum, and do not contain Fe, Mg, or Ca. The development of the formulation is discussed elsewhere [1]. While the ATS-P10 was formulated to contain 10 mass % Pu, only ~7 mass % Pu was dissolved in the glass and the remaining ~3% remained as undissolved PuO_2. In order to make a glass that completely dissolved the plutonium, a 5 mass % Pu glass, named ATS-P5, was developed.

The LaBS glass was developed at the Savannah River Site as a high temperature glass (processing temperature of 1450°C) to incorporate high levels of Pu from nuclear defense programs [3]. The LaBS glass was based upon a commercial glass developed in the 1930's for use as an optical glass [4]. The LaBS glass does not contain any alkalis and has a high concentration of lanthanides. The composition of the LaBS glass is listed in Table I.

Table I. Plutonium Glass Compositions

	Glass Compositions[a] (Element Mass %)		
Element	LaBS	ATS-P5	ATS-P10
Al	10.2	1.19	1.38
B	3.01	2.76	3.94
Cs	-	0.85	0.47
K	-	6.41	3.32
Li	-	1.82	1.58
Na	-	6.79	6.68
Si	11.7	21.2	20.0
Sn	-	1.74	1.42
Sr	2.00	-	-
Ti	-	1.24	1.32
Zr	0.89	3.66	3.70
Zn	-	2.46	-
Pu	10.5	5.00	7.5[b]
La	9.7	-	-
Nd	10.6	-	-
Gd	6.95	3.29	2.60

[a]Values in table for LaBS and ATS-P10 are based on dissolution results from ICP-MS (Inductively Coupled Plasma Mass Spectrometry). Values for ATS-P5 are the nominal composition of the starting material.

[b]The ATS-P10 glass contained a total of 10.3 mass % Pu. Approximately, 7.5 mass % is dissolved in the glass and 2.8 mass % remained as undissolved PuO_2 [1].

EXPERIMENTAL

The corrosion behavior of the LaBS, ATS-P5, and ATS-P10 glasses was evaluated by means of existing testing methods. Samples of the glasses were reacted in accordance with the MCC-1 [5]; the product consistency test (PCT) [6]; and the ANL vapor hydration test (VHT) [7]. Three-day MCC-1 tests were performed in triplicate on the ATS-P5 and LaBS glasses. The MCC-1 is a static leach test that was developed by the Materials Characterization Center to provide a comparison of high-level waste form durabilities and to bound the rate of reaction under a specific test condition [5]. The 7-day PCT performed at a surface area of glass to volume of leachant ratio (S/V) of 2,000 m^{-1} is a standard test that provides comparative information (batch to batch) required by the Waste Acceptance Product Specifications [6] used for defense high level waste glass. The PCT at a S/V of 20,000 m^{-1} is a static test in which the concentrations of glass components in solution increase rapidly and accelerate the onset of alteration phase formation. Vapor hydration tests are run under saturated vapor conditions at elevated temperatures. The VHT conditions have been shown to greatly accelerate the glass corrosion and promote the formation of secondary phases. The formation of alteration phases in the VHT has been found to correlate with the reactivity of glass under long-term PCT conditions [8]. Experimental conditions used in corrosion testing for this study are summarized in Table II. All tests on LaBS, ATS-P5, and ATS-P10 glasses were performed in stainless steel vessels with deionized water as leachant. A blank test (with no glass in test vessel) was run simultaneously under the same experimental conditions as the glass test. The S/V was adjusted with the density of the glass (4.75 and 2.88 g/cm^3 for LaBS and ATS, respectively).

Table II. Summary of Experimental Testing Conditions

Test Type	Glass Form	S/V (m^{-1})	Temperature (°C)	Test Duration (days)
MCC-1	monolith	10	90	3
PCT	-100+200 mesh glass	2,000	90	7
PCT	-100+200 mesh glass	20,000	90	3 to 728
VHT	monolith	N/A	200	7 to 56

Reacted solid samples were examined with optical microscopy to evaluate the extent of reaction; the glass surface and the cross section were examined with scanning electron microscopy (SEM). Phase identification was determined with transmission electron microscopy (TEM) and associated electron energy loss spectroscopy, x-ray spectroscopy, and electron diffraction techniques. The extent of reaction can be estimated with SEM or TEM by determining the thickness of the alteration layer.

The concentration of elements in the leachate were measured with ICP-MS. Normalized mass loss values were calculated for MCC-1 and PCT based upon the compositions given in Table I, the geometric surface area of the MCC-1 monolith sample and sieve fraction for PCT, and the solution analytical results from the terminated tests. The equation used to calculate normalized mass loss is

$$NL(i) = \frac{\left(C_i - C_i^{\circ}\right)}{\left((S/V) \bullet f_i\right)} \quad (1)$$

where NL(i) is the normalized mass loss, C_i is the concentration of element i in the leachant, C_i° is the concentration of element i in the initial solution, S is the surface area of the glass, V is the volume of the solution, and f_i is the mass fraction of element i in the glass. The mass losses for these elements are normalized to the glass surface area and the elemental content of the glass and corrected for the amount of each element in the starting leachate. The NL(i) values include a contribution from the leachate and an acid strip of the test vessel. The corrosion of glasses is typically measured by the release of boron and/or alkalis, since these elements are generally released most rapidly from the glass and are highly soluble. In addition, boron is rarely contained in alteration phases and is often a better indicator than alkalis, which are contained in zeolite and clay phases that form during glass corrosion [9]. Although silicon is the most abundant cation in the glasses and is the predominant network-forming species, the solution concentration of silicon is not a good indicator of glass reaction for two main reasons: there are other elements that are released into solution faster, and the formation of secondary phases that contain silicon is common during glass corrosion.

The dissolution rates for the glasses were calculated based on the concentration of B in solution from MCC-1 tests in which alteration phases did not form. The equation used to calculate the dissolution rate is

$$NR = \frac{\left(NL(B)_2 - NL(B)_1\right)}{\left(t_2 - t_1\right)} \quad (2)$$

where NR is the dissolution rate of the glass, and $NL(B)_1$ and $NL(B)_2$ are the normalized mass losses of boron from tests with durations of t_1 and t_2, respectively.

RESULTS AND DISCUSSION

MCC-1 Results

The data from 3-day MCC-1 tests provide valuable information regarding the initial stage of glass corrosion and the reaction rate. Normalized mass losses and reaction rates are reported in Table III. As shown in Table III, the normalized mass losses of B and Si are slightly lower in tests with the LaBS than with the ATS-P5 glass. In contrast, the normalized mass losses of Pu and Gd are lower in tests with the ATS-P5 glass than with the LaBS glass. The dissolution rate measured from the MCC-1 tests indicates that the leach resistance of LaBS glass is slightly greater than the ATS-P5 glass. In addition, the pH of the leachate from the LaBS is significantly lower than the ATS-P5 glass tests due to the lack of alkalis in the LaBS glass.

Table III. MCC-1 Results at 10 m^{-1} for Plutonium Glasses

				Normalized Mass Loss (g/m^2)[b]			
Glass	Test #	pH	NR[a] ($g/m^2 \cdot d$)	B	Si	Pu	Gd
LaBS	1	5.3	0.7	2.1	1.5	0.22	0.8
	2	4.9	0.8	2.5	1.9	0.23	1.0
	3	5.1	1.0	2.9	2.2	0.25	1.2
ATS-P5[c]	1	9.2	1.2	3.6	4.2	0.06	0.03
	2	7.4	0.9	2.7	2.2	0.10	0.05
	3	8.4	1.0	3.1	3.5	0.05	0.01
	4	7.9	1.1	3.4	3.6	0.14	0.05
	5	8.4	1.1	3.2	3.6	0.05	0.02
	6	8.6	1.2	3.5	4.6	0.14	0.06

[a]Dissolution rates based on NL(B).
[b]Tests performed at 10 m^{-1} and 90°C, and $(NL)_i$ values are calculated using the total elemental contribution released from the glass. All concentrations were blank-corrected.
[c]Tests numbered 1-3 were reported previously in [10].

PCT Results

The PCT results at 2,000 m^{-1} were conducted to monitor the corrosion progress and to compare batch to batch consistency of high-level waste glasses [6]. Normalized mass losses from PCT at 2,000 m^{-1} are reported in Table IV. As shown in Table IV, the NL(i) values of B and Si are between 10 and 30 times greater for the ATS glasses than the LaBS glass. While NL(i) values for B and Si indicate that the LaBS glass is more durable than the ATS glasses, the NL(Pu) for the glasses follows the trend: ATS-P5 > LaBS > ATS-P10, where the normalized mass loss of Pu from ATS-P5 is only slightly greater than the LaBS glass and approximately five

Table IV. PCT Results at 2,000 m^{-1} for Plutonium Glasses

		Normalized Mass Loss (g/m^2)[a]			
Glass	pH	B	Si	Pu	Gd
LaBS	8.1	0.060	0.026	0.010	0.003
	8.0	0.077	0.024	0.009	0.003
ATS-P5	11.0	0.98	0.53	0.015	0.018
ATS-P10	10.2	1.6	0.38	0.003	0.010
	10.2	1.6	0.40	0.003	0.012

[a]Tests performed at 2,000 m^{-1} and 90°C, and $(NL)_i$ values are calculated using the total elemental contribution released from the glass. All concentrations were blank-corrected.

times greater than the ATS-P10 glass. The tests results show that Gd is released at roughly the same rate as Pu from the ATS-P5, Gd is released slower than Pu from the LaBS glass, and Gd is released faster than Pu from the ATS-P10. The pH in tests with the LaBS glass remains lower than the ATS glasses, since no alkalis are present in the LaBS glass and ion exchange is not possible.

The PCT results at 20,000 m^{-1} are static tests performed at high S/V to accelerate the glass corrosion and the onset of alteration phase formation. Results from the test have previously been reported on test durations through one year for ATS-P10 and through 28 days for the ATS-P5 glass [10]. The normalized mass losses for the LaBS and ATS glasses from PCT at 20,000 m^{-1} are reported in Table V. Long-term tests are still on-going for the ATS-P5 and LaBS glasses. The NL(i) values for B and Si are at least an order of magnitude lower from the LaBS than the ATS-P5 and ATS-P10 glasses (Figure 1), although the NL(Si) is only a factor of three lower for the LaBS than that for the ATS-P10 glass. While NL(i) values for B and Si indicate that the LaBS glass is more durable than the ATS glasses, the NL(Pu) for the glasses follows the trend: LaBS > ATS-P10 ~ ATS-P5, where the normalized mass loss of Pu from LaBS is an order of magnitude greater than the ATS-P5 and ATS-P10 glasses (Figure 2). The tests results show that the normalized mass loss of Gd is approximately the same as NL(Pu) for the ATS-P5, and NL(Gd) is greater than NL(Pu) for the LaBS and ATS-P10 glasses. The NL(i) values of Pu and Gd are almost an order of magnitude higher for the LaBS than the ATS glasses. Additional long-term tests and solid analyses from are needed to understand the disposition of Pu and Gd from the LaBS glass, although early results suggest that the LaBS composition is not as leach resistant for Pu and Gd as the ATS composition.

Table V. PCT Results at 20,000 m^{-1} for Plutonium Glasses

			Normalized Mass Loss (g/m^2)[a]			
Glass	Duration (days)	pH	B	Si	Pu	Gd
LaBS	28	8.2	0.048	0.017	0.013	0.014
	98	8.8	0.072	0.025	0.010	0.016
ATS-P5	3	11.2	0.26	0.14	0.002	0.003
	28	11.3	0.48	0.26	0.002	0.003
	98	11.6	0.78	0.36	0.002	0.003
	182	11.7	0.73	0.38	0.001	0.002
ATS-P10	3	10.4	0.35	0.057	0.0009	0.003
	28	10.3	0.59	0.054	0.002	0.007
	98	10.5	0.99	0.086	0.0014	0.005
	182	10.7	1.2	0.072	0.001	0.004
	364	10.7	1.4	0.088	0.002	0.008

[a]Tests performed at 20,000 m^{-1} and 90°C, and $(NL)_i$ values are calculated using the total elemental contribution released from the glass. All concentrations were blank-corrected.

Scanning electron microscopy (SEM) of the ATS-P10 glass that underwent corrosion under PCT conditions showed that the glass appeared to be unreacted at 7, 98, and 182 days of testing. Although no alteration phases were observed with SEM, further analysis with transmission electron microscopy (TEM) on the ATS-P10 solids from a 182-day PCT showed the formation of a very thin alteration layer on the glass (layer thickness of less than 50 nm). The layer on the exterior glass surface appeared to be a thin hydration layer with a chemical composition very similar to that of the glass. Several devitrification crystallites of PuO_x were identified with TEM. The solids from the other PCTs have not yet been analyzed.

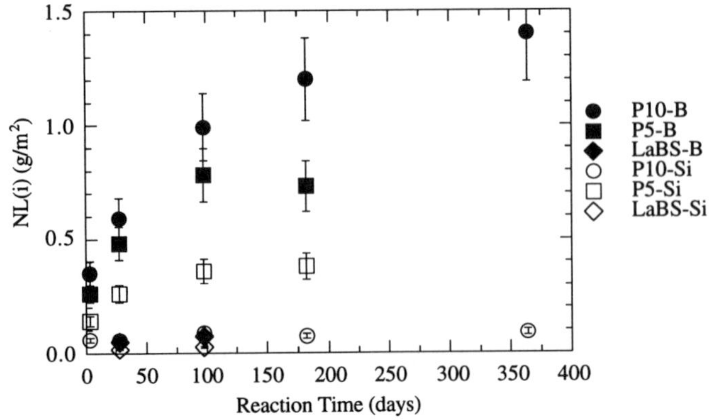

Figure 1

Normalized Mass Loss Based on Boron (B) and Silicon (Si) from LaBS, ATS-P5, and ATS-P10 Glasses at 20,000 m^{-1}.

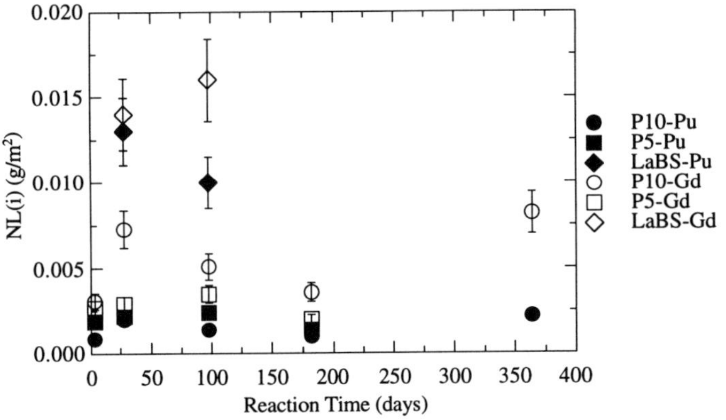

Figure 2

Normalized Mass Loss Based on Plutonium (Pu) and Gadolinium (Gd) from LaBS, ATS-P5, and ATS-P10 Glasses at 20,000 m^{-1}.

Vapor Hydration Tests

A white alteration phase was observed on all LaBS VHT monoliths (i.e., at 7, 14, 21, 35, and 56 days). The alteration phase coverage on the VHT monoliths increased dramatically for the longest VHT duration, 56 days, where the white alteration layer covered a significant portion of the LaBS sample. While analyses are ongoing for the VHT samples of the LaBS glass, TEM analysis has been completed on the 14 day VHT monolith. At 14 days, a speckling of the monolith surface with ~20 µm sized white alteration phases was observed. This alteration phase was identified as crystalline having a composition consistent with a Pu-rare earth-silicate. Enrichment of the La and Nd in the alteration phase is represented in the EDS from the alteration crystal as compared with the unaltered LaBS glass (Figure 3). The spectra were normalized to the plutonium.

An alteration layer of ~2 μm in thickness was identified with TEM at the glass surface of the 14-day LaBS VHT monolith (Figure 4). The layer thickness is at least four times less than previously reported for Pu-rich DWPF glasses reacted for 14 days [11]. The alteration layer is comprised of alternating layers of plutonium-rare earth-silicates (Regions A, C, and E in Figure 4) and aluminum silicates (Regions B and D in Figure 4), the former providing evidence for the segregation of Pu and Gd in a crystalline alteration phase.

Analysis with TEM of the ATS-P5 glass that was reacted for 21 days under VHT conditions identified a thin surface alteration layer (~100 nm) with a composition that was almost identical to the glass (Figure 5). Analysis with energy dispersive spectroscopy (EDS) determined that there was no substantial compositional difference between the surface layer and the glass.

Analysis with SEM of the P5 monoliths that were reacted for 35 and 56 days under VHT conditions showed increased corrosion of the glass compared with the 21-day test. The reacted glass contained an alteration layer which was 10- and 20-μm thick for the 35 and 56 day sample, respectively (Figure 6 shows the 56 day sample). The compositions of the alteration layer and glass were very similar, as determined with EDS. The alteration layers are porous and are probably composed of hydrated glass.

The VHT results from a 56-day test of the ATS-P10 showed a thin reacted layer on the ATS glass [11]. The layer is about 1.5 μm in thickness, compared to 20 μm for the ATS-P5 glass reacted under the same conditions. However, the LaBS glass forms an alteration layer of 2 μm at 14 days. The results indicate that the LaBS glass hydrates more rapidly than the ATS glasses, while the ATS-P5 hydrates faster than the ATS-P10 glass.

CONCLUSIONS

Glasses doped with Pu and Gd have been reacted to study the corrosion behavior of the glass compositions that can accommodate high levels of plutonium. While the normalized mass loss of boron and silicon is higher for the ATS glass compositions than the LaBS glass, the normalized mass loss of Pu from the LaBS is greater than the ATS glasses in MCC-1 and PCTs. The release behavior of the neutron absorber, Gd, from the ATS glasses is the same as the behavior of Pu during reaction. While MCC-1 and PCT results for the LaBS glass indicate that the release behavior of Pu and Gd are approximately the same, more tests are needed to determine the long-term behavior of Pu and Gd. The extent of corrosion in the Pu glasses from VHTs indicates that the hydration of the LaBS occurs more rapidly than the hydration in ATS glasses, but is slower than the hydration in the DWPF glasses. In addition, the formation of crystalline alteration phases that incorporate the Pu and rare earth elements in the LaBS VHT must be evaluated. Long-term PCT and VHT analyses must be completed before a full evaluation can be made of the reaction of the glasses to better understand the release of Pu and Gd.

ACKNOWLEDGMENTS

This work was supported by the U.S. Department of Energy, Office of Fissile Materials Disposition, under contract W-31-109-ENG-38.

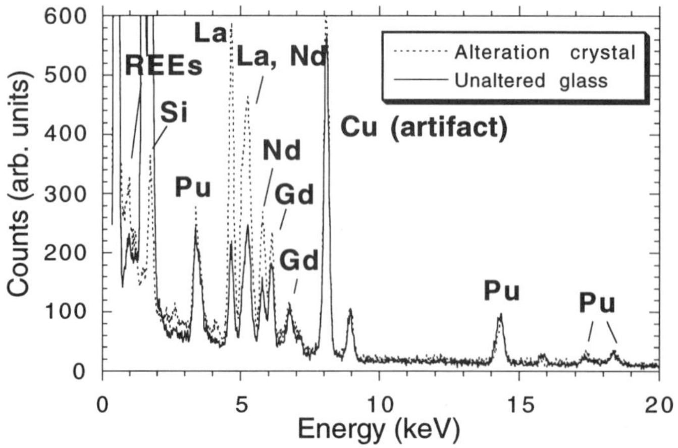

Figure 3

The EDS of an Alteration Crystal Compared with the Unaltered Glass from a 14-Day, VHT, LaBS Monolith.

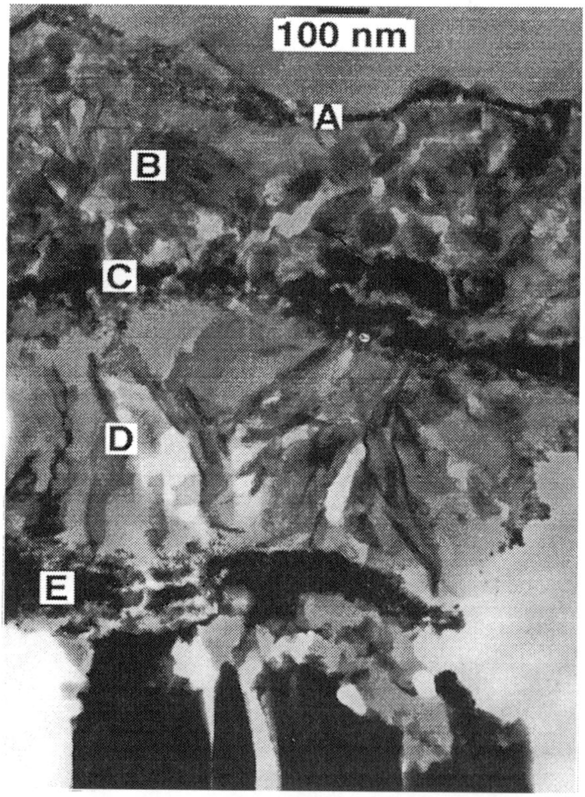

Figure 4

The TEM Micrograph of the Alteration Layer Attached to the LaBS Glass from a 14-Day VHT Monolith. The alteration layer is comprised of alternating layers of plutonium-rare earth-silicates (Regions A, C, and E) and aluminum silicates (Regions B and D).

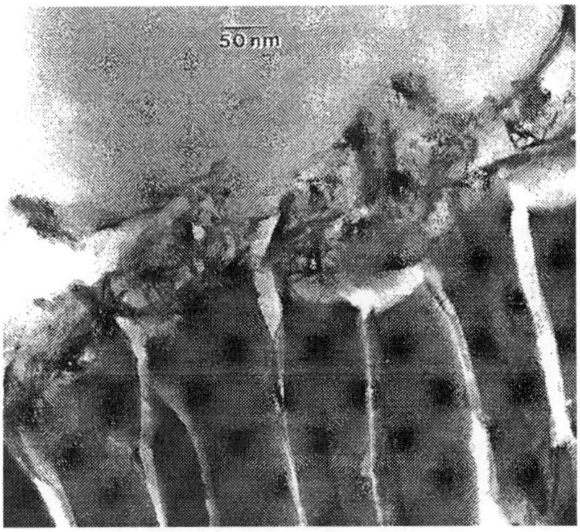

Figure 5

A TEM Micrograph of the Alteration Layer on the Surface of the P5 Glass after 21 Days of Testing under VHT Conditions.

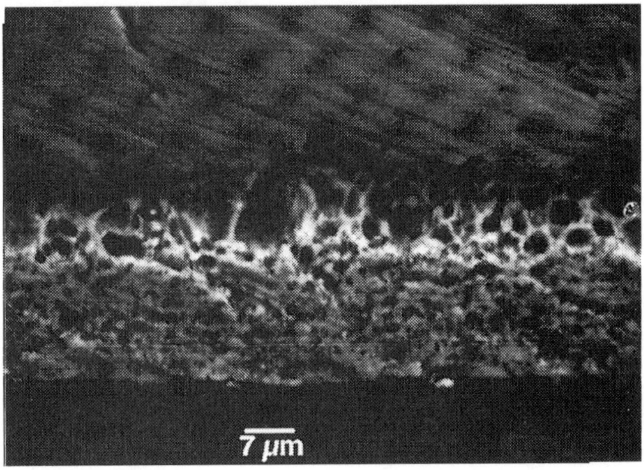

Figure 6

The Results from SEM Analysis of P5 Glass after 56 Days of Testing under VHT Conditions. The porous, alteration layer on the P5 glass is 20 μm thick.

REFERENCES

[1]J. K. Bates, A. J. G. Ellison, J. W. Emery, and J. C. Hoh, "Glass as a Waste Form for the Immobilization of Plutonium," Mater. Res. Soc. Symp. Proc. **412**, 57-64 (1996).

[2]*Management and Disposition of Excess Weapons Plutonium* Committee on International Security and Arms Control, National Academy Press, Washington, DC (1994).

[3]M. J. Plodinec, "Development of Glass Compositions for Immobilization of SRP Waste, Savannah River Laboratory Report DP-1517 (1979).

[4]J. V. Loffler, "Chemical Decolorizatrion," Glastechniche Berichte, **10**, 204-211 (1932).

[5]*Test Method for Static Leaching of Monolithic Waste Forms for Disposal of Radioactive Waste,* ASTM Standard C1220-92, ASTM, Philadelphia, PA (1992).

[6]*Standard Test Methods for Determining Chemical Durability of Nuclear Waste Glasses: The Product Consistency Test (PCT)*, ASTM Standard C1285-94, ASTM, Philadelphia, PA (1994).

[7]D. J. Wronkiewicz, L. M. Wang, J. K. Bates, and B. S. Tani, "Effects of Radiation Exposure on Glass Alteration in a Steam Environment," Mater. Res. Soc. Symp. Proc. **294**, 183-190 (1993).

[8]J. K. Bates et al., *ANL Technical Support Program for DOE of Environmental Management, Annual Report, October 1994-September 1995,* Argonne National Laboratory Report ANL-96/11 (1996).

[9]W. L. Ebert, *The Effects of Glass Surface Area/Solution Volume Ratio on Glass Corrosion: A Critical Review*, Argonne National Laboratory Report ANL-94/34 (1995).

[10]D. B. Chamberlain, J. M. Hanchar, J. W. Emery, J. C. Hoh, S. F. Wolf, R. J. Finch, J. K. Bates, A. J. G. Ellison, and D. B. Dingwell, "Development and Testing of a Glass Waste Form for the Immobilization of Plutonium," 1996 Fall Meeting of the Materials Research Society, Boston, MA, Dec. 1996.

[11]J. K. Bates, J. W. Emery, J. C. Hoh, E. C. Buck, J. A. Fortner, S. F. Wolf, and T. R. Johnson, "Reactivity of High Plutonium-Containing Glasses for the Immobilization of Surplus Fissile Materials," 1995 International High-Level Radioactive Waste Mgmt. Conf., American Nuclear Society, Las Vegas, Nevada, May, 1995.

Waste Form Property Measurement and Modeling

DOE REGULATORY REFORM INITIATIVE - VITRIFIED MIXED WASTE

Earl W. Holtzscheiter and Connie A. Cicero-Herman
Savannah River Technology Center
Westinghouse Savannah River Company
Aiken, SC 29808

Sandra J. Carroll
Environmental Protection Dept.
Westinghouse Savannah River Company
Aiken, SC 29808

James E. Flaherty
SAIC
20300 Century Blvd.
Germantown, MD 20874

ABSTRACT

In October 1995, the Department of Energy (DOE) submitted a regulatory reform proposal to the Environmental Protection Agency (EPA) relating to vitrified mixed waste forms. The technical proposal supports a regulatory strategy that would allow vitrified mixed waste forms treated through a permit or other environmental compliance mechanism to be granted an exemption from Resource Conservation and Recovery Act (RCRA) hazardous waste regulation, after treatment, based upon the inherent destruction and immobilization capabilities of vitrification technology. The vitrified waste form will meet or exceed the performance criteria of the Environmental Assessment (EA) glass[1] that has been accepted as the High Level Waste (HLW) standard for immobilizing radioactive waste components and the LDR treatment standards for inorganics and metals for controlling hazardous constituents. The proposal further provides that vitrified mixed waste would be responsibly managed under the Atomic Energy Act (AEA), while reducing overall costs. Full regulatory authority by the EPA or a State would be maintained until an acceptable vitrified mixed waste form, protective of human health and the environment, is produced.

INTRODUCTION

The DOE is charged with responsibly managing the largest volume of mixed waste in the United States. This responsibility includes managing waste in compliance with all applicable Federal and State laws and regulations, and in a cost-effective, environmentally responsible manner. Managing certain treated mixed wastes in RCRA permitted storage and disposal units (specifically those mixed wastes that pose low risks from the hazardous component) is unlikely to provide additional protection to human health and the environment beyond that afforded by managing these wastes in storage and disposal units subject to requirements for radiological control.

DOE's mixed waste are subject to dual regulation: the hazardous components are managed under RCRA, and the radioactive components are managed under the AEA. The AEA of 1954, as amended, authorizes the DOE to protect the health and safety of the public against radiation in conducting the Department's programs. Under the authority of the AEA, DOE has established a series of Orders which describe various self-imposed requirements. DOE Order 5820.2A "Radioactive Waste Management"[2], establishes policies and guidance for the management of radioactive waste and contaminated facilities.

Order 5820.2A describes a performance-based approach which allows DOE to utilize a variety of designs based upon local disposal conditions to achieve the performance criteria of protection to human health and the environment that is required under the Order. Under RCRA, mixed waste is subject to prescriptive requirements that govern the design and operation of the disposal facility, while still achieving the same goals as under DOE's Order.

A technical data package has been prepared to support a regulatory strategy which would allow vitrified mixed waste forms which have been treated through an environmental permitting process to be granted an exemption from RCRA hazardous waste regulations based upon the inherent destruction and immobilization capabilities of vitrification technology. Additionally, it proposes an alternative sampling and analysis strategy for certain highly radioactive mixed waste forms which will give the same level of confidence as sampling and analysis of the final waste form. The technical data package was submitted to the EPA in October 1995 in support of the draft Hazardous Waste Identification Rule (HWIR). This paper is a summary of the DOE's technical data package and their conditional exemption proposal for vitrified mixed waste.

WASTE STREAM OVERVIEW

Significant volumes of mixed waste have been generated within the DOE complex from various site production and maintenance activities. Process wastes that are candidates for vitrification are primarily HLW sludges and slurries stored at the Hanford Site and the Idaho National Engineering Environmental Laboratory (INEEL). These wastes are hazardous due to their corrosivity, the presence of toxic metals, and/or the presence of listed chemicals. In addition to HLW, mixed low-level wastes that are predominantly sludges, slurries, metal/metal oxides, and soils are also candidates for vitrification.

Distribution and Volume of Waste Streams that are Planned for Vitrification

Eight DOE sites have produced or will produce (over the next 75 years) wastes that are planned to be vitrified and that are potentially impacted by the proposal. They are- Battelle Columbus Lab, General Atomics in California, Knolls Atomic Power Lab in Schnectedy, NY, Rocky Flats, the Hanford Site, INEEL, the Savannah River Site (SRS) and the Oak Ridge Reservation (ORR). Total existing and projected inventory of mixed high-level and low-level wastes that are planned for vitrification at these eight sites is approximately 272,000 m^3. Approximately 74% of the waste is high-level and 26% is low-level.[3]

Waste Physical Form and Contents

The physical form of wastes planned for vitrification is varied. HLW consists predominately of aqueous liquids and inorganic sludges. The HLW at INEEL has been calcined and consists primarily of heavy metal salts and oxides. Mixed low-level waste matrices that are planned for vitrification include glass debris, waste water treatment sludges, soil, acidic waste waters, inorganic particulates, organic debris, ash, inorganic heterogeneous materials, and predominantly inorganic materials. The largest category for mixed low-level waste, "predominantly inorganic", includes filters, insulation, metallic waste and decontamination and decommissioning wastes.

VITRIFICATION TECHNOLOGY

For the purposes of the data package, vitrification refers to the process of converting waste materials into a glass or glass-like substance, typically through a thermal process. Vitrification thermally destroys the organic contaminants and stabilizes the inorganics and metals by incorporating them into the glass structure.

Vitrification Process Description

Vitrification is the thermal-chemical process whereby oxides of elemental constituents are incorporated into a solid, continuous, non-crystalline, three-dimensional network or glass structure. According to the ASTM, glass is further defined as an inorganic product of thermal fusion that has cooled to a rigid condition without crystallizing.[4] Vitrification, which occurs in a liquid mixture at an elevated temperature (nominally 1000°C to 1500°C), chemically bonds the glass elemental constituents together using oxygen to form a solution. At the required operating temperatures, organic components are either destroyed or volatilized and decompose in the off-gas treatment system and are not incorporated into the glass product.

At least one of the glass-forming elemental oxides, termed "network formers," must be present in the liquid mixture in sufficient quantity to form the glass matrix as the molten solution cools. The four primary network former oxides are those of silicon, phosphorous, boron, and germanium. Other elements break the glass-forming bonds and can lower the melt viscosity or produce other changes in the glass physical characteristics. These oxides are "network modifiers" and include the alkali metal and alkaline earth oxides. Some elements can, to a limited extent, replace the glass-formers in the glass matrix and are called "intermediate" oxides. These oxides include those of aluminum, vanadium, titanium, beryllium, molybdenum, bismuth, iron, zirconium, and others.[4] Because the network modifiers can greatly influence glass properties such as durability, viscosity, strength, and leach resistance, glass formulation development balances physical properties with processing requirements.

Gases such as radon and nitrogen are maintained in the relatively open glass network and are not chemically bonded like the metal oxides that form the glass structure. Two exceptions are fluorine and oxygen. Fluorine can be incorporated into some silicate glasses at up to 10 wt% or more. Oxygen is the bonding element in glass and is a necessary part of the glass structure.

Glass Compositions

Most waste glasses are based on the silica network because of the combination of good processing and physical characteristics. However, this requires most hazardous waste to be mixed with silica and other glass formers that are necessary for proper glass formulation. Waste glasses range from approximately 20 wt% actual waste to much higher when the waste itself contains substantial glass formers (e.g. contaminated soils and sludges). Due to the nature of the glass-making process, substantial volume reductions are common.

Table I shows several typical commercial glass compositions and the flexibility possible in formulating these glass compositions.[4] The proposed HLW glasses for Savannah River and West Valley are shown for comparison.

Table I. Compositions of Silica-Based Glasses (wt %)

Constituent	Soda-Lime Silica (Window) Glass	Borosilicate Heat-Resisting (Pyrex) Glass	Lead Glass	Nominal Savannah River	Nominal West Valley
SiO_2	70-74	80.5	56-58	49.8	41.0
B_2O_3	0-0.2	12	0	8.0	12.9
Al_2O_3	0.5-2.0	2	0-1	4.0	6.0
PbO	0	0	28-32	<0.1	--
MgO	0-4	0-1	0	1.3	0.9
CaO	5-10	0-1	0	1.0	0.5
Na_2O	12-17	4-5	0	8.8	8.0
K_2O	0	0	12-13	3.8	5.0
Other oxides	--	--	--	Remainder	Remainder

Constituent Effects

Table II lists the general effects various elements produce in a silica-based glass composition. In general, components which make the glass more processable (i.e., lower the melt temperature or reduce the melt viscosity) tend to decrease the chemical durability of the glass. The durability versus processability concern is the principal trade-off when waste glass compositions are formulated and tested.[4]

Table II. Summary of Effects of Major Glass Constituents on Glass Properties

Constituent			Chemical Durability		Liquidus*
Increased	Viscosity	Conductivity	By PCT	By TCLP	Temperature
Al	Increases	Decreases	Increases (more than TCLP)	Increases	Variable#
B	Decreases	Increases	Decreases	Decreases	Decreases
Ca	Decreases	Increases	Usually increases at low concentrations, variable at higher concentrations		Variable #
F	Decreases	Increases	Decreases	Decreases	Decreases (<10 wt%) Increases (>10 wt%)
Mg	Decreases	Increases	Similar to Ca		Variable
Na	Decreases	Increases	Decreases	Decreases	Decreases except for high levels
Si	Increases	Decreases	Increases	Increases	Increases

*The liquidus temperature is very dependent on the concentration of other glass components.

#"Variable" means the change in that specific parameter also depends on other conditions and constituent concentrations and can not be generalized independently.

Destruction of Organic Components

The DOE proposes to eliminate testing requirements for organic constituents identified in waste streams that will be vitrified because organic wastes are thermally destroyed by the vitrification process (systems typically operate above 1000°C). The destruction occurs through pyrolysis in the melt and combustion in the plenum or in the secondary combustion chamber. Studies have been performed that demonstrate the destruction and removal efficiencies (DREs) for a number of organic contaminants and compounds.[4] The DREs listed in Table III are values typical for In-Situ Vitrification systems. These values are conservative when compared with DREs for conventional melter systems with off-gas components designed to comply with (or exceed) air permit requirements. These DREs meet the requirements of 40 CFR Parts 264 and 265.

Technical Advantages of Vitrification

Vitrification has four major advantages over other methods of waste treatment. The primary advantage is the durable waste glass produced. With proper formulation, the waste glass performs exceptionally well in leach tests. The second major advantage of vitrification is the ability of the waste glass to incorporate a wide variety of contaminants and accompanying feed material in its structure without significantly affecting quality. The third advantage is that vitrification processes can accommodate various amounts of both organic and inorganic contaminants. Lastly, vitrification typically results in significant waste volume reductions, depending on the waste constituents and glass formers utilized.[4]

Vitrified waste forms possess an inherent environmental stability and will resist degradation for thousands of years, until radioactive decay lessens the threat to human health and the environment. During this decay period, the metals and inorganics remain chemically bound in the glass matrix. Due to these features, EPA has declared vitrification to be the specified treatment technology for mixed high-level waste (55 FR 22627).[5]

Table III. In Situ Vitrification Organic Destruction and Removal Efficiencies[4]

Contaminant	Initial Concentration (ppb)	Percent Destruction	Total DRE (including off-gas removal)
Aldrin	113	>97	>99.99
Chlordane	535,000	99.95	>99.999
DDD,DDE,DDT	21-24,000	99.9 - 99.99	>99.999
Dieldrin	24,000	98 - 99.9	>99.99
Dioxins	>47,000	99.9 - 99.99	>99.9999
Fuel Oils	230 - 11,000	>99	>99.999
Furans	>9,400	99.9 - 99.99	>99.9999
Glycol	NA	>90	>99.99
Heptachlor	61	98.7	>99.99
MEK	NA	>99	>99.999
PCBs	19,400,000	99.9 - 99.99	>99.9999
Pentachlorophenol	>4,000,000	99.995	>99.99999
Toluene	203,000	99.996	>99.99999
Trichloroethane	106,000	99.995	>99.99999
Xylenes	3,533,000	99.998	>99.99999

Process Limitations

Most elements can be vitrified and incorporated into glass to some extent, however some of the more volatile elements (such as cesium and the halogens) can be incorporated into glass only in very small concentrations.[4] Some metals, especially chromium and the noble metals, have limited solubility within many glass melts. However, by proper process control and design, even these volatile components can be incorporated into the glass by maintaining a melter cold cap, using reactive additives, and lowering formulation temperatures (~1050°C).

Many of the waste elements, as well as a high concentration of the network modifiers or intermediates in glass, can cause two-phase or crystalline structure distributions within the glass. Such phases can have substantial impact on the glass properties.[4] Inhomogeneities may also be present in a waste glass due to incomplete melting and homogenization of frit and waste, precipitation of crystallites, and devitrification. Any of these processes can cause the final glass to contain separate phases of various sizes and morphology. These separate phases and crystalline structures sometimes impact the corrosion and leach resistance behavior of glass. For example, studies of heat-treated SRS waste glasses have found that spinels, nepheline, and acmite devitrification products occupy 1 to 10% of the total glass volume.[4] For the SRS waste glasses, such devitrification and crystal growth did not greatly detract from the glass leach resistance when tested by the Product Consistency Test (PCT).[4] Fortunately, both crystallization and devitrification can be controlled by vitrification process parameters and final glass cooling controls.

Normally the glass-forming reaction is limited by mass transfer in the melt, so particle size control and good mixing of the solid constituents before melting is beneficial. Otherwise, non-reacted glass-coated chunks of solid material may result. This may be especially important for vitrification of large debris and may require preparation of the debris prior to vitrification. By ensuring proper glass formulation, melter residence time, thorough mixing, and overall process control, durable and acceptable glasses can be consistently produced.

OVERALL PROCESS FLOWCHART AND PROCESS CONTROL POINTS

Prior to the development of the process flow sheet and the equipment and process design, glass treatability studies are performed with either representative surrogates or the real waste depending on the radioactivity levels. In the treatability studies, the glass formulations are optimized for durability (measure of the glass capability to retain radioactive and hazardous components) and melter operating parameters (viscosity and liquidus temperature) consistent with melter capability. The durability is determined by standard leaching tests such as the PCT[6], which uses the leach rates of the more easily leached chemical components for comparison to the EA glass. In the process design phase while undergoing treatability studies, the glass would be evaluated by the Toxicity Characteristic Leaching Procedure (TCLP)[7] for hazardous metal constituents. Once the process flow sheet is defined, then the effluents and product compositions are defined and the appropriate process control strategy is put in place. The following provides a description of a typical implementation strategy of the *"Criteria For A Mixed and Hazardous Waste Process Control Program"* which is depicted in Figure 1.

1. Waste is received at the vitrification facility and held until it is fed for any pretreatment necessary.
2. The Pretreatment steps will vary depending on the nature of the feed. The pretreatment may be as coarse as chopping/sizing for heterogeneous wastes up to and including more involved processes such as mixing, neutralization, washing, dissolution, etc.
3. The Melter Feed step includes control of the waste feed to the melter, addition of glass formers to ensure meeting a predetermined glass composition and adjustment, as required, to ensure key operating parameters such as glass durability, viscosity, liquidus temperature, and/or residence time to ensure dissolution and mixing.
4. Off-Gas Treatment steps depend on the range and composition of the waste feed. Off-gas systems typically include a quenching step to drop the temperature of the off-gas exiting the melter, scrubbing of nitrogen and sulfur oxides and particulates, high efficiency mist eliminators and then high efficiency particulate air filtration as dictated by the levels of radioactivity. These systems typically produce two secondary streams: cleaned off-gas that will be defined and controlled to the applicable levels required by the air emission permits and a condensate liquid stream that will either require permitting for release, treatment prior to release, or recycle or partial recycle back to the vitrification process. The specific composition of the effluent streams will be defined in the process descriptions for various required permits.
5. Process Control is generally by one of two approaches, as shown in Figure 1:
    a. Feed forward process control systems will require characterization of key glass components prior to feeding the waste and glass formers to the melter. The acceptance of melter feed will be based on the chemical composition, or other means of product quality prediction, and key operating parameters (viscosity and liquidus temperature) required to successfully operate the melter.
    Note: The chemical composition of the melter feed must be previously determined by either surrogate or real waste testing that can be correlated with waste glass performance as determined by the PCT or by comparable durability leach tests.
    b. Feed back process control systems will typically sample and analyze the product stream (in this case the glass pour stream). The results of the analyses (either based on leach performance or another process variable correlated with leach performance) will determine whether melter feed is compositionally adjusted and/or rework of previously produced glass is required. The sampling frequency is based on the expected uniformity of the feed and the range of compositions within the process acceptance envelope.
6. Process Acceptance Envelope is the definition of the specific quality and operating parameters that the process and process control system must meet to be acceptable. The quality standard that is to be used for acceptance is the EA glass durability limit for HLW[1]. The operating parameters will be defined based on the melter and process design (e.g., viscosity required to pour glass, liquidus and operating temperature to protect the melter components and reduce volatility, residence time to assure system specific design base mixing). The operating window should be defined by the process design, expected range of waste compositions, and the glass durability. A number of the key parameters can be related to composition.

7. Sampling requirements will vary depending upon the approach used for process control.
   a. Feed forward process control will require waste or melter feed compositional knowledge based on samples. The results of the sampling will be compared to the acceptance criteria (composition, waste loading, etc. that can be related to glass durability). The pretreatment or adjustment of melter feed will be based on sample results and will control the ratio of the waste to glass formers. The extent of re-work would be related to the uniformity of the waste feed.
   b. Feed back process control would require sampling of the product stream generally as the melt exits the melter. The composition of the glass would be related to the durability or durability testing could be done directly. If the sample results did not meet acceptance requirements, the glass would have to be reworked.

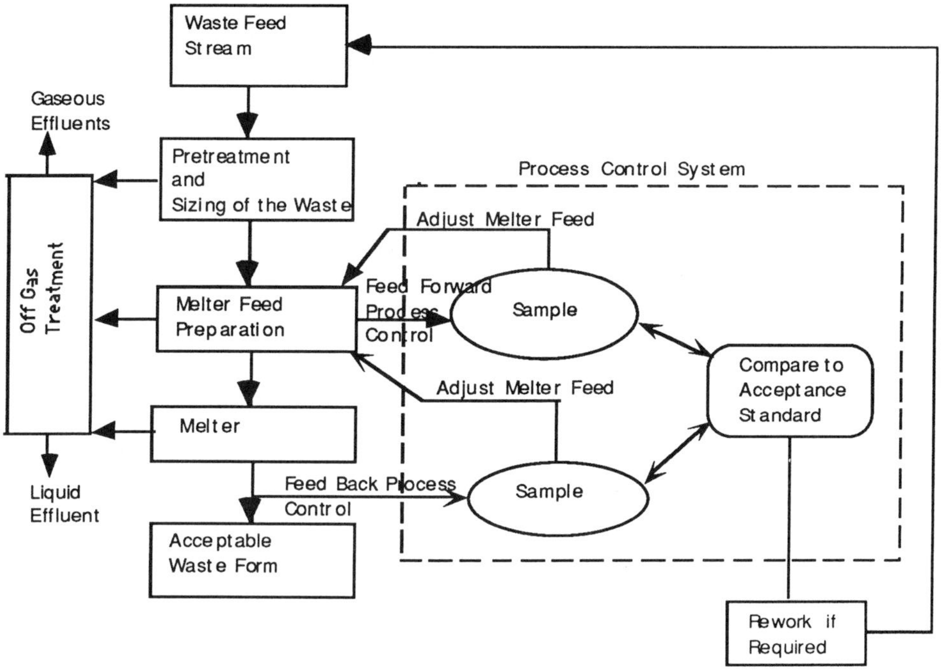

Figure 1. Typical Mixed and Hazardous Waste Process Control Flow Chart

Process Control Program

A vitrification process that will consistently provide an acceptable product requires a process control protocol for key operating parameters that yields a glass product that consistently falls within a pre-defined acceptable performance envelope. The glass formulation is defined by the treatability studies, which provide a glass formulation that will immobilize both radioactive and hazardous components. Once the right balance of variables is made to allow the production of a durable and processable glass, the range on the operating and glass composition parameters defines the performance envelope. The process control strategy is how the process is run to consistently yield the desired glass formulation (i.e. operate the process within the performance envelope). Once the process control strategy is defined and accepted as part of an environmental permitting process, the inherent properties of the final glass waste form justify no further RCRA Subtitle C control.

Criteria For a Mixed and Hazardous Waste Process Control Program:

1. The Process Control Program is consistent with "Standard Review Plan for the Review of Safety Analysis Plans for Nuclear Power Plants" Section 11.4 "Solid Waste Management Systems" NUREG 0800. It ensures the vitrification process meets applicable State and Federal regulations. It is not a set of detailed procedures, but is the bases for the detailed process control system description to be developed by the vitrification facility.
2. The Process Control Program should include, but is not limited to, the following:
   a. A commitment to a vitrification process that will satisfy State and Federal environmental regulatory requirements.
   b. A commitment that containers, shipping casks and methods of packaging meet applicable Federal Regulations, DOE Orders and DOT requirements for off-site shipment.
   c. A commitment that vitrified wastes meeting the RCRA exemption criteria will be shipped to an AEA disposal facility or Federal repository.
   d. A general description of the laboratory sampling process and its relation to the process control system.
   e. A general description of the vitrification process, process control parameters, parameter boundary conditions, waste form properties, and assurance the vitrification systems are operated within the established process parameters.
   f. A general description of the product sampling plan and the action to be taken if the sample fails to meet applicable requirements or falls outside the process control performance envelope.

Proposed Leach Testing for Determining Glass Durability

DOE will use two forms of leach tests to determine the durability of the glass. These two tests are the PCT (ASTM-C1285-94)[6] or comparable durability leach tests, and the TCLP[7]. The PCT was developed for evaluating the performance of HLW glass and will be used to determine the durability of the glass as it relates to the release of radioactive components. The TCLP test will be used to determine the leach rates for RCRA metal constituents.

The DOE proposes to pursue leach testing of the glass wasteform under an approach that considers the radiological hazard associated with testing of the final product and sampling/analytical logistics:

a. High radiation hazard wastes - Sampling and analysis will be performed on surrogate vitrified wastes that are chemically equivalent to the actual waste in cases where the radiological component renders testing of the treated form of the waste impractical/hazardous. Comparisons between radioactive SRS sludges and non-radioactive simulates have shown the simulates to be representative from both the standpoint of process parameters as well as glass performance. In these cases the development of a waste glass form that meets or exceeds the leaching characteristics of the EA glass coupled with the vitrification Process Control Program will demonstrate that an environmentally acceptable waste form will be consistently produced.
b. Low radiation hazard wastes - Sampling and analysis will be performed on the final product periodically. The period will be based on the uniformity of the feed composition. When the composition or process is altered, the process controls and performance of the final glass would have to be re-evaluated to ensure that the process produced an acceptable glass.

Environmental Assessment Glass Standard: The DOE proposes to develop vitrified waste forms with glass compositions that possess performance characteristics that meet or exceed the performance standards of the EA glass[1]. The EA glass is a borosilicate glass, and the utilization of borosilicate glass as a waste form is supported by waste form evaluation programs in the United

States and other countries. Essentially all nations now reprocessing or planning to reprocess spent nuclear fuels are either using borosilicate glass or have selected it as the preferred waste form for high-level radioactive waste. The EA glass was specifically formulated to retard the release of radioactive components in the waste. In addition, it has been shown that the EA glass retards the release of non-radioactive components. The glass has been subjected to TCLP testing for RCRA metals and the leachate concentration of these metals is well below the concentrations that EPA has proposed as generic exemption levels in the HWIR.[4]

REGULATORY CONTROL

DOE's regulatory reform initiative on vitrified mixed waste requests that EPA acknowledge that the vitrification process provides a glass final product that has inherent characteristics which provide sufficient protection to human health and the environment such that the final form should be granted an exemption from RCRA hazardous waste regulations. In order to produce an acceptable final glass product, the *"Criteria for a Mixed and Hazardous Waste Process Control Program"* must be followed. Within the context of the permitting process that controls the treatment (through vitrification) of the listed mixed/hazardous waste, the documents necessary to show that the criteria have been met will be provided for review as part of the existing permitting process. In most instances, the treatment of mixed high-level or mixed low-level waste would require a RCRA permit. For example, the RCRA permitting provisions under 40 CFR Part 264, Subpart X, would require the submission of all of the information in the Process Control Program, review and approval of all of those provisions and subsequent issuance of a permit requiring compliance with each component provision. In addition to obtaining a permit for the treatment of waste, most vitrification units will also require an air emissions permit and a wastewater treatment/effluent permit. Depending upon the actual application and the regulations in the state where the vitrification unit is to be operated, there may be other permitting requirements. In all instances, the regulatory agency retains complete control over the vitrification process and assures, through the permitting/compliance regime, the process produces a glass meeting environmentally acceptable performance characteristics. It is only after the production of glass that meets these performance characteristics (EA glass and LDR treatment standards demonstrated via inorganic analysis are the suggested performance criteria), that DOE is requesting the regulatory authority to allow the waste form to be exempt from RCRA Subtitle C control. The treated waste will continue to be managed in accordance with DOE Order 5820.2A[2].

At the request of some states that may be impacted by a proposed rule to exempt vitrified mixed waste after treatment, the DOE has recently suggested some regulatory language which could be incorporated into existing regulations. This language is subject to input from all interested parties and will likely be further refined as this initiative moves through the regulatory process. The proposed language developed to date includes an exemption in 40 CFR 261.3 (Definition of Solid Waste) to exclude mixed waste that has been treated by vitrification authorized under 40 CFR Part 270 or another environmental permitting authority acceptable by the Regional Administrator. It includes a provision that terms and conditions can be added through the permitting process to ensure the integrity of the final waste form. The proposed language also includes an addition to 40 CFR 268.40 (LDR Treatment Standards) requiring that vitrified mixed waste meet the LDR treatment standards of Subpart D and that compliance with the treatment standards will be measured by analysis for inorganic constituents listed in the Universal Treatment Standards. The proposed language does not require strict adherence to the performance standards outlined above, thus allowing for flexibility on the part of the permitting authority to set performance standards on a case by case basis.

PROGRESS AND PATH FORWARD

Since submitting this proposal to EPA in October 1995, the DOE has continued to work with the EPA to answer questions and define areas that need additional consideration. Several meetings have been held and informal comments have been made by EPA staff responsible for reviewing the proposal. In addition, shortly after the submission of the proposal, DOE began working with the National Governor's Association through an existing avenue that brought together states with an interest in DOE's mixed waste activities. Through requests from this group, DOE has produced both the proposed language mentioned above and a cost saving analysis relating the amount of money to potentially be saved if the proposal is adopted. The DOE has also initiated discussions with the National Association of Attorneys General (NAAG), and the Association of State and Territorial Waste Management Officials (ASTWMO). A Regulatory Reform newsletter outlining this and other regulatory reform initiatives has also been widely distributed. In the late summer of 1996, DOE initiated a series of meetings with key stakeholder states. These meetings were one day long technical meetings where the basis of the vitrification proposal (and another regulatory reform proposal on immobilized mixed debris) were discussed in detail, with ample time for questions from attendees. These meetings were held in Augusta GA, Las Vegas NV, Richland WA, Golden CO, Oak Ridge TN and Los Alamos NM. Attendees included representatives of EPA and state regulatory agencies, assistant attorneys general, EPA-HQ, NAAG, and ASTWMO. These meetings were well received with the technical basis for the proposals generally supported by attendees. However, a number of issues, common to most state representatives were identified. DOE has established a working group including state representatives from the technical meetings to try to identify and resolve issues that may impede adoption of the vitrification proposal.

SUMMARY AND CONCLUSIONS

In summary, the preceding discussion proposes that, based on the inherent characteristics of the vitrification process and stability of the glass waste form, the glass product made from mixed waste be exempt from RCRA control. This proposal is contingent on the required use of the Process Control Program described in this document. The Process Control Program provides the description of the key operating variables, the process stream compositions as they relate to the end product quality and the permitted emissions/effluents, the acceptable performance envelope, and the end product durability. This information not only defines the process and ensures final product performance, but is required to demonstrate that the process will meet existing federal and state environmental regulations associated with the operation of a mixed waste treatment facility. Therefore, this proposal provides a means to facilitate the use of a superior technology for responsible waste management while reducing costs and maintaining full regulatory authority until an acceptable vitrified waste form is produced.

REFERENCES

[1]U.S. Department of Energy, "Environmental Assessment Waste Form Selection for SRP High-Level Waste:, U.S. DOE Report DOE-EA-0179 (July 1982).

[2]U. S. Department of Energy Order 5820.2A, "Radioactive Waste Management", (September 1988).

[3]U. S. Department of Energy Environmental Management 1996 Baseline Environmental Management Report, DOE/EM-0290, (June 1996).

[4]S. J. Carroll, E. W. Holtzscheiter, and J. E. Flaherty, "Technical Data Package for Vitrified Waste Forms", EPA RCRA Docket No. F95-WHWO-FFFFF S0983.F, U. S. Department of Energy, (October 1995).

[5]Federal Register, "Land Disposal Restrictions for Third Scheduled Wastes, Final Rule ", 55 FR 22627 (June 1990)

[6]ASTM C 1285-94, "Standard Test Methods for Determining Chemical Durability of Nuclear Waste Glasses: The Product Consistency Test (PCT)", October 1994.

[7]Federal Register, "Method 1131 Toxicity Characteristic Leaching Procedure (TCLP)", 40 CFR 261 App. II, pp. 66-81 (1991).

CRYSTALLIZATION, REDOX, AND WASTE-LOADING EFFECTS ON PROPERTIES OF VITREOUS WASTE FORMS

P. E. Bloomer, X. Feng,* and N. Chantaraprachoom**
Pacific Northwest National Laboratory
Richland, WA 99352

* to whom correspondence should be addressed
**Office of Atomic Energy for Peace, Chatuchak, Bangkok, Thailand

ABSTRACT

A laboratory study showed that waste form durability is strongly influenced by processing conditions. A reducing melt condition had no effect on the durability of some waste forms, yet it decreased the durability of others up to 700 % when compared to the oxidized melt. One waste-form composition tested showed a 40 % increase in durability when devitrified (crystallized). Some compositions showed no change in durability after devitrification, while others showed a 200 % decrease. Increased waste loading decreased durability for one waste form, increased durability by 240 % for another, and showed no effect on a third. These responses to processing and composition variations were dictated by fundamental glass chemistry and can be adjusted to achieve maximal waste loading, acceptable durability, and desired processing characteristics with optimized formulations.

INTRODUCTION

Graphite electrode arc furnaces have been used in the steel industry since the turn of the century. A refined DC arc furnace for treating waste materials has been developed by a team from the Pacific Northwest National Laboratory (PNNL), Massachusetts Institute of Technology (MIT), and two industrial partners, T&R Associates and Electro-Pyrolysis, Inc. (EPI) [1,2].

The graphite electrode DC electric arc is used to deliver thermal energy into the material to be processed. Use of the current-controlled graphite electrode DC arc provides the high-temperature environment required for melting a range of heterogeneous waste forms. Temperatures in excess of 1700°C can be produced in the material to be processed. In addition, the high-temperature region around the DC arc will destroy organic species and vapors that evolve from the material being processed. The volume of off-gas produced in the pyrolytic DC arc treatment of solid waste is greatly reduced relative to combustion treatment—typically by more than a factor of 10.

The use of graphite electrodes and graphite crucibles in the process also provides a reducing condition for the production of the waste forms. It has been shown that the redox condition of waste-form processing may affect its chemical durability and processing properties, such as viscosity [3, 4]. The processed waste form can be poured into a large container to form a large monolith where the waste form is subjected to slow cooling, resulting in devitrification. The uncontrolled devitrification can substantially reduce glass durability by two orders of magnitude according to PCT testing [5, 6]. The controlled devitrification can in fact enhance waste-form durability substantially [7]. The waste form can also be made into small gems where glass is quenched without much devitrification, but the waste form occupies large volumes due to voids among the small gems.

It is necessary to assess how the reducing condition in a DC arc furnace and devitrification of the products affect waste-form processing characteristics and waste-form durability. This information is valuable for the design of optimal processing conditions and optimal formulations that achieve maximum waste loading, acceptable chemical durability, and proper melting temperatures and viscosities. The initial development of vitreous waste forms focuses on three separate waste streams with high metal contents.

EXPERIMENTAL

The waste forms were prepared from reagent-grade chemicals in crucibles and melted in an electrically heated resistance furnace (Deltech DT-31). The oxidizing melts were produced in air, while the reducing conditions were achieved by melting a mixture of the oxidized waste-form powders and carbon powders. The devitrified waste form is referred to as vitreous ceramic waste form and the quenched amorphous waste form is referred as glass in this report.

The chemical durability of a waste form is a measure of its capability to resist corrosion in an aqueous solution or in water vapor. The durability of the glass/vitreous ceramics was measured by two tests: the toxicity characteristic leaching procedure (TCLP) and the product consistency test (PCT). Waste-form leachates were analyzed with inductively coupled plasma–atomic emission spectrometry (ICP-AES) with an accuracy of ± 10 % for major elements. The composition of selected waste forms after melting was determined through a fusion method. Solid characterization of the waste form was performed using x-ray powder diffraction (XRD) analysis and the scanning electron microscope.

RESULTS AND DISCUSSION

Three solid mixed-waste streams were studied in this task: Metal-R, heterogeneous debris ^{238}Pu waste, and Nom-60 waste. The Metal-R waste is a waste surrogate consisting of Idaho National Engineering Laboratory (INEL) soil, lime, and metals (approximately 50 % by mass); the metals in this waste stream are almost completely carbon steel. The heterogeneous debris ^{238}Pu waste stream from the U.S. Department of Energy Savannah River site is a surrogate consisting of activated carbon, wood, PVC, neoprene, steels, glass beads, concretes, alumna crucibles, diatomaceous earth, Resource Conservation and Recovery Act metals and organics, including naphthalene, 1,2-dichlorobenzene, chlorobenzene, plutonium, and cesium. The Nom-60 waste is a waste surrogate consisting of the nominal INEL waste blended with 60 % INEL soil composition. The nominal INEL waste consists of metals, solid combustibles, silicated organics, nitrate salts, and metal hydroxide sludge.

The processing conditions and the waste loadings of each melt are listed in Table I. The target compositions are also provided in Table II. The identity of each waste form was related to the waste stream, waste loading, and processing for further clarification. Each is coded with the abbreviated name of the waste stream followed by the waste loading in mass fraction. The waste form type is abbreviated by (VC) representing a heat-treated vitreous ceramic and (G) representing a quenched glass sample. This is followed by either, (a) representing normal oxidizing melting conditions in an air environment, or (b) representing reducing melting conditions achieved with an addition of 1 wt % carbon powder mixed into the ground glass prior to melting.

Metal-R Waste

Four melts: M-.34Ga, Met-.34Gb, Met-1VCa, and Met-1VCb, were performed for this study. Met-.34Ga and Met-.34Gb are quenched glasses that contain 34 wt % of the Metal-R waste surrogate. Met-1VCa and Met-1VCb are vitreous ceramics consisting entirely of the Metal-R waste surrogate. To promote crystallization Met-1VCa and Met-1VCb were melted, then heat-treated at 1000°C for 17 hours.

Solid Characterization: Met-.34Ga was found to be fully amorphous, while Met-.34Gb contained 1 vol % silver metal. Met-1VCa was found to be 75 vol % crystalline, consisting of 80 % spinel-type, 10 % hematite, and 10 % augite-type crystals. Met-1VCb was found to be entirely crystalline, consisting of 50 vol % spinel-type and 25 % augite-type crystals. The remaining composition consists of fayalite and albite-type crystals. The results of the redox analysis are presented in Table III in a ratio of Fe(II)/(Total Fe). The concentrations of Fe(II) were found to be 3.5 % and 47 % in Met-.34Ga and Met-.34Gb respectively. This illustrates the significant reduction of Met-.34Gb. The percentage of Fe(II) in Met-1VCa is 14 % compared to 30 % Fe(II) in Met-1VCb, showing the effect of the reducing atmosphere.

Chemical Durabilities: The normalized release of silicon, sodium, and boron from the 7-day PCTs is provided in Table IV. The PCT results show that: (1) The devitrified Met-1VCa with 100 % waste loading is 44 % more durable (according to sodium release) than the glass melt of Met-.34Ga. It is known that spinel crystals are highly durable and the formation of crystalline spinel from the glass melt can beneficially alter the residual glass composition. Recalculation of the residual glass composition of Met-1VCa, subtracting 53 % of the spinel crystals, shows that the residual glass composition is enriched in SiO_2 (55 %) and Al_2O_3 (11.6 %), but low in alkalis (2.6 % Na_2O). It is now easy to understand the increased

Table I Laboratory Melting Conditions During Second Hour

Waste Form ID	Waste Loading (Wt%)	Melt Temp. (°C)	Cruc.	Atmosphere	Heat Treatment	Observed Viscosity (Pa·s)
Met-.34Ga	34	1280	Pt	Air	Quenched	12
Met-.34Gb	34	1280	Si	Reducing	Quenched	10
Met-1VCa	100	1450	Pt	Air	17hrs.@ 1000°C	5
Met-1VCb	100	1450	Al	Reducing	17hrs.@ 1000°C	
Lab-1Ga	100	1450	Pt	Air	Quenched	28
Lab-1Gb	100	1450	Al	Reducing	Quenched	25
Lab-1VCa	100	1450	Pt	Air	16hrs.@ 1000°C	28
Lab-1VCb	100	1450	Al	Reducing	16hrs.@ 1000°C	
Lab-.83Ga	83	1375	Pt	Air	Quenched	15
Lab-.83Gb	83	1375	Si	Reducing	Quenched	12
Nom-1Ga	100	1425	Pt	Air	Quenched	8
Nom-1VCa	100	1425	Pt	Air	16hrs.@ 1000°C	8
Nom-1Gb	100	1425	Si	Reducing	Quenched	5
Nom-.90Ga	90	1300	Pt	Air	Quenched	6
Nom-.90Gb	90	1300	Si	Reducing	Quenched	5

durability of Met-1VCa over Met-.34Ga by comparing the residual glass compositions. After subtraction Met-.34Ga glass is 10 times higher in Na_2O (20 %), yet has a similar amount of silica and alumina as the residual glass of Met-1VCa. (2) The reducing melting conditions decreased the chemical durability (in terms of sodium release) of glass by 400 % (Met-.34Ga versus Met-.34Gb) and of vitreous ceramics by 720 % (Met-1VCa versus Met-1VCb). The reducing condition in glass converted 44 % more ferric iron into ferrous iron. Ferric iron can serve as a network former to strengthen glass structure while ferrous iron cannot. The more severe redox effect on the durability of the vitreous ceramic is probably due to the formation of 45% silica and alumina-consuming crystals, such as albite-type ($Na[Si_3Al]O_8$), fayalite (Fe_2SiO_4), and augite ($Ca[Fe,Mg]Si_2O_6$). This results in the depletion of

Table II Target Waste Form Compositions in Wt%

Oxide	Nom-1G Nom-1VC	Nom-.90G	Met-.34G	Met-1VC	Lab-1G Lab-1VC	Lab-.83G
Ag_2O		0.000	0.199	0.584		
Al_2O_3	8.728	7.803	6.000	5.212	18.575	15.396
As_2O_3		0.001	0.002	0.006		
B_2O_3			5.917	0.000		
BaO	0.065	0.058				
CaO	21.618	19.326	2.264	6.657	3.567	2.956
CdO		0.000	0.003	0.009	0.238	0.197
CeO_2	0.428	0.383			0.296	0.245
Cl	2.566	2.294	0.006	0.016	0.498	0.412
Cr_2O_3	1.188	1.062	0.014	0.042	0.121	0.100
Cs_2O		0.001	0.060	0.177	0.355	0.294
CuO	0.276	0.247	0.001	0.003		
Fe_2O_3	20.207	18.064	20.000	58.801	24.154	20.020
K_2O	1.846	1.650			0.229	0.190
MgO	1.752	1.566	0.368	1.081	0.475	0.393
MnO	0.105	0.094	0.012	0.034	0.107	0.089
Na_2O	1.471	6.000	15.000	1.221	2.630	15.000
NiO	0.488	0.436	0.004	0.010	0.133	0.110
P_2O_5	0.106	0.095	0.045	0.133		
PbO	0.075	0.067	0.001	0.002	0.340	0.282
S	0.070	0.062			0.078	0.065
SiO_2	38.124	40.000	50.000	25.703	48.204	44.244
TiO_2	0.387	0.346	0.103	0.302		
ZrO_2	0.471	0.421	0.002	0.006		
Total	100.00	100.00	100.00	100.00	100.00	99.99

network formers in the residual glass matrix affecting the overall durability of the waste form. (3) The reducing condition increased the leachate pH in both glasses and vitreous ceramics by one pH unit within the 7 days of PCTs. Vitreous ceramics had a leach pH of one unit lower than glasses in both oxidized and reduced conditions.

The TCLP results in Table IV show that all the melts pass the Environmental Protection

Agency (EPA) land-disposal requirements. The comparison of TCLP results versus PCT results may also suggest that the TCLP results are not very sensitive to waste-form composition change and processing variations. TCLP may not be a suitable test for waste-form optimization, but is a test related to the acceptance of the waste forms developed.

Table III Results from iron II and total iron analysis

Waste Form ID	Target Fe wt%	Analyzed Fe wt%	Analyzed FeII wt%	AnalyzedFeII/Fe Ratio
Met-.34Ga	13.988	14.117	0.497	0.035
Met-.34Gb	13.988	14.101	6.663	0.473
Met-1VCa	41.125	38.711	5.416	0.140
Met-1VCb	41.125	40.187	11.904	0.296
Lab-1Ga	16.894	16.991	4.594	0.270
Lab-1Gb	16.894	17.278	10.038	0.581
Lab-1VCa	16.894	16.337	4.741	0.290
Lab-1VCb	16.894	16.590	6.219	0.375
Lab-.83a	14.003	14.170	0.984	0.069
Lab-.83b	14.003	13.942	6.009	0.431
Nom-1Ga	14.133	14.493	1.813	0.125
Nom-1VCa	14.133	11.334	0.802	0.071
Nom-1Gb	14.133	12.743	6.492	0.509
Nom-.90Ga	12.635	12.873	0.221	0.017
Nom-.90Gb	12.635	12.898	8.342	0.647

Savannah River Heterogeneous Debris ^{238}Pu Waste

Six melts: Lab-1Ga, Lab-1Gb, Lab-1VCa, Lab-1VCb, Lab-.83Ga, and Lab-.83Gb were performed for this study. Lab-1 and Lab-.83 melts contained 100 wt % and 83 wt % of the waste surrogate respectively. The glass waste forms were quenched after melting, while the vitreous ceramics were heat-treated at 1000°C for 17 hours.

Solid Characterization: Lab-1Ga was found to be 45 vol % crystalline, while Lab-1Gb contained less than 5 vol %. The crystals had a nickel-rich spinel-type structure. Lab-1VCa was 40 vol % crystalline with a spinel-type structure. Lab-1VCb was 40 vol % crystalline composed of 28 % spinel-type and 12 % albite-type $Na(Si_3Al)O_8$. Lab-.83Ga was 10 vol % crystalline and Lab-.83Gb was 5 vol % crystalline of a spinel-type structure. The results of the redox analysis are presented in Table III in a ratio of Fe(II)/(Total Fe). The ratio for Lab-1Ga is 0.27 compared to a ratio of 0.58 for Lab-1Gb. If translated to percentages, the percent of Fe(II) present under reducing conditions is about 60 %, as opposed to 30 % in the sample melted under oxidizing conditions. The percentage of Fe(II) in Lab-1VCa is also 30 %, but Lab-1VCb possesses only 40 % Fe(II). A more drastic percentage change can be seen between Lab-.83Ga, with 7 % Fe(II), and Lab-.83Gb with 40 % Fe(II). The larger percentage of Fe(II) in Lab-1Ga and 1VCa compared to that of Lab-.83Ga may be due to the higher melting temperature.

Chemical Durabilities: The pH of the PCT leachate and the normalized releases of silicon and sodium are presented in Table V. Lab-1Ga has a 100 % lower sodium release than Lab-1Gb. The pH changed from 8.16 to 8.81 when the glass was reduced. Lab-1VCa has almost

Table IV ICP Analysis Results from TCLP Testing (Mg/L)

Waste Form ID	Ag	As	Ba	Cd	Cr	Ni	Pb	Se
Met-.34Ga	<0.02	-		<0.02	<0.06	<0.06	<0.03	
Met-.34Gb	<0.02	-		<0.02	<0.06	<0.06	<0.03	
Met-1VCa	<0.02	-		<0.02	<0.06	<0.06	<0.03	
Met-1VCb	<0.02	-		<0.02	<0.06	<0.06	<0.03	
Lab-1Ga				<0.02	<0.06	0.06	<0.03	
Lab-1Gb				<0.02	<0.06	0.65	<0.03	
Lab-1VCa				<0.02	<0.06	<0.06	<0.03	
Lab-1VCb				<0.02	<0.06	<0.06	<0.03	
Lab-.83Ga				<0.02	<0.06	<0.06	<0.03	
Lab-.83Gb				<0.02	<0.06	<0.06	<0.03	
Nom-1Ga	~	~	<0.01	~	<0.06	<0.06	<0.03	~
Nom-1VCa	~	~	0.29	~	<0.06	<0.06	0.69	~
Nom-1Gb	~	~	<0.01	~	<0.06	<0.06	<0.03	~
Nom-.90Ga	<0.02	-	<0.01	<0.02	<0.06	<0.06	<0.03	-
Nom-.90Gb	0.03	-	0.03	<0.02	<0.06	0.12	<0.03	-
EPA Limit (TCLP)	**5.00**	**5.00**	**100.00**	**1.00**	**5.00**		**5.00**	**1.00**
Land Disposal Limit	**0.072**			**0.066**	**5.200**	**0.320**	**0.510**	

~element is present in small quantities in the waste stream, but was not included in the melt
-element was included in the melt, but was not detected by ICP analysis

identical sodium release as that of Lab-1Ga and the pH is also comparable at 8.26. Lab-1VCb has the highest sodium release and pH of the 100 % waste samples. This is the same trend as seen in the waste forms for Metal-R waste in that the reducing condition deteriorates waste-form durability more in vitreous ceramics than in glass. The release of sodium from L-.83Gb is 1.00 g/m^2 with a pH of 11.5. Lab-.83Ga has a sodium release of 0.30 g/m^2 and a pH of 10.8. It should be noted that Lab-.83Gb has detectable releases of the majority of the elements, including Cd, Cr, and Ni. The decreased waste loading of the Lab-.83G melts substantially decreased their chemical durability when compared with the Lab-1 series glasses with 100 % waste loading.

The results from the TCLP test are shown in Table IV. All of the waste forms except Lab-1Gb passed the land-disposal limits for all hazardous elements. In the case of Lab-1Gb, nickel was detected at a level of 0.65 mg/L, which passes the regulated disposal limits for nickel, but fails the land-disposal limit.

Table V Normalized elemental release from ICP analysis of PCT leachate

Element	Met-.34Ga	Met-.34Gb	Met-1VCa	Met-1VCb
Si	0.164	0.357	0.084	0.109
B	0.266	0.829		
Na	0.341	1.377	0.237	1.714
pH	10.310	11.285	9.305	10.305

Element	Lab-1Ga	Lab-1Gb	Lab-1VCa	Lab-1VCb	Lab-.83Ga	Lab-.83Gb
Si	0.039	0.039	0.040	0.060	0.134	0.222
Na	0.088	0.179	0.090	0.360	0.297	0.996
pH	8.155	8.805	8.255	9.650	10.840	11.505

Element	Nom-1Ga	Nom-1VCa	Nom-1Gb	Nom-.90Ga	Nom-.90Gb
Si	0.043	0.044	0.042	0.043	0.020
Na	0.129	0.392	0.097	0.123	0.117
pH	9.665	9.920	10.010	10.345	10.345

Nom-60 Waste

Five melts: Nom-1Ga, Nom-1VCa, Nom-1Gb, Nom-.90Ga, and Nom-.90Gb, were performed for this study. Nom-1Ga, 1VCa, and 1Gb were formulated to contain 100 wt % of the Nom-60 waste surrogate. Nom-.90Ga and Nom-.90Gb contained 90 wt %. The glasses were formed by quenching the melts, while Nom-1VCa was heat-treated at 1000°C for 16 hours.

Solid Characterization: Nom-1Ga was found to be approximately 25 vol % crystalline. The crystals consist of 75% magnetite ($FeFe_2O_4$) and 25% clinopyroxene Ca(Ti,Mg,Al)(Si,Al) O_6 The vitreous ceramic Nom-1VCa was found to be 100 % crystallized. The identified crystals are 50 % augite-type [$Ca(Fe,Mg)Si_2O_6$], 30 % spinel-type, 15 % wollastonite ($CaSiO_3$), and anorthite-type [(Ca,Na)(Al,Si)2Si_2O_8]. Nom-.90Ga shows much more crystallinity than Nom-.90Gb. The square-shaped augite crystals are shown clearly in Nom-.90Ga along with a well-formed triangular spinel phase. The spinel crystals are much more abundant than augite-type crystals in Nom-.90Ga. Nom-.90Gb is essentially amorphous. The redox analyses is presented in Table III in a ratio of Fe(II)/(Total Fe). The ratio for Nom-1Ga is 0.13 compared to a ratio of 0.51 for Lab-1Gb. If translated to percentages, Fe(II) formed under reducing conditions is about 51 %, as opposed to 13 % in the sample melted under oxidizing conditions. The Fe(II) in Nom-.90Ga is 2 %, whereas Nom-.90Gb possesses 65 % Fe (II).

Chemical Durabilities: The TCLP results provided in Table IV show that the heat-treated sample, Nom-1VCa, failed for lead release according to the Land Disposal Limit. The results also show that the reduced Nom-.90Gb has slightly higher releases of Ag, Ba, and Ni than those of the same composition melted under oxidizing condition (Nom-.90Gb). Such redox effects were not observed between the redox pair of Nom-1Ga and Nom-1Gb when waste loading is was at 100 wt %.

The PCT results provided in Table V are interesting and show that: (1) the 100 % waste-loading glasses have similar normalized sodium releases as the 90 % waste-loading glasses (except after heat-treatment). (2) Redox effects on glass durability are minimal in comparison of Nom-1Ga to Nom-1Gb or between Nom-.90Ga and Nom-.90Gb. (3) Crystallization decreases the glass durability by 300 %, i.e., the heat-treated Nom-1VCa has a sodium release rate three times as high as the quenched Nom-1Ga.

Comparison of Waste Forms Among the Three Wastes

A comparison of the waste-form behavior among several attributes provides insight on the relationship between waste-form performance and their compositions and processing conditions. Table VI provides a brief summary of the waste forms developed for the three waste streams. The data in Table VI indicates very different responses to each processing effect. The reducing melting condition can decrease the waste-form durability between 300 to 700 % for wastes of Metal-R and Pu^{238} debris, but essentially has no effect on waste-form durability developed for the Nom-60 waste stream. The devitrification of the waste forms increased durability for Metal-R waste, exhibited no effect for Pu^{238} debris, and substantially decreased durability for Nom-60 waste. The increased waste loading also affects waste-form behavior differently, from decreasing durability for Metal-R waste, increasing durability by 240 % for Pu^{238} debris, to no effects on Nom-60 waste. These different responses to the processing and composition variations, however, are not random and can be understood from

Table VI Waste Form Comparison

Effects	Metal-R	^{238}Pu Debris	Nom-60
Reducing Melt Atmosphere	- Decreased durability over 400-700 % - Lowered viscosity - Decreased tendency of devitrification	- Decreased durability over 300-400 % - Lowered viscosity - Decreased tendency of devitrification	- No significant effect on durability - Lowered viscosity - Decreased tendency of devitrification
Devitrification	Increased durability by about 40 %	No significant effect	Deceased durability by 200 %
Increased Waste Loading	Slightly decreased durability	Increased durability by 240 %	No significant effect

a glass chemistry point of view. The following is a brief summary of such an understanding of each effect:

Redox effects: For Metal-R waste, the oxidized (or normal melting) waste form (such as Met-1VCa) developed 90 % spinel-type crystals and 10 % augite-type crystals. The formation of spinel crystals depletes elements such as Fe, Ni, Cr, and Mn from the melt, resulting in a enrichment of silica and alumina in the residual glass. The removal of Fe(II) from the glass matrix is especially beneficial to glass durability [8]. Spinel-type crystals are themselves very stable and durable, therefore the formation of spinel usually increases the durability of the waste form. On the other hand, the formation of augite-type crystals,$Ca(Fe,Mg)Si_2O_6$, usually decreases waste-form durability because of the extraction of silica from the glassy matrix into the crystals. When the waste form with the same composition was melted under reducing conditions as in the case of Met-1VCb, the beneficial spinel-type crystals decreased to 55 % while the silica and alumina-consuming crystals of augite [$Ca(Fe,Mg)Si_2O_6$], fayalite (Fe_2SiO_4), and albite [$Na(Si_3Al)O_8$] types increased to 45 %. This is why a substantial decrease in durability for Metal-R wastes melted under reducing conditions was observed.

For ^{238}Pu debris waste, only the beneficial spinel crystals are formed in both the oxidized waste form, Lab-1Ga, and the reduced waste form, Lab-1Gb. The durability is, therefore, proportional to the amount of spinel formed in the waste form. The oxidized waste form, Lab-1Ga, formed 45 % spinel-type crystals while the reduced waste form, Lab-1Gb, formed only 5 % spinel crystals; hence the reduced waste form decreased the durability by 300-400 % due to a spinel content 9 times lower in the reduced waste form.

For Nom-60 waste, the oxidized waste form, Nom-1Ga, had about 19 % spinel and 6 %

clinopyroxene, $Ca(Ti,Mg,Al)(Si,Al)_2O_6$, which consumed silica and alumina. The reduced waste form, Nom-1Gb, had about 4 % spinel and 1 % clinopyroxene. The ratio of the beneficial crystals of spinel to the harmful crystals of clinopyroxene is similar in both oxidized and reduced waste forms. This may explain why the reducing melting conditions had no significant effects on waste-form durability.

<u>Devitrification Effects</u>: For Metal-R waste, the devitrified Met-1VCa has 90 % beneficial spinel-type crystals and only 10 % augite-type (harmful) crystals. The benefit from forming spinel outweighed the harmful effects of formation of augite. The devitrification of Met-1VCa improved its chemical durability. The glass waste form (non-devitrified) Met-.34Ga is completely amorphous. We observed an increase in waste-form durability when the waste form was heat-treated to enhance the devitrification.

For ^{238}Pu debris waste, both the heat-treated Lab-1VCa and quenched Lab-1Ga have the same amount, 40 vol %, of spinel-type crystals. Therefore the waste forms have similar chemical durability.

For Nom-60 waste, the heat-treated Nom-1VCa formed 70 % harmful crystals of augite-type $[Ca(Fe,Mg)Si_2O_6]$, wollastonite ($CaSiO_3$), and anorthite-type $[(Ca,Na)(Al,Si)_2Si_2O_8]$, which consume silica and alumina, and 30 % spinel-type crystals. The quenched Nom-1Ga had about 19 % spinel and 6 % clinopyroxene, $Ca(Ti,Mg,Al)(Si,Al)_2O_6$. Therefore, the heat-treated Nom-1VCa produced a poor durability glassy matrix after the extraction of silica and alumina from the glassy phase during the formation of the large amount of augite-, wollastonite-, and anorthite-type crystals. The quenched sample had more beneficial crystals of spinel than of the harmful clinopyroxene, which explains why devitrification of Nom-60 waste forms resulted in a 200 % decrease in durability.

<u>Waste Loading Effects</u>: For Metal-R waste, the 100 % waste loading Met-1VCb is about 20 % less durable than the 34 % waste loading Met-.34Ga. Met-1VCb had 45 % silica- and alumina-consuming crystals of augite $[Ca(Fe,Mg)Si_2O_6]$, fayalite (Fe_2SiO_4), and albite $[Na(Si_3Al)O_8]$ types, and 55 % spinel-type crystals. The formation of these crystals resulted in the deterioration of glass durability. The 34 % waste loading Met-.34Ga is essentially amorphous and its glassy matrix is not deteriorated by crystal formation as in the case of the 100 % waste loading Met-1VCb.

For ^{238}Pu debris waste, the 100 % waste loading Lab-1Ga formed about 40 % beneficial spinel-type crystals while the 83 % waste loading Lab-.83Ga contained approximately 10 % such crystals. This can explain the observed 240 % increase in durability corresponding to increased waste loading.

For the Nom-60 waste, similar chemical durabilities were observed for the 100 % waste loading Nom-1Ga and 90 % waste loading Nom-.90Ga. This similarity exists because both glasses maintained similar ratios of spinel-type crystals to silica-consuming crystals, such as augite and clinopyroxene.

ACKNOWLEDGMENT

This work was sponsored by the U.S. Department of Energy, Office of Science and Technology, Landfill Stabilization Focus Area with support from the Mixed Waste Focus Area. The work was performed by the Pacific Northwest National Laboratory which is operated by Battelle Memorial Institute. P. E. Bloomer thanks Associated Western University Inc. for an undergraduate Fellowship Appointment. Ms. N. Chantaraprachoom thanks the International Atomic Energy Agent for a trainee fellowship.

REFERENCES

[1] J.E. Surma, W.E. Lawrence, C.H. Titus, J.K. Wittle, R.A. Hamilton, D.R. Cohn, D. Rhea, P. Thomas, and P.P. Woskov, "Treatment of Simulated INEL Buried Wastes Using A Graphite Electrode DC Arc Furnace," *Proc. Spectrum '94.*

[2] J.E. Surma, D.R. Cohn, D.L. Smatlak, P.P. Waskov, C.H. Titus, J.K. Wittle, and R.A. Hamilton, "Graphite Electrode DC Arc Technology Development for Treatment of Buried Wastes," PNL-SA-21891, Pacific Northwest National Laboratory, Richland, WA, 1993.

[3] X. Feng, I. L. Pegg, E. Saad, S. Cucinell, and A. Barkatt, "Redox Effects on the Durability and Viscosity of Nuclear Waste Glasses," Nuclear Waste Management III, G. B. Mellinger, Ed. *Ceramic Transactions* **9** 165–174 (1990)

[4] A. C. Buechele, X. Feng, H. Gu, and I. L. Pegg, "Redox State Effects on Microstructure and Leaching Properties of West Valley SF-12 Glass," Nuclear Waste Management IV, G. G. Wicks, D. F. Bickford, and L. R. Bunnell, Eds. *Ceramic Transactions*, **23** 85–94 (1991).

[5] Feng, X., P.R. Hrma, J.H. Westsik, Jr., N.R. Brown, M. J. Schweiger, H. Li, J.D. Vienna, G. Chen, G.F. Piepel, D.E. Smith, B.P. McGrail, S.E. Palmer, D. Kim, Y. Peng, W.K. Hahn, A.J. Bakel, W.L. Ebert, D.K. Peeler, C. Chang, "Glass Optimization for Vitrification of Hanford Site Low-level Tank Waste," Pacific Northwest National Laboratory, PNNL-10918, Richland, Washington (1996).

[6] Kim, D., D.K. Peeler, and P. Hrma, "Effects of Crystallization on the Chemical Durability of Simulated Nuclear Waste Glasses,"*Ceramic Transactions*, **61** 177 (1995).

[7] Li, H., J.D. Vienna, P. Hrma, M.J. Schweiger, D.E. Smith, and M. Gong, "Borosilicate Based Glasses for Immobilization of Plutonium-Bearing Materials," *Ceramic Transactions*, 72, 399 (1996).

[8] Feng, X., and T. B. Metzger, "A Glass Durability Model Based on Understanding Glass Chemistry and Structural Configurations of The Glass Constituents," *Aqueous Chemistry and Geochemistry of Oxides, Oxyhydrides, and Related Materials*, **432** 27 (1997).

"EMPIRICAL MODELS" AND THERMODYNAMIC CONSTITUTIVE FUNCTIONS FOR HIGH-LEVEL WASTE GLASS PROPERTIES

P. Hrma
Pacific Northwest National Laboratory[1]
Richland, WA 99352

ABSTRACT

The "empirical model" is a statistical concept for what in physical chemistry and materials science is called "law," "equation," "constitutive assumption," or "response function." While materials scientists reserve the word "model" for theories that relate macroscopic properties to molecular or atomic characteristics, statisticians use the word "model" to refer to any mathematical expression that approximates a "true" response function. This paper attempts to clarify some misconceptions caused by this double meaning of the word "model." In particular, the concept of partial specific or molar properties of physical chemistry is applied to the response functions for glass properties represented by empirical models.

INTRODUCTION

The word "model" has been used differently by materials scientists and by statisticians. On the one hand, materials scientists reserve the term "model" for relationships between microscopic (i.e., molecular or atomic) and macroscopic properties (such as viscosity, surface tension, etc.), or between basic and complex properties (i.e., hydration energy model for glass durability). On the other hand, statisticians use the word "model" for any relationship between any measurable quantities including the relationship between material properties and state variables (temperature, pressure, composition) [1]. In physical chemistry [2] and materials science, relationships between material properties and state variables are called "laws" (i.e., Fick's first law), "equations" (such as Arrhenius equation), or, as in modern thermodynamics of continua [3], "constitutive relations, equations, or assumptions," or "response functions." These relationships are considered primitive in the structure of classical, irreversible, and rational thermodynamics. As such, they are obtained either directly from measurement or indirectly from specific molecular theories (i.e., Eyring's theory of activated processes).

The relationships between glass properties and composition are of a crucial importance for waste glass formulation and process control. This is so because of

[1] Pacific Northwest National Laboratory is operated for the U.S. Department of Energy by Battelle under Contract DE-ACO6-76RLO 1830.

the great variability of waste glass composition, the large number of chemical constituents of waste glasses, and a significant compositional dependence of the properties that define glass quality and processability [4]. The most reliable way to obtain property-composition relationships is by direct measurement, and the most economical way to conduct such measurements is based on a statistically designed test matrix that covers the composition region of interest by a network of data points, to which constitutive equations are fitted [1,4]. In the absence of a theoretical basis for a suitable form of constitutive equations/response functions, the functional form of the property versus composition relationship is chosen according to the trends suggested by the data themselves [4]. Statisticians refer to constitutive equations for property-composition relationship as "empirical mixture models" [1]. The word "model" is justified by the reasoning that the functions fitted to data are mere approximations of the true response functions, which are unknown. This terminology has caused confusion among materials scientists, who view "empirical models" with suspicion because they do not involve hypotheses based on molecular or mechanistic considerations.

This paper attempts to clarify the position of empirical mixture models within the material science field as constitutive relations, that is, relations that constitute the material (i.e., glass) behavior. In both traditional and modern thermodynamics, the dependence of material properties on composition has been expressed through partial specific or partial molar properties. The connection between response functions (empirical mixture models) and partial specific/molar properties belongs to the standard apparatus of classical thermodynamics [2, 5, 6]. These relationships are shown in the following section, which also contains examples of applying these general formulas to first- and second-order polynomial "models." These models, which can be viewed as truncated Taylor series expansions of "real" property composition functions, are commonly used for mixtures [1,4]. The discussion section stresses that empirical model coefficients are material parameters evaluated from measured data. Structural and other models based on specific hypotheses that assess the desired relationship without extensive measurements are in principle less accurate than functions derived from measured data.

PARTIAL SPECIFIC PROPERTIES

Mixture composition is conveniently expressed as a table of mass or mol fractions of mixture components. For an *n*-component mixture, only *n-1* fractions define the composition because the sum of all *n* fractions must add to 1. The array of *n-1* fractions is called the composition vector:

$$\vec{x} = (x_1, x_2, \ldots, x_{n-1}) \tag{1}$$

which represents a point in an *(n-1)*-dimensional composition space.

The dependence of a mixture property, p, on composition can be represented by a response function [3,5]

$$p = \hat{p}(\vec{x}) \tag{2}$$

where the "hat" above the symbol distinguishes a function from its value. Partial properties, p_i, are in thermodynamics defined by the equations [2,5,6]:

$$p_n = \hat{p}_n(\vec{x}) \equiv \hat{p}(\vec{x}) - \sum_{i=1}^{n-1} \frac{\partial \hat{p}(\vec{x})}{\partial x_i} x_i \tag{3}$$

and

$$p_i = \hat{p}_i(\vec{x}) \equiv \hat{p}_n(\vec{x}) + \frac{\partial \hat{p}(\vec{x})}{\partial x_i} \quad (i=1,2,...n\text{-}1) \tag{4}$$

Here, a change in i-th component occurs at the expense of a selected (preferably, but not necessarily, the most abundant) n-th component of the mixture. The partial properties are called specific or molar, dependent on whether the composition is expressed in terms of mass or mol fractions of components.

Definitions (3) and (4) allow expressing a property as products of partial properties and mass or mol fractions, summed over all mixture components:

$$p = \sum_{i=1}^{n} p_i x_i \tag{5}$$

Equations (5) and (2) are equivalent. Equation (5) expresses a property as a sum of contributions of individual components. Partial properties are functions of the state variables including composition and are uniquely defined by equations (3) and (4). Equation (5) can be rewritten as

$$p = p_n + \sum_{i=1}^{n-1} p_i' x_i \tag{6}$$

where $p_i' = p_i - p_n$ are reduced partial properties. By equation (4),

$$p_i' = \frac{\partial \hat{p}(\vec{x})}{\partial x_i} \quad (i=1,2,...n\text{-}1) \tag{7}$$

Finally, joining the first derivative of equation (5) with equation (3), we obtain the famous Gibbs-Duhem equation [2,5], which is extensively used in thermodynamics

$$\sum_{j=1}^{n} \frac{\partial p_j}{\partial x_i} x_j = 0 \quad (i=1,2,...n) \tag{8}$$

Note. Property constitutive equation (2) can be represented as a hypersurface over the composition space, and definitions (3) and (4) can be interpreted in terms of a tangential plane to this surface. Let $\vec{x} = \vec{x}_0$ be a point in composition space (such a point represents a mixture). At this point, the property p has a value $p_0 = \hat{p}(\vec{x}_0)$. Then the partial derivatives to the property surface, p, at the composition point $\vec{x}_0$ are $\partial \hat{p}(\vec{x}_0) / \partial x_i$, and the tangential plane q to the property surface p at $\vec{x}_0$ is

$$q = p_0 + \sum_{i-1}^{n-1} (x_i - x_{0i}) \partial \hat{p}(\vec{x}_0) / \partial x_i \tag{9}$$

Theorem. The tangential plane q to the property surface p at $\vec{x}_0$ is given by

$$q = \sum_{i=1}^{n} q_i x_i \tag{10}$$

where coefficients $q_i = \hat{q}_i(x_1, x_2, ..., x_{n-1})$ $(i = 1,2,...,n)$ are equal to partial specific properties at $\vec{x}_0$. Accordingly, $q_i = \hat{p}_i(\vec{x}_0)$, $(i = 1,2,...,n)$, are given by equations (3) and (4).

Proof. Using the mass balance equation in the form

$$x_n + \sum_{i=1}^{n-1} x_i = 1 \tag{11}$$

equation (10) follows from equation (9).

FIRST- AND SECOND-ORDER MIXTURE MODELS

If the response function (2) is a polynomial (or is approximated as a polynomial), its coefficients are called component coefficients for property p. If the response function (2) is linear on a composition range $\Re$, the partial specific properties are constant on $\Re$ and $p = q$. Hence, for an ideal mixture, $p_i = q_i$ for each $\vec{x} \in \Re$. A first-order polynomial function approximates the property surface by a flat plane, that is, it treats the mixture as ideal. Consequently, for a first-order response function, property coefficients represent partial specific properties of an ideal mixture. Alternatively, first-order property coefficients may be viewed as first approximations of partial specific properties over $\Re$.

Using Equations (3) and (4), partial specific properties can be obtained from empirical mixture models. Two examples shown below demonstrate this for second-order polynomial response functions.

Example 1. In a binary (n=2) mixture, the property depends on composition as $p = r_0 + r_1 x_1 + r_2 x_1^2$. Express partial specific properties.

Solution: Using Equations (3) and (4), we obtain $p_1 = p_2 + r_1 + 2r_2 x_1$, and $p_2 = r_0 - r_2 x_1^2$.

Example 2. Let $p = r_0 + r_1 x_1 + r_{12} x_1 x_2 + r_2 x_2 + r_{22} x_2^2$ for a ternary (n=3) mixture (with $r_{11} x_1^2$ term absent for simplicity). Determine partial specific properties.

Solution: $p_1 = r_0 + r_1 + r_{12} x_2 - r_{12} x_1 x_2 - r_2 x_2^2$

$$p_2 = r_0 + r_2 + r_{12} x_1 + 2r_2 x_2 - r_{12} x_1 x_2 - r_2 x_2^2$$

$$p_3 = r_0 - r_{12} x_1 x_2 - r_2 x_2^2$$

Note that for an *m*-order polynomial response function, the related reduced partial properties are *(m-1)*-order polynomials. This follows from equation (7) and is illustrated by Example 1. However, the partial properties are of the same order as the response function if this function is a second- or higher-order polynomial. This follows from equation (3) and is demonstrated by Example 2. If the response function is a first-order polynomial, both partial properties and reduced partial properties are constants.

DISCUSSION

The physical meaning of coefficients in first-order models is straightforward. As mentioned above, first-order coefficients are ideal mixture approximations of partial specific or molar properties over the composition region under study. An example is shown in Table I.

Table I. COMPOSITION REGIONS (IN MASS FRACTIONS) AND COMPONENT COEFFICIENTS (T_{Li}) FOR A FIRST-ORDER LIQUIDUS TEMPERATURE (T_L) MODEL WITH SPINEL PRIMARY PHASE [7]

Component	Range	T_{Li} °C
SiO_2	0.380-0.600	834
B_2O_3	0.000-0.120	395
Na_2O	0.080-0.200	-1826
Li_2O	0.000-0.030	-1470
MgO	0.004-0.060	2827
Fe_2O_3	0.060-0.150	2644
Al_2O_3	0.040-0.150	3307
Cr_2O_3	0.000-0.012	18864
MnO	0.000-0.040	1870
NiO	0.000-0.030	8210
Others	0.045-0.070	4419

Table I displays first-order coefficients for liquidus temperature (T_L) of Hanford high-level waste glasses that precipitate spinel as the primary crystalline phase. A close examination of the measured data reveals that the response function of T_L versus composition is nonlinear. Hence, the first order-model is an approximation, and its coefficients correspond to partial specific liquidus temperatures (T_{Li}) approximated as constant over the composition region expressed in terms of the component mass fraction ranges shown in the table.

The physical meaning of first-order coefficients is straightforward [8-10]. The difference between two first-order coefficients corresponds to a relative partial specific liquidus temperature and expresses the change in the liquidus temperature of the mixture that results from replacing one component by an equivalent fraction of another component. For example, replacing 1 wt% SiO_2 by 1 wt% Cr_2O_3 results in an increase of T_L by 189 - 8 = 181°C. Similarly, replacing 1 wt% SiO_2 by 1 wt% Na_2O results in a decrease of T_L by -18 - 8 = -26°C. The effect of an addition or a removal a of component to or from the mixture can also be easily expressed for the first-order mixture model [8-10].

The physical meaning of empirical coefficients is less clear in nonlinear models. For example, it is difficult to assess the effect of Al_2O_3 on T_L from the second-order model shown in Table II because Al_2O_3 appears in three terms: a linear term, and two second-order terms (one cross-product and one squared).

Table II. COMPONENT COEFFICIENTS (T_{Li} AND T_{Lij}) FOR A SECOND-ORDER LIQUIDUS TEMPERATURE MODEL WITH ZIRCON AND BADDELEYITE PRIMARY PHASES [11]

Term	T_{Li} °C	Term	T_{Lij} °C
SiO_2	1414	$Al_2O_3 \cdot Al_2O_3$	68867
B_2O_3	1072	$CeO_2 \cdot CeO_2$	82693
Na_2O	-3503	$Li_2O \cdot Li_2O$	69331
Li_2O	-18644	$Na_2O \cdot Na_2O$	14403
Al_2O_3	-11554	$ZrO_2 \cdot ZrO_2$	23930
ZrO_2	9613	$Na_2O \cdot Al_2O_3$	109747
Bi_2O_3	1440	$Na_2O \cdot Li_2O$	96080
CeO_2	537	$ZrO_2 \cdot Na_2O$	-84811
P_2O_5	2868		
Others	2586		

A second-order mixture model is generally of the form

$$p = \sum_{i=1}^{n} r_i x_i + \sum_i r_{ii} x_i^2 + \sum_{i,j} r_{ij} x_i x_j \qquad (12)$$

with only some x_i^2 and $x_i x_j$ terms appearing (if a non-zero squared term, x_i^2, appeared in equation (12) with a full set of cross-product terms, $x_i x_j$, the relationship would be over-parametrized). Using equations (6) and (7), we obtain the following expression for the *i*-th component partial specific property:

$$p_i' = r_i - r_n + 2r_{ii}x_i - 2r_{nn}x_n + \sum_j (r_{ij} - r_{nj})x_j \qquad (i=1,2,...n-1) \qquad (13)$$

Note that, by equation (11), $\partial x_n / \partial x_i = -1$. Thus, for Al_2O_3 replacing SiO_2 (as the *n*-th component), using the coefficient values from Table II, we obtain

$$p'_{Al_2O_3} = -12968 + 137734 x_{Al_2O_3} + 109747 x_{Na_2O}$$

This equation shows that the effect of Al_2O_3 on T_L in a glass with baddeleyite and zircon primary phases increases with increasing concentrations of Al_2O_3 and Na_2O. For a glass with 2.3 wt% Al_2O_3 and 11.4 wt% Na_2O, we have $p'_{Al_2O_3} = 2711°C$. Hence, a replacement of 1 wt% SiO_2 by 1 wt% Al_2O_3 results in an increase of T_L by

27°C. This example illustrates how obtaining partial specific (or molar) properties as functions of composition helps in understanding mixture behavior.

Both property coefficients and partial specific properties were obtained from measured experimental values of the property (i.e., liquidus temperature) itself. They represent material functions or parameters directly estimated from data by simple calculations and are considered primitive in the structure of thermodynamics.

On the other hand, structural models and models based on correlation between properties attempt to assess mixture behavior using extraneous information to avoid tedious measurements. These types of models may or may not have scientific merit in gaining insight into material behavior beyond the realm of thermodynamics. However, these models need careful checking before applying them to glass formulation or process control because they are generally less accurate than empirical models.

The accuracy of prediction is less important in situations where little risk is involved. For example, high-level waste glasses produced in the US must meet the Environmental Assessment standard [12]. Since the actual waste glasses exceed this standard by one or two orders of magnitude, the risk involved is low regardless of what kind of model is being used for checking the glass compliance with this standard. The situation becomes different when an optimized glass is produced in a melter that is sensitive to spinel precipitation. These melters require that $T_L \leq T_M - \Delta T$, where T_M is the temperature at which glass viscosity is 5 Pa·s and ΔT is approximately 100°C. Since a 10°C difference in the T_L value results in approximately 2 wt% difference in the allowable waste loading in glass, an accurate T_L model (±10°C) is required for process control. If a less accurate model is used, the waste loading in glass must be reduced or the melter will be operated at a higher risk of failure from spinel formation.

Conclusions

Partial specific properties can be derived for any response function, and hence, also for their approximations by empirical models fitted to data. For first-order models, partial specific properties are equivalent to model coefficients. For higher-order models, partial specific properties help in understanding the effects of individual components as functions of composition.

Acknowledgment

The helpful discussions of this topic with G. F. Piepel and his comments on this paper are greatly appreciated.

REFERENCES

1. J. A. Cornell, *Experiments with Mixtures: Design, Models, and the Analysis of Mixture Data*, Second Ed., Wiley, New York 1990.

2. S. R. de Groot and P. Mazur, *Nonequilibrium Thermodynamics*, North-Holland, Amsterdam 1962.

3. C. Truesdell and R. Toupin, *The Classical Field Theories*, In *Handbuch der Physik*, Vol. II/1, Springer, Berlin (1960).

4. P. Hrma, G. F. Piepel, M. J. Schweiger, D. E. Smith, D.-S. Kim, P. E. Redgate, J. D. Vienna, C. A. LoPresti, D. B. Simpson, D. K. Peeler, and M. H. Langowski, *Property/Composition Relationships for Hanford High-Level Waste Glasses Melting at 1150°C*, PNL-10359, Vol. 1 and 2, Pacific Northwest Laboratory, Richland, Washington (1994).

5. I. Samohyl, *Thermodynamics of Irreversible Processes in Fluid Mixtures*, Teubner, Leipzig 1987.

6. I. Samohyl and M. Silhavy, "Mixture Invariance and its Applications," *Arch. Rational Mech. Anal.* 109 [4] 299-321 (1990).

7. M. Mika, M. J. Schweiger, and P. Hrma, "Liquidus Temperature of Spinel Precipitating High-Level Waste Glass," *Scientific Basis for Nuclear Waste Management* (Editors W. J. Gray and I. R. Triay), Vol. 465, p. 71-78, Material Research Society, Pittsburgh, Pennsylvania, 1997.

8. P. Hrma and R. J. Robertus, "Waste Glass Design Based on Property Composition Functions," *Ceram. Eng. Sci. Proc.* 14 [11-12] 187-203 (1993).

9. P. Hrma, G. F. Piepel, J. D. Vienna, P. E. Redgate, M. J. Schweiger, and D. E. Smith, "Prediction of Nuclear Waste Glass Dissolution as a Function of Composition," *Ceram. Trans.* 61, 497-504 (1995).

10. P. Hrma, G. F. Piepel, P. E. Redgate, D. E. Smith, M. J. Schweiger, J. D. Vienna, and D.-S. Kim, "Prediction of Processing Properties for Nuclear Waste Glasses," *Ceram. Trans.* 61, 505-513 (1995).

11. J. V. Crum, P. Hrma, and M. J. Schweiger, "Liquidus Temperature Models for High-Level Waste Glasses that Precipitate Zirconium-Containing Crystalline Phases," *Ceram. Trans.,* this volume.

12. C. M. Jantzen, N. E. Bibler, D. C. Beam, C. L. Crawford, and M. A. Pickett, Development and Characterization of the Defense Waste Processing Facility (DWPF) Environmental Assessment (EA) Glass Standard Reference Material (U), *Ceram. Trans.* Vol. 39, pp. 313-322. Am. Ceram. Soc., Westerville, Ohio (1994).

PREDICTING MELT CONDUCTIVITIES OF WASTE GLASSES

Sabrina S. Fu, Vitreous State Laboratory, The Catholic University of America, Washington DC 20064

ABSTRACT

Previous modeling of the melt conductivity of waste glasses has shown that there is a good correlation between total alkali content and melt conductivity for a wide range of multi-component glasses. These data suggest that factors which often complicate the prediction of melt conductivity, such as the mixed alkali effect, do not dominate the prediction of melt conductivity of multicomponent glass systems. The correlation formed between melt conductivity and total alkali content was then used to predict new waste glass formulations.

INTRODUCTION

Melt conductivity measurements are required for processing waste glasses in Joule-heated melters. Hence, in the formulation of waste glasses, melt conductivity is a processing parameter of concern. Prediction of the conductivity of glass melts are often complicated by the mixed alkali effect[1-7] and on the dependence of electrical conductivity on melt viscosity[8-11]. Although a number of interactions undoubtedly exist in waste glasses, most of them may not be significant in the prediction of melt conductivity. This paper presents a very simple model that was then used to predict the melt conductivity of a wide range of new waste glass formulations.

EXPERIMENTAL PROCEDURE

Glasses were made by mixing the appropriate amounts of oxides, carbonates, phosphates, and sulfates. The chemicals were melted in either a quartz or a clay crucible for one to one-and-a-half hours at 1150°C to 1250°C with stirring after the first half an hour. For the actual mixed waste of Series VI glasses, dried sludge was crushed and mixed with the appropriate additives and then vitrified in a quartz crucible at 1200°C for two hours. All of the glasses examined were homogeneous, and their compositions were confirmed by direct-current plasma atomic emission spectroscopy (DCP-AES) after complete acid dissolution.

The electrical conductivity of the glass, σ, was determined by measuring the resistance of the glass melt as a function of frequency using a calibrated platinum/rhodium electrode probe attached to a Hewlett-Packard model 4194A impedance/gain phase analyzer. Salt solutions of known concentrations were used to calculated the cell constant. The resistance of the glass melt as a function of frequency was extrapolated to zero frequency to obtain the DC conductivity. These measurements were taken over a temperature range of either 900-1150°C, 1000-

1250°C, or 1100-1350°C, depending upon the particular glass series, and the liquidus temperature of the glass. No data were taken near or below the liquidus temperature. The conductivity data were interpolated to standard temperatures using the Arrhenius equation: $\ln\sigma = A/T + B$, where A and B are fitting parameters. Estimated uncertainties in the conductivity measurements are ±10-20 relative%.

The melt viscosity, η, was calculated from measurements of the torque and rotation speed of a calibrated spindle attached to a Brookfield viscometer. Measurements were performed over similar temperature ranges to those employed for the melt conductivity measurements, and the data interpolated to standard temperatures using the Volger-Fulcher equation: $\ln\eta = A/(T-T_0)+B$, where A, B, and T_0 are fitting parameters. The equipment was calibrated at room temperature using standard oils of known viscosity, and then checked at 1000°C to 1250°C using a National Institute of Standards and Technology (NIST) standard reference glass (SRM711). The precision and accuracy of the viscosity measurements are estimated to be within ±10 relative%.

RESULTS AND DISCUSSION

A total of 43 simulated and actual waste glasses were used for this study. Four criteria were used in choosing these glasses:

(1) They contain five or more cations from several different projects encompassing a wide compositional envelope.
(2) They contain <1 mole% anions other than O^{2-}.
(3) Their analyzed and calculated compositions should agree to within 10 relative%.
(4) Each glass contains only one phase.

The composition range of these 43 glasses, from six different projects is provided in Table 1. These 43 glasses range in melt conductivity from 0.006 to 1.3 S/cm and from 9 to 1550 Poise. The two order in magnitude change in melt conductivity and melt viscosity represent a large envelope of the major oxides in most waste glasses: 0-12 mole% Al_2O_3, 6-15 mole% B_2O_3, 0-35 mole% CaO, 0-10 mole% Fe_2O_3, 0-8 mole% Li_2O, 0-9 mole% MgO, 0-35 mole% Na_2O, 0-3 mole% P_2O_5, 30-51 mole% SiO_2, and 0-4 mole% ZrO_2. The six series of glasses are from six different projects: Series I and IV are glasses from systematic studies of Na-Ca exchange[12] and Al-Na exchange[13] respectively; Series II glasses are for a Savannah River study; Series III glasses are surrogate waste glasses for silo wastes at the Fernald site in Ohio[14]; Series V glasses are surrogate waste glasses for the vitrification of radioactive ion exchange resins[15]; and Series VI glasses are actual waste glasses made from Savannah River M-Area tank wastes[16]. The melt conductivity of these multi-component glasses at

Table 1. Compositional Ranges of Six Multi-Component Systems and Their Viscosity and Conductivity Ranges at 1150°C

Components (mole%)	Series I	Series II	Series III	Series IV	Series V	Series VI
Al_2O_3	7.0	5.0-6.8	6.2-10.6	0-10.7	4.3-6.8	7.0-12.4
B_2O_3	9.7	9.6-14.9	9-15.2	9.58	6.1-9.6	11.2-12.7
BaO	0.0	0.0	1.1-1.3	0.0	0.0	0.01
CaO	35-0	9.2-12.1	8.1-17.3	25.35	16.2-25.4	0.2-0.4
Fe_2O_3	0.0	0.0	1.7-4.6	0.0	0-10.1	1.4-5.8
K_2O	0.0	0.0	0.5-0.9	0.0	0.0	0.6-1.1
Li_2O	0.0	0.0	0.4-4.5	0.0	0.0	0.1-7.7
MgO	0.0	0.0	4.8-8.9	0.0	0.0	0.2-6.4
Na_2O	0-35	19.6-23.0	7.1-8.1	17.45-6.75	10.7-32.8	15-28
NiO	0.0	0.0	0.1-0.4	0.0	0.0	0.7-2.0
P_2O_5	0.0	0.0	1.1-3.3	0.0	0.0	1.2-2.1
PbO	0.0	0.0	0.7-1.8	0.0	0.0	NA
SO_3	0.0	0.0	0-1.3	0.0	0.0	NA
SiO_2	48.9	46.3-48.4	39.3-43.5	47.62	30.4-47.6	39.9-51.1
U_3O_8	0.0	0.0	0.0	0.0	0.0	0.2-0.5
ZrO_2	0.0	1.6-2.1	0-0.3	0.0	0.0	0.01-3.7
TOTAL	100.0	100.0	100.0	100.0	100.0	100.0
Viscosity Range (poise) at 1150°C	30-72	30-73	53-300	15-135	9-58	45-1550
Conductivity Range (S/cm) at 1150°C	0.006-1.3	0.24-0.76	0.02-0.08	0.08-1.09	0.08-1.1	0.17-0.62

NA = Not Analyzed

1150°C and at 1000°C as a function of mole% alkali are plotted in Figures 1(a) and 1(b), respectively. The melt conductivity appears to be well correlated to total mole% alkali. This occurs despite the more than two orders in magnitude range in melt viscosity that these glasses cover, and despite the fact that up to half of the moles of Na_2O were substituted by Li_2O for Series III and VI; hence, it appears that melt viscosity and the mixed alkali effect both play minor roles in the determination of melt conductivity for these multi-component glasses. Note that there are only four series of glasses represented in Figure 1(b) while there are six series of glasses represented in Figure 1(a); this is due to the fact that only four of these six projects required melt conductivity measurements at 1000°C.

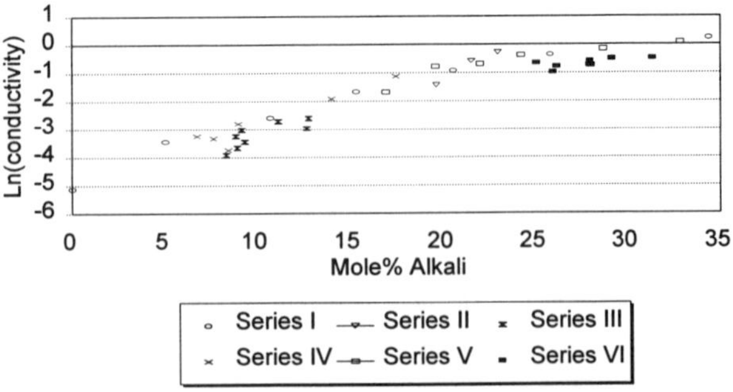

Figure 1(a). Conductivity as a function of alkali content at 1150°C.

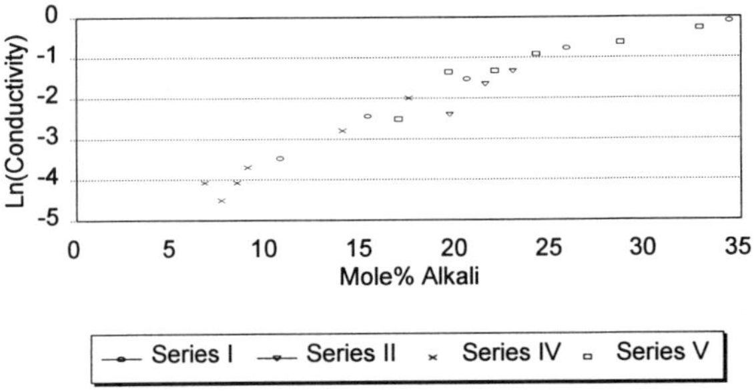

Figure 1(b). Conductivity as a function of alkali content at 1000°C

The correlation observed for melt conductivity as a function of total alkali content is in contrast to the lack of correlation between ln of melt viscosity at 1150°C and mole% alkali, as shown in Figure 2. Series IV produced a linear correlation between melt viscosity and total mole% alkali as would be expected from a systematic study of Na_2O substitution for Al_2O_3; however, there is no general correlation observed between melt viscosity and total alkali content, as expected.

The correlation between melt conductivity and total alkali content for the 43 data points observed in Figure 1(a) was then used to obtain a quadratic fit. The following equation was obtained with an R^2 of 0.94:

ln(conductivity at 1150°C)= 0.282*(mole%alkali)-0.00373*(mole% alkali)^2 -5.34.

This model was used to predict the melt conductivity of new multi-component glass formulations, and then the melt conductivity of these glasses were measured as described in the experimental section. A total of twenty new waste glasses have been tested thus far, and all those glasses which fulfill the four criteria listed above have measured melt conductivity well within 25 relative% error of the melt conductivity predicted by this simple model. The glasses, which had measured melt conductivity in disagreement with the predicted melt conductivity, were glasses which did not meet the criteria listed above. Table 2 shows the predicted and analyzed melt conductivity at 1150°C for six of these new waste glass formulations. All of the glasses are well within 25 relative% agreement with predicted melt conductivities except glass S20. However, the analyzed composition of S20 shows that the mole% alkali is about 3 mole% below the amount calculated. If the analyzed mole% alkali is used to predict the melt conductivity of this glass, the predicted melt conductivity is then within 25 relative% agreement with the measured melt conductivity, as shown in Table 2. Note that the compositions tested in Table 2 offer a wide variation in the glass components most often observed in waste glasses, with some components outside the range of the glasses used for the modeling such as an increase in the B_2O_3 content to 20 mole% and ZrO_2 content to 10 mole%. In addition, the glasses used to test the model contain components such as MnO_2 and La_2O_3, which were absent from the glasses used for forming the conductivity model.

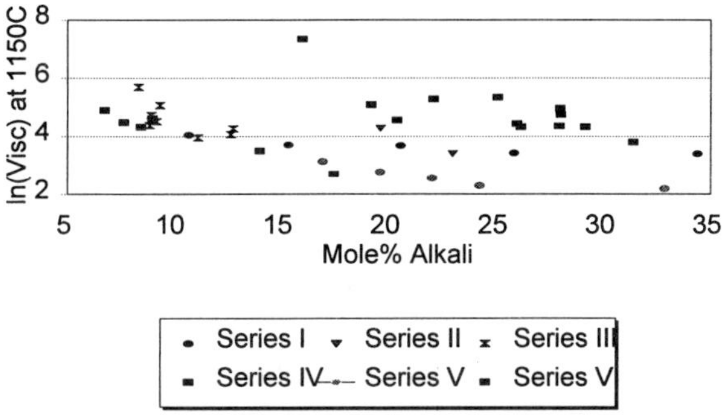

Figure 2. Lack of correlation between melt viscosity and alkali content.

When this model was used to predict glass formulations which produced two phases, such as a sodium sulfate phase on top of a glassy phase, the predicted and calculated melt conductivities did not agree, as would be expected. Previous examination of fluoride containing glasses[17] showed that fluoride containing glasses have greater melt conductivity for the same alkali content than oxide glasses and that there is greater scatter in the melt conductivity of fluoride glasses as a function of total alkali content, most likely due to the contribution of fluoride to melt conductivity[18,19]. Hence, when the above model was used to predict waste glasses with greater than 1 mole% F, the model consistently underestimated the melt conductivity.

CONCLUSIONS

We have seen that a simple correlation between the melt conductivity at 1000-1150°C and total mole% alkali exist for a wide compositional range of multi-component oxide glasses. These glasses, from six different vitrification projects, containing from five to fifteen components with large variations in the major waste glass components, produced a simple model which enabled the prediction of the melt conductivity of new waste glasses. The analyzed melt conductivity of these new waste glasses agree with the predicted values to within ± 25 relative%, often within 15 relative%. The multi-component glasses examined have up to half of their sodium atoms substituted by lithium, and spanned more than two orders in magnitude for melt viscosity, yet total mole% alkali could be used to predict melt conductivity. This suggests that both the mixed alkali effect and melt viscosity play minor roles in the determination of melt conductivity for multi-component oxide glasses in the wide compositional range examined.

ACKNOWLEDGMENTS

The data used for this paper was obtained from six different projects over the past three years at the Vitreous State Laboratory of the Catholic University of America. A number of VSL staff was involved in the glass making and glass property measurements of these glasses, and I wish to thank them all.

Table 2. Results of using Conductivity Model at 1150°C to Predict Waste Glass Melt Conductivity. Predictions are within 25 relative% except when analyzed and calculated alkali contents vary by more than 10 relative% or 2 mole% alkali. (*= at 1150°C)

Components mole%)	H13	S20		ZF31	Z32	L12	H17
		(calc)	**(analy)**				
Al2O3	2.71	8.85	9.89	6.48	0.00	4.25	3.06
B2O3	6.94	13.36	13.19	15.19	20.48	6.07	19.07
BaO	0.00	0.01	0.01	0.00	0.00	0.00	0.00
CaO	9.84	0.59	0.50	24.74	0.00	4.79	0.00
CdO	0.00	0.00	0.00	0.00	0.00	0.18	0.00
Ce2O3	0.00	0.00	0.00	0.00	0.04	0.00	0.14
CoO	0.37	0.00	0.00	0.00	0.00	0.00	0.00
Cs2O	0.04	0.00	0.00	0.03	0.05	0.00	0.04
Cr2O3	0.31	0.14	0.10	0.12	0.48	0.15	0.00
Fe2O3	10.38	0.20	0.31	0.17	0.71	2.07	12.23
K2O	**0.81**	**1.57**	**1.65**	**0.00**	**0.00**	**4.02**	**0.00**
La2O3	0.00	0.00	0.00	0.00	0.00	0.00	0.45
Li2O	**7.62**	**0.09**	**0.16**	**0.00**	**4.96**	**0.00**	**8.62**
MgO	4.11	0.24	0.23	0.00	0.00	5.23	0.00
MnO2	5.60	0.05	0.05	0.00	0.00	0.00	0.00
MoO	0.00	0.00	0.00	0.00	0.23	0.00	0.00
Na2O	**12.25**	**26.65**	**23.80**	**17.05**	**11.12**	**22.33**	**12.59**
NiO	2.08	0.22	0.21	0.50	1.93	0.13	0.19
Nd2O3	0.00	0.00	0.00	0.04	0.07	0.00	0.29
PbO	0.00	0.00	0.00	0.00	0.00	0.03	0.00
P2O5	0.00	1.94	1.99	0.00	0.00	0.85	0.00
SO3	0.00	0.00	0.00	0.00	0.00	0.43	0.00
SiO2	35.61	45.70	47.54	32.99	49.46	47.06	38.98
SrO2	1.27	0.00	0.00	0.00	0.00	0.00	1.44
TiO2	0.00	0.06	0.07	0.00	0.00	0.00	0.00
UO2	0.00	0.41	0.39	0.00	0.00	0.00	0.00
ZrO2	0.00	0.00	0.01	2.63	10.39	2.40	2.78
Total	100.00	100.00	100.00	100.00	100.00	100.00	100.00
mole% alkali	20.71	28.31	25.61	17.09	16.13	26.35	21.26
measured conductivity*	**0.34**	**0.46**	**0.46**	**0.21**	**0.17**	**0.52**	**0.34**
predicted conductivity*	**0.33**	**0.70**	**0.57**	**0.20**	**0.17**	**0.61**	**0.36**
Relative %error	**-2.85%**	**53.39%**	**24.18%**	**-3.51%**	**-1.30%**	**17.22%**	**6.16%**

REFERENCES

[1]J.O. Isard *J. Non-Cryst. Solids,* **1** 235 (1969).

[2]D.E. Day *J. Non-Cryst. Solids,* **21** 343 (1976).

[3]M.D. Ingram *Phys. and Chem. of Glasses,* **28**(6) 215 (1987).

[4]R.E. Tickle *Phys. and Chem. of Glasses,* **8**(3) 101 (1967).

[5]P.J. Hayward *Phys. and Chem. of Glasses,* **18**(1) 1 (1977).

[6]K. Matusita, S. Takayama, and S. Sakka *J. of Non-Crystalline Solids,* **40** 149 (1980).

[7]P. Hrma, G.F. Piepel, D.E. Smith, P.E. Redgate, and M.J. Schweiger, *Ceramic Transactions*, **61** 151 (1995).

[8]R. H. Doremus, *Glass Science*, 2nd ed. John Wiley & Sons Inc., New York, 1994.

[9]J.E. Davidson, M.D. Ingram, A. Bunde, K. Funke, *J. Non-Crystalline Solids*, **203** 246 (1996).

[10]D. Ravaine and J.L. Souquet, *Phys. & Chem. Of Glasses*, **18**(2) 27 (1977).

[11]H. Kahnt, *J. of Non-Crystalline Solids*, **203** 225 (1996).

[12]Q. Yan, F.C. Perez-Cardenas, W. Guo, and S.S. Fu, *Ceramics Transactions,* **72** 387 (1996).

[13]F.C. Perez-Cardenas, *Ph.D Thesis,* April 1996.

[14]S.S. Fu, K.S. Matlack, R.K. Mohr, M.C. Paul, I.L. Pegg, and P.B. Macedo, "Vitrification Testing for Fernald CRU-4 Silo Wastes," Final Report to Fluor Daniel Fernald, May 1996, Revised September 1996.

[15]M. Brandys, R. Mohr, H. Hojaji, S.S. Fu, P.B. Macedo, W. Greenman, and M. Kirche, *Waste Management*, 1995.

[16]S.S. Fu, *Characterization Studies in Support of Savannah River M-Area Engineering,* Report to GTS Duratek, May 1994.

[17]S.S. Fu, *Phys. and Chem. of Glasses* **38**(2) 100 (1997).

[18]C.C. Hunter, and M.D. Ingram, *Solid St. Ion.*, **14** 31 (1984).

[19]D.P. Button, R.P. Tandon, H.L. Tuller, and D.R. Uhlmann, *Solid St. Ion.*, **5** 655 (1981).

REDOX CHARACTERISTICS AND STRUCTURAL PROPERTIES OF IRON PHOSPHATE GLASSES: A POTENTIAL HOST MATRIX FOR VITRIFYING HIGH LEVEL NUCLEAR WASTE

G. K. Marasinghe,[a] M. Karabulut,[a] C. S. Ray,[a] D. E. Day,[a] C. H. Booth,[b] P. G. Allen,[b] and D. K. Shuh[b]

[a]*Department of Physics, Department of Ceramic Engineering, and the Graduate Center for Materials Research, University of Missouri-Rolla, Rolla, MO 65409*

[b]*Chemical Sciences Division, Lawrence Berkeley National Laboratory, University of California, Berkeley, CA 94720*

ABSTRACT

Structural features and redox equilibria in several iron phosphate glasses with batch compositions similar to $40Fe_2O_3$-$60P_2O_5$ (mol%) and those containing Na_2O and UO_2 have been investigated using Mössbauer spectroscopy, X-ray absorption spectroscopy (XAS), X-ray photoelectron spectroscopy (XPS), differential thermal (DTA) and thermo gravimetric (TGA) analysis and X-ray and neutron diffraction. Mössbauer spectra show that a redox equilibria corresponding to a Fe(II)/[Fe(II)+Fe(III)] ratio of 0.2 to 0.35 is reached when batches are melted in air. The valence state of iron ions in the glass appears to be insensitive to the oxygen content in the melting atmosphere, but the Fe(II) content can be increased even beyond the equilibrium region by increasing the partial pressure of a reducing gas in the melting atmosphere. However, the glass forming ability of iron phosphate melts decreases as the Fe(II) fraction is increased beyond that of the equilibrium region. The local structure of iron phosphate glasses appears to be related to the short range structure of crystalline $Fe_3(P_2O_7)_2$. The average coordination of iron ions appears to be distorted octahedral. Compared to other phosphate glasses, the proposed structure for iron phosphate glasses contain a smaller number of -P-O-P- bonds, a feature which is believed to be responsible for the unusually good chemical durability of iron phosphate glasses. The addition of Na_2O reduces the valence state of iron ions, drives the coordination of iron ions towards perfect octahedral symmetry, and appears to increase the fraction of bridging (taking part in -P-O-P- type bonds) oxygens. However, adding UO_2 has the opposite effect, compared to Na_2O, on the redox state and structure of iron phosphate glasses.

INTRODUCTION

Even though chemical compositions of nuclear wastes are complex and extremely diverse, only one glass material, namely borosilicate glasses, is currently approved for vitrifying high level nuclear waste (HLW). However, several of the HLWs in the U.S. contain components which make them poorly suited for disposal in borosilicate glasses. For example, the low solubility limit of phosphates and certain heavy metals in borosilicate glasses can dramatically reduce the amount of certain wastes which can be safely contained in them.[1] Recent work[2,3] at Hanford has found that the concentration of iron oxides in the waste is much higher than known previously. Such high concentrations of iron oxides also impose unfavorable waste loading restrictions on borosilicate glasses. Current plans call for vitrifying even these problematic waste feeds in borosilicate glasses after the original waste feed has been pre-processed and/or diluted to compensate for the incompatibility. However, these pre-treatment processes, as well as the larger waste volumes resulting from dilution, will significantly increase the cost of cleaning up the former nuclear weapons production facilities. Such additional costs may be avoided by developing a small number of alternative waste glasses which are suitable for vitrifying those specific waste feeds that are incompatible with borosilicate glasses.

Table 1. Selected properties of an iron phosphate base glass, three iron phosphate waste glasses containing HLWs from the Hanford site, and two borosilicate glasses.

	Base Glass[a]	Hanford Waste[b]			CVS-IS[c]	DWPF-EA[d]
		B-110	C-112	T-111		
Waste Loading (wt%)	-	35	50	35	NA	NA
Log D_R (g/cm^2/min)[e]	-9.7	-8.1	-7.5	-8.4	-7.6	-7.3
T_m (°C)[e]	1100	1150	1150	1150	NA	NA
t_m (hrs.)[e]	2	2	2	2	24	24
T_s (°C)[e]	505	494	526	512	NA	NA
T_c (°C)[e]	580	590	602	622	NA	NA
Density (g/cm^3)	3.2	3.19	3.35	3.35	NA	NA

[a] $40Fe_2O_3$-$60P_2O_5$ (mol%).
[b] B-110, C-112, and T-111 are waste tank designations at Hanford.
[c] CVS-IS is a borosilicate glass used at Hanford.
[d] DWPF-EA - Defense Waste Processing Facility Environment Assessment borosilicate glass.
[e] D_R - dissolution rate in distilled water @90°C for 16 days, T_m - melting temperature, t_m - melting time, T_s - softening temperature, T_c - crystallization temperature.

A family of iron phosphate glasses containing from 25 to about 40 mol% Fe_2O_3 appear to be a highly effective and low cost alternative to borosilicate glasses for vitrifying certain HLWs.[4,5] Compared to borrosilicate glasses,[6] iron phosphate glasses have lower melting temperatures, shorter melting times, higher waste loading, and smaller or comparable dissolution rates[7] in water. For certain HLWs, those which are incompatible with borosilicate glasses, the waste loading of iron phosphate vitrified waste forms can be as high as 60wt%.[8] Table I summarizes results from recent studies[4,5] of iron phosphate waste glasses containing several wastes from the Hanford site.

Iron phosphate glasses described herein and their nuclear waste forms have been investigated only recently, so many of their properties are not well known. For example, little is known about their atomic structure, redox equilibria, structure-property relationships, and crystallization products and characteristics. Such basic knowledge for the binary iron-phosphate glasses as well as the more compositionally complex waste forms is essential for a comprehensive scientific assessment of their use in nuclear waste disposal. In the present work, Mössbauer spectroscopy, X-ray photoelectron spectroscopy (XPS), x-ray absorption spectroscopy (XAS), and differential thermal analysis (DTA) and thermo-gravimetric analysis (TGA) have been used to study the structure and redox characteristics of several binary iron phosphate glasses and those containing up to 15 mol% Na_2O or UO_2, two common waste components.

EXPERIMENTAL

Glasses were synthesized by melting homogeneous mixtures of reagent grade chemicals in dense alumina crucibles at ~1200°C for approximately 2 hrs. Batches were melted in air or in controlled atmospheres. Table II gives the batch composition, the raw materials used, and the melting conditions for the glasses studied. Even though different raw materials were used for some of the glasses, the Fe/P atomic ratio of all the batches was approximately 0.67 which equals that in crystalline $Fe_4(P_2O_7)_3$. Each melt was quenched in air by pouring it into 1cm x 1cm x 5cm cold steel molds. The rectangular samples were immediately transferred to a furnace and annealed at 450°C for ca. 3 hrs. Parts of selected glass samples were crystallized by heating at 650°C for approximately 24 hrs. Room temperature X-ray powder diffraction (XRD) data for both glass and crystalline samples were collected on a SCINTAG XDS2000 x-ray diffractometer. High temperature XRD data was also collected for selected samples heated in air at 600 and 800°C. The Mössbauer spectra were obtained at 23°C on a ASA600 spectrometer which utilized a room temperature 50 mC rhodium matrix cobalt-57 source. The spectrometer was calibrated at 23°C with an α-iron foil and the linewidth of the α-iron spectrum was 0.27 mm/s. Mössbauer absorbers of approximate thickness 85mg/cm^2 were prepared using 200 mesh powders. The Mössbauer spectra were fit with the minimum number of broadened paramagnetic doublets necessary, usually eight, to match the absorption envelope with a misfit of less than 0.003. Details of this fitting procedure have been discussed elsewhere.[9] XPS data was collected on a Physical Electronics model 548 spectrometer. In order to minimize contamination, samples were fractured in-situ under high vacuum (~10^{-8} torr) just before analysis. Oxygen 1s scans were collected at a pass energy of 50eV using Mg Kα ($\hbar\nu$=1254.6 eV) radiation. X-ray absorption spectra were collected on BL 4-1 at the Stanford Synchrotron Radiation Laboratory (SSRL) at the Fe *K* edge using Si(220) monochromator crystals. The data was reduced and fit using standard procedures.[10,11] All absorption spectra were normalized to the absorption over a range in energy of several hundred eV above the edge. The data were fit with back-scattering amplitudes and phases calculated using the FEFF6[12] multiple scattering code for all single scattering paths up to Fe-P (~3.4 Å). Simultaneous differential thermal analysis (DTA) and thermo-gravimetric analysis (TGA) were performed in flowing air at a heating rate of 10°C/min using a Mettler type TA-1 instrument. About 300 mg of 425-500μm glass particles were used for these measurements and high purity alumina powder was used as the reference material. The thermocouples of the instrument were calibrated at the melting points of aluminum (660.4°C) and gold(1064.4°C).

Table II. Mössbauer hyperfine parameters at 23°C and the iron valence in the glass samples. Iron valence in the starting batch is also given for reference.

Sample	Batch Composition, mol%	Melting Atmosphere	Quenched State	δ (mm/s)		QS (mm/s)		Fe(II) Fraction[a]	
				Fe(II)	Fe(III)	Fe(II)	Fe(III)	Batch	Glass
A	$40Fe_2O_3$-$60P_2O_5$	Air	Glass	1.22	0.38	2.14	0.94	0	0.19
B	$31Fe_3O_4$-$69P_2O_5$	Air	Glass	1.24	0.39	2.11	0.91	0.3	0.24
C	$24FeO$-$24Fe_2O_3$-$52P_2O_5$	Air	Glass	1.24	0.39	2.15	0.95	0.5	0.34
D	$57FeC_2O_4.2H_2O$-$43P_2O_5$	Air	Glass	1.24	0.39	2.14	0.91	1	0.24
E	$31Fe_2O_3$-$23P_2O_5$-$46NH_4H_2PO_4$	Air	Glass	1.24	0.40	2.16	0.90	0	0.35
F	Same as sample A	Nitrogen	Glass	1.20	0.39	2.21	0.87	0	0.22
G	Same as sample A	Oxygen	Glass	1.23	0.41	2.11	0.80	0	0.21
H	Same as sample A	70%FG[b]-30%Air	Glass	1.24	0.40	2.14	0.92	0	0.31
J	Same as sample A	90%FG-10%Air	Glass[c]	1.23	0.40	2.18	0.94	0	0.40
K	Same as sample A	FG	Crystallized[d]					0	1
L	Same as sample E	Flowing Ar	Mixed[e]	1.23	0.39	2.2	0.94	0	0.85
M	Same as sample E	Ar	Crystallized[d]					0	1
N	$29Fe_3O_4$-$66P_2O_5$-$5Na_2O$	Air	Glass	1.24	0.39	2.13	0.91	0.3	0.23
O	$28Fe_3O_4$-$62P_2O_5$-$10Na_2O$	Air	Glass	1.22	0.40	2.17	0.90	0.3	0.27
P	$26Fe_3O_4$-$59P_2O_5$-$15Na_2O$	Air	Glass	1.23	0.40	2.16	0.88	0.3	0.28
Q	$29Fe_3O_4$-$66P_2O_5$-$5UO_2$	Air	Glass	1.21	0.38	2.13	0.91	0.3	0.17
R	$28Fe_3O_4$-$62P_2O_5$-$10UO_2$	Air	Glass	1.19	0.37	2.13	0.91	0.3	0.14
S	$26Fe_3O_4$-$59P_2O_5$-$15UO_2$	Air	Glass	1.19	0.36	2.26	0.92	0.3	0.08

[a] Fe(II)/[Fe(II)+Fe(III)] as obtained from Mössbauer spectra.

[b] Forming gas ($90N_2$-$10H_2$ at%).

[c] A trace amount of crystalline material is evident in the Mössbauer spectrum of this sample.

[d] Sample contains more than one crystalline component so Mössbauer hyperfine parameters are not reported. Mössbauer spectra show no detectable Fe(III).

[e] A thin layer on the top surface and a region at the center of the rectangular sample crystallized. Hyperfine parameters are for the glassy part of the sample.

RESULTS AND DISCUSSION

Binary Iron Phosphate Glasses

Table II gives the state of quenched samples, Mössbauer hyperfine parameters, and valence state of iron ions. The iron(II)/iron(III) ratio in a waste glass can be used to determine and control the general oxidation state of the glass during melting.[13] Controlling the oxidation state of a waste glass melt is important because extreme oxidizing conditions can facilitate the formation of foam on the melt surface which in turn diminishes the heat transfer and lowers the melting rates. On the other hand, extreme reducing conditions can cause the formation of metallic phases which can settle to the melter floor and potentially cause an electric short.[13]

The Mössbauer spectrum obtained at 23 °C for sample A, as shown in Figure 1(a), is typical of the spectra of other glasses melted in air. The valence state of iron ions in samples A through E shows that a redox equilibria corresponding to Fe(II)/[Fe(II)+Fe(III)] ratio between 0.2 and 0.35 is reached when batches are melted in air. The similarity in the iron valence of samples A, F, and G, which had the same batch composition but were melted in air, nitrogen, and oxygen, respectively, suggests that the process by which the redox equilibria is reached is independent of the oxygen content in the melting atmosphere. However, the large fraction of Fe(II) in samples H and J indicates that the Fe(II) content in the glass can be increased by melting in increasingly reducing atmospheres.

Also, the high Fe(II) content (0.34) in sample E shows that the iron valence in the glass is sensitive to the reducing ability of chemicals used in the

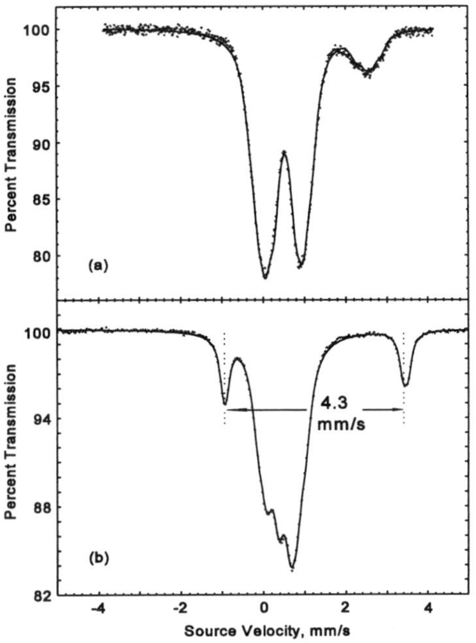

Figure 1. Mössbauer spectra obtained at 23 °C for (a) sample A, see Table II, and (b) its crystallized counterpart

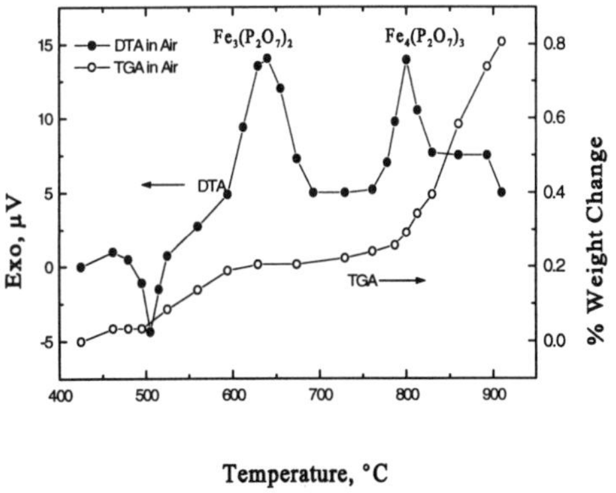

Figure 2. Differential thermal analysis (DTA) data and thermo-gravimetric analysis (TGA) data for sample A (Table II).

batch. Batch of sample E contained $NH_4H_2PO_4$ which releases NH_3 when heated and acts as a reducing agent during melting.[14]

The appearance of a trace amount of crystalline material in sample J (Fe(II) fraction = 0.4), surface and core crystallization of sample L (Fe(II) fraction = 0.85), and complete crystallization of samples K and M, both of which contains only Fe(II), suggest that the glass forming ability of melts deteriorates as the fraction of Fe(II) increases beyond that of the equilibrium region.

The DTA and TGA data shown in Figure 2 are typical for the glasses described herein. The endothermic peak at ~ 500°C in the DTA thermogram is for the glass transition and the exothermic peaks at approximately 580 and 775°C are attributed to crystallization of the glass powder. XRD data shows that the major phase crystallizing from these glasses around 580°C is the recently discovered[15] compound $Fe_3(P_2O_7)_2$, which has one Fe(II) and two Fe(III) per formula unit. $Fe_4(P_2O_7)_3$ starts to crystallize as the sample is heated above 700°C. The rate of weight gain which increases at about 800°C in the TGA curve in Figure 2 may be due to the sample reacting with oxygen from the atmosphere as some of the Fe(II) in the glass is converted to Fe(III) in $Fe_4(P_2O_7)_3$.

Knowledge of the structure of compounds which crystallize from a glass can provide clues about the structure of the parent glass. Neutron diffraction studies[9] have shown that the only crystalline component in the crystallized (at 580°C) counterpart of samples A through H is $Fe_3(P_2O_7)_2$. The structure of crystalline $Fe_3(P_2O_7)_2$, see Figure 3, is unique as it is the first time[15] that Fe(II) has been observed in a trigonal prism coordination. This unusual coordination of Fe(II) is responsible for the unusually large quadrupole splitting of 4.33 mm/s in the Mössbauer spectrum, see Table III and Figure 1(b). Each of the six-fold coordinated Fe(II) ions is sandwiched between two Fe(III) ions in distorted octahedral (Fe(III) ion is off center) coordination with oxygen ions. The $(Fe_3O_{12})^{-16}$ clusters consisting of the Fe(II) trigonal prism and the two Fe(III) octahedra are connected by two $(PO_4)^{-3}$ tetrahedra joined together to form a pyrophosphate group, $(P_2O_7)^{-4}$. An interesting feature of the structure of crystalline $Fe_3(P_2O_7)_2$ is that only one out of the eight crystallographically different oxygen ions take part in a -P-O-P- bond. Other oxygen atoms are bonded via -Fe-O-P- links. Even though the structure of the parent glass need not be identical to that of crystalline $Fe_3(P_2O_7)_2$, it is reasonable to expect general similarities between the crystal and glass structures.[16]

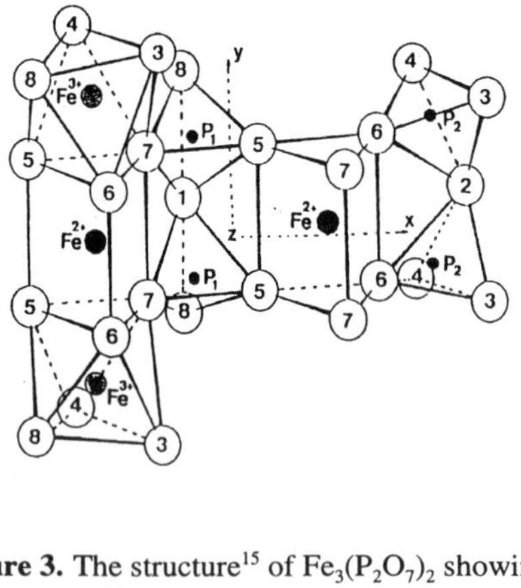

Figure 3. The structure[15] of $Fe_3(P_2O_7)_2$ showing the $(Fe_3O_{12})^{-16}$ clusters.

Recent XAS studies[9] of glass sample B and its crystallized counterpart, see Figure 4, have confirmed that the local structure around the iron ions in the glass and the crystallized sample are similar. According to XAS fit parameters, iron ions in the glass structure appear to maintain an almost six-fold coordination with near neighbor oxygen ions and second shell phosphorus ions supporting a structural model in which a majority of the oxygen atoms in the glass take part in -Fe-O-P- type bonds as is the case in crystalline $Fe_3(P_2O_7)_2$. A comparison[9] of Mössbauer hyperfine parameters obtained for glass sample B and its crystallized counterpart, see Table III, also shows that the average coordination of iron ions in the glass is most likely distorted octahedral.

The superior chemical durability of iron phosphate glasses[4,14] can be explained in terms of the structural model described above. The relatively poor chemical durability of phosphate glasses is attributed to the abundance of easily hydrated -P-O-P- bonds in these glasses.[17] Hydration of the bridging oxygen of -P-O-P- bonds results in terminal -P-OH bonds, thus destroying the cross linkage between the $(PO_4)^{-3}$ tetrahedra in what can be viewed as a depolymerization of the glass network. However, if a majority of oxygen ions in iron phosphate glasses are part of more hydration resistant -Fe(II)-O-P- or Fe(III)-O-P- bonds as proposed herein, one expects these glasses to possess a superior chemical durability like that reported.[14]

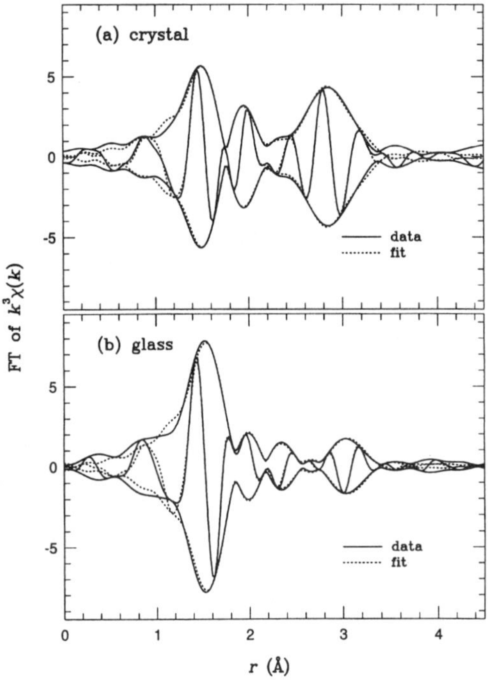

Figure 4. Fourier transform (FT) of $k^3\chi(k)$ for the glass sample B (bottom) and its crystallized counter part (top). The outer envelope is the amplitude of the transform, and the oscillatory part is the real part of the transform. Peaks correspond to Fe near-neighbors up to a phase shift. For instance, the first peak at 1.5 Å corresponds to oxygen neighbors at ~1.9 Å from the absorbing iron. Transforms are from 4.5-15 Å^{-1} (Gaussian broadened by 0.3 Å^{-1}).

Table III. Average hyperfine parameters, at 23 °C, for the crystallized counterpart of sample B (see Table II). Corresponding parameters for the parent glass sample are given within brackets for comparison.

Parameter	Fe(II)	Fe(III)
Content, at%	21{24}	79{76}
Isomer shift, mm/s	1.2{1.24}	0.42{0.39}
Quadrupole splitting, mm/s	4.33{2.11}	0.51{0.91}
Average iron valence	2.71{2.76}	

Oxygen 1s XPS studies[9] of glass samples A through E also suggest that the glass contains only a small number of bridging (-P-O-P-) oxygens supporting a glass structure which has short range similarities to that of crystalline $Fe_3(P_2O_7)_2$.

Iron Phosphate Glasses Containing Na_2O or UO_2

Mössbauer fit parameters for the iron phosphate glasses containing Na_2O (soda) or UO_2(samples N through S) are listed in Table II. Soda and UO_2 have opposite effects on the Fe(II) fraction in the glass. Mössbauer hyperfine parameters show little or no dependence on the Na_2O content indicating that structural changes introduced by the addition of Na_2O, if any, are too small to be detected by Mössbauer spectroscopy. However, the addition of UO_2 is accompanied by a small, but detectable reduction in the isomer shift, which in general is attributed to an increase in the s-electron density at the iron nucleus. Such a change in the s-electron density may occur as iron ions in the network are "squeezed" by large uranium ions.

Fe *K* near edge absorption data for sample B and for samples containing Na_2O and UO_2 is shown in Fig. 5. The feature in the pre-edge region near 7114.5 eV is from a 1s-3d transition. Studies on transition-metal compounds[18,19] have shown that such excitations are generally present, and are minimized by perfect octahedral coordination. The 1s→3d transition in the glasses shows that as UO_2 is added to the base glass, the oxygen coordination around iron becomes more distorted as compared to a perfect octahedron, yet as Na_2O is added, the iron coordination becomes more like a regular octahedron. The 15% Na_2O sample has a transition that is nearly as small as the transition previously measured[9] in crystalline $Fe_3(P_2O_7)_2$.

In considering this data, there are a few facts to keep in mind. First, the character of the 1s→3d transition does not change much across the series, even though the amplitude changes by nearly a factor of two. This means that the geometry of the distortion (away from octahedral) does not change much with UO_2 or Na_2O content. Second, the energy at the transition does not change, even though the energy of the main absorption edge shifts by ~ 0.7 eV. This indicates that the iron involved in this 1s-3d transition is in a single oxidation state. Lastly, the 1s→3d transition energy is consistent with that observed[9] for Fe_2O_3, indicating that the transition comes predominantly from iron(III) sites. This last point is consistent with the fact that the glass with the highest fraction of Fe(III) (15% UO_2) also has the largest 1s→3d transition. A detailed analysis of the XAS

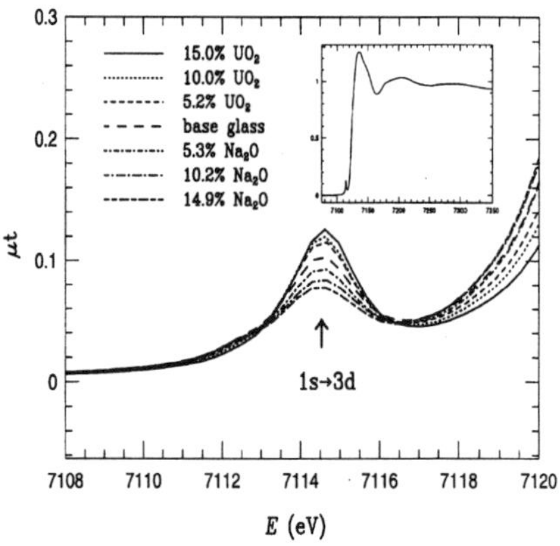

Figure 5. μt (absorption times sample thickness) vs. incident photon energy for iron-phosphate glasses. μt for each spectra is normalized to unity at the edge step for comparison, and energy is defined such that the first inflection point of an Fe foil is at 7112 eV. The main plot is focused on the pre-edge region, where the bump at ~7114.5 eV is due to a 1s→3d transition. Inset shows full edge step for the glass with 15% UO_2.

results will be published elsewhere.[20]

The photon energy required to eject a photoelectron from an absorbing atom (E_0, measured from the ½-height of the edge) is proportional to the oxidation state of the absorber. Based on measurements of E_0 for the Fe K edge from FeO and Fe_2O_3, the mean iron valence changes from +3.00 for the glass with 15% UO_2 to +2.81 for the glass with 15% Na_2O, in general agreement with Mössbauer measurements.

Table IV. O1s XPS fit parameters for samples O, B, and S listed in Table II.

Sample Composition (mol%)	% Area of Voight Peak	
	Bridgin Oxygens	Other Oxygens
(O) $28Fe_3O_4$-$62P_2O_5$-$10Na_2O$	30	70
(B) $31Fe_3O_4$-$69P_2O_5$	15	85
(S) $26Fe_3O_4$-$59P_2O_5$-$15UO_2$	12	88

Oxygen 1s XPS data was fit with two Voigt peaks. Peak corresponding to the higher binding energy is attributed to bridging oxygens taking part in -P-O-P- bonds and the other peak is attributed to the remaining oxygen ions.[21] XPS fit parameters, seeTable IV, suggest that adding Na_2O and UO_2 to iron phosphate glasses have opposite effects on the glass structure. Addition of UO_2 appears to decrease the number of bridging oxygens whereas the addition of Na_2O appears to increase it. The latter result, where the fractions of bridging oxygens appears to increase with increasing Na_2O content , is unusual and must be further investigated.

CONCLUSIONS

A redox equilibria corresponding to a Fe(II)/[Fe(II)+Fe(III)] ratio of 0.2 to 0.35 is reached when iron phosphate glasses are melted and quenched in air. The process by which this redox equilibria is reached is independent of the oxygen content in the melting atmosphere. However, the fraction of Fe(II) in the glass can be increased by increasing the reducing ability of the melting atmosphere or by using raw materials which create reducing conditions in the batch. When quenched in air as described in this paper, the glass forming ability of iron phosphate melts decreases as the Fe(II) fraction increases beyond the equilibrium region (≥0.4).

The structure of iron phosphate glasses seems to be related to the short range structure of crystalline $Fe_3(P_2O_7)_2$. The general coordination of a majority of the iron ions appears to be distorted octahedral. A majority of the oxygen ions in the glass are part of -Fe-O-P- bonds and as a result, these iron phosphate glasses contain only a small number of bridging oxygens (-P-O-P-) when compared to most other types of phosphate glasses. The addition of Na_2O reduces the valence state of the iron ions, drives the coordination of the iron ions towards octahedral symmetry, and appears to increase the fraction of bridging oxygens. On the other hand, adding UO_2 appears to have the opposite effect on the redox state and structure of iron phosphate glasses compared to that observed for Na_2O.

ACKNOWLEDGMENTS

The authors acknowledge with thanks the financial support of the Department of Energy (DOE96ER45617 and DE-AC03-76SF00098) and the University of Missouri-Research Board. The authors would also like to thank Mr. O. Ozturk, and Drs. W. J. James, D. Waddill, J. J. Bucher, and N. M. Edelstein for their valuable discussions.

REFERENCES

[1] Private communications, Bruce C. Bunker, Pacific Northwest National Laboratory, Hanford, WA.

[2] S. L. Lambert and D. S. Kim, "Tank Waste Remediation System High-Level Waste Feed Processability Assessment Report", PNNL #WHC-SP-1143, PNNL, Hanford, WA, 1994.

[3] M. J. Kipfer et al. "Best Basis for Total Inventories of Chemicals and Radionucides in Hanford Site Waste", interim report, PNNL #WHC-SD-WM-TI-740, PNNL, Hanford, WA, 1996.

[4] D. E. Day, Z. Wu, and C. S. Ray, "Chemically Durable Glasses for Vitrifying Phosphate Containing Simulated Nuclear Wastes," final report to PNNL, contract no. 276822-A-F1, December 1995.

[5] D. E. Day and M. Mesko, "Immobilization of High Level Radioactive Sludges in Iron Phosphate Glasses," final report to PNNL, contract no. 295503-A-F1, December 1996.

[6] CVS-IS borosilicate standard used at the Hanford site and DWPF-EA (Defense Waste Processing Facility Environment Assessment) borosilicate glass from Savannah River.

[7] After 16 days in distilled water maintained at 90°C.

[8] For tank C-112 sludge from Hanford.

[9] G. K. Marasinghe, M. Karabulut, C. S. Ray, D. E. Day, W. B. Yelon, C. H. Booth, P. G. Allen, and D. K. Shuh. Submitted to J. Non-Cryst. Solids.

[10] T. M. Hayes and J. B. Boyce, in *Solid State Physics*, edited by H. Ehrenreich, F. Seitz, and D. Turnbull (Academic, New York, 1982), Vol. 37, p. 173.

[11] G. G. Li, F. Bridges and C. H. Booth, Phys. Rev. B, 52, (1995) 6332.

[12] S. I. Zabinsky, A. Ankudinov, J. J. Rehr and R. C. Albers, Phys. Rev. B, 52 (1995) 2995.

[13] D. S. Goldman and D. W. Brite, J. Am. Cer. Soc., 69, 411 (1986).

[14] Y. Xiaoyan and D. E. Day, Proceedings of the 17th Intl. Congress on Glass, 2 (1995) 45.

[15] M. Ijjaali, G. Venturini, R. Gerardin, B. Malaman, and C. Gleitzer, Eur. J. Solid tate Inorg. Chem., 28 (1991) 983.

[16] D. Virgo and B. O. Mysen, Phys. Chem. Minerals, 12 (1985) 65.

[17] G. N. Greaves, S. J. Gurman, L. F. Gladden, C. A. Spence, P. Cox, B. C. Sales, L. A. Boatner, and R. N. Jenkins, Phil. Mag. B, 58 (1988) 271.

[18] J. Wong, F. W. Lytle, R. P. Messmer and D. H. Maylotte, Phys. Rev. B, 30 (1984) 5596.

[19] A. Mancau, A. I. Gorshkov, V. A. Drits, Am. Miner., 77 (1992) 1133.

[20] C. H. Booth et al. To be published.

[21] R. K. Brow, C. M. Arens, X. Yu, and D. E. Day, Phys. Chem. Glasses, 35 (1994) 132.

LIQUIDUS TEMPERATURE MODELS FOR HIGH-LEVEL WASTE GLASSES THAT PRECIPITATE ZIRCONIUM-CONTAINING CRYSTALLINE PHASES

J. V. Crum, P. Hrma, M. J. Schweiger, and G. F. Piepel
Pacific Northwest National Laboratory[1],
Richland, WA 99352

ABSTRACT

Liquidus temperature (T_L) and primary crystalline phases were determined for simulated Hanford high-level waste glasses with high concentrations of ZrO_2. Previous studies, whether varying components one at a time or statistically designed, contained insufficient number of glasses with zirconium-containing primary phases to support crossproduct terms in second-order mixture models. The data from these studies were augmented by multi-component variations of Al_2O_3, B_2O_3, Li_2O, Na_2O, and ZrO_2 to determine the interactive effects on T_L. Altogether, four second-order mixture models were fitted: two for zircon ($ZrSiO_4$) primary phase and two for zirconium-containing phases combined.

INTRODUCTION

A major concern for the vitrification of Hanford high-level waste is the precipitation of crystalline phases inside the melter during glass processing. These phases can accumulate at the melter bottom and in the glass discharge region, leading to shorter melter life. To avoid this, it is common to formulate glasses with liquidus temperature (T_L) at least 100°C below the melter operating temperature. With this constraint, understanding the effects of glass composition on T_L is crucial for obtaining maximum waste loading in a processable glass.

Zirconium-containing primary phases form in glasses with high concentrations of ZrO_2. T_L data for such glasses were reported in two previous studies [1,2]: the composition variation study (CVS) and transuranic waste study (TRU). The CVS was statistically designed to vary many components at a time, whereas the TRU study made single component changes from a baseline composition. Zircon ($ZrSiO_4$) and baddeleyite (ZrO_2) were identified in TRU glasses, and three primary phases ($ZrSiO_4$, ZrO_2, and $Na_2ZrSi_2O_7$) occurred in CVS glasses.

---

[1] Pacific Northwest National Laboratory is operated for the U. S. Department of Energy by Battelle Memorial Institute under Contract DE-AC06-76RLO 1830.

Models were fitted to different data sets according to the database (TRU or TRU+CVS) or the primary phase (zircon, baddeleyite, or sodium-zirconium silicate). First-order mixture models [3] were fitted to zirconium-precipitating glasses from CVS by Kim and Hrma [4] and from TRU by Crum et al. [2]. In addition, first- and second-order models were fitted to data from TRU and TRU+CVS by Rao et al. [5].

Unfortunately, neither the CVS nor the TRU study contains sufficient glasses with zirconium-containing primary phases to support interaction terms in second-order models. For this reason, the TRU study was augmented by nine additional glasses that allowed fitting new second-order mixture models with interactive terms.

EXPERIMENTAL

To provide maximum support for detecting interactive effects, nine additional TRU compositions were designed at or near the zircon primary phase field boundary (Table I). Components varied were Al_2O_3, B_2O_3, Na_2O, Li_2O, and Zr_2O because of their strong effects on T_L [1-4]. Components that didn't vary were kept in constant proportion with the baseline glass.

The first glass in Table I is the TRU baseline [2]. It was previously measured during the TRU single component variation study. The following six glasses vary two components at a time, and the remaining three glasses vary three components at a time. Mass fractions of the components varied were generally increased in relation to the baseline glass to provide data near the upper portion of the composition region previously studied. All but three glasses (AZ-1, BN-1, and NBZ-1) varied components that increased T_L together with components that deceased T_L. Two glass (BN-1 and NL-1) varied two components that decreased T_L, and one glass (AZ-1) varied two components that increase T_L.

Table I. Compositions (in Mass Fraction) of TRU Multi-component Variation Glasses

Glass ID	Al_2O_3	B_2O_3	Bi_2O_3	Li_2O	Na_2O	P_2O_5	SiO_2	ZrO_2	Others
TRU-BL-1	0.02296	0.08150	0.05415	0.04584	0.11392	0.00871	0.46761	0.10697	0.09834
TRU-AZ-1	**0.03000**	0.08027	0.05334	0.04515	0.11221	0.00858	0.46158	**0.11300**	0.09687
TRU-NL-1	0.02351	0.08342	0.05543	**0.01000**	**0.13000**	0.00891	0.47860	0.10948	0.10065
TRU-AB-1	**0.04000**	**0.12000**	0.05079	0.04300	0.10685	0.00817	0.43861	0.10033	0.09225
TRU-BN-1	0.02341	**0.10000**	0.05519	0.04672	**0.08000**	0.00887	0.47657	0.10902	0.10022
TRU-NZ-1	0.02211	0.07845	0.05213	0.04413	**0.13000**	0.00838	0.45013	**0.12000**	0.09467
TRU-BZ-1	0.02151	**0.12000**	0.05071	0.04293	0.10669	0.00816	0.43791	**0.12000**	0.09209
TRU-ANZ-1	**0.04000**	0.07437	0.04941	0.04183	**0.15000**	0.00795	0.42670	**0.12000**	0.08974
TRU-ABZ-1	**0.04000**	**0.10000**	0.05082	0.04302	0.10690	0.00817	0.43881	**0.12000**	0.09228
TRU-BNZ-1	0.02271	**0.06000**	0.05356	0.04534	**0.13000**	0.00861	0.46250	**0.12000**	0.09728

Others consist of the following oxides: Ag_2O, BaO, CaO, Co_2O_3, Cr_2O_3, CuO, F, Fe_2O_3, K_2O, La_2O_3, MgO, MnO, NiO, PbO, SrO, and ZnO.

Glass preparation was the same as in the previous TRU work [2]. The T_L values of the glasses were predicted using a T_L first-order mixture model for zircon phase, based on the single component variation data set [2]. These predictions were used to help speed up the measurement process.

To determine T_L, approximately 2.5 g of crushed glass was loaded into a platinum-10% rhodium rectangular crucible (12.5 mm × 12.5 mm × 8 mm) with a lid and isothermal heat-treated in a uniform temperature furnace for 24 h. Air quenched samples were examined under a low magnification (70×) metallurgical microscope to determine if crystals were present in the glass. The difference in temperature between heat-treated glass samples with and without crystals was defined to within 5°C. T_L was determined within the 5°C interval.

Selected samples were thin sectioned and examined under an optical microscope (100× to 1000×) using transmitted and reflected light. X-ray diffraction and scanning electron microscopy (SEM) with electron dispersion spectroscopy (EDS) were used to identify the crystalline phases.

Once the T_L was determined for the glasses, reduced forms of the second-order mixture model

$$T_L = \sum_{i=1}^{n} b_i g_i + \sum_{i=1}^{n} b_{ii} g_i^2 + \sum_{i=1}^{n-1} \sum_{j>i}^{n} b_{ij} g_i g_j \qquad (1)$$

were fitted to data: where b_i , b_{ii} and b_{ij} are linear and nonlinear blending coefficients, g_i is the i-th component mass fraction, and n is the number of components. Only some squared and crossproduct terms may be present in the model to avoid over-parameterization of equation (1) [3].

RESULTS

All glasses in Table I precipitated zircon as the primary crystalline phase. The measured T_L values as well as those predicted by the first-order mixture model A [2] are listed in Table II. For all but three glasses, the predicted T_L was within 20°C of the measured T_L. However, ZB-1, ANZ-1, and ABZ-1 had significantly higher measured than predicted T_L (up to 90°C); these differences indicate the existence of component interactive effects.

Second-order mixture models were fitted to the following data sets: TRU for zircon (B), TRU+CVS for zircon (C), TRU for zircon & baddeleyite (D), and TRU+CVS for all crystalline phases taken together (E). The composition regions of these data sets are summarized in Table III.

Component coefficients (Table IV) were obtained using stepwise regression software. The software was forced to select all first-order terms for components that were varied in the TRU or TRU+CVS data sets. For the TRU data, second-order terms were selected in two stages. In the first stage, squared terms (e.g., $Li_2O \times Li_2O$) were selected using stepwise regression and plots of the 27 data points in which components were varied one-at-a-time [5]. In the second stage, the 9 new glasses from Table I were added to the previous 27 glasses, and crossproduct terms (e.g., $Na_2O \times Li_2O$) were added to the model using stepwise regression. Only crossproduct terms involving components varied in the 9 new glasses were considered in this stage. This two-stage approach to select second-order terms for the TRU data was necessary because the larger component interaction effects "swamp" the smaller component curvature effects, causing some of the latter to be missed with a one-stage stepwise regression approach. For the TRU+CVS data, stepwise regression was used to select both squared and crossproduct terms in a single stage.

The number and type of second-order terms vary between the models. However, five squared terms appear in more than one model: $Li_2O \times Li_2O$ in all four models, $ZrO_2 \times ZrO_2$ in all but model C, $Na_2O \times Na_2O$ and $MgO \times MgO$ in models C and E, and $CeO_2 \times CeO_2$ and $P_2O_5 \times P_2O_5$ in models B and D. In addition, five crossproduct terms appear more than once: $Na_2O \times Al_2O_3$ and $Na_2O \times Li_2O$ in all but model C, $SiO_2 \times Al_2O_3$ and $ZrO_2 \times Al_2O_3$ in models C and E, and $ZrO_2 \times Na_2O$ in all but model E. A comparison of the models indicates that joining primary phases has less effect than combining TRU and CVS data sets.

Caution should be exercised regarding any interpretations attached to the second-order terms selected for Models C and E fitted to the TRU+CVS data. The TRU and CVS glasses have differences in components and component ranges. Merging such data sets and performing stepwise regression can cause second-order terms to be selected because of these differences, or because of correlations among second-order terms.

The predicted T_L values of the nine multi-component variation glasses by the four second-order models are shown in Table II. Model B predicts the T_L of all nine glasses within 12°C of the measured T_L, model D within 15°C, model C within 24°C, and model E within 40°C. Models B and D predicted the T_L of ANZ-1, the most underpredicted T_L value by model A, within 2°C of the measured T_L. Models B and D yield the best overall T_L predictions for the nine glasses in Table I, although for some of the glasses T_L is predicted better by model C or E.

The presence of second-order terms, especially cross-product terms, complicates understanding individual component effects on T_L. Obtaining partial specific properties for individual components allows one to determine the effect on T_L of replacing one component with another [6]. For example, using model B coefficient values (Table IV), we obtain for the effect of adding Na_2O at the expense of SiO_2:

$$\frac{\partial T_L}{\partial g_{Na_2O}} = -2705 + 127321 g_{Al_2O_3} + 50076 g_{Li_2O} - 62497 g_{ZrO_2}$$

Thus, for NL-1 glass, replacing 1 wt% SiO_2 by 1 wt% Na_2O results in a decrease of T_L by 61°C, whereas the same change in ANZ-1, which has a higher concentration of Al_2O_3, Li_2O and ZrO_2, the decrease is 3°C.

CONCLUSIONS

Liquidus temperature of glass with zirconium-containing primary phases has been expressed as a function of composition over the composition regions

Table II. Measured and Predicted T_L Values

Glass ID	Measured T_L, °C	Model A T_L, °C	Model B T_L, °C	Model C T_L, °C	Model D T_L, °C	Model E T_L, °C
TRU-BL-1	1012	1016	1013	1017	1013	1012
TRU-AZ-1	1065	1062	1071	1071	1065	1071
TRU-NL-1	1095	1093	1099	1110	1100	1119
TRU-AB-1	1056	1059	1051	1062	1056	1073
TRU-NB-1	1161	1166	1159	1157	1162	1155
TRU-ZN-1	988	978	987	1003	985	998
TRU-ZB-1	1094	1058	1093	1070	1088	1069
TRU-ANZ-1	1032	944	1030	1013	1030	1027
TRU-ABZ-1	1151	1125	1151	1156	1149	1191
TRU-NBZ-1	1010	990	998	1011	996	1011

Model A= first-order model fit to TRU single component variation data (e.g., without the nine new data points in Table I).

Table III. Component Ranges (in Mass Fractions) for Liquidus Temperature Data

	B	C	D	E
	TRU	TRU+CVS	TRU	TRU+CVS
Component	ZS	ZS	ZS&Z	ZS,Z&ZNS
Al_2O_3	0.019-0.040	0.000-0.103	0.019-0.075	0.000-0.105
B_2O_3	0.020-0.120	0.020-0.200	0.020-0.120	0.020-0.200
Bi_2O_3	0.000-0.100	0.000-0.100	0.000-0.100	0.000-0.100
CaO	0.009-0.011	0.000-0.100	0.008-0.011	0.000-0.100
CeO_2	0.000-0.030	0.000-0.030	0.000-0.030	0.000-0.030
Fe_2O_3	0.035-0.048	0.001-0.074	0.035-0.048	0.001-0.105
Li_2O	0.015-0.060	0.010-0.070	0.000-0.060	0.010-0.070
MgO	0.002-0.003	0.000-0.080	0.002-0.003	0.000-0.080
Na_2O	0.045-0.150	0.045-0.150	0.045-0.150	0.045-0.200
P_2O_5	0.007-0.050	0.000-0.050	0.007-0.050	0.000-0.050
SiO_2	0.410-0.570	0.410-0.570	0.410-0.570	0.410-0.570
ZrO_2	0.080-0.120	0.037-0.155	0.080-0.120	0.037-0.155
Others[1]	0.079-0.109		0.079-0.109	
Others[2]		0.010-0.096		0.010-0.096

ZS=zircon, Z=baddeleyite, and ZNS=sodium-zircoium-silicate
Others[1]=TRU data set (MgO, CaO, and Fe_2O_3 grouped with others).
Others[2]=TRU&CVS data sets combine (MgO, CaO, and Fe_2O_3 have individual coefficients).

Table IV. Component Coefficients for Second-Order Models

Component	Model B	Model C	Model D	Model E
	TRU	TRU&CVS	TRU	TRU&CVS
	ZS	ZS	ZS&Z	ZS,Z&ZNS
Al_2O_3	-9323	9699	-14681	30170
B_2O_3	1148	960	1114	-1498
Bi_2O_3	1519	981	1476	2566
CaO		668		1776
CeO_2	651	2161	461	3968
Fe_2O_3		2145		3851
Li_2O	-11776	-5056	-14867	-12657
MgO		1015		6645
Na_2O	-1224	-3688	-2393	-16751
P_2O_5	4338	2294	4135	4034
SiO_2	1481	1385	1436	1863
ZrO_2	2491	4949	6597	-787
Others[1]	2606		2971	
Others[2]		2042		3948
$Al_2O_3 \times Al_2O_3$			70141	
$B_2O_3 \times B_2O_3$		-2421		
$CeO_2 \times CeO_2$	81826		85074	
$Li_2O \times Li_2O$	50911	38261	64793	85368
$MgO \times MgO$		38653		-33260
$Na_2O \times Na_2O$		8108		27124
$P_2O_5 \times P_2O_5$	-26859		-24685	
$ZrO_2 \times ZrO_2$	47140		26205	13018
$Na_2O \times Al_2O_3$	127321		137617	-35923
$Na_2O \times Li_2O$	50076		67062	50227
$SiO_2 \times Al_2O_3$		-19210		-56424
$SiO_2 \times Na_2O$				17419
$ZrO_2 \times Al_2O_3$		31821		61544
$ZrO_2 \times B_2O_3$				31803
$ZrO_2 \times Na_2O$	-62497	-9243	-61691	
$ZrO_2 \times SiO_2$		7026		
R^2	.996	.990	.996	.977
# of data	36	50	40	59

ZS=zircon, ZZ=baddeleyite, and ZNS=sodium-zircoium-silicate

Others[1]=TRU data set with MgO,CaO,and Fe_2O_3 grouped into others.

Others[2]=TRU&CVS data sets combine MgO, CaO, and Fe_2O_3 have individual coefficients.

tested. T_Ls for nine glasses with components varied two- and three-at-a-time show that significant component interactions exist for T_L. Crossproduct terms added to mixture models improved the fit for each of the composition regions. However, a more systematic study is necessary to understand the interactive effects between components. Glasses precipitating baddeleyite and sodium zirconium silicate primary phases are needed to fit individual models to each primary phase. Present models can aid in formulating of high-zirconia glasses with high waste loadings.

ACKNOWLEDGMENTS

Jarrod Crum would like to thank Associated Western Universities, Inc., for his fellowship appointment at Pacific Northwest National Laboratory.

REFERENCES

1. P. Hrma, G. F. Piepel, M. J. Schweiger, D. E. Smith, D. S. Kim, P. E. Redgate, J. D. Vienna, C. A. LoPresti, D. B. Simpson, D. K. Peeler, and M. H. Langowski, *Property / Composition Relationships for Hanford High-Level Waste Glasses Melting at 1150°C* , PNL-10359, Vol. 1 and 2, Pacific Northwest Laboratory, Richland, Washington (1994).

2. J. V. Crum, M. J. Schweiger, P. Hrma, and J. D. Vienna, "Liquidus Temperature Model for Hanford High-Level Waste Glasses with High Concentrations of Zirconia," *Scientific Basis for Nuclear Waste Management Vol. XX*, edited by W. J. Gray and I. R. Triay, Vol 465, 79-86, 1996.

3. J. A. Cornell, *Experiments With Mixture: Designs, Models, and the Analysis of Mixture Data*, Second Edition, (John Wiley and Sons, New York, 1990).

4. D. S. Kim and P. Hrma, *Ceram. Trans.* 45, 327-337 (1994).

5. Q. Rao, G. F. Piepel, P. Hrma, and J. V. Crum, "Liquidus Temperatures of HLW Glasses with Zirconium-Containing Primary Crystalline Phases", *J. Non-Cryst. Solids*, (in press).

6. P. Hrma, "Empirical Models and Thermodynamic Constitutive Functions for High-Level Waste Glass Properties," *Ceram. Trans.*, this volume.

Compositional Effects on Liquidus Temperature of Hanford Simulated High-Level Waste Glasses Precipitating Nepheline ($NaAlSiO_4$)

Hong Li, Brian Jones, Pavel Hrma, and John D. Vienna
Pacific Northwest National Laboratory, Richland, Washington 99352

ABSTRACT

Borosilicate waste glasses with high concentrations of Al_2O_3 and Na_2O were shown to be susceptible to crystallization of nepheline ($NaAlSiO_4$) during slow cooling. Nepheline precipitation in high-level waste (HLW) glasses is generally detrimental to durability of the final waste form. To optimize glass formulations with achievable maximum waste loadings for Hanford HLW, a systematic study was conducted to evaluate compositional effects on liquidus temperature (T_L) of HLW glasses precipitating nepheline. A first-order mixture model was fitted to T_L data and compared with independently measured T_L data for other Hanford HLW glasses. This study showed that T_L is increased by components in the order $Al_2O_3 > Na_2O > K_2O > CaO$. On the other hand, T_L is decreased in the order $B_2O_3 > SiO_2 > Li_2O$. Fe_2O_3 and the mixture of other components had little effect on T_L within the composition region studied.

INTRODUCTION

Hanford (Richland, Washington) high-level radioactive tank wastes (HLW) generally have Al_2O_3 and Na_2O at elevated concentrations [1]. With increasing waste loading (WL) in HLW glasses to reduce total volume of the waste form for immobilization, nepheline precipitation during glass slow cooling in a canister becomes an important issue. A glass experiences a complex thermal history after it is poured, at the glass processing temperature, into a stainless steel canister. The glass cooling rate varies with its relative position in the canister; the lowest cooling rate can be expected along the canister center-line. Therefore, a heat-treatment simulating glass canister center-line cooling (CCC) is often used to study glass crystallization. A previous study showed that HLW glasses with high concentrations of Na_2O and Al_2O_3 were prone to nepheline formation during the CCC treatment [2]. The compositional effects on nepheline formation were successfully modeled using a simplified glass composition, i.e., a sub-mixture of Na_2O, Al_2O_3, and SiO_2 (NAS). This sub-mixture model [2] predicts that a HLW glass will be susceptible to nepheline formation during CCC treatment if its NAS submixture composition lies within the nepheline primary phase field defined by the Na_2O-Al_2O_3-SiO_2 phase diagram [3].

Nepheline formation in glass removes two moles of glass former, Al and Si, for every one mole of Na, which weakens the structure of the residual glassy phase resulting in lower corrosion resistance against water [2, 4-6]. For vitrification of high Na and Al HLW, several mitigation methods can be applied to suppress nepheline precipitation during canister cooling, namely, i) reducing Al by waste pretreatment, ii) decreasing waste loading, and iii) developing models for glass formulation, which can predict nepheline-forming compositional regions, concentrations of nepheline precipitated in the glass during canister cooling, and chemical durability of glass containing nepheline phase. To reduce the total costs and time for Hanford HLW cleanup, optimization of waste pretreatment and glass formulation is necessary. The composition submixture model was developed, which was capable of screening the glass composition region prone to nepheline formation [2]. Knowing the concentration of nepheline in the glass, chemical durability of the residual glassy phase was also predicted well by a second-order mixture model for glass durability [2]. The durability model was previously developed for Hanford HLW glasses [7]. To accurately predict the concentration of nepheline precipitated in the glass during cooling, both a liquidus temperature (T_L) model and a nepheline crystallization kinetics model are needed. In a crystallization kinetics model, T_L is also one of the important parameters [8]. This study focused on liquidus temperature model.

For a given CCC schedule, the effects of T_L on nepheline formation can be schematically illustrated in Figure 1. Glasses A and B are assumed to have liquidus temperatures such that $T_L^A >> T_L^B$; whereas their transition temperatures, T_g^A and T_g^B, are usually close to each other as is typically found in HLW glasses. In this case, the amount of nepheline precipitated during canister cooling should be much greater in glass A that has lower viscosity and higher T_L than in glass B that has higher viscosity and lower T_L. Therefore, it is advantageous to formulate HLW glasses with low T_L (assuming nepheline is the primary phase).

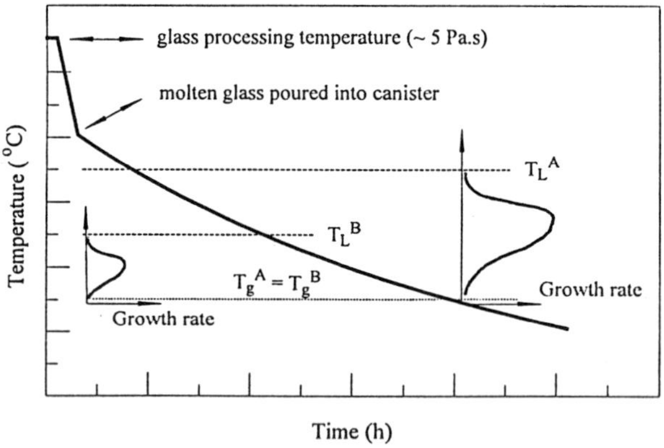

Figure 1. Schematic Diagram Showing Glass Cooling During a CCC Heat Treatment and Nepheline Precipitation in Two Glasses (A and B) with Different T_L.

In this study, 20 glasses were formulated around one baseline glass with 57 wt% simulated waste from Hanford single shell tank C106 [1]. These compositions (reported as the NP glasses here) have been previously studied for nepheline precipitation during CCC treatment [2]. The compositions of these glasses are shown in Table 1. Figure 2 shows the positions of the NP glasses in terms of the NAS submixture on the ternary Na_2O-Al_2O_3-SiO_2 phase diagram.

Table 1. Nominal Compositions of "Nepheline" (NP) Test Matrix of HLW Glasses

Glass	Components (wt%)										
NP-	SiO_2	B_2O_3	Na_2O	Li_2O	CaO	MgO	Fe_2O_3	Al_2O_3	ZrO_2	K_2O	Mix
BL	38.38	8.00	20.28	4.50	1.12	0.68	9.94	13.76	0.29	0.10	2.86
Si-3	45.31	7.10	18.00	3.99	0.99	0.60	8.82	12.21	0.26	0.09	2.54
Si-4	47.40	6.83	17.31	3.84	0.96	0.58	8.49	11.75	0.25	0.09	2.44
Al-1	41.70	8.69	22.03	4.89	1.22	0.74	10.80	6.30	0.32	0.11	3.11
Al-2	41.06	8.56	21.69	4.81	1.20	0.73	10.63	7.75	0.31	0.11	3.06
Al-3	36.64	7.64	19.36	4.30	1.07	0.65	9.49	17.68	0.28	0.10	2.73
Al-4	35.92	7.49	18.98	4.21	1.05	0.64	9.30	19.29	0.27	0.09	2.68
Na-1	41.74	8.70	13.30	4.89	1.22	0.74	10.81	14.96	0.32	0.11	3.11
Na-2	41.01	8.55	14.81	4.81	1.20	0.73	10.62	14.70	0.31	0.11	3.05
Li-1	40.19	8.38	21.23	0.00	1.17	0.71	10.41	14.41	0.30	0.10	3.00
Li-2	36.98	7.71	19.53	8.00	1.08	0.66	9.58	13.26	0.28	0.10	2.75
B-1	41.72	0.0	22.04	4.89	1.22	0.74	10.80	14.96	0.32	0.11	3.11
B-2	33.38	20.0	17.63	3.91	0.97	0.59	8.64	11.97	0.25	0.09	2.48
K-1	37.27	7.77	19.69	4.37	1.09	0.66	9.65	13.36	0.28	3.00	2.78
K-2	36.12	7.53	19.08	4.23	1.05	0.64	9.35	12.95	0.27	6.00	2.69
Ca-1	38.82	8.09	20.51	4.55	0.00	0.69	10.05	13.92	0.29	0.10	2.89
Ca-2	34.94	7.28	18.46	4.10	10.0	0.62	9.05	12.52	0.26	0.09	2.60
Fe-1	40.12	8.36	21.20	4.70	1.17	0.71	5.95	14.38	0.30	0.10	2.99
Fe-2	39.27	8.19	20.75	4.60	1.15	0.70	7.95	14.08	0.30	0.10	2.93
Fe-3	37.14	7.74	19.62	4.35	1.08	0.66	12.95	13.31	0.28	0.10	2.76

Note: Mix is composed of Ag_2O, BaO, Cr_2O_3, MnO, NiO, P_2O_5, PbO, Sb_2O_3, SeO_2, and TiO_2.

MODELING APPROACH

Many properties of HLW and silicate glasses have been successfully described as first-order interpolations. These properties include melt viscosity [7,9,10], electrical conductivity [7, 9,10], liquidus temperature of glass precipitating zirconium-containing crystals [9] or spinel [11], and glass corrosion in water [7,12]. The results of this study, which will be discussed in detail later, showed that over a narrow composition region within the nepheline primary phase field the component effects on T_L of glass precipitating nepheline were more or less of the first-order. That is, T_L changed linearly as the concentration of a component of interest varied. Therefore, first-order mixture model for T_L was adopted in this study. The model has a general form:

$$T_L = \sum_{i=1}^{n} b_i g_i \qquad (1)$$

where g_i is the i-th oxide component mass fraction in glass and b_i is the first-order coefficient for the i-th oxide.

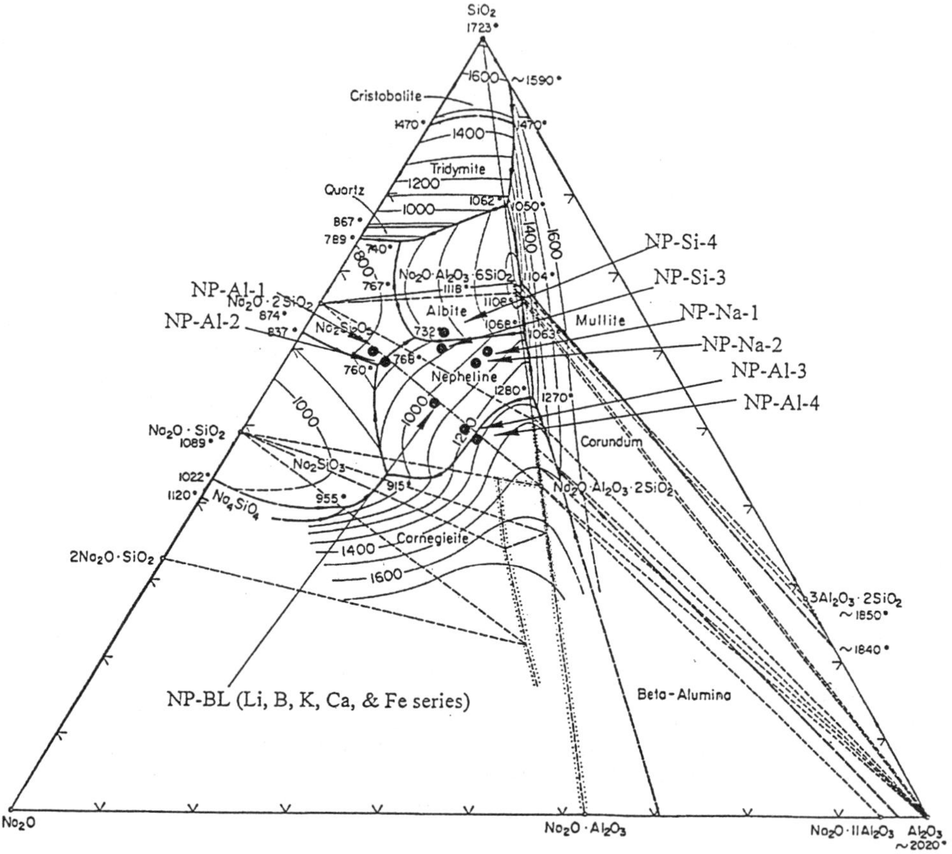

Figure 2. NAS Submixtures of NP Glasses of This Study [3].

Single component effects on T_L were evaluated by varying the concentration of that component in glass and keeping the concentrations of all other components at constant relative proportions as in the baseline glass. A specific property or the effect of a given component on glass property can be expressed as $e_i = \partial T_L(g_i)/\partial g_i$. For the existence of a first-order relationship between T_L and g_i, the effect of the i-th component on T_L of glass is described as [12, 13]:

$$e_i=[\partial T_L(g_{i,BL})/\partial g_i]=[b_i-T_L(g_{i,BL})]/[1-g_i] \qquad (2)$$

where $g_{i,BL}$ is the mass fraction of the i-th component in the baseline glass, and $T_L(g_{i,BL})$ is the estimated liquidus temperature of the baseline (BL) glass. By rearranging equation (2), the coefficient b_i can be calculated from:

$$b_i=T_L(g_{i,BL})+e_i(1-g_i) \qquad (3)$$

In Equation (3), e_i is obtained from:

$$T_L(g_i)=T_{Lo,i}+e_i g_i \qquad (4)$$

where $T_{Lo,i}$ is the intercept or the extrapolated liquidus temperature of glass without the i-th component. The coefficient, b_{Mix}, for the components not being directly measured can be calculated from Equation (1):

$$b_{Mix}=[T_L^{ave}(g_{i,BL})-\sum_{i=1}^{n-1} b_i g_{i,BL}]/g_{Mix} \qquad (5)$$

In Equation (5), $T_L^{ave}(g_{i,BL})$ is defined as:

$$T_L^{ave}(g_{i,BL})=[\sum_{i=1}^{n-1} (SD)_i^{-2}]^{-1}\sum_{i=1}^{n-1} T_L(g_{i,BL})(SD)_i^{-2} \qquad (6)$$

where $(SD)_i$ is the standard deviation with a 95% confidence interval for $T_L(g_{i,BL})$.

In addition to the above method for determining b_i values, multicomponent regression method, in a form of Equation (1), is also used. This method has an advantage of determining b_{Mix} independently, i.e., b_{Mix} is not dependent on the b_i and g_i values for other oxides. The b_i values that are determined by both methods are presented in this paper.

EXPERIMENTAL

Glass samples were from the lots made previously for the study of nepheline precipitation in CCC treated NP glasses [2] and the glass fabrication procedure is discussed in detail elsewhere [2]. To determine T_L of the NP glasses, the method of isothermal uniform temperature treatment was used. Cutlets (~1.5 g for each sample) of quenched glass were packed in a Pt-10%Rh box (12 mm × 12 mm × 6 mm) with a cover. Each heat treatment was conducted in a uniform temperature furnace for at least 24 hours at a target temperature. After heat treatment, the samples were air cooled to room temperature. Thin sections of treated samples were prepared and examined under an optical microscope. To determine the accuracy of the T_L measurements, NBS 773 glass was used as a liquidus temperature standard. The certified T_L for NBS 773 glass was 988±3°C, and the measured T_L for the same glass by the above method in our laboratory was 992.5±1.5°C. For the NP glasses, T_L was determined by the temperature that was 10°C higher than the maximum temperature at which a few nepheline crystals were found in the heat-treated glass sample. Therefore, the experimental uncertainty, as well as accuracy, of T_L in this study should be within ±5°C.

RESULTS AND DISCUSSION

Figure 3 shows the morphology of nepheline crystals found in the NP-BL glass treated at 910°C for 24 hours. The crystalline phase was previously identified by x-ray diffraction analysis and scanning electron microscopy [2]. Nepheline crystals were found in all NP glasses after 24-hour isothermal heat treatment at temperatures below liquidus. In terms of the NAS submixture, the results of this study demonstrated that the boundary lines of the nepheline primary phase field that is described by the Na_2O-Al_2O_3-SiO_2 ternary system are expanded slightly for the compositions studied. Hence, to improve the NAS model, further study is necessary to determine these boundary lines of nepheline phase field in multicomponent HLW glasses.

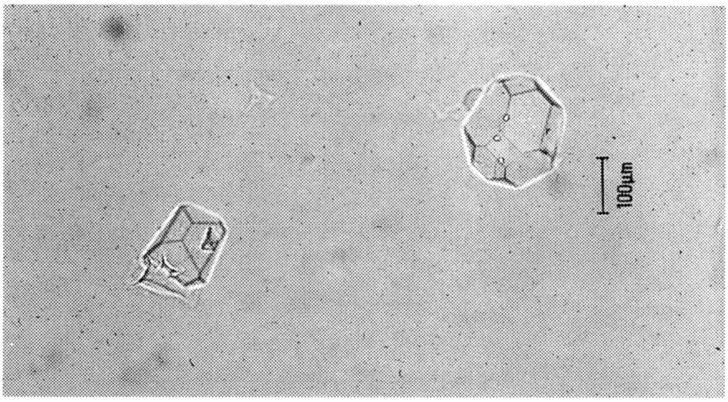

Figure 3. Transmission Optical Microscopy, Showing Nepheline Crystals in the NP-BL Glass Treated at 910°C (10°C below T_L) for 24 Hours.

The measured T_L versus the concentration of a single component of interest in glass is plotted in Figure 4. As shown in the figure, the single component effects on T_L are approximately linear under the given composition regions investigated for Na_2O, Al_2O_3, SiO_2, B_2O_3, Fe_2O_3, Li_2O, K_2O, and CaO. The linear regression analysis for each oxide of interest yields a regression coefficient, r^2, equal or greater than 0.90 with the exception of Fe_2O_3 which seemed to have little effect on T_L of the baseline glass, NP-BL.

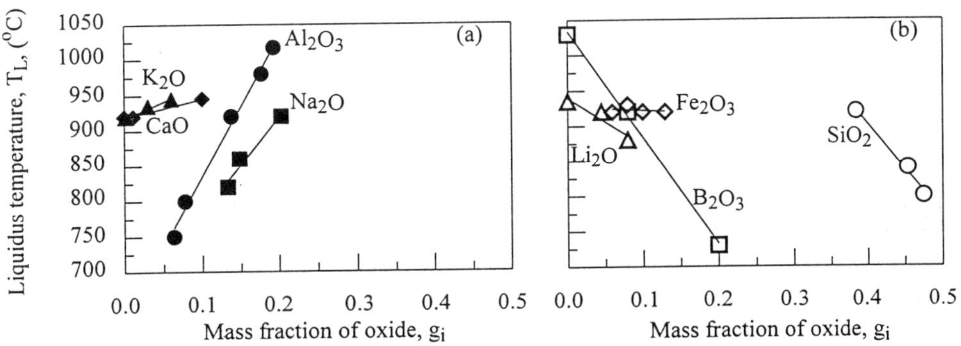

Figure 4. Liquidus Temperature, T_L, of the NP Glasses Versus Mass Fraction of (a) Al_2O_3, Na_2O, K_2O, and CaO and (b) SiO_2, B_2O_3, Li_2O, and Fe_2O_3.

Accordingly, linear regression analysis of data in Figure 4 resulted in $T_{L0,i}$ and e_i values (cf. Equation 4) for each oxide. A set of first-order model coefficients, b_i, was then calculated (cf. Equation 3 and 5). Multi-regression analysis was also performed using the same data set and a second set of the model coefficients, b_i, were obtained. Table 2 summarizes the results of regression analysis using both methods.

Table 2. Component Effects (e_i) and First-Order Mixture Model Coefficients (b_i) of Liquidus Temperature of the NP Glasses Precipitating Nepheline[(a)]

Method	Linear Regression				Multi-component Regression
Oxide	$T_{L0}(g_i)$ (10^3 °C)	e_i (10^3 °C)	r^2	b_i (10^3 °C)	b_i (10^3 °C)
Al_2O_3	0.637 ± 0.014	1.974 ± 0.099	0.993	2.611 ± 0.113	2.629 ± 0.085
B_2O_3	1.034 ± 0.007	-1.507 ± 0.057	0.999	-0.473 ± 0.064	-0.471 ± 0.074
CaO	0.919 ± 0.002	0.262 ± 0.027	0.990	1.181 ± 0.029	1.255 ± 0.130
Fe_2O_3	0.927 ± 0.011	-0.047 ± 0.113	0.079	0.880 ± 0.088	0.865 ± 0.199
K_2O	0.921 ± 0.002	0.423 ± 0.053	0.985	1.344 ± 0.055	1.496 ± 0.186
Li_2O	0.940 ± 0.012	-0.67 ± 0.229	0.896	0.269 ± 0.241	0.287 ± 0.193
Na_2O	0.650 ± 0.045	1.343 ± 0.274	0.960	1.988 ± 0.319	1.874 ± 0.111
SiO_2	1.415 ± 0.063	-1.286 ± 0.142	0.988	0.129 ± 0.205	0.191 ± 0.067
Mix	N/A	N/A	N/A	0.830	0.544 ± 1.034

(a) The values with ± sign for $T_{L0}(g_i)$ and e_i represent standard deviations with 95% confidence intervals based on statistical *t*-distribution.

Applying the first-order mixture model to the NP glass compositions in Table 1, the predictions of liquidus temperature, T_L^{P1} by the model with the first set of b_i values and T_L^{P2} by the same model with the second set of b_i values, were made. Table 3 summarizes both the measured and the predicted liquidus temperatures as well as the difference between the two. In general, the model with the second set of b_i coefficients predicted T_L slightly better than the model with the first set of b_i coefficients. The difference between T_L and T_L^{p1} was as large as 25°C, whereas the difference between T_L and T_l^{p2} was within 15°C. Therefore, the second set of b_i coefficients is recommended for the model. Using the model, glass compositional effect on T_L is illustrated in Figure 5, namely, T_L is increased by components in the order Al_2O_3 > Na_2O > K_2O > CaO, and it decreases in the order B_2O_3 > SiO_2 > LiO_2. Fe_2O_3 and the mixture of the remaining components had little effect on T_L within the composition region investigated. However, further study is necessary to assess the effects of component interactions on T_L of HLW glasses precipitating nepheline. For a wider HLW glass composition region, a second-order mixture model may be required to predict T_L more accurately.

Table 3. Liquidus Temperatures of the NP Glasses

Glass NP-	Measured T_L (°C)	Predicted T_L^{P1} (°C)	Difference (T_L - T_L^{P1})	Predicted T_L^{P2} (°C)	Difference (T_L - T_L^{P2})
BL	920	921	-1	913	7
Si-3	840	832	8	831	9
Si-4	800	805	-5	807	-7
Al-1	750	775	-25	764	-14
Al-2	800	803	-3	793	7
Al-3	980	998	-18	991	-11
Al-4	1017	1029	-12	1023	-6
Na-1	820	827	-7	828	-8
Na-2	860	848	12	847	13
Li-1	935	952	-17	942	-7
Li-2	880	897	-17	890	-10
B-1	1030	1042	-12	1033	-3
B-2	730	739	-9	732	-2
K-1	935	933	2	930	5
K-2	945	946	-1	947	-2
Ca-1	920	918	2	909	11
Ca-2	945	944	1	943	2
Fe-1	920	923	-3	914	6
Fe-2	930	924	6	916	14
Fe-3	920	921	-1	912	8

The predictability of the first-order mixture model was further evaluated using preexisting T_L data of other HLW glasses, which were determined using an isothermal gradient temperature furnace [8]. The reported T_L values were 736, 961, and 1015°C for CVS2-34, CVS2-35, and CVS2-86, respectively (CVS2 stands for composition variation

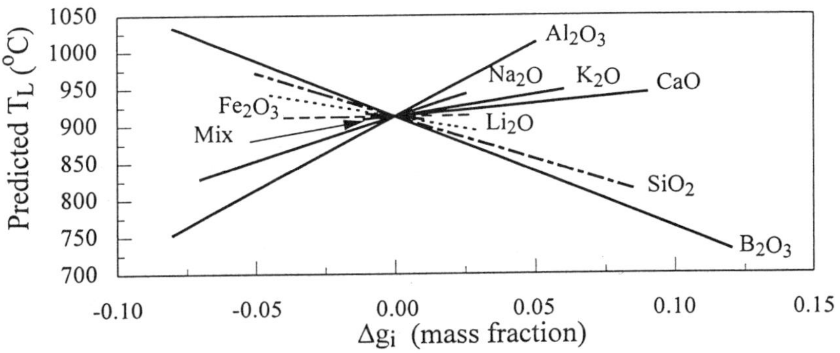

Figure 5. Glass Compositional Effects on T_L by First-Order Mixture Model (The second set of b_i values was used in the model).

study of Hanford HLW glasses in phase II). Correspondingly, the predicted values (T_L^{P2}) were 751, 956, and 935°C. The agreement is very good for two of the three samples. The data from the high temperature sample is suspected because at higher temperatures, T_L measurement is known to suffer from crystal drifting. Crystals drift from temperatures below T_L to higher temperatures and hence, the apparent liquidus temperature was higher due to slow crystal dissolution kinetics in the melt. This phenomenon is caused by surface tension driven melt convection possibly due to the evaporation of B or Na from the melt surface at the high temperature end [14,15] when the gradient temperature furnace treatment was used.

To confirm the effect of melt convection on T_L, NP-Al-2 and NP-Al-4 glasses were also heat-treated in a temperature gradient furnace for 72 hours. T_L values determined for these two glasses were 803 and >1076°C (it was the cut off working temperature of the furnace), respectively. For NP-Al-2, T_L values determined by the two methods were the same within experimental error, whereas for NP-Al-4 glass, T_L measured by the gradient furnace method was at least higher by 59°C, which confirms the effect of melt convection on T_L determined by the gradient furnace method at high temperature.

CONCLUSION

A first-order mixture model was fitted to T_L data of 20 HLW glasses precipitating nepheline. The differences between the measured and the predicted values were less than 15°C. T_L is increased by components in the order $Al_2O_3 > Na_2O > K_2O > CaO$. On the other hand, T_L is decreased by components in the order $B_2O_3 > SiO_2 > Li_2O$. Fe_2O_3, and the mixture of other components had little effect on T_L within the composition region studied.

The results of nepheline formation in the isothermal heat-treated glasses demonstrated that the previously proposed NAS model is generally applicable for screening

HLW compositions that are prone to nepheline formation. However, the boundary lines of the nepheline phase field were shifted for HLW glass compositions. The determination of these boundary lines for HLW glass compositions will further improve the NAS model predictability.

ACKNOWLEDGMENT

This work is supported by the U.S. Department of Energy under Contract DE-AC06-76RLO 1830. Discussions on statistical mixture models with G.F. Piepel and the preparation of the gradient furnace treated samples by D.E. Smith at Pacific Northwest National Laboratory are greatly appreciated.

REFERENCES

[1] S.L. Lambert, G.E. Stegen, and J.D. Vienna, WHC-SD-WM-TI-768 Rev.0, Westinghouse Hanford Company, Richland, WA, 1996.

[2] H. Li, J.D. Vienna, P. Hrma, D.E. Smith, and M.J. *Proc. Mat. Res. Soc.* Vol. 465, ed. by W. Gray and I. Triay (MRS, Pittsburgh, PA, 1997) p. 261.

[3] *Phase Diagrams for Ceramists*, Vol.1, 5th Printing, ed. by E.M. Levin, C.R. Robbins, and H.F. McMurdie (Am. Ceram. Soc. Columbus, OH, 1964) p.181.

[4] C.M. Jantzen and D.F. Bickford, *Proc. Mat. Res. Soc.*, Vol. 44, ed by C.M. Jantzen, J.A. Stone, R.C. Ewing (MRS, Boston, MA, 1980) p.135.

[5] P. Hrma and A.W. Bailey, *Rad Waste Manag. Env. Remed.* Vol. 1, ed. by S. Slate, F. Feizollahi, and J. Creer (Am. Soc. Mech. Engr., New York, 1995) p. 447.

[6] D.S. Kim, D.K. Peeler, and P. Hrma, *Ceramic Transaction*, Vol. 61, ed. by V. Jain and R. Palmer (Am. Ceram. Soc., Westville, Ohio, 1995) p.177.

[7] P. Hrma, G.F. Piepel, M.J. Schweiger, D.E. Smith, D.S. Kim, P.E. Redgate, J.D. Vienna, C.A. LoPresti, D.B. Simpson, D.K. Peeler and M.H. Langowski, PNL-10359, Pacific Northwest National Laboratory, Richland, WA (1994).

[8] J.G. Renolds and P. Hrma, *Proc. Mat. Res. Soc.*, Vol. 465, ed. by W. Gray and I. Triay (MRS, Pittsburgh, PA, 1997) p. 65.

[9] P. Hrma, G.F. Piepel, M.J. Schweiger, and D.E. Smith, *High Level Radioactive Waste Management* (Am. Nucl. Soc., La Grange Park, Illinois, 1992) p.1236.

[10] H. Scholze, Glass Nature, Structure, and Properties, Chapter 3, (Springer-Verlag, New York, 1991) p.156.

[11] J.V. Crum, M.J. Schweiger, P. Hrma, J.D. Vienna, D.E. Smith, *Scientific Basis for Nuclear Waste Management, XX*, in Proc. Mat. Res. Soc.Vol. 465, ed. by W. Gray and I. Triay (MRS, Pittsburgh, PA, 1997) p. 79.

[12] M. Mika, M.J. Schweiger, P. Hrma, and J.D. Vienna, *Proc. Mat. Res. Soc.*, Vol. 465, ed. by W. Gray and I. Triay (MRS, Pittsburgh, PA, 1997) p. 71.

[13] Cox D.R., *Biometrikc*, 58, 155-159 (1971).

[14] J.T. Wenzel and D.M. Sanders, *Phys. Chem. Glasses*, 23, 47-52 (1982).

[15] W.J. Gray, *Rad. Waste Manag.*, 1, 147-169 (1980).

High-Level Waste Processing and Vitrification

HIGH-LEVEL WASTE PROCESSING AT THE SAVANNAH RIVER SITE: AN UPDATE (U)

J.E. Marra, W.M. Bennett, H.H. Elder, E.D. Lee, S.L. Marra and P.L. Rutland
Westinghouse Savannah River Company, Aiken, SC 29808

ABSTRACT

The Defense Waste Processing Facility (DWPF) at the Savannah River Site (SRS) in Aiken, SC began immobilizing high-level radioactive waste in borosilicate glass in 1996. Currently, the radioactive glass is being produced as a "sludge-only" composition by combining washed high-level waste sludge with glass frit. The glass is poured into stainless steel canisters which will eventually be disposed of in a permanent, geological repository. To date, DWPF has produced about 100 canisters of vitrified waste. Future processing operations will be based on a "coupled" feed of washed high-level waste sludge, precipitated cesium, and glass frit. This paper provides an update of the processing activities completed to date, operational/flowsheet problems encountered, and programs underway to increase production rates.

INTRODUCTION

Approximately 130 million liters of high-level radioactive waste, currently stored in underground carbon steel tanks at the Savannah River Site (SRS), will be immobilized in stable borosilicate glass in the Defense Waste Processing Facility (DWPF). The glass is poured into stainless steel canisters for eventual disposal in a geologic repository.

Following completion of an extensive Startup Test Program, the DWPF initiated radioactive operations in 1996. The start-up program included the production of seventy-one canisters of simulated waste glass and demonstrated the DWPF's ability to comply with the Waste Acceptance Product Specifications (WAPS). Successful completion of the qualification program was instrumental in obtaining approval from the Secretary of the Department of Energy to initiate radioactive operations. Initial operations in the DWPF are based on a "sludge-only" glass composition where high-level waste sludge is mixed with glass frit and appropriate chemicals to produced the vitrified waste form. To date, over 100 canisters of radioactive glass have been produced in the DWPF.

Efforts are also underway to decontaminate high-level waste salt solutions by precipitation of Cs-137 using sodium tetraphenylborate (NaTPB). Plant operations in this area began in late-1995 and a significant decontamination factor (DF) was achieved (Cs DF~40,000). Progress in this area has been slowed by issues associated with the degradation of the tetraphenylborate (TPB^-) precipitating agent to form benzene. Extensive facility modifications are underway to address the issues associated with formation and release of benzene. In addition, a thorough laboratory test program has been initiated to better understand the mechanism(s) and catalysts for TPB^- degradation.

PROCESS/PRODUCT OVERVIEW

The radioactive waste in the SRS Tank Farms has been separated into a water soluble salt solution and saltcake, and an insoluble sludge of metal hydroxides and oxides. The salt solution and saltcake are decontaminated for disposal as low-level radioactive waste by removing the radionuclides by precipitation and sorption. This process is known as In-Tank Precipitation (ITP). The ITP facility was designed to separate the liquid waste into decontaminated salt solution (DSS) and high activity precipitate streams in preparation for treatment at downstream facilities. The ITP utilizes three underground waste tanks to separate the radioactive constituents from the remainder of the liquid waste in a combination of adsorption, precipitation, and filtration techniques.

Liquid waste, in the form of a salt solution matrix, is transferred via underground lines to the ITP reaction vessel (Tank 48). The Na^+ molarity of the salt can vary from about 6-18 M. Once inside Tank 48, this volume of waste (approximately 2.4 million liters) is diluted to about 5 M Na^+. Sodium Tetraphenylborate (STPB) is added to precipitate the ^{137}Cs and K^+. Mono-sodium Titanate (MST) removes strontium through ion exchange while adsorbing uranium and plutonium. The tank is then agitated by four long-shafted slurry pumps to fully react the contents, and then filtered to concentrate the precipitate to ~10 weight percent. This cycle is repeated until there are about 640,000 liters of the concentrated slurry. At this point the concentrated slurry is washed to reduce the Na^+ concentration to 0.2 M, a requirement of the vitrification process. After it is washed, the concentrated slurry is transferred to Tank 49 for storage until it is transferred to the Late Wash Facility for additional washing and ultimately to DWPF for immobilization.

The sludge also contains soluble components such as sodium nitrate, sodium carbonate, sodium aluminate, and sodium hydroxide in the interstitial liquid. Also contained in the sludge are most of the fission products except cesium, activation products and actinides. Cesium is quite soluble in highly alkaline solutions. The sludge portion of the waste is washed to remove the soluble salts. If necessary, insoluble aluminum is removed through high temperature caustic dissolution. This process is known as Extended Sludge Processing (ESP) and is also performed in underground carbon steel storage tanks that were originally designed for HLW storage. To remove the sludge from the waste tanks, water or dilute waste is added to the tank and the sludge is suspended using long-shafted slurry pumps. After slurrying, the slurry is pumped to one of the tanks dedicated for sludge processing. The ESP tanks have been retrofitted with mixing and transfer pumps capable of suspending and transferring the highly viscous sludge solution.

In the ESP tanks, the soluble salts (e.g., sodium nitrate and nitrite) are removed from the sludge by washing with water. This operation reduces the quantity of glass produced when the sludge is vitrified into waste glass. The ESP operates on large batches of sludge. Each washed batch supplies the DWPF for a 2-6 year period. Washing dilutes the interstitial liquid containing dissolved salts. First, washwater (with low salt content) is added to the tank and the sludge solids are suspended using slurry pumps for agitation. Then the sludge solids are allowed to settle and the diluted supernate is decanted from the tank. This cycle is repeated until the salt content of the interstitial supernate is sufficiently low. To control nitrate induced corrosion, corrosion inhibitor salts (i.e., NaOH and $NaNO_2$) must be added along with the washwater.

The allowable salt content of the sludge is limited by the DWPF feed design bases. The feed design bases ensures the molten glass has a satisfactory rheology (i.e., by control of $[Na^+]$) and meets the limits on glass insolubles and corrosive species (e.g., Cl^- and $SO_4^=$). the feed design bases also ensures that DWPF meets the air emissions permit for NO_x (i.e., NO_2^-, NO_3^-, OH^-). The allowable salt content is affected by some insoluble species present in the sludge that contribute to the total sodium concentration.

The insoluble solids content is controlled by the extent that the sludge is allowed to settle for the final decant operation. If the wt% solids is too high, then the slurry's rheology could make it difficult to pump to the DWPF. As the wt% solids decreases, then the DWPF requires additional time to process the same amount of solids (i.e., more boiling is required to remove excess water). The washing operation is accomplished by adding water inhibited with NaOH and $NaNO_2$ to the tank. Mixing is accomplished using four long-shaft slurry pumps. The wash water (containing the soluble salts) is transferred out of the tank and fresh water brought in. The washing-decant operation is continued until the concentration of soluble salts in the sludge meets the criteria established by DWPF.

Thus, the radioactive waste from the SRS Tank Farms is transferred to the DWPF in two forms: precipitate slurry and sludge slurry. The waste is then processed and blended in the DWPF before it is vitrified, poured into canisters, sealed, and placed in interim storage. Figure 1 provides a schematic of the waste treatment and vitrification process.

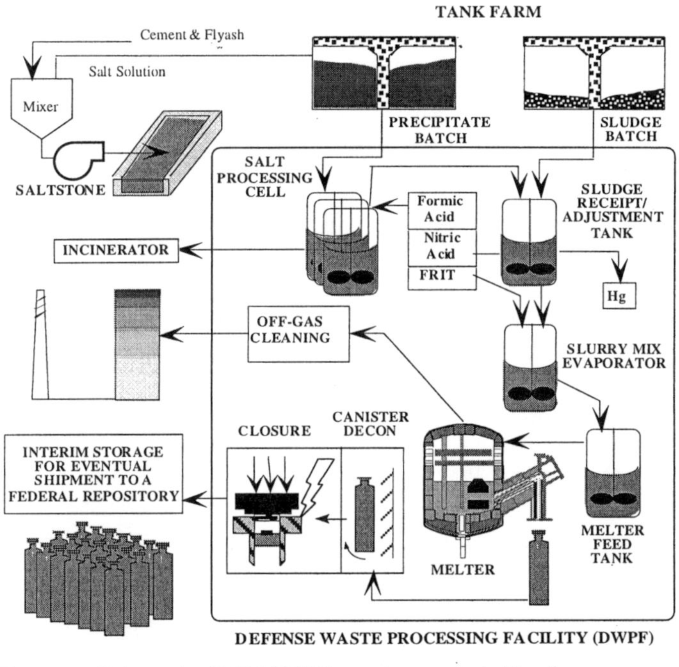

Figure 1 - Schematic of SRS HLW treatment and vitrification process.

The sludge is transferred directly into the Sludge Receipt and Adjustment Tank (SRAT) while the precipitate must first be processed in the DWPF Salt Process Cell to remove most of the organic material. The tetraphenylborate compounds in the precipitate react in the presence of formic acid and copper (II) catalyst. The products of this reaction are aromatic organic compounds (benzene, phenol, and minor amounts of higher boiling aromatics) and an aqueous phase known as Precipitate Hydrolysis Aqueous (PHA). The PHA contains the cesium, soluble formate salts, boric acid and excess formic acid.

The sludge is neutralized with nitric acid in the SRAT. The PHA is then added to the sludge (at boiling). The excess formic acid in the PHA reduces the mercuric oxide in the sludge to elemental mercury. The elemental mercury is then steam stripped from the SRAT into a holding tank from which it is later pumped and decontaminated. After the PHA and sludge are blended and processed in the SRAT, this SRAT product is transferred to the Slurry Mix Evaporator (SME) where a borosilicate glass frit is added and the slurry is concentrated to produce melter feed. The amount of sludge and PHA to be blended in the SRAT and the amount of SRAT material and frit to be blended in the SME is determined by the Product Composition Control System (PCCS).[1] The PCCS is a computer program that uses glass property models and statistical algorithms to develop blending strategies and to determine the acceptability of the melter feed before it is transferred to the melter. The PCCS uses lab analyses of tank contents, tank volumes and existing tank heel volumes to determine an appropriate tank blending region. It chooses a target that minimizes the amount of frit to be added in the SME. However, any point in the acceptable blending region can be chosen by DWPF engineers as the blending target. PCCS is also used to predict SRAT and SME compositions based on existing tank compositions, tank transfers, and the uncertainties associated with sampling and processing. These predicted compositions for the SRAT and SME can be used to diagnose processing problems.

The SME is the hold point in the process where feed acceptability is determined. The analyses of samples from the SME are used by the PCCS to determine the acceptability of the melter feed. Once the melter feed material in the SME is determined to be acceptable, it is transferred to the Melter Feed Tank (MFT) and then fed to the joule heated melter. The DWPF melter has two pairs of electrodes. The feed slurry is introduced from the top of the melter and forms a crust, or cold cap, on the surface of the melt pool as the water is evaporated and removed via the off-gas system. The cold cap melts from the bottom and forms the borosilicate glass matrix. The nominal glass melt pool temperature is 1150°C. The mixing behavior of the melter was evaluated during Waste Qualification Runs as described later. The glass is removed from the melter near the bottom through a riser and pour spout. A vacuum is drawn on the pour spout to pour the glass. A glass pour stream sample is taken occasionally during filling of a canister to confirm that the glass durability (as determined by the Product Consistency Test[2]) is acceptable. After a canister is filled, a temporary seal is installed to prevent free liquid from entering the canister during the decontamination process. Decontamination of the canister surface consists of blasting an air-injected frit slurry against the canister. The frit slurry from the decontamination process is used in the next SME batch as part of the required frit addition. The canister is then welded closed and transferred to an interim storage building. The DWPF canistered waste form contains approximately 1800 kg of glass. It is 300 cm in length and 61 cm in diameter.

STATUS OF PROCESS OPERATIONS

In-Tank Precipitation (ITP)

Initial operation of the ITP facility began on September 2, 1995 with the first addition of STPB to Tank 48. Between September 2 and September 29, 1995, ~150,000 liters of 0.55 M STPB were added to Tank 48. On September 30, 1995 8,400 liters of MST were added. The Tank 48 slurry pumps were operated and the tank was sampled to verify the decontamination of the salt solution. The decontamination was successful. The resulting Cs-137 concentration in the DSS was < 1 nCi/g.

Tests were scheduled for the ITP startup that were focused on measuring the mixing characteristics. Sodium tetraphenylborate when exposed to radiation and other factors will decompose to form benzene. This benzene presents a flammability hazard in the tanks where STPB is stored. The ITP reaction tank (48) and precipitate storage tank (49) are operated under a slight negative pressure with a nitrogen purge so that during normal operation the oxygen concentration is maintained at approximately 5 volume percent. The tests were a part of a program to reduce uncertainties associated with calculations for composite lower flammability limit (CLFL) conditions in the ITP tanks. Two vapor space sampling poles were installed in the vapor space on opposite sides of Tank 48 to sample for benzene, hydrogen, and oxygen. These poles were able to sample at 8 different levels in the tank.

During October and early November the slurry was filtered down to approximately 610,000 liters. Tests were conducted during and following the concentration. During these tests the slurry pumps were operated for several days while the Tank 48 vapor space was monitored for benzene. In all of these tests, more benzene than could be accounted for by the accepted generation mechanisms and rates of formation was evolved into the vapor space. Two subsequent loss of ventilation tests were completed in Tank 48 with the liquid phase temperatures elevated to 40 °C and approximately 52 °C. These tests were completed in late November.

On December 1, 1995 all four slurry pumps were operated in order to pull a sample of Tank 48. The pumps were run for approximately 3.5 hours. During this time a sharp rise in the vapor space benzene concentration (2000 ppm) was observed and far exceeded the values that had been seen previously during Batch 1 operations. The increase in the benzene concentration resulted in a halt to operations at ITP while the benzene generation questions are resolved. A sample of the tank pulled on December 18 indicated that over 95 % of the excess STPB (11,012 kg) added to Tank 48 had decomposed.

A program was developed to deplete the benzene inventory in Tank 48 by controlled operation of the slurry pumps. This program was completed in early May of 1996. During this time approximately 8,500 kg of benzene were removed from Tank 48.

The Savannah River Technology Center (SRTC) initiated research to investigate the unexpected rapid decomposition of STPB. The preliminary results of this investigation indicate that the decomposition reaction is a complex set of related reactions. The initial reaction of TPB^- ion forms tri-phenyl boron which subsequently degrades to Diphenyl borinic acid (DPBA). The DPBA decomposes further to phenyl boric acid (PBA) which decomposes to benzene or phenol. Copper is a known catalyst for the decomposition of TPB^-. However, the rate for decomposition of TPB^- seems to be higher than for copper alone.

In May the Defense Nuclear Facilities Safety Board (DNFSB) issued Recommendation 96-1.[3] This recommendation requires that prior to resuming normal operations at the ITP facility, a thorough understanding of the decomposition reaction kinetics be reached. A Response Plan to this recommendation was developed and has been implemented.[4] The response plan includes research and development, plant modifications, flow-sheet modifications, and in-plant testing.

One major milestone of the response plan was recently completed. A Process Verification Test (PVT) was completed that demonstrated with low amounts of excess STPB the salt solution can be decontaminated long enough to allow operation of downstream facilities. Approximately 1,200 liters of 0.59 M STPB was added to Tank 48 to precipitate the Cs-137 that had built back into the system due to lack of TPB^- in solution and radiolytic decomposition of the solid precipitate. The Cs was successfully recovered and remained at levels that would have allowed the DSS to be processed in the saltstone facility. In addition, by controlling temperature and the concentration of TPB^- in solution, only small amounts of benzene were formed in Tank 48. This testing was completed in December 1996.

Extended Sludge Processing (ESP)
As discussed above, HLW sludge is prepared for vitrification in the DWPF using the Extended Sludge Processing (ESP). This first batch of sludge was washed five times to remove the soluble salts.

Sludge Batch 1a: The first sludge sent to the DWPF, Batch 1a, consists primarily of 23 year old aged PUREX-LAW (Low Activity Waste) [approx. 93 wt%] from processing target assemblies in the F-area separations facilities and 33 year old aged HM-LAW waste (approx. 7 wt%) from the processing of fuel assemblies in H-Modified separations facilities in H-area.

In 1987, the waste was transferred from their respective tanks to Tank 51, an ESP tank. Sludge washing began in 1987 and continued during 1988. During the 1988 washes, contaminated water from the slurry pumps' bearing water column leaked on to the tank top resulting in the spread of contamination. Sludge washing was suspended between 1988 and 1993 so that additional containment could be designed and installed around the portion of the slurry pumps which extends above the tank top.

In 1993, one additional wash cycle was completed. During the 1993 wash campaign, two of the slurry pumps deteriorated to such extent that they could no longer be operated and the remaining two slurry pumps were restricted to operate at reduced speeds. In 1994 and 1995 two of the four slurry pumps that failed during the 1993 wash were replaced. The final wash for the sludge was performed during the last quarter of 1995.

The primary problems encountered during washing were the mechanical problems with the tank's four slurry pumps. These pumps are in a very difficult service partially a result of the requirement for a long vertical shaft length (45 ft), the initial sludge concentration (approx. 35 wt% insoluble solids for settled sludge) and the need for high pumping rates (> 16,000 liters/minute). Although the pumps had undergone testing at a non-radioactive test facility, a number of unforeseen problems occurred when these new pumps were used for the first time in the waste tank. Examination of one of the failed pumps as well as examination of disassembled replacement pumps, after pumping tests, revealed manufacturing and design deficiencies. The primary problem was identified to be dynamic hydraulic imbalance resulting in excessive deflection of the impeller and pump shaft. These problems have now largely been resolved by requiring extremely tight machining tolerances for the pumps' impellers and providing additional modifications to stabilize the pumps' long shafts. Now, three out of the four original slurry pumps have been replaced with modified pumps. These replacement pumps appear to be operating satisfactorily.

Final Sludge Composition: The composition of the final washed slurry for Batch 1a is shown in Table 1.

Table 1 Final Sludge Batch 1a, Major Components (> 1 wt%)

Species	Unit	Result
Fe	wt% dried solids	24.6
Na	wt% dried solids	8.74
Al	wt% dried solids	6.39
U	wt% dried solids	3.42
Mn	wt% dried solids	2.53
Ca	wt% dried solids	2.38
Mg	wt% dried solids	1.16
Insoluble Solids	wt% of slurry	12.8
Total Solids	wt% of slurry	16.8

Slurry Transfers to the DWPF: Transfers of washed sludge are performed in batches of approximately 24,000 liters through a jacketed underground transfer line. The transfer is in two steps. First the sludge in the waste tank is slurried using four slurry pumps. Then the slurry is pumped using a long shafted centrifugal pump to an intermediate transfer tank, approximately 950 meters from the waste tank. The transfer line has a minimal slope toward the intermediate transfer tank. Because of the slurry's high consistency (viscosity), the slurry does not drain from the line once the transfer pump is stopped. It has become necessary to flush the transfer line with inhibited water after each transfer, causing dilution of the slurry. From the intermediate transfer tank the slurry is pump an additional 400 meters to the DWPF.

Waste Vitrification in the DWPF

Due to SRS Tank Farm processing and blending strategies, the DWPF began radioactive operations with a "sludge-only" (no PHA) process. A frit composition higher in alkali, to compensate for the missing PHA, was developed for the initial first radioactive sludge operations. A dilute formic acid solution was added to the SRAT in place of the PHA.

Since there is no aqueous phase from the precipitate waste stream, some other source of formic acid is required to drive the reduction of certain metals to the correct oxidation state (principally, the reduction mercury to its metallic state, for removal). The dilute formic acid was close enough to the aqueous stream, that not much development work was needed to use it. Since this was adding water (which is later recycled to the waste storage), use of concentrated formic acid was desirable. After the necessary development work, the flowsheet was modified to use concentrated formic acid with sludge.

Non-Radioactive Test Production: Prior to radioactive operations several different campaigns were performed to validate the vitrification process. Cold Chemical Runs were performed to demonstrate the system chemistry using a blended simulant to represent a composite of SRS tank farm waste. Waste Qualification Runs were performed using simulated feed types representing the range of sludge compositions which would be fed to the DWPF.[5] The compositional models used to predict glass properties were verified over this range.

After this testing, additional material simulating initial radioactive feed was processed to establish inventories in the DWPF tanks to this initial feed and to increase our staff proficiency. Based on all this testing the Department of Energy (DOE) approved full radioactive operations of the DWPF in March 1996.

Radioactive Operations: The first transfer from ESP to the LPPP of approximately 3,200 liters of sludge into a receiving tank containing 14,000 liters of simulated sludge. This material was subsequently transferred to the SRAT for nitric acid addition and then to the SME for frit addition and acceptance. Transfers take place approximately every 7-10 days. Through March 1997, nineteen feed preparation cycles have been processed and one hundred fourteen canisters filled. Several problems have been encountered during the first year of operations. These included melter wicking, failure of level and density sensing equipment in the process vessels, severe foaming in the SME causing carryover of material into a condensate tank, and failure of a cooling coil in the SME due to erosion.

In vessels containing radioactive slurry level and density measurements are determined using Holledge^{TM} sensors located in the vessel. Holledge^{TM} sensors are used because a standard bubbler will become plugged when used in sludge. The Holledge^{TM} design uses a membrane sensor to measure the pressure exerted by the height of fluid covering the sensor. There have been several failures of these gauges in recent months by an unknown mechanism. Alternate designs are under evaluation to determine if a better technology is available for this application.

A recent operational occurrence was the carryover of approximately 4,800 liters of slurry from the SME to the SME Condensate Tank (SMECT). The presumed mechanism is that severe foaming occurred in the SME and the material was carried over to the SMECT before the liquid could be condensed and returned to the SME. Carryover of small amounts is not unusual and any carryover is ultimately recycled back to the tank farm for processing. However, carryovers of this magnitude are not bounded by safety analyses performed for plant operation. The SMECT is an unpurged tank and the presence of large amounts of slurry from the SME or SRAT could result from high hydrogen concentrations due to radiolytic and catalytic decomposition of water and organics. Methods of detecting imminent carryover conditions are being investigated, but in all likelihood the SMECT will require a purge to prevent accumulation of explosive gases.

In March 1997, the cooling coil in the SME failed due to erosion. This failure was not unexpected as the frit slurry in the SME is highly erosive. Such failures are expected during the life of the plant. The failed cooling coil will be replaced by a spare, and, if possible, repaired.

Glass pouring occurs by pulling a vacuum relative to the melter vapor space on the canister and pour spout. The glass level increases in the riser and overflows into the pour spout. The glass then travels vertically down the side wall approximately 38 cm where the wall of the pour spout is cut back to form a sharp 'knife edge' to allow the glass to disengage. The glass then freefalls a distance of about 60 cm through an unheated bellows and into the neck of the canister. Any glass which comes into contact with the bellows has a tendency to build-up sufficiently to completely block the flow of glass from the pour spout to the canister. Wicking is the term used to describe contact of the glass stream with the unheated bellows or lower section of the pour spout below the knife edge.

Several strategies have been employed to work around this problem while a searching for a solution. These have included studying the vacuum control loop to improve the response time to pressure transients in the melter pour spout, using a tele-robotic manipulator to physically remove any glass buildup in the pour spout and bellows, and applying heat to the lower pour spout and bellows in an effort to prevent glass sticking to the metal. All of these have had limited success in that they allowed glass pouring to continue, however, they have not solved the problem. A pour spout insert is currently in development to alter the pour spout geometry in an effort to improve the pouring characteristics of the glass stream, and studies have been undertaken to determine the most likely causes of the wicking behavior and identify solutions.

FUTURE PROCESS OPERATIONS

Future ITP Operations

The research on the decomposition reactions is continuing. Preliminary indications suggest that temperature, TPB solids concentration, oxygen, and copper are important factors in determining the rate of formation of benzene in the ITP facility and downstream processes. The Safety Analysis Report is being revised to reflect the new understanding of tetraphenylborate chemistry. Facility modifications which include a seismic inerting system for Tanks 48 and 49 are scheduled for completion in August 1997. The ITP facility is scheduled for restart in October 1997.

Additional tests will be completed during restart to verify mass transfer coefficients, reaction rate constants, and assumptions for long range operations. A phased approach will be used during restart in order to provide additional margins of safety.

Future ESP Operations

The next batch of sludge, Sludge Batch 1b, has been 90% washed. The sludge consists of 31 year old HM-HAW (high activity waste), 18 year old HM-LAW (low activity waste), and 24 year old PUREX LAW. Prior to combining with the sludge from low activity waste (LAW), the HM-HAW waste underwent aluminum dissolution in the ESP Tank. HM-HAW sludge contains high concentrations of insoluble aluminum (approximately 50 % of the HM-HAW sludge volume results from insoluble aluminum compounds). By heating the sludge in a high caustic concentration a significant amount of the aluminum is dissolved (approx. 75%). By performing aluminum dissolution less glass is required to be produced at the DWPF. There is less sludge needing to be processed and a lower frit to sludge ratio required for melting the lower wt% aluminum sludge.

In 1997, one additional wash of the batch 1b sludge in Tank 42 is planned. Then the sludge will be blended with the sludge remaining in batch 1a. The combined sludge will have a lowered concentration of sodium than the present batch. The lower sodium sludge is expected to improve conversion rates in the DWPF.

Future DWPF Operations

DWPF is processing only the insoluble sludge (from ESP) at this time. The current sludge batch is predicted to supply DWPF for approximately 3 years. As discussed above, additional sludge preparation is underway in ESP to provide sludge for future DWPF operations. Full operation, with coupled sludge (from ESP) and precipitate (from ITP) feed is expected to begin in early 1999. In addition, work is underway to address the problems encountered during process operations to date and to enhance production capability.

CONCLUSIONS

The DWPF has demonstrated the capability to safely and efficiently immobilize high-level waste in a form suitable for shipment to a geological repository. To date, over 100 canisters of vitrified waste have been produced using a feed consisting of washed HLW sludge. While certain processing problems have arisen, these problems are no different than those encountered during start-up of any complex chemical process. Efforts are underway to improve production capability. Future operations will be based on a "coupled" feed of washed high-level waste sludge, precipitated cesium, and glass frit.

ACKNOWLEDGMENT

This paper was prepared in connection with work done under Contract no. DE-AC09-96SR18500 with the U.S. Department of Energy.

REFERENCES

1. K. G. Brown, "The Product Composition Control System at Savannah River: Statistical Process Control Algorithm," **Ceramic Transactions, Volume 39 - Environmental and Waste Management Issues in the Ceramic Industry**, The American Ceramic Society (1994).

2. C. M. Jantzen, N. E. Bibler, D. C. Beam, W. G. Ramsey, and B. J. Waters, ":Nuclear Waste Product Consistency Test (PCT) - Version 7.0 (U)," **WSRC-TR-90-539, Rev. 3**, Westinghouse Savannah River Company, Aiken, South Carolina (1994).

3. J. T. Conway, "Defense Nuclear Facilities Safety Board (DNFSB) Recommendation 96-1 to the Secretary of Energy," Available at http://www.dnfsb.gov/recommend/96-1.html.

4. "Defense Nuclear Facilities Safety Board (DNFSB) Recommendation 96-1 Implementation Plan, US Department of Energy.

5. S. L. Marra, D. E. Snyder, H. H. Elder, and J. E. Occhipinti, "The DWPF: Results of Full Scale Qualification Runs Leading to Radioactive Operations," **Proceedings - Waste Management '96**, CD-ROM Session 8-2.

COLLOIDAL AND RHEOLOGICAL CHARACTERIZATION OF HIGH-LEVEL NUCLEAR WASTE

PA Smith, PJ Bruinsma, DR Rector, A Shekarriz, and KP Brooks
Pacific Northwest National Laboratory, Richland, Washington

ABSTRACT

An understanding of the transport and solid/liquid separation properties of high-level nuclear waste is required for successful processing. This study focuses on the colloidal properties of several Hanford tank wastes and the rheology of wastes in tanks 241-SY-102 and C-107. The experimental results elucidate the relationship between the microscopic features of the suspension: particle size, morphology, chemical speciation and zeta potential; and macroscopic behavior: rheology and sedimentation.

INTRODUCTION

Suspensions of radioactive waste are temporarily stored in 177 underground storage tanks at the U. S. Department of Energy's Hanford Site located in southeast Washington. In the current scenario, wastes will be retrieved from tanks by sluicing and pumped through pipelines to a centralized processing plant for separation of radionuclides and vitrification. Typically, these wastes are concentrated suspensions of colloidal aggregates with non-Newtonian rheology. To ensure success in implementation of the processing scheme, it is critical to predict the behavior of the suspensions under a variety of conditions, especially shear history.

This study points to some potential problems which could be catastrophic. During transport of the suspension down 6 mi of pipeline, the rheological behaviors observed suggest that the viscosity could increase to an extent that results in a clogged pipe. Boehmite suspensions which are good simulants of C-107 are known to gel under static conditions. Gelation following an interruption of pumping could permanently block a pipeline. This phenomena has probably occurred. Several transfer pipeline at Hanford have been clogged over the years. However, the shear thinning property of the waste suspensions in pipeline transport may be advantageous because the suspension of small particles would help carry the larger particles and agglomerates that would otherwise settle to the bottom of the pipe. This strategy has been used for decades in petroleum production; i.e. clay suspension drilling muds are used to transport drill debris from the well base.

EXPERIMENTAL

The crystallinity, morphology, chemical composition, and crystalline phases of samples of the SY-102 and C-107 sludge were studied using transmission electron microscopy (TEM), electron energy dispersive spectroscopy (EDS), and electron diffraction. In summary, the process involved dispersing a small quantity of the sample (~1 mg) in a solution of methanol (<5 ml) and placing a small drop of the solution on a TEM copper grid. The morphology, distribution, and size of primary particles composing the agglomerates were obtained from electron imaging. Elemental composition of the particles was identified by EDS. The crystal structure was obtained from electron diffraction patterns and compared with the JCPDS-EDD™ database published by the JCPDS International Center for Diffraction Data.

Particle size distributions were measured by the Microtrac full range analyzer (FRA) and photon correlation spectroscopy (PCS) (Brookhaven ZetaPlus). Samples were prepared in three conditions: distilled deionized water, 0.1 M $NaNO_3$, and 1 M $NaNO_3$. These conditions encompassed the range of electrolyte conditions that can affect flocculation and measured agglomerate size. The particle sizes obtained are not of the primary particles; these measurements represent the size of the flocs or aggregates present in the waste as exists in the particle size analysis system.

Zeta potential is a measure of the surface charge density of particles. Zeta potential measurements were performed on dilute suspensions of waste in distilled deionized water using electrophoretic light scattering (Brookhaven Zetaplus). The pH was varied with nitric acid and sodium hydroxide.

The rheology experiments were performed on a rotational viscometer, Haake CV-20 rheometer, with a PK 45/4 cone-and-plate sensor geometry. Sample volumes were approximately 0.20 mL. For flow curve determination, the shear rates were varied between 0 and 300 s^{-1}. For time-dependent experiments, shear rates employed included 0.06, 0.1, 10 and 100 s^{-1} for approximately 30 minutes continuously. For the latter two shear rates, shear on-shear off-shear on experiments were performed with 30 minute intervals. These experiments were performed to determine the time evolution of the viscosity under steady shear conditions, targeted to understand material rheopexy or thixotropy.

Samples of the SY-102 and C-107 waste were placed in a test tube and subjected to centrifugation. The revolutions per minute (rpm) was converted to acceleration. Sediment volume was measured by a calibrated test tube. The acceleration obtained is an average value because the acceleration varies in the test tube as a function of position.

RESULTS

The morphology of the SY-102 is dominated by the large-size (2-5 mm x 10-30 mm) NaOH rods, which are accompanied by smaller particles with different contrasts, sizes, and shapes. Figure 1a shows a mm-sized sharp-edged γ-Al_2O_3 particle surrounded by dark contrast amorphous FeO(OH) agglomerates containing fine particles. In other analyses, needle-like FeOOH particles, irregular U-containing agglomerates, $Al_2Si_4O_{10}$ particles, Al/Cr-containing particles and amorphous silica particles were obtained.

Figure 1b shows the general view of the phases present in the C-107 tank waste. Similar to the SY-102, the samples contain primarily FeOOH particles as well as alumina and aluminosilicate phases. In addition, zirconia is observed. Additional TEM inspection showed the presence of distinct U-containing particles, and P-containing particles.

Overall, the EDS showed that the SY-102 waste is primarily composed of Na, Si, Al, Fe, Cr and U. The C-107 EDS analysis showed Na, Fe, Al, Si, Zr, U, P and Pb. The EDS results corresponded well with the wet chemical analysis (see Table 1).

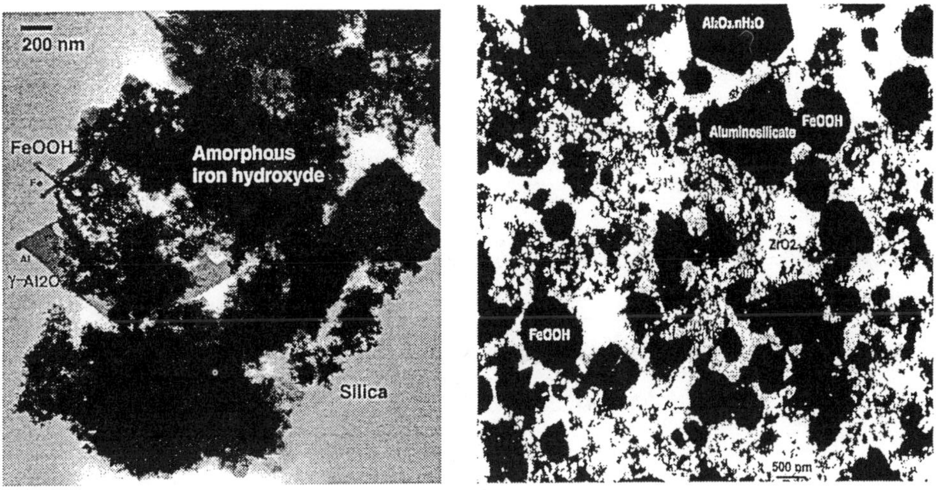

(a) (b)

Figure 1. Transmission electron micrograph (TEM) of the SY-102 (a) and C-107 (b) wastes showing the microstructure of the primary particles that form the agglomerates.

Table 1. Chemical Analysis of C-107 and SY-102

Component	C-107 wt%	SY-102 wt.%	Component	C-107 wt%	SY-102 wt.%
Al	17.87	14.02	Ag	0.12	0.06
Ba	0.15	0.02	Na	33.85	22.78
Cd	0.03	0.14	Sr	0.06	0.04
Cr	0.31	2.62	Zn	0.05	0.23
Cu	0.04	0.09	Zr	0.52	0.16
Ca	0.41	Not Analyzed	Pb	1.16	Not Analyzed
Bi	3.35	Not Analyzed	Tl	0.08	Not Analyzed
Fe	24.1	9.20	U	3.35	0.36
Pb	1.16	0.54	NO_3^-	Not Analyzed	10.46
Mg	0.20	0.54	Cl	Not Analyzed	0.21
Mn	1.17	2.81	F	Not Analyzed	0.52
Ni	0.97	0.21	CO_3^{2-}	Not Analyzed	5.25
P	6.15	13.96	Ttl Org. Carbon	Not Analyzed	1.54
K	0	0.54	NO_2^-	Not Analyzed	2.34
Si	5.39	4.73	SO_4^{2-}	Not Analyzed	0.70

A compendium of the particle sizes characterized at the Hanford site is provided in Table 2. The Microtrac data are described in number-averaged (*mn*) and volume-averaged (*mv*) format. In general, the PCS values correspond to the FRA *mn*.

Examination of the FRA *mn* and PCS effective diameters and the corresponding TEM data from SY-102 and C-107 show that the system is colloidal. Specifically, the waste is composed of individual phases ranging from the nanometer to micron length scales. In addition, the primary particles are arranged in larger aggregates. The FRA volume-averaged sizes show the larger particles/aggregates are present. Since the larger particles/aggregates are weighted more in volume-averaging, it is understandable that the average size is greater than the FRA *mn* and PCS values.

Table 2. Results of Particle Size Analysis Under Different Solvent Conditions

Name	condition	FRA *mn*	FRA *mv*	PCS
BY 110C	distilled water	0.521	6.15	
BY 104 C	distilled water	0.68	10.51	
BY 104 H	distilled water	0.833	1.958	0.624
SX108	distilled water	0.61	12.84	
C106-H	distilled water	0.243	5.579	1.84
	0.1M $NaNO_3$	0.223	8.009	1.2
C-107	distilled water	0.3	10	0.617
	1M $NaNO_3$	0.29	12.3	1.63
S104-H	distilled water	0.313	5.214	
	0.1M $NaNO_3$	0.318	3.495	1.02
SY-102	distilled water	0.417	5.95	0.565

ZETA POTENTIAL

The zeta potentials for the SY-102 and C-107 waste sample as a function of pH are shown in Figure 2. The isoelectric points are approximated to be 3.2 and 3.6 for the SY 102 and C-107, respectively. Parks (1965) has shown that the isoelectric points of metal oxide/hydroxide phases vary according to the binding force and correspondingly to the phase. Specifically, MO_2 phases such as silica and uranium possess isoelectric points typically in the pH range of 2 to 4. Clearly, the silica phase is controlling the surface charge development of the SY-102 waste as indicated by the zeta potential. This is consistent with the TEM results that show a presence of the colloidal amorphous silica in the aggregates.

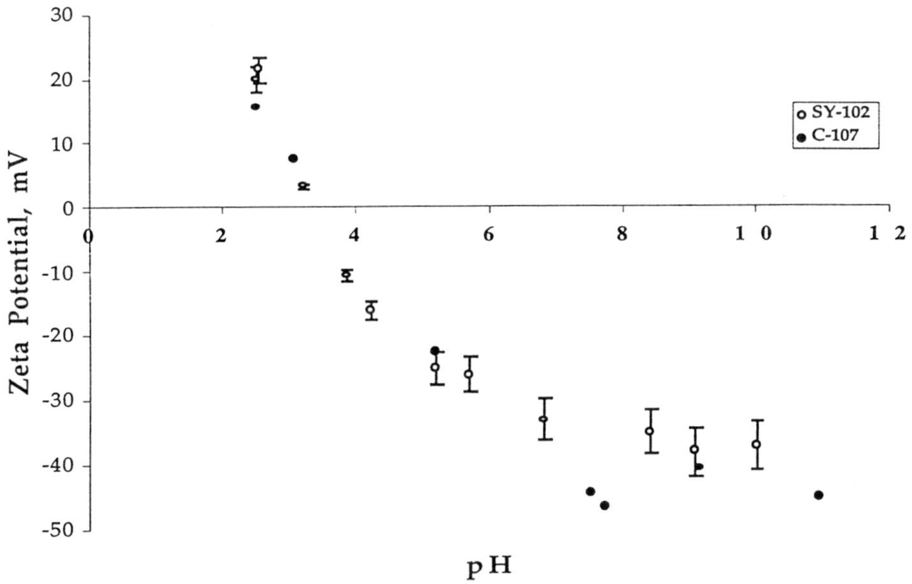

Figure 2. Zeta Potential as a Function of pH for Tank SY-102 Waste Slurry

RHEOLOGY

Figure 3 shows the viscosity behavior for the shear rates of 0.06 through 100 s^{-1}. The viscosity increases as a function of time for shear rates of 0.06 and 0.2 s^{-1} i.e., a rheopectic or antithixotropic behavior is observed. The figure shows an incubation period before the onset of the rheopexy. To obtain rheopexy, the microstructure of the suspension must be altered so that particles are aggregated. Therefore, the incubation is a result of the time required for particle collisions to be enacted by the small applied strain rates. Figure 4 also presents the results of tests in the shear on-shear off-shear on mode for strain rates of 10 and 100 s^{-1}. In the 10 s^{-1} test, the suspension shows thixotropic behavior. Examination of the initial curve at the terminal region and the data produced after the 30 min static period shows that the structure did not rebuild. The difference in the curves suggests that the system behaves thixotropically in this region and an area estimate of thixotropy (8029 mPa-s) was obtained. At 100 s^{-1}, the aggregates in the suspension are not destroyed any further, and time effects are absent.

A viscosity versus shear rate plot was produced for both the SY-102 and C-107 suspensions (Figure 4). The SY-102 waste has a classic behavior which may be represented as a Bingham plastic fluid over the range of 0 to 300 s^{-1} shear rates. This is one of the many models that can be used to describe the material behavior. The constitutive rheological model of this fluid can be presented as

$$\tau = 2.7\ \text{Pa} + 0.005\,\dot{\gamma} \tag{1}$$

where, the value of 2.7 Pa is the apparent yield stress of the material.

The C-107 sample was characterized for its viscosity behavior as a function of shear rate (see Figure 4). The figure shows plastic behavior followed by a transition region and finally, a terminal region which can be described by the power law:

$$\tau = K\dot{\gamma}^{n} \tag{2}$$

where τ is the shear stress, K is the consistency,$\dot{\gamma}$ is the shear strain rate and n is the power law exponent. A single fit was made for the C-107 sample and the consistency was 66.2 Pa·s and the shear thinning exponent was 0.135. The plasticity observed is characteristic of a flocculated suspension with all of the particles in a continuous network. The behavior in the rheometer shows that this solid-like structure is not deformed for shear rates below 0.1 s-1. At ~ 0.1 s-1, the

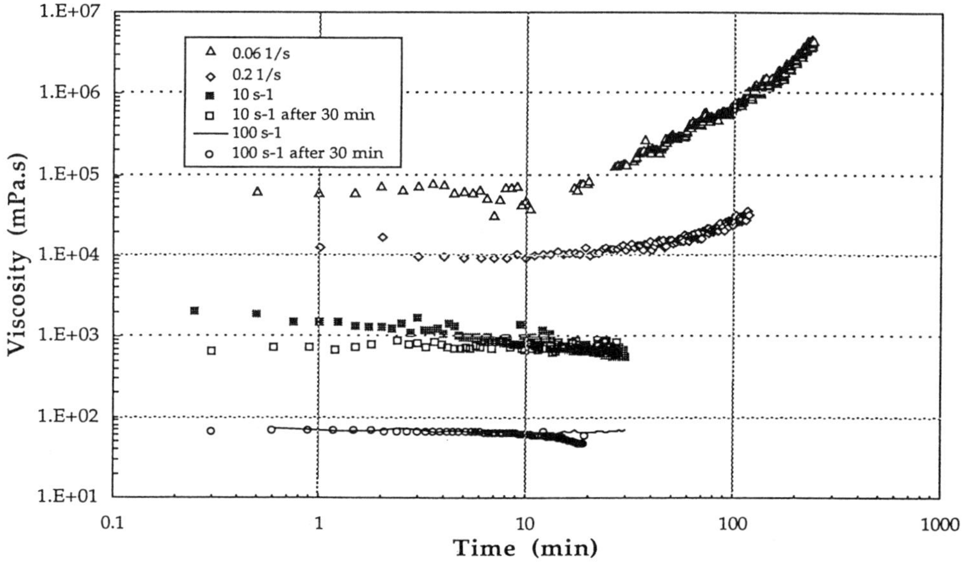

Figure 3. Time-dependent viscosity of tank 241-SY-102 waste showing both rheopectic and thixotropic behavior under different operating conditions. At low shear rates, aggregation is promoted and as a result the viscosity of the suspension increases.

sample in the rheometer begins to behave as a liquid. The transition region in the viscosity versus shear rate curve between 0.1 and 1 s^{-1} is analogous to a phase transition [Macosko (1994)].

Typically, nuclear waste suspensions have been simulated with chemical analogs (Chang, 1996). This data represent the first simulant which closely matched the waste behavior over 4.5 orders of magnitude. The boehmite analog reproduces the flow conditions with a single phase colloidal material that can be manipulated via electrolyte and salt conditions. While the size and phase chemistry do not *vis a vis* match the C-107, it is clear that variation of the physical properties (i.e matching particle size) and solution conditions are crucial for analog transport phenomena.

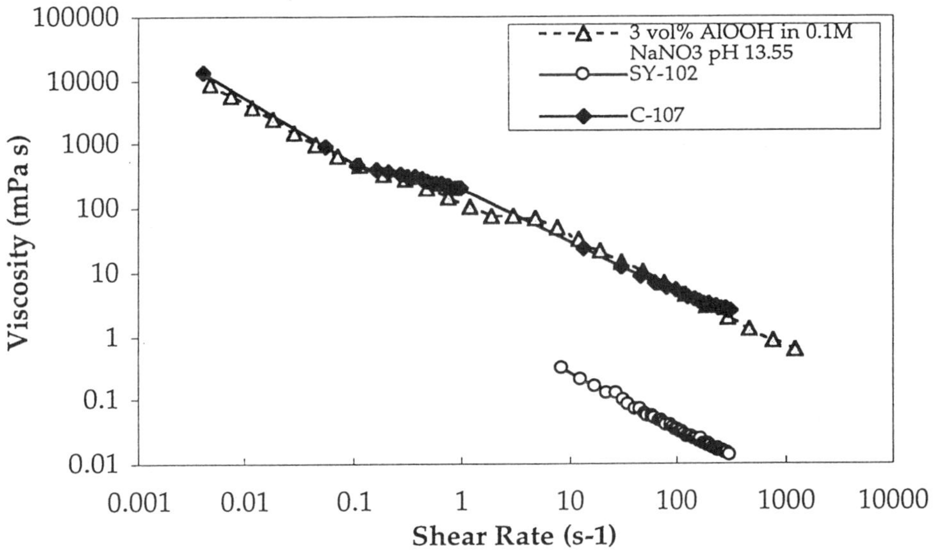

Figure 4. Viscosity versus shear rate for SY-102 and C-107. The viscosity behavior of a proposed analog is overlayed with the boehmite data.

COMPRESSIVE YIELD

The parameters for the power law curves may be determined using equilibrium sediment height data. Compressive yield data describes the sediment microstructure as a function of position and provides the understanding for engineering scale up. The compressive yield stress, $P_y(\phi)$, is an implicit function of many variables, including the size, shape, composition and relative number of particles involved, and the interparticle forces (which in turn depends on the solution chemistry). This form can be divided into two regions, with the boundary defined by the gel point volume fraction, ϕ_g. Below this value, the aggregates are not connected and only interact hydrodynamically. At the gel point, these aggregates become interconnected throughout the container. Above the gel point, the compressive yield stress is typically modeled using power law curves of the following type (Landman, White, and Buscall 1988):

$$P_y(\phi) = c\left[\left(\frac{\phi}{\phi_g}\right) - 1\right]^m \qquad \phi > \phi_g \tag{3}$$

with m varying between 4 and 10.

From a sedimentation/centrifugation curve, expressions predicting the compressive yield stresses for the C-107 and SY-102 sediment was developed. For the SY-102 waste, the compressive yield stress (gm/cm^2) is given by the expression

$$P_y(\phi) = 162.0\left[\left(\frac{\phi}{0.27}\right) - 1\right]^{5.3}, \quad \phi > 0.27 \tag{4}$$

and for C-107

$$P_y(\phi) = 41.937\left[\left(\frac{\phi}{0.079}\right) - 1\right]^{4.0}, \quad \phi > 0.079 \tag{5}$$

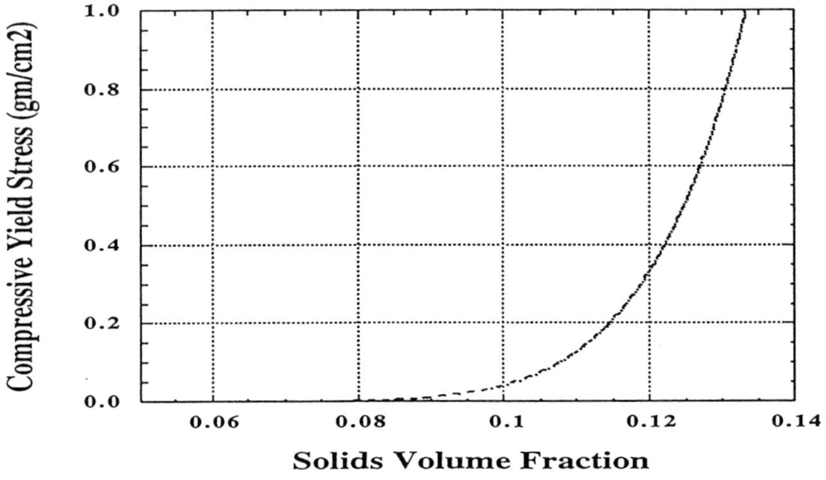

Figure 5. Compressive Yield Stress as a Function of Solids Volume Fraction for C-107.

The compressive yield stress curve as a function of solids volume fraction is shown in Figure 5. As the weight of sediment increases, either through the addition of solid material or increasing the gravitational force, the solids volume fraction also increases. Equivalent effects are observed in pressure filtration. If the solids were resuspended, the resulting sediment density profile is presented in Figure 6. The compressive strength of the sludge at the bottom of the SY 102 tank is estimated to be approximately 7.9 kPa. This value is significantly higher than the yield stress (obtained with curve fitting of the rheology data), which implies that, depending on what failure theory is used, the compressive strength might be the dominant resistive strength of the material.

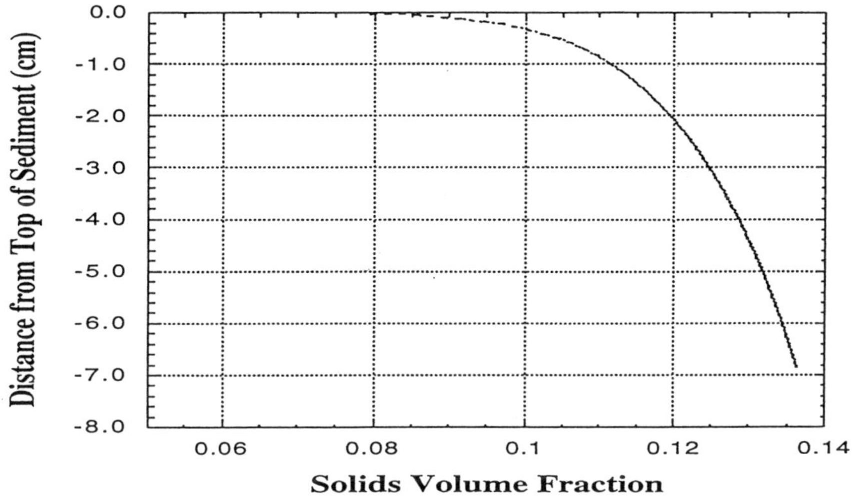

Figure 6. Solids Density Profiles versus Distance From the Top of the Sediment

DISCUSSION

The observation of rheopexy, thixotropy, and shear thinning behaviors are complex. However, the knowledge provided by the particle size analysis, TEM results, zeta potential measurement, and the literature allows some intelligible statements to be made. Clearly, there is an appreciable quantity of silica in both the C-107 and SY-102 systems of colloidal particles. In both samples, beyond 10 s^{-1} the agglomerates are significantly broken up due to the action of shear. Essentially the system contains a mixture of particles which would aggregate due to heterocoagulation. More importantly, the system has a high electrolyte content which allows the particles to be attracted to one another. Thus, the behavior of shear thinning (or alternatively Bingham plasticity) is understood in terms of aggregate disruption. Thixotropy was observed due to the time-dependent nature of the aggregate destruction.

The observation of rheopexy is quite different. Essentially, at low shear rates, the collisions between particles lead to aggregation. Simply, the silica renders the particles sticky. Analogous behavior has been observed in sol gel films (Brinker and Scherer 1990) and simulated nuclear waste suspensions containing frit (a silica source) particles (Chang and Smith 1996).

Some qualitative observations may have broad implications on waste processing. For particle density determination, suspensions were dried inside of a glove box at room temperature. Following drying, the particles were found to be dispersed over an area of a 1 m^2. It is our interpretation that under alpha/beta bombardment, secondary-electron emission self-charges the particles. When dry, charge dissipation is less effective, causing charge build-up and, eventually, repulsion. Under both laboratory and processing conditions, this self-charging is a dangerous dispersion mechanism that would not be anticipated from studies of non-radioactive simulants.

Above the isoelectric point, electrophoretic light scattering gave sharp, consistent peaks (in the power versus frequency spectrum) indicating that the particles have uniform surface charge densities. Though the suspensions contain several particle phases, the particle interfaces have had years to come to an equilibrium with the basic solution phase. Below the isoelectric point, the electrophoretic light scattering data was more scattered, suggesting a multiplicity of surface charge densities. Therefore, particle-particle separation based on chemical differences will require suspension acidification or alternative means of removing the interfacial layer.

SUMMARY

The colloidal and rheological aspects of two Hanford tank wastes, SY-102 and C-107, were successfully characterized and related. In addition, the characterization of a variety of tank wastes showed that the tanks contain similar sized microstructural features (with respect to transport). The rheology was characterized as a function of time (at fixed shear rates), as a function of shear rate and in compression. The material displayed a variety of rheological phenomenon: shear thinning, yield, thixotropy, and rheopexy. Specifically, the aggregated nature of both wastes produced rheology behavior at 10 and 100 s-1 which showed shear thinning. Chemical and TEM data showed the presence of silica which provided the parameters necessary for a phenomenological explanation of low shear rate rheopexy. For the SY-102 data, viscometric measurements suggested the possibility of a yield stress of 2.7 Pa. In compression, a yield stress of ~ 7.9 kPa was observed and suggested that the dominant failure mode may be compressive. For C-107, the first simulant for transport applications was developed on a physical basis that accounts for colloidal principles. The success of the analog was demonstrated over 4.5 orders of magnitude in viscosity.

The characterization of these properties is crucial to the effort to predict transport behavior, for engineering scale-up and for safety considerations. The rheological data presented represent only two waste compositions. Fortunately, particle size data shows that continued developments in colloidal rheology will be applicable to the aggregated submicron primary particles contained in Hanford tank wastes.

ACKNOWLEDGMENT

The authors would like to acknowledge the technical contributions made by Dr. Jun Liu for preparation of the TEM images. Though critical to the full-scale implementation of waste processing, rheological studies of real wastes using sample volumes on the order of tenths of ml are now nearly impossible because of a recent lowering of the radioactivity limits set internally by PNNL. Pacific Northwest National Laboratory is operated for U. S. Department of Energy by Battelle Memorial Institute under contract DE-AC06-76RL01830.

REFERENCES

Brinker, C. J. and G. W. Scherer. *Sol-Gel Science The Physics and Chemistry of Sol-Gel Processing.* Academic Press, New York. 1990.

Chang, C. and P. A. Smith. "Flow-induced Microstructures in a System of Nuclear Waste Suspensions." *Rheologica Acta.* 35. 1996.

Chang, C. P. A. Smith, and E. Tracey, "Rheological Characterization of Hanford Neutralized

Current Acid Waste Simulant Slurries " *J. Rheol.* July/Aug., 483-495, 1996.

Landman, K. A., L. R. White, and R. Buscall. "The Continuous-Flow Gravity Thickener: Steady State Behavior." *AIChE J.*, No. 34, p. 239. 1988.

Macosko, C. *Rheology Principles Measurements and Applications.* VCH Publishers, New York. 1994.

Parks, G. A. The Isoelectric Point of Solid Oxides, Solid Hydroxides and Aqueous Hydroxyl Complex Systems. *Chem. Rev.* **62**, 2, 177-198. 1965.

Tracey, E. M., P. A. Smith, and E. V. Morrey,"Rheology of Concentrated, Heterogeneous Slurries Containing > 1M Electrolyte - A Case Study in Waste Suspensions" *J. Nuc. Mater.*,19-35, **230**(1), 1996.

WEST VALLEY NUCLEAR SERVICES CO., INC. FLOWSHEET FOR TANK 8D-1 AND TANK 8D-2 RESIDUAL WASTE REMEDIATION USING OXALIC ACID

H. D. Smith, G. L. Smith, R. L. Russell, and G. K. Patello
Pacific Northwest National Laboratory[a], P.O. Box 999, Richland, WA 99352-0999

ABSTRACT

Following an initial vitrification phase at the West Valley Demonstration Project (WVDP), residual tank waste, consisting of cesium (Cs)-loaded zeolite in Tank 8D-1 and a mixture of neutralized PUREX sludge, THOREX waste and Cs-loaded zeolite remaining in Tank 8D-2, may require additional processing so that the waste can be removed from the tanks and vitrified in preparation for site closure. The objective of the treatment would be to increase the mobility of radionuclides by eluting Cs from the zeolite in 8D-1 and by dissolving the sludge in 8D-2. Laboratory-scale tests at Pacific Northwest National Laboratory (PNNL) with simulated waste have focused on using oxalic acid to facilitate the transfer of radionuclides from the residual waste to the Vitrification Facility for treatment through the melter. In order to develop these processes for the residual waste in Tanks 8D-1 and 8D-2, an integrated Cs elution/sludge dissolution flowsheet involving oxalic acid was evaluated. Effects of agitation, presence of corrosion products, and pH adjustment of the oxalic acid to pH 5 with 10M NaOH after sludge dissolution were also evaluated. Iron oxide added to simulate tank rust slightly decreased the effectiveness of oxalic acid on Cs elution. Adjusting the pH to 5 with 10M NaOH resulted in a decrease in solubility for iron, nickel, manganese, strontium and calcium.

---

a. Pacific Northwest National Laboratory is operated for the US Department of Energy by Battelle Memorial Institute under Contract DE-AC06-76RLO 1830.

INTRODUCTION

The work reported here is part of the West Valley Support Project (WVSP) conducted by the Pacific Northwest National Laboratory (PNNL) and is in support of residual waste removal during high-level waste (HLW) transition activities to be performed by the West Valley Demonstration Project (WVDP). The high-level waste (HLW), approximately 2 million liters, originated from a former commercial nuclear fuel reprocessing plant at West Valley, New York, and was originally stored in two tanks at the site of the WVDP[1]. The waste consisted of an alkaline sludge and supernatant in Tank 8D-2 and an acidic THOREX solution in Tank 8D-4. Subsequent processing has resulted in cesium-loaded zeolite which was stored in Tank 8D-1.

In preparation for the vitrification phase of the West Valley Demonstration Project waste remediation, the THOREX solution was transferred to tank 8D-2, and a major portion of the zeolite was transferred from 8D-1 to 8D-2, leaving residual cesium-loaded zeolite behind which may require treatment to remove the cesium in preparation for site closure. The other tank, 8D-2, will have residual waste consisting of waste from the PUREX and THOREX process and cesium-loaded zeolite that can not be pumped out. The residual waste remaining in 8D-2 may also require treatment to dissolve the waste and transfer the radionuclides to the Vitrification Facility.

Substantial work has been completed at PNNL to evaluate processes for cesium elution from the residual zeolite in tank 8D-1 and sludge dissolution of the residual waste in tank 8D-2[2,3]. This work was designed to develop treatment processes and give information on process limits, conditions, and controls for the individual treatment steps for removal of 8D-1 and 8D-2 tank residual wastes as an interim step to decontamination and decommissioning of the West Valley nuclear processing facility. The work reported in this paper focused on testing an integrated "Cs Elution/Sludge Dissolution flowsheet," i.e., a series of individual processing steps to produce a final result, the tank residual waste remediation. Several factors such as the presence of Fe_2O_3, the lack of slurry agitation, and the results of in-tank neutralization of oxalic acid after the sludge dissolution step were evaluated as part of this integrated flowsheet.

EXPERIMENTAL PROCEDURES

A WVNS flowsheet for residual zeolite/sludge oxalic acid washing tested in this study is described below (Note: 1.5 L/kg is not the baseline, rather, the "real" baseline is 20 L/kg in Tank 8D-2. Additional testing was run, but not included in this report, on the baseline flowsheet (@ 20 L/kg)):

1) The zeolite in contact with a 0.045 M NaOH heel at 4 L/ kg in tank 8D-1 will be contacted with an 8 wt% oxalic acid solution under ambient (50°C) conditions at a solution volume to zeolite ratio of 40 L/kg. The contact time, during which the mixture will be agitated, is set at 20 hours since previous testing [2] has shown that this contact time is well past the saturation time for Cs elution under these conditions.

2) Supernatant will be transferred from Tank 8D-1 to Tank 8D-2; which contains sludge wastes (i.e., PUREX sludge, THOREX, and zeolite) having an approximate 12 wt% solids loading. The supernatant/sludge mixture will be agitated for 20 hours under 50°C conditions. The ratio of 8D-1 eluent to 8D-2 sludge will be about 1.5 L eluent/kg sludge based on the total sludge (water plus dissolved and suspended solids).

The laboratory-scale tests followed the basic two stage testing flowsheet with the following exceptions besides system size: the zeolite elution period was 28 hours and one test was not agitated.

A total of four experiments were conducted (Table I): 1) control; 2) no iron compound; 3) no agitation during sludge dissolution; 4) neutralization. The experiments were performed in sealed 500 mL or 2000 mL vessels depending on the required volume. To provide the presence of iron oxide solids representing tank steel corrosion products, 4 grams of ferric oxide were added to the tests along with the zeolite and 2 grams of ferric oxide were added to the tests along with the sludge for Tests 1, 3, and 4. The only iron in Test 2 came from the initial zeolite elution and subsequently from added washed sludge. Sampling times were: 28 hrs. for the cesium elution stage, 15 min., 30 min., 1 hr., 2 hrs., 4 hrs., 10 hrs., 20 hrs., 28 hrs., and 50 hrs for the oxalic acid-sludge contact step, and 15 min., 30 min., 1 hr., 2 hrs., and 16 hrs for the adjusted pH stage. These conditions and sampling times were found to be most favorable based on previous testing experience[2].

Table I. Experimental test matrix

Test Condition:	Test 1	Test 2	Test 3	Test 4
	Control	No Iron	No Agitation	pH Adjust
Standard Conditions - Cs Elution (a) ("a" is test stage designator)				
Added ferric oxide with 10 g zeolite	4 g	0 g	4 g	4 g
Standard Conditions - Oxalic / Sludge Slurry (b) ("b" is test stage designator)				
Added ferric oxide with 240 g sludge	2 g	0 g	2 g	2 g
Adjustment to pH 5	•			•

During the tests, the temperature was maintained at 50°C in sealed vessels. Each of the agitated tests was stirred with a propeller stirrer on a stainless steel shaft at a rate which was thought to be enough to maintain the solids in a suspended condition. Each test was sampled through a port in the vessel lid using a pipette with a rubber suction bulb on the end. Attempts to sample just the supernatant from each test were not very successful because it was difficult to control the depth of penetration of the pipette tip because visual access was from the top only and a thin layer (< cm) of solids free supernatant developed after five minutes of non-agitation. Two samples were taken at each sample time; one for oxalate analysis by ion chromatography (IC) and one for multi-element analysis by inductively coupled plasma-mass spectroscopy (ICP-MS).

The pH adjustment step was tested as an option for the WVNS baseline flowsheet. It would be included after the sludge dissolution in tank 8D-2. It was performed as follows: first a small quantity of slurry (22 g) was removed from Test 4b to be pH adjusted. This slurry was titrated from pH ~2 to pH 8.0 in small steps using 10 M NaOH solution. Based on the amount of 10 M NaOH used to adjust the slurry sample pH to 5, the amount of 10 M NaOH needed to adjust the full laboratory-scale quantity was calculated. The oxalic acid containing mixture of sludge and supernatant for Test 1b and Test 4b was adjusted to a pH of about 5 ± 0.5 using 10 M NaOH solution at a rate of approximately 2% per minute of the calculated total (8.7 g) needed to accomplish the pH adjustment. The calculation of the amount of 10M NaOH needed to reach pH 5 ± 0.5 was based on a titration performed at ≈ 23°C. The error incurred by not performing the titration at test temperature (50°C) was ≈ 0.18 pH units[4]. After the pH adjustment step, samples were taken at 15 min., 30 min., 1 hr., 2 hrs., and 16 hrs.

Sludge dissolution and oxalate salt formation were evaluated by analyzing the sludge and the supernatant using ICP-MS, and IC. The amount of zeolite remaining after contact with the oxalic acid solution was determined by retrieving, drying, and weighing the zeolite. Data reproducibility was tested by running Tests 1 and 4 in parallel and comparing test results.

RESULTS AND DISCUSSION

Results include visual observations, IC analyses for oxalate, ICP-MS for the dissolved elements and the washed slurry composition. The appropriate data have also been graphed for each element as a function of contact time. Error bars have not been assigned to the data because the data scatter may be related to both analytical error and actual concentration fluctuations. The most accurate analytical results are probably ± 5 wt% with an average ±10-20 wt%, the error increasing as the detection limit is approached. Some elements such as silicon are difficult to accurately analyze because its oxide does not dissolve easily. Other elements such as iron and potassium can be a problem that is indicated by analytical values which vary by as much as a factor of five when there is no reason for such a variation. When this is observed, the samples are rerun with additional precautions taken and for this work the analytical values were observed to become consistent within analytical error.

Test Observations and pH

The test stages and pH are summarized in Figure 1. In Figure 1, the pH of the supernatant for each test system is shown at each sampling stage. The final pH of the cesium elution solution for Tests 1a through 4a was taken as time zero. An arbitrary time interval of one hour was included between the final pH reading of the cesium elution solution and the initial pH measurement of the mixture of the cesium elution solution and sludge simulant so that the initial data points would not overlap on the Y-axis. The caption "Sludge added to Cs Elution Solution" indicates the beginning of the time line for sludge mixed with the elution solution (Tests 1b through 4b) which continued for 50 hours at a pH that resulted from the mixing of 240 g sludge simulant and about 360 g of the supernatant solution each from Tests 1a through 4a. At the end of 50 hours Tests 1b and 4b were adjusted with NaOH to pH 5 over a period of about one hour (This hour is not included in Figure 1). After pH 5 was achieved, testing continued another 16 hours.

No dramatic changes in appearance were observed during these tests. Even though it was not possible to view the acid-slurry mixture directly, every time a sample was pipetted out of the reaction vessel, color and clarity were noted. Most samples contained solids (a few appeared to be clear) and therefore most supernatant samples were in constant contact with solids. The presence of solids in the samples is a concern because this condition allows the supernatant and solids to continue to react after removal from the test system. This might be the reason for the consistency of concentrations of species between long and short term samples as observed later. This possibility could be checked by repeating some of the tests
and immediately isolating (such as by centrifuging the sample to quickly settle the solids) the supernatant from any solids.

The pH of the cesium elution solution was uniformly acidic for all of the tests. When the sludge simulant was added, all of the systems except Test 2b (no iron added) displayed a pH maximum before settling down to equilibrium values. Test 3b, the non-agitated system, took the longest to come to a "constant" value, which might be expected based on the following rationale. The initial pH variations might relate to the relative rate of buffering reactions involving oxalic acid and the various metal hydroxides in the sludge simulant. The only difference known between Test 2b and the other tests was the lack of added iron oxide suggesting that the presence of iron oxide may temporarily influence pH. In Tests 1b and 4b the pH was increased to 5 with the addition of 10 M NaOH solution at a rate of about 2 wt% of the estimated total per minute until a pH of 5 was attained. Almost immediately it was noticed that the pH began to decrease with time. It may be similar to the pH variation (buffering) observed when the sludge simulant was added to the oxalic acid solution. As shown in Figure 1, the pH stabilization time was not determined since a sequence of constant values was not observed. The trend suggests that the pH will stabilize, but additional data points before and after 16 hours are needed to establish this.

Cesium Elution and Zeolite Mass Change

The cesium elution data are consistent with previously obtained values for eluting cesium- loaded zeolite[3]. During the cesium elution, iron was dissolved into solution. Table II presents the iron mass balance for each test system during the 28 hour cesium elution. The iron has been recalculated as elemental iron and consists of 0.245 g per 10 g of zeolite and 2.798 g per 4 g of iron oxide. (The control was a sample of the zeolite material exposed to water for 24 hours and then dried which illustrates that wetting and drying is not the reason for weight loss observed for the test solids.) Considering Test 2a first, it is observed that all of the iron has been removed out of the zeolite and resides in solution. The same thing could be true of the other tests but no measurements have been made to determine this.

Table II. Solids mass changes during the 28 hour cesium elution from zeolite

Test Identification:	1a	2a	3a	4a	control
Amount of Cs Loaded Zeolite (g)	10	10	10	10	5
Finely Divided Iron Oxide (g)	4	0	4	4	0
Final Dried Weight Solids (g)	7.74	6.88	8.91	9.59	4.94
Net Weight Loss (g)	6.26	3.12	5.09	4.42	0.06
Total Elemental Iron in Test (g)	3.043	0.245	3.043	3.043	-
Total Elemental Iron in Solution (g)	1.838	0.262	1.450	1.051	-
Percent Total Elemental Iron in Solution	60.4	106.9	47.7	34.5	-
Final pH of Eluent	0.358	0.360	0.319	0.341	-

Tests 1a, 3a, and 4a had the same amount of zeolitic iron plus added iron, but had considerable differences in the amount of iron in solution. It was expected that Tests 1a and 4a would give the same result and that Test 3a would be lower, because it was not stirred. It is not known why there is a discrepancy in the results but there are two factors that may have influenced these results. First it is not clear that the stirring action used for these tests was energetic enough to agitate all of the solids into a suspended state. Hence, iron oxide and other sludge components could have partially caked on the bottom of the vessels and not be in a condition to rapidly equilibrate with the supernatant. Since only final iron concentrations were determined, it is not known if the measured values were equilibrium values or if the system was still approaching equilibrium, i.e., if the sludge components were caked to various degrees, it is possible that equilibrium wasn't reached and actual numbers would be different if equilibrium had been achieved. Since the agitation was gentle, it is believed that different degrees of solids caking is the most likely explanation for the observed differences in analytical results.

Supernatant Concentrations of Selected Sludge Components in the Presence of Oxalic Acid

During the contact with the oxalic acid solution at pH $\sim$ 2.5, the solution appears to have come into equilibrium with the sludge solid almost immediately based on the fact that the analytical values change little from the 15 minute sample to the 50 hour sample such as indicated in Table III for nickel and iron (See Figures 2 and 3 relating to iron). Other elements such as strontium displayed a similar pattern (Figure 4). Considering the fine grain size of the sludge, this may not be surprising. However, the concern with respect to suspended solids in the samples needs to be considered here because continued contact with suspended solids in the sample bottle may have allowed the supernatant composition to be altered by continuing reaction.

The presence of additional iron oxide appears to have decreased the relative solubility of nickel as well as iron itself (Table III). Manganese and chromium also exhibit the same behavior, but their data is not shown. Elements other than Ni, Fe, Mn, or Cr did not show this trend. As shown in Table III, nickel and iron have a distinctly higher percentage in solution in the system with the lowest iron content (Test 2b). This trend is also true for iron where Test 2b has the highest supernatant iron content of all the test supernatants.

Table III. Weight percent of total Ni and Fe dissolved in solution

time (hours)	Weight Percent of Total Ni				Weight Percent of Total Fe			
	1b (control)	2b (no Fe)	3b (no stir)	4b (neutral)	1b (control)	2b (no Fe)	3b (no stir)	4b (neutral)
0.25	41.86	61.96	48.67	46.61	20.65	43.17	26.49	26.56
0.5	39.64	63.34	46.45	54.49	22.98	40.87	24.77	27.02
1	39.24	61.03	42.28	52.11	24.48	38.16	22.90	25.92
2	43.9	63.09	43.25	53.28	23.13	41.68	23.02	20.55
4	46.17	49.49	41.53	50.81	22.38	38.92	24.81	23.40
10	38.84	68.49	41.33	59.79	23.26	43.56	24.57	24.06
20	47.58	59.23	45.28	49.35	23.31	42.22	24.13	21.98
28	53.48	60.49	48.52	48.14	22.15	44.28	23.79	27.36
50	51.79	53.32	52.16	37.32	20.82	42.35	22.21	23.89
Adjusted to pH 5.0								
50.25	12.28			11.24	12.38			9.75
50.5	12.23			12.18	11.03			12.05
51	12.31			12.27	10.07			11.35
52	12.01			12.2	11.51			12.10
66	12.06			12.2	12.35			10.99

As indicated in Figure 2 and Table III, a higher percentage of iron is in solution in Test 2b than for the other tests. Also iron is more concentrated in the supernatant of Test 2b ≈ 4500 μg/mL versus ≈3700 μg/mL for each of the other tests (see Figure 3). Recall that in Figure 1, it appears that each test, except Test 2, experienced some degree of pH maximum over the subsequent equilibrium values when the sludge component was first added to the oxalic acid eluent solution. The pH peak lasted over an hour and was over a pH unit higher than the equilibrium range at it maximum extent. Since it was subsequently found that during and after the neutralization phase increased pH lowered the iron concentration in Tests 1 and 4 (see following section), it is possible that the lowered iron concentrations during and after addition of the sludge component were related to the pH peaking but the mechanism is not clear since the test pH values become equal again. The only chemical difference known between Test 2 and the other three tests was that Test 2 did not contain additional iron (2 grams ferric oxide added in addition to 240 grams of sludge). Our conclusion is that the solubility (or rate of solubility) of iron is being moderated by the presence other iron containing phases. Performing additional tests where the pH is cycled up and down several times under otherwise indentical conditions may help to clarify this issue.

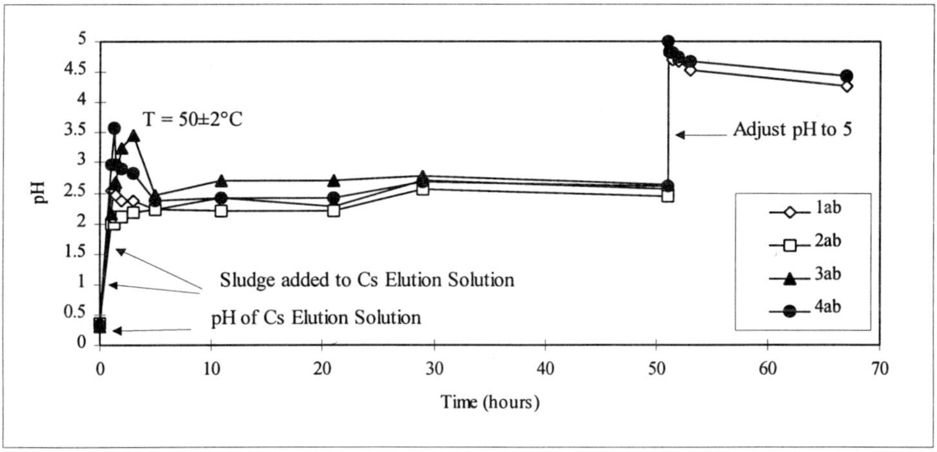

Figure 1. System pH as a Function of Flowsheet Stage as Reproduced by this Study

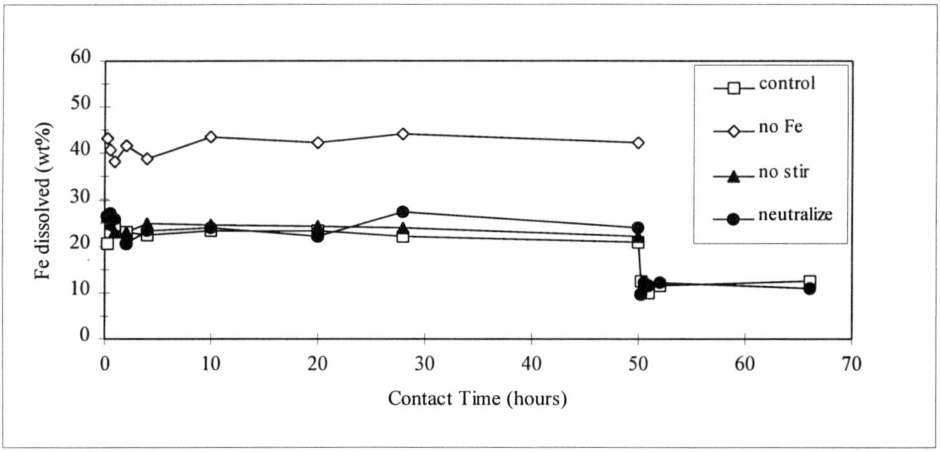

Figure 2. Weight Percent of Fe Dissolved in Each of the Four Test Systems as a Function of Contact Time

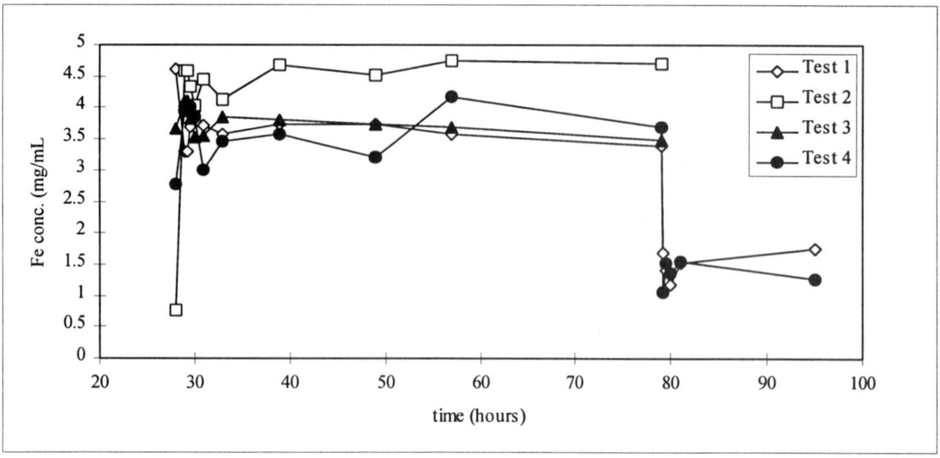

Figure 3. Measured Iron Concentration in the Supernatant of Each Test as a Function of Time

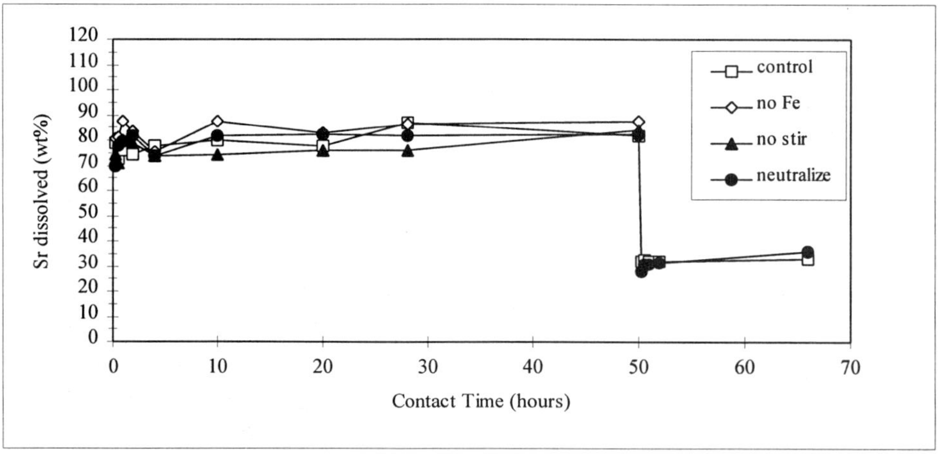

Figure 4. Percent Strontium Dissolved as a Function of Time

Supernatant Concentrations of Sludge Components after System Neutralization

After adjusting the pH to 5.0, elements which form more insoluble oxalates such as iron (II), nickel, manganese, calcium, and strontium show a significantly reduced concentration. Aluminum concentration has been reduced to a smaller but still detectable amount. It probably also precipitates as an oxalate. Silicon concentration may also have been reduced in solution precipitating along with the oxalates. An example of this change in concentration is shown in Figure 4 (for strontium). Sodium which can precipitate as sodium oxalate also shows a steplike decrease in concentration. The diminution of concentration of these elements appears to be complete by the time the first sample has been taken since the concentration (wt% dissolved) does not change significantly over a period of 16 hours. Again the concern of "supernatant in contact with sludge solids re-equilibrating in the sample bottle waiting for analysis" potentially affecting the results needs to be considered here because the pH was observed to shift over the 16 hour period suggesting that the chemical environment was changing during that period (see Figure 1).

CONCLUSIONS

The results of the tests provided the basis for the following conclusions.

Effect of Iron Addition

During zeolite treatment, when no iron was added, the zeolite was completely stripped of iron (100% based on the supernatant analysis and the reported iron content of the zeolite). Also, the measured cesium eluted was 100 wt% (again based on the supernatant analysis and the reported loading of the zeolite). When additional iron was added, 90 wt% of the cesium was eluted which was consistent with past testing results. The iron analytical values for the cesium elution solutions to which iron additions were made displayed considerable scatter which may be the result of incomplete suspension of solids due to too little agitation. The average value was similar to that previously reported[3].

During sludge treatment, added iron appeared to reduce the solubility of iron and other transition elements. The reason might be that the equilibrium solubility of the added iron phase is lower than that of the iron phases contained in the slurry simulant. The reduced solubility of the other transition elements is assumed to be due to coprecipitation with iron, substituting for iron in the phase precipitated. The coprecipitation is presumably due to the chemical similarity of the transition elements, but since the actual phases involved were not identified more cannot be said.

pH Adjustment

Adjusting the pH from ≈ 2.5 to 5.0 with 10 M NaOH over a period of about one hour causes a significant decrease in the solubility of iron, nickel, manganese, strontium, and calcium (50 - 80%) probably due to oxalate precipitation. Equilibrium concentrations of all of the elements appeared to have been attained by the end of the pH adjustment period and the solubility of the other sludge components was not changed by the pH adjustment.

REFERENCES

[1]L.M. Cadoff, "Waste Compliance Plan for the West Valley Demonstration Project," WVNS-WCP-001, REV 3. West Valley Nuclear Services Co., West Valley, New York, 1991.

[2]J.A,. Sills, M.R. Elmore, G.K. Patello, W.G. Richmond, J.S. Roberts, R.L. Russell, and K.D. Wiemers. "Fiscal Year 1995 Laboratory Scale Studies of Cs Elution in Tank 8D-1 and Sludge Dissolution in Tank 8D-2," PNNL-10945, Pacific Northwest National Laboratory, Richland, Washington, 1996.

[3]R.L. Russell, G.K. Patello, J.A. Sills, "Fiscal Year 1995 Laboratory Scale Studies of Cs Elution in Tank 8D-1," PNNL-11274, Pacific Northwest National Laboratory, Richland, Washington, 1996.

[4]Omega Engineering, *The pH and Conductivity Handbook and Encyclopedia,* Vol. 28, Omega Engineering, Stamford, Connecticut, 1992.

FABRICATION AND CHARACTERIZATION OF A FULLY LOADED HIGH LEVEL WASTE GLASS

G.L. Smith and H.D. Smith, Pacific Northwest National Laboratory [a], Richland, WA 99352, 509-372-1957 and 509-376-3588, gl_smith@pnl.gov and hd_smith@pnl.gov

ABSTRACT

The West Valley Demonstration Project (WVDP) is incorporating high-level waste (HLW) into a borosilicate glass to produce a stable waste form for placement in a nuclear waste repository. Fabrication of approximately 200 grams of fully loaded HLW glass has been accomplished using actual radioactive sludge from the WVDP located in West Valley, New York. The furnace was designed to simulate cold-crown melting and was used in a hot cell. The density of the black glass produced is 2.785 g/cc and the glass redox, ferrous to ferric ratio, is 0.083. Beta-gamma autoradiography indicated that the glass is homogeneous with respect to the distribution of the fission products from the HLW. Radiochemical analyses indicate that all of the radionuclides originally placed into the glass batch were incorporated into the final glass except technetium. Optical microscopy and XRD evaluations of the glass indicate a very low level of crystallinity. The preliminary conclusion is that a waste glass has been produced which is suitable for a durability testing glass standard and may be qualifiable as a nuclear waste analytical glass standard.

INTRODUCTION

The Materials Characterization Center (MCC) at the Pacific Northwest National Laboratory (PNNL) has assisted the West Valley Demonstration Project (WVDP) in qualifying its waste glass for geologic disposal by obtaining baseline leach data on simulated West Valley glass and by providing assistance to West Valley related to analytical laboratory methods validation and quality control. The MCC has fabricated and is characterizing a radioactive glass made from actual spent fuel reprocessing waste which was shipped to the West Valley Support Project (WVSP) at PNNL by West Valley Nuclear Services Company, Inc. The WVSP at PNNL received radioactive samples of West Valley Tank 8D-2 caustic washed sludge and THOREX acid waste from Tank 8D-1 and combined these wastes with Cs-137-

a. Pacific Northwest National Laboratory is operated for the U.S. Department of Energy by Battelle Memorial Institute under Contract DE-AC06-76RLO 1830.

loaded zeolite using a process that mimics the WVDP flowsheet. The WVSP performed characterization and blending of these wastes, added appropriate glass component materials to achieve a representative melter feed sample (final glass composition is similar to that produced at West Valley) and analyzed this final mixture before its transfer to the MCC for vitrification. Approximately 376 grams of dried/sieved melter feed were delivered to the MCC for melting. The melter feed consisted of about 152 grams of high-level waste (a mixture of caustic washed sludge, THOREX acid waste, and Cs-loaded zeolite) plus approximately 230 grams of glass additives. It was estimated that about 6 grams of melter feed was lost during processing. The average cation analytical results of the final melter feed composition in weight percent oxide and the radiochemical results normalized to the final glass composition are provided in Tables I and II (see column labeled "FEED"), respectively, for comparison to the WVSG-2 results. The radioactive waste glass, West Valley Sludge Glass-2 (WVSG-2) was vitrified and archived at PNNL; some of the WVSG-2 samples have been shipped to Argonne National Laboratory-East for testing. The "Waste Form Compliance Plan for the West Valley Demonstration Project High-Level Waste Form," Rev. 10, document number WVDP-185, provides a good description of the above processes and systems [1].

The purpose of this work was to control the melting process so that the glass redox state, as determined by the iron redox state, was close to a Fe^{+2}/Fe^{+3} ratio of 0.1, the glass was to be homogeneous with respect to chemical and radiochemical constituents, and the glass was to contain little or no secondary phase precipitates. The intended use of the glass produced is to be both short- and long-term glass durability samples and possibly as a future high-level nuclear waste analytical glass standard. This paper describes the results of the fabrication and characterization activities. A large portion of the WVSG-2 glass has been shipped to ANL-East for short- and long-term glass durability testing.

EXPERIMENTAL

Fabrication of West Valley Sludge Glass-2 (WVSG-2)

The vitrification of the radioactive waste feed was completed using a hydrogen/argon gas mixture (the 4% hydrogen/96% argon gas mixture facilitated the production of a homogeneous glass melt and necessary reduction of the glass) which was bubbled through the melt at 100 cc/min for 24 hours while at a temperature of 1175°C. The same gas mixture and was also used as a flush gas above the melt. Approximately 376 grams of melter feed was melted in a platinum crucible. This amount of melter feed was estimated to yield approximately 276 grams of glass. The actual amount of glass recovered was on the order of 200 grams. The unrecovered glass remained attached to the furnace and furnace hardware largely due to foaming of the melter feed during vitrification and hang-up in the melter throat. The resultant WVSG-2 was annealed at 500°C in the platinum crucible used to melt the radioactive waste feed. A cylinder of glass was core drilled from the crucible using a diamond core drill and cooling water. The right circular cylinder, "hockey puck", obtained from the platinum crucible was approximately 5.8 cm in diameter and 2 cm in height. The weight of the WVSG-2 "hockey puck" was 114.1 grams. The weight of the broken-up WVSG-2 that was attached to the platinum crucible after the "hockey puck" was removed was 63.5 grams.

The "hockey puck" was sectioned into three discs, top (T), middle (M), and bottom (B), with approximate thickness of 0.7 cm, 0.6 cm, and 0.5 cm, respectively (see Figure 1). Half of each disc was archived. The other half of each disc was cut into five pieces per the characterization plan. The number "1" and "2" samples from each half of the disc (three discs) were weighed and dimensions and dose rate measured. Part of each number "3" sample, approximately 1 gram, was crushed and ground to -200 mesh for use in chemical, radiochemical, X-ray diffraction, and iron redox analyses. The approximately 1 gram, -200 mesh radioactive powdered glass samples are designated as T3A, M3A, and B3A.

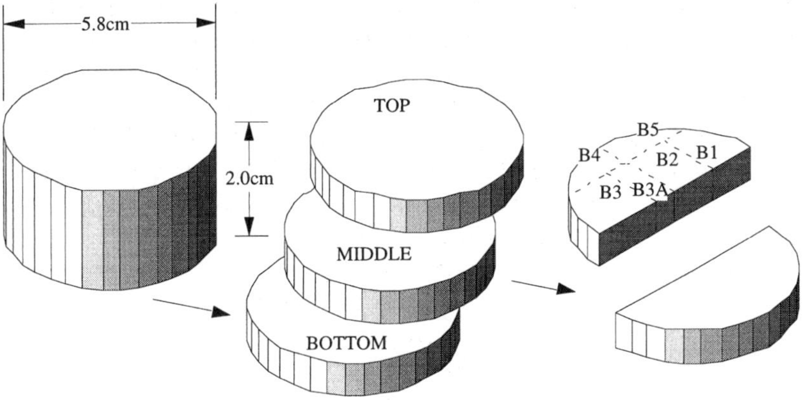

Figure 1 Sectioning Diagram of WVSG-2 Core Drilled Glass Cylinder

The six monolithic samples were mounted in resin after density measurements were taken and ground and polished to be used for optical microscopy, alpha and beta-gamma autoradiography, and scanning electron microscopy (SEM) examination. Alpha autoradiography and SEM examinations are still pending. The six monolithic samples are designated as T1, T2, M1, M2, B1, and B2.

Analytical Characterization of WVSG-2: Samples T3A, M3A, and B3A

Duplicate analyses of the three individual WVSG-2 radioactive powdered glass samples were analyzed for cation composition using inductively coupled argon plasma atomic emission spectrometry (ICP-AES). Two fusion/dilution procedures were used for each sample: 1) a sodium peroxide, Na_2O_2, fusion in zirconium crucibles; and 2) a potassium hydroxide, KOH, fusion in nickel crucibles. The overall dilution after fusions was approximately 2000. Preparation blanks, analytical spikes, and laboratory control samples, Analytical Reference Glass-1 (ARG-1) [2], were used to check the accuracy of the results.

Various radiochemical analytical methods were performed on the powdered glass samples to provide an inventory of radionuclides with half lives greater that 10 years and concentrations greater than 0.05 percent of the total radioactive inventory. The Waste Compliance Plan for the West Valley Demonstration Project [1] was used to specify the radionuclides for analysis that were likely to be in the vitrified melter feed which met the above criteria along with various radionuclides of interest. The glass samples in solution form were analyzed by the appropriate counting techniques for the following radionuclides: ^{60}Co, $^{59,63}Ni$, ^{90}Sr, $^{93m,94}Nb$, ^{93}Zr, ^{99}Tc, ^{107}Pd, ^{126}Sn, $^{133-135,137}Cs$, ^{151}Sm, $^{154,155}Eu$, ^{227}Ac, ^{231}Pa, $^{236-237}Np$, $^{233-236,238}U$, $^{238-242}Pu$, $^{241,242m,243}Am$, and $^{242-244}Cm$. Additionally, total alpha and beta were measured.

The three individual WVSG-2 radioactive powdered glass samples were examined using X-ray diffraction (XRD) to check for crystalline phases which may have formed during glass fabrication. The samples were prepared as a thin sample on a glass slide. Each powdered sample, limited to 5 mg because of radioactivity, was slurried using a 5% collodion/amyl acetate solution. The sample slurry was coated onto an adhesive tape covered glass slide and dried. A Scintag PAD V X-ray diffractometer, using a copper anode, set at 45 KV and 40 mA was used to scan over the 2-theta range of 5 to 65 degrees. The step size was 0.02 degrees with a counting time of 25 seconds per step. The sensitivity was limited to about 5 vol% to 10 vol% by the amount of material that could be used in each mount.

Duplicate ferrous to ferric ratio (Fe^{+2}/Fe^{+3}) measurements were conducted on WVSG-2 samples T3A, M3A, and B3A to determine if there was a redox difference in the original sample when measured from top to bottom. The ferrous to ferric ratio is used as an indicator of the glass oxidation state. The Fe^{+2}/Fe^{+3} digestions were performed in a hot cell and the determinations of reduced iron (Fe^{+2}) and total iron in each glass were performed in the laboratory. The 1,10-phenanthroline spectrophotometric procedure with a diode array spectrophotometer had an accuracy of ±7% based on parallel runs with a NIST standard [3-5]. The amount of ferric iron (Fe^{+3}) was calculated by difference.

Analytical Characterization of WVSG-2: Samples T1, T2, M1, M2, B1, and B2

The dose rate of the six WVSG-2 samples was measured at contact and 30 cm using an Eberline, Model R0-7 ion chamber dose rate meter. Measurements were taken in the air lock of the hot cell.

Density was measured on six of the WVSG-2 samples via an immersion liquid displacement method similar to ASTM C693-84 [6]. The density measurements were done in a shielded hot cell using an inert, fluorinated, high wetting factor liquid, Fluorinert FC-43, manufactured by 3M Company. The WVSG-2 samples were equilibrated in the Fluorinert FC-43 at 20°C.

Beta-gamma autoradiography of each of the six WVSG-2 samples was performed by placing the ground and polished surface in contact with Kodak SO-343 high resolution film for exposure times of 10 minutes each. The film was developed and photographic prints made for each sample.

RESULTS AND DISCUSSION

In Table I, the melter feed (see column labeled "FEED") and final glass analyzed compositions are presented. The composition of the WVSG-2 waste glass agrees quite well with the analyzed feed composition for all of the major elements. These element concentrations are also observed to be consistent from place to place in the vitrified waste glass. The values for Nd_2O_3, NiO, and P_2O_5 are also fairly consistent between the melter feed and resultant glass except for one or two values which may be flyers. Both ThO_2 and ZrO_2 values are not consistent, neither are volatile compounds during glass melting, and there were no other observations that would suggest a reason for a nonhomogeneous distribution and therefore suggests analytical problems for these elements.

Radiochemical analyses of the melter feed (see column labeled "FEED") and final glass are presented in Table II. The analytical values for the feed have been adjusted to represent activity per gram of glass produced from the feed, i.e. 376 grams of melter feed produced 276 grams of glass so the analytical values for the feed were increased by a factor of 1.362 to place them on the same basis. Comparing feed analytical values with those for the glass, it appears that there was good retention of most of the elements in the glass. The only exceptions were technetium and perhaps cesium. Most of the isotopes appear to be very uniformly distributed top to bottom in the glass. Some isotopes (i.e. ^{155}Eu and ^{90}Sr) were found to have higher analytical values in the glass than in the feed. Since the elements were not created by melting, the analytical value for the feed was probably low and the contamination with specific elements is unlikely. The over all results indicate that the radioactive elements (transuranic elements and fission products except for technetium) are completely captured and uniformly distributed throughout the glass.

Results from X-ray diffraction measurements indicate that all three samples, T3A, M3A, and B3A, are almost completely amorphous. No detectable crystalline peaks were found in samples T3A and M3A with only one faint crystalline peak near 9 degrees 2-theta found in sample B3A. The crystalline phase volume percent was too low to make an accurate determination of the crystalline phase(s) present. However, optical micrographs indicated that a crystalline phase was present at the ~1 vol% level and homogeneously distributed so the lack of XRD sensitivity is not believed to be a serious deficiency.

The ferrous to ferric ratio (Fe^{+2}/Fe^{+3}) duplicate results are summarized in Table III. The redox values from the top(T), middle(M), and bottom(B), samples are very close in value as expected because of the hydrogen/argon purge of the glass during melting. The values for Fe^{+2}/Fe^{+3} in Table III are reported to have a ±7% error. Based on these results and reported accuracy the glass appears to have a uniform redox state. It is possible that the surface of the glass melt (T3A sample) was slightly more oxidized than the bulk of the glass (i.e. the Fe^{+2}/Fe^{+3} is outside the ±7% envelope). This is consistent with the potential access of air to the melt surface by entrainment with the flush gas which provided an overall positive pressure to the furnace melting chamber. However, the redox condition appears to have been well controlled by the approach taken. The data below provide sufficient evidence that bubbling a reducing gas through a small glass melt can produce a

Table I. WVSG-2 oxide composition using inductively coupled plasma atomic emission spectrometry (ICP-AES)

Oxide	Weight Percent Oxide (wt% oxide)								
	FEED	T3A	T3A-Dup.	M3A	M3A-Dup.	B3A	B3A-Dup.	(Average)	(%RSD)
Al_2O_3	**5.69**	5.77	5.81	5.66	5.67	5.79	5.77	5.74	1.1
B_2O_3	**11.30**	11.98	12.09	11.71	11.73	11.83	11.77	11.85	1.3
BaO	**0.15**	0.15	0.15	0.15	0.15	0.15	0.15	0.15	1.6
CaO	**0.53**								
Ce_2O_3	**0.16**								
Cr_2O_3	**0.17**	0.19	0.20	0.19	0.19	0.19	0.19	0.19	2.4
CuO		0.08	0.08	0.08	0.07	0.08	0.07	0.08	2.7
Fe_2O_3	**12.03**	12.67	12.79	12.38	12.40	12.63	12.55	12.57	1.3
K_2O	**4.38**	4.43	4.37	4.24	4.23	4.33	4.22	4.30	2.0
Li_2O	**3.31**	3.31	3.35	3.22	3.24	3.24	3.28	3.27	1.5
MgO	**0.79**	0.88	0.89	0.86	0.86	0.87	0.86	0.87	1.4
MnO	**0.94**	0.96	0.97	0.95	0.93	0.99	0.95	0.95	1.8
Na_2O	**6.67**	7.03	7.06	6.91	6.76	6.97	6.87	6.93	1.6
Nd_2O_3	**0.14**	0.16	0.18	0.11	0.09	0.10	0.09	0.12	32.0
NiO	**0.27**	0.42	0.79	0.39	0.40	0.40	0.39	0.46	34.0
P_2O_5	**0.87**	1.13	0.64	0.72	0.77	0.72	0.82	0.80	22.0
SiO_2	**40.98**	39.52	39.07	38.40	38.07	38.67	38.25	38.67	1.4
SrO	**0.03**	0.03	0.03	0.03	0.03	0.03	0.03	0.03	0.2
ThO_2	**3.21**	2.34	2.81	1.08	0.96	0.84	0.99	1.50	56.0
TiO_2	**0.73**	0.76	0.76	0.74	0.75	0.77	0.77	0.76	1.3
UO_3		1.30	1.27	1.20	1.12	1.15	1.09	1.19	7.0
ZnO	**0.06**	0:05	0.06	0.05	0.05	0.05	0.05	0.05	5.2
ZrO_2	**1.08**	1.02	0.42	0.46	0.42	0.47	0.60	0.57	41.2

Table II. WVSG-2 radiochemical contents

Analysis/Analyte		Feed	T3A		T3A-Dup.		M3A		M3A-Dup.		B3A		B3A-Dup.	
		μCi/g	μCi/g	%Err ±	μCi/g	%Err ±	μCi/g	%Err ±	μCi/g	%Err ±	μCi/g	%Err ±	μCi/g	%Err ±
GEA	Co-60	**4.48E0**	4.59E0	10	4.46E0	8	3.44E0	15	5.32E0	8	4.71E0	11	4.18E0	10
	Cs-134		<6E0		<4E0		<7E0		<6E0		<6E0		<6E0	
	Cs-137	**1.22E4**	7.59E3	5	7.68E3	5	8.05E3	5	8.28E3	5	7.93E3	5	8.03E3	5
	Eu-154	**1.06E2**	1.02E2	3	1.00E2	3	7.85E1	4	7.69E1	4	8.44E1	4	8.27E1	3
	Eu-155	**1.20E1**	2.37E1	20	1.92E1	17	<3E1		<2E1		<3E1		<2E1	
	Am-241	**1.29E2**	1.14E2	7	1.15E2	6	1.01E2	8	1.27E2	7	1.06E2	8	1.23E2	7
AEA	Np-237	**6.44E-2**	6.73E-2	13	7.01E-2	12	6.89E-2	12	8.11E-2	12	6.26E-2	16	5.53E-2	14
	Pu-238		1.42E1	3	1.42E1	3	1.47E1	2	1.53E1	3	1.49E1	3	1.59E1	3
	Pu-239/240	**6.49E0**	6.18E0	3	5.78E0	4	6.60E0	3	6.59E0	3	6.65E0	3	6.9E0	3
	Am-241		1.11E2	2	1.05E2	2	1.15E2	2	1.16E2	2	1.08E2	2	1.11E2	2
	Cm-242		7.34E-1	9	7.44E-1	9	5.00E-1	12	5.08E-1	10	4.45E-1	12	6.19E-1	10
	Cm-243/244		1.89E1	3	1.98E1	3	1.26E1	3	1.23E1	3	1.34E1	3	1.18E1	3
Other	Ni-59		1.7E-1	20	9.5E-2	21	1.0E-1	28	9.3E-2	23	1.1E-1	20	1.0E-1	22
	Ni-63		2.27E1	8	2.18E1	8	2.28E1	8	2.30E1	8	2.08E1	8	2.30E1	8
	Sr-90	**8.69E3**	1.15E4	8	1.16E4	8	1.19E4	8	1.15E4	8	1.16E4	8	1.23E4	8
	Tc-99	**3.55E-1**	<6E-2		<6E-2		<7E-2		<5E-2		<6E-2		<5E-2	
	Sn-126		<1.7E-1		<1.4E-1		<1.7E-1		<1.3E-1		<1.7E-1		<1.8E-1	
	Pa-231		<1.8E0		<1.0E0		<1.3E0		<1.0E0		<1.1E0		<1.3E0	
	Np-236		<1.0E-1		<8.6E-2		<4.5E-2		<3.7E-2		<9.8E-2		<7.9E-2	
	Total Alpha	**1.50E2**	1.54E2	5	1.82E2	5	1.73E2	5	1.78E2	5	1.52E2	5	1.69E2	5
	Total Beta	**3.13E4**	3.21E4	4	3.29E4	4	3.32E4	4	3.06E4	4	3.28E4	4	3.16E4	4
		μg/g	μg/g	%Err ±	μg/g	%Err ±	μg/g	%Err ±	μg/g	%Err ±	μg/g	%Err ±	μg/g	%Err ±
From GEA	Cs (Tot.)		3.70E2	5	3.74E2	5	3.99E2	5	4.05E2	5	3.89E2	5	3.96E2	5
From GEA/MS	Cs-133		2.30E2	5	2.32E2	5	2.50E2	5	2.53E2	5	2.44E2	5	2.48E2	5
From GEA/MS	Cs-135		5.3E1	5	5.3E1	5	5.6E1	5	5.7E1	5	5.5E1	5	5.6E1	5
From GEA/MS	Cs-137		8.7E1	5	8.8E1	5	9.3E1	5	9.5E1	5	9.1E1	5	9.2E1	5

Table II. WVSG-2 radiochemical contents (cont.)

Analysis/Analyte		T3A		T3A-Dup.		M3A		M3A-Dup.		B3A		B3A-Dup.	
		μCi/g	%Err ±	μCi/g	%Err ±	μCi/g	%Err ±	μCi/g	%Err ±	μCi/g	%Err ±	μCi/g	%Err ±
RADIOCHEM	Zr-93	3.2E-1	15	1.7E-1	15	1.9E-1	15	1.4E-1	22	2.2E-1	15	1.9E-1	15
	Nb-93m	<8E-1		<6E-1		<7E-1		<7E-1		<7E-1		<6E-1	
	Nb-94	<1.9E-1		<8.1E-2		<1.0E-1		<1.1E-1		<1.3E-1		<1.2E-1	
	Pd-107	<2.6E-2		<2.3E-2		<2.7E-2		<2.1E-2		<2.6E-2		<2.1E-2	
	Sm-151	1.5E2	14	1.4E2	11	9.9E1	17	8.4E1	16	1.0E2	18	1.2E2	13
	Ac-227	<4.4E-1		<3.4E-1		<4.2E-1		<3.8E-1		<4.2E-1		<5.3E-1	
	Am-242m	7.3E-1	9	7.4E-1	9	5.0E-1	12	5.1E-1	10	4.5E-1	12	6.2E-1	10
	Am-243	2.0E0	19	1.8E0	18	1.5E0	22	2.1E0	18	1.3E0	18	1.3E0	25

		T3A	T3A-Dup.	M3A	M3A-Dup.	B3A	B3A-Dup.
TIMS	U-233 μg/g	3.45E0	3.37E0	3.18E0	2.97E0	3.05E0	2.89E0
	U-234 μg/g	2.16E0	2.11E0	1.99E0	1.86E0	1.91E0	1.81E0
	U-235 μg/g	1.89E2	1.85E2	1.75E2	1.63E2	1.68E2	1.59E2
	U-236 μg/g	1.86E1	1.81E1	1.71E1	1.60E1	1.64E1	1.56E1
	U-238 μg/g	1.06E4	1.04E4	9.79E3	9.13E3	9.38E3	8.89E3
	Pu-238 μCi/g	1.50E1	1.41E1	1.61E1	1.61E1	1.62E1	1.64E1
	μg/g	8.74E-1	8.23E-1	9.4E-1	9.38E-1	9.47E-1	9.6E-1
	Pu-239 μCi/g	3.60E0	3.36E0	3.84E0	3.83E0	3.87E0	4.01E0
	μg/g	5.81E1	5.42E1	6.19E1	6.18E1	6.23E1	6.46E1
	Pu-240 μCi/g	2.58E0	2.42E0	2.76E0	2.76E0	2.78E0	2.89E0
	μg/g	1.13E1	1.07E1	1.22E1	1.22E1	1.23E1	1.27E1
	Pu-241 μCi/g	1.24E2	1.15E2	1.33E2	1.35E2	1.34E2	1.39E2
	μg/g	1.20E0	1.12E0	1.29E0	1.31E0	1.30E0	1.35E0
	Pu-242 μCi/g	3.01E-3	2.81E-3	3.21E-3	3.17E-3	3.26E-3	3.45E-3
	μg/g	7.66E-1	7.15E-1	8.17E-1	8.08E-1	8.30E-1	8.78E-1

GEA = gamma energy analysis
AEA = alpha energy analysis
GEA/MS = combination of gamma energy analysis and mass spectroscopy
TIMS = thermal ionization mass spectrometry

homogeneous glass with respect to the redox condition.

Table III. WVSG-2 iron redox as a function of depth

WVSG-2 ID	%Fe^{+2}	%Fe^{+3}	Fe^{+2}/Fe^{+3}
T3A	0.627	7.95	0.079
T3A Duplicate	0.519	7.92	0.066
M3A	0.677	7.76	0.087
M3A Duplicate	0.706	7.69	0.092
B3A	0.667	7.59	0.088
B3A Duplicate	0.657	7.70	0.085

Table IV provides the dose rate at contact and 30 cm and the density for all six WVSG-2 samples measured. The dose rate is a combination of alpha, beta, and gamma radiation. At contact all three contribute. The large scatter of the contact reading for the samples is due to sample self adsorption of alpha and beta radiation coupled with the odd shapes of the samples. At a distance of 30 cm from the sample gamma radiation is the principle radiation component measured and is combined with background gamma radiation in the hot cell. This imprecise evidence is consistent with a uniform distribution of radionuclides in the glass. As can be seen from the data, the density measurements indicate that the glass produced is homogeneous. The average density of the six samples is 2.785 g/cm^3. No radial differences or density settling is evident from the six density measurements.

Table IV. WVSG-2 dose rate and density as a function of radial position and depth

WVSG-2 ID	T1	T2	M1	M2	B1	B2
Dose @ contact (R/hr)	9.5 ± 3					
Dose @ 30 cm (mR/hr)	700					
Density (g/cm^3)	2.7856	2.7853	2.7843	2.7863	2.7845	2.7853

The MCC has conducted a number of interlaboratory test programs (Nuclear Waste Analytical Round Robins) that have focused on analysis of glasses pertinent to nuclear waste repository programs [7]. These round robins have shown that more than one or two reference glasses will be required to appropriately cover the range of compositions of waste glasses [8]. As WVSG-2 is a homogeneous glass and well characterized, especially with respect to both cation and radiochemical composition, it would be a suitable candidate for inclusion in a Nuclear Waste Analytical Round Robin testing program. Interlaboratory testing would provide sufficient analytical data to qualify WVSG-2 as a consensus analytical reference glass.

CONCLUSIONS

A fully loaded HLW glass has been produced using actual waste from the West Valley Demonstration Project. This glass, West Valley Sludge Glass-2 (WVSG-2), has a uniform composition with respect to its redox state, density, crystallinity, elemental composition of major glass components, and radionuclides. This glass meets the objectives of the work and is suitable for use as short- and long-term glass durability samples. In addition, it would make a good high-level nuclear waste analytical glass standard, although the amount of material available after testing may be too small to accommodate this purpose.

ACKNOWLEDGMENTS

The authors gratefully acknowledge W. M. Gerry for his extensive help in fabricating WVSG-2; B. O. Barnes for his assistance with data preparation and reliability and Quality Assurance aspects of the project; the Analytical Chemistry Laboratory and the Postirradiation Testing Laboratory for help in gathering data; and financial support by the U.S. Department of Energy through a contract with West Valley Nuclear Services, Inc. The Pacific Northwest National Laboratory is operated for DOE by Battelle Memorial Institute under Contract DE-AC06-76RLO 1830.

REFERENCES

1. L. H. Cadoff, “Waste Form Compliance Plan for the West Valley Demonstration Project High-Level Waste Form,” WVDP-185, Rev. 10, West Valley Nuclear Services Co., Inc., West Valley, New York, August 2, 1996.
2. G. L. Smith, “Characterization of Analytical Reference Glass-1 (ARG-1),” PNL-8992, Pacific Northwest National Laboratory, Richland, Washington, December 1993.
3. M. C. Burt, “Spectrophotometric Determination of Reduced and Total Iron in Glass with 1,10 Phenanthroline” PNNL Procedure No. PNL-ALO-255, Rev. 1, Pacific Northwest National Laboratory, Richland, WA, 1991.
4. D. R. Jones IV, W. C. Janheski and D. S. Goldman, Analytical Chemistry, **53**, 923, 1981.
5. L. T. Begheijn, Analyst (London) 1070. 1055-1061.
6. American Society for Testing and Materials (ASTM) C 693 - 84 (Reapproved 1988) “Standard Test Method for Density of Glass by Buoyancy,” ASTM, 100 Barr Harbor Drive, West Conshohocken, Pennsylvania, 19428-2959.
7. G.L. Smith and S.C. Marschman, "Nuclear Waste Analytical Round Robins 1-6 Summary Report," in Scientific Basis for Nuclear Waste Management XVII, Vol. 333, ed. by R. V. Konynenburg and A. Barkatt, pp 461-472, Materials Research Society, Pittsburgh, Pennsylvania, 1994.
8. G.L. Smith, D.L. Eggett, and H.D. Smith, "Waste Glass Analytical Bias Correction Using a Reference Standard Glass," in Scientific Basis for Nuclear Waste Management XX, Vol. 465, ed. by W.J. Gray and I. Triay, pp 173-180 Materials Research Society, Pittsburgh, Pennsylvania, 1997.

WEST VALLEY HIGH-LEVEL WASTE GLASS CRYSTALLIZATION IN CANISTERS

M. J. Schweiger, M. W. Stachnik, and P. Hrma
Pacific Northwest National Laboratory,[1] Richland, WA 99352

ABSTRACT

Simulated West Valley high-level waste glass samples from core-drilled canisters were examined by optical microscope. Spinel was the major crystalline phase. Both spatial and size distribution of spinel crystals exhibited pronounced nonuniformity. Altering layers of glass with high and low crystal number density produced a marbling pattern. Spinel crystals were present in two distinct sizes, 7-15 μm and 0.5-2 μm. Microscopic examination of canister and laboratory samples has shown that the marbling pattern results from spinel agglomeration in a "turbulent" flow. Steep velocity gradients cause spinel agglomeration, which result in segregation into layers. The bimodal distribution of crystal size is caused by the temperature history of the glass: larger crystals formed as glass cooled slowly after pouring, and massive numbers of the smaller crystals precipitated as fresh hot glass reheated cooler regions.

INTRODUCTION

Segregation of spinel crystals into densely packed areas is a common feature observed in high-level waste glass. Spinel crystals are arranged in streaks, striae, layers, or islands, often forming a marble-like pattern. Another typical feature of spinel in waste glass is a bimodal size distribution of the crystal size. This paper aims to demonstrate these phenomena on West Valley core-drilled canister samples from non-radioactive testing, previously characterized by Jain and Barnes [1] as well as samples produced in the laboratory, and to offer their plausible explanation.

WEST VALLEY CORE-DRILLED GLASS

Canister core-drilled samples of simulated high-level waste glass were obtained from West Valley Nuclear Services. Thin sections of these samples were examined by optical microscopy and X-ray diffraction analysis. A typical example is shown in Figure 1a. Spinel crystals are arranged in alternating layers of glass with high and low crystal number density, producing a marbling pattern. The layers are

[1] Pacific Northwest National Laboratory is operated for the U.S. Department of Energy by Battelle under Contract DE-ACO6-76RLO 1830.

typically 150 μm thick with a fine structure of 20 μm sublayers, as can be seen at higher magnification (Figure 1b). In Figure 1b, two distinct sizes of spinel crystals are prominent: larger crystals (7-15 μm) and substantially smaller crystals (0.5-2 μm) with virtually no crystals of intermediate size.

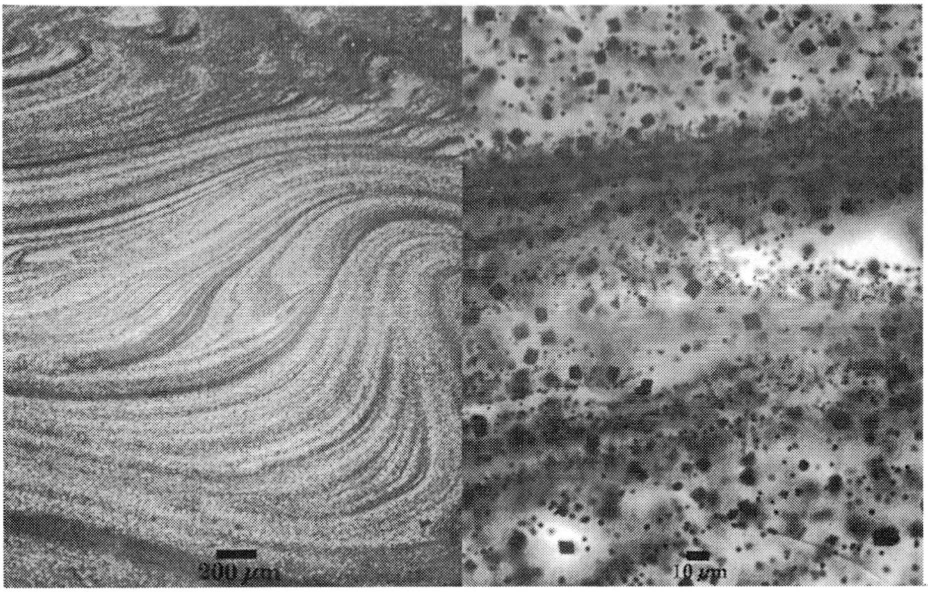

(a) (b)

Figure 1. Spinel Distribution in a West Valley Core-drilled Glass Sample: (a) the Marbling Pattern, (b) a Magnified View Showing the Fine Structure of Striae

MARBLING PATTERN

When reasoning about the origin of spinel segregation into layers, the first thing that comes to mind as a possible cause is glass inhomogeneity. The temperature at which spinel begins to precipitate during cooling depends on glass composition. As Table I shows, increasing concentration of chromium and nickel strongly increases liquidus temperature, whereas increasing concentration of alkali oxides decreases it [2]. Thus, one can reason, depletion of alkalis or local fluctuation of Ni and Cr can lead to irregularities in spinel crystal distribution within the glass.

Inhomogeneities can result from imperfect mixing of batch materials, segregation during early stages of melting due to buoyancy, melt volatilization, and melt interaction with the crucible. Alkali evaporation decreases the ability of molten glass to dissolve spinel components and results in an enhanced precipitation. Alumina enrichment, which can result from crucible wall dissolution if glass is prepared in an alumina crucible, also promotes spinel precipitation. Although

TABLE I. COMPONENT COEFFICIENTS (IN °C) FOR LIQUIDUS TEMPERATURE WITH SPINEL PRIMARY PHASE [2]

COMPONENT	T_L, °C	COMPONENT	T_L, °C	COMPONENT	T_L, °C
SiO_2	834	MgO	2827	MnO	1870
B_2O_3	395	Fe_2O_3	2644	NiO	8210
Na_2O	-1826	Al_2O_3	3307	Others	4419
Li_2O	-1470	Cr_2O_3	18864		

dealkalization and enrichment by alumina promote spinel precipitation, these effects are strictly localized and cannot be responsible for spinel segregation in a large-scale melter. Also, when glass is carefully homogenized by crushing, mixing, and remelting, the pattern reappears. Moreover, concentration fluctuation can hardly account for segregation of spinel in layers of regular thickness. Clearly, inhomogeneity plays a secondary role, responsible for only a small fraction of the nonuniformity in spinel distribution.

As a recent study [3] showed, spinel nucleation in a waste glass is nearly instantaneous. No measurable incubation time has been obtained from data analysis, and the rate of growth is high at temperatures up to 200°C below liquidus. Because of this ease of nucleation in a homogeneous melt, spinel precipitates virtually uniformly. Therefore, spinel segregation takes place only after the crystals were nucleated.

It is conceivable that segregation of spinel crystals into parallel layers can occur as periodic precipitation. Liesegang [4] observed such precipitation in photographic emulsion. This effect was explained by Ostwald [5] who attributed it to crystallization in an advancing concentration gradient, from which the diffusing component is periodically depleted. This theory was reexamined and found satisfactory by Stern [6]. An alternative explanation is a nonlinear behavior leading to spatial and temporal periodicity due to autocatalytic growth coupled with diffusion [7,8]. One can reason that the Liesegang effect can operate also in glass as a result of advancing temperature gradient. The interplay between oversaturation from cooling and diffusion can lead to periodic precipitation. Steep temperature gradient exists when glass is poured into laboratory molds or into vitrification plant canisters. Jain and Barnes [1] suggest that local temperature variations caused spinel agglomeration into layers in the West Valley canistered glass. However, close observation of laboratory samples provides little evidence that the Liesegang phenomenon is responsible for the marbling pattern. Rather, it seems that it is related to the flow of molten glass.

Figure 2 shows a thin section through the meniscus area of a SP-1 glass [2] sample heat treated at 873°C for 24 h in a small platinum crucible. The flow cell marked by the layers of spinel is driven by a surface tension gradient that is produced by evaporation of alkali borates. Spinel is segregated into curved layers that follow the convection pattern. Dense spinel-filled areas are separated by glass with little spinel in it, as is obvious from the transmitting light micrograph. The streaks are directed diagonally from the meniscus area toward crucible centerline, where they turn upward and horizontally back along melt surface toward the meniscus.

Figure 2. Spinel Distribution Showing a Flow Cell in the Meniscus Area (note the curved top surface of glass in upper left and a crack separating the meniscus tip in upper right)

The evaporation of alkali borates leads to a depletion of surface-active oxides within the thin meniscus layer, resulting in a surface tension increase. This promotes "interfacial turbulence," a flow of melt from the central area where the surface tension is low because the loss of alkali borates is replenished from the bulk glass.

Within the meniscus, the velocity gradients are extremely steep. Because the surface tension gradient is a powerful pump, most of the crucible content circulates though the meniscus within a short time. As was demonstrated in the case of RuO_2 [9], small solid particles such as spinel crystals agglomerate within the meniscus area where the high velocity gradient brings them into contact. This phenomenon, also called shear flocculation, was studied by Smoluchowski (see [10]). The agglomerates tend to grow and eventually create islands of densely packed spinel, which gradually merge into spinel sludge layers. Spinel sludge has a high viscosity, two orders of magnitude higher than glass, and thus resists flow. As a result, layers of viscous spinel sludge form "channels," through which the less viscous molten glass flows. This configuration of spinel in melt is relatively stable.

It is difficult to prepare high-level waste glass that has a uniform distribution of spinel crystals within its volume. However, as has been demonstrated by LaMont [11], spinel segregation can be prevented and its uniform distribution can be preserved for an indefinite period of time if the meniscus convection is inhibited. Hence, convection is the most probable cause of spinel marbling.

During pouring high-level waste glass into canisters, steep velocity gradients are present in the pour spout, through which a thin stream of glass continuously flows, and extremely high gradients occur where the stream of glass impacts solidified glass in the canister and flows radially towards canister walls. Figure 3a shows an area of a core-drilled sample in which segregation between spinel and glass was extremely severe. A detail, shown in Figure 3b, makes an impression, in transmitted light, of an almost solid mass of spinel with a sharp boundary between this mass and the melt in with only sporadic crystals are suspended.

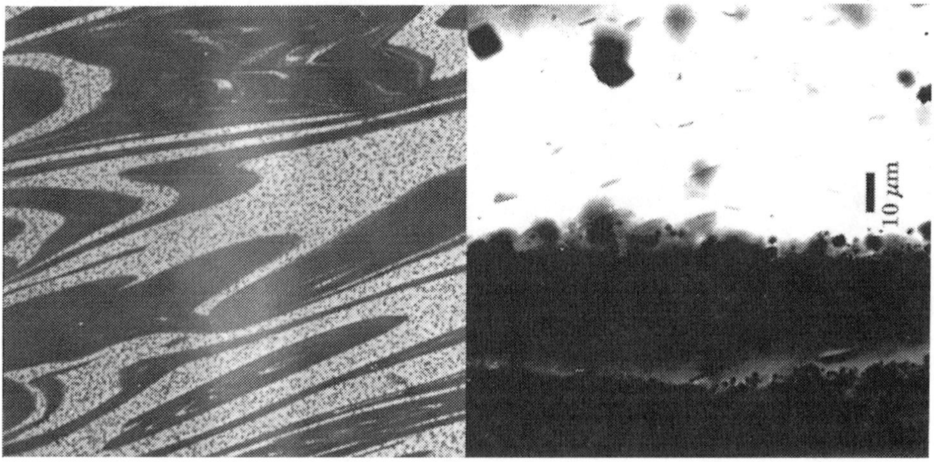

(a) (b)

Figure 3. (a) Spinel Segregation in a Core-drilled West Valley Glass Sample under a Low Magnification (secondary crystallization occurred in the glass probably during reheat); (b) Boundary Region under a High Magnification

In laboratory crucibles, spinel layers, which are heavier than the spinel-free melt, tend to sink to the bottom if the temperature of the melt is kept constant for a sufficiently long time. In the canister, where glass cools relatively quickly, and glass is poured over a long period of time, no settling can occur.

An interference of spinel layers with gas bubbles is frequently observed in both laboratory crucible and plant canister. A bubble deforming spinel layers in a crucible is captured in Figure 4a. A severe distortion of spinel layers in a core-drilled sample, most likely caused by bubbles, can be seen in Figure 4b. Occasionally, bubbles punch holes in spinel layers and bend them up. Bubbles may also drag chunks of spinel with them and lose them in their wake. Continuous interaction between bubbles and melt flow create complex patterns in spinel distribution, which are difficult to decipher. Spinel foam created by flotation of spinel by rising bubbles was observed in the laboratory by LaMont [11].

Bimodal Size Distribution

Bimodal size distribution of spinel is seen in Figures 1b and 5. In laboratory melts, large crystals result from slow cooling or heat treatment of molten glass, the small

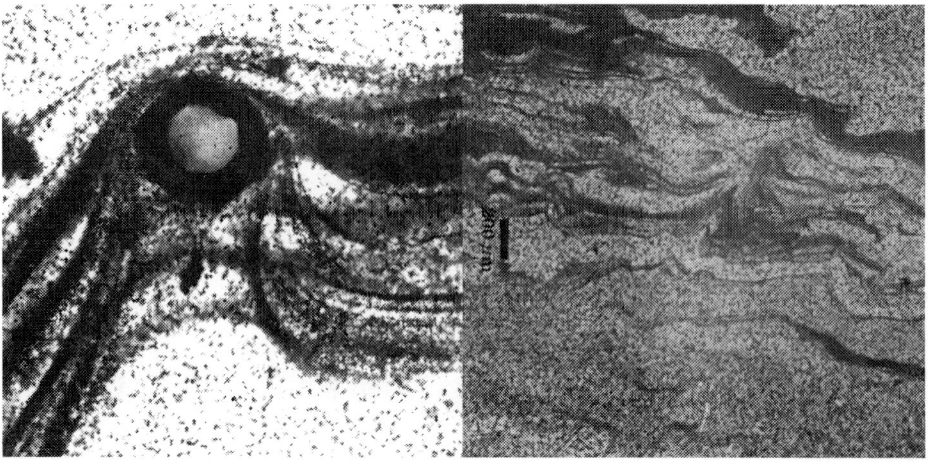

(a) (b)

Figure 4. (a) A Laboratory Example of a Bubble Penetration through a Layer of Spinel; (b) Distortion of Spinel Layers in a West Valley Core-drilled Sample, Probably Caused by Rising Bubbles

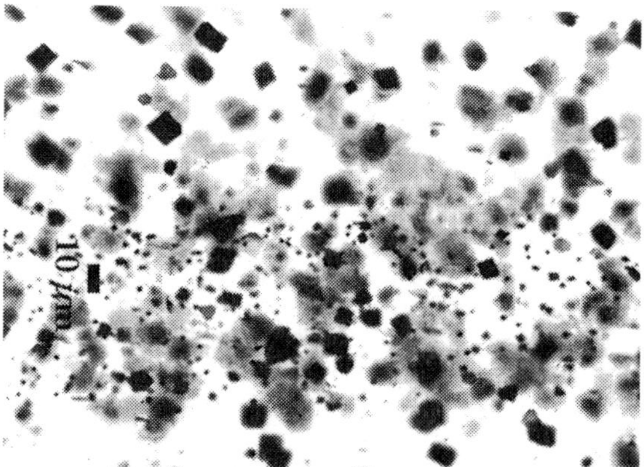

Figure 5. A Layer of Secondary Microcrystalline Spinel in West Valley Core-drilled Glass

ones form during rapid cooling to the ambient temperature. Tiny crystals precipitate on rapid cooling because of the oversaturation at lower temperatures and the ease of spinel formation mentioned above. Fast nucleation and growth makes it difficult to quench glass rapidly enough to avoid spinel precipitation. In glass canisters filled by pouring a narrow stream of glass, the top surface of glass is periodically reheated [1] as the fresh glass flows over it from the central region to which the glass jet is directed. This reheat above T_g produces additional crystallization.

Conclusion

Spinel, a ubiquitous and harmless crystalline form present in high-level waste glasses, is typically distributed nonuniformly with respect to space and size. The nonuniformity of spinel spatial distribution is caused by agglomeration associated with high velocity gradients. The bimodal spinel distribution results from the response of spinel formation to the glass temperature history.

ACKNOWLEDGMENT. The authors are grateful to Scott Bulkley for his help in preparation of this paper.

References

1. V. Jain and S. M. Barnes, "Effect of Glass Pour Cycle on the Crystallization Behavior in the Canistered Product at the West Valley Demonstration Project," *Ceram. Tans.* 23, 239-249 (1991).
2. M. Mika, M. J. Schweiger, P. Hrma, and J. D. Vienna, "Liquidus Temperature of Spinel Precipitating High-Level Waste Glass," *Scientific Basis for Nuclear Waste Management* (Editors W. J. Gray and I. R. Triay), Vol. 465, Material Research Society, Pittsburgh, Pennsylvania, 1997 (in press).
3. J. G. Reynolds and P. Hrma, "The Kinetics of Spinel Crystallization from High-Level Waste Glass," *Scientific Basis for Nuclear Waste Management* (Editors W. J. Gray and I. R. Triay), Vol. 465, Material Research Society, Pittsburgh, Pennsylvania, 1997 (in press).
4. R. E. Liesegang, *Naturwiss. Wochenschr.* 11, 353 (1896).
5. W. Ostwald, *Lehrbuch der allgemeinen Chemie*, p. 778, Engelman, Leipzig (1897).
6. K. H. Stern, "The Liesegang Phenomenon," *Chem. Rev.* 54, 79-99 (1954).
7. S. Kai, S. C. Muller, and J. Ross, "Measurements of Temporal and Spatial Sequences of Events in Periodic Precipitation Processes," *J. Chem. Phys.* 76 [3] 1392-1404 (1982).
8. S. C. Muller, S. Kai, and J Ross, "Periodic Precipitation Patters in the Presence of Concentration Gradients," *J. Phys. Chem* 86, 4078-4087 (1982).
9. W. T. Cobb and P. Hrma, "Behavior of RuO_2 in a Glass Melt," *Ceram. Trans.* 23, 233-237 (1991).
10. R. D. Vold and M. J. Vold, *Colloid and Interface Chemistry*, Addison-Wesley, London, p. 281-282 (1983).
11. M. J. LaMont and P. Hrma, "A Crucible Study of Spinel Settling in a High-Level-Waste Glass," *Ceram. Trans.*, this volume.

A Crucible Study of Spinel Settling in a High-Level Waste Glass

Micheal J. LaMont and Pavel Hrma
Pacific Northwest National Laboratory,[1] Richland, Washington

Abstract

The settling of spinel crystals in a melter for high-level waste vitrification can cause an excessive accumulation of sludge and a blockage of the pour spout. This study elucidates the complex nature of spinel settling in molten glass. The mechanism and kinetics of spinel settling in high-level waste glass were studied in a laboratory crucible at temperatures 16°C and 49°C below the liquidus. Precautions were taken to eliminate melt-line convection and bubble generation, thereby minimizing agglomeration and avoiding "spinel foam." The settling depth in the quiescent melt was found to be a function of the radial distance from the crucible centerline and was 4 to 5 times larger at the centerline than predicted by Stokes law. This difference was attributed to a convective cellular flow driven by settling.

Introduction

Because most Hanford high-level wastes are high in iron and contain chromium and nickel, they generally precipitate spinel crystals, possibly during glass processing at high waste loading [1,2]. The settling of spinel results in the formation of a sludge layer at melter bottom, which may hinder melter performance and shorten the melter life [3]. This study was undertaken to observe the settling rate of spinel along with various other related phenomena such as bubble/spinel interaction. The focus was to test the ability of Stokes law to predict the settling rate of spinel in molten glass. Accurate modeling of spinel settling will allow a more realistic prediction of melter performance at an increased waste loading.

Theory

Settling of an individual solid particle placed in a large volume of liquid can be described by Stokes law [4]

$$v = k\frac{g\Delta\rho r^2}{\eta} \qquad (1)$$

---

[1] Pacific Northwest National Laboratory is operated for the U.S. Department of Energy by Battelle under Contract DE-AC06-76RLO 1830.

where k is a constant, v the settling velocity, g gravity, $\Delta\rho$ density difference between solid and melt, r effective particle radius, and η viscosity. The constant value is $k = 0.205$ for cubic particles [5], such as spinel crystals.

The Reynolds number associated with a single spinel crystal in molten glass is very low, $Re \approx 10^{-7}$. Because of the high viscosity (typically 10 Pa.s), the velocity field associated with the falling crystal extends to the crucible walls; thus the assumption of the walls being "far" is not met, and deviation from Stokes law can be expected. In molten glass, spinel crystals that precipitate as a result of a temperature decrease are approximately 40 μm apart from each other. Therefore, they settle as a swarm of crystals rather than independent spinel particles. One can anticipate the development of one or several flow cells within the melt. Their velocity field will accelerate settling of some crystals and reduce the settling rate of others.

EXPERIMENTAL

A spinel-precipitating glass of the following composition (in mass fractions of oxides) was used: SiO_2 (0.4400), B_2O_3 (0.0680), Na_2O (0.1580), Li_2O (0.0291), MgO (0.0058), Al_2O_3 (0.0777), Fe_2O_3 (0.1500), Cr_2O_3 (0.0024), MnO (0.0035), and NiO (0.0051). The remainder (0.0603) consisted of 24 minor components that included Ag_2O (0.0006) and RuO_2 (0.0003). The glass batch was mixed in a tungsten carbide grinding cell with an Angstrom® disk mill for 8 min and melted in a covered platinum crucible for 1 h at 1150°C. The temperature was raised to 1220°C for approximately 20 min to dissolve any traces of spinel that formed during the initial stage of melting.

To observe settling, an experimental setup was developed that minimized the interference of bubbles and surface-tension-induced convection. Trials were made with silica, alumina, and platinum crucibles. In a silica crucible, bubbles were generated at the bottom and disturbed settling. In alumina and platinum crucibles, bubbles did not form to an intolerable extent, but a strong convective cell driven by the surface tension gradient in the meniscus stirred the melt. The final setup consisted of an alumina crucible (21 mm in diameter and 30 to 38 mm high) positioned inside a silica crucible and completely immersed in the melt.

The double crucible assembly was preheated at 1150°C for 10 to 15 min. The glass melt was poured into the inner crucible and allowed to run over the edges until the top of the inner crucible was just covered with melt. The crucible assembly was then placed back into the furnace at 1150°C. After 10 min, a preheated silica lid was placed on the silica crucible to minimize evaporation. Twenty minutes later, the temperature was lowered to the test temperature. Settling was observed in three tests, at 1050°C for 24 h, at 1050°C for 48 h, and at 1083°C for 24 h. These temperatures were 49°C and 16°C below the liquidus temperature (1099°C). At the end of the heat treatment, the sample was allowed to cool to approximately 850°C to increase melt viscosity before the transport to the annealing furnace, and annealed at 482°C for 4 h. Vertical thin sections were prepared to investigate the area inside the alumina crucible with the Olympus PMG3 Microscope with Leco 2001 Image Analyzer.

Spinel Settling

In all three tests, the settling line of spinel (defined by the position of initially topmost particles) had a jagged U shape (Fig. 1). The upper region was not crystal free due to crystal nucleation and growth upon cooling. In the tests conducted at 1050°C for 48 h and at 1083°C for 24 h, the settling line was marked by a change in the volume fraction of crystals (Fig. 2), whereas in the test conducted at 1050°C for 24 h, the volume fraction of crystals was nearly the same in the upper and lower regions, but the lower region was marked by tiny crystals that were absent above the settling line. At the crucible centerline, the distance of the settling line from the top surface of the melt was 7.2 mm for the 1083°C specimen, and 12 mm for both specimens treated at 1050°C.

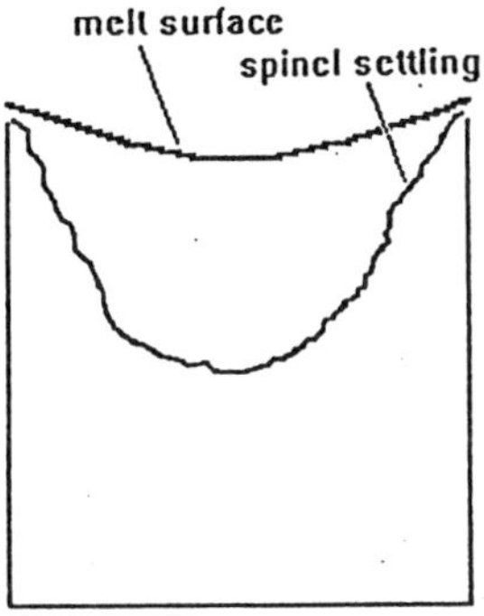

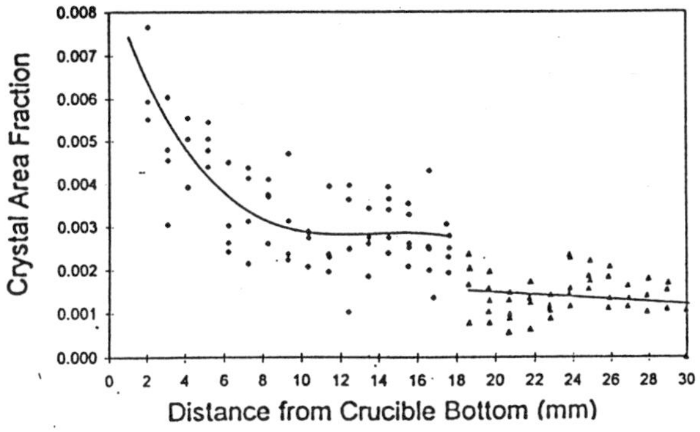

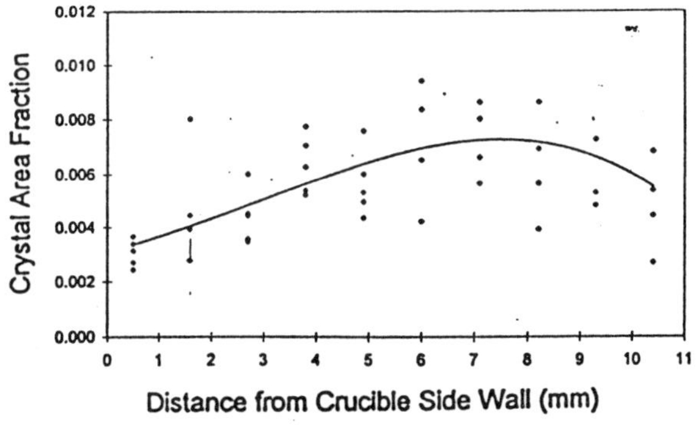

Figure 1. Spinel Settling Line

Figure 2. Spinel Fraction vs. Distance from Crucible Bottom (1050°C/24 h sample at the centerline)

Figure 3. Spinel Fraction vs. Distance from Crucible Sidewall (1083°C/24 h sample, approximately 15 mm above the crucible bottom)

To calculate the Stokes settling velocity, the following property values were used. Spinel precipitating from high-level waste glasses is a solid solution of simple spinels $NiCr_2O_4$, $FeCr_2O_4$, $NiFe_2O_4$, and $FeFe_2O_4$, the density of which ranges from 5.05 to 5.27 g/cm^3. The composition of spinel in our glass was estimated using measured compositions of spinel formed in similar glasses [2]. A mass averaged value was 5.12 g/cm^3. Using empirical models for the room-temperature density[2] [6] and thermal expansion coefficients [6,7], the glass density was estimated as 2.62 g/cm^3. The average crystal size, measured by the optical microscope, was 9.9 μm for the 24-h test at 1083°C, 12.5 μm for the 48-h test at 1050°C, and 15.3 μm for the 24-h test at 1050°C. The measured glass melt viscosities were 12.0 Pa.s and 8.6 Pa.s at 1050°C and 1083°C, respectively.

By Eq. (1), the settling distances of a single crystal in a large volume of melt are 1.3 mm (1083°C, 24 h), 3.1 mm (1050°C, 48 h), and 2.5 mm (1050°C, 24 h). These distances are 4 to 5 times smaller than the measured settling distance at the crucible centerline.

The observed settling behavior is a result of the convection cell produced by the settling crystals, which force the melt to flow down along the cell axis and upward along its perimeter, i.e., the cylindrical walls of the crucible. The mixture velocity and the crystal velocity are superimposed, resulting in the U-shaped settling line. Because the flow and settling velocities are not generally parallel, the radial velocity component affects the distribution of crystal volume fraction within the crucible. The crystal concentration increased with the radial distance from the centerline, reaching a maximum at about two thirds of the crucible radius (Figure 3).

Spinel Agglomeration

Agglomeration was observed in the initial tests. It was caused by convection driven by the surface tension gradient generated in the meniscus by melt evaporation and alumina dissolution. The meniscus convection created a flow cell below the top surface [8], producing dark streaks of agglomerated spinel separated by almost clear melt. Meniscus convection was effectively inhibited in a silica crucible, where the viscosity of the meniscus melt was increased by dissolving silica.

Gas bubbles can also interfere with settling. They are nucleated on crucible walls, grow, break away, and ascend. In alumina crucibles, the rate of bubble formation was too low to affect spinel settling, though the settling line was jagged by passing bubbles. However, bubbles rapidly formed at the silica crucible bottom. These bubbles dragged with them a high-viscosity silica-rich melt. As they passed through the layer of settled crystals, a mass of crystals was raised on each bubble (Figure 4a), dropping off clusters of spinel in its wake (Figure 4b). The remaining spinel was deposited at the top surface where crystals, bubbles, and high-viscosity melt created a persistent foam. The formation of bubbles in the silica-rich melt under the layer of spinel sludge can be attributed to a shift in the iron redox ratio toward a more reduced state. Spinel agglomeration was also assisted by the needle-like crystals of noble metals, on which spinel crystals were nucleated (Figure 4c). This secondary effect did not interfere with settling.

Possible Effects of Spinel on Melter Operation

Formulation of glasses that precipitate spinel is desirable because these glasses can immobilize a substantially higher amount of waste per unit volume of the waste

---

[2] P. T. Fini, Partial Specific Volumes of Oxides in High-Level Waste Glasses, private communication.

form. Precipitation of spinel in the melter may not lead to an increasing accumulation of spinel sludge if the crystals are removed from the melter with the glass. This may be achieved by preventing settling or by the removal of settled spinel from the melter, either continuously or periodically. To predict spinel behavior in a large-scale melter, spinel interaction with bubbles and with thermal convection must be considered. Spinel agglomerates may settle more rapidly than individual crystals. As this study shows, hydrodynamic interaction of individual crystals results in an accelerated settling.

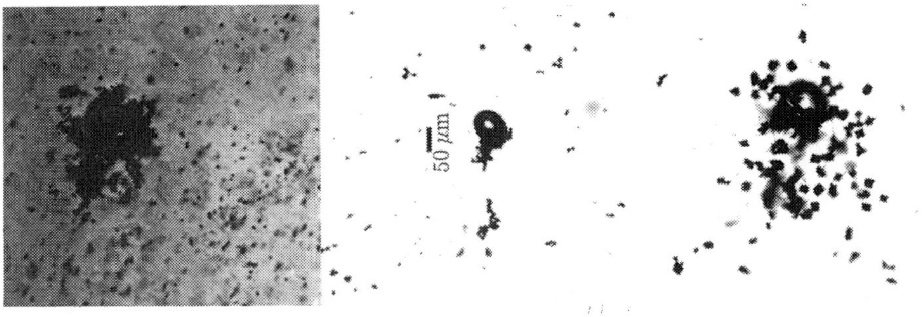

(a) (b) (c)

Figure 4. A Spinel-coated Gas Bubble (a), a Bubble Losing Spinel Clusters (b), and a RuO_2-bonded Spinel Agglomerate (c).

A buildup of spinel on the bottom of the melter can shorten the melter lifespan. Also, precipitation of spinel in the pour spout can hinder or stop melter discharge. In addition, rising bubbles can create "spinel foam" that may harden into a spinel crust, which is extremely difficult to remove. The melter must be stopped or fed with a low-viscosity frit until the foam collapses and the refractory crust dissolves [9]. However, this may offset the benefit of increased waste loading.

Glass formulation to maximize waste loading while minimizing the risk of melter problems requires understanding of spinel behavior, that is, 1) the equilibrium concentration of spinel as a function of glass composition and temperature, 2) the agglomeration and settling rate of spinel in a large-scale melter, and 3) the rheology of spinel sludge. The database of the equilibrium concentration of spinel will consist of the liquidus temperature and the fraction of spinel below the liquidus, both as a function of glass composition. This study shows that a quantitative prediction of spinel settling in a melter is possible if spinel crystals are treated as swarms that exchange momentum with molten glass. Preliminary experiments[3] indicate that spinel sludge behaves as a pseudoplastic rheopectic liquid.

---

[3] M. Mika, private communication.

Conclusions

This study demonstrates the complex nature of spinel settling in molten glass that involves hydrodynamic interaction of settling crystals that may be complicated by agglomeration. Spinel settling in glass is not described by Stokes law. The behavior of melt-spinel mixture can be adequately described as a two-phase system, in which the phases exchange momentum. Hence, it is conceivable that spinel behavior can be incorporated into mathematical modeling of flow and heat transfer in the melter. Also, glass formulation can be optimized to maximum waste loading while minimizing the risk for the melter. Finally, melter construction can be modified to process melts with spinel while maintaining a steady state during the melter campaign.

Acknowledgment

Authors are grateful to Michael Schweiger, John Vienna, Don Smith, Jarrod Crum, Michael Elliott, and Reza Shekarriz for helpful discussions. They are also grateful to Scott Bulkley for scanning the micrographs. Micheal LaMont would like to thank the SERS program, Associated Western Universities, Inc., for the fellowship at the Pacific Northwest National Laboratory.

References

1. J. G. Reynolds and P. Hrma, "The Kinetics of Spinel Crystallization from High-Level Waste Glass," *Scientific Basis for Nuclear Waste Management* (Editors W. J. Gray and I. R. Triay), Vol. 465, p. 65-70, Material Research Society, Pittsburgh, Pennsylvania, 1997.
2. M. Mika, M. J. Schweiger, P. Hrma, and J. D. Vienna, "Liquidus Temperature of Spinel Precipitating High-Level Waste Glass," *Scientific Basis for Nuclear Waste Management* (Editors W. J. Gray and I. R. Triay), Vol. 465, p. 71-78, Material Research Society, Pittsburgh, Pennsylvania, 1997.
3. L. R. Bunnell, *Laboratory Work in Support of West Valley Glass Development*, PNL-6539, Pacific Northwest Laboratory, Richland, Washington (1988).
4. J. Happel and H. Brenner, *Low Reynolds Number Hydrodynamics*, Martinus Nijhoff, Hague, Netherlands (1973).
5. E. A. Pettyjohn and E. B. Christiansen, *Chem. Eng. Progress*, 44, 157 (1948).
6. H. Scholze, *Glass, Nature, Structure, and Properties*, Springer, New York (1991).
7. P. Hrma, G. F. Piepel, M. J. Schweiger, D. E. Smith, D.-S. Kim, P. E. Redgate, J. D. Vienna, C. A. LoPresti, D. B. Simpson, D. K. Peeler, and M. H. Langowski, *Property/Composition Relationships for Hanford High-Level Waste Glasses Melting at 1150°C*, PNL-10359, Pacific Northwest National Laboratory, Richland, Washington (1994).
8. W. T. Cobb and P. Hrma, "Behavior of RuO_2 in a Glass Melt," *Ceram. Trans.* 61, 187-193 (1995).
9. G. P. Abrigo and D. A. Lamar, *SSHTM-1 and SSHTM-2 Melter Performance Test*, PVTD-T3C-95-138, Pacific Northwest National Laboratory, Richland, Washington (1995).

Hazardous Waste Stabilization and Alternative Waste Forms

COMPARISON OF BENCH AND PILOT PLANT RESULTS ON THE VITRIFICATION OF U.S. AIR FORCE HAZARDOUS WASTE STREAMS

P. Nevatia* and C.H. Drummond, III, The Ohio State University, Columbus, OH 43210; R. Richards, Seiler Pollution Control Inc., Dublin, OH 43017; C.R. Burnett, McClellan AFB, Sacramento, CA 95652.

ABSTRACT:

The ever increasing disposal costs as well as stringent environmental regulations have necessitated alternative ways of handling hazardous waste. Vitrification technology presents a product based recycling approach to the problem. Five U.S. Air Force waste streams were characterized and a bench scale feasibility study was conducted. Four formulations were developed and subsequently processed through a pilot system. The mass balance and incorporation of hazardous metal from the Seiler high temperature vitrification pilot system have been reported and critically evaluated. Glass-ceramic non-hazardous products were developed with spinel and augite phases in a glass matrix. The products had critical performance properties. The effect of salt content and waste variability have been studied and some solutions have been recommended.

INTRODUCTION:

Worldwide concern about the safety and wisdom of traditional disposal methods of hazardous wastes and the ever increasing costs associated with them have caused governments and industry to recognize the imperative need for a different and better approach to hazardous waste management. The United States generates over 300 million tons of hazardous waste each year and the worldwide generation is unknown. Currently, hazardous wastes are disposed by incineration, deposited in surface impoundments, placed in landfills and sometimes stored in abandoned mines. All of these methods require close and continuous monitoring of the disposal sites to ensure their security and safety. A much superior technique is to use a recycling approach to make non-hazardous commercially viable products using the hazardous wastes as raw materials. Glass and glass-ceramic products are ideally suited for the purpose as the glass incorporates the heavy metal ions into its structure making them non-susceptible to chemical and physical attack[1]. Products like medium grade abrasives for blasting applications and roofing granules have been developed[2] using the Seiler high temperature vitrification (HTV) system.

The Seiler HTV system comprises of a preheater operated at 800°C and a vitrifier along with air and water treatment units. The vitrifier is an oxygen-gas fired furnace with a product burner flame that operates at temperatures of 1700°C - 2000°C and a reservoir for homogenization of the product at temperatures around 1500°C. The vitrified product exits the furnace and can be processed to obtain the final product. The volatalized material is exhausted through a baghouse filter where it can be condensed and collected. The pilot plant has a throughput of 150 kg/h and the commercial system has a throughput of 500 kg/h .The typical residence time in the system is 10-15 minutes.

Five waste streams generated at the McClellan and Tinker U.S. Air Force bases were studied on a bench scale to determine their feasibility for product development using the Seiler HTV system. Table I lists the waste streams and their disposal costs. The products developed were medium grade abrasives and roofing granules. The performance properties of roofing granules are consistent color and sizing. Medium grade abrasives should have acceptable hardness (Vickers hardness 600-800 kgf/mm^2) and consistency. Critical property required of the products is incorporation of the RCRA8 hazardous metals (Ag, As, Ba, Cd, Cr, Hg, Pb, Se). Also the products must be non-leachable and in compliance with the EPA Toxicity Characteristic Leaching Procedure (TCLP). The waste streams were characterized and formulations were developed, using the waste streams and waste additives, such as foundry sand in crucible melts. The products obtained were characterized and accepted if they met the requisite performance and critical properties. These formulations were subsequently processed through the pilot plant to gauge the feasibility of installing a commercial system for processing these wastes and also to evaluate the products generated in large quantities. The waste streams processed through the pilot, the products obtained, as well as the baghouse dust collected, were characterized and their properties were compared with the bench results. Mass balance and metal incorporation calculations were conducted and the differences due to the scale-up were evaluated.

Table I: U.S. Air Force waste streams and their disposal costs

Waste streams	**Disposal costs**
	($/ton, dry basis)
A- Garnet blasting media	483
B- IWTP sludge	1833
C- IWTP sludge	697
D- Oil water separator sludge	1845
E- Plastic blasting media	NA

EXPERIMENTAL:

The waste streams and products for bench and pilot studies and the baghouse dust from the pilot run were characterized in a similar manner. Percent solids, percent loss on ignition (LOI), elemental constituents using SEM/EDS and trace heavy metal content were measured. In addition, for the products optical microscopy and XRD were used to characterize the microstructure. The microhardness was measured using Vicker's Microhardness analyzer. The color was determined using Munsell color charts and the specific gravity was obtained using a pychnometer.

The waste streams were mixed with foundry sand and melted in alumina crucibles in an electric furnace at 1500^{0}C for two hours. The melts were furnace cooled to produce glass-ceramic samples. The pilot operation in Leibstadt, Switzerland consisted of a preheater at 800^{0}C to evaporate the moisture and volatiles and an oxygen-gas burner with a flame temperature of 2000^{0}C and a reservoir at 1500^{0}C. The product then exited the system and the volatiles were condensed and captured in a baghouse as dust. The air was exhausted through an air pollution control system and was released through the stacks. Representative samples were taken from all streams for characterization.

% Solids and % LOI

Samples of the waste streams and baghouse dust were dried in an electric oven at 105^0C for 4 hours and the weight loss recorded to calculate the % solids. These samples were further fired at 800^0C for 4 hours and weight loss recorded to calculate the % loss on ignition (LOI).

SEM/EDS

Pellets of the waste stream and baghouse dust samples already fired at 800^0C were made and coated with colloidal graphite. The product samples were mounted in Al filled conductive bakelite and polished to 0.5 microns using standard metallographic polishing techniques. These samples were viewed under an optical microscope. The polished samples were carbon coated. Using SEM and EDS on a Phillips XL30, the sample constituents were identified.

X-ray Diffraction (XRD)

The product samples were crushed into fine powder for X-ray diffraction testing to identify crystal phases present, using a Scintag PAD-V X-ray diffractometer. The samples were irradiated with Cu-Kα radiation with a scan rate of 0.5 degree/min over 20 to 80 degrees. The low scan rate was chosen to increase the signal over the background to account for fluorescence due to iron in the samples. The crystal phases were identified using the JCPDS database.

Microhardness and Color

A Buehler Micromet® II Digital Micro Hardness Tester was used to measure the Vicker's hardness of the samples. An indentation load of 200g and a dwell time of 30 seconds was used. Multiple values were recorded and the mean values and standard deviation were reported. The color of the products was determined by comparing them to standard Munsell color charts[3].

Trace Metal Analyses and Toxicity Characteristic Leaching Procedure (TCLP)

The waste, product and the baghouse dust were analyzed for the RCRA8 heavy metal content . The metals being arsenic, barium, cadmium, chromium, lead, mercury, selenium and silver. The toxic content leached out of the products was determined by the TCLP as per the EPA procedure SW846. The analyses was conducted at Wilson Environmental Laboratories Inc.[4]

RESULTS:

The major constituents of the waste steams as determined by EDS are listed in Table II. In addition, all the wastes had other oxides like Na_2O, MgO and K_2O present. Waste E also had some TiO_2 present. The toxic metals of concern were cadmium, chromium and lead. Their concentration is shown in Table III. The quantities present are well in excess of the EPA guidelines. The waste were then mixed in appropriate ratios to produce four formulations for glass-ceramic products. The third formulation was characterized twice to insure quality control. Formulations were consistent with the relative quantities of the waste at the separate facilities.

Table II: Major constituents (wt %) of waste streams used in bench study

Oxide	A	B	C	D	E
Fe_2O_3	36.15	2.96	44.40	7.76	33.26
Al_2O_3	20.27	7.7	4.19	11.09	7.76
SiO_2	37.5	33.58	10.85	50.65	44.34
CaO	1.56	49.70	27.28	13.86	3.32
Cr_2O_3	0.05	0.74	4.19	6.28	2.21
other		5.32	9.09	10.36	9.11

Table III: Toxic metals (mg/kg) of concern in waste streams used in bench study

Elements	A	B	C	D	E
Cd	16.7	130	199	353	46.1
Cr	118	2180	7600	13,800	259
Pb	389	851	85.8	275	228
org. carbon	< 1621	16,800	47,400	46,100	41,800

The formulations for both bench level and pilot run are shown in Table IV on an organic flashed basis. The percent solids and percent residue on ignition of the formulations for the bench scale and pilot run are shown in Table V. The values obtained were vastly different. The feed for the pilot run had a much higher moisture and volatile content resulting in huge differences on the relative ratios of the waste streams as compared to the formulations on the bench scale. The product formulation 4 had a residue on ignition of 800°C of 8.6 % as opposed to 58.27 % on the bench studies. The difference is manifested in the formulations shown in Table IV. The formulations for both bench and pilot runs were the same on a wet basis.

Table IV: Formulations on an Organic Flashed Basis (wt %)

Waste	Formula 1		Formula 2		Formula 3		Formula 4	
Stream	pilot	bench	pilot	bench	pilot	bench	pilot	bench
A	82.8	78.2	66.3	61.8	0	0	0	0
B	16	19.8	22.3	27.2	0	0	0	0
C	0	0	0	0	67.8	66.5	66.7	42.7
D	0	0	0	0	0	0	1.8	35.8
E	1.2	2	0	0	0	0	0	0
Sand	0	0	11.4	11	32.2	33.5	31.5	21.5
Total	100	100	100	100	100	100	100	100

Table V: Solid and Volatile content of the Feed

Sample	% Solids		Residue on ignition (%) @ 800°C	
	Pilot	Bench	Pilot	Bench
Formula 1	85.12	66.29	68.74	79.83
Formula 2	74.45	69.90	67.06	90.62
Formula 3	44.88	24.82	30.72	56.56
Formula 4	14.77	19.60	8.60	58.27

The difference in feed compositions led to a large variation in the product compositions and hence on the product properties. The values for TCLP on products from both bench and pilot runs were well within EPA requirements as shown in Table VI. Therefore, the product was able to incorporate the toxic metals into its structure inspite of the variations in composition. The chromium, in particular, was incorporated into the highly durable spinel phase which is present in these products. The values for the Vicker's hardness of the products are listed in Table VII. It was found that the hardness values were in the same range for all the four products. Formulation 1 and 2 were developed for medium grade abrasives and the hardness values in the pilot were lower but still high enough to be used as an abrasive media. All the products were very uniform as indicated by the low standard deviation values. The products were homogeneous and consistent. The specific gravity values were in the range 2.9-3.3 These values are higher than typical glasses due to the presence of the denser spinel phase. The color was consistent and the values are shown in Table VIII. Thus, the products can be used as roofing granules.

Table VI: TCLP Data (mg/l) on Products

Elements	Formulation 1	Formulation 2	Formulation 3	Formulation 4
Cd	<0.02	<0.02	<0.02	<0.02
Cr	0.04	<0.03	<0.03	<0.03
Pb	0.019	0.002	0.003	0.021

Table VII: Vicker's Hardness of Product

Sample	Hardness, kgf/mm^2 (mean, standard deviation)	
	Pilot	Bench
Formula 1	611, 14	719, 66
Formula 2	615, 30	657, 28
Formula 3	637, 34	600, 13
Formula 3b	624, 22	600, 13
Formula 4	629, 14	626, 18

Table VIII: Color and Specific Gravity of Products

Sample	Munsell color code, Hue Value/Chroma	Specific gravity	
		Pilot	Bench
Formula 1	N 2/0	3.00	2.92
Formula 2	N 2/0	2.92	3.05
Formula 3	N 1/0	3.07	2.91
Formula 3b	N 1/0	3.11	2.91
Formula 4	5BG 3/1	3.25	3.29

The x-ray diffraction pattern was compared to JCPDS files and all the products showed the presence of spinel, believed to be a spinel solid solution and augite [$Ca(Mg,Fe)(SiO_3)_2$]. A typical microstructure for the medium grade abrasive products is shown in Figure 1. It shows presence of dendritic spinel and some equiaxed spinel phase and few augite crystals in a glass matrix.

Mass balance calculations were conducted on the feed and product streams in the pilot run. An estimate of the product expected was made by using the data on weight loss on vitrification obtained during the bench runs. The mass balance is summarized in Table IX. It was found that 87.2% of the expected quantity of the product was obtained. It is believed that the difference in quantities was due to the change in atmosphere from electric to gas fired furnaces and also due to scale up from crucible quantities to several pounds in the pilot. Also, the volatalization was greater in the pilot plant due to higher flame temperatures. The effect of temperature on weight loss from these melts was studied on formulae 3 and 4. It was found that the weight loss increased with temperature in a near linear manner as shown in Figure 2.

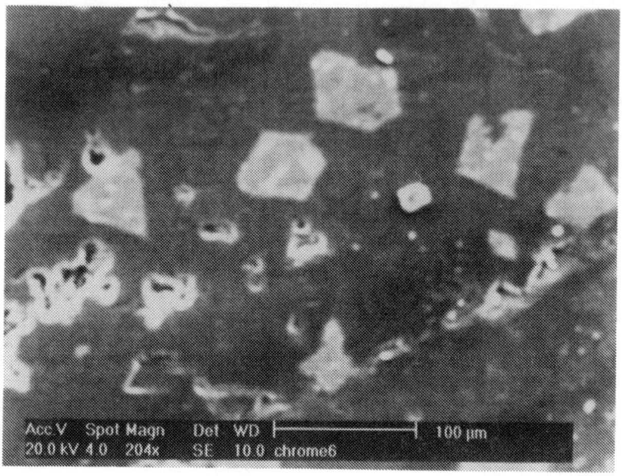

Figure 1: Microstructure of a Typical Medium Grade Abrasive Product

Table IX: Mass Balance Data on Pilot Run

Sample	Feed, kg	Product, kg		Baghouse dust, kg
		Actual	Expected	
Formulation 1	231	67.4	151.9	2
Formulation 2	476	270	305.7	6
Formulation 3	660	189	179.7	5
Formulation 4	609	48	48.8	2
Total	1976	574.4	686.1	15

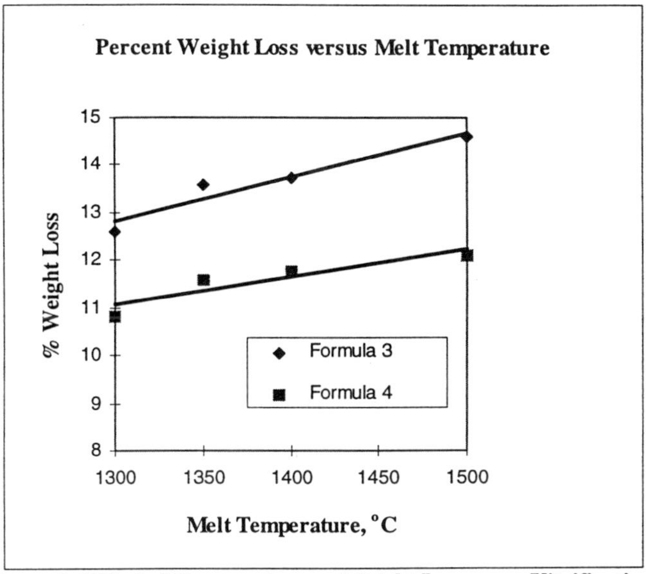

Figure 2: Effect of Temperature on Weight Loss upon Vitrification.

The incorporation of the toxic metals into the product was studied. The metal incorporation data is summarized in Table X. The incorporation was high for formula 1 and 2 on bench runs. There was relatively poor incorporation in run 3 and 4 on the bench scale, however run 3 exhibited high incorporation on the pilot scale. The incorporation of cadmium was found to be a problem. The incorporation of lead was high as it has easy acceptance in the glass structure. The chromium solubility in the glass is limited and it was found that all the chromium partitioned to the spinel phase. This is beneficial from a chemical durability standpoint as the spinel phase is more durable than glass. In the pilot runs the overall incorporation of metals was found to be high except in formula 4. The incorporation of cadmium and lead was low. In some cases the amount of chromium in the product was found to be more than the amount in the feed. This chromium is believed to be coming from refractory corrosion. In such cases the incorporation has been reported as 100% as the product was capable of incorporating all the chromium in the feed.

Table X: Metal Incorporation Data of Toxic Metals

Sample	% Lead		% Cadmium		% Chromium		Total	
	pilot	bench	pilot	bench	pilot	bench	pilot	bench
Formula 1	12	100	3	81.7	82	100	67	98
Formula 2	32	100	4	100	100	91.6	86	99
Formula 3	23	85.4	6	3.9	93	39.1	90	40
Formula 4	5	68.5	2	2.74	28	28.5	26	29

The molecular state of the metal ions has a drastic effect on the vapor pressure of the metal or its salt. This is shown in Figure 3. The change in vapor pressure greatly affects the volatilization of the metal and its incorporation. In general, the vapor pressure of the oxide is much lower than the metal or its salt. Thus, presence of the metal in its oxide form should result in higher rates of incorporation. This is particularly true for cadmium. The cadmium metal volatilizes under standard pressure at 763°C whereas cadmium oxide volatilizes at 1550°C. Thus, change in the salt content of the waste should result in different metal incorporation rates.

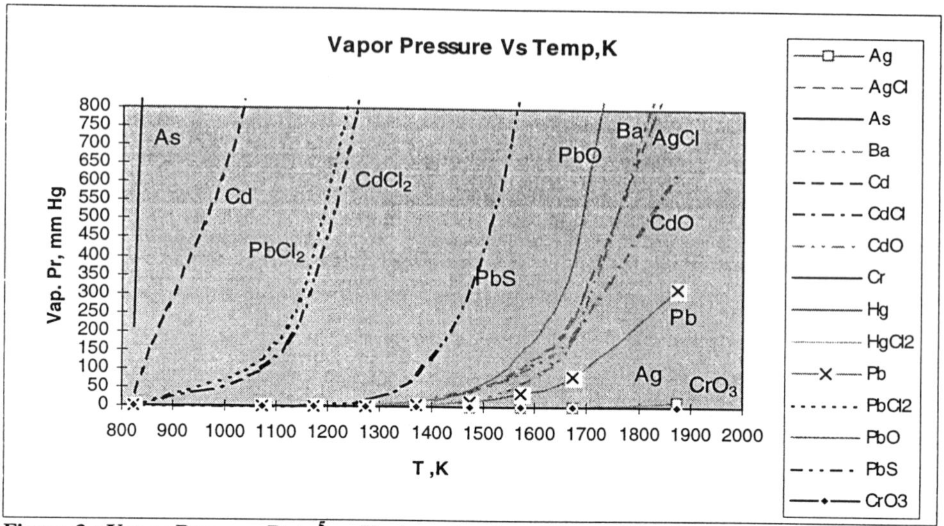

Figure 3 : Vapor Pressure Data[5] on Various Species of Heavy Metals

Table XI: Relative anion content of the waste streams (Elemental Weight Percent)

Anions	A	B	C	D	E	Sand
O	98.81	96.98	77.99	83.87	97.63	100.00
P	0.00	0.75	2.75	8.22	0.09	0.00
S	0.62	1.98	3.11	4.43	2.28	0.00
Cl	0.56	0.29	16.16	3.49	0.00	0.00

Table XI shows the analyses of the relative anion content of the waste streams used in bench experiments. The data suggests that the McClellan wastes A, B and E, had a high percentage of oxides and very little P, S, or Cl was detected. The formulations utilizing the Tinker wastes, C and D, showed a significant content of anions other than oxygen. This suggests that a significant weight fraction of these wastes contained metal chlorides, sulfides or sulfates, and/or phosphates. The poor metal incorporation in formulae 3 and 4 which utilized waste stream C and D can thus be attributed to a higher salt content such as chlorides, sulfides, sulfates and phosphates.

SUMMARY:

Glass-ceramic products were developed for applications, such as blasting media and roofing granules, from five U.S. Air Force hazardous waste streams. The waste streams were characterized on a bench scale and four formulations were developed from crucible melts. These formulations were run on a Seiler high temperature vitrification pilot system in Switzerland. The waste streams, products and baghouse dust from the pilot run were characterized. Significant differences in percent solids and percent loss on ignition at 800°C were found in the waste streams for the pilot run as compared to the bench waste streams. It is recommended that the waste streams be characterized for percent solids and percent loss on ignition on line and the formulation be kept constant on a volatile free basis. The products however were consistent and had the intended properties. In particular, the products passed the US EPA Toxicity Characteristic Leaching Procedure and were able to incorporate significant quantities of hazardous metals. The microstructure of the products consisted of dendritic and equiaxed spinel and augite in a glass matrix. The spinel phase was found to enhance the chromium incorporation in the product. The products from formula 4 had a low hazardous metal incorporation (26%). This is believed to be due to high salt content in the waste stream. The vapor pressure data indicates that the hazardous metals can be better incorporated in an oxide form. Further processing may be done to convert salts into oxides for higher hazardous metal incorporation, in waste streams with high salt content. The mass balance data indicated that 87.2% of the expected product was obtained. Higher volatilization rates in the pilot due to high flame temperature and significantly higher volatile content of the feed decreased the amount of material produced. The pilot demonstrated the feasibility of using varied hazardous waste streams to make non-hazardous glass ceramic products for industrial applications.

ACKNOWLEDGMENT:

This work was supported by the U.S. Air Force (McClellan and Tinker bases) through a contract with Radian Corporation and a sub-contract with Seiler Pollution Control Inc.

REFERENCES:

1. D. F. Bickford and R. Schumacher, "Vitrification of Hazardous and Radioactive Wastes", Westinghouse Savannah River Technology Center, Aiken SC 29808, (1995).
2. R. D. Blume, C. H. Drummond, III and A. B. Sarko, "High Grade Abrasive Product Development from Vitrified Industrial Waste", pp 229-239 in *Environmental Issues and Waste Management Technologies in the Ceramic and Nuclear Industries II*, Ceramic Transactions Series, Vol 72. Edited by V. Jain and D. Peeler, (1996).
3. A.H. Munsell, *Atlas of the Munsell Color System*; Malden, Mass., Wadsworth, Howland & Co., Inc., (1915).
4. Wilson T. Walker, Wilson Environmental Laboratories Inc., 401 Venture Drive, Suite C, Westerville Ohio 43081.
5. C. L. Yaws; Appendix C, pp 352-357 in *Handbook of Vapor Pressures, Vol. 4*, Gulf Publishing Company, Houston TX, (1995).

PHASE DEVELOPMENT IN HEAT-TREATED VITRIFIED INDUSTRIAL WASTE

Zheng Gao and Charles H. Drummond, III, Department of Material Science and Engineering, The Ohio State University, Columbus, OH 43210

ABSTRACT

Differential thermal analysis (DTA), X-ray diffraction and scanning electron microscopy were used to study nucleation and crystal growth behavior of vitrified industrial waste containing electric arc furnace (EAF) dust and foundry sand. Three crystalline phases were identified in the devitrified product: spinel, augite and feldspar solid solutions. Differential thermal analysis of the vitrified and heat-treated samples established that the DTA exothermic peak at 1060-1085°C was associated with the crystallization of feldspar and iron-rich spinel solid solutions. The endothermic DTA peak at 1180-1220°C was associated with the melting of a metasilicate (augite) and a tectosilicate (feldspar). Simulated melts of the augite composition based on EDS analysis confirmed that augite melted within a temperature range that was consistent with the DTA endothermic peak position. It was established that increasing melting time resulted in decreased formation of augite. This was due to increased vaporization of alkali and alkaline-earth constituents from the melt which are necessary for the formation of augite. An increased cooling rate inhibited the crystallization of both feldspar and augite to a greater extent than was observed for spinel which indicates that the silicate phases are kinetically slower to crystallize.

INTRODUCTION

Vitritication of hazardous industrial waste into durable commercial glass or glass-ceramic products has been given more attention than traditional waste treatment and landfilling because of both environmental and economic benefits. The market of ceramic products expands every year in the following areas: abrasives, refractories, engineering and technical ceramic products and ceramic fibers. Abrasive applications[1] share 23% of the multibillion industrial ceramic market. Among them, the market value for high grade abrasives (Vickers hardness value (VHN)>1000 Kg/mm^2) is approximately $250/ton which is up to 10 times higher than that of the medium grade abrasive (750-1000 Kg/mm^2). Previous work[2] by Blume and Drummond had produced an medium grade abrasive product utilizing industrial waste (EAF dust and foundry sand) as the primary constituents. This product has high VHN value (936±124 Kg/mm^2), good mechanical durability, and also incorporated hazardous components, such as Cr, Cd and Pb into a chemically durable matrix.

It has been shown[2] that the formation of equiaxed spinel is necessary to increase VHN for high grade abrasive applications. The understanding and control of the crystallization behavior in this system can lead to optimization of product characteristics resulting in a higher value-added product. Previous work[3] has determined two crystalline phases are present in this system: spinel solid solution of type $(Fe^{+2}, Cr^{+6}, Zn^{+2}, Mg^{+2}, Si^{+4})(Fe^{+3}, Cr^{+3}, Al^{+3}, Fe^{+2})_2O_4$ and augite solid solution of type $(Fe^{+2}, Ca^{+2}, Zn^{+2}, Mg^{+2}, Mn^{+4})[(Al^{+3},Si^{+4})_2O_3]_2$. The temperature of the maximum nucleation rate for the spinel phase is 635°C. The VHN value for spinel phase is over 1000 Kg/mm^2. Achievement of desired properties strongly depends on a high volume fraction of spinel. In order to maximize spinel in the product, the quenched sample was heat-treated at the nucleation temperature (635°C) and the crystal growth temperature (835°C). The morphology of as-quenched sample with subsequent heat-treatment is shown in Figure 1. Curve 1 of the DTA scan (Figure 2) shows two exothermic crystalline peaks. The first crystalline peak corresponds to formation of both spinel and augite. Curve 2 shows complete crystallization of the spinel as indicated by the absence of the first crystallization phases peak. The crystalline phases formed at second DTA peak is much more complicated. The formulation used in this study is quite similar to compositions of natural basaltic melts. Work by Bowen[4] and Osborn [5], on the crystallization sequences in basaltic magmas, has established that magnetite and feldspar (anorthite and albite) phases will crystallize from the melt at higher temperatures.

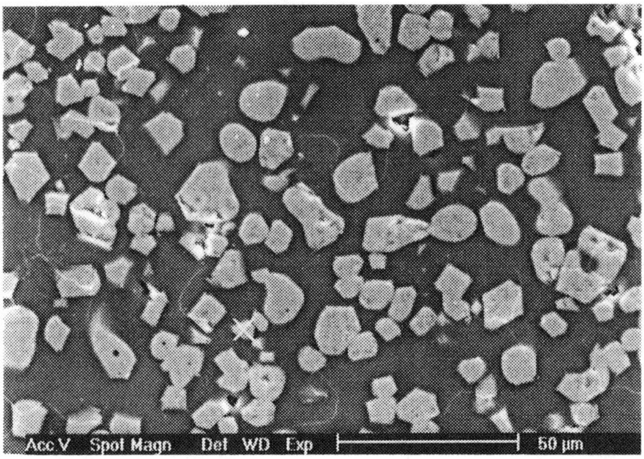

Figure 1: As-quenched sample heat-treated at nucleation temperature (635°C) and crystalline growth temperature (835°C)

In addition to composition and heat-treatment, process parameters such as melting temperature[3], time and cooling rate will also affect the product properties and morphology. Increased melting time alters the chemical composition[6] due to loss of volatile constituents and corrosion of refractory materials. Glasses containing high proportions of alkali or alkaline-earth oxides are very corrosive.

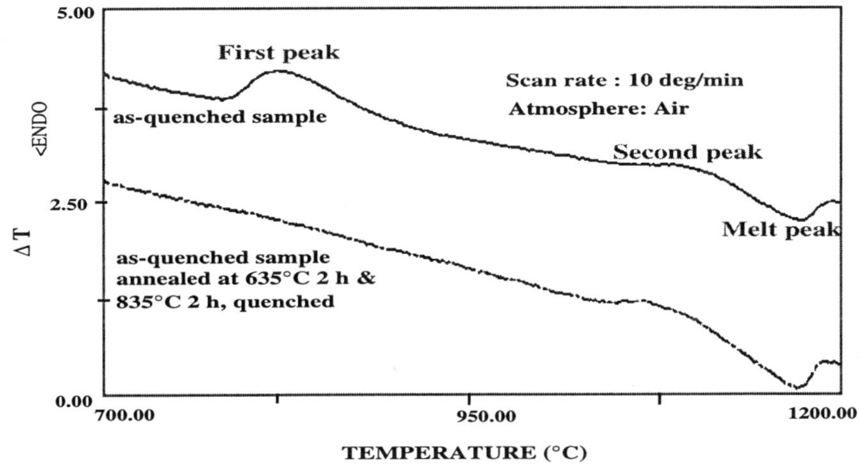

Figure 2: Comparison DTA scans of as-quenched sample and sample annealed at 635°C for 2 h and 835°C for 2 h

EXPERIMENTAL

Sample Preparation

Samples of 40 grams were placed in a Al_2O_3 crucibles and heated in electric furnace with a heating rate of 10°C/min. Samples were held for 0.5 , 2 and 6 h at 1500°C and then furnace cooled or quenched to room temperature. Tables I and II list the sample formulation on a dry basis and on a flashed oxide basis. Table III lists the composition of industrial waste (EAF dust and foundry sand) used in this formulation.

Sample Characterization

Simulated augite composition was produced using pure oxides based on EDS analysis of the augite phase. Approximately 30 mg of a furnace-cooled augite sample was ground to -150 mesh and placed in a 100 mm^3 Al_2O_3 crucible for subsequent thermal analysis utilizing a Perkin-Elmer DTA 1700. DTA scans were performed in an air atmosphere with a heating rate of 10°C/min.

Crystalline phases were identified by X-ray diffractometer (Scintag PAD-V diffractometer). Powder samples (-150 mesh) were characterized with Cu-Kα radiation at a scan rate 0.5 deg/min. The scan angle (2θ) varied from 20° to 80°.

Crystal morphology and composition were investigated by SEM/EDS (Philips XL30 FEG). Samples were mounted in epoxy or bakelite, polished to 0.3μm and coated with colloidal graphite to reduce surface charging.

Table I: Sample Composition (wt.%) on Dry Basis — A Medium Grade Abrasive Formulation

Component	Weight Percent (%)
Armco EAF Dust	53.46
Duriron Foundry Sand	19.66
Kaolinite Clay	15.31
$CaCO_3$ (>99.5%)	7.00
Al_2O_3 (>99.0%)	3.06
$NaHCO_3$(>99.5%)	1.52
Total	100.01

Table II: Sample Composition on a Flashed (500°C) Oxide Basis — A Medium Grade Abrasive Formulation

Component	Weight Percent (%)
SiO_2	33.36
Fe_2O_3	24.25
MgO	11.45
Al_2O_3	9.81
CaO	9.01
ZnO	6.59
Cr_2O_3	3.58
PbO	1.21
Na_2O	0.63
MnO	0.10
Total	99.99

Table III : Composition of EAF Dust and Foundry Sand (wt. %)

Component	Electric Arc Furnace Dust *	Foundry Sand**
Al_2O_3	0.7	1.0
CaO	6.7	0.0
Cr_2O_3	4.2	0.0
Fe_2O_3	29.3	0.1
K_2O	0.0	0.8
MgO	23.0	0.0
MnO	1.9	0.0
PbO	1.4	0.0
Na_2O	14.7	0.2
SiO_2	10.2	97.8
ZnO	7.9	0.0
Total	100.0	99.9

* from Armco Corporation, Middletown, OH ** from Duriron Foundry, Dayton, OH

RESULTS AND DISCUSSION

Phase Identification

Figure 3 shows the comparison of XRD scans of samples under different heat-treatments. The as-quenched sample was almost completely amorphous. Spinel and augite formed when the as-quenched sample was annealed at nucleation (635°C) and crystal growth (835°C) temperatures[3]. Further investigation indicated that in addition to formation of dendritic spinel[3], feldspar also crystallized in the same high temperature range (1150-1185°C). The morphology of dendritic spinel is shown in figure 4(a). The structure and orientation of the dendritic phase is also shown in Figure 4(b) at higher magnification. Feldspar solid solution can be expressed as $(Ca,Na)(Al,Si)_4O_8$. Formation of feldspar in this system is consistent with right branch of Bowen's reaction series. Formation of iron-rich dendritic spinel (magnetic solid-solution) is consistent with middle branch of Osborn's reaction series.

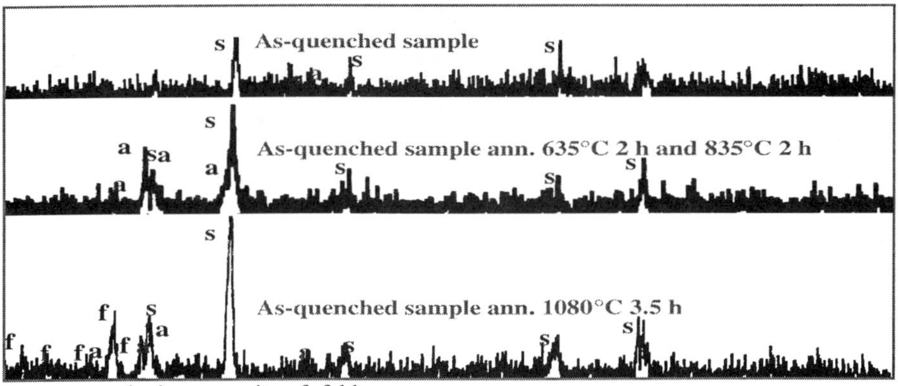

s: spinel a: augite f: feldspar

Figure 3: Comparison of XRD data of abrasive samples with different heat-treatments

Figure 4: (a) As-quenched sample annealed for 2 h at 1075°C, quenched

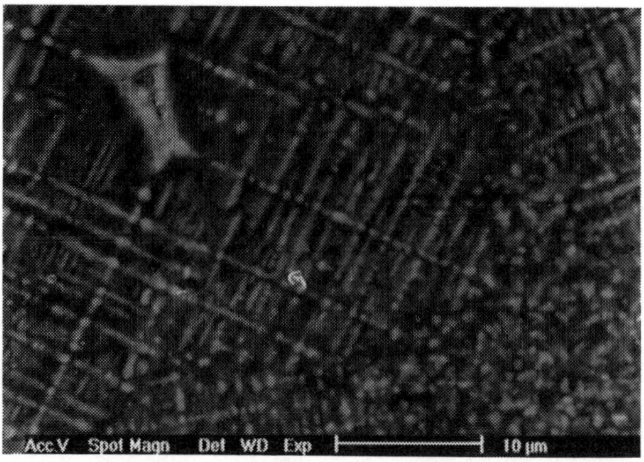

Figure 4: (b) As-quenched sample annealed for 5 h at 1065°C, quenched

Melting Phase Analysis

In order to prove that the endothermic peak is due to the melting of augite, an augite composition was synthesized (Table IV). The simulated augite sample, designated as augite 1, was melted at 1400°C for 2 h and then furnace cooled. However, as indicated by XRD data (Figure 5), the furnace-cooled sample did not crystallize and resulted in a glass of augite composition. Based on the DTA scan of augite 1 (Figure 6, curve 1), a sample, designated as augite 2, was nucleated at 720°C (70-85% of the crystallization peak from DTA scan, curve 1) for 3 h and grown at crystallization peak temperature 906 °C for 2 h. XRD pattern showed crystallization of augite in the augite 2 sample. A DTA scan of sample augite 2 (Figure 6, curve 2) indicated completely crystallization. The morphology of dendritic augite is shown in Figure 7. The endothermic peak of DTA scan indicated that the melting temperature of this phase was between 1180-1200°C which is consistent with the melting temperature range (Figure 2) of the abrasive product. Previous work[3] established that there was only spinel after an as-quenched abrasive sample was annealed at a temperature above the endothermic melting peak. Thus, the melting phase in the abrasive product is augite. The crystallized feldspar in the abrasive product may also melt in this temperature range since it has a melting point below 1200°C. Albite ($NaAlSi_3O_8$), a typical feldspar, has a melting point of 1100°C[7].

Table IV. Augite formulation based on EDS data

wt.%	Al_2O_3	$CaCO_3$	Fe_2O_3	MgO	$NaHCO_3$	SiO_2	ZnO	Total
augite	8.43	25.35	12.78	7.61	2.86	40.74	2.24	100.00

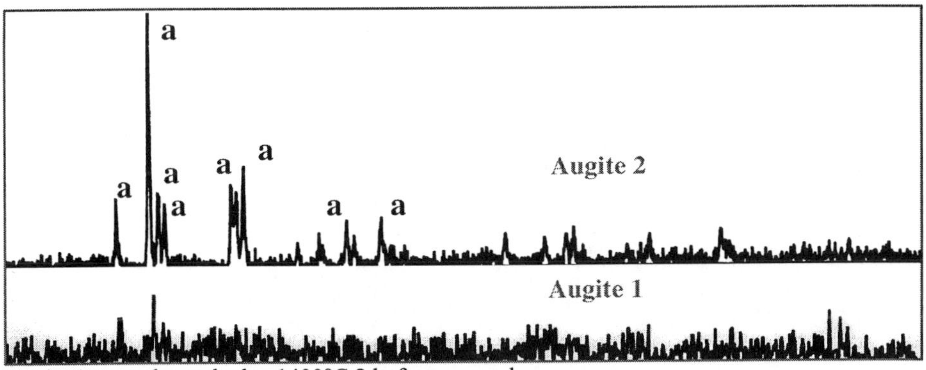

Augite 1: powder melted at 1400°C 2 h, furnace cool
Augite 2: augite 1 annealed 720 °C 3 h and 906°C 2 h , furnace cooled

Figure 5: Comparison (XRD) of samples augite 1 and augite 2

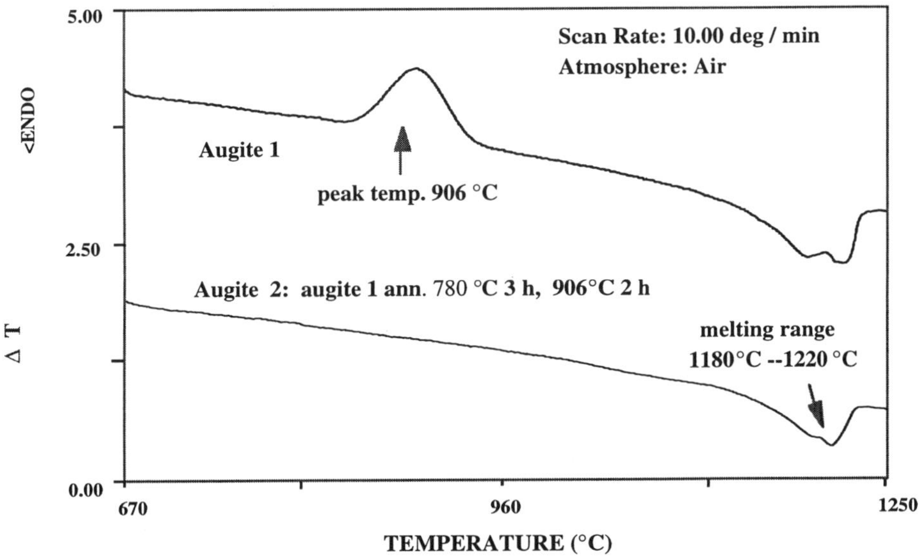

Figure 6: Comparison (DTA) of samples augite 1 and augite 2

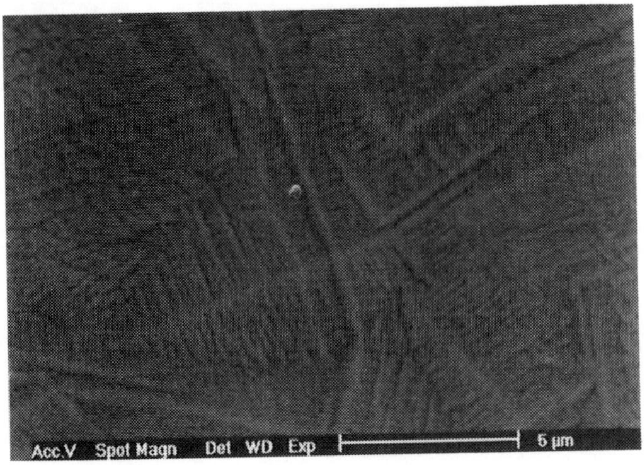

Figure 7: Morphology of augite 2

Effect of Melting Time and Cooling Rate

Melting times greater than 6 h resulted in no crystallization of augite due to volatilization of alkali and alkaline-earth metals such as Na, Mg and Ca. These cations are necessary constituents for the formation of augite. As shown in Figure 8, no augite peaks are observed for the sample heated for 6 h at melting temperature. During a long melting time, Al also diffuses into melt which can be observed by SEM at crucible / melt interface. The refractory was corroded due to the large amount of alkali and alkaline-earth metal present.

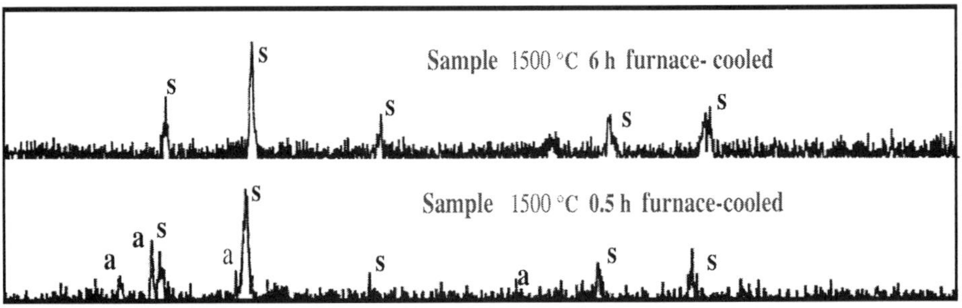

a: augite s: spinel

Figure 8: Comparison of abrasive samples with different melting times

The cooling rate affected the phases crystallization as shown in Table V. The crystallization kinetics of augite and feldspar were both much slower than the formation of the spinel. Quenched samples only formed spinel, furnace-cooled samples formed spinel and augite but no feldspar, only with additional heat-treatment was feldspar crystallized.

Table V. Phase formation and cooling rate

Sample	Spinel	Augite	Feldspar
Quenched	✓	—	—
Furnace-cooled	✓	✓	—
Heat-treated	✓	✓	✓

— not present ✓ present

CONCLUSIONS

Dendritic iron-rich spinel and feldspar solid solution were formed at temperatures between 1060-1085°C in heat-treated samples. Silicate phases augite and feldspar solid-solutions were found to melt in a temperature range 1180-1220°C. Samples melted at 6 h did not contain augite and feldspar due to volatilization of the cations, such as Na, Mg, and Ca. Rapid cooling also prevents formation of augite and feldspar but not spinel in the abrasive sample. Feldspar was found to be much harder to form under slow furnace-cooled conditions.

REFERENCES

1. Breckling, J. J., "Industrial Ceramics, Current Market and Future Trends", April, 1988.
2. Russell D. Blume and Charles H. Drummond, III, " High Grade Abrasive Product Development from Vitrified Industrial waste", in *Environment Issues and Waste Management Technology in Ceramic and Nuclear Industries II, Ceramic Transaction Series*, Vol. 72. pp. 229-239, Edited by V. Jain and D. Peeler, 1996.
3. Zheng Gao and Charles H. Drummond, " Thermal Analysis of Nucleation and Growth of Crystalline Phases in Vitrified Industrial Wastes" accepted for Ceramic Society Proceedings.
4. Bowen, N. L., "The Reaction Principle in Petrogenesis", *Journal of Geology*, **30**: 177-198, (1922)
5. Osborn, E. F., "Reaction Series for Subalkaline Igneous Rocks based on Different Oxygen Pressure Conditions", *The American Mineralogist*, **47**: 211-226 (1962)
6. P. W. Mcmillan, *Glass-Ceramics* , pp. 57-88, Academic Press, London, 1979.
7. C. T. Lynch, *CR. Handbook of Chemistry and Physics*, pp. 4:96, 73rd ed., CRC Press, London, 1993.

THE EFFECT OF PO_2 AND COOLING RATE ON THE CRYSTALLIZATION AND PROPERTIES OF VITRIFIED INDUSTRIAL WASTE FOR ABRASIVE APPLICATIONS

Russell D. Blume*, Charles H. Drummond, III, Department of Material Science, The Ohio State University, Columbus OH 43210, and Alan B. Sarko, Seiler Pollution Control Systems, Dublin OH 43017

ABSTRACT

The crystallization sequences in abrasive formulations utilizing electric arc furnace (EAF) dust and foundry sand appear to be quite similar to the crystallization sequences observed in natural basaltic melts. Oxidizing conditions have been observed to promote spinel formation, while reducing conditions tend to promote the crystallization of an iron rich metasilicate (augite) solid solution. Spinel compositions were observed to vary continuously from a magnetite to a (Mg,Zn) ferrite solid solution under oxidizing conditions. Under reducing conditions, the spinel compositions were observed to vary continuously from the (Mg,Zn) ferrite to an aluminous chromite solid solution. Cr was observed to partition exclusively to the spinel phase. An increase in Vicker's hardness of approximately 25% was observed between the most reduced sample which was water quenched and the most oxidizing sample which was furnace cooled. Increased cooling rate led to decreased augite formation but increased spinel formation at constant pO_2 conditions. Measurements of pO_2 during melting and during cooling of the melt offer potential insights into the melt reactions as well as shifts in iron redox during cooling. This technique also offers a sensitive measurement of glass transition temperature for transition cation containing melts.

INTRODUCTION

Previous work[1] has led to the development of medium grade abrasives utilizing electric arc furnace (EAF) dust and foundry sand as the primary waste constituents with a Vicker's hardness of 936 ± 124 KgF/mm^2, acceptable processing parameters(melt viscosity and liquidus temperatures), and superior resistance to particle size deterioration upon high energy impact during blasting applications as compared to commercially available Al_2O_3 and garnet blasting media. In addition to high hardness, the developed formulations must also incorporate the hazardous components (Ag, As, Ba, Cd, Cr, Pb, Se) into the final product such that the hazardous components impart functional characteristics to the final product application. It was found that the formation of spinel solid solutions crystalline phase in the final product is necessary to develop the required hardness in the final product as typical glass and silicate crystalline phases have hardness less than 800 KgF/mm^2. The formation of a dendritic metasilicate (augite) crystalline phase is thought to increase the fracture toughness of the product by a crack deflection mechanism and/or by the formation of residual compressive stresses within the glass matrix due to mismatches in thermal expansion coefficients. Energy dispersive spectroscopic (EDS) analysis of the spinel, augite and glass phases indicate that Cr preferentially segregates to the chemically durable spinel phase, in which the high field strength of the Cr cation likely imparts increased hardness. The presence of Cr within the silicate melt has been shown to increase the tendency for spinel formation due to both it's low solubility in siliceous melts and the fact that both Cr^{+3} and Cr^{+6} can be incorporated into the octahedral and tetrahedral sites respectively within the spinel crystal lattice[2]. These facts establish that the Cr, the predominate hazardous component of concern, is incorporated into the final product in a manner which imparts functional characteristics to the final product application.

It was desirable to develop formulations that were compatible with the processing constraints of the Seiler system. The Seiler system is a high throughput, oxy-gas fired vitrification system comprised of a pre-heater, a vitrification unit, and a product reservoir. The pre-heater typically operates at 800°C and is used to remove moisture and organics from the waste feed. Vitrification of the waste feed occurs within the oxy-gas fired flame at temperatures which typically are between 1700 - 2000°C. After vitrification, the waste resides within a reservoir in which homogenization of the product occurs at approximately 1500°C. Typical residence time of the Seiler system ranges between 10 - 15 minutes and the commercial system has a throughput of approximately 500 kg/hr.

The utilization of an oxy-gas fired vitrification systems allows for the control of the partial pressure of oxygen (pO_2) within the vitrification unit by adjustment of the gas-air ratio in the burner. The control of pO_2 during vitrification has the potential of impacting both the incorporation of hazardous metals into the glass-ceramic product as well as product characteristics such as hardness which depend on subsequent crystallization. The phases which form as well as their volume fractions in the resultant product are likely effected by the redox state of the transition metal cation constituents. For this reason it is desirable to determine the relationship between pO_2 and the crystallization sequences in vitrified industrial waste. In addition to the control of pO_2 during vitrification, control of the cooling rate of the glass melt as it exits the vitrification system will likely affect the resultant product morphology and properties. Understanding the role of pO_2 and cooling rate on product characteristics can increase the versatility of waste vitrification systems in producing a wider range of potential product applications which will likely make the system more economically viable.

As can be seen by inspection of **Table I** below, the composition of the developed formulation, henceforth designated as the Base Formulation, is quite similar to natural basaltic magma compositions[3]. Osborn[4] investigated the role of pO_2 on the crystallization of subalkaline igneous basaltic melts and found that these systems crystallize by means of a coupled reaction series. Under oxidizing conditions, these melts crystallized by coupled reaction series involving a set of continuous reaction series of feldspar and magnetite solid solutions and a discontinuous reaction series in which orthosilicate (olivine) phases form at high temperatures and then react incongruently with the remaining liquid phase to form a silicate phase (meta- and tectosilicates) lower in the series as the melt cools. The residual liquid formed upon continued fractionation of the melt becomes enriched in SiO_2 relative to the initial homogenous melt. The crystallization reaction series under oxidizing conditions is shown in **Figure 1** below.

Under reducing condition, Osborn found that basaltic melts crystallize by a similar reaction series with the exception that the magnetite series is replaced with an iron rich olivine series and that the residual liquid is enriched in FeO rather than SiO_2. The reaction series under reducing conditions is shown in **Figure 2** below.

As both spinel and metasilicate (augite) phases occur in both of these reaction series, it is likely that crystallization occurs by a similar reaction series in this system and that the formation of the spinel phase will likely be affected by the pO_2.

Table I: Composition of base formulation and typical basaltic magmas (wt.%)

	SiO_2	Al_2O_3	Fe_2O_3	CaO	MgO	Na_2O	Other
Base	33.9	10.3	16.7	8.0	13.1	9.0	8.9
Basalt[4]	47.1	15.9	12.1	10.7	6.1	2.9	5.2

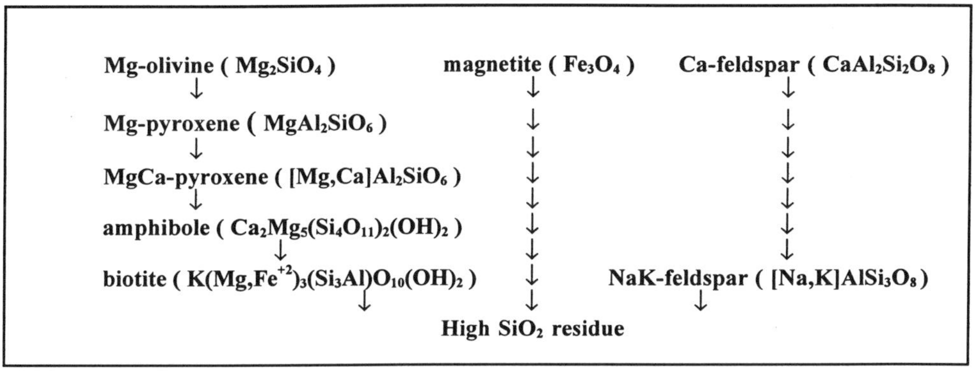

Figure 1: Osborn's reaction series at high *pO_2's*

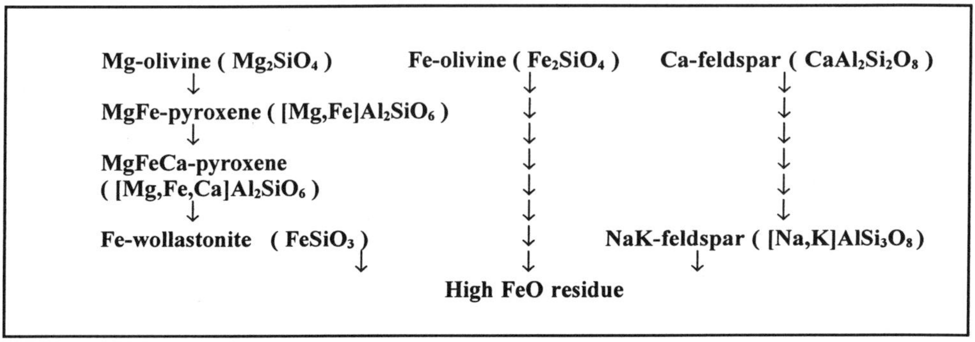

Figure 2: Osborn's reaction series at low *pO_2's*

EXPERIMENTAL:

In order to simulate the effect of changes in pO_2 of the melt atmosphere during vitrification, oxidizing and reducing agents were added to the base formulation such that the bulk composition of the base formulation remained invariant and the decomposition of the reducing and oxidizing agents occurred at a temperature at which a liquid phase is present in the system

Several oxidizing agents were considered including ammonium nitrate (NH_4NO_3), aluminum nitrite $(Al(NO_3)_3{\bullet}9H_2O)$, calcium nitrate $(Ca(NO_3)_2)$, and sodium nitrate $(NaNO_3)$. Although all of the above oxidizing agents could be substituted in a manner which would keep the base composition invariant, only calcium nitrate (m.p. =561°C)[5] has the potential of decomposing when a liquid phase is present in the system. $Ca(NO_3)_2$ undergoes a decomposition reaction[6] and liberates oxygen according to the reaction: $Ca(NO_3)_2 \leftrightarrow CaO + NO + NO_2 + O_2$. $Ca(NO_3)_2{\bullet}4H_2O$ was used in this study to simulate oxidizing conditions.

Graphite was used as a reducing agent due to the fact the graphite additions would not alter the base formulation, it is stable to high temperatures, and that it is known to be an effective reducing agent. **Table II** below summarizes the formulations used in this study to simulate 5 levels of pO_2 in the glass melt.

Table II: Invariant Base Compositions with Oxidizing and Reducing Additions (wt. %)

	Base + 5% Graphite	Base + 2.5% Graphite	Base Formulation	Base + 2.5% Oxidizer	Base + 5.0% Oxidizer
EAF Dust	50.79	52.12	53.46	52.69	51.91
Sand	18.68	19.17	19.66	19.37	19.09
Kaolinite	14.54	14.93	15.31	15.09	14.87
$CaCO_3$	6.65	6.83	7.00	5.84	4.68
$NaHCO_3$	1.44	1.48	1.52	1.50	1.48
Al_2O_3	2.91	2.98	3.06	3.02	2.97
$Ca(NO_3)_2{\bullet}4H_2O$	0.00	0.00	0.00	2.50	5.00
Graphite	5.00	2.50	0.00	0.00	0.00
Sum	100.01	100.01	100.01	100.01	100.00

Test Matrix: *pO_2* and Cooling Rate

It was decided that the experimental design would consist of 3 levels of cooling rate: furnace cool, air cool, and water quench, and 5 levels of pO_2. The 5 levels of pO_2 are simulated by 2.5% and 5.0 wt% of graphite and $Ca(NO_3)_2{\bullet}4H_2O$ additions to the base formulation. The order of the experiments were randomized to prevent systematic errors from compounding the statistical analysis of the data. This experimental design consist of 15 melts with no replicates which allows for a total of 7 degrees of freedom for subsequent Analysis of Variance (ANOVA).

The above melts were conducted in an electric furnace at 1500°C for 1 hours and then quenched according to the desired quench rate. The heating rate was kept constant at 10°C/min and sample mass was fixed at 40.0 ± 0.1 grams.

Product Characterization:

Samples were mounted in Al filled conductive Bakelite and polished to 0.5 microns using standard metallographic polishing methodology. Vicker's hardness were measured using a Buehler Micromet® II Digital Micro Hardness Tester with an indention load of 200 g and a dwell time of 30 seconds. To insure representative values, at least 15 measurements were conducted.

The samples were crushed into fine powder for powder X-Ray diffraction testing with a Scintag PAD-V X-ray diffractometer. These samples were characterized with Cu-Kα radiation with 2θ ranging between 20 and 80° at a scan rate of 0.5 deg/min. This scan rate was chosen to increase the signal to background ratio which is low at high scan rates due to X-ray florescence of the Fe within the samples. Crystalline phases were then identified using JCPDS data.

Mounted and polished specimens were carbon coated for SEM/EDS characterization. All EDS data was obtained with an accelerating voltage of 25kV. The average bulk composition was determined by operating the EDS in survey mode at 200X magnification at 3 randomly chosen areas. The composition of each phase present including all crystalline and glass phases were determined by taking 3 measurements at higher magnifications (500X - 10000X) as indicated by the relative sizes of the phase.

Iron Redox Determination:

Samples of the furnace annealed melts were submitted for wet chemical analysis to Monarch Analytical Laboratories, Inc[7]. Although the exact dissolution and analytical methodology is proprietary, the technique involved sample dissolution with hydrofluoric and perchloric acids under an argon atmosphere to minimize the potential for redox changes during dissolution. Total iron was then determined by titration with potassium dichromate and fusion of the residue with lithium carbonate/ sodium borate. Reduced iron was determined by back titration after sample dissolution with HF and H_2SO_4 acids in the presence of excess cerric sulfate. Total iron was reported as weight percent Fe_2O_3 while ferrous iron was reported as weight percent FeO. Each value represents the average of two replicate measurements.

Melt pO_2 Measurements:

10 ml capacity yittria (9.2 wt%) stabilized zirconia cylindrical crucibles (height: 33 mm, O.D. : 22mm, wall thickness: 2mm) were obtained from the Coors Ceramic company[8] (Catalog#65533-ZIR) for subsequent fabrication of O_2 sensors. Both the inner and outer floor surfaces were coated with Pt-paste[9] and fired at 1200°C for 12 hours to insure bonding of the Pt electrodes to the crucible. Electrical contact with the outer electrode was provided by placing a coil of Pt wire on a zirconia disk and then utilizing the weight of the charged crucible to provide mechanical contact between the Pt electrode and Pt wire. The zirconia disk was used as a base in order to prevent chemical reactions between the crucible and the Al_2O_3 refractory lining the furnace. The electrical contact to the inner electrode was provided by threading a Pt wire through a 8mm diameter Al_2O_3 tube and allowing the weight of the tube to provide a mechanical contact between the inner electrode and the Pt wire.

Approximately 5 grams of each formulation was charged into each crucible and the Emf of the inner (sample) electrode was measured relative to the outer (reference) electrode using a Strawberry Datatree data acquisition system as a function of melt temperature and time. In accordance to the standard sign convention adopted for the operation of solid-electrolyte O_2 sensors, the sample electrode was connected to the positive terminal and the reference electrode was connected to the negative terminal of the data acquisition system. The data acquisition system was configured to sample the Emf of the cell every 10 seconds to a resolution of $\pm 1\ \mu V$.

The samples were fired at a heating rate of 10°C/min from 23°C to 1500°C, held at 1500°C for 1 hour, and then allowed to furnace cool under ambient conditions. The furnace temperature was recorded every 5 minutes during heat-up and during the isothermal hold at 1500°C. The temperature was recorded every minute during cooling to obtain an accurate cooling curve.

Emf(V) and temperature(K) data was converted to the melt pO_2 by utilizing the Nernst equation[10]:

$$\log(pO_2) = 20155.3596 * [\ Emf(V) / T(K)\ ] - 0.678 \qquad (1)$$

RESULTS AND DISCUSSION:

X-Ray Diffraction

Comparison of peak intensity versus 2θ plots with JCPDS cards indicates the presence of two crystalline phases: a spinel solid solution; donathite $(Fe, Mg)(Cr, Fe)_2O_4$ and an augite; $Ca(Mg,Fe)(SiO_3)_2$ crystal phase.

Lattice parameters were calculated from the 2θ position of the most intense spinel peak which corresponds to the {311} type reflection using Bragg's Law:

$$\lambda = 2d_{hkl} \sin\theta_{hkl} \qquad (2)$$

with λ = 1.5406 Å. These lattice parameters varied from 8.29 to 8.35Å and fall within the ranges of lattice parameters with pure spinel minerals[11]: $MgAl_2O_4$ (8.0831 Å), $ZnAl_2O_4$ (8.0970 Å), $ZnCr_2O_4$ (8.3275 Å), $MgCr_2O_4$ (8.333 Å), $FeCr_2O_4$ (8.3790 Å), $MgFe_2O_4$ (8.392 Å), Fe_3O_4 (8.396 Å), and $ZnFe_2O_4$ (8.4411 Å).

The spinel {311} peak intensity as a function of cooling rate and pO_2 were examined in order to determine the effects of these variables on spinel formation. Both the furnace cooled and air cooled samples showed an increase in spinel peak intensity with increased melt pO_2. The furnace cooled samples had spinel peak intensities of 70, 88, 115, 124, 155 cps respectively from the most reduced sample to the most oxidizing sample. The air cooled samples showed a similar trend with spinel peak intensities of 35, 139, 159, 174, 189 cps respectively. The water quenched samples showed no definite trends with spinel peak intensities of 68, 196, 180, 211, 196 cps respectively. ANOVA established that both pO_2 and cooling rate are statistically significant at a confidence level greater than 95%. It is also interesting to note that the spinel peak intensity increases with increased cooling rate.

The intensity of the most intense augite peak, the { $\underline{2}21$ } reflection, as a function of cooling rate and pO_2 was likewise examined. No apparent trend in augite peak intensity with pO_2 is evident; however, the effect of cooling rate is significant in that samples which were water cooled had no detectable augite phase present. Furnace cooled augite peak intensities were 96, 52, 57, 69, 75 cps while the air cooled samples had augite peak intensities of 84, 109, 62, 81, 58 cps respectively.

Spinel Compositions:

A plot of $Fe^{+3}/(Al+Cr+Fe^{+3})$ versus $Fe^{+2}/(Mg+Zn+Fe^{+2})$ for the spinel phases present in these samples is shown in **Figure 3** below. The diagram can be divided into 4 regions in which $(Mg,Zn)Fe_2O_4$, $(Mg,Zn)(Al,Cr)_2O_4$, Fe_3O_4 and $Fe(Al,Cr)_2O_4$ are the predominate spinel solid solution components. As can be seen by inspection, samples fired under ambient to oxidizing conditions (◆) have spinel compositions which vary continuously from a predominately magnetite (Fe_3O_4) solid solution to a predominately $(Zn,Mg)Fe_2O_4$ solid solution. Samples fired under reducing conditions (●) have spinels which vary continuously from a $(Zn,Mg)Fe_2O_4$ solid solution to a $Fe(Al,Cr)_2O_4$ solid solution. This result is consistent with the fact that the ion is predominately ferric iron under oxidizing conditions while under reducing conditions, the iron redox ratio shifts towards the ferrous state. EDS analysis indicates that Cr is partitioned exclusively to the spinel phase under both oxidizing and reducing conditions.

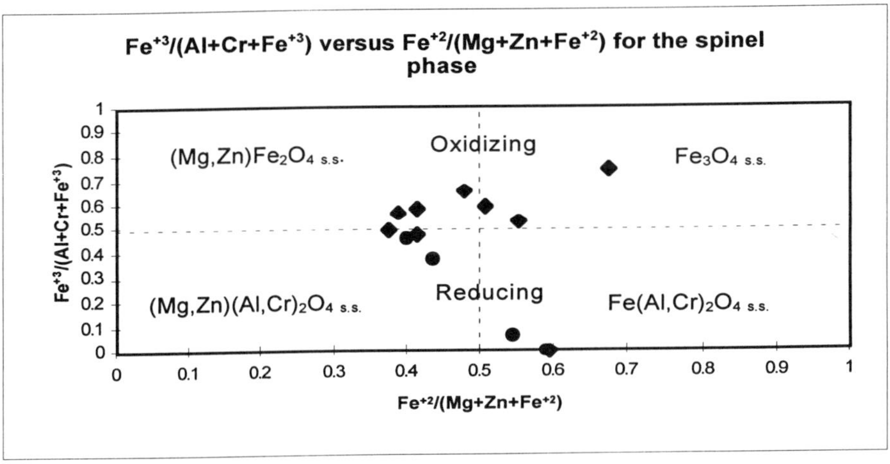

Figure 3: Fe^{+3} / $(Al+Cr+Fe^{+3})$ versus Fe^{+2} / $(Mg+Zn+Fe^{+2})$ for the spinel phase
◆ = oxidizing conditions, ● = reducing conditions

Vicker's Hardness:

The Vicker's hardness values were observed to change throughout the tested areas of the samples due to changes in morphology and phase distribution. A total of 15 measurements were conducted on each sample in randomly selected areas. **Table III** summarizes the average Vicker's hardness and standard deviation of the samples as a function of pO_2 and cooling rate.

Table 3: Vicker's hardness as a function of pO_2 and cooling rate

(kgF/mm^2)	**Furnace Cool**	**Air Cooled**	**Water Quenched**
5.0% $Ca(NO_3)_2•4H_2O$	777, 46	741, 25	657, 48
2.5% $Ca(NO_3)_2•4H_2O$	695, 45	733, 45	636, 40
Base Formulation	712, 42	687, 52	647, 55
Base + 2.5% Graphite	676, 51	725, 53	652, 11
Base + 5.0% Graphite	634, 50	696, 53	608, 16

ANOVA indicates that cooling rate is statistically significant at a confidence level greater than 95% while pO_2 is statistically significant at a confidence level of 93%. It can be seen that, in general, oxidizing conditions and slower cooling rates are associated with increased average Vicker's hardness. This observation is further emphasized by the fact that the highest hardness (777 kgF/mm^2) is associated with furnace cooling and 5% $Ca(NO_3)_2$ additions while the lowest hardness (608 kgF/mm^2) is associated with water cooling and 5% graphite additions. This represents an approximate 28% increase in Vicker's hardness.

Iron Redox:

The weight percent of FeO from the most reduced sample to the most oxidized sample was measured to be 19.87, 8.35, 6.04, 6.12, and 5.40 weight percent. The weight percent of Fe_2O_3 was measured to be 4.83, 17.55, 21.06, 20.58, and 21.00 wt.%. Converting these weight percents to moles of Fe^{+2} and Fe^{+3} , the Fe^{+2} /Fe^{+3} ratio was calculated to be 8.50, 0.55, 0.33, 0.34 and 0.29 from the most reduced to the most oxidized samples respectively.

As can be seen by inspection of the above data, $Ca(NO_3)_2$ additions did not significantly affect iron redox state of the melts whereas graphite additions were effective as a reducing agents in these melts. It is thought that the relatively low decomposition temperature of $Ca(NO_3)_2$ is the primary reason that this additive was ineffective as an oxidizing agent as it decomposes at a temperature at which no liquid phase is present and will therefore interact with the transition metal cations quite sparingly. It is also likely that the samples were initial quite oxidized as is seen in base formulation data and that further oxidation of the iron is inhibited by the preferential oxidation of MnO to Mn_2O_3 or Cr_2O_3 to hexavalent chromium.

Melt pO_2 Measurements:

The electromotive force (Emf) of the oxygen sensor relative to the reference atmosphere at a pO_2 of 0.21 atm was obtained as a function of time and furnace temperature for the 5% $Ca(NO_3)_2 \bullet 4H_2O$ sample. **Figure 4** below presents the calculated melt pO_2 as a function of temperature and time for the 5% $Ca(NO_3)_2 \bullet 4H_2O$ sample.

Examination of **Figure 4** suggest that the sensor is recording changes in pO_2 during melting, at the melt temperature, and during cooling. Of particular interest are the two sharp rises in pO_2 at 439 and 539°C which is most likely due to the stepwise decomposition of the $Ca(NO_3)_2$ additions. A shallow decrease followed by an increase in pO_2 is observed at approximately 1398°C which most likely corresponds to the liquidus temperature for this melt. The subsequent increase in pO_2 is likely attributed to the entrapment and finning of air bubbles in the newly formed liquid. It should be noted that the activity of O_2 in the melt is initially supersaturated with respect to atmospheric pO_2 upon reaching 1500°C. With subsequent equilibration during the hold time, the pO_2 drops to a value just below atmospheric pO_2. During cooling the pO_2 is observed to decrease by many orders of magnitude and this effect is attributed to the oxidation of FeO to Fe_2O_3 with decreasing temperature.

At a temperature of approximately 647°C, the pO_2 is observed to increase by many orders of magnitude to a value approaching 0.21 atm at room temperature. This behavior is not consistent with the established redox models and another mechanism must be operating upon further cooling. It is proposed that this behavior occurs at the glass transition temperature for this system at which point, O_2 can no longer diffuse into the glass melt from the walls of the zirconia crucible. As long as the net flux of O_2 into the crucible from the atmosphere is equal to the net flux of O_2 into the melt from the crucible, a quasi-steady state condition can be maintained and the sensor can register changes in melt pO_2. At the glass transition temperature(T_g) the impedance at the glass/crucible interface to O_2 diffusion approaches infinity. As O_2 can no longer exit the solid electrolyte, the activity gradient which is present at the T_g acts as a driving force for equilibration of O_2 activity with the atmosphere. This results in a decrease in the slope of the activity gradient as O_2 continues to diffuse into the electrolyte from the atmosphere. The equilibration process, reduces the magnitude of the Emf of the cell and results in an apparent increase in pO_2 of the melt. This apparent increase in pO_2 does not correspond to further changes in pO_2 within the glass, as such process are kinetically inhibited.

This technique appears to offer a quite sensitive measure of glass transition temperature, though further work is indicated to correlate experimentally determined T_g (DTA) to the T_g indicated by this method. This technique is likely only applicable to glass melts containing transition metal cations, as the cooling process will not have as dramatic effect on melt pO_2 in the absence of the applicable redox reactions.

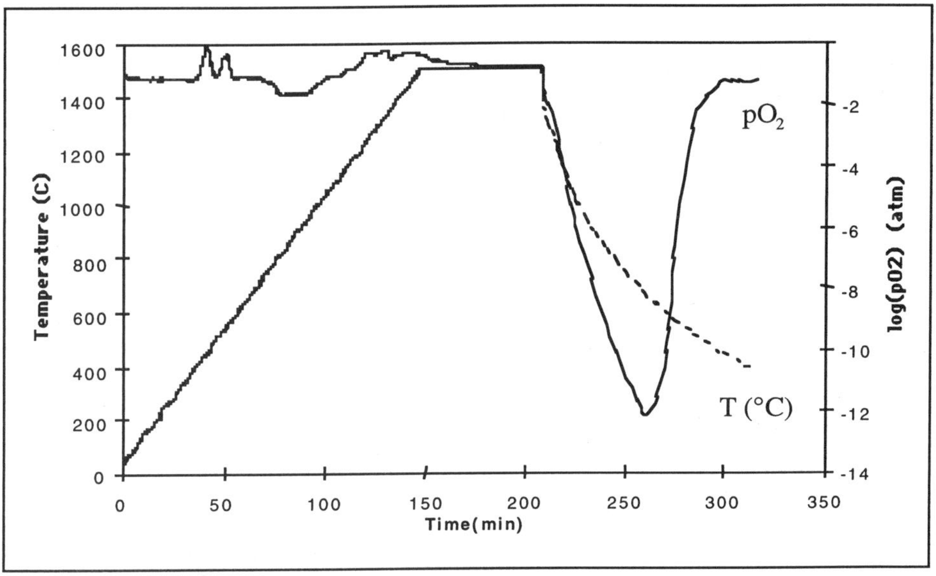

Figure 4: Log(pO_2) as a function of time and temperature for furnace cooled, 5.0% $Ca(NO_3)_2$ Melt

CONCLUSIONS:

Both cooling rate and changes in melt pO_2 have been observed to affect the crystallization and properties of abrasive product obtained by the vitrification of industrial waste. Oxidizing conditions and slow cooling rates tend to promote the formation of equiaxed spinel grains and dendritic augite phase, both of which are desired for increased Vicker's hardness and increased mechanical durability. An increase in Vicker's hardness of approximately 25% was observed between the water quenched, 5% graphite addition sample and the furnace cooled, 5% $Ca(NO_3)_2 \bullet 4H_2O$ sample. Spinel composition was observed to vary continuously between a magnetite and a Mg-Zn ferrite solid solution under oxidizing conditions. Under reducing conditions, the spinel composition varied continuously between the Mg-Zn ferrite and an aluminous chromite solid solution. Cr was observed to partition exclusively to the spinel phase in which its high field strength promotes increased hardness of the chemically durable spinel phase.

It is proposed that this system crystallizes similarly to natural basaltic melts and the reaction series consist of a continuous spinel series with spinels rich in high field strength cations forming at high temperatures. At lower temperatures, the spinel compositions approach the magnetite composition. Under reducing conditions, the iron is predominately ferrous and the increased solubility of ferrous iron in the silicate liquids favors the formation of iron rich metasilicate phases. Faster cooling rates result in decreased volume fraction of the metasilicate phase with a corresponding increase in spinel formation. The spinel which forms under faster cooling rates have a dendritic morphology and have been shown to be less effective in promoting increased hardness than do equiaxed grains of spinel.

The experimental technique utilized in this study to measure melt pO_2 during firing and during cooling appears to offer potential insights into the melt reactions which occur during firing and the changes in iron redox state upon cooling of the melt. This technique also appears to measure the T_g of the glass to a high sensitivity, though comparison with standard thermal (DTA) techniques is indicated to establish the accuracy of this measurement.

Acknowledgments

This work was supported by the Edison Materials Technology Center (EMTEC) through a sub-contract with Seiler Pollution Control Systems, Inc.

References

1. R.D. Blume, C.H. Drummond III, and A.B. Sarko, "High Grade Abrasive Product Development From Vitrified Industrial Waste", pp.229-239 in *Environmental Issues and Waste Management Technologies in the Ceramic and Nuclear Industries II*, Ceramic Transactions Series, Vol 72. Edited by V. Jain and D. Peeler, 1996
2. M.B. Volf, "Chromium"; pp.336-338 in *Chemical Approach to Glass*, Elsevier, Amsterdam, 1984.
3. R.F. Fudali, "Oxygen Fugacity of Basaltic and Andesitic Magmas", *Geochimica et Cosmochimica Acta*, **29**: 1063-1075 (1965).
4. E.F. Osborn, "Reaction Series for Subalkaline Igneous Rocks Based on Different Oxygen Partial Pressure Conditions", *The American Mineralogist*, **47**: 211-226 (1962).
5. C.T. Lynch, *CRC Handbook of Chemistry and Physics*, pp. 4:36-114, 73rd ed., CRC Press, 1993.
6. O. Knacke, *Thermochemical Properties of Inorganic Substances*, Springer-Verlag, Berlin, 2nd Edition, 1991, p. 360
7. J.P. Hojnicki, Monarch Analytical Labortories, Inc., 349 Tomahawk Dr., Maumee, OH 43537-1696, (419)897-9000
8. B. Johnson, Coors Ceramic Company, Structural Products Group, 16000 Table Mountain Parkway, Golden Colorado, 80403, (303)277-4922
9. Platnium Ink, part# A3788, Engelhard, 1 W. Central Ave., E. Newark, NJ
10. K.S. Goto, *Solid State Electrochemistry and Its Applications to Sensors and Electronic Devices, Material Science Monographs*, Elsevier, Amsterdam, 1988, Vol. 45, pp. 314-315
11. Henk Colijn, Powder Diffraction File; HTTP://kcgl23.eng.ohio-state.edu/pdf/

ION EXCHANGE IN A ZEOLITE-MOLTEN CHLORIDE SYSTEM

R. H. Woodman and C. Pereira
Chemical Technology Division
Argonne National Laboratory
9700 South Cass Avenue
Argonne, Illinois 60439

ABSTRACT

Electrometallurgical treatment of spent nuclear fuel results in a secondary waste stream of radioactive fission products dissolved in chloride salt. Disposal plans include a waste form that can incorporate chloride forms featuring one or more zeolites consolidated with sintered glass. A candidate method for incorporating fission products in the zeolites is passing the contaminated salt over a zeolite column for ion exchange.

To date, the molten chloride ion-exchange properties of four zeolites have been investigated for this process: zeolite A, IE95®, clinoptilolite, and mordenite. Of these, zeolite A has been the most promising. Treating zeolite 4A, the sodium form of zeolite A, with the solvent salt for the waste stream–lithium-potassium chloride of eutectic melting composition, is expected to provide a material with favorable ion-exchange properties for the treatment of the waste salt. We constructed a pilot-plant system for the ion-exchange column. Initial results indicate that there is a direct relationship between the two operating variable of interest, temperature, and initial sodium concentration. Also, the mass ratio has been about 3-5 to bring the sodium concentration of the effluent below 1 mol%.

INTRODUCTION

Between 1963 and 1993, Argonne National Laboratory (ANL) operated a nuclear reactor facility known as the Experimental Breeder Reactor II (EBR-II).[1] Like most nuclear reactor facilities in the United States, EBR-II accumulated an inventory of spent fuel. However, unlike most spent fuel, the EBR-II inventory has properties that make it undesirable to dispose by repository burial. Consequently, research and development has been conducted for treating fuel to produce more stable chemical and physical forms for disposal is underway.[2,3]

Electrometallurgical treatment involves dissolution of the spent fuel in an electrorefiner-a large electrochemical cell, with chopped fuel pins in an anode basket and a cathode selected to allow precipitation of desired materials. The electrolyte is a lithium-potassium chloride salt of eutectic melting composition (59.5 mol% LiCl, 40.5 mol% KCl), with a melting point of approximately 350°C.[4]

Certain fission products, notably the alkalis, the alkali earths, and the rare earths, remain dissolved in the electrorefiner salt. Eventually, either the heat load of the accumulated fission products or the accumulation of hypereuctectic sodium-lithium-potassium chloride will render this salt no longer serviceable.[5] Figure 1, the equilibrium phase diagram of the ternary system, reveals the sharp increase in liquidus temperature that occurs when sodium is present in amounts which exceed the ternary eutectic composition.[6]. Figure 1 also reveals the relatively low sodium content required to exceed the euctectic sodium content. Once the salt bath becomes unserviceable, the process flowsheet calls for one of two options: disposal of the salt or treating it to remove sodium and fission products.

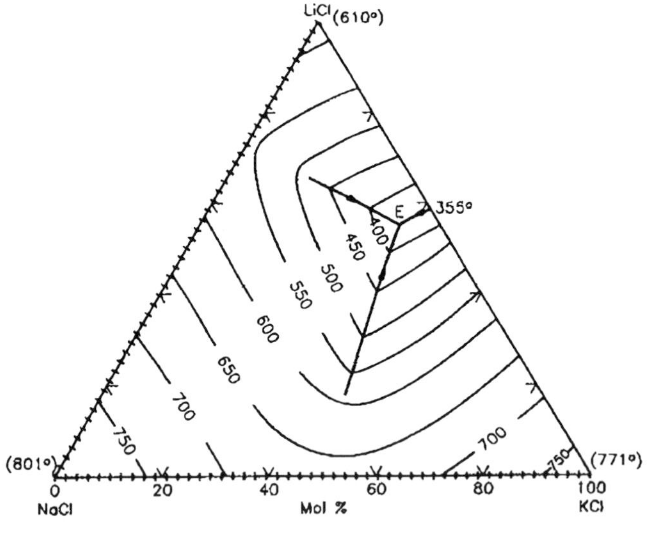

Figure 1. Equilibrium Phase Diagram for the Ternary System Lithium-Potassium-Sodium Chloride, (Ref. 6).

Disposal plans include a waste form that can incorporate the fission products as chlorides. Certain mineral zeolites, such as sodalite, have chlorides in their structure, and may be able to accommodate some or all of the fission products.[7, p. 41] Certain synthetic zeolites, such as Linde Type A (often referred to as zeolite A), can, after dehydration, occlude salt within their structural cavities.[7, p. 585] Waste forms featuring one or more such zeolites, consolidated by sintering with a glass frit, are under development.[8]

Alternatively, it has been established since the late 1960s that zeolite A, when occluded with salts, exhibits cation exchange.[7,p.587-8, and 9-13] Hence, the electrorefiner salt might be treated by exchange with zeolite A. Fission products and sodium would be incorporated with the exchange bed and, ideally, a eutectic lithium-potassium chloride salt would be returned to the electrorefiner.

Research and development at ANL has been directed towards establishing the feasibility of treating spent electrorefiner salt in this manner. Initial testing established that, of zeolite A, clinoptilolite, mordenite, and IE95®, zeolite A was by far the most effective ion-exchange medium.[8]

APPARATUS

A pilot-scale system for testing the ion-exchange column has been constructed. A schematic appears as Figure 2. The system is contained within a heated enclosure. The enclosure is maintained at temperatures ranging between 450 to 550°C. Because the chloride salt is hygroscopic, and the zeolites are dehydrated, the apparatus is housed in a glovebox with a helium atmosphere. The helium is circulated through absorbent beds to maintain the water vapor and oxygen gas levels below 1 ppm.

The primary feature of the system is the pair of stainless steel tanks, each capable of holding approximately 20 kg of salt. The tanks can be made gastight and pressurized to 2 to 3 psig. Each tank is also fitted with a cylindrical, annular cup that is filled with a braze. A cylindrical part that mates to the annular cup can be inserted. When the braze melts and freezes, a seal called freeze seal is formed.

Salt is transferred from one tank to the other by fitting a pipe to the freeze seal of the feeding tank and pressurizing the tank to force the salt through the pipe to the receiving tank. The lower end of the pipe is below the level of the molten salt. When the pipe is packed with beads of zeolite particles, it becomes an ion-exchange column. Because successful ion exchange requires a slow transfer rate, an electronic mass flow controller regulates the pressure.

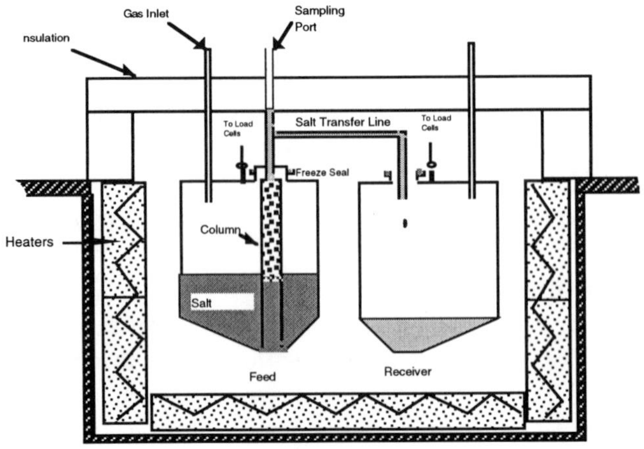

Figure 2. Schematic of Ion-Exchange Column Test Apparatus

The tanks themselves are mounted on piezoelectric load cells. The flow rate of the salt during the process can be measured by recording the weight change of the tanks over time.

The transfer pipe is fitted with a tube which permits sampling of the effluent stream from the column exit. The elemental composition of the effluent salt is determined by instrumental analysis techniques, typically inductively coupled plasma–atomic emission spectroscopy (for most species), inductively coupled plasma–mass spectroscopy (for ions, such cesium, with emissions outside the detection window of the emission spectrometer), or ion chromatography (for chloride ion).

DISCUSSION AND RESULTS

Beaded particles of zeolite 4A are available in industrial quantities. While such beads exhibit ion-exchange behavior that is promising for treating fission product-laden molten chlorides, its physical and chemical form require modification to enhance the exchange. First, the beads must be dehydrated. Second, the channels need to be occluded with chloride salt. Third, sodium needs to be removed from the zeolite (for the same reasons that sodium should not accumulate in the electrorefiner).

The beads are dehydrated by implementing of a controlled heating schedule under a flow of dry nitrogen.[14] At present, three methods for sodium removal and chloride occlusion are under investigation: (1) immersion of beads in molten lithium-potassium chloride, (2) aqueous exchangeof sodium with lithium or potassium chloride salts followed by salt occlusion, or (3) treatment of the dehydrated material in a column, using dehydrated beads as the column packing and lithium-potassium chloride eutectic as the feed salt. We explored the latter option.

Two control variables are of interest: temperature and residence time in the column. Two output variables are of interest: the concentration of sodium in the effluent salt over time and the mass of feed salt required to bring the sodium concentration of the effluent to a desirable level, such as 1 mol %, for a given mass of beads..

Figure 3 illustrates the change in composition of the effluent salt over the course of a representative column operation. As can be seen there, the initial effluent has the highest sodium concentration, and thus the greatest risk of the liquidus exceeding the operating temperature of the system. The concentration of sodium falls as more salt is transferred.

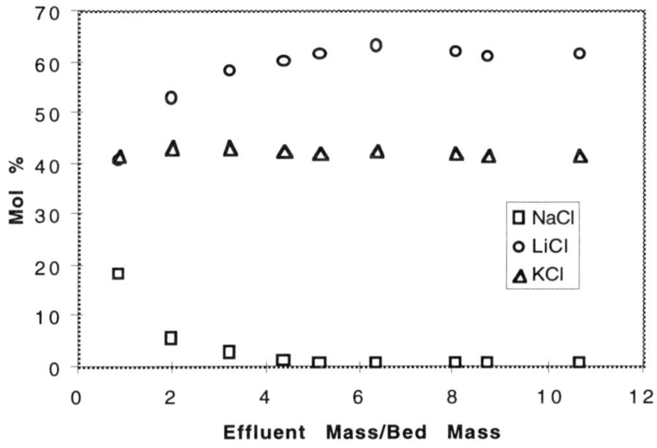

Figure 3. Composition of the Effluent Salt versus Amount of Salt Transferred for a System of Dehydrated Zeolite 4A Beads and a Lithium-Potassium Chloride Eutectic Feed Salt. Test was conducted at 550°C (825 K) and a residence time of 25 minutes.

At present, we do not have definitive data that describes the relationship between the operating variables. However, there does appear to be a direct proportion between residence time and the initial sodium concentration. Also, the mass ratio to bring the sodium concentration of the effluent below 1 mol % is typically between 3 and 5.

SUMMARY

A pilot-scale system for investigating ion-exchange behavior between molten chloride and zeolite A has been developed. To date, most operations of the column system have been devoted to treatment of dehydrated zeolite 4A beads to a chloride-loaded form with minimal sodium. While the relationship between the operating variable is not definitive, initial sodium concentration and residence time appear to be directly linked. The mass ratio is about 3 to 5 for optimum sodium concentration in the effluent.

ACKNOWLEDGMENTS

Many people contributed to the work described. The authors gratefully acknowledge Don Gracyk, Alice Essling, Ed Huff, Donna Huff, Paul Johnson, and Florence Smith for chemical analyses. John Ackerman provided us with much-needed guidance and direction, and Larry Putty with invaluable technical service. Our special thanks to Terry Johnson, Michele Lewis, and their co-workers, who started the work in this area. We also acknowledge Debbie Rausch for her contributions in preparing the manuscript.

REFERENCES

[1] Stevenson, C.E., The EBR-II Fuel Cycle Story, American Nuclear Society, LaGrange Park, IL, 1987.

[2] Ackerman, J.P., and McPheeters, C.C., "Advanced Waste Forms from Spent Nuclear Fuel", presented at GLOBAL '95 Int. Conf. on Evaluation of Emerging Nuclear Fuel Cycle Systems, Versailles, France, Sept. 11-14, 1995.

[3] Johnson, T.R., Lewis, M.A., Newman, A.E., and Laidler, J.J., "Treatment of High-Level Wastes from the IFR Fuel Cycle," in Chemical Pretreatment of Nuclear Waste for Disposal, Schulz, W.W., and Horwitz, E.P., eds., Plenum Press, New York, 1994.

[4] Janz, G.J., Tomkins, R.P.T., Allen, C.B., Downey, J.R., Gardner, G.L., Krebs, U., and Singer, S.K., Journal of Physical and Chemical Reference Data, Vol. 4, No. 4, 1975, p. 1044.

[5] Ackerman, J.P., Ind. Eng. Chem. Res., Vol. 30, No. 1, 1991, pp. 141-145.

[6] Sangster, J.M., and Pelton, A.D., "Critical Coupled Evaluation of Phase Diagrams and Thermodynamic Properties of Binary and Ternary Alkali

Salt Systems," Special Report to the Phase Equlibria Program, American Ceramic Society, Westerville, Ohio, pp. 33-35, 1987.

[7] Breck, D.W., Zeolite Molecular Sieves: Structure, Chemistry, and Use, John Wiley & Sons, Inc., New York, 1974.

[8] Lewis, M.A., Fischer, D.F., and Smith, L.J., J. Am. Ceram. Soc., Vol .76, No. 11, pp. 2826-32, 1993.

[9] Rabo, J.A., "Salt Occlusion in Zeolite Crystals," in Zeolite Chemistry and Catalysis, Rabo, J.A., ed., American Chemical Society, Washington, D.C., 1976, pp. 334-336.

[10] Liquornik, M., and Marcus, Y., Israel J. Chem., Vol. 6, 1968, pp. 115-121.

[11] Liquornik, M., and Marcus, Y., J. of Phys. Chem., Vol. 75, No. 16, 1971, pp. 2523-2525.

[12] Liquornik, M., J. Inorg. Nucl. Chem., Vol. 33, 1971, pp. 283-4.

[13] Susic, M.V., Petranovic, N.A., and Mioc, D.A., J. Inorg. Nucl. Chem., Vol. 33, 1971, pp. 2667-2675.

[14] Hash, M.C., personal communication, 1997.

A GENERAL METHOD FOR THE CONVERSION OF FLY ASH INTO ZEOLITES FOR IMMOBILIZING RADIOACTIVE WASTES

H.-L. Chang* and W.-H. Shih, Department of Materials Engineering, Drexel University, Philadelphia, PA 19104

ABSTRACT

A general method that can convert various coal ashes into zeolites was studied. Fly ashes were reacted with fused sodium hydroxide at 550°C and ambient pressure followed by dissolution in aqueous solution and heat treatment. The faujasite type zeolite and zeolite P were precipitated from the solution of fused powders cured at 60°C and 90°C, respectively, and ambient pressure. Those zeolites converted from coal ashes were characterized by the X-ray diffraction method and Scanning Electron Microscopy. The ion exchange capacities with Cesium ion of zeolites converted from fly ashes were obtained by measuring the increase of sodium ion concentration in solution. When faujasite type zeolite converted from fly ash was heated at different temperatures, the thermal stability of faujasite was studied by XRD and nitrogen gas adsorption analysis. The result indicates that the crystal structure of faujasite was thermally stable at 540°C. The fusion method is critical to convert fly ashes with different chemical compositions into faujasite. With a higher concentration of the soluble aluminosilicate in the fused powders, the yield of faujasite increased.

INTRODUCTION

The amount of waste ash released by factories and thermal power stations which use coal as fuel has been increasing all over the world. This combustion by-product consists mostly of aluminum-silicon-oxygen compounds mixed with smaller amounts of iron, sodium, potassium, calcium, and titanium oxides. Fly ash is collected by cyclones, electric precipitators, and/or bag filters from the flue gas at the top of the burner. Each year near 850 million tons of coal are used in the United States to generate electricity (1). Depending on the coal sources, different amounts and chemical compositions of fly ashes, which are the inorganic part of the coal, will be generated. In our previous study, zeolite P, zeolite A and faujasite have been synthesized from fly ash (2, 3). However, the direct use of as-received fly ash to generate zeolites without any pretreatment is only possible for faujasite type zeolite using fly ash supplied by Eddystone electric power plant, PECO Energy Inc. The amount of soluble aluminosilicate from fly ash is limited by the crystalline phases of fly ash, which are mullite and quartz. These crystalline phases are difficult to dissolve in water. As a result the synthesis of zeolites from fly ash varies according to the chemical composition and crystalline phases of different coal fly ashes. Recently, Shigemoto *et al* .(4) found the fusion of fly ash with NaOH powder prior to hydrothermal processing increased the yield of Na-X zeolite converted from coal fly ash. It was proposed by Shigemoto *et al.* that the fly ash particles reacted with sodium hydroxide producing silicates and aluminates that are more soluble than mullite and quartz. This is consistent with the results of Weldes *et al.*(5) that the silicates containing a higher amount of sodium have the higher solubility in water. As a result, the yield of Na-X zeolite converted from fly ash increased because more silicon and aluminum could be consumed during hydrothermal processing. To investigate whether the fusion process is useful for the conversion of zeolites from different fly ashes, three fly ashes, supplied by Eddystone, Goudey and Conemiaugh electric power plants PECO Energy Inc., were used as the raw materials of aluminosilicate. It was found that the faujasite and zeolite P can be converted from those fly ashes with the fusion process followed by curing at different temperatures. The ion exchange capacities of zeolite P and faujasite with cations showed that the exchange capacity of zeolite P is lower than that of faujasite due to the smaller pore size of zeolite P compared to that of faujasite. For the applications in ion exchange with radioactive ions, the synthesis of faujasite will be discussed in detail. In addition, the ion exchange capacity with cesium ions of faujasites converted from three fly ashes were measured and found to be similar. The thermal stability of converted faujasite is very important for the applications in nuclear waste clean-up. Therefore, the nitrogen gas adsorption isotherms and the XRD patterns of converted faujasites, that have been heated to different temperatures, were measured and the results indicate that the crystal structure of faujasite remains at 540°C.

EXPERIMENTAL PROCEDURES

Three class F fly ashes from Eddystone, Goudey and Conemiaugh electric power plants were used. The chemical compositions of the fly ashes were analyzed by EDXS (Energy Dispersive X-ray Spectroscopy) in SEM (Scanning Electron Microscopy , Amray 1830D4) and were obtained by the average of different spherical particles with a spot size of electron beam of 0.5 μm. The fly ash particle size was measured with a Horiba Laser Scattering Particle Size Distribution Analyzer (LA 900).

As-received fly ash was mixed with sodium hydroxide powder with a weight ratio of 1 : 1.2. The powdered mixture was heated at 550°C in air for 1 hour and ground by mortar. The melting point of sodium hydroxide is 318°C. The alkali fusion method has been used to decompose species of silicon and aluminum in the inorganic materials (6). After fusion, the fused powder was dissolved in distilled water and aged with agitation for one day at room temperature and ambient pressure. The aged fly ash solution was cured at 60°C and 90°C for various times at ambient pressure. The precipitated solids at various curing times during the hydrothermal process were collected by centrifuge, and then washed 2 times with distilled water followed by centrifugation to remove the residue of sodium hydroxide and soluble impurities. The solid powder was dried in air for 12 hours at 80°C. The crystalline phase of solid powder was determined by XRD (X-ray diffraction, Siemens D500). For the test of thermal stability, the faujasite, converted from Conemiaugh ash, was kept in a crucible plate and heated in air at 540°C and 740°C for 7 hours, respectively. The heated samples were characterized by X-ray diffraction and nitrogen gas adsorption (Quantachrome NOVA 2200).

The ion exchange capacities of faujasites converted from the three fly ashes were tested for cesium ion at room temperature. The sodium ion in the faujasite is easily exchanged with cesium ion in the solution so that the concentration of sodium ion in the solution will increase and eventually reach equilibrium. The sodium ion concentration of the solution was measured by a sodium probe (Orion, Inc.). 0.9 g of faujasites converted from the fly ash were mixed in 38 ml distilled water and stirred for three hours at room temperature. 2 ml of 0.1 N CsCl solution were added to the faujasite solution. At various reaction times, the concentration of sodium ion and the pH value in the solution were measured until equilibrium occurred. For the purpose of comparison, the ion exchange capacities of zeolite P converted from fly ash with fusion was also measured.

RESULTS OF EXPERIMENT

The chemical compositions of the three fly ashes are shown in Table I. The cation elements of fly ash usually exist in the oxide phases. The contents of silicon and alumina are much larger than other cations. The chemical composition of fly ash depends on the source of coal. The mean particle sizes of the three ashes are also shown in Table I. Due to the similar particle size of Eddystone, Goudey and Conemiaugh ashes, the effects of the particle size on the solubility of aluminosilicate in fly ash should be small.

Table I. Chemical Compositions and Mean Particle Size of Untreated Fly Ashes

Chemical Compositions (wt%)	Goudey Ash	Eddystone Ash	Conemiaugh Ash
Si	26.97 $\pm$ 0.46	28.90 $\pm$ 1.07	27.09 $\pm$ 1.11
Al	14.82 $\pm$ 2.42	15.34 $\pm$ 2.03	14.00 $\pm$ 1.50
Fe	3.89 $\pm$ 0.25	1.43 $\pm$ 0.23	4.81 $\pm$ 1.18
Na	0.22 $\pm$ 0.21	0.46 $\pm$ 0.11	0.17 $\pm$ 0.04
K	2.59 $\pm$ 0.26	2.11 $\pm$ 0.55	2.17 $\pm$ 0.52
Ca	2.01 $\pm$ 0.86	1.43 $\pm$ 0.02	0.93 $\pm$ 0.21
Mg	ND	ND	0.15 $\pm$ 0.07
Ti	1.62 $\pm$ 0.84	0.15 $\pm$ 0.04	0.92 $\pm$ 0.16
Mean particle size (μm)	0.76	0.69	0.73

ND : not determined

The X-ray diffraction pattern of untreated fly ash, for example, Goudey ash, is shown in Fig.1. The main crystalline phases of fly ash are mullite and quartz. The amount of crystalline phases will depend on the combustion temperature during coal burning. Because mullite and quartz peaks disappeared after fusion, it indicated that mullite and quartz have reacted with sodium hydroxide during fusion to form silicates and amorphous aluminates. The formation of soluble phases were very important to the increase of the yield of faujasites in the zeolite conversion. In our previous paper (3), when the as-received fly ash was directly used as the raw material without fusion, the peaks of mullite and quartz decreased but still existed during hydrothermal processing indicating only small amount of amorphous aluminosilicate was consumed to form zeolites.

The X-ray diffraction patterns of faujasites converted from Eddystone, Goudey and Conemiaugh ashes are shown in Fig.2, 3 and 4, respectively. The intensity of these faujasites is higher than that of faujasite converted from fly ash without fusion (3). This means that the yield of faujasite converted from fly ash with fusion is higher and all three fly ashes can be used for the formation of faujasite. The XRD intensity of faujasites converted from the three ashes are different. As the cation amount in the three ashes varies, the converted faujasites could contain different amounts and species of cations. The XRD peak intensities of the zeolites have been known to change with different cations contained in the crystal structure (7).

Fig.5 shows the variation of sodium ion concentration in faujasites and zeolite P solutions at different reaction times during ion exchange tests. The pH value of the solution is above 10. The sodium ion concentration in solution

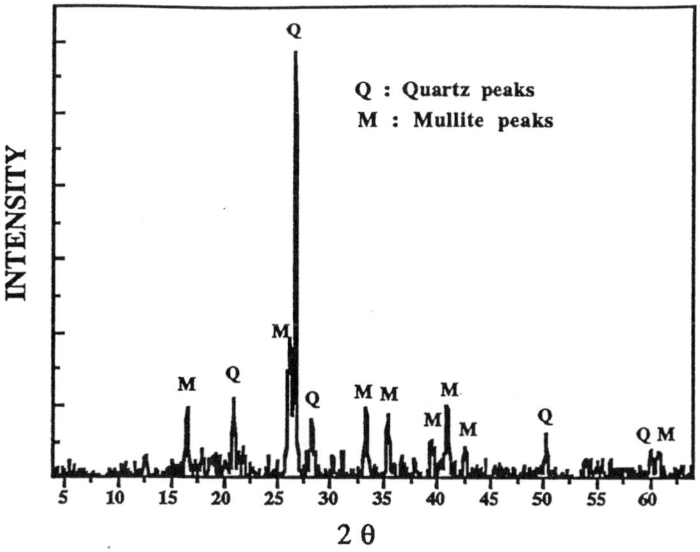

Figure 1. X-ray diffraction pattern of untreated fly ash (Eddystone)

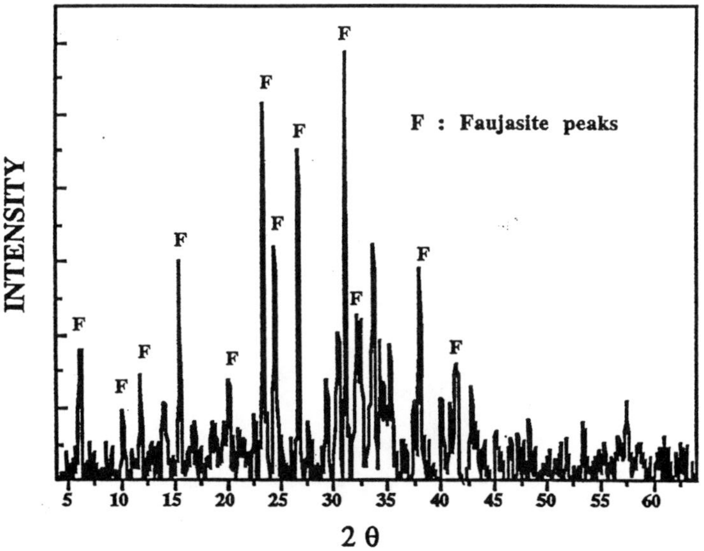

Figure 2. X-ray diffraction pattern of faujasite converted from Eddystone ash

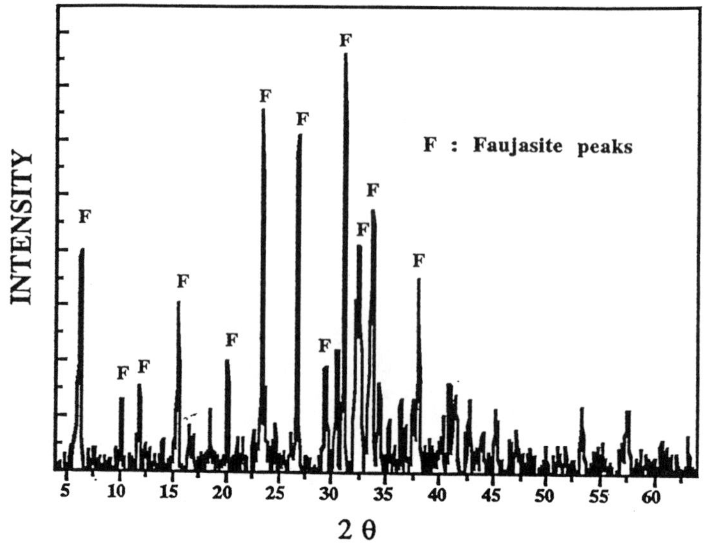

Figure 3. X-ray diffraction pattern of faujasite converted from Goudey ash

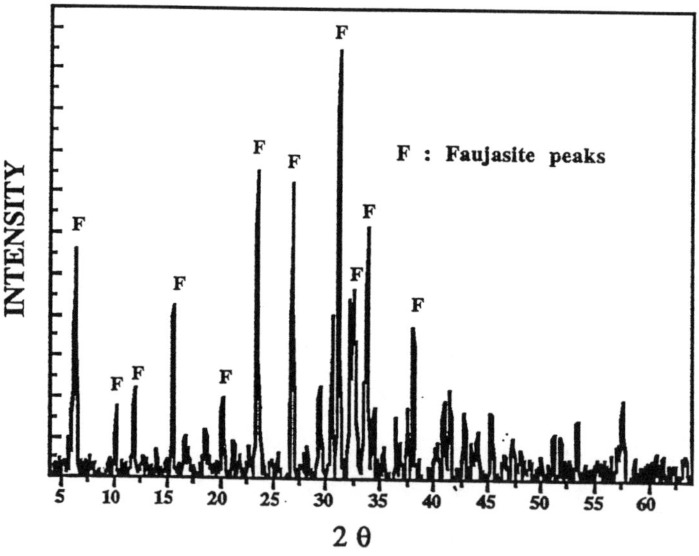

Figure 4. X-ray diffraction pattern of faujasite converted from Conemiaugh ash

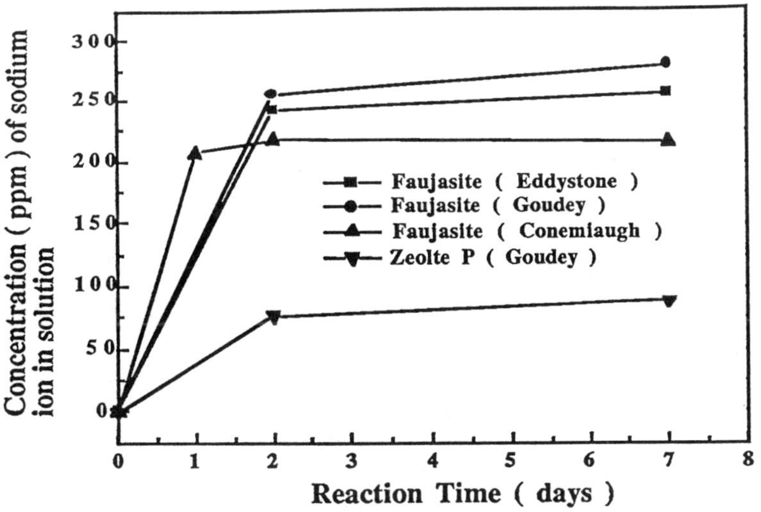

Figure 5. Concentration of sodium ion in 38 ml of water with 0.9 g of zeolites as a function of reaction time after 2 ml of 0.1 N CsCl were added.

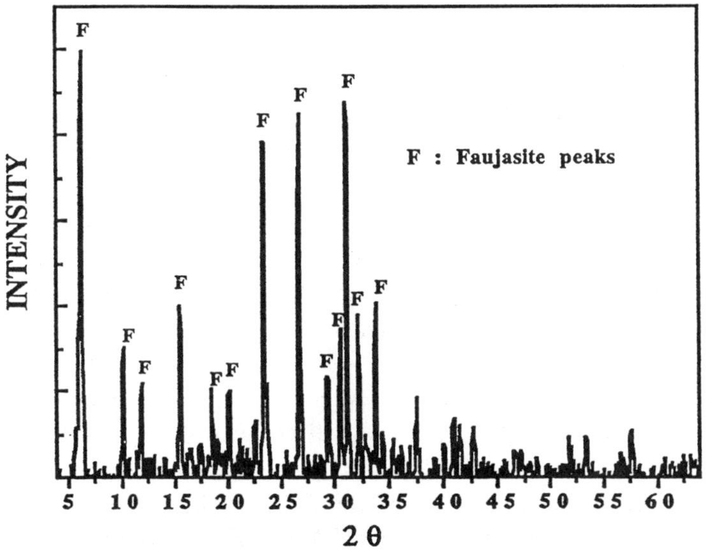

Figure 6. X-ray diffraction pattern of faujasite converted from Goudey ash after an ion exchange with Cesium ion

increases rapidly at the initial stage of the test and reaches equilibrium after 2 days. The ion exchange capacity with cesium ion of the faujasite converted from fly ash is higher than that of zeolite P from fly ash. This result correlates to the different crystal structure of zeolite P (single 4-ring silicate) and faujasite (double 6-ring silicate and cage), and the number of exchangeable sodium ions in those structures (7). The higher pore size and larger quantity of exchangeable sodium ions in the faujasite type zeolite compared with that of zeolite P gives rise to the larger ion exchange capacity of faujasite. On the other hand, the ion exchange capacities of faujasites converted from the three fly ashes are similar. The X-ray diffraction pattern of faujasite converted from Goudey ash after an ion exchange test is shown in Fig. 6. The crystalline phase of faujasite after the ion exchange test is similar to that before the test although the intensity of the peaks are slightly changed. In the SEM micrograph, Fig.7, the morphology of treated fly ash is different from typical fly ash particles. Typically, the fly ash particles are spherical with a wide range of diameters. On the other hand, faujasite crystals are octahedral and tend to agglomerate together.

The thermal stability of faujasite converted from Conemiaugh ash was tested by heating faujasite powder in air at 540°C and 740°C for 7 hours, respectively. The nitrogen gas adsorption isotherms of the heat-treated faujasite powder are shown in Fig.8. The isotherm of the sample heated at 540°C is similar to that of the as-synthesized faujasite powder although the amount of nitrogen gas adsorbed is slightly less. On the contrary, the isotherm of the sample heated at 740°C significantly differs from those of the sample heated at 540°C and the as-synthesized faujasite powder. It is conspicuous that there is little of the nitrogen gas adsorbed by the sample heated at 740°C. This adsorption behavior implies the pores of faujasite have collapsed due to the change of crystal structure. The X-ray diffraction pattern of faujasite converted from Conemiaugh ash heat treated at 740°C is shown in Fig.9. The peaks of faujasite disappeared indicating that the thermal stability of faujasite converted from fly ash can persist at 540°C.

CONCLUSION

Fly ashes with a wide range of chemical compositions have been converted to zeolites with fusion pretreatment. The fusion of NaOH with the fly ash decomposes the silicon and aluminum compounds to form soluble aluminosilicate species. Higher amounts of aluminosilicate species in the alkali solution increases the yield of converted zeolites. The ion exchange capacities of faujasites converted from different ashes with cesium ions are similar. The crystalline structure of the faujasite was thermally stable at 540°C, but broken down at 740°C. Converted faujasites may be a better ion exchanger for radioactive ions than organic resin.

Figure 7. SEM micrograph of Goudey faujasite after ion exchange test

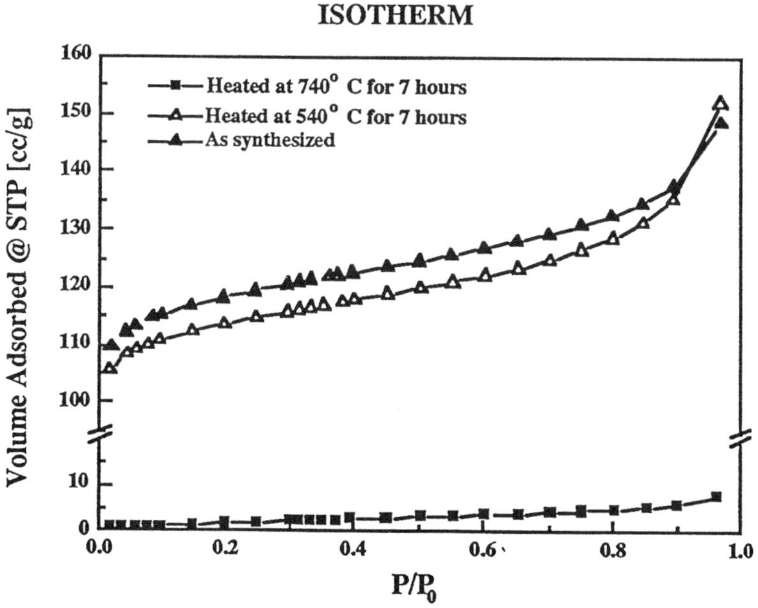

Figure 8. Nitrogen gas adsorption isotherms of Conemiaugh faujasites after heat treatment at different temperatures

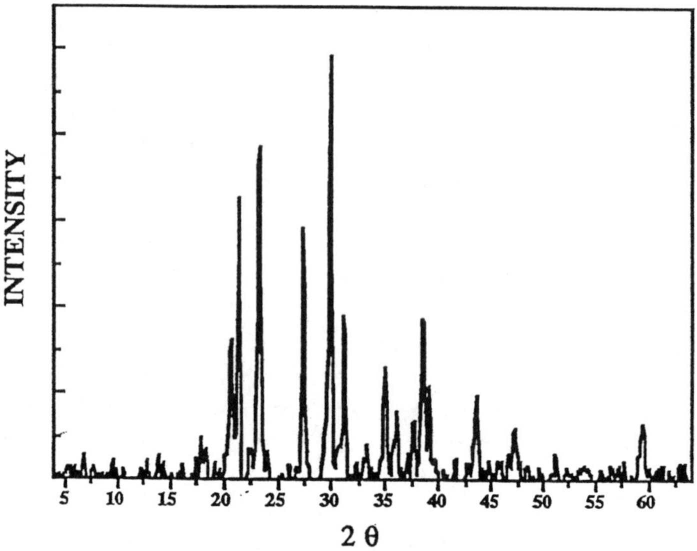

Figure 9. X-ray diffraction pattern of faujasite converted from Conemiaugh ash after heated in air at 740°C for 7 hours

REFERENCES

1. L.A. Ruth, Chemtech, June, 33-39 (1993)
2. W.H. Shih, H.L. Chang, and Z. Shen, " Conversion of class F fly ash to zeolites ", Mat. Res. Soc. Symp. Proc. Vol. 371, p39-44, (1995)
3. W.H. Shih and H.L. Chang, " Conversion of fly ash into zeolites for ion-exchange applications ", Mat. Lett., 28, p263-68, (1996)
4. N. Shigemoto, H. Hayashi, and K. Miyaura, " Selective formation of Na-X zeolite from coal fly ash by fusion with sodium hydroxide prior to hydrothermal reaction ", Mat. Sci., 28, p4781-4786, (1993)
5. H.H. Weldes and K.R. Lange, " Properties of soluble silicates ", Ind. & Eng. Chem., 61(4), p29-44, (1969)
6. W.F. Hillebrand, G.E.F. Lundell, H.A. Bright, and J.I. Hoffman, Applied Inorganic Analysis, 2nd ed., Wiley, NewYork, 1962
7. D.W. Breck, Zeolite Molecular Sieves, John Wiley & Sons, New York, 1974

PREPARATION TECHNIQUES FOR CERAMIC WASTE FORM POWDER

M.C. Hash, C. Pereira, V.N. Zyryanov, M.A. Lewis, G.L. Burns, J.J. Smith, J.P. Thalacker, and J.P. Ackerman
Argonne National Laboratory, Chemical Technology Division, Nuclear Technology Program, 9700 South Cass Avenue, Argonne, IL 60439

ABSTRACT

The electrometallurgical treatment of spent nuclear fuels results in a chloride waste salt requiring geologic disposal. Argonne National Laboratory (ANL) is developing ceramic waste forms which can incorporate this waste. Currently, zeolite- or sodalite-glass composites are produced by hot isostatic pressing (HIP) techniques. Powder preparations include dehydration of the raw zeolite powders, hot blending of these zeolite powders with the chloride waste salt, and subsequent dry mixing with various glass powders and secondary additives. Various approaches are being pursued to achieve adequate mixing, and the resulting powders are being characterized by quality assurance methods. These powders have been HIPed and characterized for leach resistance, phase equlibria, and physical integrity.

INTRODUCTION

A ceramic waste form is being developed at ANL to immobilize chloride waste salt generated during the electrometallurgical treatment of spent nuclear fuel.[1] This waste form comprises two phases: zeolite or sodalite and glass. The zeolite or sodalite phase, both aluminosilicates, incorporates the waste salt into its crystalline lattice or cage structure.[2,3,4] The glass phase binds the crystalline phase together lending structural integrity.[4] Hot isostatic pressing (HIP) techniques are used in fabrication of these composite waste forms.

Hot isostatically pressed composites show adequate leach performance for consideration as a spent nuclear waste form surpassing that of comparable hot uniaxially pressed composites.[4,5] The overall mass loss during leach testing is relatively low, the normalized release rate is less than 1 g/m^2d in 28-d leach tests at 363 K (90°C) in deionized water, and this rate decreases with time.[5] Mass loss was measured by a modified MCC-1 leach test procedure (ASTM C 1220). Normalized release rates were determined from analysis of leachates.

Fabrication of these ceramic waste forms includes a number of powder processing techniques. The raw materials used are glass powder, simulated waste salt, and zeolite-A powder. The simulated waste salt is prepared as a fine (-400 mesh) powder. The zeolite-A powder is dehydrated, blended with the simulated waste salt, combined with the glass, and then prepared for HIP cycling. Each stage of the process depends on the success of the previous stage. As is the case for all spent nuclear waste forms, leach behavior is of prime concern. Therefore, knowledge of how variations in intermediate characteristics using powder preparation affect the leach behavior of the final composite is essential.

Most of our work has focused on the blending stage. Previous studies suggested that this stage of our powder preparation has the greatest effect on the composite leach behavior.[4] However, due to the interdependence of the stages we will discuss processing before and after blending in detail. Also, several quality assurance (QA) methods are being developed to provide response variables which may predict leach behavior. The most useful of these methods appears to be our characterization of the blended powders.

EXPERIMENTAL

Raw Materials

Three raw materials are used in the fabrication of our ceramic waste forms: glass powder, simulated waste salt, and zeolite-A powder. Glass powders obtained from Bayer (Baltimore, MD) or Corning (Corning, NY) are typically -325 mesh and are used as-received (stored in air).

The simulated waste salt is prepared in a purified argon atmosphere glovebox. This salt composition is given in Table I. This mixture of chlorides is melted and quenched, yielding a uniform, multiphase salt that represents an average composition of what is expected from the electrometallurgical treatment of spent nuclear fuels. The salt is crushed by means of a porcelain mortar and pestle to a fine powder, typically -400 mesh. All handling and storing of salts is done in a purified argon atmosphere glovebox.

Zeolite-A powders are obtained from UOP, Molecular Sieves Division (DesPlaines, IL). Two types are used in this study: 4A, which is a sodium aluminosilicate, and 5A, which is a calcium/sodium aluminosilicate. Particle size analysis of these powders is done using a light scattering technique. A typical particle size distribution is shown in Figure 1. Raw zeolite powders are stored in air.

Table I. Nominal Simulated Waste Salt Composition

Compound	Weight %
LiCl-KCl Eutectic	89.3
CsCl	2.3
$BaCl_2$	1.0
$SrCl_2$	0.4
$CeCl_3$	2.0
$LaCl_3$	0.6
$NdCl_3$	3.2
$PrCl_3$	0.8
YCl_3	0.1
KI	0.3

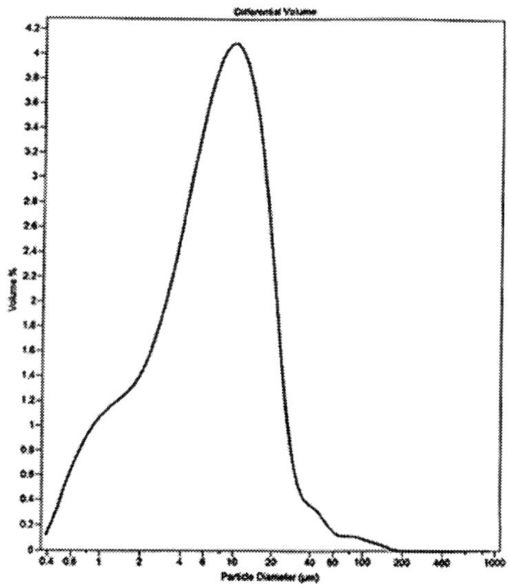

Figure 1. Typical Particle Size Distribution of Zeolite-A Powder

Dehydrating Zeolite Powders

Zeolite powders may contain as much as 22 wt% water.[6] Our dehydration process removes most of this water, yielding an activated powder suitable for blending with simulated waste salt. The dehydrator, shown below in Figure 2, passes nitrogen (99.999%) gas through a loosely packed zeolite powder, which is heated under the thermal cycle shown in Figure 3. This thermal cycle removes most of the water from the zeolite without damaging the open, cage structure desired for inclusion of the simulated waste salt.

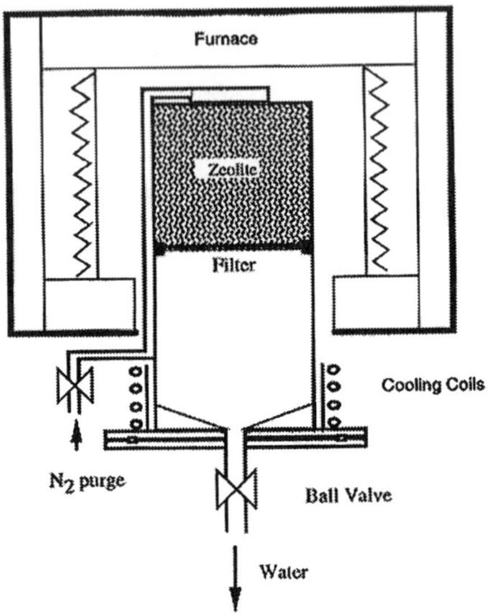

Figure 2. Schematic of Dehydration Vessel

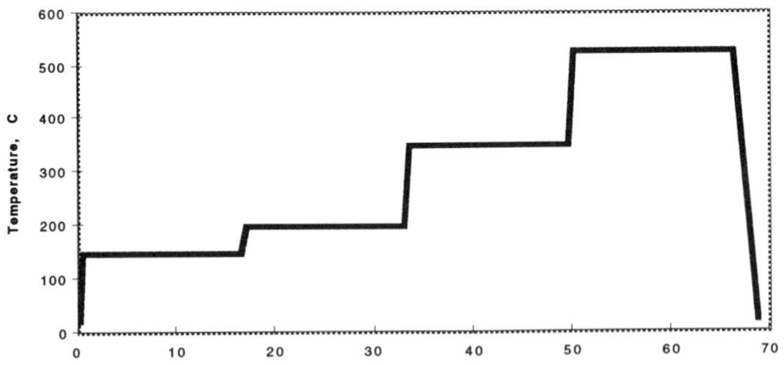

Figure 3. Dehydration Thermal Schedule

Blending Dehydrated Zeolite and Simulated Waste Salt

Dehydrated zeolite powder is combined with simulated waste salt powder in a heated, tumbling vessel known as a "blender" (shown in Figure 4).

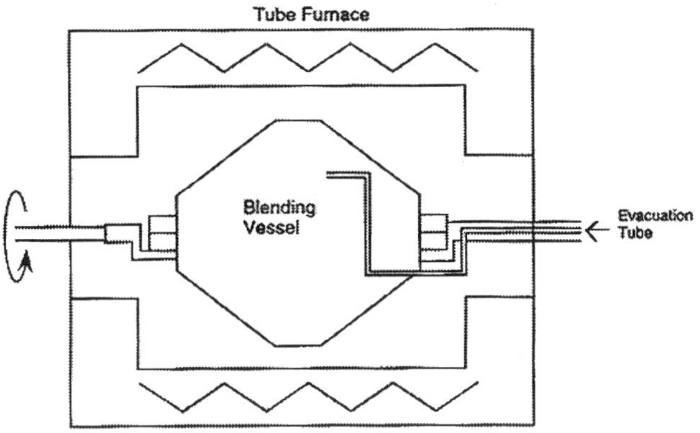

Figure 4. Blending Vessel Used to Combine Dehydrated Zeolite and Salt

Desired proportions of the powders are loaded into the blender in a purified argon atmosphere glovebox. The blender is sealed and removed from the glovebox. The blender is then located within a large tube furnace, which is used to heat the whole vessel to 500°C. Prior to heating, the vessel is evacuated using a mechanical vacuum pump. The vessel is rotated slowly on its major axis while heating to the desired temperature at a rate of 5-10°C/min. Once at 500°C, the vacuum is re-established (if necessary) prior to a 20-h isothermal soak with rotation. After this soak, the vessel is furnace cooled and returned to the argon atmosphere glovebox for unloading and storage of the blended zeolite.

Mixing Blended Zeolite and Glass

The powder product from the blender, referred to as "blended zeolite", is combined with glass powder in a purified argon atmosphere glovebox. Desired proportions are mixed dry using a rotating-blade mill. Powder batches of 50-100 g each are milled together for 4 minutes. This powder is stored in the glovebox until loaded into HIP canisters.

Hot Isostatic Pressing of Zeolite-Glass Powders

The combined zeolite-glass powder is loaded into canisters, which serve as dies for HIPing. The canisters are 1-in. (2.54 cm) diameter x 3-in. (7.62 cm) tall right cylinders with a wall thickness of 0.020-in. (0.05 cm) and are made of Type 304 stainless steel. Uniaxial cold-pressing techniques are employed for loading the powder in layers directly into the canisters until full. This technique yields green densities of 55-60% solids. Lids that include a 1/8-in. outer diameter tube used for evacuating the canisters are then welded into place.

The canisters are evacuated using a mechanical vacuum pump. Once a vacuum has been achieved, the canisters are heated to 500°C and held at this temperature for 16 h. The canisters are then cooled rapidly under vacuum. When cool, the evacuation tubes are crimped and welded to ensure closure.

The canisters are commonly HIPed using the cycle depicted in Figure 5. This thermal profile yields high-density zeolite-glass ceramics having greater than 99% theoretical density.

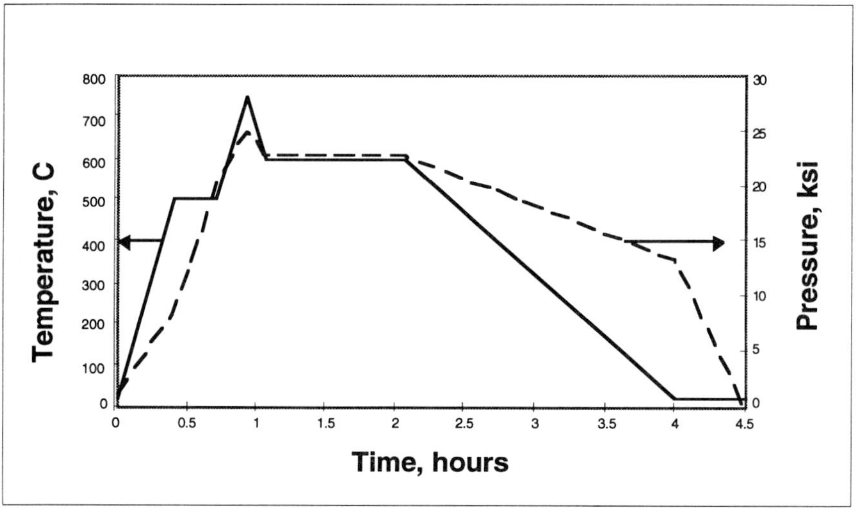

Figure 5. Typical HIP Cycle Used for Densification of Zeolite-Glass Composites

RESULTS AND DISCUSSION

Our ultimate concern is how variations at the many stages of this process affect the leach behavior of the resulting composite ceramic. The response variable we have chosen to study is cesium release. Cesium has a high level of radioactivity and one of the longest half-lives of all the radionuclides contained in the waste salt. Cesium is also the most difficult to retain. Cesium can, therefore, be considered the weak link in a Weibull analysis (performance of the system or material being studied cannot surpass its weakest component). The cesium fractional mass release is determined by comparing the initial cesium content in the composite to that found in the leachant after MCC-1 leach tests.

The cesium content is determined using inductively coupled plasma-mass spectroscopy (ICP-MS). Where appropriate, therefore, cesium retention data are presented to provide insight as to how variations in our powder preparations affect leach behavior.

Dehydrating Zeolite Powders

The relative humidity of the off-gas and the centerline temperature are monitored during dehydration. The relative humidity serves as a control parameter. Since we are not merely driving off all of the water, which would destroy the zeolite-A structure, it is necessary to carefully define conditions under which the dehydration cycle is considered complete. The temperature profile for this process includes four isotherms (Fig. 3). Using the relative humidity signal, we can discern whether the powder is actively releasing water or has come to equilibrium with the nitrogen gas at each temperature plateau. The centerline temperature is watched to ensure that the powder bed has come to thermal equilibrium.

Dehydrated zeolite absorbs water very readily, making analysis in air difficult. In response to this, we are developing a QA method to characterize the dehydrated product without exposure to moisture. Thermogravimetric analysis (TGA) of the resulting powders under well-defined operating parameters is a promising technique. Operating parameters are being refined as we build a suitable database of trends.

Blending Dehydrated Zeolite and Simulated Waste Salt

The QA method used at this stage of our powder preparation was developed by us. After blending of dehydrated zeolite and simulated waste salt, samples of the product are washed with water. This washing method provides a means of detecting readily dissolved or non-occluded chloride salts. High chloride values indicate zeolites which were not properly dehydrated or have been exposed to moisture during storage.

Variables at this stage are many: zeolite type (4A or 5A), salt-to-zeolite ratio (or "salt loading"), and process duration. We are currently investigating two zeolite-A raw materials: 4A, which is a sodium aluminosilicate, and 5A, which is a calcium/sodium aluminosilicate. Cesium release data for comparable samples show that composites made with either 4A or 5A zeolite exhibit similar leach behavior.

The amount of salt blended and potentially occluded by the zeolite is expressed as the number of chloride species per psuedo unit cell of zeolite (Cl^-/u.c.). We have produced comparable zeolite-glass composites with zeolites blended at 3.2, 7.5, 9.5, and 10.0 Cl^-/u.c. Cesium fractional mass release data for leach tests of 3 and 28 days with these composites are represented in Figure 6. Unfortunately, 28-d leach testing for the sample containing zeolite blended at 10.0 Cl-/u.c. is not yet available. The available release data after 3-d leach tests show a relatively flat response with respect to salt loading. However, the 28-d leach tests show cesium fractional mass release to be directly proportional to the amount of salt incorporated in the zeolite. The 28-d data suggest that the release of cesium is driven by the concentration gradient that exists between the composite and the deionized water. Assuming this to be the major driving force, cesium release normalized to leach testing duration should continously decrease with time.

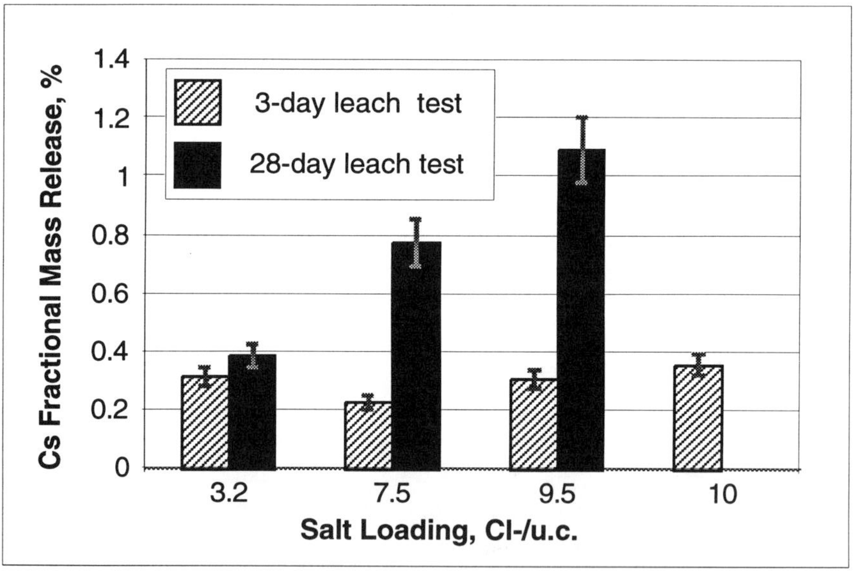

Figure 6. Cesium Fractional Mass Release (Cs FMR) at Various Salt Loadings in Zeolite-Glass Composites

The error bars shown in Figure 6 represent variance associated with our method of analysis (ICP-MS). Thje individual data points represent either an average of several tests or a single test: at 7.5 Cl-/u.c. the 3-d data point is an average of samples ($\sigma = 0.02$) and the 28-d data point is an average of 4 samples ($\sigma = 0.27$), at 9.5 Cl-/u.c. the 3-d data point is an average of 2 samples ($\sigma = 0.03$), all other data points represent single samples.

As previously discussed, our blending process is achieved in a heated, rotating vessel. Tests were conducted to see if we had achieved a homogeneously blended zeolite. Initially, while heating the zeolite-salt mixture to 500°C, the salt melts incongruently. The dehydrated (activated) zeolite rapidly sorbs the molten salt, leading to a powder with no liquid phase present. At this point, any macroscopic mixing occurs according to dry powder mixing kinetics. Dry powder mixing is slow and limited in the degree of mixing that can be achieved. This is especially true in our system since the zeolite tends to form agglomerates.

Using our free chloride QA method we studied the kinetics of one of our blending runs. The thermal cycle was interrupted several times for sampling of the powders. All the liquid was, indeed, absorbed in less than 2 h at processing temperature. The free chloride level was measured at 0.04% after 2 h of blending. After 4 h the free salt remained at 0.04%. This level of free salt was constant thereafter out to a blending time of 24 h after which the experiment was halted. The observation of the free salt level reaching 0.04% in only 2 h suggests only that macroscopic mixing may be complete. This result should not be confused with the time necessary to fully distribute the ionic species within the zeolite structure. During blending beyond the time at which our QA method can detect change, the salt occluded zeolite may thermodynamically equilibrate yielding the most stable phase possible. Experimentation to study the effect of blending duration on leach behavior has been initiated.

CONCLUSIONS AND FUTURE WORK

Zeolite powder preparation techniques currently being employed seem appropriate for the production of leach-resistant ceramic waste forms. Quality assurance methods still require refinement such that scale-up activities will not be hindered by mis-characterization of intermediate powder products.

To date, we have not attempted tailoring particle size distributions to enhance mixing and flow characteristics. The zeolite powder is as fine as it can be made. Particle size reduction of this powder would damage the zeolite crystals.

However, our development of these ceramic waste forms will eventually include agglomerated zeolite pellets, which may induce particle size design efforts.

ACKNOWLEDGMENTS

The authors gratefully acknowledge the sample preparation rendered by A. Candiotti (University of Illinois-Chicago) and M. Stanley (Eastern Illinois University). Also gratefully acknowledged are B. Tani, P. Johnson, and J. Crain (ANL, Analytical Chemistry Laboratory).

REFERENCES

[1] J. P. Ackerman, T. R. Johnson, and J. J. Laidler, Waste Removal in Pyrochemical Fuel Cell Processing for the Integral Fast Reactor, *Actinide Processing, Methods and Materials,* pp. 261-266, Eds., B. Mishra and W. A. Averill, The Minerals, Metals, and Materials Soc., Warrenville, PA (1994)

[2] M. A. Lewis, L. J. Smith, and D. F. Fischer, Salt-Occluded Zeolites as an Immobilization Matrix for Chloride Salt Waste, Am. Ceram. Soc. 76 (11), 2826-2832 (1993)

[3] M. A. Lewis, D. F. Fischer, and C. D. Murphy, Densification of Salt-Occluded Zeolite A Powders to a Leach Resistant Solid Monolith, *Scientific Basis for Nuclear Waste Management*, Eds., A. Barkatt and R. Van Konynenburg, Vol. 333, pp. 277-284, Materials Research Society, Pittsburgh, PA (1994)

[4] M. C. Hash, C. Pereira, M. A. Lewis, R. J. Blaskovitz, and J. P. Ackerman, Hot Isostatic Pressing of Glass-Zeolite Composites, *Environmental and Waste Management Issues in the Ceramic Industry II*, pp. 135-143(1996)

[5] M. A. Lewis, D. F. Fischer, and C. D. Murphy, Properties of Glass-Bonded Zeolite Monoliths, *Environmental and Waste Management Issues in the Ceramic Industry II*, pp. 277-286 (1994)

[6] Donald W. Breck, *Zeolite Molecular Sieves*, New York, John Wiley & Sons, Inc., 1974.

LOW TEMPERATURE SETTING IRON PHOSPHATE CERAMICS AS A STABILIZATION AND SOLIDIFICATION AGENT FOR INCINERATOR ASH CONTAMINATED WITH TRANSURANIC AND RCRA METALS.

P. G. Medvedev, J. Macheret*, M. Hansen, E. L. Wood, S. M. Frank, R.W. Sidwell, J. J. Giglio and S. G. Johnson

Argonne National Laboratory-West
P.O. Box 2528
Idaho Falls, ID 83403

*Dept. Of Energy, Idaho Office
Idaho Falls, ID 83415

ABSTRACT

Incineration of combustible Mixed Transuranic Waste yields an ash residue that contains oxides of Resource Conservation and Recovery Act (RCRA) and transuranic metals. In order to dispose of this ash safely, it has to be solidified and stabilized to satisfy appropriate requirements for repository disposal. This paper describes a new method for solidification of incinerator ash, using room temperature setting iron phosphate ceramics, and includes fabrication procedures for these waste forms as well as results of the MCC-1 static leach test, XRD analysis, scanning electron microscopy studies and density measurements of the solidified waste form produced.

INTRODUCTION

The goal of this work is investigation and development of a new method for solidification of incinerator ash containing oxides of RCRA and transuranic metals using low temperature setting iron phosphate ceramics. More specifically, this work seeks to elucidate the means in which the reaction rate of the setting process can be controlled, easily and economically, while still yielding a waste form with acceptable characteristics and performance. Iron phosphate ceramics represent a large group of acid-base cements. Their properties as well as some practical applications are well described in the literature (1-5). Iron phosphate ceramics are themselves a sub-group of the larger body of ceramics that include other types, such as magnesia based ceramics, that have been investigated for immobilization of transuranic wastes (6).

Solidification of the ash waste using this process results not only in micro- and macro-encapsulation of contaminants but also in chemical binding of TRU and RCRA metals in the form of phosphates. Phosphates are known as host minerals for some actinides and heavy metals (7). This process has several distinct advantages over other methods: 1) it is mechanically very simple requiring minimal equipment, 2) it does not produce secondary waste streams such as free standing liquids or off-gases, 3) it proceeds at room temperature which eliminates the need for off-gas treatment and 4) the treatment is cost-effective from a processing standpoint.

EXPERIMENTAL

Waste stream

Solidification experiments were conducted using simulated incinerator ash. The surrogate ash was prepared and supplied by Dr. Wagh of Argonne National Laboratory East, Energy Technology Division, Ceramic Section. Composition of the waste stream is given in Table 1.

Table 1. Composition of simulated ash waste

Components	Concentration (g/kg)
Class F fly ash	387.96
Coal bottom ash	320.07
Vermiculite	193.98
Activated carbon	48.49
Cr_2O_3	14.18
NiO	6.16
PbO	5.22
CdO	5.54
CsCl	6.15
Ce_2O_3	5.70
$HgCl_2$	6.56

As indicated in Table 1, oxides and chlorides of various metals were added to the ash to simulate actual waste ash. Cr_2O_3, PbO, $HgCl_2$ and CdO represented compounds of RCRA metals. Ce_2O_3 was used to simulate plutonium oxide. Natural CsCl was used to simulate fission product cesium.

Solidification Experiments

Solidification experiments were conducted in 50 ml plastic beakers. Ash waste was mixed with magnetite powder to form a homogeneous mixture. The powder mixture was slowly added to orthophosphoric acid forming a slurry. The slurry was blended thoroughly using a spatula. Monolithic specimens produced as a result of solidification were cured for 3-4 weeks at room temperature. The flow chart for the solidification process is shown in Fig. 1.

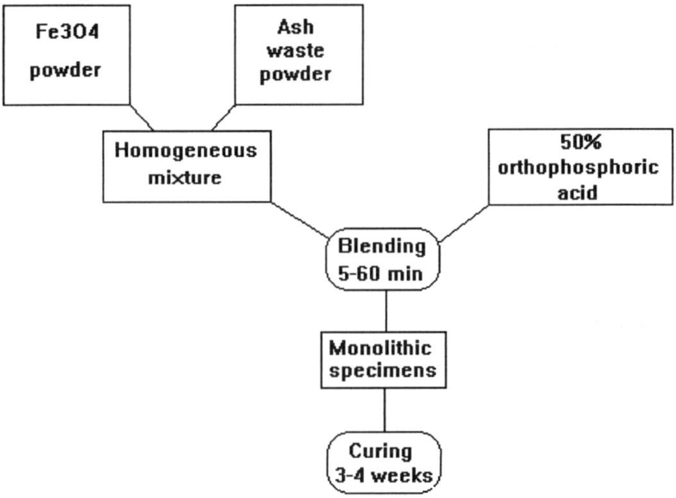

Figure 1. Flow chart for the process of ash waste solidification

The ratios of the constituents and the acid concentration were determined by an iterative method based on the composition diagram shown in Fig. 2. The key parameter used to narrow the area of interest in Fig. 2 was a visible inspection of the structural integrity of the waste form produced. A large number of cracks or a "crumbly" waste form was regarded as a non-attractive composition. The shaded area on the composition diagram (Fig. 2) corresponds to the waste form compositions for which solidification occurs. Investigations presented in this paper were focused on the compositions with a waste loading of 30 and 40%. The waste loadings cited in this paper are calculated by dividing the mass of waste by the total mass of waste, binder and phosphoric acid.

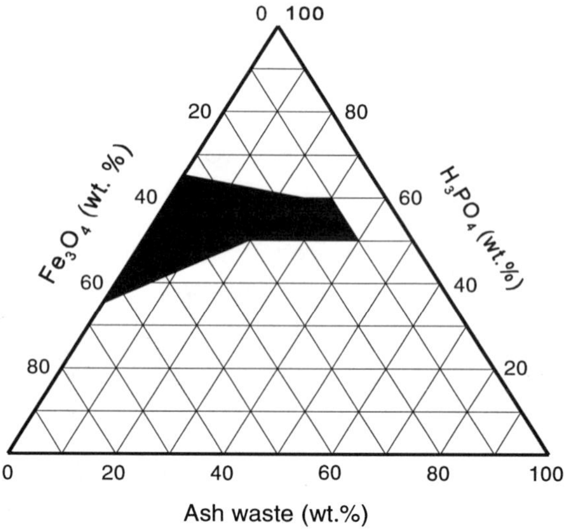

Figure 2. Composition Diagram for the Solidification Process.

Leach Tests

The leaching properties of the fabricated waste forms were investigated by conducting the MCC-1 static leach test (ASTM C1220-92). The procedure described in the MCC-1 method was modified in regard to the sample preparation steps. The monolith utilized was obtained via core drilling and the surface finish was 600 grit. The (surface area/volume) ratio was 10 m^{-1}. The leach tests were run for 7 days at 90 °C.

The normalized release rate (NRR) was evaluated using the equation:

$$NRR(i) = C_i V / f_i A d \qquad (1)$$

where C_i is the concentration of the element i in the leachate in g/ml, V is the volume of leachate in ml, f_i is the fraction of the element i in the solid initially, A is the macroscopic area of the solid in m^2, and d is the duration of the test in days. The surface area was calculated from the linear dimensions of the cylindrical samples. The monolithic specimens, obtained as a result of solidification, were cured for 21 days at room temperature before they were submitted for the leach test.

Instruments Used for Analysis

1. Density measurements

Density of the samples was measured using Ultrapycnometer-1000 (UPY-2, Quantachrome Inc, Boyton Beach, FL).

2. Microscopy studies and EDS analysis

An ISI SS440 scanning electron microscope supplied by Topcon Technologies (Paramus, New Jersey) with a Kevex 7070 EDS system (Kevex Fisons Instruments, San Carlos, CA) was used for microscopy studies and EDS analysis.

3. Analysis of leachates

Leachates were analyzed using an ICP system that included a plasma unit consisting of a 40 MHz, 1600 W generator and a torch box (ICP 16L, RFPP, New Jersey, USA). The torch box was fitted with a demountable Fassel type torch. A 1.5 path length monochromator, fitted with a 2400 lines per mm holographic grating (THR 1500, ISA Inc. Metuchen, NJ, USA) was used to detect analyte emissions from ICP. The monochromator stepper motor was controlled via a Spectrolink control unit, (ISA Inc., Metuchen, NJ, USA).

4. X-ray diffraction analysis

XRD analysis was performed in order to identify the crystalline phases that are present in the ceramics. A Scintag X1 Powder X-ray Diffractometer (Scintag, Inc., CA) with a Theta-Theta type goniometer, Cu K-alpha x-ray tube with normal focus, and top loading powder sample holder was used for this purpose.

RESULTS AND DISCUSSION

The methodology pursued for this study revolved around first investigating the appropriate binder compositions to provide the desired setting kinetics, not too slow or too fast. Once the approximate composition was determined the waste loading was optimized and a final fine tuning of the components was attempted. The waste forms produced were characterized by means of SEM, XRD, density and setting times. The durability of the waste forms produced was assessed by the MCC-1 test.

Estimation of Reaction Rates and Setting Times

The quick setting behavior of the Fe_3O_4-H_3PO_4 system is caused by the rapid formation of normal and acidic iron phosphates (2). Since this reaction is exothermic, its rate can be monitored via the temperature rise of the slurry. As can be seen in Fig. 3, the reaction rate is controllable at several waste loadings of interest by the simple addition of a single agent. The reaction retardant is dissolved in phosphoric acid before the acid is reacted with the powder mixture of ash waste and magnetite. By varying concentration of the retardant in the acid from 0.025 g/ml to 0.1 g/ml, we were able to fabricate ceramic waste forms with the setting time from 5 minutes to 2 hours. The degree of flexibility that this affords the process in the field can not be over-stated.

Density

Results of density measurements are summarized in Table 2. The true density of the fast setting compositions is less than that of the slow setting. A difference between the true and bulk density is explained by the presence of air voids trapped in the matrix. Bulk density was determined using the macroscopic volume and mass. The higher the density, the lower the porosity and this should be reflected in better normalized release rates from the MCC-1 test. A method of removing this variable from the leach test would be to apply the product consistency test (PCT), ASTM 1285-94, which utilizes a crushed sample and not a monolithic one. Providing the compressive strength measurments are acceptable the PCT test may be the method of choice in the immediate future for our research needs.

Table 2. Densities and setting times of selected waste form compositions

Parameters		Fast setting compositions		Compositions with extended setting time							
Waste loading, %		30	40	30				40			
Amount of retardant added, g/ml H_3PO_4		0	0	0.025	0.05	0.075	0.1	0.025	0.05	0.075	0.1
Setting time, min		5	5	8	30	60	90	8	30	60	90
Density, g/cm^3	true	2.68	2.50	2.62	2.78	2.80	2.90	2.63	2.77	2.83	2.67
	bulk	1.46	1.57	1.37	1.58	1.67	1.63	1.79	1.57	1.67	1.68

Static Leach Results

The results of a 7 day MCC-1 test are presented in Table 3. These are normalized release rates for the cations of interest including Fe. Concentrations of Ce, Cs and RCRA metals in the leachate solutions were below detection levels. Relatively high weight losses of the samples can be explained by leaching of unreacted phosphoric acid from the samples. This hypothesis is supported by the relatively low pH of the leachates, pH<4.

Table 3. MCC-1 test results

Parameters		Fast setting compositions		Compositions with extended setting time				
Waste loading, %		30	40	30		40		
Amount of retardant added, g/ml H_3PO_4		0.00	0.00	0.075	0.1	0.05	0.075	0.1
Weight loss, wt %		5.95	7.10	2.02	1.95	2.32	2.13	1.48
pH of the leachate solutions		3.47	3.32	3.36	3.33	3.51	3.55	3.37
Normalized release rate, $g/m^2/day$	Cr	<0.049	<0.037	<0.049	<0.051	<0.036	<0.037	<0.036
	Pb	<0.294	<0.220	<0.294	<0.305	<0.217	<0.219	<0.219
	Ce	<0.049	<0.037	<0.049	<0.051	<0.036	<0.037	<0.036
	Fe	0.042	0.038	0.018	0.039	<0.016	0.036	0.039

Scanning Electron Microscopy Analysis

Scanning electron microscopy combined with EDS analysis was used to characterize different phases present in monolithic waste forms. The samples were found to consist of a bulk matrix phase with smaller phases present in suspension. The bulk phase consisted of Al, Si, P and Fe (our EDS is incapable of detecting oxygen). The bulk phase appears as a dense amorphous material; its amorphoric character was confirmed by XRD analysis. The suspended particles present in the matrix were of non-uniform compositions consistent with the encapsulated nature of the waste form. It was found that major constituents of the ceramic matrix are Al, Si, P and Fe. SEM images of selected monolithic waste forms are presented in Fig. 4. Fig. 4a) and 4b) correspond to the fast setting compositions with the waste loading of 30 and 40% respectively. Fig. 4c) and 4d) correspond to the slow setting compositions with the same waste loading. It was observed that slow setting compositions have fewer pores and cracks compared to fast setting, which suggests that slow setting is more favorable for the solidification process.

XRD Results

XRD analysis was performed to identify the crystalline phases present in the solidified product. Crystalline phases include those identified in the surrogate waste starting material (i.e., silicate or aluminum-silicate phases, carbon, mullite), unreacted magnetite (Fe_3O_4) and possibly an iron phosphate phase ($Fe_3H_9(PO_4)_6$ $6H_2O$). It appears the fast setting composition has a greater amorphus content than the slow setting material.

CONCLUSIONS

The preliminary study presented in this paper using surrogate fly ash has demonstrated the potential of the iron phosphate ceramic system for application to actual TRU-containing waste. Furthermore, the results show promise and the need for further work in this area using both surrogates and actual TRU containing fly ash. The following points deserve emphasis as significant findings of this work:

1) High, 40 wt%, waste loadings of surrogate fly ash containing Cs, Ce and RCRA metals have been successfully immobilized using an iron phosphate ceramic waste form.
2) The ceramic waste form shows preliminary normalized release rates for the RCRA metals that are promising.
3) The setting time of the waste form compositions is easily controlled in a wide range using an easily obtained, economical retardant.
4) The method outlined in this work is quick, effective, economical and displays a significant waste loading.

ACKNOWLEDGMENTS

Argonne National Laboratory is operated for the U.S. Department of Energy by the University of Chicago. This work was supported by the U.S. Department of Energy, Reactor Systems, development and Technology, under Contract W-31-109-ENG-38. The Plutonium Focus Area, Dept. Of Energy, Idaho Office is gratefully acknowledged for supporting one of us, P. G. M., for this work. Dr. Arun Wagh is gratefully acknowledged for providing the surrogate ash used for this study and for helpful discussions with P. G. M.

REFERENCES

1. L. I. Turkina, L. G. Sudakas, V. A. Paramonova, and A.A. Chernikova, "Phase Composition and Strength of Iron Phosphate Binders," *Izvestiya Akademii Nauk SSSR, Neorganicheskie Materialy, Vol.26, No. 9, pp. 1967-1972, September, 1990.*
2. V. V. Gerasimov and S. P. Sheptinskii, "X-Ray Phase and Electron Microscopic Studies of Pyrite Phosphate Mixtures", *Izvestiya Akademii Nauk SSSR, Neorganicheskie Materialy, Vol.22, No. 1, pp. 120-125, January, 1986*
3. S. S. Zhukovskii, Yu. M. Yunovich, and O.E. Nevskaya," Phosphate-Bonded Cold Setting Sands for Steel and Iron Casings Production," *Liteinoe Proizvodstvo, No. 4, pp. 19-20, 1987.*
4. L. I. Turkina et. al., "Principles of Cold-Setting Sand Preparation Using Iron Phosphate Binders," *Liteinoe Proizvodstvo, No. 4, p.5, 1989*
5. M. M. Sychev and V.G. Komlev "Setting of Iron, Cobalt, and Nickel Phosphate Cements," *Izvestiya Akademii Nauk SSSR, Neorganicheskie Materialy, Vol. 7, No. 9, pp. 1612-1615, 1971.*
6. Jeong, S.; Wagh, A.S.; Singh, D., *"Chemically bonded phosphate ceramics for stabilizing low-level radioactive wastes", April 1996, NTIS Order Number DE96009457 .*
7. *"Handbook of Geochemistry," edited by K.H. Wedepohl, Berlin, Heidelberg, New York, Springer, v.2, pp. 82B9-89B11, 92D4-92D5,1969.*

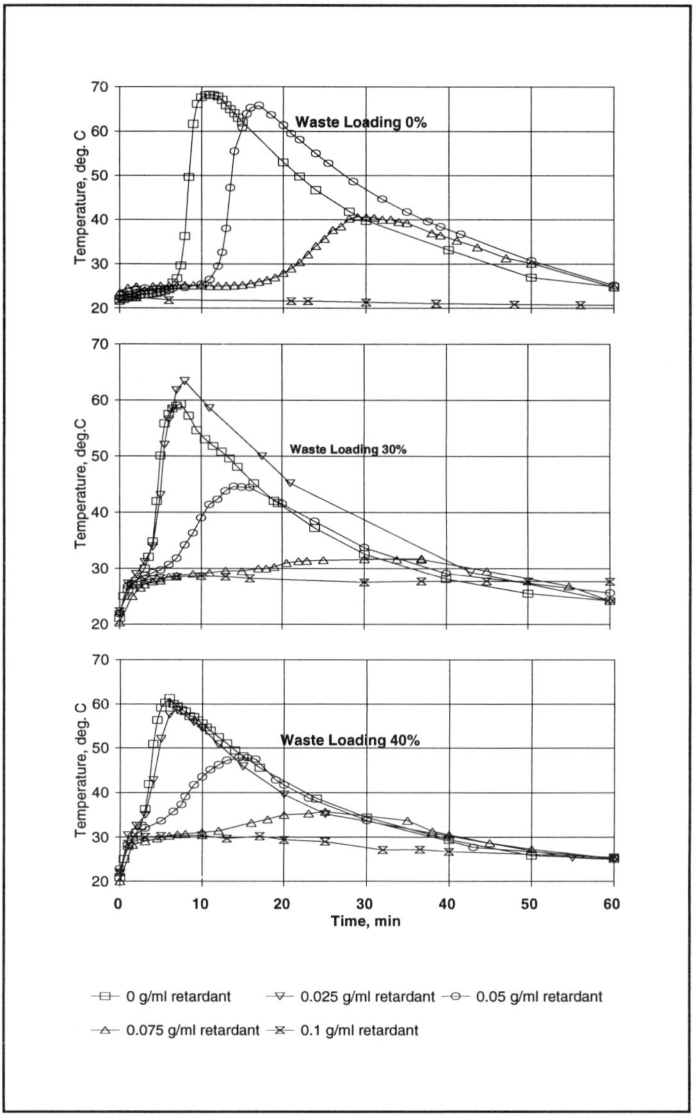

Figure 3. Temperature rise during setting of iron phosphate ceramics. Amount of the retardant added is in grams per milliliter of phosphoric acid.

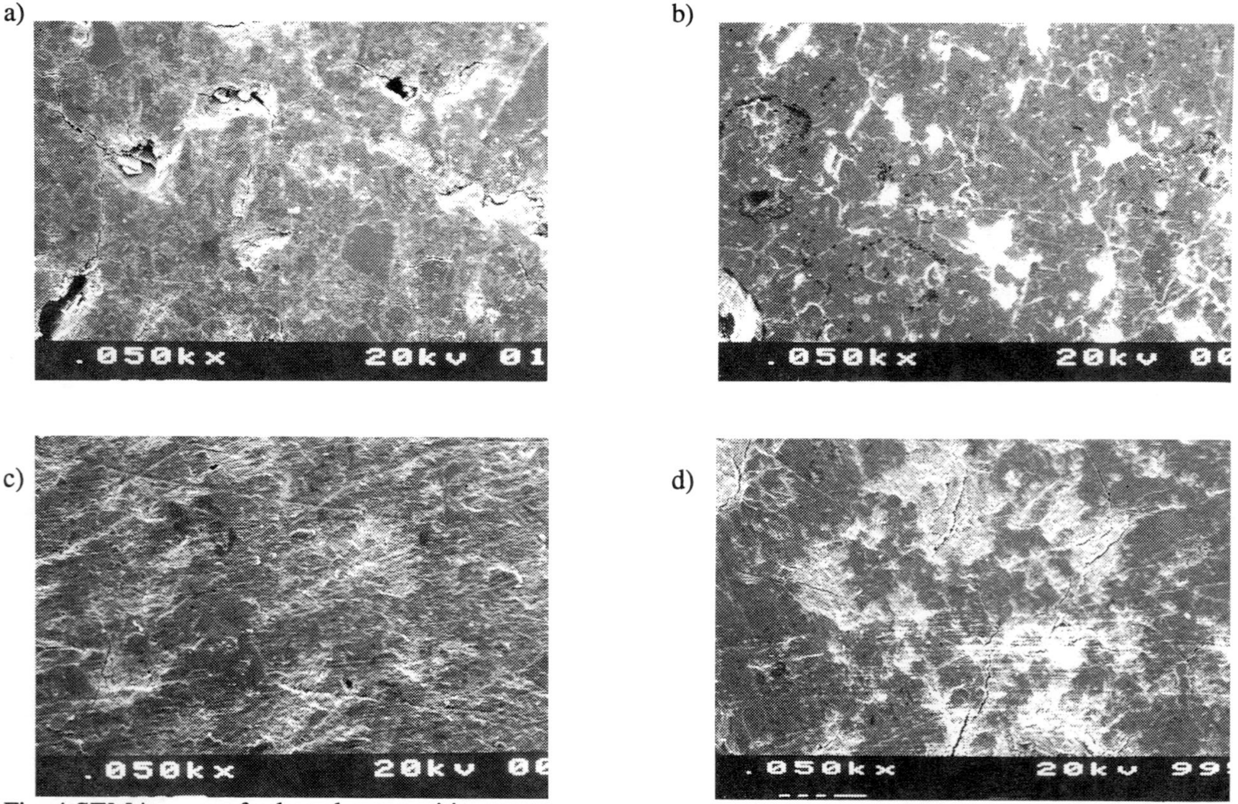

Fig. 4 SEM images of selected compositions
a) fast setting composition, waste loading 30%; b) fast setting composition, waste loading 40%; c) slow setting composition, waste loading 30%; d) slow setting composition, waste loading 40%

Leaching, Durability and Performance Assessment of Waste Forms

ALTERATION LAYERS ON GLASSES AFTER LONG-TERM LEACHING

A.C. Buechele, S.-T. Lai, and I.L. Pegg, Vitreous State Laboratory, The Catholic University of America, Washington, D.C. 20064

ABSTRACT

Multicomponent glasses subjected to PCT leaching for periods up to 8 years show widely differing behaviors depending on composition variations and test S/V ratio. Samples of the reacted glasses with their associated alteration layers have been subjected to microstructural analyses using SEM-EDS in conjunction with XRD in order to determine the morphology of the structures formed and the identity of the principal phases. Replicas taken from selected modified areas have been examined using TEM to look for liquid-liquid phase separation. Changes in glass composition between high-iron borosilicate HLW glass and high-sodium borosilicate low-level tank waste glasses lead to distinct differences in alteration layers. Analysis of leachates sampled periodically frequently reveals accelerations in leach rates whose time of onset can vary considerably with composition, S/V ratio, and other factors. The observed effects of glass composition, test conditions, and test duration on the alteration layer morphology and phases that are formed are discussed. The results are compared to the predictions made by geochemical codes based on solution phase compositions.

BACKGROUND

Over the past 18 years, a large inventory of glasses that are undergoing long-term leach testing has been built up at the Vitreous State Laboratory of The Catholic University of America. This includes a very wide range of glass compositions, most from actual mixed wastes, and many natural analogs. There is likewise great variety in the test protocols (e.g., PCT, ANSI, TCLP, IAEA, Flow, Soxhlet, MCC1, MCC3, DIN) and test conditions. Twenty different leachants including groundwaters, humic acids and pH buffers are being used. Periodic sampling with fluid replacement is used to monitor the progress of these ongoing tests. Occasionally some of the glass is removed from the leaching vessels to observe the modified layers on the glasses and to identify reaction products. Some of the phenomena observed in these tests have been reported previously [1-3].

The glasses selected for this study were a U- and Th-containing high-iron borosilicate high-level waste (HLW) glass that had been under test for 2900 days at three different S/V ratios (600, 2000, and 10000 m^{-1}) and three members of a family of Tc-99 containing, high-sodium, low-level tank waste (LLTW) glasses that had been under test for 722 and 904 days. It is frequently noted that during leaching there will be a sudden change in the behavior of the system which is manifested in the accelerated rate of release of certain components from the glass [4-9]. The likely cause of this effect is the evolution of the solution phase composition into a region in which the formation of a new alteration phase is favored; as this phase forms and depletes the solution, additional glass must dissolve to maintain a near-zero reaction affinity. However, other mechanisms, including alteration layer spallation, may also contribute [3,5]. Figure 1 shows the normalized boron release using the

PCT procedure (ASTM C 1285) for the HLW glass on a log scale. The effect of increasing S/V in promoting higher rates of boron release is immediately evident, but far more striking is the sudden acceleration in the rate for the S/V=10000 m^{-1} test between 700 and 1400 days. Figure 2 shows the sodium PCT release rate for the three LLTW glasses selected for this study. It can be seen that all three have a relatively constant release rate up to about 275 days; at that point the sodium release rate of HAW5 begins to increase sharply. After 350 days the sodium release rate from HAW4 increases, although less sharply, and after 550 days a very slight increase is apparent for HAN38. As noted above, these LLTW glasses belong to a family of similar compositions. Sodium release rates for several other glasses of this family show accelerations with onsets at times ranging from 50 to 400 days, emphasizing that such accelerations arise often in long-term testing and that they are influenced by changes, sometimes minor, in the composition of the glass (see, e.g., ref. 4). Indeed, data from our test inventory show this to be a rather common phenomenon if tests are extended to sufficiently long times.

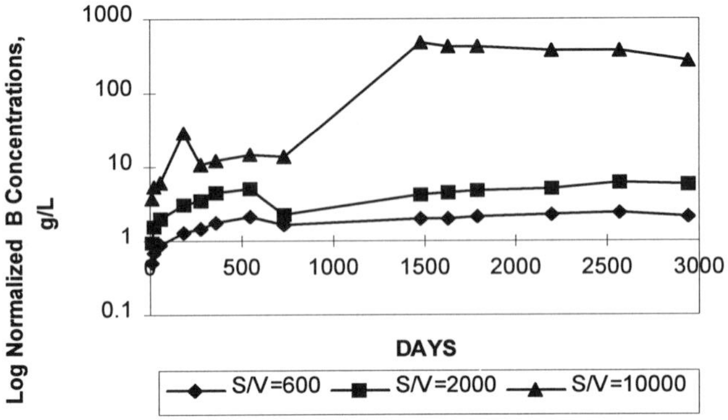

Figure 1. Log normalized boron release versus S/V for HLW glass.

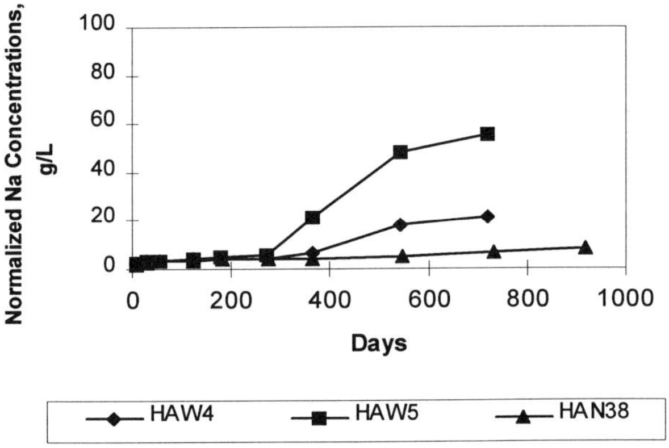

Figure 2. Sodium release rates for LLTW glasses selected for this study.

METHOD

The stainless steel leaching vessels containing the glasses under test were opened and the contents were temporarily transferred to a polypropylene beaker. A small amount of the glass and its leaching liquid were placed in a glass sampling vial which was then capped. The remainder of the glass and its leaching liquid were then returned to the leaching vessel, sealed and placed back in the 90°C test chamber.

The leached glass grains to be examined in the scanning electron microscope (SEM) were embedded in epoxy blocks and sectioned by polishing on diamond lapping films. A low-viscosity resin (Spurr [10]) was used and two different techniques were employed to prepare the grains for embedding. In the case of the HLW glasses, the grains were first placed in acetonitrile, which displaces the water in the leached layer and mixes well with the mounting resin. Because acetonitrile caused clumping of the LLTW glass grains, a second mount was made in which the grains were simply dried first in a 55°C oven and then vacuum impregnated with the resin. Despite the clumping in the one case, no difference was seen in the SEM between the two methods of mounting. Specimens were carbon coated before examination in the SEM, and energy dispersive x-ray spectroscopy (EDS) was used to assist in characterization of leach modified areas and secondary phases. In one case, where phase separation was suspected, the polished surface of the leached grains was replicated using acetate film and transmission electron microscopy (TEM) was used to examine the final extracted, shadowed replica.

X-ray diffraction was used to attempt to identify the secondary phases produced during leaching in the most severely corroded glass of each group (HLW glass $S/V=10000\ m^{-1}$, and LLTW glass HAW5). The glass samples were ground with a mortar and pestle while being kept wet with acetonitrile to which a small amount of a binder had been added. The ground samples were mounted on glass slides and scanned in an x-ray powder diffractometer. Patterns were matched with the PDF-2 Database [11] using JADE [12] software.

Leachate solution analysis data were used in conjunction with the EQ3/6 geochemical reaction modeling program [13] to predict phases which might be present, and the results were compared with SEM/EDS and XRD data.

RESULTS

HLW Glasses

Figures 3 through 6 are micrographs of the HLW leached grains. The leached layer is actually more evident on the grains from the test at $S/V=600\ m^{-1}$ than at $S/V=2000\ m^{-1}$. In the first case, the apparent layer thickness is about 5 μm and in the second, it is about 2 μm. No substructure could be discerned in these layers using the SEM. At $S/V=10000\ m^{-1}$, Figure 5 shows that there is considerable secondary phase material in the debris accompanying the grains taken from the leaching vessel. For the most part, there was no adherent layer present on the grains, but the bulk of the grain was found to be modified to its full depth when examined using EDS. The modification appeared in the EDS spectrum as a severe depletion in Na, Al, P, and K, and slight depletion in Ca. The elements Li and B cannot be detected by EDS. Figure 6 shows a closeup of some of the secondary phases present in the debris. The acicular spray shows the presence primarily of Ca and P, and the other gray material in the micrograph contains predominantly Na, Al, Si, and K. The XRD pattern for this sample shown in Figure 7 includes a few very broad lines characteristic of variable planar spacing and solid solution found in clay-type minerals and zeolites, which commonly form when alkali alumino-silicate glass compositions are leached. Matching this pattern unambiguously proved to be impossible, but a number of the phases that showed reasonable matching were consistent with the SEM/EDS observations. Phillipsite, a mineral of the zeolite group and intermediate between solid solution end members analcime and wairakite [14], was one of the better matches and was consistent with the gray material present in Figure 6 and referred to above. There were three lithium compounds which showed very good matching with portions of the pattern:

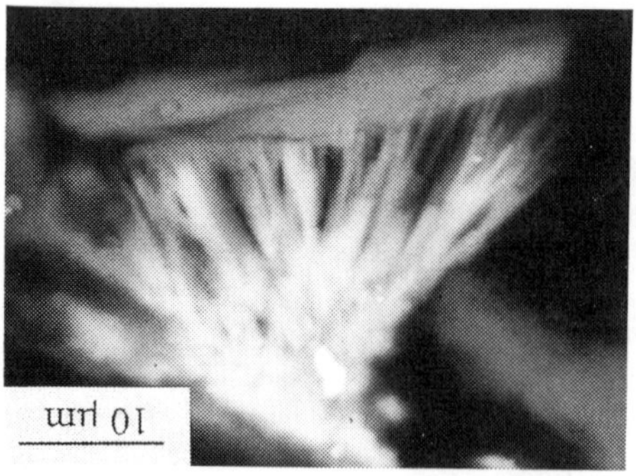

Figure 3. Leached grains, HLW glass, 2900 days, S/V = 600 m^{-1}.

Figure 4. Leached grains, HLW glass, 2900 days, S/V = 2000 m^{-1}.

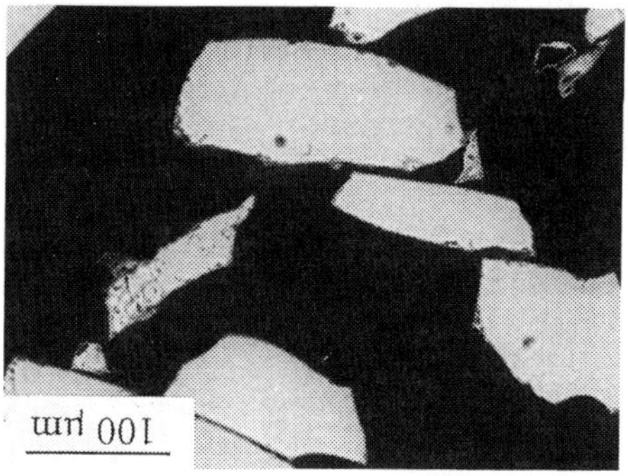

Figure 5. Leached grains, HLW glass, 2900 days, S/V = 10000 m^{-1}.

Figure 6. Calcium phosphate phase (acicular spray) and Na-K-alumino-silicate phase (grey platelet) from HLW glass, 2900 days, S/V = 10000 m^{-1}.

$Li_2Al_2Si_2O_8 \bullet 0.18H_2O$ and its dehydrated form, and $Li_3HP_2O_7$. No calcium phosphate phases were among the best matches, but by running the search match with a chemical filter limited to Ca, P, O, and H, three phases were identified that were consistent with portions of the pattern: $Ca_2P_2O_7$, $Ca_3H_2(P_2O_7)_2 \bullet H_2O$, and $Ca_2P_6O_{17}$. Although hydroxylapatite has been reported by Ogino and Hench [15] as a reaction product of water with P-containing glasses (there is about 3 wt% P_2O_5 in this glass), no credible matching for it was found here.

Figure 8 shows selected element concentrations in the leachate of the HLW glass at S/V = 10000 m^{-1}, normalized to that of boron, before and after the acceleration in leach rate. Significant differences are found in several elements, including lithium, phosphorus, and silicon; these correlate well with the alteration phases observed in that test. The EQ3/6 program identified 23 phases as likely on the basis of leachate chemistry. Of these, only tridymite was in common with the leading phases returned by the JADE search/match. However, Ca-, K-, Mg- and Na-beidellite were among the 23 likely phases, and although our edition of the PDF-2 did not include these phases, they are dioctahedral smectites like the montmorillonites, some of which showed a reasonable degree of matching. The EQ3/6 program also listed 77 pure mineral phases for which the solution was supersaturated. Since the solution was found supersaturated for analcime and unsaturated for wairakite, an intermediate like phillipsite (which was not in the EQ3/6 list of possible phases) is quite likely. The solution was supersaturated for both the calcium phosphate phases listed (whitlockite and hydroxylapatite) but these phases did not show good matching to the pattern, and the calcium phosphate phases for which good matching was found were not listed by EQ3/6.

The full-depth modification of the grains in the S/V=10000 m^{-1} test suggested the possibility of a phase separated glass, so a replica of the cleaned polished surface was made, shadowed, and examined in the TEM. No structure was visible in the replica to the resolution limit of our TEM (about 2-3 nm). Therefore, we feel confident in saying that the full-depth modification of the grains is a consequence of the solution chemistry resulting from the high S/V ratio.

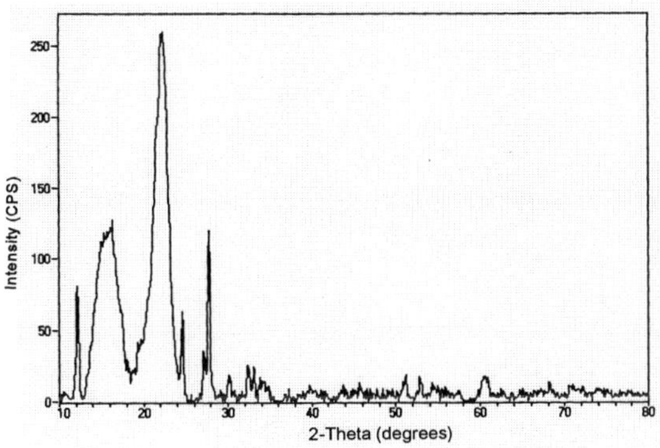

Figure 7. Background subtracted XRD pattern for HLW glass, 2900 days, S/V =10000.

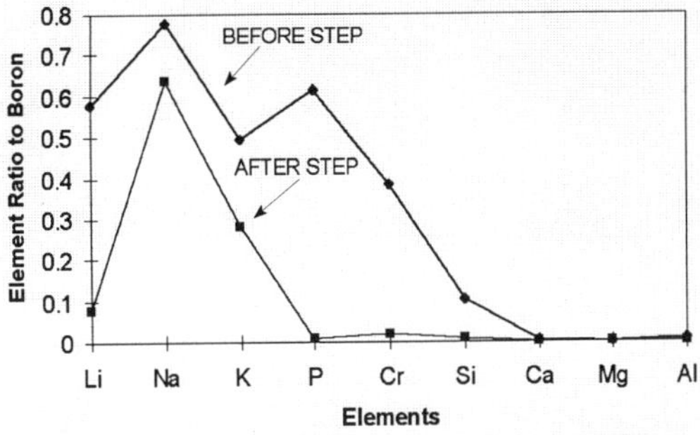

Figure 8. Elemental concentrations normalized to boron in leachate of HLW glass, S/V = 10000 m^{-1}, before and after step in leach rate.

LLTW Glasses

Micrographs of the leached LLTW grains are shown in Figures 9 through 12. The glass HAN38 pictured in Figure 9 is the most durable as judged by sodium release and displays the smallest apparent layer thickness, at least on the larger grains. Next in order of decreasing durability is HAW4 whose thicker leached layers are visible in Figure 10; some small amounts of secondary phases are also present, as well as evidence of variable layer thickness. Finally, the least durable glass of those selected, HAW5, (Figure 11) shows an apparent leached layer thicknesses of about 30 μm, with some internal structure and the presence of secondary phases. The grain cores of HAW5 are sometimes modified to full depth, and at other times there is unmodified glass remaining, depending on the size of the grain. The modified areas show depletion in Na, Al, and Ca. Figure 12 is a slightly higher magnification of the secondary phases found in the HAW5 specimen. The nearly round, gray, slightly faceted crystals are a variant of analcime ($NaAlSi_2O_6 \cdot H_2O$) containing a small amount of Fe. The identity was confirmed both by EDS analysis and by XRD. The other major phase, which was characterized by EDS analysis, was similar in composition to analcime with partial substitution of K and Ca for Na. This latter phase would again be consistent with phillipsite, and the XRD pattern shown in Figure 13 supports this. But, as before, the pattern contains some broad lines and indications that there are other phases present as well. Furthermore, the morphology of some of the leaching products suggests that they are probably not single-phase material. Modeling of the leachate via EQ3/6 produced a list of six phases for which the solution was approximately saturated. Of these, only albite ($NaAlSi_3O_8$) showed good matching to the XRD pattern. The solid solution albite-anorthite ($CaAl_2Si_2O_8$) was also indicated by EQ3/6 as approximately saturated and this is consistent with the EDS observations of the secondary phases. Fourteen pure mineral species were identified by EQ3/6 as supersaturated for the leachate chemistry. The solution was supersaturated for analcime and unsaturated for wairakite, although to a different degree than for the HLW glass, so the presence of phillipsite is likely.

Figure 9. Leached grains, LLTW glass HAN38, 904 days, S/V = 2000 m^{-1}.

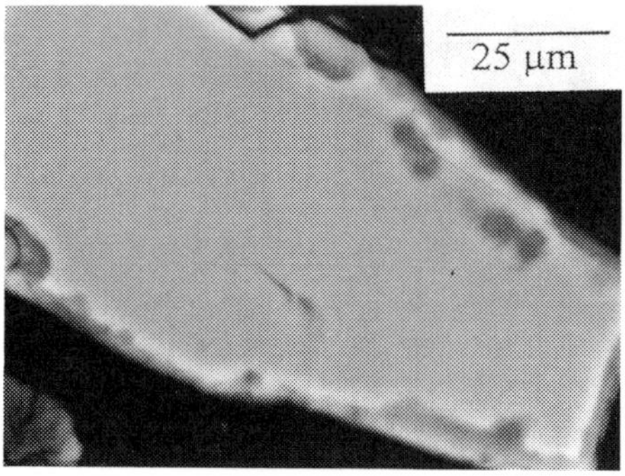

Figure 10. Leached grains, LLTW glass HAW4, 722, S/V = 2000 m^{-1}.

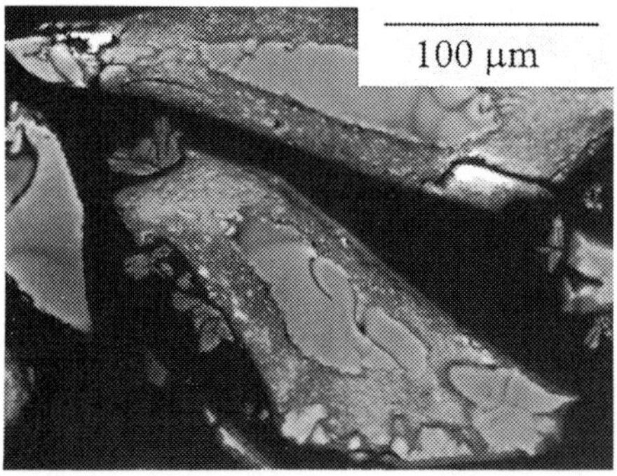

Figure 11. Leached grains, LLTW glass HAW5, 772 days, S/V = 2000 m^{-1}.

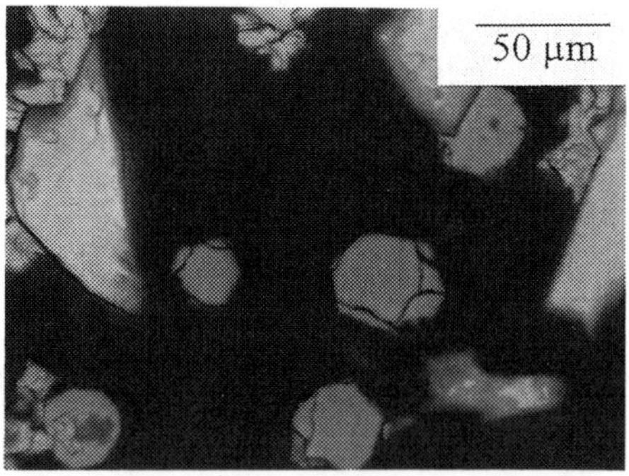

Figure 12. Analcime (rounded, dark grey) and Na-K-Ca-alumino-silicate (irregular, medium gray) on leached grains of LLTW glass, HAW5.

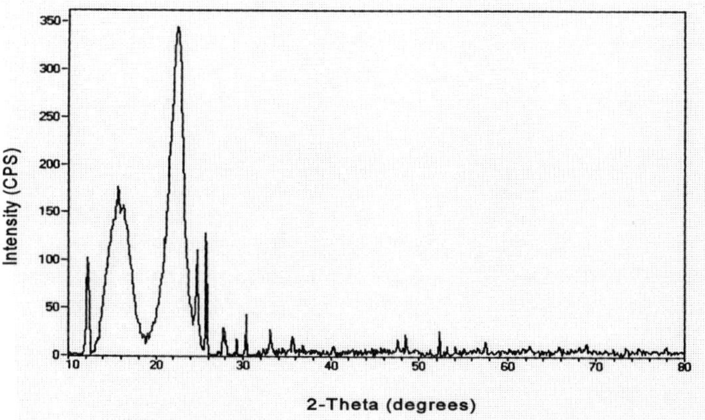

Figure 13. Background-subtracted XRD pattern for HAW5.

DISCUSSION AND CONCLUSIONS

The results presented above show that the observed accelerations in leach rates are influenced by several factors including time, S/V, and glass composition; other data from our test inventory show similar dependencies on temperature and leachant composition (including pH). Furthermore, this phenomenon is actually rather common, and particularly so under highly interactive conditions (e.g. high S/V, high temperature, long times). The results presented here appear to be generally consistent with the picture of the acceleration arising as a result of the formation of a new alteration phase which thereby increases the reaction affinity.

There appears to be little need to invoke a layer-spallation mechanism to explain the data for these two families of glasses. In fact, such a mechanism seems particularly unlikely in the case of the HLW glass at S/V = 10000 m^{-1}, since the glass grains had essentially their original dimensions but were reacted to their full depth.

In principle, an accurate kinetic model coupled with a geochemical code using complete and reliable thermodynamic data could predict the *potential* onset of such accelerations. However, the *actual* onset can be kinetically hindered by the nucleation and growth rates of such controlling phases, which is very difficult to predict. These issues highlight the care that must be exercised in projecting long-term behavior on the basis of short-term test data.

ACKNOWLEDGMENTS

We gratefully acknowledge the help of M.C. Paul in manuscript preparation.

REFERENCES

[1]E. Wang, S.B. Xing, S.S. Fu, I.L. Pegg and A.C. Buechele, "Effects of Liquid Phase Immiscibility on the Leaching of Fluorosilicate Waste Glasses," *Ceramic Transactions, Volume 45: Environmental and Waste Management Issues in the Ceramic Industry II,* (Edited by D. Bickford, S. Bates, V. Jain and G Smith) The American Ceramic Society, Westerville, Ohio, 1994, pp. 209-218.

[2]S.B. Xing, S.S. Fu, A.C. Buechele, Z.B. Djordjevic and I.L. Pegg, "Non-Linear Effect of Glass Composition on Glass Dissolution in both Fluorosilicate and Borosilicate Waste Glasses," *Ceramic Transactions, Volume 45: Environmental and Waste Management Issues in the Ceramic Industry II,* (Edited by D. Bickford, S. Bates, V. Jain and G. Smith) The American Ceramic Society, Westerville, Ohio, 1994, pp. 177-185.

[3]I.S. Muller, A.C. Buechele, I.L. Pegg and P.B. Macedo, "Non-Linear Effects of Glass Composition on Chemical Durability: Physical Stability of Surface Layers," *Mat. Res. Soc. Symp. Proc.,* **257**, 91-98 (1992).

[4]I.S. Muller, H. Gan, A.C. Buechele, S.T. Lai, and I.L. Pegg, "Development of the Vitrification Compositional Envelope to Support Complex-Wide Application of MAWS Technology," DOE/CH-9601, 1996.

[5]Aa. Barkatt, S.A. Olszowka, W. Sousanpour, M.A. Adel-Hadadi, R. Adiga, Al. Barkatt, G.S. Marbury, and S. Li, "Leach Rate Excursions in Borosilicate Glasses: Effects of Glass and Leachant Composition," *Mat. Res. Soc. Symp. Proc.,* **212**, 65-76 (1991).

[6]J. Patyn, P. Van Iseghem, and W. Timmermans, "The Long-Term Corrosion and Modelling of Two Simulated Belgian Reference High-Level Waste Glasses - Part II," *Mat. Res. Soc. Symp. Proc.,* **176**, 299-307 (1990).

[7]M. Adel- Hadadi, R. Adiga, Aa. Barkatt, et al. "Preliminary Results of Durability Testing with Borosilicate Glass Compositions," DOE/NE/44139-34, 1987.

[8]X. Feng, Ph. D. Dissertation, The Catholic University of America, 1988.

[9]X. Feng, J.K. Bates, C.R. Bradley, and E.C. Buck, "Does Fully Radioactive Glass Behave Differently Than Simulated Waste Glass?" *Mat. Res. Soc. Symp. Proc.,* **294**, 207-214, (1993).

[10]A.R. Spurr, "A Low-Viscosity Epoxy Resin Embedding Medium for Electron Microscopy," *J. Ultrastruc. Res.,* **26**, 31-43, (1969).

[11]*PDF-2 Database*, JCPDS-International Center for Diffraction Data, Swarthmore, PA, 1992.

[12]*JADE+ SEARCH/MATCH, Version 2.0,* Materials Data, Inc., Livermore, CA, 1992.

[13]T.J. Wolery, "EQ3NR, A Computer Program for Geochemical Aqueous Speciation-Solubility Calculations: Theoretical Manual, User's Guide and Related Documentation (Version 7.0)", Lawrence Livermore National Laboratory Report UCRL-MA-110662 PT III, 1992.

[14]D.M. Moore and R.C. Reynolds, Jr., *X-ray Diffraction and the Identification and Analysis of Clay Minerals,* Oxford University Press, New York, 1989, pp.232-234.

[15]M. Ogino and L.L. Hench, "Formation of Calcium Phosphate Films on Silicate Glasses," *J. Non-Cryst. Solids,* **38-39**, 1487-1510 (1991).

THE INCORPORATION OF TECHNETIUM INTO A REPRESENTATIVE LOW-ACTIVITY WASTE GLASS

W. L. Ebert, A. J. Bakel, D. L. Bowers, E. C. Buck, and J. W. Emery
Chemical Technology Division
Argonne National Laboratory
9700 South Cass Avenue
Argonne, IL 60439

ABSTRACT

A glass that has been extensively tested to understand the corrosion behavior of waste glasses with high soda contents for immobilizing Hanford incidental low-activity wastes has been doped with technetium by melting crushed glass with either TcO_2 or $NaTcO_4$ at temperatures between 1100 and 1300°C. The Tc-doped glass will be used in dissolution tests. Incorporation of technetium in the glass was affected by solubility or kinetic effects. Metallic technetium inclusions formed in all the glasses with added TcO_2. Inclusions also formed in glasses with added $NaTcO_4$ that were melted at 1100°C, but a glass melted at 1200°C did not contain detectable inclusions. A glass that was free of inclusions was desired because the presence of Tc-bearing inclusions will complicate measurement of the release behavior of technetium from dissolution tests. A glass containing about 0.15 mass % Tc dissolved in the glass and without inclusions will be used in dissolution tests to study the release behavior of technetium. While this technetium content is much higher than that expected in actual waste forms, it facilitates measurement of the disposition of technetium during glass corrosion.

INTRODUCTION

Incidental wastes recovered during remediation of Hanford tanks will be vitrified for final disposal. The resulting waste forms will have low levels of radioactivity, referred to here as "low-activity waste" (LAW), and will be disposed in a near-surface disposal site. However, before LAW can be disposed, it must be demonstrated that the disposal system will adequately contain radionuclides and prevent them from contaminating the surrounding environment. Assessments conducted to provide guidance to the Storage and Disposal Project for Hanford LAW have identified Tc-99 to be the major dose contributor (about 70% of the peak dose) during the first 30,000 years after disposal [1]. The next highest contribution to this peak is from Se-79 (about 20%). These calculations assume that the glass will contain 82% of the Tc-99 inventory, 100% of the Se-79 inventory, 6% of the U (all isotopes) inventory, and 5% of the Np-237 inventory in the tanks. Doses at times beyond 30,000 years are due primarily to uranium isotopes and Np-237. Under the oxidizing conditions of anticipated disposal sites, the Tc released from the waste form will dissolve as the highly soluble pertechnetate ion, TcO_4^-. This species is stable under anticipated disposal conditions. Reduced forms are stable only under strongly reducing conditions. Since the transport of pertechnetate away from the disposal site will likely not be restricted by interactions with minerals in the surrounding geology, the waste form will provide the primary containment for technetium.

The work described in this paper is part of a program that is underway at Argonne National Laboratory to study the release behavior and disposition of technetium and other radionuclides from low-activity waste glasses similar to those that may be produced to immobilize Hanford LAW. Two main issues related to the management of Tc-bearing wastes are (1) how technetium is incorporated into a LAW glass and (2) the process by which technetium is released from the glass. We are addressing these issues by studying the incorporation of technetium into and the release of technetium from LD6-5412 glass, which is a glass that has been studied extensively in recent tests [2-6]. The LD6-5412 glass was developed in a program to study the relationship between composition and various properties of glasses for immobilizing low-activity waste streams that will be generated during the treatment of Hanford tank wastes.

Two aspects of the corrosion behavior of LD6-5412 glass are expected to impact the release and disposition of radionuclides. The first is that ion exchange to release sodium from the glass continues for very long time periods at the low temperatures relevant to LAW disposal systems; dissolution tests being conducted at 20°C have shown sodium to be released from the glass preferentially to B and Si in tests conducted longer than two years [2]. This behavior differs from the transient nature of ion exchange observed in tests with other glasses conducted at higher temperatures. This may mean that components that are weakly bound in the glass structure, possibly including technetium, may be leached from the glass at a rate that is much higher than the rate at which the glass matrix dissolves. The second is that relative release rates of Na, B, and Si are different in tests that maintain dilute solutions than in tests in which dissolved glass components accumulate. Boron is released congruently with sodium and faster than silicon in very dilute solutions, but is released congruently with silicon and slower than sodium in more concentrated solutions. Therefore, tests will be conducted to determine if the release of either Na or Si can be used to approximate the release of technetium.

Several researchers have investigated the dissolution of Tc into silicate glasses, and thorough reviews have been published by Vida [7] and by Darab and Smith [8]. In fact, Darab and Smith themselves doped LD6-5412 glass with technetium. The vitrification of technetium is strongly affected by the high volatilities of most Tc-bearing species. Technetium has been added to glasses as Tc(VII) in a pertechnetate salt (e.g., $NaTcO_4$ or NH_4TcO_4), and it has also been reduced to Tc(IV) and added as the oxide $TcO_2 \bullet 2H_2O$. Several studies have shown that a greater percentage of technetium is retained in glasses made with TcO_2 than with pertechnetate salts [8]. However, $TcO_2(s)$ is known to sublime at temperatures above 900°C and disproportionate at temperatures above about 1100°C to generate metallic Tc and gaseous Tc_2O_7 [7].

In this paper, we discuss the production of Tc-doped glasses by remelting a dry mixture of crushed LD6-5412 glass with either TcO_2 or $NaTcO_4$. While the primary purpose of this work was to produce a glass for use in laboratory dissolution tests, the results are also relevant to the vitrification of actual waste streams. Because of their high volatilities, large amounts of Tc-bearing species will need to be recovered from off-gas systems during vitrification. Separation of technetium from the waste stream into the off-gas will provide an opportunity for developing a treatment to increase the amount retained in the waste during possible recycle into the LAW waste form, or its incorporation into the high-level waste form. The present study may also provide insight into whether or not reduction to Tc(IV) will increase the retention of technetium in either low-level or high-level silicate waste glasses.

EXPERIMENTAL

Several glasses were prepared by remelting crushed LD6-5412 glass with various amounts of TcO_2 or $NaTcO_4$. The composition of the LD6-5412 "base" glass was measured at ANL and is given in Table I. The original LD6-5412 glass was obtained from Pacific Northwest National Laboratory, where the glass had been prepared by melting oxide components at about 1400°C [4]. We desired to remelt the Tc-doped glasses at as low a temperature as possible to decrease the amount of technetium lost to volatilization. Therefore, tests were conducted by melting small amounts of the LD6-5412 glass at temperatures of about 900, 1000, 1100, 1200, and 1300°C for about 1 hour then quenching the samples in air to (1) determine the temperature at which the glass could be remelted and (2) determine if insoluble phases were exolved during remelting. The resulting glasses were examined with a scanning electron microscope (SEM) to determine if the glass had completely vitrified and to detect and identify any exolved phases.

Table I. Measured Composition of LD6-5412 Glass

Oxide	mass %	Oxide	mass %	Oxide	mass %
Al_2O_3	12	F	0.31	Na_2O	20
B_2O_3	5	Fe_2O_3	0.005	NiO	<0.03[a]
CaO	4	I	0.14	P_2O_5	0.19
Cl	0.36	K_2O	1.51	SO_3	0.22
Cr_2O_3	0.04	MgO	0.003	SiO_2	55.65
Cs_2O	0.15	MnO	0.002	SrO	0.12
CuO	<0.03[a]	MoO_3	0.16		

[a]Below detection limit.

Aliquots of crushed LD6-5412 glass were then doped with small amounts of TcO_2 or $NaTcO_4$. The TcO_2 was prepared by first dissolving 0.8 g of NH_4TcO_4 in 30 g H_2O plus a few drops of concentrated NaOH. The TcO_4^- was then reduced by adding 5 mL of a 5% aqueous hydrazine solution. A finely divided, black precipitate formed, probably $TcO_2 \cdot 2H_2O$. The precipitate was removed from suspension by filtration through a bed of crushed (-200 +325 mesh) LD6-5412 glass. A bed of crushed glass was used because filtration of the precipitate alone had clogged 0.4 μm filters. The bed was washed with additional hydrazine solution then removed from the filter. The resulting mixture of $TcO_2 \cdot 2H_2O$ and crushed glass was allowed to air dry at room temperature. The amount of $TcO_2 \cdot 2H_2O$ recovered was calculated from the mass of the LD6-5412 bed before and after filtration. The $TcO_2 \cdot 2H_2O$ /glass mixture was transferred into a glovebox, and an aliquot was mixed with more crushed LD6-5412 glass in a platinum crucible to achieve the desired loading of technetium. The amount of technetium added to the glass was calculated by assuming homogeneous distribution of technetium both in the filter bed and in the mixture with the added glass.

The $NaTcO_4$ was prepared by first acidifying NH_4TcO_4 with HNO_3 and taking the resulting solution to dryness. The white crystalline residue was dissolved in about 10 mL water, to which 0.2 mL conc. NaOH was then added to precipitate $NaTcO_4$. This was filtered through a bed of crushed LD6-5412 glass, mixed with more crushed glass, and melted following the same procedure as described above.

The crucibles for all glasses were covered with a loose-fitting lid to prevent splattering of the molten glass. The lid did not restrict the movement of air into or out of the crucible. The crucible was placed into a furnace that had been preheated to the desired temperature and left undisturbed for the scheduled melting time. A bottom loading furnace was used to melt the glass. After the desired melting time, glass was poured from the crucible into platinum loaf pans that had been preheated to about 500°C. The glass in the loaf pans was then annealed at about 500°C. After about 2 hours, the annealing oven was turned off and allowed to cool to room temperature with the loaves of glass still in the oven. The glass was then broken out of the loaf pans and small chips were removed for analysis.

Samples were prepared for SEM examination by potting small pieces of glass in epoxy and polishing to a 600-grit final surface finish with carborundum paper and water lubrication. These samples were examined with an SEM to determine if inclusion phases had formed in the glasses. Samples of some glasses were also examined with analytical electron microscopy (AEM), which is a transmission electron microscope with associated energy dispersive x-ray spectroscopy (EDS) and electron energy loss spectroscopy (EELS).

Small amounts of these glasses were crushed and dissolved for analysis. The glasses were dissolved overnight in HCl and HF at 200°C in a sealed Teflon vessel. Insoluble material remained after dissolution of one of the glasses. This material was dissolved after the addition of more HNO_3. The technetium concentrations were measured with inductively coupled plasma-mass spectrometry (ICP-MS).

The specific procedures followed for making the various glasses with TcO_2 or $NaTcO_4$ and the results of the examinations are summarized below.

RESULTS

Undoped LD6-5412: Glasses made by remelting crushed LD6-5412 glass at temperatures of 1200°C and below were all opaque. Examination of these glasses with SEM revealed the presence of various small (<0.5 μm) inclusions that contained Cu, Ni, Fe, and/or Cr. As given in Table I, the original glass contained small amounts of Cr and Fe, but Cu and Ni were not detected. Small Cr-bearing inclusions had been detected in LD6-5412 glass previously. Included phases were not detected in the glass melted at 1300°C. It was transparent and lime green, and appeared identical to the original glass that had been crushed. Analysis of the glasses with EDS indicated that the compositions of the glasses remelted at 1200°C and below were the same as that of the original glass. Remelting at 1300°C led to a very small decrease in the amount of sodium--from 15.0 ± 0.5 mass % Na in the original glass to about 14.3 ± 0.2 mass % Na in the remelted glass.

Glasses Made with $TcO_2 \cdot 2H_2O$

Glass 1: A mixture of 0.108 g $TcO_2 \cdot 2H_2O$ was mixed with 7.0 g glass (total of glass in filter bed and glass added to mix) then melted for 60 minutes at about 1300°C. The molten glass was poured into loaf pans and annealed at 500°C for 2 hours, then slowly cooled to room temperature. The glass was visibly dark and opaque. A small amount of this glass was dissolved for chemical analysis; it was found to contain 0.20 mass % Tc; this includes technetium present in inclusions and dissolved in the glass. The as-mixed fraction of technetium was about 1.5 mass %, so about 87% of the added technetium was volatilized. Examination of polished cross sections with SEM revealed the presence of small (1-5 μm) particles distributed in the glass. Analysis with EDS in the SEM indicated that the particles contained technetium, but the amount of technetium dissolved in the glass was below the detection limit. An SEM image of one large and two small Tc-bearing inclusions in a polished cross-section is shown in Figure 1.

Glass 2: About 40 g of Glass 1 was crushed and mixed with about 70 g of LD6-5412 glass to dilute Glass 1. The mixture is calculated to contain 0.07 mass % Tc. This glass was melted at 1300°C for one hour, then heated to 1400°C for another hour. The molten glass was poured into loaf pans and annealed at about 500°C for about 2 hours. Examination of the resulting glass with SEM/EDS revealed the presence of small (about 1 μm) Tc-bearing inclusions similar to those found in

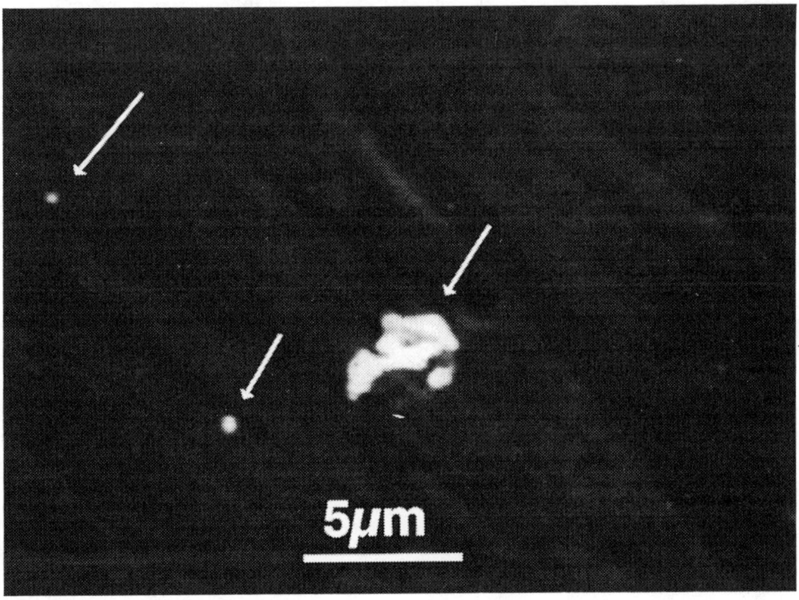

Figure 1. Polished Cross-Section Showing Tc-Bearing Inclusions in Tc-Doped LD6-5412 Glass

Glass 1, except they were smaller and less abundant. Dissolution and ICP-MS analysis of Glass 2 indicated that the glass contained about 0.01 mass % Tc total in inclusions and dissolved in the glass. Similar to Glass 1, about 15% of the technetium was retained after remelting. Given the size and abundance of the inclusions, it is likely that most of the technetium retained in Glasses 1 and 2 is present in the inclusions.

The inclusions were examined with AEM to determine if they were undissolved TcO_2 or metallic technetium. Several samples were prepared for AEM examination from small chips of Glass 2 that contained inclusions visible in an optical microscope. Sections containing the included phase were prepared with an ultramicrotome. The chemical compositions of the inclusion and the adjacent glass were analyzed with an ultra-thin window detector. The spectrum for the inclusion is shown in Fig. 2a. The spectrum is dominated by the peak at about 2.4 keV, which is due to the Tc $\beta_1(L_{II}M_{IV})$ and $\alpha_1(L_{III}M_V)$ transitions at 2.54 and 2.42 keV, respectively. A strong peak was also observed at about 18.4 keV, which is due to the Tc $\alpha_2(KL_{II})$ and $\alpha_1(KL_{III})$ transitions (not shown in Fig. 2a). Essentially no oxygen was detected in the analysis, which indicates that the technetium in the inclusion is metallic. This is consistent with the deformation and curling of the inclusion that were observed during sectioning. We have found these attributes to be characteristic of sectioned metallic samples. Technetium was not detected during analysis of the glass next to the inclusion, as shown by the absence of a peak near 2.4 keV in Fig. 2b. Evaluation of other sections revealed that some inclusions contained technetium plus small amounts of nickel; inclusions containing nickel, chromium, and iron had been detected after remelting the stock LD6-5412 glass.

Additional evidence that the inclusion phases are metallic technetium is provided by electron energy loss spectroscopy (EELS). In the EELS spectrum of an inclusion phase shown in Fig. 3, the $M_{4,5}$ line of technetium is clearly seen, while no peak is seen at the O-K edge at 532 eV, which is usually seen as a strong, broad peak in oxides. (The C-K edge was measured for a region free of inclusions and has been subtracted from the spectrum shown in Fig. 3.)

These results suggest that disproportionation of Tc(IV) leading the generation of Tc_2O_7 and metallic technetium has occurred during melting. The metallic inclusions probably formed from small clusters of TcO_2 that were not well mixed in the glass. If the TcO_2 had been better mixed, the inclusions may have been too small to be detected.

Glasses Made with $NaTcO_4$

Glasses 3 and 4: These glasses were made by mixing 0.09 g $NaTcO_4$ with 6.82 g LD6-5412 glass (1.4 mass % Tc as-mixed) and by mixing 0.009 g $NaTcO_4$ with 7.18 g LD6-5412 glass (0.13 mass % Tc as-mixed), respectively. These glasses were heated in quartz crucibles for about 1 hour at 1100°C. The molten glasses were not annealed, rather they were quenched in air. Examination of the resulting glasses with SEM revealed the presence of small Tc-bearing inclusions. Examination of the inclusions with AEM has not yet been completed. They are probably undissolved $NaTcO_4$, since Tc(VII) does not disproportionate. Chemical analysis of Glasses 3 and 4 showed that they contained a total of about 0.15 and 0.05 mass % Tc, respectively. This includes both dissolved and precipitated technetium.

a.

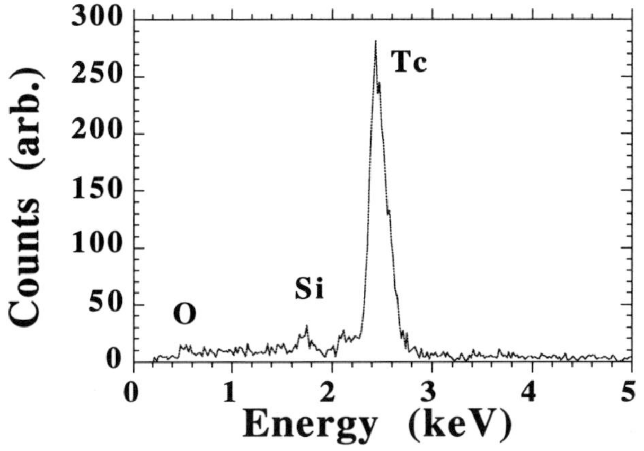

b.

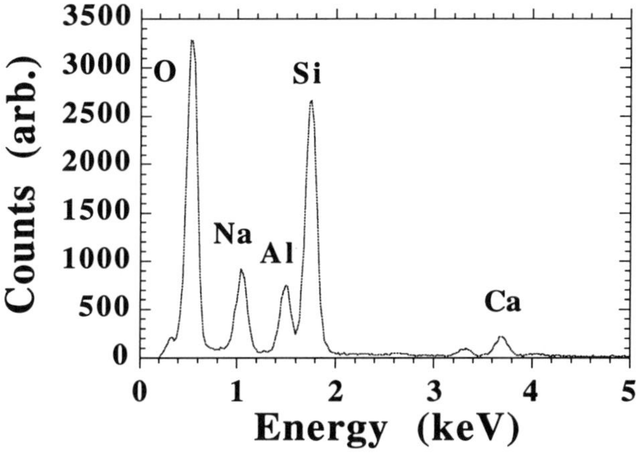

Figure 2. Compositional Analysis of Glass 2: (a) Included Phase and (b) Adjacent Glass

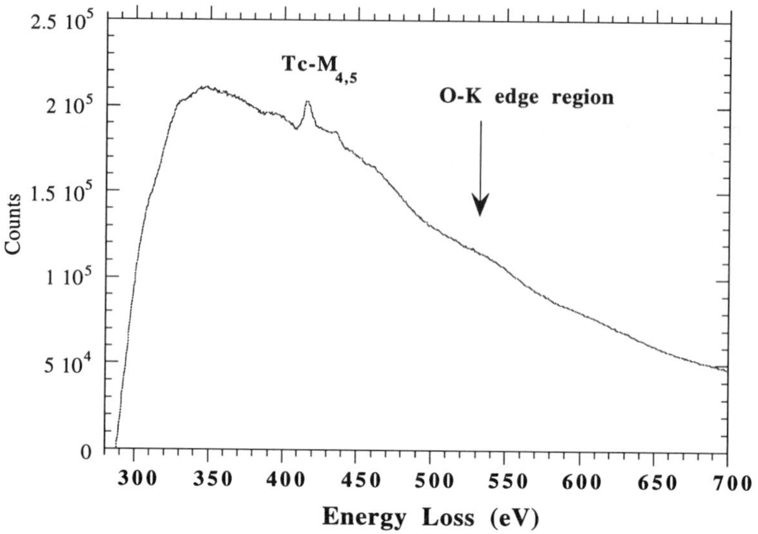

Figure 3. Electron Energy-Loss Spectrum of the Inclusion Phase in Glass 2.

Glasses 5 and 6: These glasses were made with mixtures of LD6–5412 glass and $NaTcO_4$ that were the same as those used for Glasses 3 and 4. Glasses 5 and 6 were melted in platinum crucibles for 1 hour at 1200°C, whereas Glasses 3 and 4 were melted at 1100°C. No inclusions were found during examination of Glasses 5 and 6 with SEM. Chemical analyses of these glasses show that they contained 0.18 and 0.06 mass % Tc, respectively.

Glass 7: A large batch of glass was made under the same conditions used to make Glasses 5 and 6. About 120 g of a mixture of $NaTcO_4$ and crushed LD6-5412 glass was melted in a platinum crucible at 1200°C for about 1 hour. The mixture contained about 1 mass % Tc. The molten glass was poured into platinum loaf pans and annealed at 500°C for about 2 hours, then slowly cooled. Examination of a chip of the resulting glass showed it to contain inclusions. Therefore, the glass was crushed and remelted at 1300°C for 1 hour, annealed, then allowed to cool to room temperature. No Tc-bearing inclusion phases were observed in this glass. Some of the glass was dissolved in HF and HCl and analyzed with ICP-MS. The glass was found to contain 0.15 mass % Tc. Samples of Glass 7 have been prepared for use in corrosion tests.

DISCUSSION

These results indicate that, while the retention of Tc(IV) that is added to waste glasses may be greater than the retention of Tc(VII), disproportionation of the Tc(IV) results in the formation of metallic inclusions. Reduction of technetium prior to recycling would lead to several disadvantages in vitrification. The formation of metallic inclusions in glass melts is detrimental to Joule-heated melters, and phase separation in the final waste form complicates its qualification. From the stoichiometry of the disproportionation reaction, only one-third of the added Tc(IV) is reduced to the metal. The rest is lost as the highly volatile heptoxide. Analysis of the glasses made with TcO_2 indicates that less than 20% of the added technetium is retained. The behavior of the metallic inclusions during glass corrosion remains to be determined.

Technetium added as $NaTcO_4$ did dissolve in the glass that was melted at 1200°C, but did not fully dissolve in glasses melted at lower temperatures for one hour. It is not clear whether the inclusions in glasses melted at temperatures below 1200°C remained because of solubility or kinetic limitations. However, the amount of technetium that was fully dissolved in Glass 7 is less than that reported by Darab and Smith [7] for glasses made with NH_4TcO_4. They did not report if the glasses were analyzed to verify that the technetium was completely dissolved in the glass and not present as undissolved inclusions.

The presence of technetium in the glass and in inclusions complicates the characterization of the release of technetium with laboratory tests. This is because technetium released as the glass dissolves may sorb onto the included phase and not be detected by solution analysis. Alternatively, the simultaneous dissolution of the glass and inclusion phases will prevent measurement of the rate at which technetium is released from the glass. That is why it was important to verify that included phases were not retained in the glass to be used in the corrosion tests.

CONCLUSION

Small amount of TcO_2 or $NaTcO_4$ were mixed with a representative waste glass for Hanford incidental low-activity wastes and melted to produce glasses doped with technetium. Examination with SEM and AEM of the TcO_2-doped glasses melted at 1300°C revealed the formation of small inclusions of metallic technetium. This is probably a result of the disproportionation reaction in which Tc_2O_7 and metallic technetium form. The size of the inclusions is probably related to the size of the agglomerates of TcO_2 that were added to make the glass. The amounts of technetium dissolved in the glasses doped with TcO_2 were below the detection limit of EDS, and most of the technetium that was retained in these glasses was probably present in the inclusions. Therefore, treatment to reduce pertechnetates in the waste stream or collected as off-gas will not facilitate the dissolution of technetium into this glass, although the total amount retained in the glass, including metallic inclusions, may increase.

Small Tc-bearing inclusion phases were also observed in glasses doped with $NaTcO_4$ that were melted at temperatures of about 1200°C and below; we assume the inclusions are undissolved $NaTcO_4$, but this remains to be verified. No inclusions were observed in glasses melted at about 1300°C.

ACKNOWLEDGMENTS

Samples for AEM examination prepared by J. K. Bates and N. L. Dietz. Technical assistance by L. Hafenrichter and M. Surchik. Work supported by the U.S. Department of Energy through Lockheed Martin Hanford Corporation Tank Waste Remediation System and Richland Operations Office, and conducted under contract W-31-109-ENG-38.

REFERENCES

[1]F. Mann, *Impacts of Disposal System Design Options on Low-Level Glass Waste Disposal System Performance*, Westinghouse Hanford Company report WHC-EP-0810 Revision 1, (Sept. 1995).

[2]A. J. Bakel, W. L. Ebert, and J. S. Luo, "Long-Term Performance of Glasses for Hanford Low-Level Waste," Ceram. Trans. Environmental and Waste Management Issues in the Ceramics Industry, **61**, pp. 515-522 (1995).

[3]B. P. McGrail and D. K. Peeler, *Evaluation of the Single-Pass Flow-Through Test to Support a Low-Activity Waste Specification*, Pacific Northwest Laboratory Report PNL-10746 (1995).

[4]X. Feng et al. *Glass Optimization for Vitrification of Hanford Site Low-Level Tank Waste*, Pacific Northwest National Laboratory Report PNNL-10918 (1996).

[5]W. L. Ebert, A. J. Bakel, and N. R. Brown, "Measurement of the Glass Dissolution Rate in the Presence of Alteration Phases," Proceedings International Topical Meeting on Nuclear and Hazardous Waste Management, Spectrum '96, Seattle, WA, August 18-23, 1996, pp. 569-575 (1996).

[6]B. P. McGrail, W. L. Ebert, A. J. Bakel, and D. K. Peeler, "Measurement of Kinetic Rate Law Parameters on a Na-Ca-Al Borosilicate Glass for Low-Activity Waste," J. Nucl. Mat. (in press).

[7]J. Vida, *The Chemical Behavior of Technetium during the Treatment of High-Level Radioactive Waste*, Institute for Hot Chemistry, Karlsruhe Nuclear Research Center, KfK-4642 (1989). Translated by J. R. Jewett, Westinghouse Hanford Company PNL-TR-497 (1994).

[8]J. G. Darab and P. A. Smith, "Chemistry of Technetium and Rhenium Species during Low-Level Radioactive Waste Vitrification," Chem. Mater., **8**(5) 1004-1021 (1996).

CHARACTERIZATION OF SAVANNAH RIVER GLASSES BURIED IN BOOM CLAY - THE HADES TESTS AT MOL, BELGIUM

R.L. Schulz, D.E. Clark - University of Florida, Gainesville, FL 32611
G.G. Wicks - Westinghouse Savannah River Tech. Center, Aiken, SC 29802
A.R. Lodding - Chalmers University of Technology, 41296 Gothenburg, Sweden
P. Van Iseghem - SCK-CEN, 2400 Mol, Belgium

ABSTRACT

Several glass compositions originally designed to immobilize the high-level radioactive waste generated at the Savannah River Site (SRS) were buried for a period of 2 years in Boom clay during the HADES (high activity disposal experimental site) burial study at Mol. The simulated waste glass burial samples were characterized using secondary ion mass spectroscopy (SIMS) and scanning electron microscopy (SEM/EDS). Fourier transform infrared reflection spectroscopy (FT-IR) was performed before and after testing. Data from these characterization studies will be presented for the SR-165 waste glass, along with a compilation and comparison of SR-165 of glass performance from other field burial studies .

INTRODUCTION

A number of burial studies have been conducted in a variety of geologic media to determine the chemical durability of a wide range of proposed vitreous waste forms and package components (e.g., canister and overpack metals, backfill materials). The studies were carried out at both ambient temperatures and at 90°C. The results discussed in this paper include data from the Materials Interface Interaction Tests conducted at the Waste Isolation Pilot Plant in New Mexico, US (salt), the Stripa burial study conducted in Sweden (granite), the Ballidon study in the United Kingdom (limestone), and finally, the Boom clay study carried out at Mol, Belgium. The proposed glass waste form developed for the high level radioactive waste stored at the Savannah River Site was placed and monitored in each of these environments (in contact with the geology and other barrier materials) for periods up to 5 years. The composition of the SR-165/TDS glass will be emphasized in this discussion, as shown in Table1[1].

BACKGROUND - BURIAL STUDIES

HADES Burial Study (Boom Clay - Belgium)

The HADES burial program was initiated in 1985 [2,3]. The study was developed and managed by the Studiecentrum voor Kernenergie/Centre d'Etude de L'Energie Nucleaire (SCK-CEN). The SRS glass samples discussed in this paper were entombed in 1986 and were retrieved

Table 1. Composition of the SR-165/TDS Simulated Nuclear Waste Glass [1].

COMPONENT	FRIT	SIMULATED WASTE	WT%
SiO_2	X		47.7
Na_2O	X		9.1
B_2O_3	X		7.0
Li_2O	X		4.9
MgO	X		0.7
ZrO_2	X		0.7
Fe_2O_3	X		13.4
MnO_2		X	3.9
Zeolite		X	2.9
Al_2O_3		X	2.7
NiO		X	1.6
CaO		X	1.0
Coal		X	0.7
SiO_2		X	1.2
Na_2O		X	0.9
Na_2SO_4		X	0.2
Cs_2CO_3		X	0.1
$SrCO_3$		X	0.1
U_3O_8		X	1.1

in 1988. Samples were placed approximately 224 meters below the surface in the Boom clay geology (see Figure 1). Glass specimens (two or three specimens of each) were mounted in specially designed stainless steel assemblies, as shown in Figure 2 and were maintained at 90°C for the two-year burial period. Tests were also conducted at ambient temperatures (16°C).

Important characteristics of the Boom clay including composition, water content and mineralogy are presented in Table 2 [4].

Other Burial Studies of SRS Glasses

Ballidon (Limestone - UK). The Ballidon burial study was initiated in 1970 by the British Glass Industry Research Association and involved the burial of archaeological and modern glass specimens in carboniferous limestone at Ballidon in Derbyshire UK [5-8]. In the years following the onset of the study, samples were retrieved and others were placed in the vacated positions. In 1986, after an extraction of samples within the burial mound, Savannah River waste glass compositions along with several other relevant glass compositions were placed in the vacated positions. Samples were to be retrieved at 1, 2, 5 and 16 year intervals. The specimens were buried at a depth of approximately 1 meter in alkaline soil (pH = 9.5-9.7) and were maintained at ambient temperatures (0.7-18.9°C). The 1, 2 and 5 year samples have been excavated and characterized. The 16-year samples are set to be retrieved in 2002.

Stripa (Granite - Sweden). The Stripa burial study was conducted in a granitic environment within an abandoned iron mine in Sweden under the direction of the Swedish

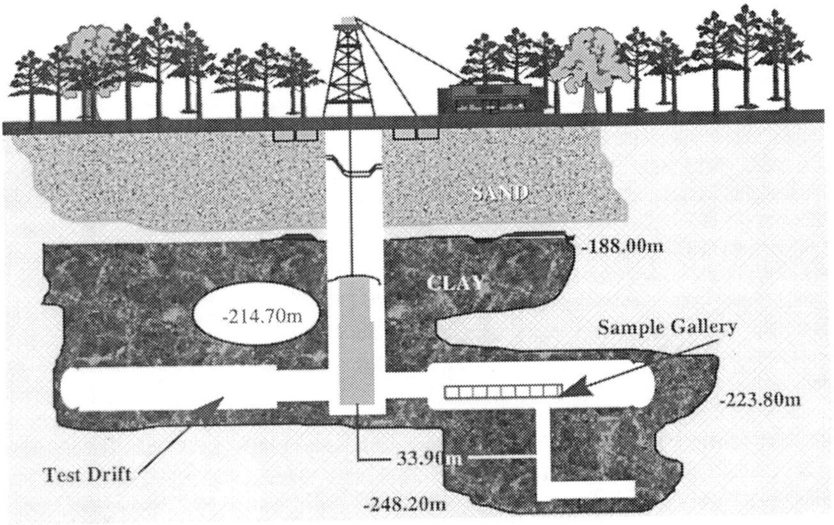

Figure 1. Schematic of the underground test facility at Mol, Belgium.

Figure 2. Photo showing the glass samples on the stainless steel assembly.

Table 2. Characteristics of the Mol Boom Clay [4].

Chemical Composition (Dry)	64% SiO_2, 14% Al_2O_3, 5.3% Fe_2O_3, 2.2% K_2O, 1.4% Na_2O, 0.6% CaO, 0.5% TiO_2, 0.7& MgO
Minerals Present	25% Illite, 20 % Smectite, 30% Vermiculite, 15% Illite-Montmorillonite interstratified, 10% Chlorite + Chlorite vermiculite interstratified, ~ 1% pyrite
Organic Material (wt%)	1-5%
Sulfates (wt%) Sulfides (wt%) Chlorides (wt%) Fluorides (wt%) Phosphates (wt%)	0.26 - 4.58% 0.0103 - 0.0225% 0.04 - 0.24% 0.03% 0.01 - 0.1%
Natural Water (wt%)	~.79 - 26%

Nuclear Fuels Safety Division/Nuclear Fuel Supply Co. (SKBF/KBS). This international three-phase burial program was established in 1980 and included participants from 9 countries. Three US SRS waste glass compositions (SR-131/29.8% waste, SR-131/35% waste and SR-165/ 29.8% waste) were added to the study in 1982. Samples were stacked on stainless steel rods in a "pineapple" slice configuration and were positioned in boreholes approximately 350 meters beneath the surface. The pineapple slice configuration permitted many different types of interfaces to be examined (e.g. glass/glass, glass/metal, glass/granite, etc.). Ground water was allowed to flow freely through the boreholes during the course of the study. Some boreholes were maintained at ambient mine temperatures (8-10°C, groundwater pH = 7.8-9.0), but most were held at 90°C. The SR glasses were extracted at 1, 3, 6, 12 and 24 month intervals. Samples were characterized using SEM/EDS, FT-IR, SIMS. The information gained from this burial study is well documented in the literature [9-11] and demonstrated that a systematic burial study was useful in providing valuable data on the leach performance of waste package materials.

Materials Interface Interaction Test (MIIT- Salt - US). The MIIT burial study was conducted in salt deposits of the Salado Formation at the Waste Isolation Pilot Plant in New Mexico, US. This international program was initiated in 1986 and involved the placement of over 900 waste form samples as well as a large variety of potential canister, overpack, backfill and geological specimens. Samples were stacked on Teflon® assemblies with heater rods placed in the center of the unit. The assembly was then lowered into a pre-drilled borehole approximately 655 meters below the surface. Samples were retrieved at 0.5, 1, 2 and 5 year intervals. The 5 year samples were extracted in 1991. All samples were characterized prior to burial using FT-IR. Upon removal from the test samples were further characterized with SEM/EDS, SIMS, FT-IR, wide angle x-ray diffraction (WAXD), electron microprobe, and transmission electron microscopy. Characterization data from the MIIT study, as well as comparison data from laboratory and other burial studies have been presented in the literature [9,12-15].

RESULTS

Samples extracted from the Mol burial study have been characterized using FT-IR, SEM/EDS and SIMS. The data from these analyses appear to indicate little surface alteration. However, mass loss data (Table 3) suggest that portions of the surface may have been removed or exfoliated either during testing, sample removal, or analytical preparation processes. A representative schematic of pre- and post-burial FT-IR data for the SR-165/TDS composition is shown in Figure 3. For this specimen, there is little evidence of surface alteration as the pre-burial data indicate peak positions ranging from 1006-1010cm^{-1} and post burial spectra had peak positions ranging from 1007-1009cm^{-1}, depending on the location scanned. The decrease in peak intensity is attributed primarily to surface attenuation involving either roughening or precipitation.

Table 3. Mass Loss Data [16].

SAMPLE	BURIAL TIME/TEMP	MASS LOSS (g/m^2)
SR-165/TDS	2 years, 90°C	253
SR-131/TDS	2 years, 90°C	264
SR-131/35% TDS	2 years, 90°C	252

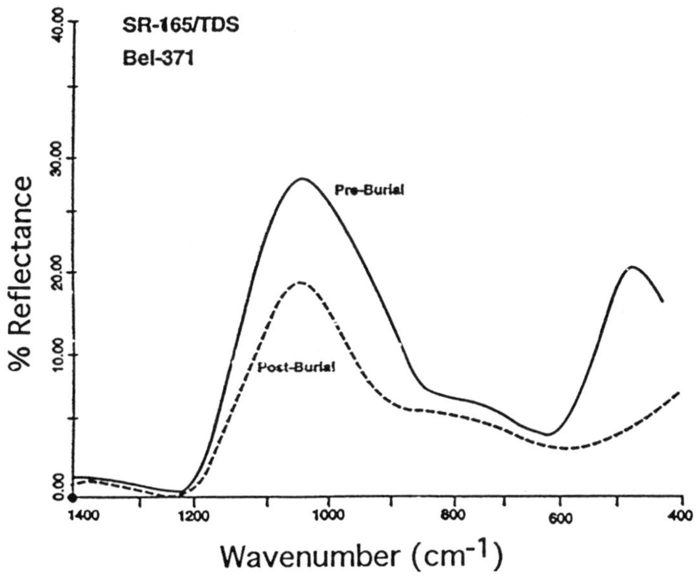

Figure 3. Schematic representation of FT-IR pre- and post-burial spectra for the SR-165/TDS glass sample.

The SEM micrograph shown in Figure 4 and the accompanying EDS spectra show a portion of the surface of the glass which contains a surface layer where magnesium and chlorine were present in measurable amounts. These two components are found in the Mol clay, but were not part of the glass composition. Small amounts of titanium and chromium were also detected. In contrast, the SEM/EDS data for the SR-131 glass samples show no surface layers.

Analyses by secondary ion mass spectroscopy may offer a possible explanation for the apparent lack of surface alteration seen on some of the glass samples, as well as the relatively large mass losses. The SIMS data for the SR-165/TDS glass sample are shown in Figure 5A,B. These spectra exhibit data for the two types of glass leaching behavior that are thought to occur during burial in the clay geology. The first spectra was taken in an area that was smooth and lustrous. In the "lustrous areas," congruent dissolution (Mol-C, Figure 5A) is thought to have occurred, where all elements found in the glass are removed at similar rates. Figure 5B (Mol-S), shows the spectra taken from the glasses in "cloudy" patches where selective dissolution was believed to occur, perhaps as a precursor to the congruent dissolution. Leaching depths in the selectively leached areas were in the range of 0.07 - 2.75 microns. However, taking into account the mass loss data and the congruent dissolution SIMS spectra, it is possible that dissolution on the order of 90 microns occurred. These data are consistent with several other glass systems exposed to clay environments. For example, some of the reference and precursor Belgian high-level borosilicate waste glass compositions also experienced an increase in corrosion. The SAN 60 composition was reported to have a leach depth of ~ 25 µm/year and the SM 58 glass had a leaching depth of ~120µm/year after two years burial in the Boom clay environment [2,17]. While the leachability of many of these systems in a clay environment may be described as relatively low, the leaching of glass in this environment is larger than observed in other burial studies. The effects of clay on glass performance have been observed in other efforts. For example, in glasses buried in the granite site at Stripa, glass corrosion was seen to increase when the glasses were placed in contact with some clay backfill materials [18-20]. Also, data from laboratory studies show that glasses in contact with some clay compositions exhibited greater leaching than glasses in contact with other repository/engineered barrier materials [21]. In Figure 6, the leaching depths of SR-165/TDS glass are summarized and compared as a function of time, after burial in granite, salt, limestone and clay. Where these depths can be described as relatively low, glass leaching in clay is slightly more than an order of magnitude greater than observed in the other geologic media.

SUMMARY

In summary, the following points can be made:

- ➢ The leaching performance of the SR-165/TDS glass buried in Mol exhibited residual surfaces with selectively leached layer depths of less than 3µm. However, data (mass loss and SIMS), indicate that congruent dissolution of the glass may have been dominant with actual leaching depths as large as 90µm. The rate of such congruent attack appears about the same for all three compositions of SRS glasses tested.

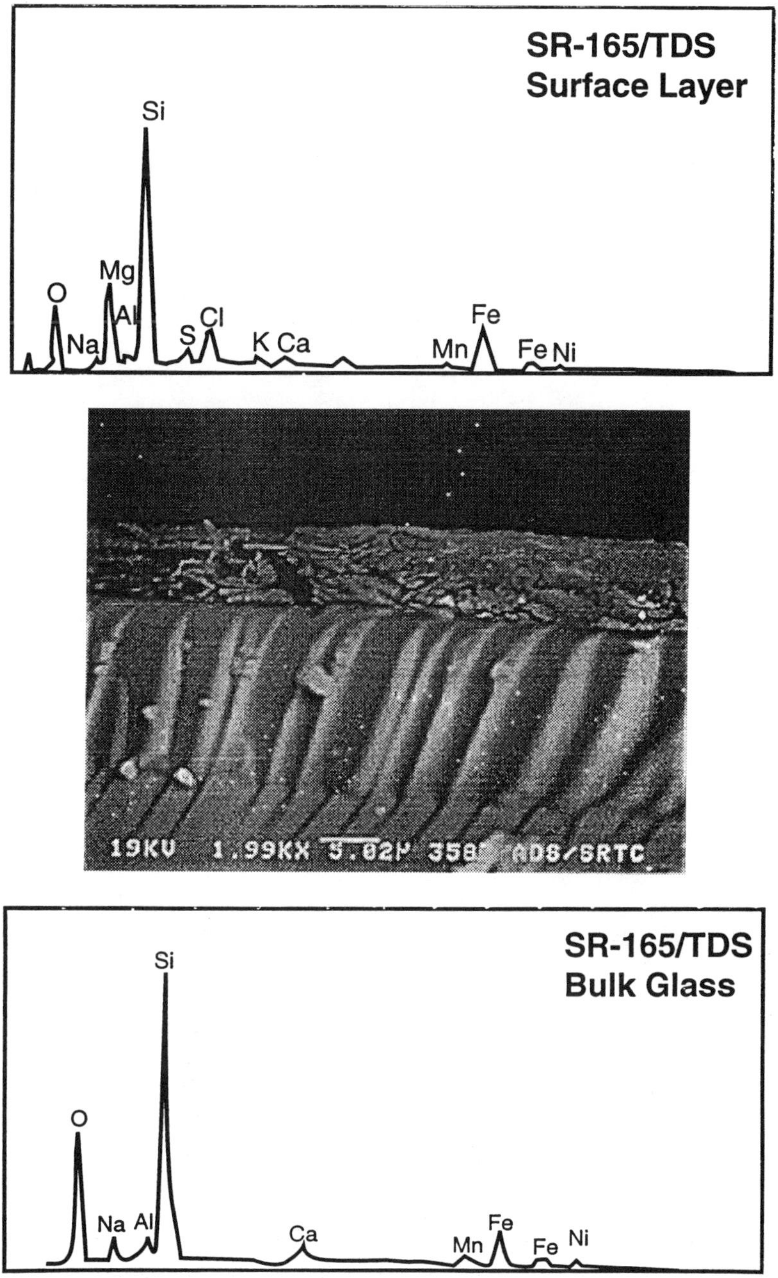

Figure 4. Scanning electron micrograph of the SR-165/TDS glass sample with accompanying EDS spectra.

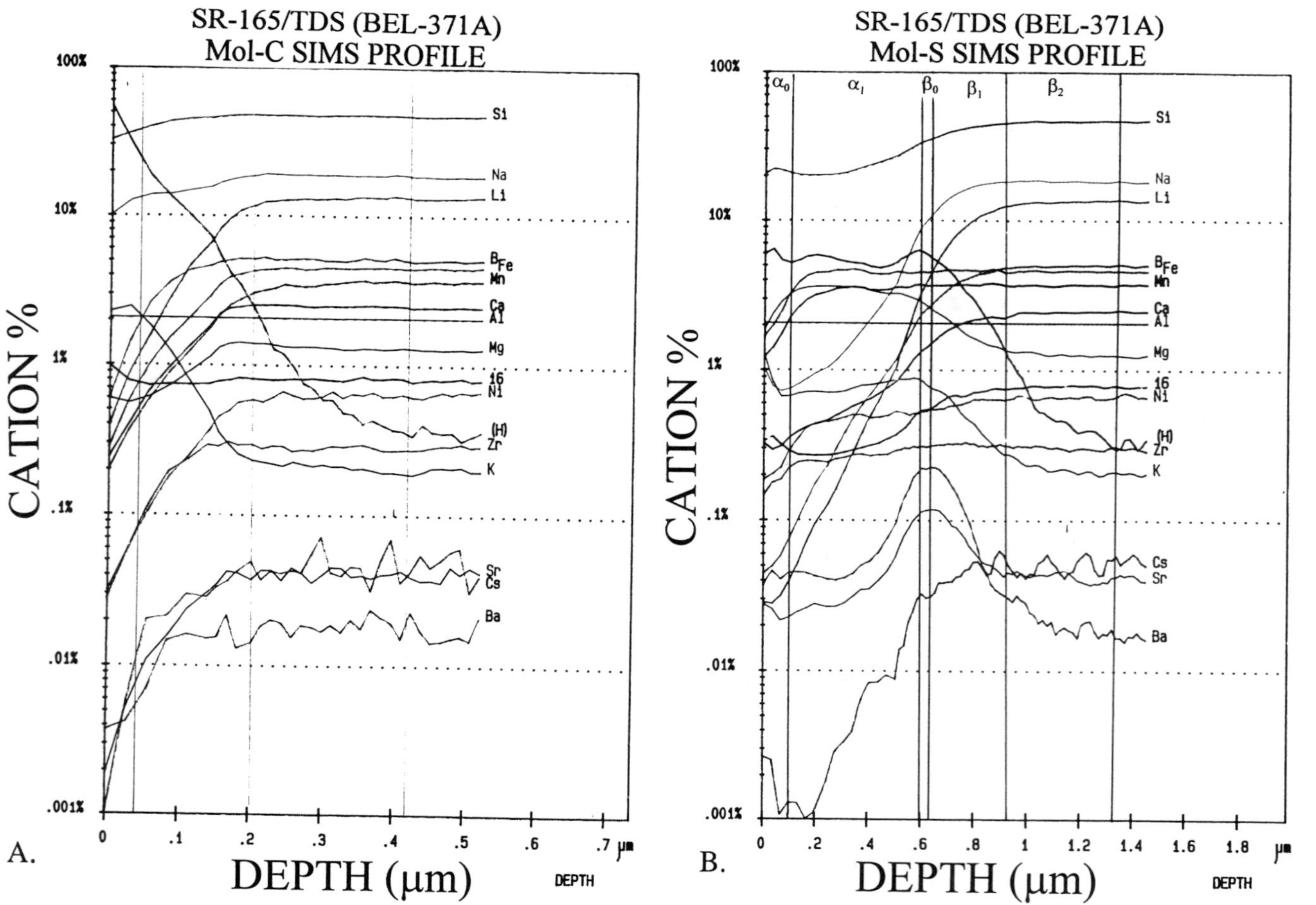

Figure 5. A). Mol-C SIMS profile showing congruent dissolution; B) Mol-S SIMS profile showing selectiive dissolution.

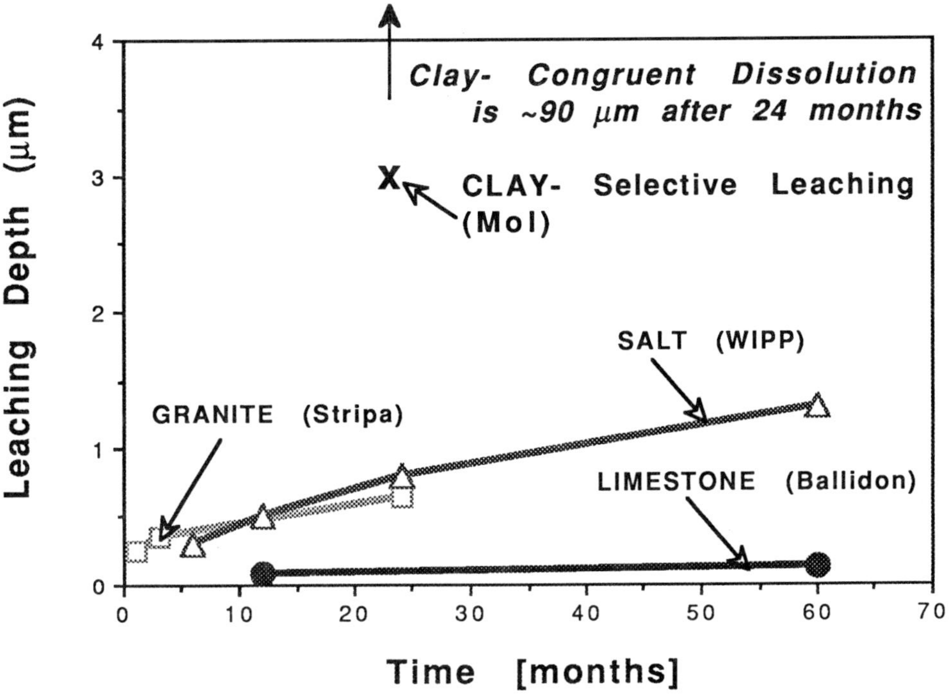

Figure 6. Leached layer depth versus time for the SR-165/TDS glass in granite, limestone, salt and clay.

- While overall leaching depths of glasses buried in clay were less than about 50µm/ year (i.e., low from an application point of view), they were slightly more than an order of magnitude greater than those observed for glasses buried in granite, salt and limestone geologies.

ACKNOWLEDGEMENTS

The authors thank Westinghouse Savannah River Technology Center for partial financial support during this work under Subcontract AB-46359-0

REFERENCES

1. D.E. Clark, B-F. Zhu, L.L. Hench, G.G. Wicks and L. Werme, Journees Sur le Verre et L'eau, Revista della Stazione Sperimentale del Vetro, n.5 (1984).
2. B. Kursten, B. Cornelus, S. Labat and P. Van Iseghem, Completion of the Corrosion Programme in Boom Clay - In-Situ Experiments, EUR 17105 (1997).
3. J. May, M. Odom and G.G. Wicks, "Analyses of SRS Waste Glass Samples Buried in Boom Clay in Mol, Belgium," WSRC-RP-93 Report (1993).

4. F. Casteels, H. Tas, M. Brabers, and R. Heremans "Corrosion of Materials in a Clay Environment," presented at the **Proceedings and International Symposium on Chemistry and Process Engineering for High-Level Liquid Waste Solidification,** Julich, Germany (1981).
5. R.G. Newton and D. Fuchs, *Glass Tech.,* 29[1], pp. 43-48 (1988).
6. R.G. Newton, *Glass Tech.* 33[5], pp. 179-180 (1992).
7. C.G. Namboodri, Jr., S.L. Namboodri, G.G. Wicks, A.R. Lodding, L.L. Hench and R.G. Newton, Nuclear Waste Management IV, (G.G. Wicks, D. Bickford and L. Bunnell, eds.), American Ceramic Society, Westerville, OH, pp. 653-661 (1991).
8. R.L. Schulz, D.E. Clark, G.G. Wicks, A.R. Lodding, R.G. Newton and K. Welcher, Environmental Issues and Waste Management in the Ceramic and Nuclear Industries, *Ceramic Transactions*, Vol. 61 (V. Jain and R. Palmer, eds.), American Ceramic Society, Westerville, OH, pp. 569-576 (1995).
9. G.G. Wicks in, Corrosion of Glass, Ceramics and Ceramic Superconductors (D.E. Clark and B.K. Zoitos, eds.), Noyes Publications, Park Ridge, NJ, pp. 218-268 (1992).
10. L.O. Werme, H. Odelius and A.R. Lodding, in, In-Situ Testing of Radioactive Waste Forms and Engineered Barriers, Proceedings of the CEC, SCK/CEN and U.S. DOE Workshop, CEC-EUR 15629-EN Corsendonck, Belgium (1993).
11. B-F. Zhu, D.E. Clark, L.L. Hench, G.G. Wicks and L. Werme in, Scientific Basis for Nuclear Waste Management VIII (C.M. Jantzen, J.A. Stone and R.C. Ewing, eds.), Vol. 44, pp. 187-194 (1985).
12. D.E. Clark, R.L. Schulz, B.K. Zoitos, G.G. Wicks and A.R. Lodding in, In-Situ Testing of Radioactive Waste Forms and Engineered Barriers, Proceedings of the CEC, SCK/CEN and U.S. DOE Workshop, CEC-EUR 15629-EN Corsendonck, Belgium (1993).
13. G.G. Wicks, A.R. Lodding, P. Macedo and D.E. Clark in, Scientific Basis for Nuclear Waste Management XV, Vol. 257 (C.G. Sombret, ed.), Materials Research Society, Pittsburgh, PA, pp. 119-126 (1992).
14. R.L. Schulz, D.E. Clark, A.R. Lodding and G.G. Wicks in, Scientific Basis for Nuclear Waste Management XV, Vol. 257 (C.G. Sombret, ed.), Materials Research Society, Pittsburgh, PA, pp. 65-72 (1992).
15. D.E. Clark, R.L. Schulz, G.G. Wicks and A.R. Lodding in, Scientific Basis for Nuclear Waste Management XVII, Vol. 333 (A.A. Barkatt and R.A. Van Konynenburg, eds.), Materials Research Society, Pittsburgh, PA, pp. 107-122 (1994).
16. Personal Communication - P. Van Iseghem (1992).
17. A.R. Lodding and P. Van Iseghem, Mat. Res. Soc. Symp., Vol. 412, pp. 229-238 (1996).
18. L.L. Hench, A.R. Lodding and L. Werme in, Nuclear Waste Management, *Advances in Ceramics*, Vol. 8, (G.G. Wicks, W.A. Ross, eds.), American Ceramic Society, Westerville, OH, pp. 310-323 (1984).
19. D.E. Clark, B-F. Zhu, R.S. Robinson and G.G. Wicks in, Nuclear Waste Management, *Advances in Ceramics*, Vol. 8, (G.G. Wicks, W.A. Ross, eds.), American Ceramic Society, Westerville, OH, pp. 324-336 (1984).
20. A.R. Lodding, E.U. Engstrom, D.E. Clark, L.O. Werme and G.G.Wicks in, Nuclear Waste Management II, *Advances in Ceramics*, Vol. 20, (D.E. Clark, W.B. White and A.J. Machiels, eds.), American Ceramic Society, Westerville, OH, pp. 565-581 (1986).
21. G.G. Wicks, B.M. Robnett and W.D. Rankin in, Scientific Basis for Nuclear Waste 0Management V (W. Lutze, ed.), North-Holland, NY, pp. 15-24 (1982).

AN AEM STUDY OF GLASS-PALAGONITE INTERFACES FORMED IN HAWAIIAN BASALT GLASSES

J. S. Luo, W. L. Ebert, N. L. Dietz, and J. K. Bates
Argonne National Laboratory
9700 South Cass Avenue
Argonne, IL 60439-4837

ABSTRACT

Basalt glass has been used as a natural analogue of high-level waste glass because both types of glass have similar Si contents. Most important for identifying the dominant long-term corrosion process is the exact chemical nature of the glass-palagonite boundary. In this paper, we present the results of a detailed study of alteration interfaces formed in subaerially altered Hawaiian basalt glasses (400 to 700 years old) and analyzed with high resolution transmission electron microscopy (TEM) and nanometer-size energy dispersive x-ray spectrometric (EDS) probe. Our study reveals a sharp contact between fresh glass and alteration products at nanometer scale. The alteration products, which were typically several nanometers in size, were found to be iron oxide hydroxide (goethite), aluminum hydroxide hydrate (scarbroite), and an Si-Fe-Al rich amorphous phase. The EDS line profile across the boundary showed a zone of 60-130 nm thickness that is depleted in alkali elements but exhibits no variation for silicon. These results point to a dominant dissolution-precipitation mechanism plus a limited ion-exchange process under subaerial conditions.

INTRODUCTION

Basalt glasses have been found in a wide range of geological environments, including deep-sea, subglacial, and subaerial environments.[1] The physical and chemical nature of glass alteration in these environments has been investigated extensively.[2-4] In all cases, the alteration of basaltic glasses with water can be characterized by the formation of a palagonite layer followed by the formation of secondary mineral phases on the outer surface of the palagonite.

Basalt glass has a similar range of compositions to the waste form borosilicate glasses selected for the high-level waste form and, therefore, has been utilized as a natural analogue for nuclear waste glasses to elucidate their long-term reaction mechanisms.[1] Understanding the processes by which natural glasses are hydrated and react can assist in understanding the complex interactions of nuclear waste glass reaction. The dominant process by which a basalt glass hydrates is still disputed, although it certainly includes a combination of molecular water diffusion, dealkalization, network hydrolysis, and network dissolution.[2,5]

Analytical electron microscopy (AEM) represents a powerful investigative technique for determining the reaction pathway of glass corrosion. Because the crystal sizes and/or amounts of the alteration products in palagonite are usually rather small, conventional techniques such as electron microprobe analysis and x-ray powder diffraction cannot be used to identify these alteration products rigorously. Zhou and others[6] examined with transmission electron microscopy (TEM) the palagonite formed on a 13-million-year-old basalt glass from a deep sea site and found fibrous and lathy particles, both of which were interpreted as precursors of smectite clay minerals. Crovisier and others[2] analyzed a sea-water-altered basalt glass recovered from the Contrada Acqua Amara, near Palagonia, and reported that the palagonite is mainly composed of poorly crystallized smectites. However, there have been few previous AEM studies on basalt glasses subaerially altered by meteoric water and humidity. Such an environment may be more representative of the conditions expected during the alteration of nuclear waste glasses in an unsaturated repository setting, such as that which characterizes the proposed Yucca Mountain nuclear waste repository site.

The purpose of this work is to understand the reaction process that controls the corrosion of Hawaiian basalt glass in an unsaturated environment. The alteration phases that were formed in the vicinity of the interface were identified and analyzed using a combination of high-resolution TEM lattice imaging, selected area electron diffraction, and energy dispersive x-ray spectroscopy (EDS). Identification of the alteration products is important because it helps to elucidate the solution chemistry that existed during the formation of these phases. The interface between the glass and palagonite was also examined, and a compositional profile across the boundary was made with a focused electron beam and EDS. These results will determine the possible presence of a leached zone on the surface, which is indicative of whether the corrosion is controlled by a congruent or non-congruent dissolution process. We also discuss the implication of this study to the projection of long-term corrosion behavior of nuclear waste glasses.

SAMPLES

Ten samples of volcanic basaltic glass were collected by Bishop Museum staff from two archaeological sites of Oahu, Hawaii. These samples were located 5 to 85 cm below the ground level in generally moist and cool environments. Although the exact humidity and temperature values for the Hawaiian samples are not known, the relative humidity has been reported to be close to 100% in most soil environments.[7] The ages of the samples were measured to range from 480 to 720 $\pm$ 80 years by radiocarbon dating of the occupation layer of charcoal. The glasses have been naturally altered by meteoric water, ground water, and a humid atmosphere. The palagonite layer formed on the surface is yellowish. Where the layer has spalled off, the glass surface beneath can clearly be seen with etching pits. Two glass samples from which the palagonite layers had completely spalled off were dissolved and chemically analyzed with inductively coupled plasma/atomic emission spectroscopy (ICP/AES), and their compositions are listed in Table I. The compositions of the two samples are nearly identical within the 3% uncertainties of the ICP/AES analyses.

Table I. Chemical compositions of major elements determined by ICP/AES for two Hawaiian basalt glasses

Oxide	Glass 1 (wt.%)	Glass 2 (wt.%)
Al_2O_3	14.4	14.5
CaO	9.71	9.86
Fe_2O_3	11.9	12.2
K_2O	0.51	0.53
MgO	6.72	6.54
MnO	0.13	0.17
Na_2O	2.82	2.87
SiO_2	51.6	51.2
TiO_2	2.06	2.07

EXPERIMENTAL

Each sample was first examined with a Wild MPS 52 optical microscope. Typical areas of palagonite were selected and mounted in a plastic holder for ultramicrotomy. The samples for AEM analysis were prepared with ultramicrotomy by impregnation with epoxy resin that had been poured onto the surface of the specimens in a rough vacuum (~10^2 torr) to ensure penetration of the resin through the surface layer and eliminate of air bubbles. Specimens were then cut to a thickness of about 50 nm by a Reichert-Jung Microtome. Transmission electron microscopy was performed with a JEOL 2000FX unit operating at 200 kV and equipped with an ultra-thin window energy dispersive x-ray spectrometer. Selected area electron diffraction was employed along with bright and dark field imaging. The chemical analyses by EDS were performed with a focused electron beam of about 15 nm, which corresponds roughly to the resolution area of EDS. Unreacted fresh glass, whose composition had been analyzed with ICP/AES, was used as a standard for the quantitative determination of major components in the alteration products by means of the well-established Cliff-Lorimer method.[8] The crystallinity of precipitates in the palagonite layer was determined with selected area electron diffraction and the alteration phases were identified by direct imaging and by matching of the measured composition and diffraction patterns against reference materials.

RESULTS AND DISCUSSION

Microstructure of Palagonite

Figure 1 shows a typical TEM image of palagonite, displaying a multiphase mixture of a greyish matrix, various dark precipitates, and several layered or fibrous phases. The majority of the palagonite seems to consist of hemispherical masses, which were determined to be goethite [FeO(OH)] by electron diffraction. Analysis of this phase with EDS revealed the presence of a small amounts of Ti and Cr. The formation of goethite is consistent with the unsaturated conditions under which the transition metals such as Fe, Ti, and Cr are insoluble and precipitate, whereas other species do not reach their solubility limits and therefore remain in solution. Many of the other crystalline precipitates have a layered or fibrous texture, which is characteristic of smectite type clays.[2-4,6]

Palagonite/Glass Interface

Figure 2 is a TEM image of the glass/palagonite interface, which generally exhibits distinct changes across the boundary between the glass and palagonite. This observation is consistent with the theory that the basalt glass undergoes a congruent dissolution upon contact with water.[2-4,6] In many areas, palagonite may have been ripped or spalled off from the glass surface during the sample preparation because of its different texture and loose adherence to the glass.

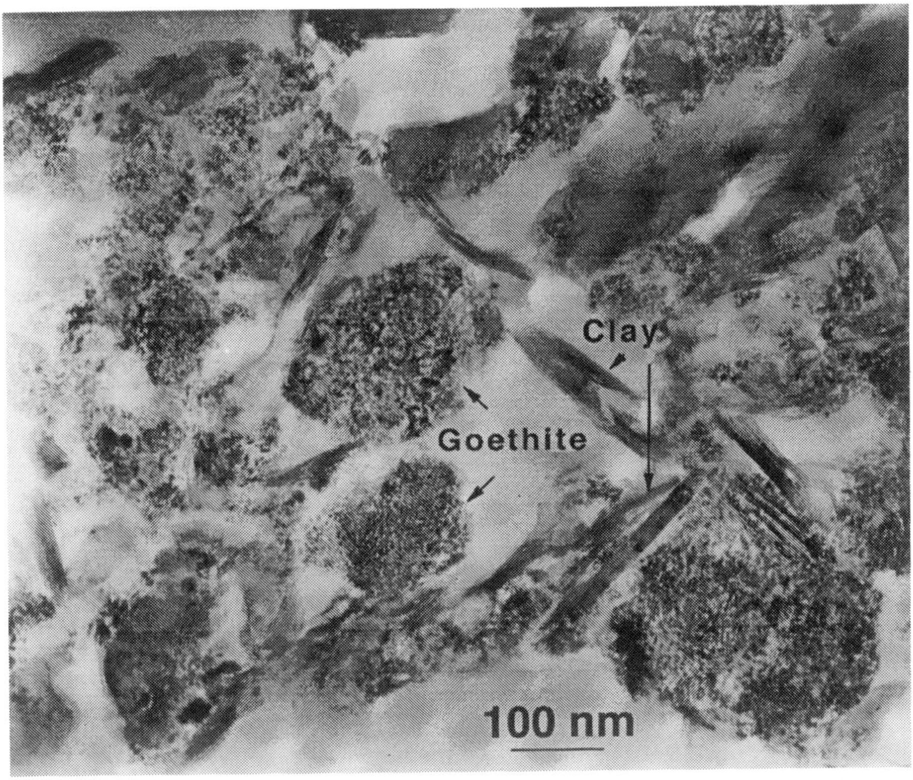

Fig. 1. Transmission electron micrograph of a typical section of palagonite, which mainly consists of hemispherical masses of goethite and layered smectite clays.

The alteration products formed in the vicinity of the palagonite/glass interface were also identified. The goethite had the characteristic hemispherical texture and was easy to distinguish from other phases (see Fig. 2). Figure 3 is a higher magnification image of the interface. In addition to the hemispherical masses of goethite, a layered phase, with different d-spacings from those smectites, was imaged and identified as scarbroite [$Al_2(OH)_6{\bullet}H_2O$] with electron diffraction and EDS. An amorphous phase, rich in Si, Al, and Fe, was also present among the goethite and scarbroite particles.

Figure 4 is a line profile of the major constituents determined with quantitative EDS from the glass surface to the central core. The profile revealed the extent of the dealkalized zone in which soluble elements such as Ca and Mg were depleted

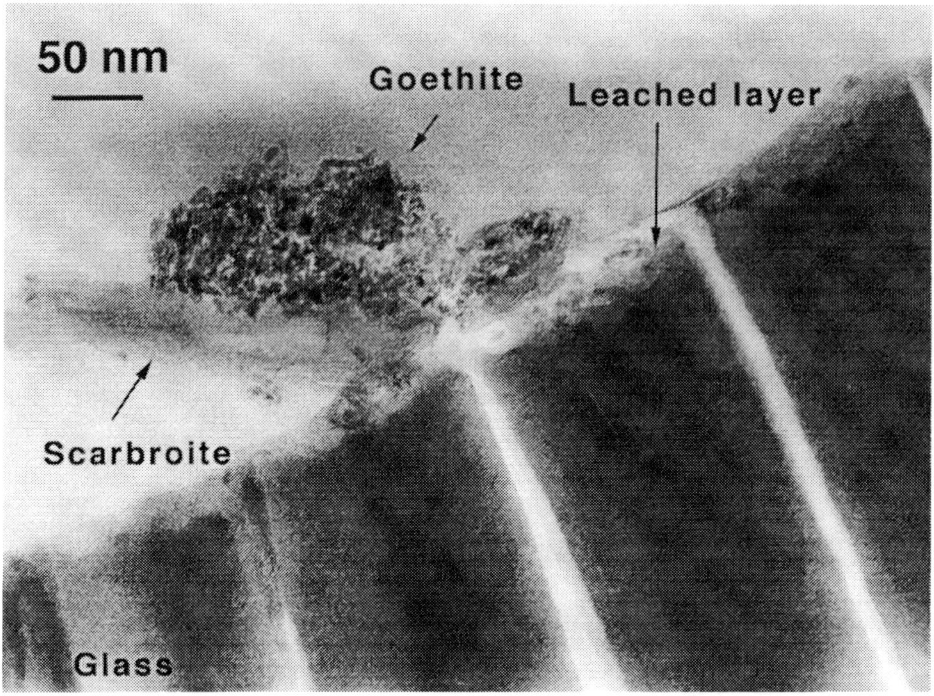

Fig. 2. Transmission electron micrograph of the palagonite/glass interface. The alteration products in the vicinity of the interface were identified to be geothite, scarbroite, and a Si-Fe-Al rich amorphous phase.

from the glass (the Na content is the glass was too low to be accurately profiled with EDS). As can be seen in Fig. 4, Mg began to become depleted starting from a depth of about 130 nm from the surface. Calcium was also depleted in this area. However, its concentrations increased at 60 nm, which may be due to inhomogeneity of the sample, and then decreased at the surface (interface). The concentrations of Si was virtually unchanged within the experimental uncertainty, but Al and Fe displayed enrichment between 60 and 90 nm from the interface region. Ti also displayed an increase within this region, however, the observed change is less than the experimental uncertainty. This observation suggests that elements such as Mg and Ca were preferentially leached out from the glass prior to a variable degree of network dissolution of the leached zone. This leached zone was best seen from Fig. 2 as the whitish layer formed on the glass surface.

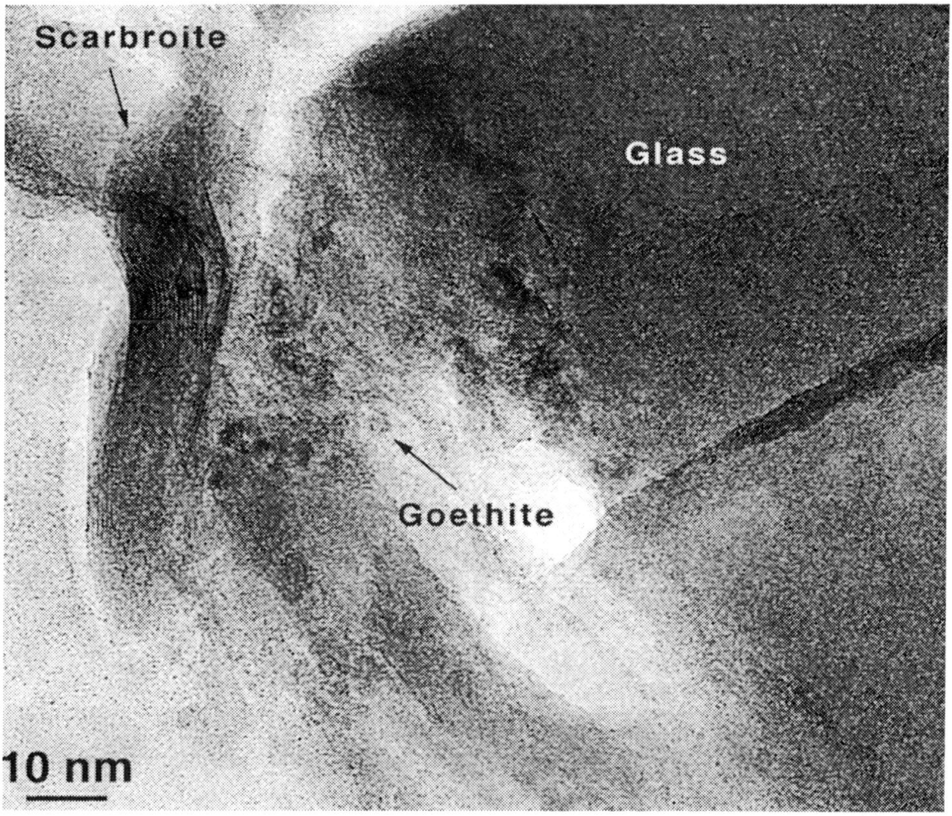

Fig. 3. High-resolution TEM image of the palagonite/glass interface showing the precipitate of goethite and scarbroite at the surface.

CONCLUSIONS

We have analyzed the alteration interfaces formed in subaerially altered Hawaiian basalt glasses (400 to 700 years old) with high-resolution TEM and EDS. Our study revealed (1) the presence of a leached zone of about 100 nm between palagonite and fresh glass and (2) a different texture and a relatively sharp contact between glass and palagonite. The alteration products present in the palagonite layer were determined to be iron oxide hydroxide (goethite), aluminum hydroxide hydrate (scarbroite), and a Si-Fe-Al rich amorphous phase. These

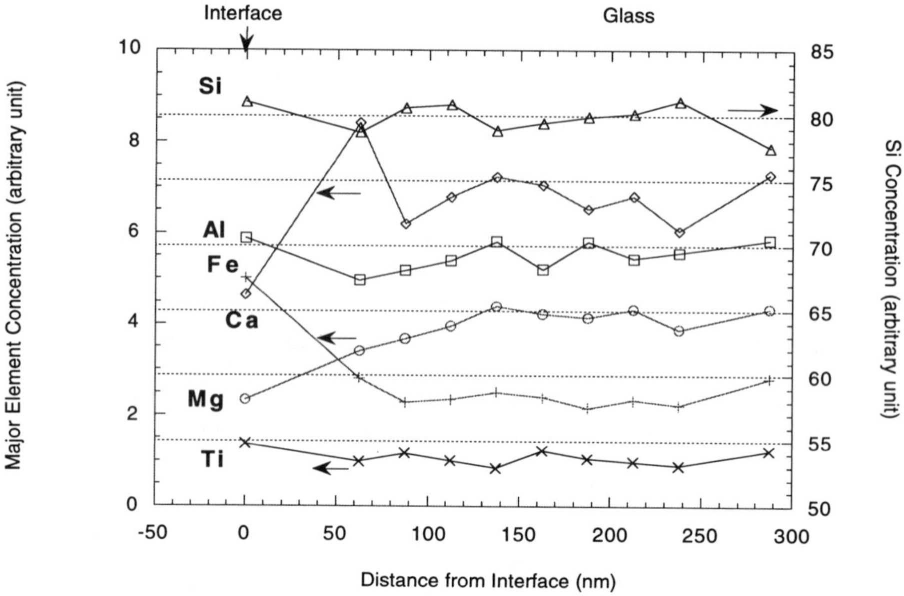

Fig. 4. Profile of major elements in the glass from core to the palagonite/glass interface revealing the presence of a 60-130 nm thick zone in which Ca and Mg were depleted, Al, Fe, and Ti enriched, and Si unchanged.

results are consistent with a dominant dissolution-precipitation mechanism plus a limited ion-exchange process during subaerial alteration of basalt glass.

IMPLICATIONS FOR WASTE MANAGEMENT

Basalt glasses are analogues for the long-term behavior of nuclear waste glasses, because of the similar silica contents between the two types of glasses. A natural analogue study could provide confirmation of the thermodynamic and kinetic models used to extrapolate the long-term corrosion behavior of nuclear waste glasses.[1] In this work, we characterized the palagonite layer and the palagonite/glass interface in the subaerially altered glasses, which in turn sheds light on the important alteration processes involved. The information obtained from this work can then be compared to alteration of basalt and waste glasses that occurs in laboratory tests. Work is ongoing in our laboratory to characterize the basalt and waste glasses reacted under unsaturated vapor conditions. If it can be demonstrated that (1) the long-term alteration products formed by natural processes can be reproduced by laboratory experiments and (2) the experimental method developed to simulate naturally occurring alteration produces similar reactions in waste glass, then the kinetic models that have been developed based on short-term experiments can be applied to calculate the long-term corrosion of waste glass with more confidence.

ACKNOWLEDGMENT

The authors thank D.M. Strachan, E.C. Buck and D.J. Wronkiewicz for their critical reading of the manuscript and R. Riel and J. Harmon for their editorial assistance. The work at ANL was sponsored by the U.S. Department of Energy, Office of Environmental Management, under Contract W-31-109-ENG-38.

REFERENCES

[1]R.C. Ewing, "Natural Glasses: Analogues for Radioactive Waste Forms"; pp. 57-68 in *Scientific Basis for Nuclear Waste Management*, Edited by G.J. McCarthy et al. Plenum Press, New York, 1978.

[2]J.L. Crovisier, J. Honnores, and J.P. Eberhart, "Dissolution of Basaltic Glass in Seawater: Mechanism and Rate," *Geochim. Cosmochim. Acta*, **51** 2977-2990 (1987).

[3]H. Furnes, "Element Mobility during Palagonitization of a Subglacial Hyaloclastite in Iceland," *Chem. Geol.*, **22** 249-264 (1978).

[4]M.J. Jercinovic, K. Keil, M.R. Smith, and R.A. Schmitt, "Alteration of Basaltic Glasses from North-Central British Columbia, Canada," *Geochim. Cosmochim. Acta*, **54** 2679-2696 (1987).

[5]T.A. Abrajano, Jr., J.K. Bates, and J.J. Mazer, "Aqueous Corrosion of Natural and Nuclear Waste Glasses: II. Mechanisms of Vapor Hydration of Nuclear Waste Glasses," *Journal of Non-Crystalline Solids*, **108** 269-288 (1989).

[6]Z. Zhou and W.S. Fyfe, "Palagonitization of Basaltic Glass of DSDP Site 335, Leg 37: Texture, Chemical composition, and Mechanism of Formation," *Am. Miner.*, **74** 1045-1053 (1989).

[7]I. Friedman, F.W. Trembour, F.L. Smith, and G.I. Smith, "Is Obsidian Hydration Dating Affected by Relative Humidity?" *Quaternary Research*, **41** 185-190 (1994).

[8]J.S. Luo, S.F. Wolf, W.L. Ebert, and J.K. Bates, "Quantitative SEM/EDS Analysis of High-Level Waste Glasses"; in *Proceedings of the 29th Annual Conference of the Microbeam Analysis Society*, Edited by E.S. Etz, VCH Publishers Inc., New York, 1995.

CORROSION AND LONG-TERM STABILITY OF VITRIFIED E-FILTER DUST FROM MUNICIPAL WASTE INCINERATOR PLANTS

N. H. Menzler, H. Mörtel
University of Erlangen-Nuernberg
Department of Materials Science, Chair of Glass and Ceramics
91058 Erlangen, Martensstr. 5, Germany
Phone : ++49-9131-857551, Fax : ++49-9131-858311
e-mail : moertel@ww.uni-erlangen.de

ABSTRACT

On behalf of the Bavarian Ministry for Development and Environment, our institute has a research project called "Optimization of the long-term stability of vitrified e-filter dust of municipal waste incinerator plants." The project has two major goals: First, the optimization of the incorporation in glasses of residues from waste incinerators which are not reusable nowadays; and second, the characterization of the corrosion behavior of the resulting glass. Parallel to this work we examined granules of three different high-temperature waste incinerator plants. The granules are the residue of the melting chamber or the gasification reactor of the incinerator plants. They are cooled down by falling in a water bath.

The leaching results of the vitrified e-filter dust was compared to the leaching results of the granules.

The two most important results will be discussed. We were successful in preparing a durable glass by combining different technical and materials parameters. Additionally the amount of heavy metals incorporated in the glass was comparatively high. Also, we found that the products of various

high-temperature waste incinerator plants were not glasses. The granules of these processes were partially crystallized and had many "Schlieren," other inhomogeneities, and bubbles. Only the matrix is glassy.
Therefore, we examined different test methods for glasses or amorphous materials to find whether or not it is possible to describe the corrosion behavior of the granules using these methods.

1 CHARACTERIZATION OF THE "RAW MATERIALS"

By "raw materials" we mean the residues of the waste incinerator plants (e.g. flue-gas scrubber dust, e-filter dust, fly ash, slag, bottom ash ...) and the granules of the high-temperature waste incinerator plants. We obtained e-filter dust from one Bavarian special waste incinerator plant and two batches of e-filter dust and flue-gas scrubber dust from two different plants. The e-filter dust is a residue who is separated in a waste incinerator plant via electrostatic precipitators and is normally containing a high amount of heavy metals.
We determined their chemical composition, mineralogical phases, grain size distribution, and thermal behavior. We found the major elements are aluminum, silicon, calcium, iron, potassium, sodium, chlorine, and sulfur (for chemical composition of e-filter dust see table I).

Table I : Chemical composition of the e-filter dust

Elements/compounds (mg/kg)	e-filter dust
Aluminum	33,000
Barium	5,100
Lead	9,800
Calcium	55,500
Chromium	2,700
Iron	40,800
Potassium	41,200
Copper	3,200
Magnesium	13,100
Manganese	1,300
Sodium	152,600
Silicon	40,400
Titanium	31,300
Vanadium	1,200
Zinc	32,200
Chlorine	158,000
Fluorine	2,000
Phosphate (P2O5)	4,000
Sulfate (SO3)	140,500

The amount of heavy metals varies between 4.8 and 13 wt. (counted as elements). The x-ray diffraction analysis shows alkali and alkaline-earth chlorides, sulfates and sulfides, as well as silica and silicate compounds. The grain size distribution shows that the amount of fine particles in the pure e-filter dust is higher (90 % < 48µm) than in the mixes (60-80 % < 128 µm). Thermal analysis shows high mass losses up to 1200°C (batch 1, 20 wt., batch 2, 40 wt., e-filter dust, 30 wt.).
The granules were also characterized with respect to their chemical composition, mineral phases, and microstructure, as well as their density. Very informative was the optical microscopy of thin sections.

The major oxides in the granules were SiO_2, Fe_2O_3, Al_2O_3, CaO and P_2O_5. The amount of heavy metals varies between 1.1 and 4.1 wt. (counted as oxides!).

The microscopic analysis showed many inhomogeneities: "Schlieren," bubbles, and cracks. Additionally, there are many different crystalline particles present.

2 MELTING OF RESIDUES

The aim of the vitrification was to formulate a glass having long-term durability. Various parameters must be considered :

- glass composition (high amount of SiO_2 $\rightarrow$ high durability of the glass; high amount of alkalis $\rightarrow$ reduction in durability),
- amount of residue (too high $\rightarrow$ no glass or less durable glass; too low $\rightarrow$ only dilution and an uneconomical procedure),
- melting parameters (temperature $\rightarrow$ influence on viscosity and evaporation; time $\rightarrow$ influence on evaporation, crystallization, and segregation),
- melting technology (melting furnace $\rightarrow$ "cold top" or "hot top"; oxidizing or reducing atmosphere; separation of heavy metals or immobilization in the glass network),
- others (costs, quantities and availability of admixtures, types of admixtures).

Keeping these parameters in mind, a series of crucible melt experiments were carried out using different compositions, temperatures, and times. Finally, the corrosion experiments were done on one glass with the following composition:

50 wt. SiO_2 + 12 wt. Na_2O + 38 wt. e-filter dust,

which was melted for 1 hour (at maximum temperature) at 1200°C.

3 RESULTS USING GERMAN DIN 38414, "SWISS TEST", ISO 720

3.1 German DIN 38414

In Germany, the DIN 38414 is the stipulated method for describing the suitability of waste materials for being placed in deposits, and is the only

standard for defining classes of waste for disposal. 100 g of waste was dispersed in 1000 ml distilled water for 24 hours in 2000-ml polymer bottles. After this 24 hours, the solution is examined. There are two different threshold values for waste disposal.
If the amount of each component lies below the maximum threshold value, the material may be disposed, otherwise the waste must be treated.
The results for the granules are as follows: The characterization of their corrosion behavior via DIN 38414 is not satisfactory. The reasons are first, an inexact description of the sampling of the materials (quantitative, representative); second, the inhomogeneity of the material; and third, the absence of any treatment or homogenization before testing. Treatment especially includes grain size separation. The material should be tested as it is delivered (statement of the method). No requirement exists for separating the fine particles or for grinding the material. Only if the particles are too large should they be crushed (statement of the method). Thus, if we tested the same material three times, we got three different results. For the glass, the test method is not satisfactory because of its short duration and low temperature (room temperature). Figure 1 shows one test result for the glass and the three granules. In this test, most of the results are below or at the threshold values, respectively.

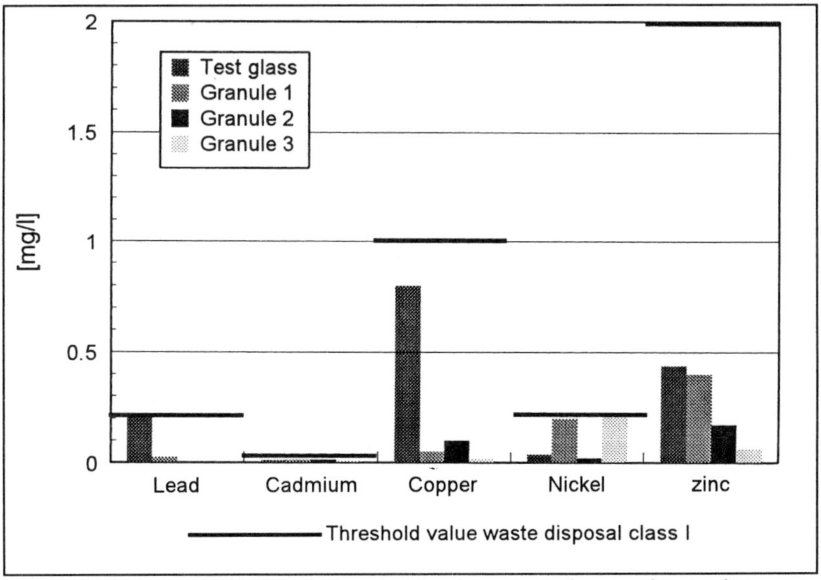

Figure 1 : German DIN 38414; threshold values; test glass and granules

3.2 "Swiss test"

The 'Swiss test," comparable to the German DIN 38414, is prescribed in Switzerland. The ratio of waste material to solution is also 1:10. The difference is that this test method works with CO_2-saturated distilled water. This means that the pH decreases to 4 or 4.5. The duration is two 24-hour periods. After the first 24 hours, the solution is renewed. For the whole time (48 h) CO_2 bubbles flow through the solution.

Table II : Swiss test; comparison between the results of the granules, the test glass, and the threshold value for inert waste disposal class

Elements	Granule 1	Granule 2	Granule 3	Test glass	Threshold value inert waste disposal
Aluminum	1.49	2.16	0.48	0.26	1.0
Arsenic	br	br	0.04	nd	0.01
Lead	< 0.01	< 0.01	< 0.01	0.028	0.1
Cadmium	< 0.01	< 0.01	< 0.01	br	0.01
Chromium-VI	br	br	nd	nd	0.01
Copper	br	0.52	0.94	0.41	0.2
Nickel	0.028	0.29	0.17	br	0.2
Zinc	1.10	0.30	0.53	0.09	1.0

all values (mg/l); br = measured value comparable to blank reading;
nd = not determined

For the results of the "Swiss test," the same statements and conclusions could be drawn as with the DIN test (see table II). Additionally, two granules and the glass exceed the threshold value for heavy metals in the dry starting material. If this first value is exceeded, the material is not disposable in an ordinary waste deposit. It would be hazardous waste (special waste burial with top and bottom seals).

The authors have the opinion that this threshold value for dry material is questionable. We were successful in producing a durable glass from a dry material with a high amount of heavy metals, but we would not be allowed to deposit this material (in Switzerland). We think this indicates problems with the present test that should be discussed.

3.3 ISO 720

The ISO 720 is a short-term test method to describe the corrosion behavior of durable glass grains. The advantage of this method is that the particle size is defined. Accordingly we had a calculable surface area of the material, and it is possible to get normalized leach rates (NL_i, especially for Si).The grain size is 300 - 425 µm. The grains have to be cleaned with acetone before testing. The running time of the test is half an hour in an autoclave at 121°C (1 bar pressure). The particles lie in Erlenmeyer flasks (5 g in 50 ml distilled water). The leach rate for the glass was lower than rates for the granules. The optimized glass therefore had a higher durability than the granules. Using the ratios Si/Mg, Si/Al, Si/Ca and Si/Na in solution and in the unreacted bulk glass (granule), we could explain which corrosion mechanism is predominant (leaching or network dissolution). If all ratios in the bulk glass and in solution are comparable, network dissolution is predominant, otherwise leaching is predominant or other reactions like precipitation take place.
At these short testing times, leaching is always the predominant mechanism. The granules are, in fact, not glass, but rather are glassy materials with crystalline inhomogeneities.

4 MCC-3S TEST

The MCC-3S test is a long-term test method that uses glass grains of specific fractions at five different temperatures, four different solutions, and three times. It was originally developed for testing nuclear waste glasses. The crucible is made of Teflon. Each test was performed in triplicate.

With these numerous parameters, we were able to detect time and temperature influences on the glass corrosion. We found that the influence of temperature on glass corrosion is higher than that of time. In distilled water and a silicate/bicarbonate solution, the glass behaved as expected. The behavior is an exponentially increasing function of corrosion (Si in solution) with temperature.

The limiting factor is the saturation of the solution. Figure 2 represents a typical result (each test result is displayed). The exponential increase is present up to 150°C. At 190°C, the solution is saturated or over-saturated (because of different evaporations the measured values differ strongly). The solubility of SiO_2 was reduced in the salt solution. The organic solution (not stipulated) shows that not only glass corrosion takes place but complexing and precipitation, respectively. For the granules, the test results again agree with the granules consisting of a glassy material with crystalline particles.

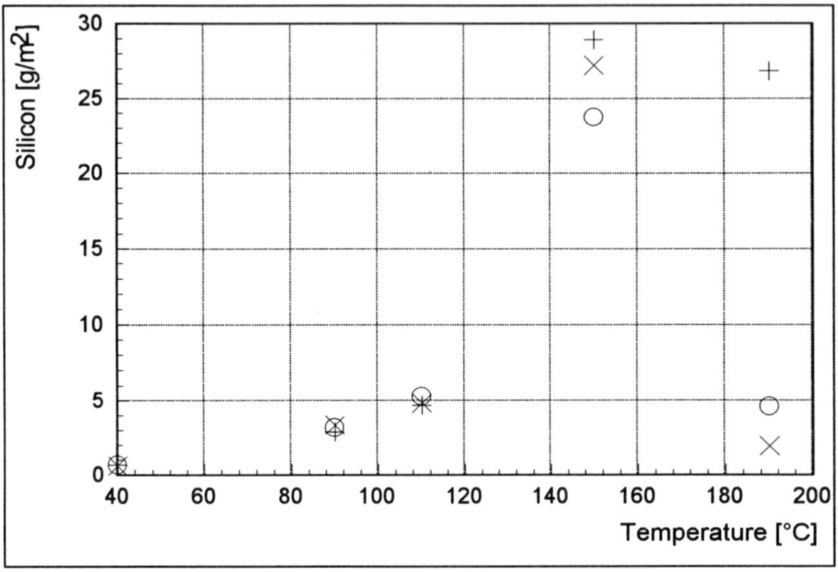

Figure 2 : MCC-3S test at 28 days in distilled water, test glass

5 SUMMARY

The results of three different short-term test methods, one long-term test method, XRD data, and optical microscopy show that the granules produced in high-temperature waste incinerator plants are not glasses. Therefore, it is obvious that these leach test methods are insufficient for describing the behavior of these granules in contact with liquid media.

We showed that it is possible to vitrify e-filter dust and batches of e-filter dust and flue-gas scrubber dust independent of their chemical composition. We were successful in formulating a durable glass by using admixtures and a maximum melting temperature of 1200°C; at this temperature the losses due to evaporation are less than those at higher temperatures (e.g. 1400°C) or longer melting times. The melting process yields a short-term and long-term durable glass with an acceptable amount of strongly bonded heavy metals.

The influence of temperature, time, and ratio of surface area to solution volume on the corrosion behavior on the glass was described.

6 LITERATURE

[1] **Adams P.B.** : Glass corrosion. J.Non-Cryst. Solids (1984), 193-205

[2] **Conradt R., Roggendorf H., Scholze H.** : A contribution to the modelling of the corrosion porcesses for HLW glasses. Mat.Res.Soc.Symp.Proc. Vol. **44** (1985), 155-62

[3] **Feng X., Pegg I.L.** : Effects of salt solutions on glass dissolution. Phys. and Chem. Of Glasses Vol. **35** No. 2 (1994), 98-102

[4] **Grambow B.** : A general rate equation for nuclear waste glass corrosion. Mat.Res.Soc.Symp.Proc. Vol. **44** (1985), 15-27

[5] **Hench L.L**. : Characterization of glass corrosion and durability. J. Non-Cryst. Solids **19** (1975), 27-39

[6] **Iler R.K.** : The chemistry of silica. John Wiley and Sons, New York, (1979)

[7] **Jantzen C.M.** : Prediction of glass durability as a function of glass composition and test conditions: thermodynamics and kinetics. In Proceedings - First international conference on advances in the fusion of glass. Amer.Ceram-Soc. Westerville, USA, Chapter 24

[8] **Macedo P.B. Barkatt A., Gibson B.C., Montrose C.J.** : Long-term release rates of borosilicate glass waste forms. Nucl.Tec. Vol **73** (1986), 199-209

[9] **Newton R.G.** : The durability of glass - a review. Glass Tec. Vol **26** No. 1 (1985), 21-38

[10] **Scholze H.** : Chemical durability of glasses. J.Non-Cryst. Solids **52** (1982), 91-103

[11] **Van Iseghem P. Cantale C., Coquerelle M., Dussossoy J.L., Malow G., Roggendorf H. :** Corrosion mechanisms of vitrified high level waste. Proc. 3rd Europ.Comm.Conf.Rad.Waste (1990), 275-86

METHODS AND CRITERIA DEVELOPMENT RELATED TO CONSISTENCY FOR QUALIFICATION OF THE GLASS-ZEOLITE CERAMIC WASTE FORM

L. J. Simpson, D. J. Wronkiewicz, and S. F. Wolf
Chemical Technology Division
Argonne National Laboratory
9700 South Cass Avenue
Argonne, IL 60439

ABSTRACT

A ceramic waste form composed of glass-bonded zeolite is being developed by the Argonne National Laboratory's Electrometallurgical Treatment Program for the disposition of radionuclides associated with the U.S. Department of Energy's spent nuclear fuel conditioning activities. Methods and criteria have been evaluated that can be applied towards the process of qualifying this material for acceptance into the Civilian Radioactive Waste Management System. A range of conditions, simulating potential repository environments, were evaluated to determine the bounding parameters appropriate for the corrosion testing of the ceramic waste form. Consistency, sample preparation, and acceptance test procedure issues related to the ceramic waste form have been investigated. Results from the Product Consistency Tests with the ceramic waste form in deionized water indicate that the Sr, Ba, and rare earth simulated fission products were preferentially retained, with respect to the singly ionized cations, in the zeolite. Glass matrix dissolution and high-ionic-strength brines affect the fission product release. With slight modifications to sample preparation procedures, the American Society of Testing and Materials standard Product Consistency Test procedure provided appropriate data that may be used to develop acceptance criteria for the ceramic waste form with regards to consistency. This investigation is limited to the 7 day Product Consistency Test, but other test methodologies may also provide useful results and should not be excluded.

INTRODUCTION

The Electrometallurgical Treatment Program at Argonne National Laboratory (ANL) is developing conditioning processes for the potential treatment of a diverse inventory of U.S. Department of Energy (DOE) spent nuclear fuel that may not be amenable for direct geologic disposal [1]. The electrometallurgical treatment program uses molten anhydrous LiCl-KCl salt in an electrorefining process that accumulates Sr, Cs, Ba, Ce, La, Nd, Y, Ag, Ru, and Pd fission products. The anode residue is incorporated in a metal waste form [2] and fission products that accumulate in the salt are loaded into zeolites through an ion exchange process [1]. Equal amounts by weight of salt-loaded zeolite and Bayer "57" glass are mixed together in an industrial mill, packed into 304 stainless steel canisters, and hot isostatically pressed to produce the ceramic waste form of glass-bonded zeolite (Table I) [3-7]. The samples used in this independent evaluation of the ceramic waste form are similar to the "baseline" material produced by ANL's Ceramic Waste Form Development Group.

Table I. Elemental compositions of the 57 glass, salt-loaded zeolite, and ceramic waste form

Elements	Li	B	Na	Al	Si	Cl	K	Ca	Y	Sr	Cs	Ba	La	Ce	Nd	O
57 glass, wt%		4.9	5.0	3.8	28.6		0.5	0.8		0.01						56.4
SLZ, wt%	1.4		4.3	15.2	15.8	12.8	5.4	7.5	0.01	0.05	0.4	0.1	0.07	0.2	0.5	36.3
CWF, wt%	0.7	2.5	4.6	9.5	22.2	6.4	2.9	4.2	0.01	0.03	0.2	0.07	0.04	0.1	0.2	46.3

SLZ: salt-loaded zeolite; CWF: ceramic waste form.

To be accepted into the DOE Civilian Radioactive Waste Management System, the ceramic waste form must be reviewed and deemed "acceptable" with the identification and development of the appropriate qualification methods and criteria following the guidelines presented in the Waste Acceptance System Requirements Document (WASRD) for spent nuclear fuel and high-level waste [8,9]. Consistency, sample preparation, and test procedures are specific issues that must be addressed for this qualification process and were investigated in this study.

The effects of leachant composition and S/V ratios on the test results are also investigated in an attempt to determine the release mechanisms and other parameters that are important for the acceptance of the ceramic waste form.

The salt-loaded zeolite has a "cubic" structure with edge lengths of up to 4 μm. The relatively small size, the ion exchange properties, and the crystallinity of the zeolite are extremely important parameters that must be considered when evaluating the different test methodologies. Detailed investigations by Simpson *et al.* [3,4] and Lewis *et al.* [5] of the ion exchange behavior of the salt-loaded zeolite without the glass binder have been performed with an extensive matrix of modified Product Consistency Tests (PCTs) [3-7]. The results from these experiments that used salt-loaded zeolites without the glass binder indicated that:

1. The preferential cation release order in deionized water (DIW) at 90°C was: Li > Na ≥ K > Cs > Al > Si > rare earths (RE) ≥ Sr ≥ Ba.
2. The Si release displayed a significant pH dependence in buffered solutions.
3. Solution Ca and Mg species ion exchanged with the Na and K in the zeolite.
4. Rare earth release rates substantially increased in brines.

The PCT is an American Society of Testing and Materials (ASTM) approved test procedure [10] that uses granulated samples. The 7 day PCT has been approved as an acceptance test method for determining the consistency of high-level borosilicate waste glass samples [8] through a comparison with the environmental assessment (EA) benchmark glass [10]. However, there are extensive material property differences between the ceramic and glass waste forms that must be considered when evaluating acceptance qualifications that may include comparisons with the EA glass. Longer duration PCTs may also provide information about the mechanistic release behavior of the ceramic waste form.

For disposition in a geologic repository additional durability testing will be required to fulfill the "repository-relevant performance assessment" criteria. This may require investigations with durability tests that use unsaturated conditions at elevated temperatures. Ultimately, the short term results from a consistency test may need to be related to the longer term repository relevant behavior. The primary goal of the present investigation is to provide the information that may be used to evaluate the utility of the PCT procedure for determining the criteria that the ceramic waste form will need for acceptance into the Civilian Radioactive Waste Management System. However, the search for acceptance tests is not limited to the PCT and these results have also been used to investigate the mechanistic behavior of the ceramic waste form.

EXPERIMENTAL METHODS AND DEVELOPMENT

The samples used in these experiments were provided by the Ceramic Waste Form Development Group of the ANL Electrometallurgical Treatment Program [6,7]. The ceramic waste form samples were cut from their hot isostatically pressed containers with a diamond saw without the use of lubricants. Samples were crushed and sieved according to the ASTM protocols [10], except ethanol was used in place of DIW for rinsing the fines from the crushed material [3,4,9-11]. This change was imposed to limit the alkali metal and chloride loss from the ceramic waste form material during sample preparation. Preliminary analysis of the ethanol rinse solutions indicates that less than 1% of the total salt in the crushed ceramic waste form material was released. However, the salt fraction released during the ethanol rinse may be significant and may require the addition of rinse solution analysis in the test protocols for the ceramic waste form.

Approximately 20% of the crushed ceramic waste form material for the PCTs had the correct size distribution (-100 to +200 mesh; 75 to 150 μm). Comparison of the data from energy dispersive spectroscopy indicate that the chemical composition of the -100 to +200 mesh material is nominally the same as the pre-crushed material and the material that passed through a 325 mesh screen. Also significant amounts of free zeolite particles were not observed in any of the crushed material. This suggests that to within the sensitivity of the energy dispersive spectroscopy technique (±10%) the zeolite and glass were sufficiently bonded to allow the ceramic to be ground for the PCTs without disproportionately separating the crystalline and glass phases of the samples.

The PCTs were performed in Teflon vessels at 90°C. The tests were conducted in duplicate or triplicate to provide a statistical basis and to provide backup tests in the event that an individual test mechanically failed. In accordance with the ASTM procedures, the solution volumes used in the tests were adjusted to the geometric surface areas of the samples to provide a S/V ratio of 2000 or 20000 m^{-1}. Deionized water, equilibrated Yucca Mountain groundwater (EJ-13) [12,14], and an ASTM Mg-Na-K-Cl brine [11] were used as leachants in the tests. The effect of brines on the ceramic waste form may be important when considering disposition in an unsaturated repository where fluids that condense on the waste form surface react with the glass matrix, thereby creating a concentrated brine. The composition of such a fluid is likely to be controlled by the composition of the solid waste form under high solid/liquid ratio conditions.

Leachate analyses were performed with a Fison Instruments Plasma Quad inductively coupled plasma mass spectrometer (ICP/MS) for the elemental cation analysis, ion chromatography (IC) for anion analysis, a Brinkmann 605 pH meter, and a Dohrmann Carbon Analyzer. The leachate cation, anion, and carbon analyses are accurate to within ±10% for elements present above the level of quantitation. The pH measurements are accurate to within ±0.2 pH units. The analyses were performed at room temperature, and the cation solutions were passed through a 0.45 μm filter to remove large sample residue and acidified to prevent precipitate formation.

The leachate analyses provide information about the concentration of a specific ion in the solution (C_i). The leachate concentrations can be converted to normalized elemental mass loss (NL_i, g/m^2) defined as: $NL_i = (C_i - C_{is})\, V_s / f_i\, A$; where, C_i - concentration of *i*th element in the leachate (mg/L), C_{is} - starting concentration of *ith* element in the leachant (mg/L), V - volume of the leachate (m^3), f_i - fraction of *ith* element in the sample, A - surface area of the sample (m^2). The NL_i provides a measure of the amount of released solids, accounting for the surface area of the sample interacting with the leachant and the weight fraction of the element present in the solid. The surface areas are estimated for crushed samples based on density and assuming spherical particles with mean diameters of 75 to 150 μm. The surface areas are based purely on geometry and do not account for surface roughness, fractures, or the internal surfaces of the zeolites.

RESULTS AND DISCUSSION

Product consistency tests were initiated with the ceramic waste form in several leachants to investigate the size-fraction release and solubility of specific components. The incorporation of different fluids into the acceptance tests provided information about the bounding parameters and mechanistic release scenarios appropriate for the corrosion testing of the ceramic waste form and about the behavior of the waste form under accelerated testing conditions. Tests were performed on samples that were held at 90°C for 7 days. Leachate aliquots were removed and analyzed upon termination of the tests, and the solids were retained for microscopic analysis.

Seven individual ceramic waste form samples [3-7] were crushed and sized separately for use in evaluating the PCT protocols. Prior to use in a PCT, small samples of the crushed material were examined by scanning electron microscopy and energy dispersive spectroscopy analysis. These studies indicated that the fines were sufficiently removed from the 75 to 150 μm particles by the ethanol rinse process. However, NaCl and other "salt" crystals formed as the materials were dried in a 90°C oven. Three samples were dried at ambient temperatures and salt crystal formation was not observed. The ethanol appears to aid the salt transport to the surface at elevated temperatures. A modified protocol recommendation may be the addition of a room temperature drying step to evaporate most of the ethanol, before the final drying in a 90°C oven.

While the initial compositions and hot isostatic pressing conditions of the seven samples were nominally the same, the crushed material from each of the seven samples was evaluated individually with the PCT protocol [10] to determine the sample-to-sample variability. The average and standard deviation from all the samples were calculated for each element (Table II). In general, deviations of ±10 to 15% in the data might be attributed to analytical and systematic errors in the leachate analysis. Statistically, this analytical error for each of the samples will translate into a corresponding ±10 to 15% normalized average deviation in the leachate results (Normalized Average Deviation = $\Sigma_i\ |x_i - \langle x\rangle|/n\langle x\rangle$; where, x_i - the *ith* number in a series, $\langle x\rangle$- the average of a series of numbers, n - the total number in a series). Analysis of the PCT results indicate that the deviation between the samples for the release of Cl, Li, Na, K, Al, and Si is within the expected deviation of the analyses from the PCTs with the 7 individual samples (Figure 1). Larger deviations were associated with the release of Cs, Ca, Sr, Ba, Nd, Ce, La, Y, and B that may be related to the sample-to-sample variations in release behavior. No individual sample could be identified as the cause of the sample-to-sample variations for all of the elements. The sample-to-sample variations in the release of simulated radionuclide elements, including Cs, may be important for waste form development as well as waste form acceptance evaluations.

The results from these PCTs with the seven individual samples, along with the statistical derivation of the average and standard deviation (Table II), also provide for a meaningful evaluation of the consistency issues associated with the acceptance qualifications. The specification for borosilicate glass product consistency requires a demonstration of waste form production control by comparison of the corrosion behavior of waste form samples with the EA glass [13,14]. For acceptance of borosilicate glass, the mean concentrations of Li, Na, and B in the leachate after normalizing for the initial compositions in the glass must each be less than those of the EA benchmark glass. One acceptable method of demonstrating that the acceptance criteria is met, is to ensure that the mean PCT result for each radionuclide element is at least two standard deviations below the mean PCT results of the EA glass.

While the salt dissolution and preferential exchange release mechanisms for Li and Na from the ceramic waste form are significantly different from that of borosilicate glass [3,4], a direct comparison between the release of Li, Na, Si, and B from the ceramic waste form with the EA glass (Table II, Figure 2) indicates that the ceramic waste form meets the borosilicate glass acceptance criteria with regards to consistency for the release of Na, Si, and B. The release of Li from the ceramic waste form is on average less than from the EA benchmark, but with these samples the Li release is within 2 standard deviations from the EA benchmark. Since the preferential release of Li from the ceramic waste form is significantly different from glass [3-12] and improvements in the overall durability of the ceramic waste form is being investigated, these results should only be used as a guide for further waste form development efforts.

Table II. Average leachate NL_i results and st. dev. for 7 day, 2000 m^{-1} (h-20000 m^{-1}) PCTs in 90°C DIW, EJ-13, and Mg-Na-K-Cl brine for the ceramic waste form and other waste glasses in DIW

	7PGZD	PGZD	PGZDd	PGZDh	PGZEJ	PGZB	EA [13]	EA [15]	57 Glass	NCAW [15]	SRL-202U [14]	SRL-131A [14] EJ-13	SRL-165F [16]
Cl	3.1 ±0.5	3.7 ±0.3	3.0	3.3	3.6								
Li	3.9 ±0.6	2.9 ±0.06	2.5	2.7	2.9	3.5	4.8±0.4	4.3		0.74	0.41	1.0	
Na	1.5 ±0.3	1.4 ±0.02	9.6E-1	1.2	1.1		6.7±0.5	7.3	1.8	0.77	0.42	0.95	
K	1.9 ±0.8	1.9 ±0.03	1.5	1.6	1.8				0.17		0.28	0.56	
Cs	0.60 ±0.2	0.40 ±0.01	3.3E-1	2.7E-1	3.8E-1	3.4				0.05			
Ca	0.062 ±0.03	0.067 ±0.002	4.6E-2	1.7E-1	2.4E-2	1.6			0.31		0.02		0.008
Sr	0.015 ±0.005	0.0062 ±0.0003	6.2E-3	6.3E-3	>0.001	3.1E-1			0.10				
Ba	0.0083 ±0.002	0.0038 ±0.0003	3.5E-3	1.2E-3	2.0E-3	3.0E-2			0.04		0.01		
Nd	0.003 ±0.002	0.0008 ±0.0002	8.9E-4	3.1E-5	4.0E-4	2.3E-1							
Ce	0.005 ±0.003	0.0015 ±0.0003	1.9E-3	7.3E-5	5.9E-4	2.0E-1							
La	0.006 ±0.002	0.0019 ±0.0004	2.3E-3	1.0E-4	8.6E-4	2.8E-1							
Y	0.010 ±0.004	0.009	4.4E-1	8.7E-2									
Al	0.064 ±0.009	0.051 ±0.003	5.6E-2	1.2E-3	1.7E-4	<.0001			0.01		0.13	0.042	0.10
B	0.38±0.1	0.24 ±0.03	2.2E-1	6.5E-2	3.9E-1	6.4E-1	8.3±0.6	10.0	3.4	0.96	0.39	1.1	0.46
Si	0.041 ±0.009	0.030 ±0.002	3.1E-2	2.0E-3	8.6E-3	5.1E-2	2.0±0.2	2.0	0.02	0.27	0.26	0.30	0.20
pH	9.0	9.1	9.1	8.9	9.3	5.6	11.8	11.8	8.8	10.2	10.0	10.3	10.2

7PGZD: 2000 m^{-1} PCTs with the 7 different ceramic waste form samples in DIW.
PGZD: 2000 m^{-1} PCTs with the combined ceramic waste form samples in DIW.
PGZDd: 2000 m^{-1} PCTs with the DIW washed ceramic waste form material in DIW.
PGZDh: 20000 m^{-1} PCTs with the combined ceramic waste form samples in DIW.
PGZEJ: 2000 m^{-1} PCTs with the combined ceramic waste form samples in EJ-13.
PGZB: 2000 m^{-1} PCTs with the combined ceramic waste form samples in Mg-Na-K-Cl brine.

The limitations on sample availability required modification of the test procedure by the use of only 0.5 g (5 g) of crushed material per 2000 m^{-1} (20000 m^{-1}) PCT. Furthermore, to eliminate the sample-to-sample variability observed in the above tests, crushed material from all 7 samples was thoroughly mixed together to provide enough homogeneous material for the remainder of the PCTs. This mixture of samples will be designated as the "combined" ceramic waste form material. Three PCTs (designated PGZDd in Table II) used crushed ceramic waste form material that was rinsed according to the ASTM PCT standard with DIW and ethanol. The remaining PCTs used material rinsed only in ethanol. The leachate results are reported in Table II and Figure 2.

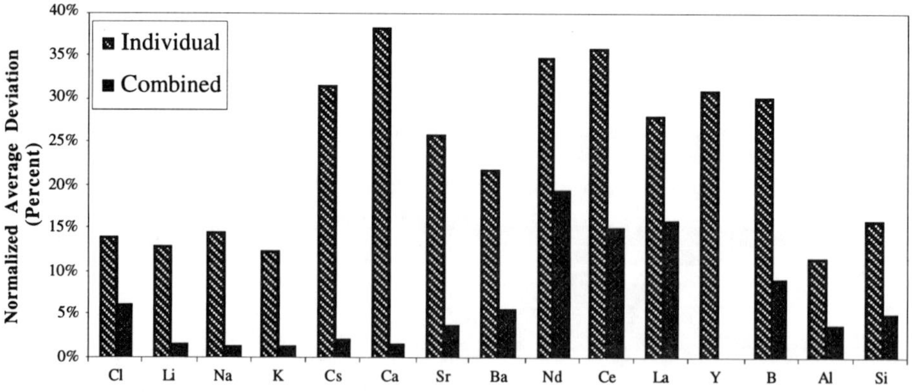

Figure 1. Normalized average deviations of leachate results from ceramic waste form samples in 2000 m^{-1} 7 day PCTs with 90°C DIW.

The relatively high alkali metals and chloride release from the ceramic waste form (Table II) would suggest that interactions between the leachant and the zeolites located at the surface of the ceramic waste form material may significantly influence the acceptance test results. Comparison of the average results reported in the PGZDd (DIW washed) column of Table II with similar tests that used material washed with only ethanol indicates that approximately 1.4% of additional salt from the total crushed ceramic waste form material was removed during the DIW washing process. This corresponds to a 20% decrease in normalized release numbers. These results indicate a high degree of alkali metal release from the zeolites located at the surface of the crushed sample particles in DIW and reinforce the premise that use of aqueous solvents should be limited or eliminated in the sample preparation procedures.

The normalized average deviations associated with the results from the PCTs with the combined ceramic waste form samples in DIW varied from 1 to 9% for all of the elements except the REs (Figure 1). The relatively low leachate concentrations and low initial compositions of the RE elements in the ceramic waste form result in large uncertainties that create 15 to 20% analytical deviations. An element-by-element comparison indicates that the deviations from the PCTs with the combined ceramic waste form samples are lower than the deviations observed with the PCTs that used the seven different samples separately and indicates that the analytical and systematic errors inherent in these type of tests is below 10% for elements that are present at or above the level of quantitation in the leachates. These data suggest that at least ±10% of the statistical uncertainties with the individual sample PCTs may be attributed to sample-to-sample variations in release behavior.

On average, the release of Cs, Li, Sr, Ba, Nd, Ce, La, and B from the PCTs with the combined ceramic waste form samples was lower than the average of the PCTs with the seven individual samples. This suggests that sample-to-sample variation may significantly affect consistency issues related to the testing of the ceramic waste form (Figure 2).

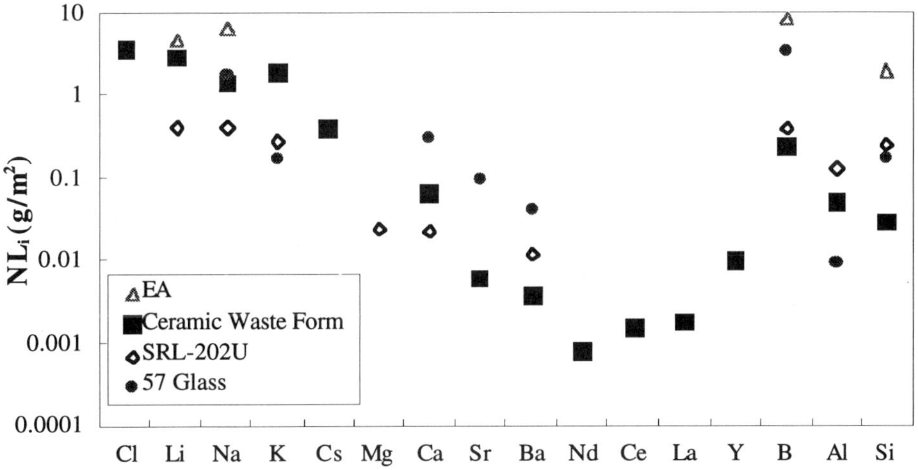

Figure 2. Comparison of leachate NL_i results from the ceramic waste form, EA glass, 57 glass, and SRL-202U glass in 2000 m^{-1}, 7 day PCTs with 90°C DIW. The standard deviation associated with each data point is within the symbol size, except for the RE data from the ceramic waste form, which have errors between 15 and 20%.

Table II contains similar data for other waste glasses that may be compared to the data from the PCTs with the ceramic waste form. A direct comparison between NL_i leachate results for the ceramic waste form and the EA glass [13], SRL-202U [14] glass, and the pure "57" glass matrix material used in the ceramic waste form is illustrated in Figure 2. The alkali metals and chlorides were preferentially released from the ceramic waste form, as compared to the Si and Ca dissolution of the zeolite framework ions or the B dissolution of the glass (Table II, Figure 2). The RE, Sr, and Ba simulated fission products were preferentially retained in solid phases in the PCTs with DIW or EJ-13 (Table II, Figures 2 and 3). The Sr and Ba fission product retention by the ceramic waste form was as good or better than that of waste glasses like SRL-202U and the pure 57 glass (Table II, Figure 2). This result and the observed adsorption of Sr and RE ions from the EJ-13 leachant into the remaining solids indicate that the zeolitic component of the waste form is retaining most of the higher-valence fission products as well as actively absorbing components from the leachants (Figure 3). However, enhanced dissolution of Ba, Sr, and RE elements from the ceramic waste form was observed in the PCTs with the Mg-Na-K-Cl brine (Table II, Figure 3). The increased release of Ba, Sr, and RE ions in the presence of high-ionic-strength fluids may be important should brine-like fluids contact the waste form in the repository [17].

These observations are all consistent with previously reported results from similar tests that used only the salt-loaded zeolites without the glass matrix binder [3-7]. The release and adsorption of specific élements to and from the leachate was explained by the preferential exchange of ions between the zeolite and the leachant [3,4]. In this previous work with the salt-loaded zeolites [3-7], the release order was correlated to ionic radius and charge, with a dependence on the solubility and the intrinsic zeolite adsorption preference of the individual elements.

Analysis of the residual solid material indicated that relatively minor compositional and structural changes occurred in the crushed ceramic waste form material during these PCTs [4]. The only notable differences were the incorporation of Mg from the brine leachant into the ceramic waste form and the presence of particles much smaller than 75 μm in the remaining solid material. These small particles may be related to the fracturing that was observed with the monoliths used in MCC-1 and vapor hydration tests [3,4], due to the release of alteration products, and or due to colloid agglomeration. The similar compositions of particles trapped on a 0.45 μm filter and an alteration phase observed on the solid material after the PCTs suggests that the small material may be related to alteration product formation.

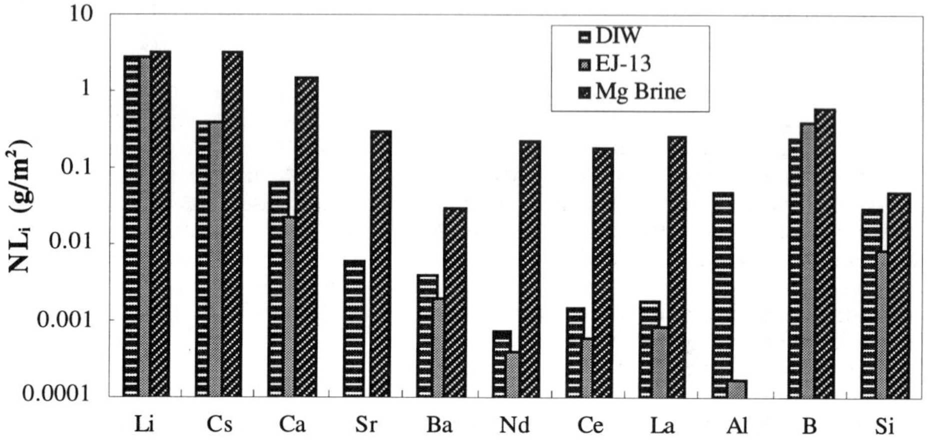

Figure 3. Comparison of leachate NL_i results from the ceramic waste form in 2000 m^{-1} 7 day PCTs with 90°C DIW, EJ-13, and Mg-Na-K-Cl brine. The K, Na, and Cl results from the brine leachant tests are not reported due to matrix interferences.

Colloid formation was identified through the observation of 0.5 to 5 μm size filter residue particulates and light scattering measurements of the PCT leachates that used DIW and EJ-13 leachants, but not in the Mg-Na-K-Cl brine leachate. The formation of colloids may be an important issue related to the transport of relatively insoluble radionuclides in a repository environment. Compositional as well as geometrical characterization will be required to address some of these radionuclide transport issues. The lack of colloids in the PCTs with the Mg-Na-K-Cl brine leachant is consistent with the effect of brines discharging the zeta potential of the colloids and causing particle agglomeration, producing relatively large particulates that settle out of solution quickly.

Except for the preferential adsorption of Ba and Sr from the EJ-13 leachant and the overall slower ceramic waste form dissolution, the results from the PCTs with EJ-13 leachant are similar to those with DIW. Compared to the results from PCTs with DIW, the lower Al and higher Sr, Ba, and RE leachate concentrations from the PCTs with the Mg-Na-K-Cl brine indicate an enhanced formation of alteration phases and preferential ion release, respectively (Table II, Figure 3). The increase in glass matrix dissolution in the brine leachants, as indicated by boron dissolution, does not completely account for the increase in fission product release. These results indicate that the higher ionic strength of the brine leachant modified the ion retention conditions in the zeolites (Figure 3).

CONCLUSIONS

The ceramic waste form has a two-component structure of salt-loaded zeolite in a glass matrix that incorporates a relatively large amount of diverse fission products and waste materials associated with DOE nuclear fuel conditioning activities. The potential utility of PCTs for elucidating the criteria for acceptance of the ceramic waste form into the DOE Civilian Radioactive Waste Management System was evaluated in this study. The two main advantages of performing PCTs with ceramic waste form samples is the quantitative measure of elemental release, which may be directly compared with results from other waste forms, and the ability to measure the consistency of the samples being produced. The crushing process did not disproportionately separate the ceramic waste form phases and should not adversely affect the PCT results. The results from these initial tests indicated that the glass and zeolite matrix dissolution was significantly lower than the preferential release of alkali earth and chloride elements. The Sr, Ba, and RE fission products had a lower release than the matrix dissolution of Si, B, or Ca.

The univalent ions are preferentially released from the zeolite relative to the divalent and trivalent ions in DIW and EJ-13 leachant. The order of cation release from the salt-loaded zeolites is affected by the leachant solubility and zeolite retention preference of specific ions. Thus, the addition of specific ions to the leachant may influence the ion exchange properties of the zeolite. This was most evident for samples tested in the Mg-Na-K-Cl brine. The high-ionic-strength brines changed the ion retention conditions of the zeolites such that the release of the alkaline earth and the RE ions was dramatically increased and Mg was preferentially ion exchanged in the zeolites.

These results indicate that tests in both DIW, which affects the dissolution of the glass matrix and zeolite framework of the waste form, and other leachants, which specifically affect the ion exchange behavior of the radionuclides, are required to fully assess the durability of the ceramic waste form. Furthermore, to avoid premature dissolution of alkali metal and chloride ions, the use of water in sample preparation should be limited. Ethanol rinses followed by room temperature drying was found to be an adequate cleaning process that minimized salt transport and removal from the ceramic waste form material.

The trends observed in these experiments suggest that the zeolite in the ceramic waste form should preferentially retain divalent and trivalent fission products, except in the presence of a brine, where retention of these ions may be more dependent on the leach resistance of the glass matrix. The univalent ion retention is dependent to a large extent on the glass matrix durability in the presence of all leachants. Thus tests like the PCT that artificially expose a greater fraction of the zeolites to the leachant and allow solution interactions with more of the interior surfaces of the zeolites disproportionately affect the release of components with a high affinity for ion exchange compared to monolithic tests that have a small fraction of zeolites exposed to the leachant.

ACKNOWLEDGMENTS

This work was supported by the U.S. Department of Energy, Nuclear Energy Research and Development Program, under Contract No. W-31-109-Eng-38. Valuable discussions with Drs. J. K. Bates, D. M. Strachan, M. Lewis, C. C. Pereira, and J. P. Ackerman are greatly appreciated. Valuable technical support was provided by M. T. Surchik, Dr. J. S. Luo, L. D. Hafenrichter, D. L. Bowers, T. DiSanto, J. W. Emery, and J. C. Hoh.

REFERENCES

[1]C. C. McPheeters, J. P. Ackerman, G. K. Johnson, "Electrometallurgical Treatment of Spent Nuclear Fuels for Repository Disposal," *1st Seminar on Molten Salts in Nuclear Technologies*, Dimitrovgrad, Russia, June 19-23 (1995).

[2]D. P. Abraham, J. W. Richardson Jr., and S. M. McDeavitt, "Formation of the $Fe_{23}Zr_6$ Phase in an Fe-Zr Alloy," accepted by *Scripta Materialia* (1997).

[3]L. J. Simpson and D. Wronkiewicz, "Evaluation of Standard Durability Tests Towards the Qualification Process for the Ceramic Waste Form," *Mat. Res. Soc. Symp. Proc.* **465** (1997).

[4]L. J. Simpson, Argonne National Laboratory, personal communication (1997).

[5]M. A. Lewis, D. F. Fischer, and L. J. Smith, "Salt-Occluded Zeolites as an Immobilization Matrix for Chloride Waste Salt," *J. Am. Ceram. Soc.* **76**(11), 2826 (1993).

[6]M. A. Lewis, D. F. Fischer, and C. D. Murphy, "Properties of Glass-Bonded Zeolite Monoliths," *Ceramic Trans.* **45**, 277 (1994).

[7]M. A. Lewis, Argonne National Laboratory, personal communication (1997).

[8]U.S. Department of Energy, Office of Civilian Radioactive Waste Management, "Waste Acceptance System Requirements Document," Report No. DOE/RW-0351, Rev. 2, 1996.

[9]G. B. Mellinger, "Standardized Test Methods for use in Waste Compliance Testing in the Department of Energy's Waste Acceptance Process," *Mat. Res. Soc. Symp. Proc.* **84**, 483 (1987).

[10]C. M. Jantzen, N. E. Bibler, D. C. Beam, W. G. Ramsey, and B. J. Waters, "Nuclear Waste Glass Product Consistency Test (PCT) version 5.0(T)," Westinghouse Savannah River Co. Report WSRC-TC-90-539, Rev. 2 (1992). *American Society of Testing and Materials*, "Annual Book of ASTM Standards" **12.01**, C1285-94 (1996).

[11]*American Society of Testing and Materials*, "Annual Book of ASTM Standards" **12.01**, C1220-92 (1996).

[12]"High-Level Waste Borosilicate Glass, A Compendium of Corrosion Characteristics (Vol. 2)", DOE Office of Waste Management, DOE-EM-0177 (1993).

[13]C. M. Jantzen, N. E. Bibler, D. C. Beam, C. L. Crawford, and M. A. Pickett, "Characterization of the Defense Waste Processing Facility (DWPF) Environmental Assessment (EA) Glass Standard Reference Material (U)," Westinghouse Savannah River Co. Report WSRC-TC-92-346, Rev. 1 (1993).

[14]W. L. Ebert, "The Effect of the Glass Surface Area/Solution Volume Ratio on Glass Corrosion: A Critical Review," Argonne National Laboratory Report ANL-94/34 (1995).

[15]P. R. Hrma, G. F. Piepel, M. J. Schweiger, D. E. Smith, D. S. Kim, P. E. Redgate, J. D. Vienna, C. A. LoPresti, D. B. Simpson, D. K. Peeler, and M. H. Langowski, "Property/Composition Relationships for Hanford High-Level Waste Glasses Melting at 1150°C," Pacific Northwest Laboratory Report No. 10359, Vol. 2 (1994).

[16]D. M. Strachan, "Leach Testing and Evaluation of WV-205 Glass from a Repository Perspective," Appendix in L. R. Brunnell report on "Laboratory Work in Support of West Valley Glass Development," Pacific Northwest Laboratory, Report Number PNL-6539/UC-510 (1988).

[17]J. C. Walton, "Effects of Evaporation and Solute Concentration on Presence and Composition of Water in and Around the Waste Package at Yucca Mountain," *Waste Man.* **13**, 293 (1993).

CHEMICAL DURABILITY OF A Nd DOPED OXYNITRIDE GLASS

G. LETURCQ*[(1)], G. BERGER[(2)], T. ADVOCAT[(1)], F. DELAGE[(1)], J. ROCHERULLE[(3)] and E. VERNAZ[(1)]

(1)Commissariat à l'Energie Atomique (CEA) Centre d'Etudes de la Vallée du Rhône, DCC/DRDD, Service de Confinement des Déchets, BP 171, F-30207 Bagnols-sur-Cèze Cedex, France

(2)Université Paul Sabatier UMR 5563, 38 rue des 36 Ponts, F-31400 Toulouse, France

(3)Université de Rennes 1, Laboratoire de Chimie des Matériaux, Campus Beaulieu, F-35042 Rennes, France

ABSTRACT

The chemical durability of a SiAlON glass highly doped with Nd (53.8 oxide wt%) was investigated experimentally between 90 and 200°C under flow-through conditions using deionized water. Although the proportions of the major glass network formers, Si and Al, were near those found in tholeitic basalt glass, the initial dissolution rates in pure water were up to two orders of magnitude lower than those observed for tholeites, in which the initial alteration kinetics are controlled by hydrolysis of the Si-O and Al-O bonds, and were dependent on the experimental conditions. Initial rates comparable to those of the tholeites were obtained using a 10^{-1} M sodium sulfate solution (a Nd complexing agent). These results indicate that in the absence of a complexant, an insoluble surface complex (< 100 nm) forms almost immediately, resulting in passivation. Additional experiments were carried out in static media at a high (20000 m^{-1}) SA/V ratio for three months at 90°C to simulate the long-term behavior of the matrix; the alteration rate measured in a confined environment was closed to those of the tholeites.

INTRODUCTION

This study is part of a research curriculum on specific containment of actinides (Am, Np, Cm, Th ...) coming from nuclear spent fuel reprocessing. Many studies on alteration of aluminosilicate glasses have allowed to define a range of composition for glasses able to contain all radioelements in nuclear waste (for example the french nuclear waste form : R7T7 glass). Assuming that nitrogen addition increases significantly the mechanical properties of aluminosilicate glasses such as Vickers hardness and Young's modulus (ROCHERULLE *et al.*, 1988 [1]), it seems interesting to study the ability of this anion to potentially improve the chemical durability. Neodynium is used in this study to simulate the trivalent actinides. We conducted two types of experiments: dynamic leach tests to determine initial dissolution rates at three temperatures, 90, 150 and 200°C and the activation energy; static leach tests during three months to simulate long term behavior : saturation state and long-term alteration rate.

METHODS

Starting Glass

Starting materials are commercial products AlN(Starck), MgO(Rhône-poulenc), Al_2O_3 (Johnson Matthey), SiO_2(Quartz et Silice), Y_2O_3(Rhône-poulenc). The glass was prepared in a dry box by heating up to 1750K the powders inside of a molybdenum crucible under a nitrogen atmosphere to prevent oxidation. Then, the melts were poured into a molybdenum mould at room temperature. The glass composition is given in table I.

Table I. Chemical composition of SiAlON glass (analyzed wt %)

SiO_2	Al_2O_3	Nd_2O_3	N	Σ
32.3	11.4	53.8	1.7	99.2

The glass was crushed and sized (125-250 μm and 40-50 μm) and cleaned ultrasonically to eliminate the ultrafine particles produced by the grinding. The specific surface area of those fractions were measured using the BET method with krypton adsorbate gas and were found to be 0.028 and 0.094 m^2/g respectively.

Dynamic Leach Tests.

Dynamic leach tests were conducted in a mixed flow reactor. Solutions (pure water and Na_2SO_4 $10^{-1}M$) were continuously injected into a titanium reactor, containing the immersed glass powder specimen, and were maintained at a fixed temperature. The solution was homogenized by stirring. A regulator maintained the pressure at 10.3 MPa in the reactor by means of a leakage flow to compensate for the inflowing solution under flow rates ranging from 1 to 9 ml/min.

Static Leach Tests.

For the static leaching experiments, a powder specimen was placed in Teflon® containers with 5 ml of pure water to obtain a glass-surface-area-to-leaching-solution-volume ratio of 200 cm^{-1}. The test containers were then placed in flasks with a small quantity of water to minimize leakage by balancing the pressures inside and outside the containers. The flasks were then placed on a stirring device in a temperature-regulated chamber at 90 ± 0.1°C during one week to three months.

Analytical measurements :

The nitrogen content of the glass have been measured with an O_2-N_2 analyzer (LECO -USA). Leachates have been analyzed by colorimetry (Technicon Analyzer II), emission/absorption spectrometry (Perkin-Elmer Zeemann 5000), ICP-MS (Perkin-Elmer Elan 5000). pH of leachates have been measured at 25°C using a method that minimized atmospheric CO_2-leachate exchanges. Altered surfaces have been observed by XR diffraction (Phillips PW 1130), SEM (JEOL 6400) with EDS/WDS, STEM (Philips 120kV) with EDS and XPS (ESCALAB NK2).

RESULTS AND DISCUSSION

Dynamic leach test

Dissolution rates, measured in pure water at 200°C, were not reproducible. So, many further tests were conducted at this temperature to allow a better understanding of this variation. The lowest values at 200°C, about 0.05 g(glass)/m^2/day, were up to two orders of magnitude lower than those observed for basaltic and nuclear R7T7 glasses (BERGER et al., 1994 [2]; DELAGE, 1992 [3]). On the other hand, higher dissolution rates are obtained with higher flow rates and lower surface area. Figure I, shows a linear dependence between the dissolution rate at 200°C and a factor quantifying both the undersaturation of the leaching medium (the flow-rate-to-surface-area ratio) and the possibility of creating a concentration gradient at the reaction interface (the stirring speed). The higher this factor, the greater the dilution of the solution at the reaction interface, and thus the closer we approach the conditions to measure the dissolution rate constants. This dependence suggests that SiAlON glass dissolution at 200°C in pure water is controlled by the diffusion of aqueous species near the reaction interface. Other published studies (GRAMBOW, 1985 [4] ; BERGER, 1994 [2]) on the dissolution of tholeitic and R7T7 glasses in similar condition have suggested that the rate is depending on the silica concentration of leachate. In this scheme, low concentrations in the bulk solution (< 2 ppm Si) compared to the apparent "solubility" of the other silicate glasses, require a strong chemical gradient near the reaction interface.

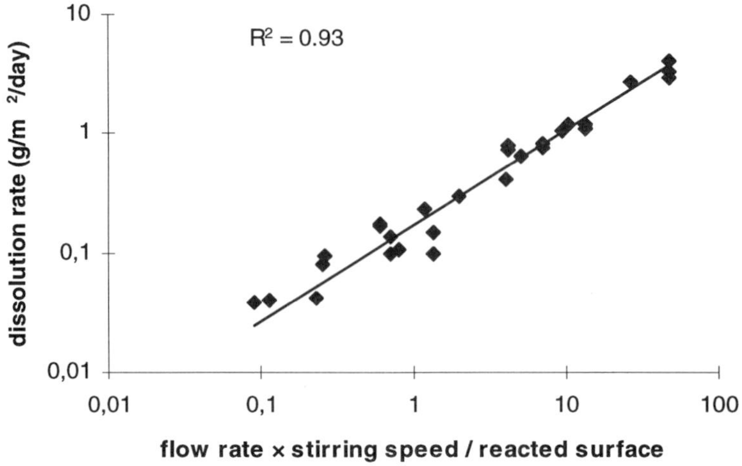

Figure I : SiAlON glass dissolution rate in pure water at 200°C versus experimental conditions.

Amorphous layer:

Table II. XPS analysis

	Pristine glass	Alteration layer (200°C)
Al/Si	0.27	0.35
Nd/Si	2.79	8.73

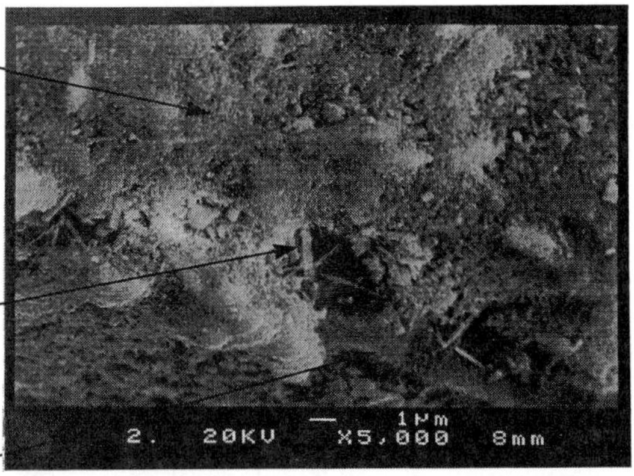

Figure II. SEM observations on SiAlON glass altered in renewed pure water at 200°C for 161 hours.

SEM observations of altered grains revealed a thick alteration layer (Figure II), shown by XPS analysis to be enriched in aluminum and neodymium, which could cause such a gradient. X-ray diffraction analysis indicated that the needles are hydrated neodymium carbonate crystals (ICDD : 27-1296).

In order to avoid the precipitation of neodynium carbonates, we altered SiAlON glass in decarbonated water, but we again obtained the same results as before. If a gradient exists, it is thus not caused by the neodymium carbonate layer. Another experiment using a glass in which the nitrogen was replaced by lithium resulted in the same rates as for the nitride glass under the same experimental conditions. Nitrogen is not the cause either. The most likely reaction which may control the dissolution rate of SiAlON glass seems to be related to the presence of neodymium in the glass. The development of an Nd-enriched and lowly reactive superficial skin could explain the low measured rates by a "passivation" effect. This assumption was tested by performing experiments with a sodium sulfate solution, which forms complexes with the rare earth elements. The observed dissolution rates under these conditions were 1) much higher, 2) reproducible and 3) very near to those measured for other aluminosilicate glasses, as shown on Figure III.

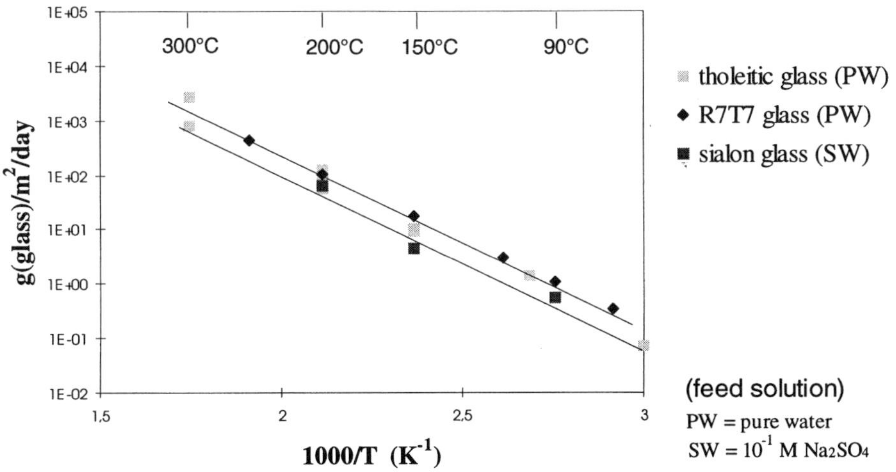

Figure III. Arrhenius plot of initial dissolution rate of SiAlON glass compared with tholeitic and R7T7 glass.

The differences between our data and the other published values may be attributed to the different procedures for determining the reaction surface area: in the present study we used the BET areas, whereas the points shown for the other two glass compositions correspond to geometric surface areas and are therefore underestimated. However the rate activation energy is the same (60 kJ/mol), suggesting that the initial SiAlON glass dissolution is controlled by the same reactions as those controlling hydrolysis of basalt glass and R7T7 glass: the rupture of Si–O and Al–O bonds. The hydrolysis of Nd-O bonds could produce a protective residual layer (Nd-OH ?) when Nd is not complexed by an appropriate solution.

Static leach test

The tests were conducted for periods of about a hundred days. The solutions very quickly reached a steady state concentration in silica about 50 ppm (Figure IV). Nitrogen increases with time and neodynium is more scattered. Three of the leachate samples, marked "UF", were ultrafiltered (10 000 D): in each case, the neodymium concentration dropped considerably, showing that it was present in particle form.

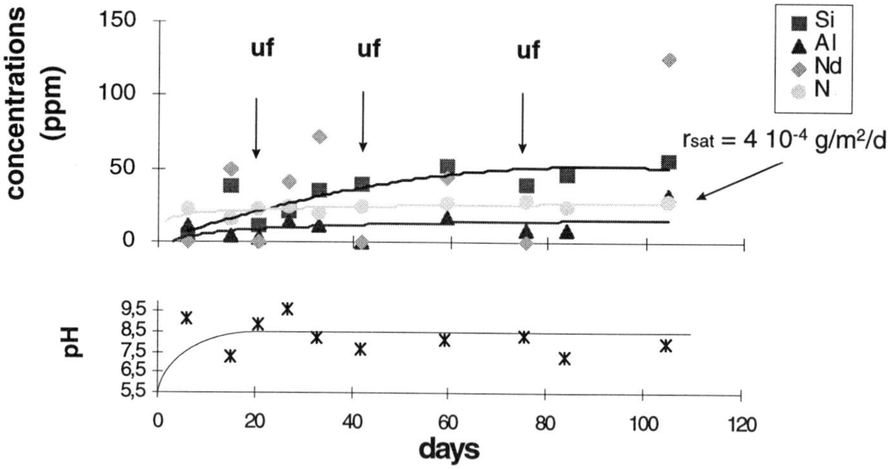

Figure IV. Alteration in closed medium at 90°C in initially pure water

Using nitrogen as a corrosion tracer, we determined that the dissolution rate at saturation was 4×10^{-4} g(glass)/m^2/day, comparable to the rate observed with other aluminosilicate glass compositions. The pH increased initially as a result of nitrogen release in the form of proton-consuming ammonia in solution. These low rates are generally due to the development of a protective alteration layer on the glass surface. In our case, this layer is difficult to observe even by scanning transmission electron microscopy. If it exists, it is less than 50 nm thick. However, the surface area of the glass powder increased by a factor of 3 after alteration and XPS measurements show that altered surface is enriched in Al and Si, suggesting the existence of such a layer.

If there was an alteration layer, it was less than 50 nm thick.

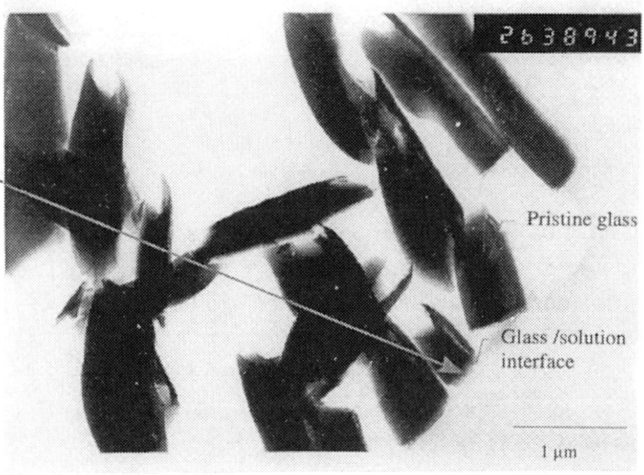

Figure V. STEM observations on SiAlON glass altered in initially pure water at 90°C, in closed medium.

CONCLUSIONS

The alteration of a SiAlON glass highly doped with Nd (53.8 oxide wt%) was investigated experimentally between 90 and 200°C. The nitrogen concentration tested in the glass had no observable effect on the chemical durability of the aluminosilicate glass. In pure renewed water, a protective effect was observed that may be a sort of neodymium passivation. After neutralizing this phenomenon, by complexing neodymium by sulfates, we observed practically the same alteration rates as in basalt glass and nuclear R7T7 glass whose hydrolysis was controlled by the rupture of Si-O and Al-O bonds. Finally, the long-term behavior of this glass seems to be the same as basaltic glasses.

REFERENCES

[1]J. Rocherullé, P. Verdier and Y. Laurent, "Preparation and Properties of Gadolinium Oxide and Oxynitride Glasses", *Materials Science and Engineering*, **B2** 265-268 (1989).

[2] G. Berger, C. Claparols, C. Guy and V. Daux, "Dissolution rate of a basalt glass in silica-rich solutions: Implications for long-term alteration", *Geochimica et* Cosmochimica *Acta*, **58** [2] 4875-86 (1994).

[3] F. Delage "Etude de la fonction cinétique de dissolution d'un verre nucléaire" *PhD thesis, Université des Sciences et Techniques du Languedoc, Montpellier, France,* 178p. (1992).

[4] B. Grambow "A general rate equation for nuclear waste glass corrosion" *Mat. Res. Soc. Symp. Proc.* Scientific Basis for Nuclear Waste Management VIII, **44** 15-21 (1985).

THERMODYNAMIC APPROACH TO GLASS ETCHING BY HF

M. J. Plodinec
Diagnostic Instrumentation and Analysis Laboratory
Mississippi State University
Post Office Drawer MM
Mississippi State, MS 39762-5932

ABSTRACT

In theory, the hydration thermodynamics approach used for control of glass production in the Defense Waste Processing Facility should be applicable to virtually any situation in which glass is allowed to react with an aqueous medium. In this paper, the general approach has been modified for application to glass etching by aqueous HF. Using measurements made by Spierings, who etched a wide variety of silicate and borosilicate compositions in 2.9 M HF, excellent agreement is found between the theory and the experimental measurements.

INTRODUCTION

The Department of Energy (DoE) currently has 130 million liters of high-level radioactive waste (HLW) in storage at the Savannah River Site. Late in 1996, the onerous task of converting the highly mobile alkaline HLW slurries to a stable borosilicate glass was begun in the Defense Waste Processing Facility (DWPF).

The borosilicate glass product must satisfy a set of specifications intended to ensure suitability for acceptance into DoE's system for transportation, storage and disposal of HLW.[1] The most important of these requires that HLW glass must be consistently less reactive with water than a benchmark glass, with high confidence, using a standard leachability test.[2]

To ensure a consistent product (consistently better than the benchmark glass), waste glass producers need a tool with which to control the production process. In the DWPF, the most important process variable is the composition of the batch being fed to the melter. The chemical composition of each batch varies, resulting in variations in the chemical composition of the glass. This implies that the process control tool must be able to correlate glass composition with the results of the standard leach test. An approach based on hydration thermodynamics has proven useful for control of HLW glass production in the DWPF. In this approach, the glass composition is used to calculate a free energy of hydration of the glass, using the reactions expected between water and the glass. The free energies of hydration of a large set of waste glasses have been calculated, and correlated with the results of the standard leach test. This correlation is currently being used to ensure the consistency of the waste glass product.

This same general approach has been applied to the results of a wide variety of leach tests in aqueous media over a wide range of aqueous conditions in the pH range 6-14. The approach has been used to explain the effects of radiation, glass oxidation state, and pressure on the reaction between glass and water. These results have shown that the same general approach can be used to explain both the results of short-term leach tests and the results of burial tests in various media extending over several years. However, the approach has not previously been applied to acidic media, nor to reactive species (e.g., HF) in solution. In addition, most of the previous work has concentrated on nuclear waste glasses.

Because of its basis in fundamental thermodynamics, the approach should be applicable to virtually any situation in which glass is allowed to react with an aqueous medium, as long as the set of reactions used accurately reflects the interactions between the glass and the aqueous solution. The objective of the work reported here was to verify that, in fact, this approach could be applied to the interactions between glasses and an acidic medium containing a reactive component. The general hydration thermodynamics approach was modified by developing a set of reactions specific to the etching of glass by aqueous HF. These were then used to calculate the free energies of reaction for a set of glasses for which the etching rates in 2.9M HF had previously been determined.[3] None of the glasses tested were waste glasses, but they represented a rather wide compositional range, thus offering a rigorous test of the approach. If successful in explaining this type of reaction, the general usefulness of the approach would be further validated.

HYDRATION THERMODYNAMICS APPROACH

A model of glass durability based on hydration thermodynamics was originally developed by Newton and Paul to explain the hydration behavior of antique glasses in such diverse environments as medieval churches and clay burial sites.[4,5,6] In its simplest terms, the interaction between glass and an aqueous solution can be thought of as a binary reaction:

$$\text{Glass} + \text{Solution} \rightarrow \text{Aqueous product species}$$

If we define the progress of the reaction at some time, t, to be ξ, then

$$\xi = \int_0^t k \cdot a_{glass}(t) \cdot a_{water}(t) \cdot dt \qquad (1)$$

where "k" is the rate constant, a_{glass} indicates the activity of glass (assumed to be proportional to the surface area), and a_{water} denotes the activities of water and other species in the water (e.g., HF). If we assume that the activities of the glass and solution are constant over the period of the test, then equation (1) implies that k is directly proportional to ξ (Note that the assumption of constant activity of the glass, while reasonable, and sufficient for the following development, is probably not absolutely necessary. As will be noted later, this assumption can introduce error into the correlation.). In the case of the data to be examined later, ξ is the measured depth of penetration of the etchant solution into the glass surface.

It is generally assumed that any chemical reaction proceeds through an intermediate state, called the activated complex. This complex may be an association of the reactants (in this case, glass and solution), or it may more closely resemble either the reactants or the products. As a useful approximation, it is generally assumed that the reactants are in a pseudo-equilibrium with this activated complex, which then "decays" to the products. The rate constant is then treated as the equilibrium constant for the pseudo-equilibrium. Thus, the rate constant, k, depends on the free energy of the pseudo-equilibrium ($\Delta G_{activation}$), according to:

$$k \propto \exp\left(-\frac{\Delta G_{activation}}{RT}\right) \tag{2}$$

where T is the temperature, R is the gas constant, and $\Delta G_{activation} \approx$ G(activated complex) – G(reacting species).

In the case of reactions between glass and a solution, it is reasonable to assume that the activated complex is very similar to the final products. Based on the pseudo-equilibrium,

$$\Delta G_{activation} \approx G(\text{product species}) - G(\text{glass}) - G(\text{solution}) \tag{3}$$

However, the quantity on the right hand side of the equation is simply the free energy of reaction of the glass. Since the reaction progress is directly proportional to the rate constant, equations (2) and (3) imply that the logarithm of the reaction progress is proportional to the free energy of reaction of the glass. Conversely, these equations suggest that a much weaker relationship should exist between the activation energy of the reaction (e.g., determined by an Arrhenius plot) and the free energy of reaction. This is because the energy of activation is an enthalpic term only, and does not include the entropy of activation. Thus, the strength of the correlation between the free energy of reaction and the activation energy can be viewed as a sort of measure of the relative importance of the enthalpic and the entropic components of the free energy.

In order to calculate the free energy of reaction, the glass is treated as an ideal mixture (in the thermodynamic sense) of simple components. The free energy of the glass, G(glass), is represented by

$$G(glass) = \sum_i n_i \cdot \mu_i$$

where n_i are the number of moles of species "i" which react, and μ_i is the free energy of formation of species "i".

The free energy of the product species, G(product species), is represented by

$$G(product\ \ species) = \sum_j n_j \cdot \mu_{p,j}$$

where n_j are the number of moles of the product species "j" which are produced by the glass-solution reaction, and $\mu_{p,j}$ is the free energy of formation of the product form of species "j".

The free energy of reaction of component "i" in the glass, ΔG_i (i.e., the free energy of the reaction between the component and water), is thus assumed to be the sum of the free energies of formation of the pure component "i" and of the solution subtracted from the free energy of the product species containing "i," all in their standard states.

$$\Delta G_i = \mu_{p,i} - \mu_i - \mu_{solution}$$

The stability of the glass toward reaction with the solution is then evaluated as the sum of these free energies of reaction, each weighted by the amount of that oxide in the glass (in mol%), n_i. In terms of the measure of reaction progress (penetration into the glass surface),

$$\log \xi \approx a_1 + a_2 \cdot \sum_i n_i \cdot \Delta G_i \tag{4}$$

where the a-values are constants.

Thus, for a set of glasses and a given type of leach test (i.e., constant reaction conditions), a correlation between the results of the leach test and glass composition can be obtained. The correlation should be applicable to any glass, with the following limitations:

- The activity of each of the glasses used to develop the correlation must either remain constant over the time of reaction, or at least must vary in the same manner. Otherwise, the correlation may be significantly in error. Thus, if the activity of the glass is proportional to the surface area, then care must be taken to ensure that the surface area is constant for each of the glasses used to develop the correlation.
- The activity of the solution must remain constant, or at least must vary in a consistent manner, for the set of glasses used to develop the correlation. In tests in deionized water, for example, some glasses may release so much hydroxide to solution that it cannot be ignored as a reactant. If this does not happen for all glasses of a given set, then significant error may result (However, if the change in solution activity with time is known, then the set of ΔG_i's can be modified to take this change into account.).
- The set of components (both for the glass and the solution) used in developing the correlation must accurately reflect the reactions actually taking place. As an extreme case, a correlation for borosilicate glasses cannot generally be used for phosphate glasses.
- The methods used to determine the glass composition (which is used to calculate the free energy) must be consistent. Different laboratories usually have different capabilities in this regard. This has been a significant source of error in previous attempts to apply the approach from one laboratory to another.
- The glass must not contain second phases (either crystalline or as the result of amorphous phase separation). The approach is predicated on the assumption that the measured glass composition represents the glass phase which is actually being leached. It is well-established that the presence of either crystalline material or

amorphous second phases can affect the reactivity of glass by changing the composition of the material actually reacting.

• The reaction must be carried out for all the glasses in the same manner, and the reaction progress estimated in the same way.

SPIERINGS' EXPERIMENTS

Spierings has examined the effects of glass composition on the etching of multicomponent silicate glasses by HF solutions.[3] The compositions of the glasses tested (based on the as-batched compositions) are given in Table I. A chromium metal mask was placed on each glass surface, which exposed part, but not all, of the glass surface. These samples were exposed to 500 mL of 2.9 M HF at temperatures from 1 to 50°C, and for times varying from 120 to 900 s. The depth of the etched region was determined by profilometry. Besides facilitating measurement, the mask also tended to ensure that the glass surface area was maintained constant.

Spierings found that the depth of the etching was directly proportional to time, indicating that the activities of the glass and the solution were constant over the course of the experiment. For glasses 3 and 21, a slight initial surface reaction evidently took place, because the depth vs time curves were slightly offset from the origin. For all of the other glasses, the depth vs time lines passed through the origin.

For all of the glasses, Spierings calculated the rate of reaction by plotting the depth of etching vs. time, and calculating the slope. Thus, the rates calculated by Spierings are directly proportional to the extent of reaction, implying that the log (calculated rate) should also be proportional to the free energy of reaction of the glass, by equation (4).

In order to determine the free energy of the reaction, a "basis set" of reactions are required. These were chosen in the following manner.

• Both water and HF were considered as reactants. This means that at least two sets of reactions were considered for each glass component.

• For Na_2O, K_2O, CaO, MgO, and ZnO, reactions involving both the oxides and the monosilicate of each were considered. Thus, for these components, four different reactions were considered.

• For each of the reactants and products, standard free energy of formation data were used to calculate μ_i and $\mu_{p,i}$.[7]

• All of the possible combinations of reactants were used to calculate free energies of etching for the set of glasses.

• When the reactants included the simple silicates, the amount of silica tied up in each silicate (e.g., Na_2SiO_3) was subtracted from the total amount of silica to determine the contribution of silica to the free energy of etching, i.e., n_{SiO2} = mol% SiO_2 – mol% Na_2SiO_3.

• Each set of free energies was then correlated with Spierings' log (rate) data.

• The correlation coefficient and the standard error of the correlation were used as measures of how accurately each reaction set represented the etching reaction. The reaction set with the highest correlation coefficient and the lowest standard error is given in Table II. For this set, the following correlation was found:

$$\log(rate) = 0.289 - 0.202 \cdot \sum_i n_i \cdot \Delta G_i \qquad (5)$$

The correlation coefficient for equation (5) is 0.876, and the standard error of the fit is 0.1736. The data and the strong correlation between the log (rate) and ΔG(etching) are shown in Figure I.

Table I. Compositions of glasses tested by Spierings (in mol%)

Glass	SiO_2	Na_2O	Al_2O_3	B_2O_3	CaO	K_2O	MgO	ZnO
1	1	0	0	0	0	0	0	0
2	0.8	0.2	0	0	0	0	0	0
3	0.7	0.3	0	0	0	0	0	0
4	0.7	0.25	0	0	0.05	0	0	0
5	0.7	0.2	0	0	0.1	0	0	0
6	0.7	0.15	0	0	0.15	0	0	0
7	0.7	0.25	0.05	0	0	0	0	0
8	0.7	0.2	0.1	0	0	0	0	0
9	0.7	0.25	0	0.05	0	0	0	0
10	0.7	0.2	0	0.1	0	0	0	0
11	0.7	0.15	0	0.15	0	0	0	0
12	0.7	0.1	0	0.1	0.1	0	0	0
13	0.7	0.1	0.025	0.1	0.075	0	0	0
14	0.7	0.1	0.05	0.1	0.05	0	0	0
15	0.7	0.1	0.1	0.1	0	0	0	0
16	0.7	0.1	0	0.1	0	0	0	0.1
17	0.7	0.1	0.025	0.1	0	0	0	0.075
18	0.89	0.055	0	0.055	0	0	0	0
19	0.56	0.22	0	0.22	0	0	0	0
20	0.625	0.125	0.125	0	0	0	0	0.125
21	0.7	0	0	0	0.1	0.2	0	0
22	0.7	0.2	0	0	0	0	0.1	0
23	0.7	0.2	0	0	0	0	0	0.1

There are several interesting features of the reactions in Table II. Most importantly, the alkaline and alkaline earth cations are each paired with silica, as the monosilicates, producing the corresponding fluorides, and silicic acid. The set of reactions for the oxides going to the same products gave a correlation coefficient of 0.676, with an error of 0.2804. This significantly poorer representation of the data is primarily due to the alkalis. If the only modification to the best reaction set is use of Na_2O or K_2O instead of the corresponding silicate, then the correlation coefficient drops below 0.4 and 0.6, respectively. In the case of the alkaline earths, use of the oxides instead of the silicates coarsens the correlation only a small amount. For all of these, treating the etching reaction as simple hydrolysis (going to the hydrated cation) rather than leading to fluoride formation greatly worsens the correlation.

Table II. Reaction set which gave the best correlation with the data. Free energy values calculated from those of the reactants using the data from reference 3.

REACTANTS		PRODUCTS		$\Delta G°$ (kcal/mol)
Al_2O_3	$6H^+$	$2Al^{3+}$	$3H_2O$	-23.68
B_2O_3	$3H_2O$	$2H_3BO_3$		-10.47
$CaSiO_3$	2HF	CaF_2	H_2SiO_3	-22.66
$MgSiO_3$	2HF	MgF_2	H_2SiO_3	-19.90
K_2SiO_3	2HF	2KF	H_2SiO_3	-34.79
Na_2SiO_3	2HF	2NaF	H_2SiO_3	-21.79
ZnO	$2H^+$	Zn^{++}	H_2O	-15.76
SiO_2	H_2O	H_2SiO_3		4.05

Silica and B_2O_3 are also interesting. For both of these, the hydrolysis reactions produced much better fits than any others. Use of any other reactions which produced either silicon or boron fluorides or ionic species greatly worsened the correlation ($R^2 < 0.4$). This was especially true for boron ($R^2 \approx 0$).

For both alumina and ZnO the hydrolysis reactions produced slightly better fits to the data. The correlation coefficient for the reaction producing AlF_3 was 0.749, and the standard error was 0.2464. For the reaction producing ZnF_2, $R^2 = 0.790$, and the standard error was 0.2249. The results were virtually the same for both $ZnSiO_3$ and ZnO as reactants. Since there was a slightly better correlation when ZnO was assumed to be the reactive Zn species, it is shown in Table II.

Using the measured rates as a function of temperature, Spierings also calculated energies of activation for each glass. These varied from 35 to 40 kcal/mol, for all of the glasses except the two which had an initial surface reaction. As can be seen in Figure II, there is virtually no correlation between the free energies calculated using Table II, and the activation energy for the reaction (It should be noted that the

Figure I. Correlation between calculated free energy of reaction and rate of etching.

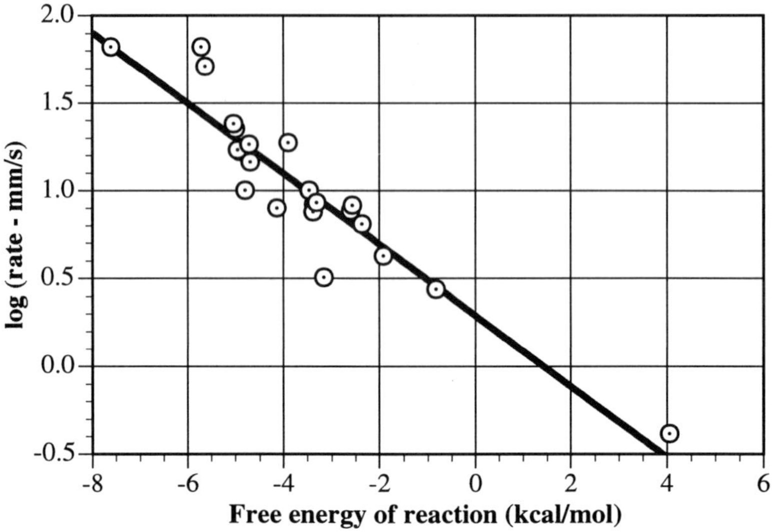

Figure II. Correlation between calculated free energy of reaction and energy of activation.

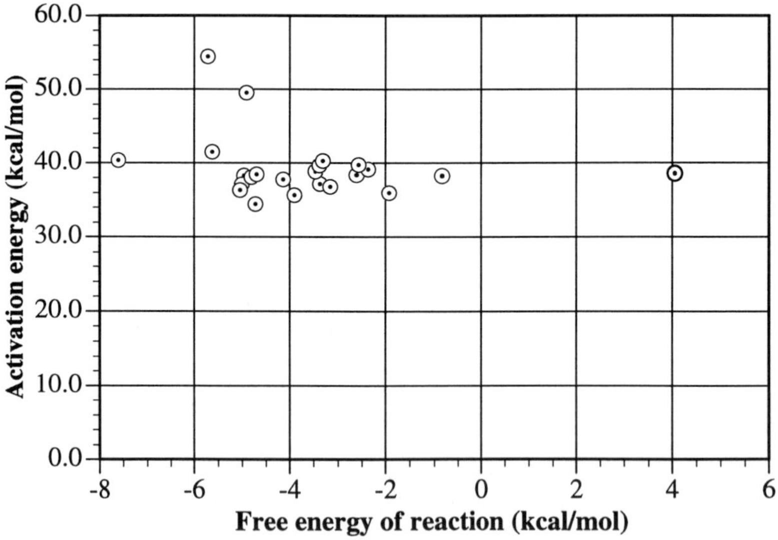

two points signficantly above the others in Figure II are those which underwent an initial surface reaction - glasses 3 and 21). As noted above, this probably indicates the relative importance of the entropic term to the overall free energy of activation.

CONCLUSIONS

The key parameter in the thermodynamic approach is the free energy of reaction. Theoretically, the logarithm of the extent of reaction at constant time for a given set of glasses should be proportional to the free energy of reaction. For Spierings' data, the log(reaction progress) is directly proportional to ΔG(etching), $R^2 = 0.876$. In this case, the calculated rates ranged over more than two orders of magnitude. Conversely, the energy of activation for the reaction is virtually independent of the free energy of reaction.

The excellent agreement between theory and experiment clearly demonstrates that this simple thermodynamic approach can be generalized to include acidic media with reactive solution constituents. A key to its successful application is clear identification of the "basis set" of reactions expected to occur, and investigation of alternatives. In this case, it appears that treating the glass reactant as a mixture of simple silicates, boria, and silica gives better results than treating the entire glass as a mixture of oxides. In addition, the data indicate that fluoride is the active component in the aqueous reactant, for most of the glass components, but apparently not for boron oxide or silica.

REFERENCES

1. Office of Environmental Management, *Waste Acceptance Product Specifications for Vitrified High-Level Waste Forms, Revision 1,* USDOE Document EM-WAPS, U. S. Department of Energy, Germantown, MD (1995).

2. C. M. Jantzen, N. E. Bibler, D. C. Beam, W. G. Ramsey, B. J. Waters, *Nuclear Waste Product Consistency Test - Version 7.0,* USDOE Document WSRC-TR-90-539, Revision 3, Savannah River Technology Center, Aiken SC (1994).

3. G. A. C. M. Spierings, "Compositional effects in the dissolution of multicomponent silicate glasses in aqueous HF solution," *Journal of Materials Science*, **26**, 3329-36 (1991).

4. A. Paul, *J. Materials Science, 12,* 2246 (1977).

5. R. G. Newton, A. Paul, *Glass Technology, 21,* 307 (1980).

6. A. Paul, *The Chemistry of Glasses,* Chapman and Hall, chapter 4 (1984).

7. Donald D. Wagman, et al., *The NBS tables of chemical thermodynamic properties,* American Chemical Society (1982).

Actinide Stabilization II

MICROSTRUCTURE AND PHASE FORMATION IN A 17 WEIGHT PERCENT PLUTONIUM OXIDE DEVITRIFIED WASTE GLASS

Mitchell K. Meyer, Stephen G. Johnson, Thomas P. O'Holleran, and Steven M. Frank
Argonne National Laboratory-West
P.O. Box 2528
Idaho Falls, ID 83404

ABSTRACT

Plutonium containing ceramic waste forms have been prepared by dry mixing and melting glass frit, simulated zirconia high level waste calcine from chemical reprocessing, samarium oxide, titanium metal, and plutonium oxide. Materials were produced using melt times of 4 and 12 hours at 1450°C followed by a thermal anneal at 500°C. Complex materials with a substantial volume fraction of crystalline phases were the result. The principle plutonium bearing phase was identified as a fluorite structured plutonium-zirconium-samarium phase of variable stoichiometry. This high plutonium phase preferentially segregated to the bottom of the waste form. A waste form was also melted using metallic plutonium in a quantity equivalent to 15 wt% plutonium oxide. XRD results indicate that the metal was completely oxidized on melting.

INTRODUCTION

Dissolution of plutonium in a silicate glass spiked with fission products sufficient to provide a lethal radiation dose is a long-term disposition option recommended by the National Academy of Sciences.[1] This approach puts weapons grade material in a form similar to spent nuclear fuel.

The objective of this project is to explore the feasibility of preparing a plutonium bearing waste form containing calcined high level waste from reprocessing, a matrix glass, and a neutron poison. One possible method of preparing such a waste form is by dry mixing and crucible melting of the constituent materials. The melting step is followed by a devitrification heat treatment to produce a product with a large volume fraction of crystalline phases. This method provides the possibility of partitioning plutonium and the neutron poison into durable crystalline host phases.

EXPERIMENTAL

Powdered glass frit, surrogate zirconia calcine, samarium oxide, and titanium metal were dry mixed with either powdered plutonium oxide, or no plutonium. The compositions of the glass frit and surrogate calcine are given elsewhere.[2,3] The waste form batch and net composition are given in Tables I and II. The batch was melted in an alumina crucible in air at 1450°C for 4 hours, then cooled at 350°C/hr to 500°C and held for 4 hours. The resulting product was sampled by core drilling. The remainder of the sample was remelted at 1450°C for 12 hours and annealed using the same cycle. Another melt was made by dry mixing the ceramic components with millimeter size and smaller pieces of metallic plutonium in an amount equivalent to 15 weight percent plutonium oxide. This batch was heated to 1050°C for 48 hours, furnace cooled to 500°C and held for 4 hours to promote nucleation, then ramped to 700°C at 60°C/hr. The sample was then held at 700°C for 140 hours. Samples for analysis were obtained by core drilling vertically through the melt, then axially sectioning and polishing samples. Waste form microstructures were studied using SEM/EDS and XRD. The plutonium bearing samples provided some analytical difficulties in determining quantitative compositions by EDS. EDS results, although reproducible, should not be considered quantitative due to a high background at low energy, peak broadening, and energy shift due to emission of radiation from the samples. Fluorine and oxygen cannot be detected with the EDS spectrometer used in analyzing the plutonium bearing samples, making differentiation of fluoride and oxide phases difficult. EDS data were analyzed assuming that the elements were present as oxides to allow for an absorption correction more representative of a nonmetallic matrix. EDS data are thus presented in terms of weight percent of cation, with the balance of the analysis being oxygen. A batch of the calcine, frit, samarium, and titanium (sans plutonium) was prepared and melted at 1450°C for 4 hours in order to study phase formation, particularly partitioning of the fluoride anions, in a non-radioactive sample.

Table I. Waste form batch

Component	Wt%
Zirconia Calcine	51.9
127 Glass Frit[2]	20.8
Sm_2O_3	6.2
Ti (as metal)	4.1
PuO_2	17

Table II. Net composition of PuO_2 waste form

Component	Wt%	Component	Wt%
CaF_2	17.7	Ti (metal)	4.1
PuO_2	17.0	Na_2O	2.8
SiO_2	14.7	B_2O_3	1.9
CaO	13.4	Li_2O	1.3
ZrO_2	11.8	CuO	0.4
Al_2O_3	7.5	SrO	0.4
Sm_2O_3	6.2	Ni,Cr,Cs,Fe oxide	0.8

WASTE FORM MATRIX

A low magnification image of the microstructure of the waste form (sans plutonium) is shown as Figure 1. The melt shows distinct bands of coloration, changing from black in the lower region of the melt to off white at the top. Melt depth was 20 mm. Metallic copper, iron, and tin were detected in the waste form by SEM/EDS 3-4 mm above the bottom of the crucible (the tin is present as an impurity). The presence of these elements in the metallic state indicates that oxygen fugacity is low in the lower region of the melt and that redox reactions with more stable species are occuring. Higher magnification images of typical areas in the upper and lower regions of the melt are shown in Figure 2. It is clear from these micrographs that stratification of phases has occurred. The upper region of the melt is rich in corundum and relatively zirconium poor. The lower region of the melt is rich in zirconia. The approximate composition of the sample at locations shown in Figure 2 was measured by EDS analysis and is given in Table III. X-ray diffraction patterns taken from the top and bottom regions of the melt indicate that although amounts differ, the major phases present in the two regions are the same: CaF_2, monoclinic zirconia, a cubic zirconia solid solution, and corundum. In addition, there are several unidentified reflections that may correspond with the Al-Ti-O and Zr-Ti-Sm-Al-Ca-O phases shown in Figure 2. EDS analysis indicates that zirconia occurs in solutions with titania, and calcia and forms in a spherical morphology in the lower sample region. Near the top of the sample the zirconium rich phase has a dendritic morphology and occurs with samaria, alumina, calcia and titania. CaF_2 was the only fluorinated phase present, indicating that CaF_2 is stable in

Figure 1. Low magnification image of undoped waste form showing stratification.

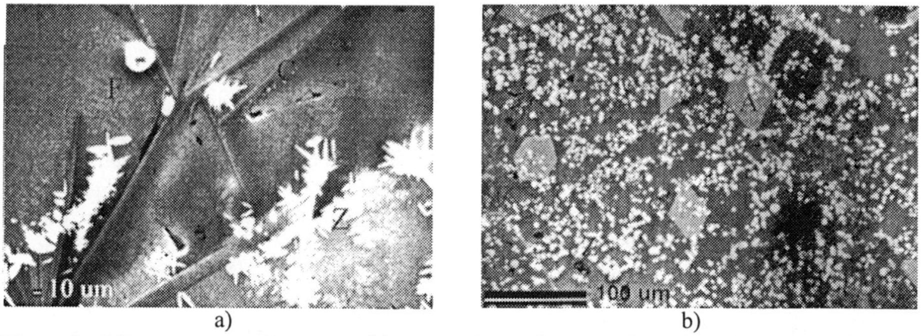

Figure 2. Microstructure of upper and lower regions of waste glass sans plutonium. a) upper region is corundum (C) rich. Bright dendritic phase is zirconia (Z) and Zr-Ti-Sm-Al-Ca-O phase of unknown structure. CaF_2 (F) is present as darker dendritic phase. b) Lower region is zirconia rich and contains an Al-Ti-O (A) phase of unknown crystal structure.

contact with the oxide melt under these processing conditions.

Table III. Approximate composition of undoped waste form in regions shown in Figure 2. Balance O and F.

Location	Al	Si	Ca	Ti	Zr	Sm
Top (2a)	37.8	28.0	19.6	4.1	6.4	4.1
Bottom (2b)	30.4	17.7	16.8	5.0	30.2	1

PLUTONIUM OXIDE WASTE FORM

After heating at 1450°C for 4 hours, the product was complex and multi-phase. Twelve different phases (+ matrix phase) were identified by SEM/EDS contrast and chemical analysis, and are listed in Table IV. The matrix, presumably glassy, has variable composition. Some of these phases may be variations in chemistry among minerals with ranges of solid solution. Phases 1 and 2 are the only phases that contain plutonium in detectable quantities. Phase 1 is a high plutonium content phase that also contains zirconium, samarium, and calcium. The plutonium content of this phase ranges from 20-50 wt%. The morphology is roughly spherical. PuO_2 was not identified by SEM/EDS as a phase in the waste form. The absence of PuO_2 was confirmed by XRD. Instead, a series of four fluorite structured phases was identified in the diffraction pattern, with lattice parameters ranging from 5.11 to 5.31 Å. The lattice parameter for ZrO_2, PuO_2, and Sm_2O_3 are 5.09, 5.40, and 10.93 Å respectively.[4] These phases likely represent solid solutions of the three oxides with varying stoichiometry.

Table IV. Approximate chemical analysis of phases (wt%). Balance O and F.

Phase	Al	Si	Ca	Ti	Zr	Sm	Pu
1	-	-	3-6	<1	20-44	5-8	23-52
2	2.8	-	8.1	16.3	27.3	11.1	5.7
3	15.9	18.7	11.7	<1	5.2	4.8	-
4	21.7	17.8	6.1	<1	7.6	1.5	-
5	18.5	19	8.2	<1	8.6	<1	-
6	35.4	8.6	2.2	1	6.1	1.3	-
7	36.5	6.9	2.7	1	5.6	2.8	-
8	15.5	12.7	11.6	3.1	8.4	9.3	-
9	24.5	15.6	6.4	<1	7.6	-	-
10	28.1	10.9	6.4	1	7.8	2.1	-
11	-	-	100	-	-	-	-
12	13.5	11.4	27.7	1.2	4.7	2.7	-

*Phases 6,7 and 3,8 have similar cation analysis but differ in BSE contrast.

Phase 2 is a lower plutonium content phase containing about 6 wt% plutonium. In contrast to the high plutonium phase, this phase has a fixed composition. Phase 2 has a high samarium content, about 11 wt%. Other elements present are zirconium, titanium, calcium, and aluminum. Table V gives the bulk chemistry of the core sample at top, center, and bottom locations. (Note that this analysis represents trends in composition only). SEM/EDS indicates that the dense plutonium-zirconium rich phase (Phase 1) is settles to the bottom of the crucible, causing the compositional gradient. Another phase clearly identified by XRD is CaF_2. As mentioned above, XRD patterns are complex and solubility ranges over a large extent for these phases. More refinement is required to gain a better understanding of the crystal structures of the phases in Table IV.

Table V. Approximate bulk chemistry of the sample (wt%). Balance O and F.

	Al	Si	Ca	Ti	Zr	Sm	Pu
Bottom	17.1	9.8	14.3	1.6	10.1	2.4	6.7
Center	17.5	10.9	14.3	1.8	8.8	3.6	4.1
Top	20.1	10.7	14.2	2.7	7.6	3.9	<1

Remelting

In order to study the formation of phases nearer to equilibrium conditions, the waste form was remelted for 12 hours at 1450°C. A low magnification micrograph of the sample is shown as Figure 3. There is a sharp transition between the upper and lower regions of the glass at roughly one half of the melt depth. The higher density plutonium rich phases settle to the bottom of the crucible. The lower, plutonium rich region of the melt contains large pores on the order of 0.1 to 1 mm in size. A plot of the EDS chemical analysis as a function of vertical height in the crucible is shown as Figure 4. The average plutonium oxide content of the sample differs by approximately a factor of six in traversing from the bottom to the top of the sample.

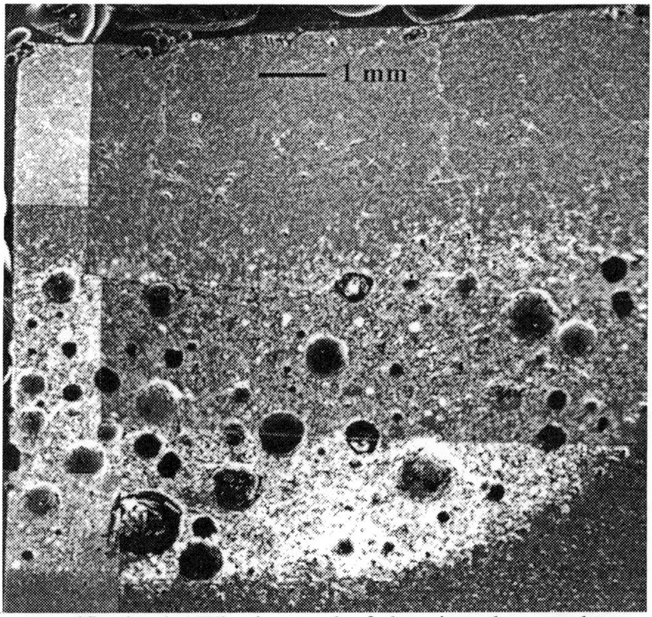

Figure 3. Low magnification (~15X) micrograph of plutonium glass remelt.

The sample is silicon and aluminum rich near the surface. An abrupt change in composition occurs near the transition zone. The sample contains three major phases in addition to the matrix. These phases are described in Table VI. (In addition, there were 3 minor phases amounting in total to less than 10 volume percent.) Plutonium is detectable only in phases R1 and R2. Phase R1 contains only plutonium, zirconium, and samarium and is found primarily near the bottom of the crucible, although it can also be found in much lesser volume above the transition zone. A micrograph of an area near the bottom of the crucible containing phases R1

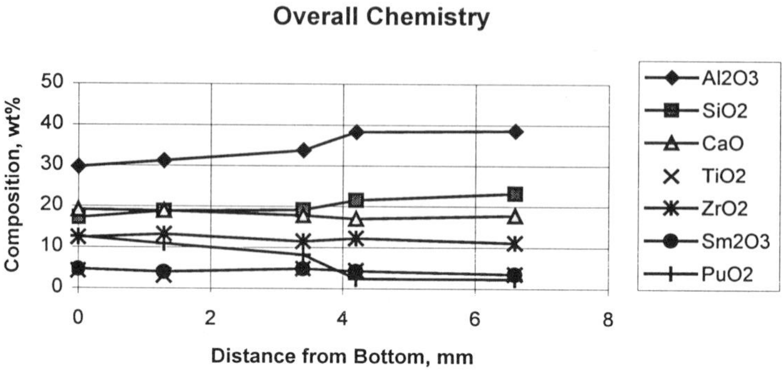

Figure 4. EDS analysis of sample as a function of location.

TableVI. Composition of major phases after remelting for 10 hours at 1450°C

Phase	Al	Si	Ca	Ti	Zr	Sm	Pu
R1	-	-	-	<1	19-40	5-12	27-60
R2	0-11	-	3-9	0-3	33-42	4-22	11-42
R3	6-12	0-6	16-18	13-22	0-3	13-21	-

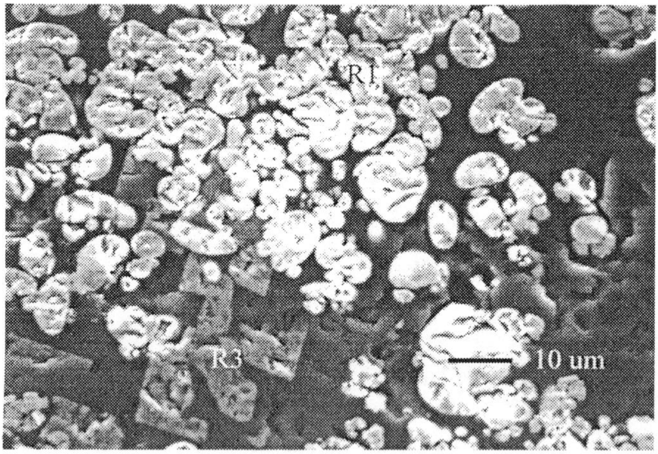

Figure 5. Micrograph of area near bottom of crucible containing phase R1 (bright) and phase R3 (gray) in a glassy matrix.

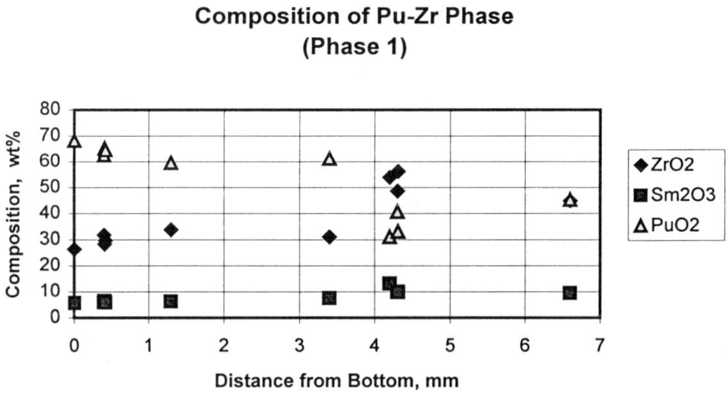

Figure 6. Composition of phase R1 as a function of distance from the bottom of the melt.

and R3 is shown in Figure 5. EDS spectra for phase R1 were taken as a function of location; analysis are shown in Figure 6. The plutonium content of phase R1 in general decreases with distance from the bottom of the crucible, although this decrease is not monotonic. A discontinuity appears near the transition zone, where the plutonium content of the phase decreases sharply, then rises again near the surface.

Phase R2 contains zirconium, plutonium, samarium, calcium, and sometimes aluminum and titanium. It is found above the transition zone near the top of the crucible, and was never identified below the transition zone. Silicon was the only major component present in the batch not identified in phase R2 in more than trace amounts. EDS analysis of phase R2 as a function of location shows a varying composition with no clear spatial dependence.

Phase R3 always contains aluminum, silicon, calcium, titanium, zirconium, and samarium. There is a sharp discontinuity in composition at the transition zone. In contrast to the first melt, the matrix glass is of nearly constant composition with respect to the cations analyzed. There is, however, a discontinuity in composition at the transition zone. Sm and Pu were not detected in the matrix glass.

Since the waste form segregated into two visibly distinct top and bottom layers, x-ray diffraction patterns were recorded separately on each layer. CaF_2 was not found by XRD or SEM/EDS in the remelted sample. The phase(s) in which the fluorine anions reside is unknown. The bottom layer was found to contain three major phases. Two fluorite structured phases with lattice parameters of 5.29 and 5.24 Å were identified. The phase with a_0=5.29 is presumably R1, the major phase in this layer. A phase that fit well to an orthorhombic $CaTiO_3$ type perovskite with a=5.37, b=7.63, and c=5.42 Å was also identified. The most notable difference between the top and bottom layers of the sample is the absence of strong reflections due to phase R1 in the top layer, which is consistent with SEM/EDS results.

WASTE FORM INCORPORATING METALLIC PLUTONIUM

After melting, the waste form was cored in an area containing an obvious heterogeneity. A low magnification image of a cross section of this core is displayed in Figure 7. The sample a large feature (phase M1) that appears by EDS to be nearly pure plutonium oxide. There are no peaks assignable to metallic plutonium in the x-ray diffraction pattern. Three other major

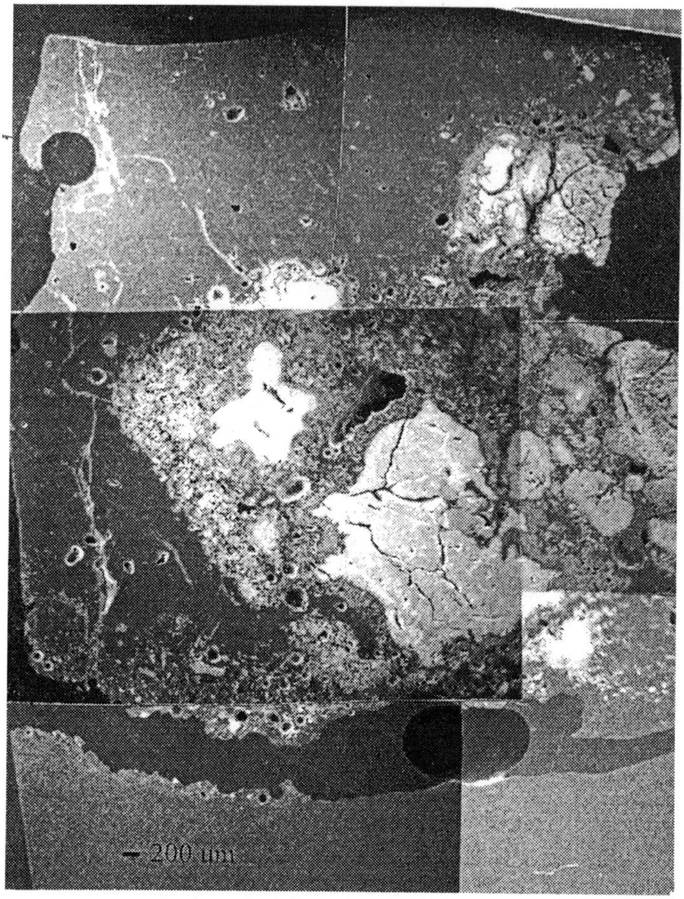

Figure 7. Low magnification image of a core sample from the plutonium metal/glass/calcine melt

phases were found, as described in Table VIII and shown in Figure 8. These phases and the matrix varied in composition with distance from the plutonium rich area. The bright blocky phase (M2), such as the triangular grain in the lower left of Figure 8, was the only phase (other than the large feature described above) found to contain plutonium. This phase is similar in composition to a fluorite structured Pu-Zr phase found in other melts. The bright 'fishbone' phase (M3) contains the majority of the samarium. The phase that appears light gray (M4) in Figure 8 is rich in calcium and silicon. Zirconium was sometimes found in this phase. The matrix was aluminum and silicon rich, and contained most of the sodium.

Table VII. EDS analysis of phases in waste from made with metallic plutonium

Phase	Na	Al	Si	Ca	Zr	Sm	Pu
M1						tr.	95+
M2					10-20		80-90
M3		tr.	20-33	13-18	9-12	39-48	
M4		tr	40-63	36-45	0-15		
M5			~20				~80
matrix	6-9	15-20	54-68	2-5	0-20		

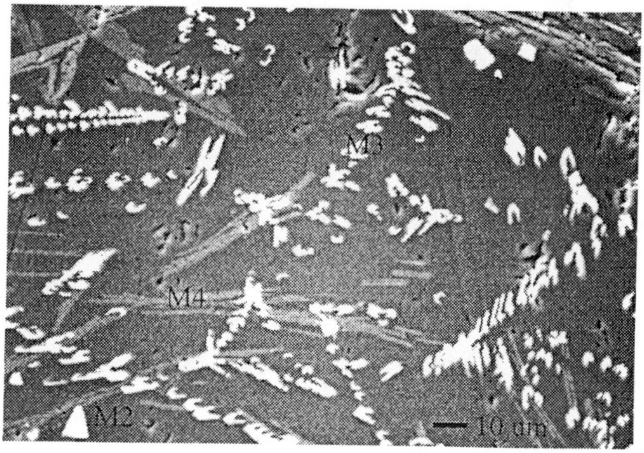

Figure 8. Microstructure of waste form made with plutonium metal away from M1, showing phases M2, M3 and M4.

Other interesting features in this melt were the formation of a Pu-Si phase (M5) at the interface between phase M1 and the matrix. This phase may be an intermediate in the dissolution of plutonium and subsequent precipitation of Pu-Zr phases, as it is not found at other locations away from the interface.

SUMMARY

A melting route was used in the synthesis of plutonium bearing waste forms made with simulated zirconia calcine, glass frit, titanium, and samarium. The process produces a waste form with durable[5] zirconate host phases and a high capacity for plutonium. The primary plutonium bearing phase was a zirconium-plutonium-samarium solid solution with the cubic fluorite structure. The stoichiometry of this phase varies with depth in the melt. Another plutonium bearing phase containing calcium, titanium, zirconium, aluminum, and samarium of unknown crystal structure was identified. Samarium is present as a neutron absorber in both phases.

It was found that the plutonium rich phases segregate to the lower half of the melt, and that this segregation becomes worse at longer melt times, introducing a compositional gradient into the waste form. Segregation is apparently due to gravitational settling of the dense phases in the melt. The segregation might be controlled by adjusting the melt viscosity through a change in processing temperature or by mechanical agitation of the melt.

When plutonium is introduced into the waste form in metallic form and processed at a temperature of 1050°C, the plutonium is completely oxidized. It appears that larger pieces of the plutonium oxide react slowly to form Pu-Zr phases under these conditions.

ACKNOWLEDGEMENTS

Argonne National Laboratory is operated for the U.S. Department of Energy by the University of Chicago. This work was supported by U.S. D.O.E. Reactor Systems, Development and Technology under contract W-31-109-ENG-38. The authors gratefully acknowledge E.L. Wood, M. Hansen, and D.A. Knecht for their contributions to this project.

---

[1] "Management and Disposition of Excess Weapons Plutonium," Committee on International Security and Arms Control, National Academy of Sciences, National Academy Press, 1994.

[2] B.A. Staples, H.S. Cole, and D.A. Pavlica, "Properties of Formula 127 Glass Prepared with Radioactive Zirconia Calcine," pp. 125-134 in *Scientific Basis for Waste Management II*, (Mater. Res. Soc. Symp. Proc **2**), Edited by V.G. Brookins. Materials Research Society, Pittsburgh, PA, 1979.

[3] T.P. O'Holleran, S.G. Johnson, S.M. Frank, M.K. Meyer, M. Noy, E.L. Wood, D.A. Knecht, K. Vinjamuri, and B.A. Staples, "Glass-Ceramic Waste Forms for Immobilizing Plutonium," to be published in in *Scientific Basis for Waste Management XX*, Edited by W. Gray and I. Triay. Materials Research Society, Pittsburgh, PA, 1997.

[4] JCPDS-ICDD cards 27-997, 6-360, and 43-1029. International Center for Diffraction Data, Swarthmore, PA.

[5] T.P. O'Holleran, S.G. Johnson, P.C. Kong, B.A. Staples, "Plutonium Host Phases Derived from High Level Waste at the Idaho Chemical Processing Plant," this proceedings.

DEVELOPMENT OF APATITE-BASED GLASS CERAMICS FOR ACTINIDE IMMOBILIZATION

Florence Bart, Vincent L'Hermite, Stéphane Houpert, Catherine Fillet, François Pacaud and Noël Jacquet-Francillon
Commissariat à l'Energie Atomique (CEA), Centre d'Etudes de la Vallée du Rhône, DCC/DRDD/SCD, BP 171, 30207 Bagnols-sur-Cèze, France

ABSTRACT

Crystallization of glasses was studied in the $MgO-CaO-P_2O_5-SiO_2$ system with 5 wt% Nd_2O_3. Depending upon the composition and the heat treatment, glass ceramics containing apatite and whitlockite were observed. Whitlockite phase was obtained on cooling samples containing MgO; on heating the glass, following a nucleation-growth scenario, apatite was crystallized. For $CaO-P_2O_5-SiO_2-Nd_2O_3$ glasses, the crystallization on cooling led to a non-stoichiometric apatitic structure. Nd partitioned equally between the glass and crystalline phases when apatite crystallized in $CaO-P_2O_5-SiO_2-Nd_2O_3$ glass ceramics. When MgO was added to the melt, the Nd concentration was higher in the crystals (apatite or whitlockite) than in the residual glass matrix.

INTRODUCTION

The use of glass based materials, principally borosilicate compositions, to immobilize high-level liquid waste is clearly established as the preferred waste form in most countries (France, UK, USA). However, an international research effort is underway to develop improved actinide waste forms, mainly in the field of crystalline materials[1]. Most of this work is devoted to a few crystalline families: silicates (e.g. zircon), titanates (e.g. sphene, perovskite), zirconates (e.g. zirconolite) and phosphates (monazite, apatite).

Another development is glass-ceramic materials for incorporation of the long-lived elements. This work is based on two ideas:

- from a material standpoint, glass ceramics provide a double barrier for radionuclides: the first is the crystalline phase in which the radionuclide is located; the second is the highly durable glassy phase in which crystals are embedded;
- from a technical standpoint, it is advantageous to use the well developed technology of glass fabrication.

The first studies of glass ceramic materials were performed in 1976 on the crystallization of celsian ($BaAl_2Si_2O_8$) in borosilicate melts[1]. Following this pioneering work, several laboratories have developed a variety of compositions[2]. In recent years, the Commissariat à l'Energie Atomique (CEA) has studied three kinds of glass crystalline phases: sphene, zirconolite[3] and apatite, considering their ability to incorporate the actinide elements.

The use of apatite as an actinide host phase has been proposed in the past[4]. Many studies have been carried out on radiation effects in silicated apatite such as $Ca_2Nd_8(SiO_4)_6O_2$[5] and in France on the development of a ceramic apatitic material for iodine containment[6,7]. In addition, Wronkiewicz *et al.*[8] recently developed some apatite- and monazite-bearing glass-crystal composites for the immobilization of low-level nuclear and hazardous wastes.

This paper discusses the synthesis and characterization of apatite-based glass ceramics in the MgO-CaO-SiO_2-P_2O_5-Nd_2O_3 system.

BRIEF STRUCTURAL DESCRIPTION OF APATITES

The apatite structure defines a large family of crystalline phases, made more complex by the large number of sites available for ionic substitutions.

Fluoroapatite $Ca_{10}(PO_4)_6F_2$ has hexagonal symmetry with the space group $P6_3/m$, where the unit cell contains two formula units. The structure consists of a column of fluorine ions on the hexagonal axis parallel to the c axis. The 10 Ca ions are distributed among two nonequivalent crystallographic sites:

- CaII (6h) are symmetrically located about the 6_3 axis (coordination 7)
- 4 CaI (4f) are also in the unit cell, surrounded by 3 oxygen atoms, one at the same height, one above and one below (coordination 9).

Various literature sources[9] indicate that apatite $A_{10}(XO_4)_6Y_2$ may accept a large variety of substitutions as follows

- A site : vacancy, Na, K, Cs, Ca, Mg, Sr, Pb, Zn, rare earth elements, Ce, actinides
- X site : Si, P, V, As
- Y site : vacancy, O, OH, F, Cl, I.

Replacement within this structure often employs a charge-balanced coupled substitution. For example, the substitution of Si^{4+} for P^{5+} may be compensated by substitution of REE^{3+} for Ca^{2+}. It is interesting to note that this structure can accept actinides and REE (Rare Earth Elements which are used in this study to simulate actinides).

FABRICATION AND CHARACTERIZATION OF THE INITIAL GLASSES

Sample Casting

Recent papers have explored the crystallization of MgO-CaO-SiO_2-P_2O_5 glasses, in the context of biomedical applications[10]. Depending on the composition, it has been shown that these glasses develop some apatite crystallization, after suitable additional heat treatment. Based on these observations, we studied some glasses having the eutectic composition: 33.3 wt% SiO_2, 18.3% P_2O_5, 37.5% CaO and 11.2% MgO, to which we added 5 or 10% of neodymium oxide.

The compositions are given in **Table I**. NSVA is the initial Non-Substituted Vitro-Apatite. SVA-Mg5 means 5 wt% Nd_2O_3 Substituted Vitro-Apatite. In SVA-Ca, all the MgO is replaced by CaO (on a molar basis).

Table I. Compositions of the precursor glasses (wt%)

Sample	SiO_2	P_2O_5	MgO	CaO	Nd_2O_3
NSVA	33.3	18.3	11.2	37.5	
SVA-Mg5	31.6	17.4	10.6	35.6	5.0
SVA-Mg10	29.4	16.2	9.9	32.8	11.7
SVA-Ca 5	30.0	16.5		47.6	5.8

Well-mixed powders containing suitable amounts of SiO_2, $NH_4H_2PO_4$, Nd_2O_3, CaO and MgO were melted in zirconia crucibles for 3 hours at 1450°C. The melts were then quenched in air. Except for the calcium-rich melt (SVA-Ca5), the initial violet glasses were clear, but SVA-Ca5 was partly crystallized in the bottom of the crucible. Crystallization of the glasses was conducted by isothermal treatments at temperatures determined from differential thermal analysis results.

Differential Thermal Analysis Results

DTA was performed on each sample, in order to determine the glass transition (Tg) and crystallization (Tc) temperatures on heating from the solid state, and the devitrification temperature (Td) on cooling from the liquid state

Heating and cooling were performed at 10°C/minute on finely ground samples. The curves corresponding to the SVA-Ca5 and SVA-Mg10 glasses are given **Figure 1**. Note the very similar shapes for each sample. The first endothermic phenomenon corresponds to the glass transition. The next exothermic peaks are related to crystallization phenomena (Tc). Finally the last endothermic peaks indicate that melting of crystallized species has occurred (Tm). On cooling the sample, in each case a single crystallization peak, designated Td, is observed. **Table II** summarizes the measured temperatures for all the studied glasses.

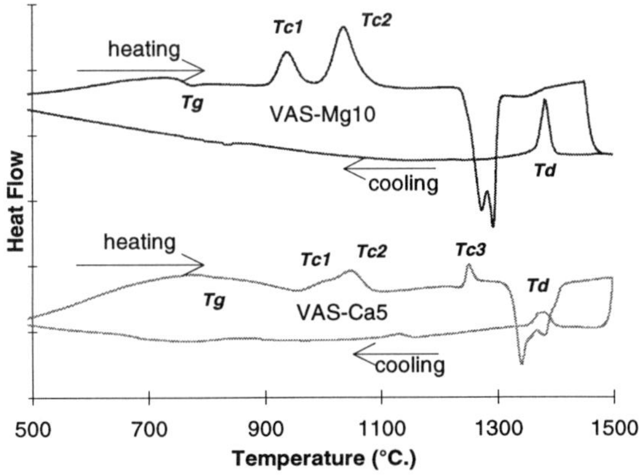

Figure 1. DTA curves for the initial glasses

Compared with the initial 5% Nd_2O_3 doped eutectic composition (SVA-Mg5), a larger quantity of CaO increased the glass transition temperature by about 50°C. The same trend was observed for crystallization and melting temperature. Another effect of the substitution of the whole quantity of MgO by CaO was the appearance of a third crystallization peak that was absent for SVA-Mg5 and 10.

Table II. DTA measured temperatures (°C) (n.m = not measured)

Sample	Tg	Tc_1	Tc_2	Tc3	Tm_1	Tm_2	Td
NSVA	748	955	970	X	1274	X	n.m.
SVA-Mg10	746	928	1034	X	1270	1290	1381
SVA-Mg5	742	942	972	X	1267	1290	n.m.
SVA-Ca 5	791	1006	1046	1250	1350	1378	1377

FABRICATION AND CHARACTERIZATION OF GLASS CERAMICS

According to previous studies[10], the first crystallization peak corresponds to the apatitic phase, and the second to the eutectic components, i.e. a mixture of β-$Ca_3(PO_4)$ (β-C3P) and diopside $CaMgSi_2O_6$ (CMS2). When it contains Mg, β-C3P is isostructural[11] with whitlockite $Ca_{18}Mg_2Ca_1\nabla_1(PO_4)_{14}$ (∇=vacancy). In order to compare the crystallization trends, we used different heat treatments for each sample, as indicated in **Table III**.

Table III. Heat treatments used to obtain glass ceramics
(TSVA: Treated Substituted Vitro-Apatite)

Sample	Scenario	Temperature	Time
TSVA-Mg5	heating	792°C	2 hours
		942°C	2 hours
TSVA-Mg10	cooling	1300°C	2 hours
TSVA-Ca 5	cooling	1380°C	2 hours

After polishing, the treated samples were analyzed using a scanning electronic microscope (JEOL JSM 6300), equipped with an EDS system (Oxford ISIS), and with a Cameca electron probe. Additionally, the crystalline phases were analyzed in the powders, using a SEIFERT X-ray diffraction apparatus.

Magnesium Glass Ceramics (TSVA-Mg5 and TSVA-Mg10)

TSVA-Mg5: Nucleation and Growth Process:

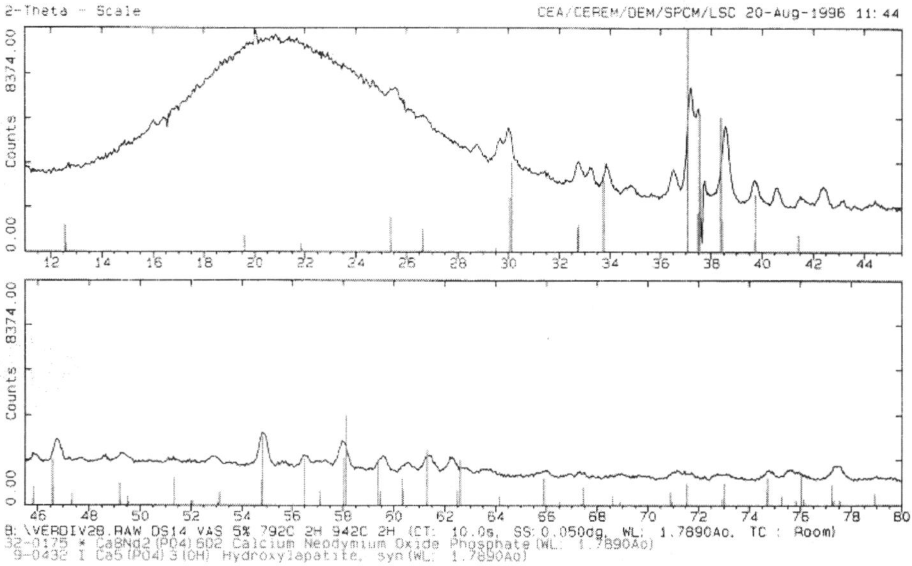

Figure 3. X-ray diffraction pattern of TSVA-Mg5

Figure 2 shows a micrograph of TSVA-Mg5, as observed with backscattered electron detection. The X-ray diffraction pattern (**Figure 3**) establishes that the main crystalline phase belongs to the apatite family. The clearly hexagonal morphology of the crystals is very close to the apatite shape observed without neodymium[12].

The size and morphology of crystals did not allow us to perform a quantitative analysis of a single crystal; however, it can be seen in **Figure 2** that Nd is more concentrated in the crystals than in the residual glass matrix, since the crystals are whiter than the glass matrix. The glassy phase was observed to comprise mainly Si, Mg, and the crystalline phase contained mainly Ca, P.

Figure 2. Micrography of TSVA-Mg5 (backscattered electrons)

TSVA-Mg10: Devitrification during Cooling:

Figure 4 shows the morphology of TSVA-Mg10 (SEM). The crystals are quite different from previous sample, forming elongated, irregular needles, and the typical hexagonal shape is not observed. An X-ray partition map of P (Kα) and Nd (Lα) reveals that the crystals contain P, Nd, Ca (little Si or Mg) and the glass matrix is enriched in Si, Mg, Ca with little P or Nd (**Figure 4**). Nevertheless, a more detailed quantitative analysis of both phases is necessary to describe their composition.

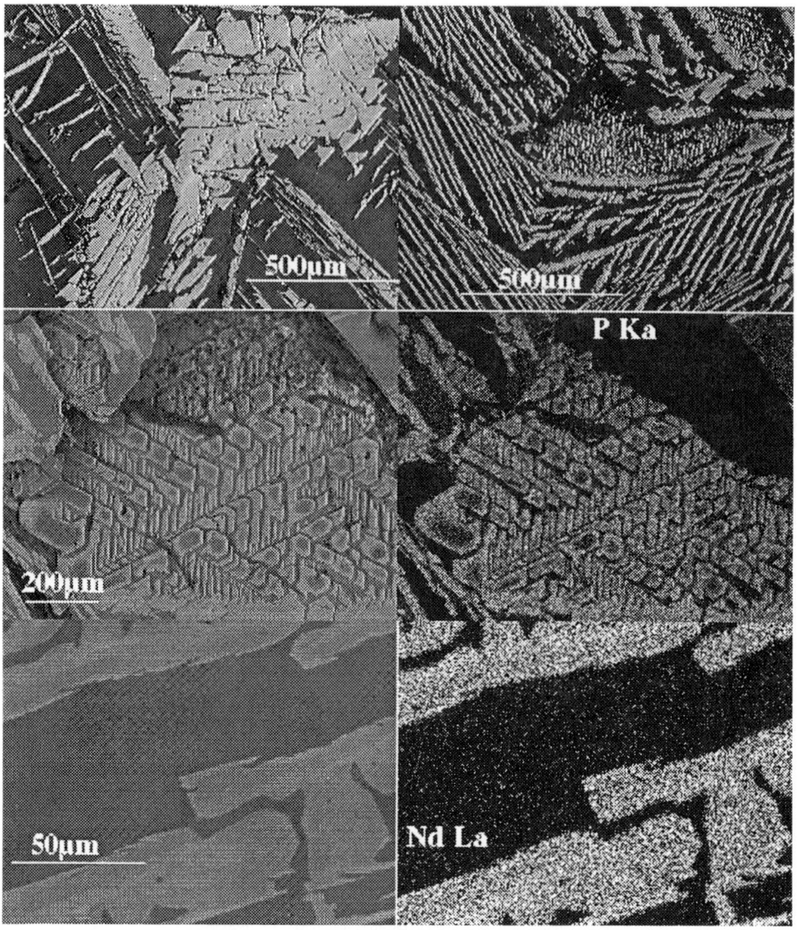

Figure 4. Micrograph of TSVA-Mg10
Upper, middle left and lower left: backscattered electrons.
Middle right: phosphorus partition (Kα).
Lower right: Neodymium partition (Nd Lα)

The X-ray spectrum shows that the structure of crystalline phase is related to the whitlockite structure, such as $Ca_8MgNd(PO_4)_7$ (**Figure 5**).The crystallization of Rare Earth Elements containing whitlockite has been described in lunar rock assemblages rich in REE elements[13] where it was shown that total concentrations of REE oxides in whitlockite range from 9 to 13 wt%. In some samples, this study demonstrates that the introduction of REE in whitlockite, into a Ca site, can be accompanied by the substitution of Si for P. The generic formula for such substituted whitlockite is $Ca_{19-x}Mg_2Nd_x(PO_4)_{14-x}(SiO_4)_x$.

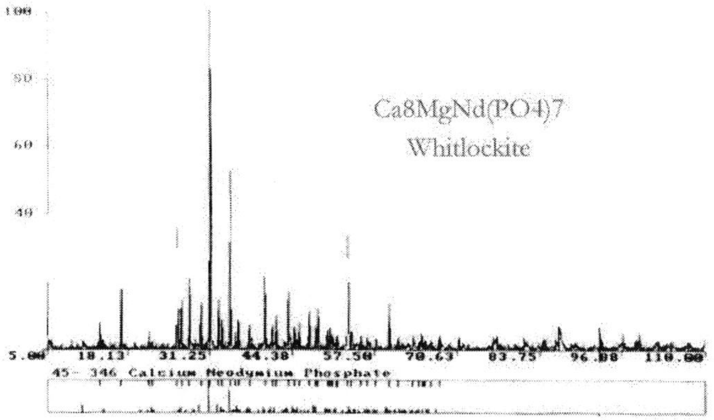

Figure 5. X-ray spectrum of TSVA-Mg10

The comparison between TSVA-Mg10 (crystallized on heating) and TSVA-Mg5 (on cooling) establishes that the first crystallization peak corresponds to the crystallization of apatite, and the devitrification peak to whitlockite. We know also that the second crystallization peak is related to whitlockite[10]. It is thus not surprising that the phase formed on cooling corresponds to the crystalline phase with the higher melting temperature. Jolliff *et al.*[13] also observed that for SiO_2-Al_2O_3-FeO-MgO-CaO-P_2O_5 melts, whitlockite crystallization may have preceded that of apatite.

Calcium Glass Ceramics (TSVA-Ca5)

The X-ray diffraction spectrum of TSVA-Ca5 is shown in **Figure 6**. All the peaks are related to an apatitic structure, since the spectrum is close to that of the oxyapatite $Ca_8Nd_2(PO_4)_6O_2$. **Figure 7** shows the morphology of TSVA-Ca5. The X-ray map of O, Si, Ca, P and Nd is shown in the lower images. The crystalline phase is enriched in Ca and P, and the glassy

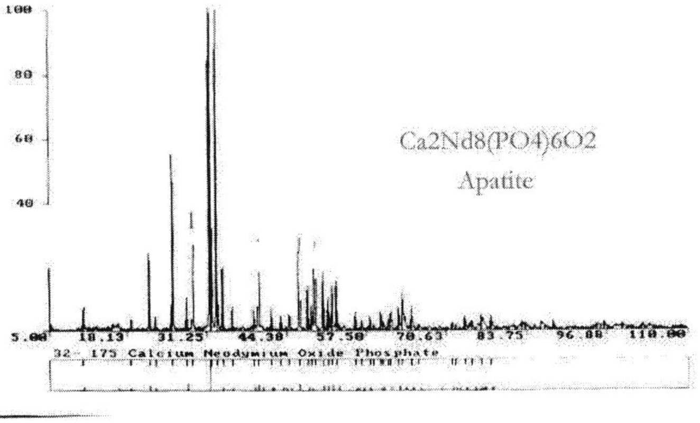

Figure 6. X-ray diffraction spectrum of TSVA-Ca5

phase is enriched in Si. The Nd distribution is nearly identical between the two phases. This is confirmed by the quantitative EPMA analysis of both phases (**Table IV**), which establishes that large quantities of P and Nd still remain in the glass phase.

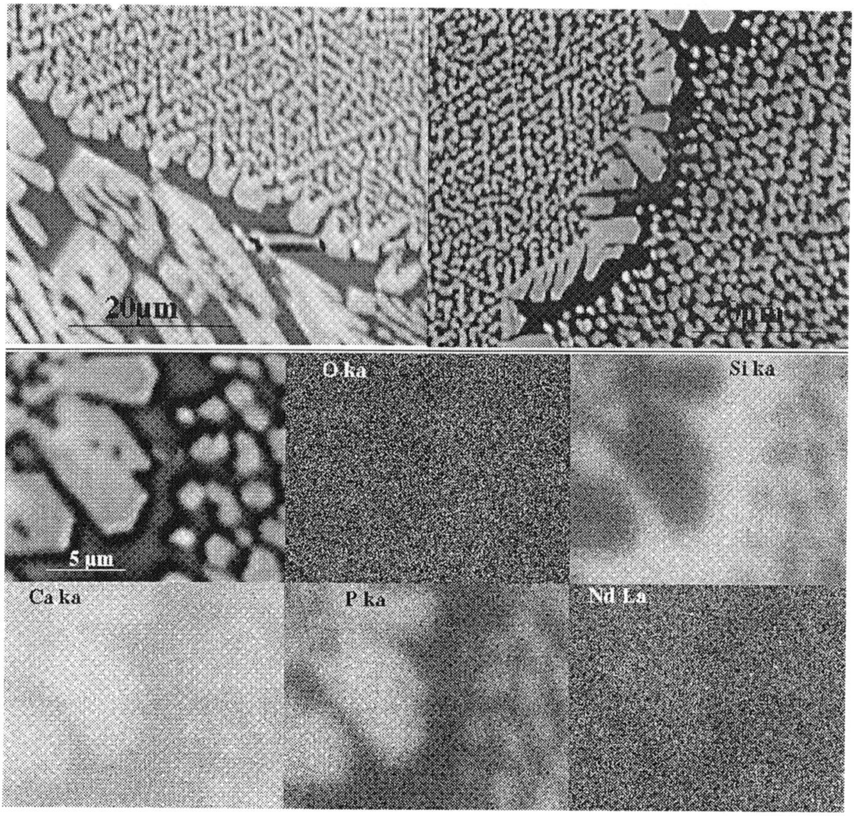

Figure 7. Morphology of TSVA-Ca5 (backscattered electrons and X-ray partition images)

Table IV. EPMA quantitative analysis of TSVA-Ca5 (oxygen was not measured)

Element atomic %	Glassy phase	Crystalline phase
O	64.9	63.2
Si	17.7	6.9
P	1.6	8.5
Ca	15.5	20.4
Nd	0.6	1
(Ca+Nd)/(Si+P) ratio	0.8	1.4

Analysis of these data shows that the (Ca+Nd)/(P+Si) ratio in the crystalline phase is less than 1.66, which is the theoretical value for a stoichiometric apatitic structure. This may be attributed to the presence of vacancies on the Ca site.

CONCLUSION

Crystallization of glasses in the MgO-CaO-P_2O_5-SiO_2 system with 5 wt% of Nd_2O_3 was studied using differential thermal analysis. Depending upon the composition and the heat treatment, apatite- and whitlockite-forming compositions were observed and characterized using SEM-EDS, XRD and EPMA. Two different crystallization processes exist for samples containing MgO: on cooling the glass, a whitlockite phase is obtained, while on heating after a nucleation-growth scenario near the first crystallization temperature, apatite is crystallized.

For CaO-P_2O_5-SiO_2-Nd_2O_3 glasses, crystallization on cooling leads to a non-stoichiometric apatitic structure. Nd partitions equally between the glass and crystalline phases when apatite is crystallized in CaO-P_2O_5-SiO_2-Nd_2O_3 glass ceramics. When MgO is added to the melt, the Nd concentration in the crystals (apatite or whitlockite) is higher than in the residual glass matrix.

Further studies are underway to improve apatite crystallization, in order to place all the neodymium in the crystalline phase. Leaching tests will subsequently be carried out to evaluate the durability of apatite-based glass ceramics.

Acknowledgments

We are grateful to J.M. Boubals, P. Pérouty and S. Pugnet for their assistance with the SEM, XRD and EPMA analyses.

REFERENCES

1. W. Lütze and R.C. Ewing, *Radioactive Waste Forms for the Future*, North Holland (1988).
2. P.J. Hayward, "The use of Glass Ceramics for Immobilizing High Level Wastes from Nuclear Fuel Recycling", *Glass Technology* 29(**4**), 122-136 (1988).
3. C. Fillet, J. Marillet, J.L. Dussossoy, F. Pacaud, N. Jacquet-Francillon and J. Phalippou, "Sphene and Zirconolite Glass Ceramics for Long-Lived Actinide Immobilization" (*this conference).*
4. W.J. Weber, R.P. Turcotte, L.R. Bunnell, F.P. Roberts and J.H. Wetsik, "Radiation Effects in Vitreous and Devitrified Simulated Waste Glass", pp. 294-299 in *Ceramics in Nuclear Waste Management*, edited by T.D. Chikalla and J.E. Mendel (1979).
5. W.J. Weber, "Radiation-Induced Swelling and Amorphization in $Ca_2Nd_8(SiO_4)_6O_2$", *Radiation Effects*, **77**, 295-308 (1983).
6. J. Carpéna, F. Audubert and J.L. Lacout, "Method of Packaging Radioactive Iodine, in Particular Iodine 129, Using Apatite as the Confinement Matrix", Pat. No. WO 96/18196, Nov. 6, 1995.

7. J. Carpéna and J.L. Lacout, "Method for Encapsulating Radioactive Waste Using Phosphosilicated Apatites as the Confining Matrix", Pat. No. WO 95/02886, July 12, 1994.
8. D.J. Wronkiewicz, S.F. Wolf and T.S. DiSanto, "Apatite- and Monazite-Bearing Glass-Crystal Composites for the Immobilization of Low-Level Nuclear and Hazardous Wastes", in *Material Research Society Symposium Proceedings*, vol. 412, pp. 345-352 (1996).
9. D.M. Roy, L.E. Drafall and R. Roy, "Crystal Chemistry, Crystal Growth and Phase Equilibria of Apatites", in *Material Science and Technology*, vol. V, Academic Press, pp. 185-239 (1978).
10. J. Shyu and J. Wu, "Growth Kinetics of Spherulitic Apatite in Some MgO-CaO-SiO_2-P_2O_5 Glasses", *J. Mater. Sci.* **29**, 3167-3171 (1994).
11. J.C. Elliot, "Structure and Chemistry of the Apatites and Other Calcium Orthophosphates", Studies in Inorganic Chemistry 18, Elsevier 1994.
12. J. Shyu and J. Wu, "Crystallization of MgO-CaO-SiO_2-P_2O_5 Glass", *J. Am. Ceram. Soc.* 73 1062-1068 (1990).
13. B.L. Jolliff, L.A. Haskin, R.O. Colson and M. Wadhwa, "Partitioning in REE-Saturating Minerals : Theory, Experiment, and Modeling of Whitlockite, Apatite and Evolution of Lunar Residual Magmas", *Geochim. and Cosmochim. Acta* 57 4069-4094 (1993).

GLASS-CERAMIC FORMS FOR IMMOBILIZING PLUTONIUM USING CERIUM AS SURROGATE

Dieter A. Knecht, Krishna Vinjamuri, Jon Grandy, Swami Raman
Idaho National Engineering and Environmental Laboratory
Lockheed Martin Idaho Technologies Co.
P. O. Box 1625
Idaho Falls, ID 83415.

ABSTRACT

New glass-ceramic forms have been developed that may be suitable for immobilizing plutonium, based on tests using cerium as a surrogate for plutonium. The glass-ceramic waste forms were formed with cerium oxide loadings equivalent to 15 wt% of plutonium oxide and using surrogate compositions of Idaho Chemical Processing Plant (ICPP) high-level waste (HLW) calcines either by melting at 1450 °C and atmospheric pressure or by sintering at 1050 °C and 138 MPa in a hot isostatic press (HIP). Calculations show that if radioactive calcine were used, these materials would result in a dose considered by DOE Order to be self-protecting and greater than 200 rem at 1 m, assuming the dimensions of a HLW glass canister. X-ray diffraction and scanning electron microscopy showed that durable host crystalline phases for cerium were formed, including fluorite, zirconolite, and zirconia. Durability tests using the Product Consistency Test (PCT) at 90 °C and 7 days at a surface-to-volume ratio of 500 m^{-1} resulted in undetectable levels of Ce and Sm, less than 0.02 g/m^2, and other components detected at comparable levels to Hanford standard ARM-1 glass and from 20-200 times lower levels than for the Savannah River Site standard EA glass.

INTRODUCTION

The end of the Cold War has resulted in a large amount of surplus weapons-usable fissile materials both in the United States and the former Soviet Union. In addition, the shutdown of the weapons production left a wide variety of plutonium-containing scrap and residue materials in the manufacturing "pipeline." A report by the National Academy of Sciences (NAS)[1] presented two alternatives for disposition of residual plutonium by (1) using fuel in existing or modified reactors and (2) vitrification in a mixture with radioactive waste. In January 1997 the *Record of Decision for Storage and Disposition of Weapons-Usable Fissile Materials Final Programmatic Environmental Impact Statement*[2] selected these two options for plutonium disposition including immobilizing of Pu in a mixture with radioactive wastes as a glass or ceramic. Some of the waste forms that have been shown to be good candidates for immobilizing plutonium include crystalline zircon, zirconolite and zirconia ceramics and glass ceramics consisting of a durable glass phase embedded with some of these ceramic phases.[3,4,5,6] The potential benefit of glass-ceramics is a result of the compositional flexibility of a glass as combined with the high durability of a ceramic for plutonium and actinides, especially if applied to plutonium residue materials which include other impurities.

Glass-ceramics, that contain crystalline phases such as zircon and zirconia, have been developed as waste forms for Idaho Chemical Processing Plant (ICPP) high-level wastes.[7,8] The major components in HLW calcine, formed from high-level liquid radioactive waste by heating at

500 °C, include alumina, fluorite, zirconia, calcia, alkali oxides, borate, and cadmium oxide.[9] Fission product content is typically less than 1 wt%. Durable glass-ceramic forms were developed for zirconia and alumina calcine with additives including silica, alkali, and yttria. Detailed microstructure analysis of these waste forms revealed the presence of actinide host phases, including fluorite, stabilized cubic zirconia, zircon, zirconolite, perovskite and amorphous aluminosilicate glass.[7,8]

This paper describes compositions and properties of glass-ceramic materials developed at the Idaho National Engineering and Environmental Laboratory (INEEL) for immobilizing plutonium. Glass-ceramic waste forms that have been tested using cerium as a surrogate for plutonium[5] include zirconium and titanium-based glass ceramics formed from surrogate HLW calcine and additives in a crucible melt or hot isostatic press.

SELF-PROTECTING RADIATION DOSE

The NAS study recommends that if plutonium is to be immobilized by vitrification, it should be as protected from potential diversion by a high radiation field similar to that in spent fuel until disposal.[1] Doses at 1 m from commercial spent fuel run typically in the 1000 rem/hr range. DOE Order 5633.3B, "Control and Accountability of Nuclear Materials," defines highly radioactive special nuclear materials as those that, unshielded, emit a radiation dose measured at 1 m that exceeds 100 rem/hr. For these doses from highly radioactive materials, the time to receive an LD_{50}, or fatality to one-half of the expose population, is approximately 3 hr. The self protecting dose is thus desirable at a high level during the anticipated time of storage of the immobilized plutonium.

Since one of the options for producing a self-protecting dose involves vitrification of DOE HLW either with added plutonium mixed in the waste solution or by encapsulating smaller cans of immobilized plutonium within the canisters of HLW glass, hypothetical doses were calculated for anticipated doses resulting from ICPP HLW. Doses at 1m were calculated[10] using a computer code, "Microshield," for glass or glass-ceramics containing ICPP HLW zirconia and alumina calcines containing approximately 20,000 and 47,000 Curies per cubic meter, respectively. The stainless steel canister was assumed to have a 66-cm (26-inch) diameter with 0.32-cm (1/8-inch) wall thickness. For a glass containing 33 wt% zirconia calcine, the calculated dose at 1m was 169 rem/hr; for a glass-ceramic containing 70 wt% zirconia calcine, the dose was 284 rem/hr; for a glass-ceramic containing 50 wt% alumina calcine, the dose was estimated at 300 rem/hr.

EXPERIMENTAL PROCEDURE

The simulated calcines that were used in this study were prepared in fluidized-bed pilot-plant calciner units and consist of Run 80 zirconia and Run 17 zirconia-sodium blended calcines. The compositions of the calcines and frits used in these experiments are shown in Table I. Glass-ceramic samples were prepared from the simulated calcines, additives, and ceria as a surrogate for plutonium by melting in crucibles or HIPing. The crucible melting was run by heating at 1450 °C for 4 hr followed by annealing at 500 °C for 4 hr and furnace cooling to room temperature. HIPed samples used surrogate calcine that had been heat treated for about 30 hr at 600 °C to decompose and release nitrates. The calcine additive mixture was uniaxially precompacted at 3.4 MPa (500 psi) in a 2.5-cm diameter by 7.5-cm long stainless-steel cans and out-gassed at 200 °C. The cans were evacuated, sealed and HIPed for 4 hr at 1050 °C and 138 MPa (20,000 psi).

The waste forms were characterized for density, chemical durability (leachability), microstructure, and glass and crystalline phase composition. Density measurements used a helium gas displacement pycnometer. Chemical durability was determined by a 7-day product consistency test (PCT) in which 1 g of the crushed sample in a 100-200 mesh fraction was contacted by 40 mL deionized water for 7 days at 90 °C. The particle surface area to water volume ratio was 500 m^{-1} for all waste forms. Elemental composition in the leachate was analyzed using inductively coupled plasma-atomic emission spectroscopy (IP-AES). Total release was also estimated by measuring the electrical conductivity of the leachate solution. Powder X-ray diffraction (XRD) analysis used powdered samples with particle size smaller than 100 mesh. The diffractometer used K-alpha copper radiation by scanning from diffraction angles of 10-75° at 2° per minute. The XRD spectra was indexed using Joint Committee on Powder Diffraction Standards (JPDS) data and computer search/match software to identify the crystalline phases. Scanning electron microscopy with associated energy dispersive X-ray fluorescence spectrometry (EDS) was used to identify predominant crystalline and amorphous phases in the samples.

Table I. Normalized composition in wt% of simulated zirconia (Run 80) and zirconia-sodium (Run 17) calcines and frits.

Component	Zirconia Calcine	Zirconia-Na Calcine	Frit-127	Frit-KV	Frit-Soil
Al_2O_3	14.5	12.0	0	0	12.5
SiO_2	0	0	70.4	81.0	65.4
B_2O_3	0.3	3.5	8.5	7.0	0
CaO	26.1	10.4	0	0	9.6
CaF_2	34.3	40.7	0	0	0
Na_2O	0.3	5.4	12.6	9.0	1.5
Li_2O	0	0	6.2	3.0	0
K_2O	0.0	1.1	0	0	2.9
Cs_2O	0.3	0.0	0	0	0
CdO	0.0	6.0	0	0	0
Ce_2O_3	0.1	0.0	0	0	0
Cr_2O_3	0.7	1.6	0	0	0
Fe_2O_3	0.2	0.5	0	0	4.8
MnO_2	0.0	0.2	0	0	0
NiO	0.1	0.2	0	0	0
SrO	0.8	0.1	0	0	0
ZrO_2	22.8	19.7	0	0	0
TiO_2	0	0	0	0	0.7
CuO	0	0	2.3	0	0
MgO	0	0	0	0	2.5

RESULTS

Glass-ceramic formation conditions and compositions of representative formulations are shown in Table II. See Table I for the compositions of the calcines and frits. Experiments CWF-1 and -3 formed the glass-ceramic samples by melting at 1450 °C in a crucible, and experiments GC-71 and -75 used a HIP at 1050 °C and 130 MPa. Table III gives the compositions of the final resulting forms.

Table II. Formation conditions and compositions of representative glass-ceramic formulations (wt%)

Experiment	CWF-1	CWF-3	GC-71	GC-75
Conditions	Melt - 1450°C	Melt - 1450 °C	HIP - 1050 °C	HIP - 1050 °C
Calcine	[Zirconia] 62	[Zirconia] 62	[Zirconia] 62	[Zirconia-Na] 51
Frit	[127] 25	[Soil] 25	[127] 25	[KV] 25
Ti Metal	5	5	5	5
Ce_2O_3	7	7	7	9
Sm_2O_3	1	1	1	10

Table III. Compositions of waste forms (wt%)

Experiment	CWF-1	CWF-3	GC-71	GC-75
Al_2O_3	8.9	12.1	8.9	6.0
SiO_2	17.6	16.4	17.6	20.2
B_2O_3	2.3	0.2	2.3	3.5
CaO	16.1	18.5	16.1	5.2
CaF_2	21.1	21.1	21.1	20.2
Na_2O	3.3	0.6	3.3	4.9
Li_2O	1.6	0	1.6	0.8
K_2O	0	0.7	0	0.6
Cs_2O	0.2	0.2	0.2	0
CdO	0	0	0	3.0
Cr_2O_3	0.4	0.4	0.4	0.8
Fe_2O_3	0.1	1.3	0.1	0.3
MnO_2	0	0	0	0.1
NiO	0.1	0.1	0.1	0.1
SrO	0.5	0.5	0.5	0.1
ZrO_2	14.1	14.1	14.1	10.0
Ti	5	5.2	5	5
CuO	0.6	0	0.6	0
MgO	0	0.6	0	0
Ce_2O_3	7.0	7.0	7.0	9.5
Sm_2O_3	1.0	1.0	1.0	9.7
TOTAL	100.0	100.0	100.0	100.0

Table IV. XRD analysis results for crystalline phases of calcine-bearing Pu-surrogate glass-ceramics.

Glass-ceramic	Calcine	Major crystalline phases	Possible minor crystalline phases
CWF-1	zirconia	CaF_2	$CaZrTi_2O_7$, (Zr, Ce)O_2, Al_2O_3, Ce metal
CWF-3	zirconia	CaF_2	$CaZrTi_2O_7$, ZrO_2, $ZrTiO_4$, Al_2O_3, Ce metal
GC 71	zirconia	CaF_2	ZrO_2, $CaAl_2Si_2O_8$, $Ca_2Al_2SiO_7$, Ca_2SiO_4
GC 75	zirconia-Na	CaF_2	ZrO_2, $ZrSiO_4$, CDs, $CaTiO_3$, $CaZrTi_2O_7$, Sm_2O_3

The X-ray diffraction analyses are summarized in Table IV. In all samples, fluorite (CaF_2) is the major phase. Minor phases include zirconolite ($CaZrTi_2O_7$), zirconia, alumina, and other silicates. Scanning electron microscopy with X-ray fluorescence also inferred these phases including an alumino-silicate glass. Figures 1-4 show the SEM micrographs for CWF-1, CWF-3, GC-71, and GC-75, respectively. Table V summarizes the results of the SEM-EDS analyses. Figure 1 shows the crystal phases of zirconia, fluorite, alumina, and sphene formed in CWF-1; Figure 2 shows zirconia, fluorite, zirconolite, and alumina formed in CWF-3; Figure 3 shows zircon-type structure, fluorite, and zirconia formed in GC-71; Figure 4 shows Cd metal, sphene, zirconia, and fluorite formed in GC-75. In all cases, the glass phase contained about one mole per cent cerium oxide and one-half mole per cent samarium oxide.

The durability was measured using the 7-day PCT at 90 °C at a surface to volume ratio of about 500 $m,^{-1}$ and the results are summarized in Table VI. In all cases, the normalized elemental releases for Ce and Sm were below detectability of 0.05 μg/mL and 0.02 μg/mL, respectively. The other components were comparable to the Hanford standard ARM-1 glass, indicating a high durability. The leach rates were much lower than for the Savannah River Site standard EA glass when run under the same surface-to-volume conditions.

Table VI. Normalized 7-day PCT elemental release, g/m^2, for melted and HIPed glass-ceramics.[a]

Element	Al	B	Ca	Ce	Cs	Na	Si	Sm
CWF-1	0.24	0.34	0.07	<0.001	<0.02	0.12	0.08	<0.002
CWF-3	0.12	-	0.09	<0.001	<0.02	0.10	0.08	<0.002
GC 71	0.24	1.31	0.11	<0.001	0.12	0.50	0.20	<0.002
GC 75	0.05	0.69	0.30	<0.001	-	0.19	0.02	<0.0002
EA glass		18.14				14.31	4.16	
ARM-1 glass		0.56				0.60	0.31	

[a] At 90 °C and surface area to volume ratio of 500^{-1}

CONCLUSIONS

A number of candidate glass-ceramic materials capable of immobilizing plutonium were prepared using cerium as a surrogate for plutonium and surrogate zirconia-based ICPP calcines. In the zirconia calcine glass-ceramic produced by devitrifying a high-temperature glass formed at 1450°C, the predominant phases hosting plutonium and neutron absorber samarium were zirconia and zirconolite with cerium oxide loadings up to 7 wt% (plutonium oxide equivalent of 12 wt%). Similar results were observed using cerium as a plutonium surrogate in a HIP at 1050° C and 138 MPa and up to 9 wt% (plutonium oxide equivalent of 15 wt%). In the HIP experiments, a zircon-like phase was formed in addition to zirconia, sphene, and zirconolite as cerium and samarium host phases. Leach rates were low, in the range observed for the most durable HLW glasses at other sites, including the ARM-1 glass.

ACKNOWLEDGMENTS

This work is sponsored by the U. S. Department of Energy, Idaho Operations Office, under contract DE-AC07-94ID13223. The authors would like to thank Bill Robson and Doug Wenzel for calculating the anticipated doses at 1 m of immobilized ICPP HLW.

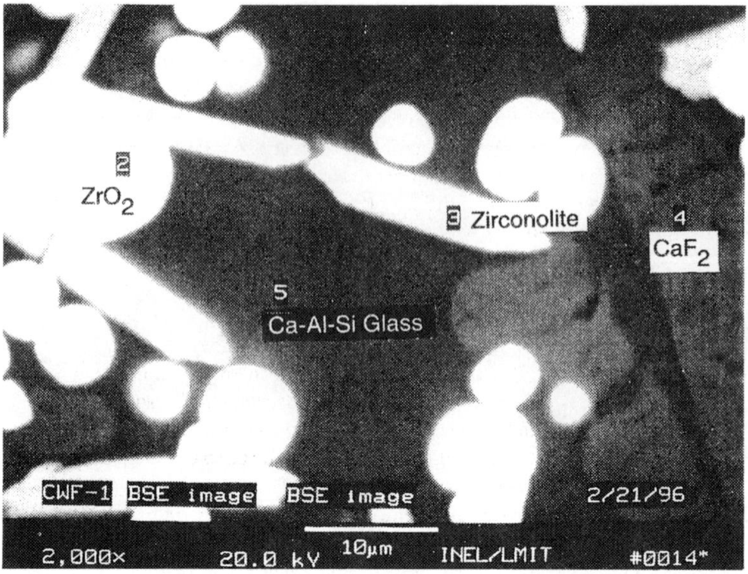

Figure 1. SEM back scattered electron image of glass ceramic sample CWF-1 formed in a high-temperature melt

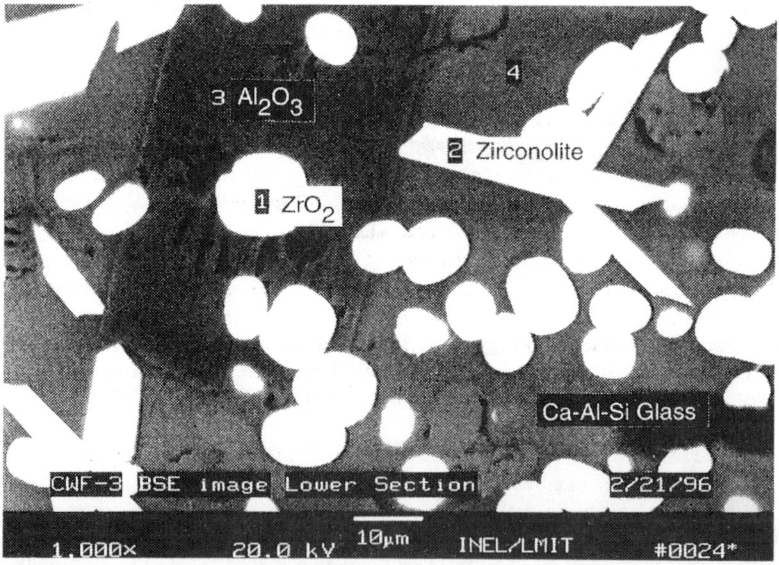

Figure 2. SEM back scattered electron image of glass ceramic sample CWF-3 formed in a high-temperature melt

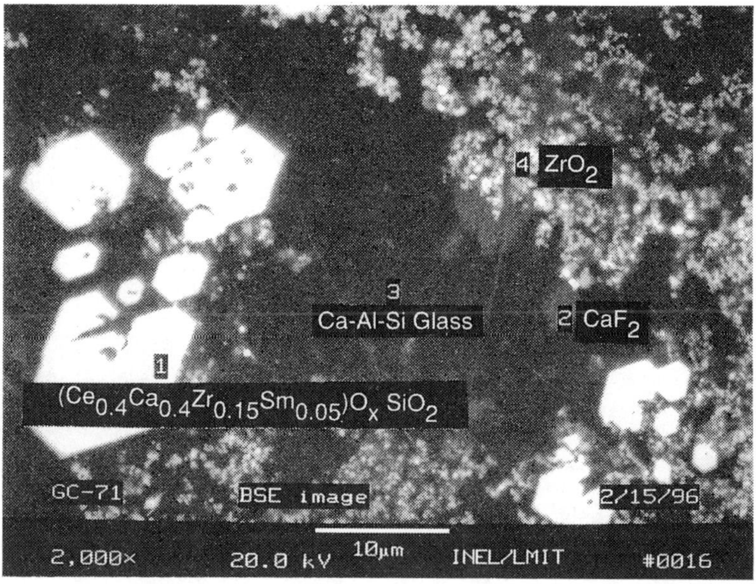

Figure 3. SEM back scattered electron image of glass ceramic sample GC-71 formed in a HIP

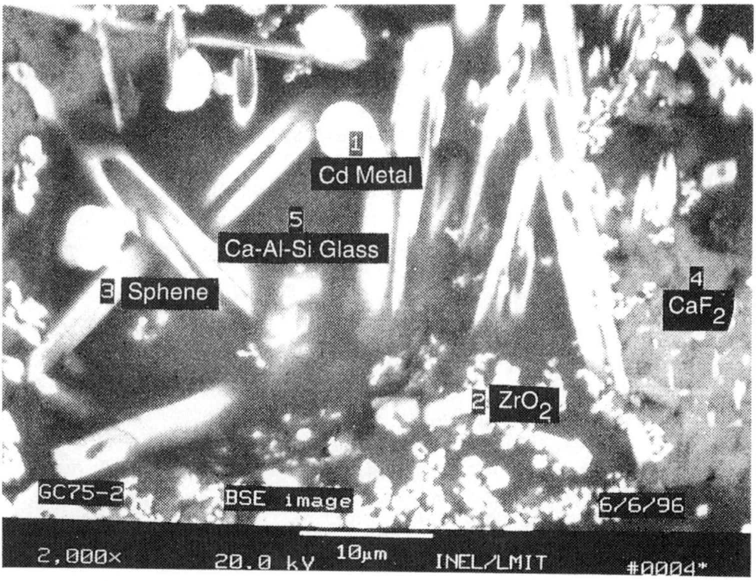

Figure 4. SEM back scattered electron image of glass ceramic sample GC-75 formed in a HIP

Table V. SEM/EDS analysis of crystalline and glass phases in high temperature and HIPed Zr calcine-bearing Pu-surrogate glasses CWF-1, CWF-3, GC-71, and GC-75

CWF-1	Component in mole %
Spot 5 (Glass)	$40SiO_2$, $27Al_2O_3$, 23CaO, $5Na_2O$, $2ZrO_2$, $1CeO_2$, $1TiO_2$.
Spot 2	100 ZrO_2.
Spot 3[a]	$35TiO_2$, $34ZrO_2$, 20CaO, $6Al_2O_3$, $4CeO_2$, $1Sm_2O_3$
Spot 4	CaF_2
CWF-3	**Component in mole %**
Spot 1	$96ZrO_2$, $3TiO_2$, $1Al_2O_3$, $1CeO_2$
Spot 2	$35TiO_2$, $35ZrO_2$, 19CaO, $6Al_2O_3$, $3CeO_2$, $2Fe_2O_3$, $1Sm_2O_3$
Spot 3	$80Al_2O_3$, 11CaO, $3TiO_2$, $2SiO_2$., $2CeO_2$., $1Fe_2O_3$., $1Sm_2O_3$
Spot 4	$77Al_2O_3$, 10CaO, $4TiO_2$, $3SiO_2$, $2CeO_2$, $2ZrO_2$, $1Fe_2O_3$, CaF_2
Glass (not shown)	$40SiO_2$, 31CaO, $21Al_2O_3$, $5TiO_2$, $2ZrO_2$, $1CeO_2$.
Spot 9 (not shown)	$36SiO_2$, $34ZrO_2$, 20CaO, $5Al_2O_3$, $3CeO_2$, $2Fe_2O_3$, $1Sm_2O_3$
GC 71	**Component in mole %**
Spot 1 (Crystal)	$41SiO_2$, $18CeO_2$, 17CaO, $9CaF_2$, $6ZrO_2$, $3Sm_2O_3$, $2TiO_2$.
Spot 2 (Crystal)	$69CaF_2$, 23CaO, $5SiO_2$, $2Al_2O_3$, $1TiO_2$, $1ZrO_2$, $1Na_2O$
Spot 3 (Glass)	$51SiO_2$, $16Al_2O_3$, 14CaO, $7Na_2O$, $6CaF_2$, $3TiO_2$, $2ZrO_2$, $1CeO_2$
Spot 4 (Crystal)	$53ZrO_2$, $23SiO_2$, 10CaO, $7Al_2O_3$, $5TiO_2$, $3Na_2O$
GC 75	**Component in mole %**
Spot 1 (Metal)	Cd metal
Spot 2 (Crystal)	Si, Zr, Ca, Al, Ce, Sm.
Spot 3 (Crystal)	$34SiO_2$, $28TiO_2$, 13CaO, $7CeO_2$, $7Sm_2O_3$ $5Al_2O_3$, $2Na_2O$, $2ZrO_2$, $1K_2O$
Spot 4 (Crystal)	CaF_2 with Ce and Sm.
Spot 5 (Glass)	$58SiO_2$, 20CaO, $13Al_2O_3$, $3Na_2O$, $2ZrO_2$, $1TiO_2$, $1CeO_2$, $1Sm_2O_3$
Spot 5 (Glass)	$52SiO_2$, $23Al_2O_3$, $9Na_2O$, 7CaO, $3K_2O$, $3TiO_2$, $2ZrO_2$,. $1CeO_2$

REFERENCES

[1]National Academy of Sciences, *Management and Disposition of Excess Weapons Plutonium*, National Academy Press, Washington, D.C. (1994).

[2]U. S. Department of Energy Record of Decision for the Storage and Disposition of Weapons-Usable Fissile Materials Final Programmatic Environmental Impact Statement, January 14, 1997.

[3]B. B. Ebbinghaus, R. A. Van Konynenburg, E. R. Vance, A. Jostsons, R. G. Anthony, C. V. Philip, and D. J. Wronkiewicz, "Status of Plutonium Ceramic Immobilization Processes and Immobilization Forms," *Proc. Pu Stabilization & Immobilization Workshop*, Washington, D.C., December 12-14, 1995, DOE (1995).

[4]R. C. Ewing, W. Lutze, and W. J. Weber, "Zircon: A Host-Phase for the Disposal of Weapons Plutonium," *J. Mater. Res.u*, **10**, 243-46 (1995).

[5]E. B. Anderson, B. Burakov, B. Galkin, V. Starchenko, and V. Vasilev, *Crystalline Phases for Plutonium and Actinide Immobilization, Proc. Pu Stabilization & Immobilization Workshop*, Washington, D.C., December 12-14, 1995, DOE (1995).

[6]R. B. Heimann and T. T. Vandergraaf, "Cubic Zirconia as a Candidate Waste Form for Actinides: Dissolution Studies," *J. Mat. Sci. Lett.*, **7**, 583-586 (1988).

[7]D. A. Knecht, P. C. Kong and T. P. O'Holleran, "Proposed Glass-Ceramic Waste Forms for Immobilizing Excess Plutonium," *Proceedings of the Embedded Topical Meeting on DOE Spent Nuclear Fuel and Fissile Material Management*, ANS, La Grange Park, IL, pp. 183-191 (1996).

[8]K. Vinjamuri, "Candidate Glass-Ceramic Waste Forms for Immobilization of the Calcines Stored at the Idaho Chemical Processing Plant," *Environmental Issues and Waste Management Technologies in the Ceramic and Nuclear Industries*," V. Jain and R. Palmer, eds., *Ceramic Transactions*, **61**, p.439 (1995.)

[9]J. R. Berreth, *Inventories and Properties of ICPP Calcined High-Level Waste*, WINCO-1050 (1988).

[10]W. G. Robson and D. R. Wenzel, Private Communication, Lockheed Martin Idaho Technologies Co., P.O. Box 1625, Idaho Falls, ID 83415.

SPHENE AND ZIRCONOLITE GLASS CERAMICS FOR LONG-LIVED ACTINIDE IMMOBILIZATION

C. Fillet[1], J. Marillet[2], J.L. Dussossoy[1], F. Pacaud[1], N. Jacquet-Francillon[1], J. Phalippou[2]

[1]*Commissariat à l'Energie Atomique (CEA), Centre d'Etudes de la Vallée du Rhône, DCC/DRDD/SCD, BP 171 30207 Bagnols-sur-Cèze, France*

[2]*Laboratoire des Verres (UMR 5587), Université de Montpellier 2, CC069, Place Eugène Bataillon, 34095 Montpellier Cedex 5, France*

INTRODUCTION

The investigations now in progress in France on enhanced reprocessing of spent fuel, and notably on separating the actinides from the other fission products, have led to the development of specific matrices for conditioning these long-lived radionuclides. Glass ceramic materials provide a means for double containment: primary containment in a stable crystalline phase with proven resistance to leaching and to irradiation damage, and secondary containment in a vitreous phase with high chemical durability containing little or no radioactive species.

The crystalline phases investigated in this study are sphene ($CaTiSiO_5$)[1,2,3,4] and zirconolite ($CaZrTi_2O_7$)[1,5,6,7,8]. The expected composition for the residual vitreous phase is a calcium aluminosilicate. Neodymium was selected to simulate the trivalent actinides. Tests are conducted in the SiO_2-Al_2O_3-CaO-TiO_2-ZrO_2-Nd_2O_3 system to determine the composition range best suited to crystallization of the host phases by devitrification of the parent glass. The objective is to obtain a glass ceramic material using a vitrification process in which oxides are melted in a cold crucible – a technique demonstrated at industrial scale – followed by devitrification.

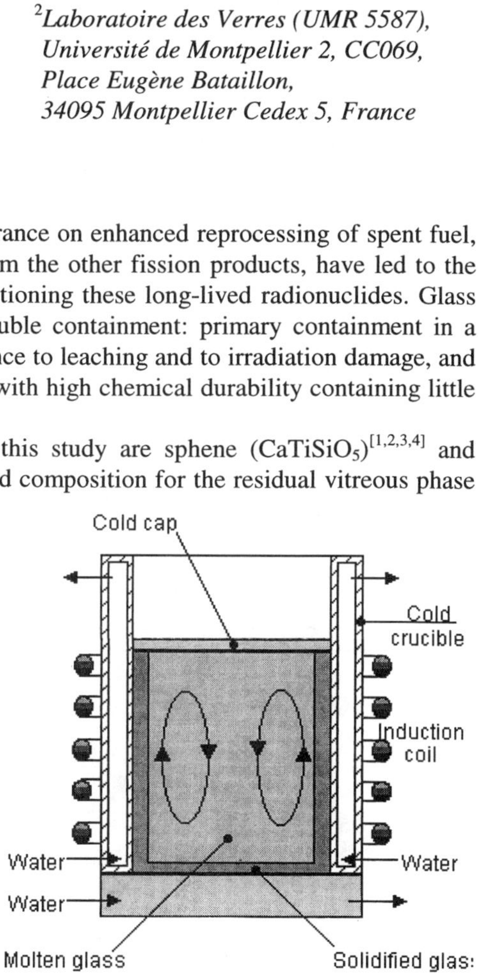

Figure 1. Induction-heated cold crucible

Cold Crucible Melting Technique

The induction-heated cold crucible melter process is illustrated in **Figure 1**. It has three significant advantages : the melt forms a solid crust along the cooled crucible wall, effectively protecting it from corrosion by the molten material; the process is relatively simple and easy to implement; and the "cold cap" of unmelted material at the surface diminishes the process volatility[9].

EXPERIMENTAL PROCEDURE

The test materials were initially fabricated in quantities of about 50 grams by melting at 1400–1500°C in alumina crucibles, refining for two hours and quenched in air. Devitrification of the parent glass is induced by isothermal heat treatment.

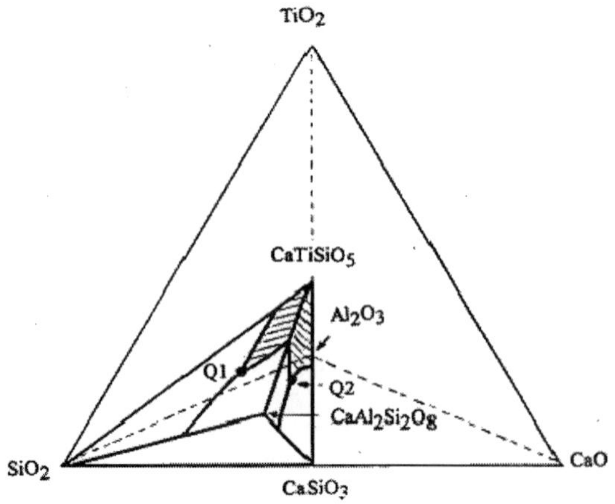

Figure 2. Sphene domain in the SiO_2–Al_2O_3–CaO–TiO_2 quaternary

The sphene-based compositions (**Table I**) are situated in the SiO_2–Al_2O_3–CaO–TiO_2 quaternary (**Figure 2**). The stability domain for sphene can be situated based on published data concerning the ternary SiO_2–CaO–TiO_2[10] and the pseudo-ternaries SiO_2–$CaAl_2Si_2O_8$–$CaTiSiO_5$[11] and $CaSiO_3$–$CaAl_2Si_2O_8$–$CaTiSiO_5$[12].

The compositions investigated were selected from two eutectics Q1 and Q2 (respectively SiO_2 rich and CaO rich) existing in the pseudo-ternaries, and then by mixing Q1 and Q2 with TiO_2 addition to be inside the sphene stability domain. Na_2O and

Table I. Glass compositions investigated (wt%)

Sample	SiO_2	Al_2O_3	CaO	Na_2O	B_2O_3	TiO_2	ZrO_2	Nd_2O_3
Q1	57.46	15.50	16.16			10.88		
$Q1_L$	54.72	14.77	15.39	0.95	3.81	10.36		
$Q1_L$-Nd	52.70	14.22	14.82	0.97	3.67	9.97		3.70
$Q1_H$	52.23	14.10	14.69	1.82	7.27	9.89		
Q2	43.25	14.14	32.58			10.03		
$Q2_H$	39.32	12.86	29.62	1.82	7.27	9.11		
T_L-Nd	39.59	11.66	19.16	0.79	3.14	12.35		13.31
Z_L-Nd	40.53	11.94	19.62	0.80	3.22	12.64	4.44	6.81
T-Nd	44.54	13.11	21.55			13.67		7.13
Z-Nd	40.75	12.00	19.72			12.51	8.50	6.52

B_2O_3 were added to some compositions in order to extend the meltable composition domain to 1400°C. Tests were also conducted by adding 10 wt%, 20 wt% and 40 wt% of $Nd(NO_3)_2{,}6H_2O$. Zirconolite precipitation was tested by adding 5 wt% and 10 wt% ZrO_2 to the T_L and T compositions.

The parent glasses and the glass ceramics were characterized by differential thermal analysis, X-ray diffraction and scanning electron microscopy coupled with X-ray analysis, and by Soxhlet-mode leach testing.

RESULTS AND DISCUSSION

Differential Thermal Analysis (DTA)

DTA was used to assess the devitrification properties of the parent glass. Test materials were pulverized with a mortar and pestle, and placed in 6 µl platinum cups. Temperature difference between sample and alumina reference was recorded from room temperature to 1150°C with a heating rate of 10°C per minute.

The glass transition temperatures (Tg) were similar between the Q1 and Q2 poles, but diminished with the use of melting additives (**Figure 3**); similarly, the exothermic peak corresponding to crystallization shifted toward lower temperatures when additives were used. The decrease in the crystallization temperature was more pronounced for Q1 (about 100°C) than for Q2 (less than 50°C) despite the apparently lower exothermic effect.

As for Q1 and Q2, the glass transition temperatures and the crystallization

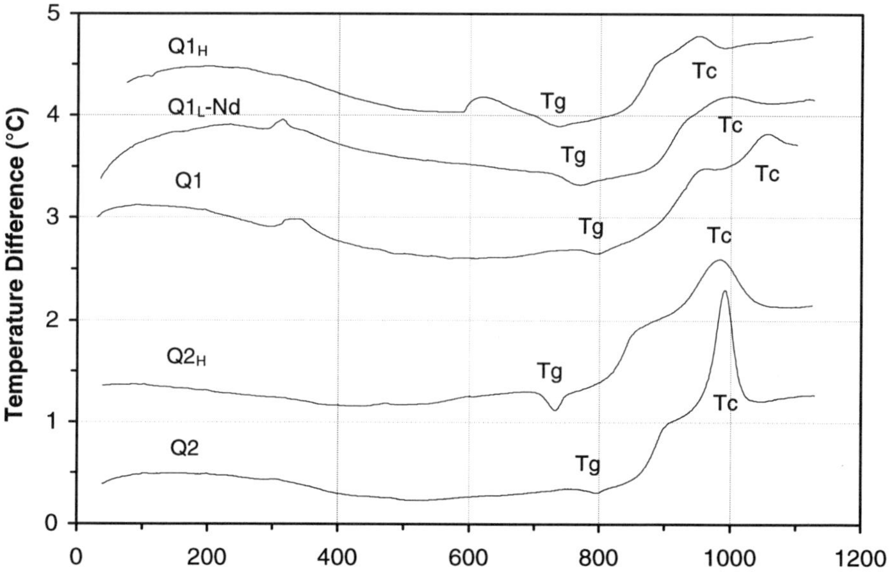

Figure 3. Differential thermal analysis (DTA) of vitrified compositions based on eutectics Q1 and Q2

temperatures for the T and Z compositions decreased with Na_2O and B_2O_3 additions (**Figure 4**).

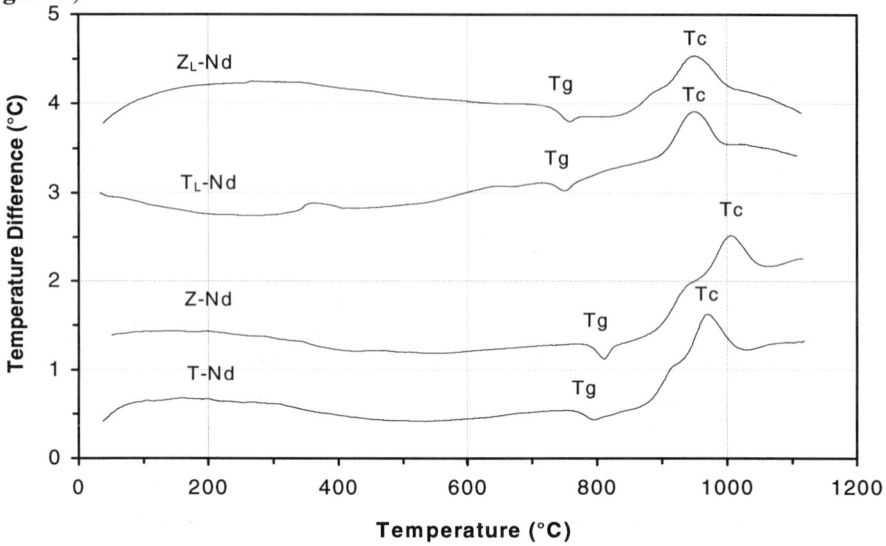

Figure 4. Differential thermal analysis (DTA) of vitrified compositions based on mixtures with equal weight of eutectics Q1 and Q2

For the compositions designed to induce zirconolite precipitation, the glass transition and crystallization temperatures were unaffected by adding 5 wt% ZrO_2 (composition Z_L-Nd) but increased with 10 wt% ZrO_2 (composition Z-Nd).

Identification of Phases (XRD, EDX-SEM)

The glass samples were heat treated at 1050°C for 1 hour (Q1, Q2, T_L-Nd and Z_L-Nd

Table II. Identification of phases by X-ray diffraction

Sample	Primary phases		Secondary phases				
	Sphene	Zirconolite	Apatite	Anorthite	Cristobalite	Perovskite	Wollastonite
Q1	■	–	–	–	■	–	–
$Q1_L$	■	–	–	–	■	–	–
$Q1_L$-Nd	■	–	–	–	–	–	–
$Q1_H$	■	–	–	–	–	–	–
Q2	■	–	–	■	–	■	■
$Q2_H$	■	–	–	■	–	■	–
T_L-Nd	■	–	■	■	–	–	–
Z_L-Nd	■	–	–	■	–	–	–
T-Nd	■	–	–	–	–	–	–
Z-Nd	–	■	–	–	–	–	–

compositions) or for 2 hours (T-Nd and Z-Nd compositions). The resulting phases were identified by XRD (**Table II**) and by SEM-EDX.

For the SiO_2-rich pole (Q1 composition), the presence of melting additives appeared to inhibit crystallization of secondary phases (cristobalite) and promoted sphene crystallization. Adding Nd_2O_3 favored sphene crystallization.

For the CaO-rich pole (Q2 composition) a perovskite secondary phase was observed with or without additives.

Three phases precipitated in the Nd-rich composition(T_L-Nd) (**Figure 5** and **6**) : sphene, as gray, comb-shaped crystals a few tens of micrometers long, enriched in Si, Ca, Ti, Al,Nd $[(Ca^{2+}+Nd^{3+})(Ti^{4+}+Al^{3+})SiO_5]$; an apatitic structure $(Ca_2Nd_8(SiO_4)6O_2)$ containing neodymium in the form of white needle-like or hexagonal crystals measuring 5 to 10 μm; and a dark phase containing Si, Al, Ca, Na, corresponding to a disordered anorthite $(Ca,Na)(Si,Al)_4O_8$.

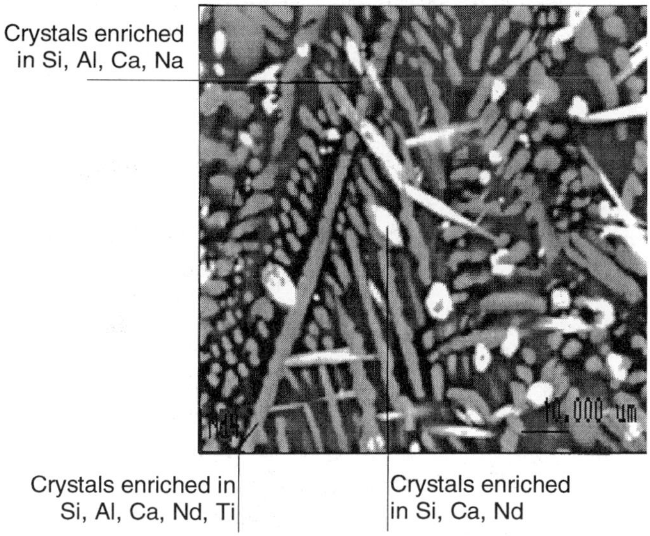

Figure 5. TL-Nd glass ceramic material observed by SEM-EDX

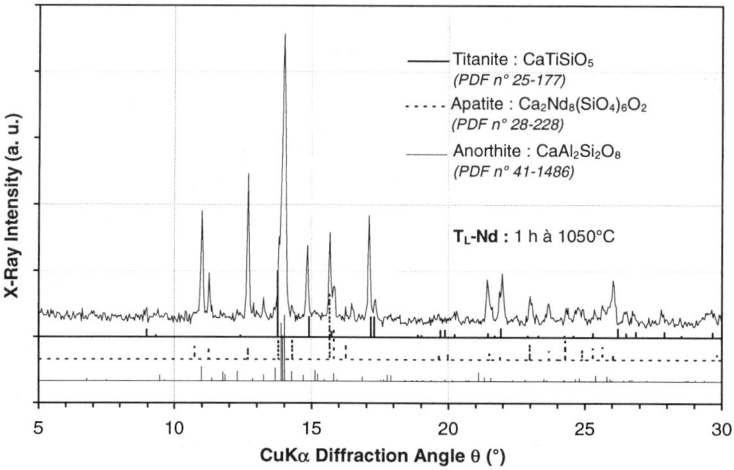

Figure 6. X-Ray Diffraction (XRD) pattern of T_L-Nd sample treated 1 hour at 1050°C.

Adding 5 wt% ZrO_2 to the T_L-Nd composition (Z_L-Nd) did not result in the formation of zirconolite.

The T-Nd composition yielded a sphene phase in the form of comb-like crystals a few tens of micrometers long.Adding 10 wt% ZrO_2 (Z-Nd) resulted in the precipitation of zirconolite in the form of sub-micron needles forming a mosaic pattern (**Figure 7**).

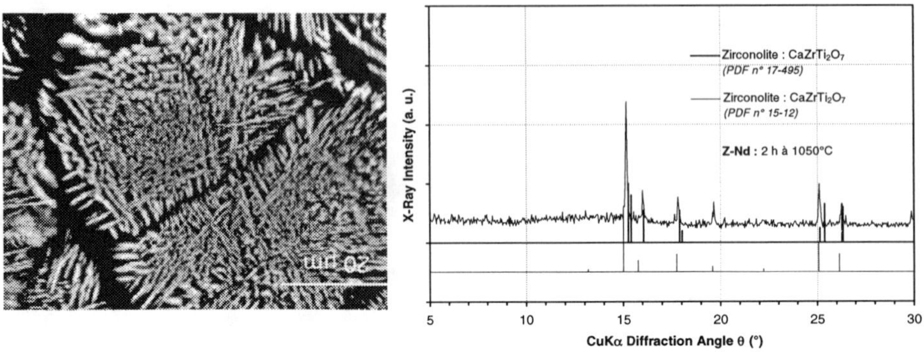

Figure 7. Z-Nd glass ceramic material

Soxhlet Leaching Behavior: Measuring the Initial Rate at 100°C

The inherent chemical durability of each material was determined by Soxhlet-mode leach tests for 28 days at 100°C with coupons 3 mm thick and 30 mm in diameter. The initial mean mass leaching rate *LM(m)* was determined by the difference between the test coupon mass before and after leaching, according to the following relation:

$$LM(m) = \frac{m_{28days} - m_0}{S \times t}$$

where m_0 is the initial coupon mass (grams), m_{28days} is the coupon mass (grams) after 28 days of leaching, S is the coupon surface area (m^2), and t is the test duration (days).

The leachates were also analyzed by ICP-AES to determine the mean element leaching rates *LM(i)* from the following formula:

$$LM(i) = \frac{C_i \times f}{x_i \times \frac{S}{V} \times t}$$

where C_i is the concentration of element i in the leachate, f is the element-to-oxide mass conversion factor, x_i is the oxide weight percentage of element i, S is the coupon surface area, V is the leachate volume, and t is the test duration.

The leach rates for the glass ceramics are overall values that do not discriminate between the inherent properties and chemical durability of the residual vitreous phase and those of the crystalline phases. Nevertheless, the are indicative of any gain in chemical durability of the devitrified material with respect to the totally amorphous parent glass. It is important to note that the leach rates determined here are the maximum rates at the test temperature of 100°C. Finally, only one test was carried out for each material, and the reproducibility results cannot be considered here. The test results are indicated in **Table III** and in **Figure 8**.

Table III. Mean mass leaching rate *LM(m)* and mean element leach rates *LM(i)* in $g \cdot m^{-2}d^{-1}$ for parent glasses and for glass ceramics (designated Tt)

	LM(m)	*LM(Si)*	*LM(Al)*	*LM(Ca)*	*LM(Na)*	*LM(B)*	pH
Q1	0.56	0.55	0.60	0.65			7.88
Q1H	0.55	1.22	1.16	1.16	0.95	0.91	9.12
Q2H	8.52						11.13
TL-Nd	2.03	2.21	2.33	2.32	3.60	2.84	10.00
TL-Nd (Tt)	0.97	1.41	1.63	1.03	3.22	3.36	9.06
ZL-Nd	1.45	1.76	1.79	1.65	2.54	2.14	9.63
ZL-Nd (Tt)	1.70	2.00	2.67	2.07	4.70	9.17	9.48
T-Nd	1.17	1.36	1.40	1.32			10.13
T-Nd (Tt)	0.65	0.79	1.05	0.63			9.63
Z-Nd	0.94	1.20	1.26	1.14			9.90
Z-Nd (Tt)	0.80	1.18	1.26	0.87			9.48

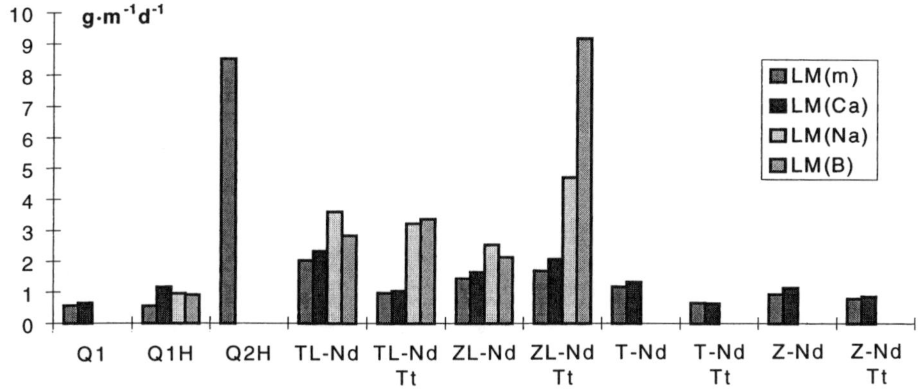

Figure 8. Initial alteration rates ($g \cdot m^{-2}d^{-1}$) of materials measured after Soxhlet-mode leaching for 28 days at 100°C

Q1 is the pole richest in silica, and can thus be expected to exhibit a slightly lower leach rate than glass $Q1_H$, containing Na and B to lower the melting point, and much lower than glass $Q2_H$, corresponding to the calcium-rich glass with the same additives.

With or without heat treatment, the T_L-Nd and Z_L-Nd materials (intermediate compositions between Q1 and Q2) exhibited leach rates two to three times higher than for Q1, despite the addition of refractory oxides (Ti, Nd, Zr) and a concentration of melting additives lower by a factor of 2 than for $Q1_H$. Heat-treated T_L-Nd had slightly lower *LM(m)* and *LM(Ca)* values than the parent glass, while the *LM(B)* and *(LM(Na)* values were unaffected; the fact that sodium and boron shall remain in the residual glass may account for these differences. Similar behavior could have been expected for the Z_L-Nd materials. In fact, *LM(m)* and *LM(Ca)* did not decrease after heat treatment, while *LM(B)* and *(LM(Na)* increased significantly (by nearly a factor of 4); this behaviour is not explained.

T-Nd and Z-Nd materials contain neither boron nor sodium, but similar or higher amounts of silicon and zirconium, and thus exhibit slightly greater chemical durability than TL-Nd and ZL-Nd. Moreover, the T-Nd and Z-Nd glass ceramics had slightly lower leach rates than their parent glasses.

CONCLUSION

Composition studies within the SiO_2–Al_2O_3–CaO–TiO_2 quaternary allowed us to obtain sphene precipitation; depending on the composition, minor phases were also obtained that may diminish the properties of the residual vitreous phase. Precipitation of zirconolite was obtained by adding 10 wt% of zirconium oxide. The T and Z materials were found to have the most interesting properties: precipitation of the desired phase, leach rates not exceeding 1.2 $g{\cdot}m^{-2}d^{-1}$ at 100°C. The subsequent investigations will focus on compositions of this type. One of the next steps will be to quantify the containment properties with respect to the elements simulating the actinides, as well as the chemical durability of the glass ceramics under leaching conditions more representative of long-term behavior, i.e. in static media and at lower temperatures.

REFERENCES

1. W. Lutze and R.C. Ewing Editors, *Radioactive Waste Forms for the Future* (1988).
2. P.J. Hayward and E.V. Ceccheto, "Development of Sphene-Based Glass Ceramics Tailored for Canadian Waste Disposal Conditions", *Scientific Basis for Nuclear Waste Management IV*, Mat. Res. Symp. Proc., Vol. 6 (1982).
3. E.R. Vance, P.J. Hayward, C.D. Cann, S.L. Mitchell, M.A.T. Stanchell and D.J. Wren, "Crystallization of Uranium-Bearing Sphene Glass Ceramics for the Immobilization of Nuclear Waste", *Glass Technology*, 25 [5] 232-9 (1984).
4. P.J. Hayward, "Review of Progress in Development of Sphene-Based Glass Ceramics", Scientific Basis for Nuclear Waste Management XV, Mat. Res. Symp. Proc., Vol. 257 289-98 (1992).
5. A.E. Ringwood, S.E. Kesson, NG. Ware, W. Hibberson and A.Major, "Immobilisation of High-Level Nuclear reactor wastes in Synroc", Nature, 278 219-23 (1979).
6. A. Jotsons, "Making Steady Progress with Synroc Research", Nucl.Eng.Int. 37 (2) 36-8 (1992).
7. E.R. Vance, B.D. Begg, R.A. Day and C.J. Ball, "Zirconolite-Rich Ceramics for Actinide Wastes", Scientific Basis for Nuclear Waste Management, Mat. Res. Symp. Proc., Vol. 353 (1995).
8. F.W. Clinard, "Review of Self-Irradiated Effects in Pu-Substituted Zirconolite", *Am. Ceram. Bull.* **65**, 1181-87 (1986).

9. A. Jouan, JP. Moncouyoux, S. Merlin, *Proceedings of Waste Management 95*, Feb 26-March 2,1995,Tucson
10. R.C. Devries, R. Roy and E.F. Osborn, *J. Am. Ceram. Soc.*, 38 [5] 161 (1955).
11. Y.M. Agamawi and J. White, *Trans. Brit. Ceram. Soc.*, 53 [1] 23 (1954).
12. Keizo Iwasé and Usaburô Nisioka, *Science Reports*. Tôhoku Imp. Niv., Ser.1, Honda Anniv., pp. 1-10 (1936).

PLUTONIUM HOST PHASES DERIVED FROM HIGH-LEVEL WASTE AT THE IDAHO CHEMICAL PROCESSING PLANT

Thomas P. O'Holleran and Stephen G. Johnson
Argonne National Laboratory-West
P. O. Box 2528
Idaho Falls, ID 83403-2528

Peter C. Kong and Bruce A. Staples
Idaho National Engineering and Environmental Laboratory
P. O. Box 1625
Idaho Falls, ID 83415-2210

ABSTRACT

The National Academy of Sciences recommended dissolution in a silicate matrix, with fission products to provide a diversion-resistant radiation field, as one option for dispositioning excess weapons-grade plutonium. Candidate materials and processing conditions have been developed to pursue this option using high-level waste stored at the Idaho Chemical Processing Plant. Devitrification of glassy host materials achieves increased plutonium loading by partitioning plutonium into durable crystalline host phases. Results of devitrification experiments are summarized, and several unique plutonium host phases are reported. These phases were initially synthesized and characterized using cerium as a plutonium surrogate, to simplify operational requirements. Tests using plutonium were performed to validate results obtained with surrogate materials. Characterization and leach test results are reported.

INTRODUCTION

As the Cold War came to an end, the United States and Russia agreed to reduce their nuclear arsenals. As a result, the United States declared 38.2 metric tons of weapons grade plutonium "excess material" (1). For political reasons, this excess material must be dispositioned in such a way that it becomes unattractive for use in nuclear weapons. In their report on dispositioning of excess weapons grade plutonium (2), the National Academy of Sciences recommended two options:

1. Fabricate the material into mixed-oxide (MOX) fuel and burn it in reactors, and

2. Incorporate the material into a borosilicate glass along with sufficient fission products from high-level waste (HLW) to make the plutonium no more attractive from a proliferation standpoint than the plutonium in spent fuel from commercial reactors.

The second option is referred to as the "spent fuel standard." The Department of Energy (DOE) recently issued a Record of Decision (ROD) as the culmination of the process established by the National Environmental Policy Act (NEPA) selecting a hybrid approach using both the MOX fuel approach and the spent fuel standard for dispositioning of appropriate portions of the Nation's excess weapons-grade plutonium (3).

The Idaho Chemical Processing Plant (ICPP) at the Idaho National Engineering and Environmental Laboratory (INEEL) reprocessed highly enriched reactor fuel, primarily from naval propulsion reactors, to recover the unburned highly enriched uranium. The reprocessing operation produced an aqueous HLW stream that was calcined in a fluidized bed calciner to reduce volume. The solidified HLW calcine (approximately 3800 m^3) (4) is currently stored at the ICPP in shielded underground facilities called binsets. A variety of process and waste forms have been investigated as candidates for preparing the calcine for ultimate disposition in a geologic repository. As part of this effort, borosilicate glass formulations were developed that were tailored to the unique calcine chemistry (5,6). The unique chemical nature of the calcine results from the dissolution step in the reprocessing operation. Unlike other reprocessing operations, the operation at the ICPP did not chop the fuel to dissolve the fuel meat. Instead, the entire fuel element was dissolved. As a result, elements from the cladding are included in the HLW. Consequently, much of the calcine stored at the ICPP is rich in zirconium.

This work was initiated to demonstrate the feasibility of using HLW stored at the ICPP to "denature" excess weapons grade plutonium to meet the spent fuel standard. Initial efforts focused on vitrification technology, which has a proven track record world-wide for immobilizing HLW. In the conceptual stages of the project, it was recognized that the unique chemistry of the HLW at the ICPP, when combined with borosilicate glass, had the potential for yielding plutonium host phases that would not only be more chemically durable (and therefore proliferation resistant) than borosilicate glass, but that might also achieve overall plutonium loadings higher than solubility limits would permit in a conventional borosilicate glass. For example, it was initially proposed that plutonium might be partitioned into a known durable transuranic host like zircon (7-9) formed from zirconia in the HLW and silica in the glass-forming additives under suitable processing conditions. The plutonium-bearing zircon would be microencapsulated in a glass matrix containing the fission products that provide the diversion-resistant radiation field to meet the spent fuel standard.

TECHNICAL APPROACH

Initial experiments used glass formulations developed in the late 1970s and early 1980s to immobilize zirconia-rich calcine, and investigated processing conditions intended to devitrify the glass in an attempt to produce durable crystalline phases (10). Cerium was used as a surrogate for plutonium in these initial experiments, and surrogate (non-radioactive) calcine that was used during development of waste forms for HLW was substituted for the radioactive material. Samarium was added to simulate candidate neutron absorbers. Neutron absorbers would be added to materials destined for geologic disposal for criticality control. It is desirable for neutron absorbers to partition into the same phases as the plutonium to provide long term criticality control by, presumably, being released into ground water at the same rate as the plutonium. After initial success developing formulations and processing conditions based on HLW glass forming technology, other waste forms and related processes (e.g. ceramics from cold pressed and sintered powders, and hot isostatically presses ceramics) were also investigated (11). All experiments initially used cerium as a surrogate for plutonium, and samarium as the generic neutron absorber. Verification experiments using plutonium were performed after promising formulations and processing conditions were identified with surrogates.

EXPERIMENTS TO ACHIEVE THE SPENT FUEL STANDARD

The first experiments used a simulated zirconia calcine called Run 80 (for pilot plant run 80) and a glass additive called Frit 127. The compositions of each of these materials is given in Table I. Frit 127 and Run 80 calcine were mixed in a 7/3 weight ratio. This was the basic formulation developed to immobilize zirconia calcine (5). To this mixture was added 6 weight percent Sm_2O_3 and 9.5 weight percent CeO_2 (equivalent to 15 weight percent PuO_2 on a molar basis).

Table I. Compositions of Run 80 simulated calcine and Frit 127 glass

Run 80 Calcine		Frit 127	
Compound	wt. %	Compound	wt. %
Al_2O_3	14.4	SiO2	70.3
B_2O_3	0.3	Na_2O	12.8
CaO	25.9	Li_2O	6.2
Ce_2O_3	0.1	B_2O_3	8.5
Cr_2O_3	0.7	CuO	2.1
Cs_2O	0.3		
CaF_2	34.1		
Fe_2O_3	0.2		
Na_2O	0.3		
NiO	0.1		
SrO	0.8		
ZrO_2	22.8		

This mixture was melted at 1050°C in an alumina crucible using an electric resistance furnace. After four hours at 1050° C, the furnace temperature was reduced to 500° C for four hours, then raised to 700° C at 2° C per minute. The material was held at 700° C for 144 hours, then furnace cooled. The 500° C hold was intended to promote crystal nucleation, and the 700° C hold was to promote crystal growth (devitrification). The microstructure of the resulting devitrified glass is shown in Figure 1. The bright phases, one of which is labeled 1, was identified as a cerium-zirconium-samarium oxide containing 35 mole percent ZrO_2, 47 mole percent CeO_2, and 7.5 mole percent Sm_2O_3. The phase labeled 2 is a sodium calcium aluminosilicate of somewhat variable composition, typically containing a few atom percent cerium and samarium. No cerium or samarium was detected in either the glassy matrix (3 in Figure 1) or the calcium silicate phase (2 in Figure 1). Interpretation of X-ray diffraction results was difficult, but eventually a pattern very similar to cubic zirconia was identified, along with an anorthite-like pattern, leaving a number of unidentified peaks. The expected zircon phase was never detected in these first experiments.

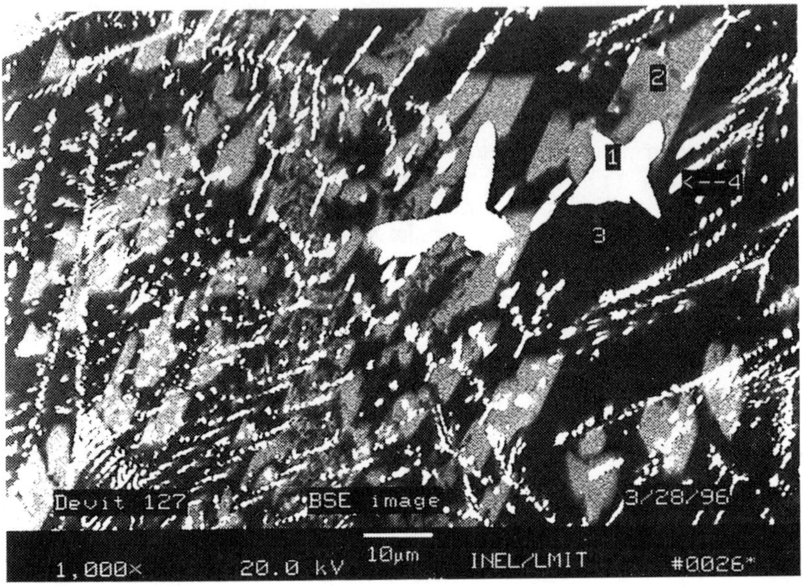

Figure 1: Scanning electron micrograph of an early sample of devitrified Frit 127-based glass containing cerium as a surrogate for plutonium. Brighter areas represent regions of higher average atomic number in this backscattered electron image. The area labeled 1 is the ceria-zirconia-samaria phase, 2 is primarily a calcium silicate phase also containing some zirconium, the bright area labeled 3 is the glassy matrix, and 4 is the sodium calcium aluminosilicate phase.

As surrogate experiments continued, other formulations and processing conditions were tested, including cold pressing and sintering, and hot isostatic pressing. When zirconium was present in the starting mixture, cerium was often found in apparent solid solution with zirconium (with or without samarium) in the face centered cubic phase identified in the initial devitrification experiments. Even when zircon was identified in a sample, little of the cerium was found partitioned into the zircon. Rather, the cerium was mostly partitioned into the cubic zirconia phase, which was also present. Ceria was usually present in the zirconia phase at concentrations from about 40 to 60 mole percent, with Sm_2O_3 at concentrations from 0 to nearly 10 mole percent. Subsequent experiments with plutonium revealed similar behavior, validating the use of cerium as a surrogate for plutonium in these experiments.

Several experiments were conducted to gain insight into the crystallization kinetics that produced the first devitrified glass samples. Batches of the starting mixture were prepared (with cerium), and melted at various temperatures for various times. Samples of the melt were quenched by pouring into a graphite mold. Up to a melt temperature of 1150° C, the quenched samples consisted of sub-micron CeO_2 crystals in an otherwise homogeneous glass matrix. Very little cerium was dissolved in the glass matrix. When re-heated to 700° C for 144 hours (the devitrification schedule used in the original experiments), the quenched samples devitrified, but the cubic ceria/zirconia phase did not form. Instead, the CeO_2 crystals remained embedded in the devitrified glass. These results suggest that the cubic ceria/zirconia phase forms at higher temperatures as the melt is cooling, with the ceria crystals serving as nuclei for the apparent solid solution phase. Unfortunately, operational restrictions prevented similar experiments from being performed with plutonium oxide.

SYNTHESIS OF THE PREFERRED PLUTONIUM HOST PHASE

Actinide-bearing zirconias have been identified in the solidified material that formed after the Chernobyl accident, and have been suggested as durable actinide host phases (9). Since a cubic zirconia phase heavily substituted with plutonium (or cerium) seemed to form readily in many of the candidate dispositioning materials tested, an effort was initiated to synthesize the material in pure form. Energy dispersive X-ray spectroscopy (EDXS) was used to acquire chemical data during scanning electron microscopic examinations of these materials. The EDXS analysis of the ceria-samaria-zirconia phase identified as spot 1 in Figure 1 is shown in Table II. This data was used to formulate the starting mixture to synthesize the pure ceria-samaria-zirconia phase.

Table II. Results of EDXS analysis of the ceria-zirconia-samaria phase identified as spot 1 in Figure 1. Weak signals from Na, Si, and Ca likely originate from the surrounding matrix material due to the small size of the phase identified as spot 1.

Element	weight %
O	10.1
Na	0.2
Si	0.3
Ca	0.3
Zr	23.4
Ce	48.9
Sm	16.8

The X-ray spectra for sodium, aluminum, silicon, and calcium may have originated from surrounding material. Nevertheless, they were included in the starting formulation in case they had actually been incorporated into the crystal structure. The sodium and calcium were incorporated into the starting mixture as nitrates, all other cations were incorporated as oxides. The molar ratio of cerium to zirconium was 1.36, and Sm_2O_3 was incorporated at 8.5 mole %. The starting ingredients were mixed and placed in an alumina crucible as a loose powder. All firings were performed in an electric furnace with air atmosphere and $MoSi_2$ heating elements. The first firing was at 1500° C for 24 hours, followed by a firing at 1600° C for 24 hours, and a final firing at 1625° C for 17 hours. The progress of the reaction was followed by X-ray diffraction after each firing. The X-ray diffraction patterns of the starting mixture and samples of material after each firing are shown in Figure 2. Progress of the reaction is most easily followed by observing the shift and initial broadening, then sharpening, of the reflections assigned to the face-centered cubic (fluorite) structure. The final ceria-zirconia-samaria phase exhibits this structure, as does CeO_2 in the starting material. The reaction was judged essentially complete after the three higher-angle peaks at 70.6°, 78.3°, and 80.8° 2Θ had become sharp. The two small unidentified peaks at 47.2° and 56.1° 2Θ are probably due to a small amount of unreacted material, possibly resulting from the small amounts of Na, Ca, Al, and Si included in the starting mixture.

After the initial synthesis experiment, another batch of the ceria-zirconia-samaria phase was prepared in a single step by firing the starting mixture at 1675° C for 24 hours. Complete reaction using this firing schedule was confirmed by comparing the X-ray diffraction pattern to the pattern in Figure 2d. Reflections from this pattern also matched reflections in the complex patterns from devitrified materials, indicating that synthesis of the ceria-zirconia-samaria phase observed in the devitrified material had been successful.

Using the experience gained synthesizing the cerium surrogate material, two batches of plutonium-containing material were synthesized, one with samarium and one without. The starting materials were mixtures of PuO_2 and ZrO_2 powders, with or without Sm_2O_3 powder. The molar ratio of plutonium to zirconium was 1.36 (duplicating the Ce/Zr ratio of the surrogate material), with 8.5 mole % Sm_2O_3 added to one sample. The starting powders were placed in alumina crucibles. and fired simultaneously in an electric furnace with air atmosphere and $MoSi_2$ heating elements. The materials were fired for 48 hours at 1650° C to ensure complete reaction. Synthesis was successful, as shown by the X-ray diffraction patterns in Figure 3. The X-ray diffraction pattern of pure PuO_2, which also exhibits the fluorite structure, is shown for reference.

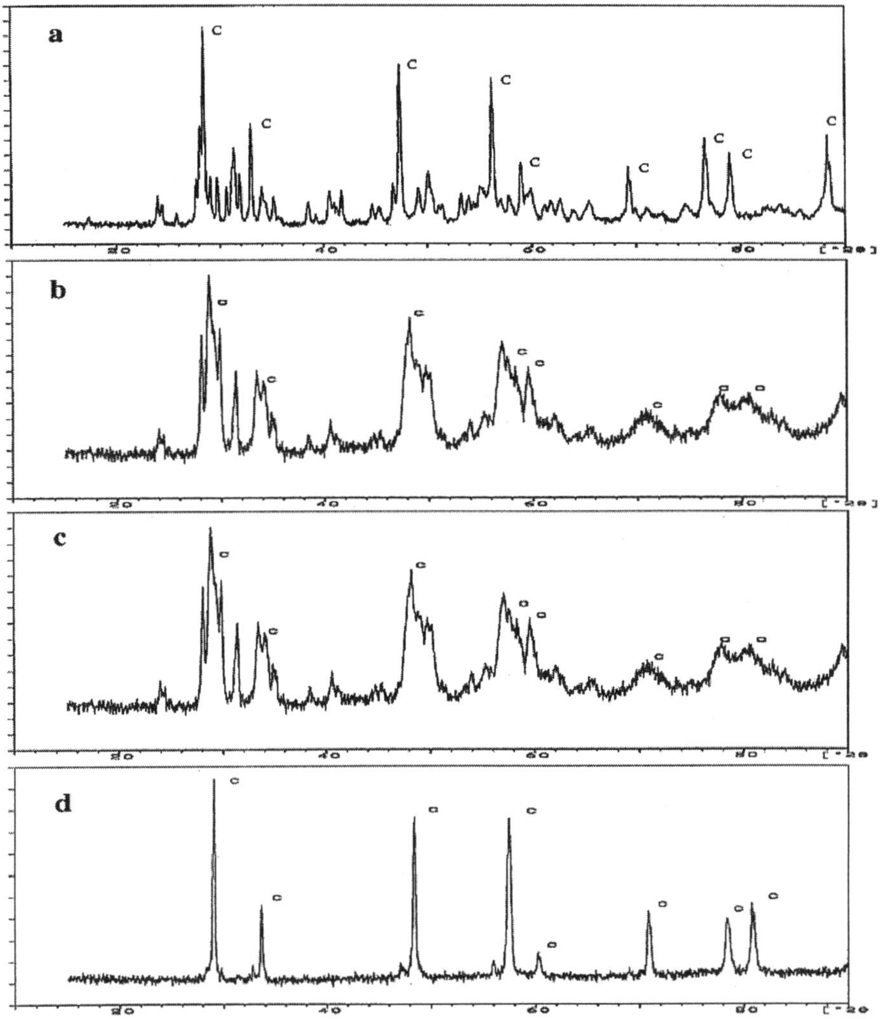

Figure 2: X-ray diffraction patterns of a) the starting mixture, b) the material after 24 hours at 1500° C, c) after an additional 24 hours at 1600° C, and d) after an additional 17 hours at 1625° C. Reflections from a face-centered cubic (fluorite-type) structure are labeled with "C." The final ceria-zirconia-samaria phase exhibits this structure, as does CeO_2 in the starting mixture. The ZrO_2 in the starting mixture was monoclinic, as determined from the diffraction pattern.

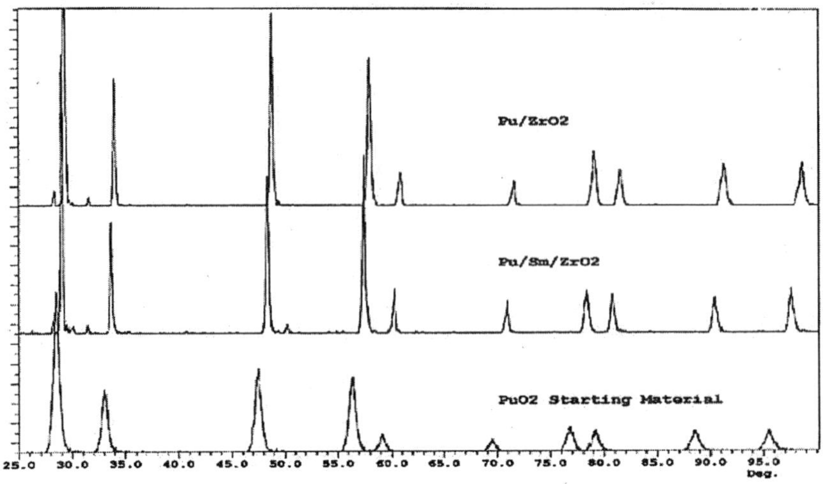

Figure 3: X-ray diffraction patterns of plutonia-zirconia phases, (top) without Sm_2O_3 (lattice parameter 0.528 nm), (middle) with Sm_2O_3 (lattice parameter 0.533 nm), and (bottom) pure PuO_2 for reference.

Normalized plutonium release rates for the zirconates with and without samarium were measured using the seven day Product Consistency Test (12). Plutonium concentrations were calculated from measurements of plutonium activity in samples of the leachate. Normalized plutonium release rates for the plutonium zirconate and plutonium zirconate with samarium were calculated at 4.26×10^{-6} and 1.3×10^{-5} $g/m^2/d$ respectively. These numbers compare to an "acceptable" release rate of 1 $g/m^2/d$ for HLW glass. Leachate pH for the samarium-containing material was 7.58, and conductivity was 0.049 milli-Siemens/cm. These numbers compare to a pH of 6.03 and conductivity of 0.010 mS/cm for control blanks. Unfortunately, insufficient leachate was available to perform these measurements on the plutonium zirconate leachate.

CONCLUSIONS

A durable zirconia-based host phase for plutonium immobilization has been identified and synthesized in phase-pure form. The material can be easily prepared, incorporating neutron absorbers for criticality control, if desired. The material can be prepared by devitrification of appropriate glass formulations, by solid state reaction in powder compacts, or by solid state reaction in loose powders. The ease of preparation and durability of this material make it attractive for a variety of plutonium immobilization applications. To meet the spent fuel standard for dispositioning of weapons grade plutonium, the ability to form or incorporate the material in either a glass or ceramic matrix allows the plutonium to be intimately mixed with fission products or other materials to generate a diversion-resistant radiation field.

SUGGESTIONS FOR FUTURE WORK

The chemical stability of the material suggests other possibilities for plutonium immobilization not previously considered. These possibilities offer opportunities for additional process development work. For example, excess plutonium could be reacted with zirconia and neutron absorbers to form a powder feedstock that could be introduced into HLW immobilization processes in a variety of ways. Plutonium immobilized with neutron absorbers in the zirconate phase could be mixed with molten cesium chloride from the stockpile at Hanford, and cast into sealed containers for dispositioning. This approach would provide much higher fields to deter diversion, and follow a dispositioning route already proposed for the cesium chloride inventory. Other applications might include stabilization of plutonium "residuals" (non-weapons grade plutonium-containing materials). For example, plutonium nitrate solution could be mixed with zirconia and an appropriate neutron absorber, dried, calcined and fired to immobilize the plutonium. The resulting material could be safely handled and stored either as a powder, or encapsulated in an appropriate matrix such as polyethylene or cement to prevent inadvertent dispersion.

ACKNOWLEDGEMENTS

The authors gratefully acknowledge the contributions of M. Hansen, B. Scholes, B. Boyle, T. Morris, M. Hankins, S. Frank, E. Wood, and M. Meyer. This work was performed for the U. S. Department of Energy, Assistant Secretary for Environmental Management, Office of Nuclear Energy under DOE Idaho Operations Office Contract DE-AC07-94ID13223.

REFERENCES

(1) Office of Fissile Materials Disposition, United States Department of Energy, "Technical Summary Report for Surplus Weapons-Usable Plutonium Disposition," Rev. 1, DOE/MD-0003 (1996).

(2) Committee on International Security and Arms Control, National Academy of Sciences, "Management and Disposition of Excess Weapons Plutonium," National Academy Press (1994).

(3) Office of Fissile Materials Disposition, United States Department of Energy, "Record of Decision for the Storage and Disposition of Weapons-Usable Fissile Materials Final Programmatic Environmental Impact Statement," (1997).

(4) J. R. Berreth, "Inventories and Properties of ICPP Calcined High-Level Waste," WINCO-1050, Idaho National Engineering Laboratory, U. S. Department of Energy (1988).

(5) B. A. Staples, D. A. Pavlica, and H. S. Cole, "Properties of Formula 127 Glass Prepared with Radioactive Zirconia Calcine," pp. 125-134 in *Scientific Basis for Nuclear Waste Management II* , Edited by D. G. Brookins. Materials Research Society, Pittsburgh, PA, 1979.

(6) R. S. Baker, B. A. Staples, and and H. C. Wood, "Development of a Ceramic-Based Waste Form to Immobilize ICPP HLW," pp. 924-934 in

Proceedings of Spectrum '86 . American Nuclear Society, La Grange Park, IL, 1987.

(7) R. C. Ewing, W. Lutze, and W. J. Weber, "Zircon: A Host Phase for the Disposal of Weapons Plutonium," *Journal of Materials Research* , **10** 243-46 (1995).

(8) W. J. Weber, R. C. Ewing, and W. Lutze, "Performance Assessment of Zircon as a Waste Form for Excess Weapons Plutonium Under Deep Borehole Burial Conditions," pp. 25-32 in *Scientific Basis for Nuclear Waste Management XIX* , Edited by W. M. Murphy and D. A. Knecht. Materials Research Society, Pittsburgh, PA, 1996.

(9) B. E. Burakov, E. B. Anderson, V. S. Rovsha, S. V. Ushakov, R. C. Ewing, W. Lutze, and W. J. Weber, "Synthesis of Zircon for Immobilization of Actinides," pp. 33-39 in *Scientific Basis for Nuclear Waste Management XIX* , Edited by W. M. Murphy and D. A. Knecht. Materials Research Society, Pittsburgh, PA, 1996.

(10) D. A. Knecht, P. C. Kong, and T. P. O'Holleran, "Proposed Glass-Ceramic Waste Forms for Immobilizing Excess Plutonium," pp. 183-191 in *Proceedings of the Embedded Topical Meeting on DOE Spent Nuclear Fuel and Fissile Material Management* . American Nuclear Society, La Grange Park, IL 1996.

(11) D. A. Knecht, S. Johnson, T. P. O'Holleran, S. Frank, K. Vinjamuri, S. V. Raman, and B. A. Staples, "Glass and Glass Ceramic Waste Forms Developed for Immobilizing Actinides," in *Scientific Basis for Nuclear Waste Management XX* . Materials Research Society, Pittsburgh, PA, in press.

(12) Annual Book of ASTM Standards, **12.01**, pp. 797-814, ASTM, Philadelphia, PA 1995.

CORROSION OF A Pu-DOPED ZIRCONOLITE-RICH CERAMIC

A. J. Bakel, E. C. Buck, S. F. Wolf, B. B. Ebbinghaus,* D. B. Chamberlain, and J. K. Bates
Argonne National Laboratory
9700 South Cass Avenue
Argonne, IL 60439
*Lawrence Livermore National Laboratory
P. O. Box 808
Livermore, CA 94550

ABSTRACT

As part of a large Pu disposition program, a zirconolite-rich titanate ceramic is being developed at Lawrence Livermore National Laboratory (LLNL) as a possible immobilization material. This same material is being tested at Argonne National Laboratory (ANL). The goal of this study is to describe the corrosion behavior of this ceramic, particularly the release of Pu and Gd, using results from several static corrosion tests (MCC-1, PCT-A, and PCT-B). The release of relatively large amounts of Al, Ba, and Ca in short-term tests (3 day MCC-1 and 7 day PCT-A) indicates that these elements are released from grain boundaries or from highly soluble phases. Results from long-term (28, 98, and 182 day) PCT-B show that the releases of Al, Ba, and Ca decrease with time, the releases of U and Zr increase with time, and that the releases of Cs, Gd, Mo, and Pu remain fairly constant. Formation of alteration phases may lead to the decrease of Ba and Ca in leachate solutions. Due to the heterogeneous nature of the material, the formation of alteration phases, and the inherently low solubility of several elements, no element(s) could be recommended as good markers for the overall corrosion of this ceramic. Our data show that, due to the complex nature of this material, the release of each element should be considered separately.

INTRODUCTION

Due to the ratification of several nuclear nonproliferation treaties, large quantities of weapons-grade Pu, contaminated Pu stock, and Pu scrap existing in the United States have been declared surplus and suitable for disposition. Several options are currently being considered by the US Department of Energy, Office of Materials Disposition, including immobilization. This option involves incorporating Pu into a glass or ceramic material [1]. As part of this effort, titanate-based ceramics are being developed at LLNL as materials for immobilizing Pu.

Studies of the corrosion of titanate ceramics and natural titanate minerals show that these materials are very resistant to aqueous corrosion [2-5]. Studies of SYNROC (a titanate ceramic designed to stabilize nuclear wastes) show that its corrosion is characterized by a relatively large initial release of some elements (Ba, Cs, and Mo), presumably from grain boundaries and amorphous material [5-7]. In this case, the corrosion rates observed in short-term tests (e.g. MCC-1) are not necessarily representative of longer-term corrosion rates of the same material. The corrosion rate might also be affected by the formation of a Ti-rich layer on the surface [8], which could retard the interdiffusion of water and Ca^{+2} [9].

The purpose of this study is to describe the corrosion behavior of a Pu-doped, zirconolite-rich ceramic material. Test results will be used to evaluate the resistance of the ceramic to aqueous corrosion, and to compare its corrosion behavior to other candidate immobilization materials. In particular, we are interested in the release of Pu and Gd. Since the goal of this program is to immobilize Pu, its release from the material is an important criterion for deciding on a final immobilization material. Gadolinium will be present in any Pu immobilization material as a neutron absorber to prevent a criticality event. The corrosion data presented here will be used, along with microcharacterization of the ceramic previously reported [10], to provide a comprehensive description of the corrosion of the materials.

EXPERIMENTAL

Preparation of the Pu-Doped Ceramic

The PuO_2 used to prepare the ceramic was made at LLNL from a low temperature oxidation of $PuH_{2.7}$ using 4%O_2/Ar. The temperature never exceeded 200°C during the oxidation process. The hydride was formed from actual pit material at about 100°C.

Samples of four ceramics were prepared at LLNL and sent to ANL for testing. Details of the preparation of these ceramics are presented elsewhere [10]. The target chemical compositions (as batched) of each ceramics is given in Table I. The target phase composition of these ceramic is 90% zirconolite ($CaZrTi_2O_7$) with the balance comprised of perovskite ($CaTiO_3$), hollandite $\{(Ba_xCs_y)[(TiAl)_{2x-y}]O_{16}\}$, and rutile ($TiO_2$).

Table I. The Compositions (As Mixed) of Ceramics Used in this Study

Oxide	Ceramic A	Ceramic B	Ceramic C	Ceramic D	PCT Mixture*
CaO	7.35	6.55	8.07	8.07	7.38
ZrO_2	16.2	14.4	17.7	17.7	16.2
TiO_2	38.6	34.4	40.4	40.4	40.0
Al_2O_3	10.9	9.71	4.71	4.71	7.99
BaO	2.99	2.67	3.37	3.37	3.04
PuO_2	14.4	19.3	16.9	16.9	17.1
Gd_2O_3	9.61	12.9	8.75	8.75	10.4

*A mixture of 26.4 mass % of ceramic A, 32.9 mass % of ceramic B, 20.2 mass % of ceramic C, and 20.5 mass % of ceramic D was used in PCT.

Corrosion Test Procedures

Several types of tests were performed to highlight different aspects of the corrosion process.

The MCC-1 test was originally developed to directly compare the relative durabilities of different materials. The test method has been standardized in ASTM C1220 [11]. Monoliths (1 cm in diameter, 1 mm thickness) were cut from ceramic B (Table I) and polished to a 240-grit finish. Tests were conducted by sealing a monolith in a stainless steel vessel with deionized water (18 g) for three days at 90°C. After completion of the test the vessel was filled with 2% HNO_3, and placed in a 90°C oven for at least 8 hours. This "acid strip" solution contains any material that was fixed on the vessel wall. The leachate and acid strip solutions were analyzed with inductively coupled plasma-mass spectrometry (ICP-MS). This test provides valuable information regarding the corrosion behavior of the material in dilute solutions. It has been shown that the corrosion rate of a material is highest when the contacting solution is dilute.

The Product Consistency Test (PCT) was originally developed to monitor the consistency of high-level radioactive waste glasses. It has been standardized in ASTM C1285 [12]. The test is performed with crushed material (-100+200 mesh) and deionized water in stainless steel vessels. Due to the small amount of each of the four ceramic samples, we were unable to test each of them separately. Ceramics A, C and D, along with the portion of ceramic B remaining after monolith preparation were crushed and sieved to produce -100+200 mesh powder. The composition of the mixture used for the PCT is also given in Table I.

Tests were conducted under two sets of conditions: PCT-A, in which one gram of test material (ρ=4.5 g/cm^3) is sealed in the vessel with 10 g of water [giving a solid surface area to solution volume ratio (S/V) of about 1200 m^{-1}] for 7 days at 90°C; and PCT-B in which 5 g of test material is sealed in the vessel with 5 g of water (giving an S/V of about 12,000 m^{-1}) for various times at 90°C. After completion of the test the vessel was filled with 2% HNO3, and placed in a 90°C oven for at least 8 hours. This "acid strip" solution contains any material that was fixed on the vessel wall. The leachate and acid strip solutions were analyzed using ICP-MS. These tests provide valuable information regarding the corrosion behavior of the material in concentrated solutions. The corrosion rate of a material typically decreases as the concentration of certain components increases in the contacting solution.

RESULTS AND DISCUSSION

Short-Term Test Results

Table II shows the amounts of several elements released from the ceramic during the 3 day MCC-1 and 7 day PCT-A. Six replicates of the MCC-1 and three replicates of the PCT-A were performed. The values presented in Table II are averages of the replicates ± the standard deviation and have been background-corrected. The normalized mass loss values shown in Table 2 are calculated using the total surface area of the ceramic in the test. These values are maxima, since any one element is released from only a portion of the total surface area. The background values for each element were calculated from results of a series of blank tests. Some of the elements found in the test solutions (U, Cs, and Mo) were not reported to ANL as added components. These elements are presumably contaminants from the feedstock material or from processing.

The results from these short-term tests (3 day MCC-1 and 7 day PCT-A, Table II) can be used to understand the corrosion behavior of the ceramic material. The total surface area available in a PCT-A (1.2x10^{-2} m^2) is much larger than that available in a MCC-1 test (1.9x10^{-4} m^2) and the results of these tests must be interpreted with this in mind. The amount of Al, Ca and Ba released from the ceramic in PCT-A is much larger than is released in the MCC-1 tests. These elements are probably incorporated into grain boundaries or phases that are

Table II. Results From Static Corrosion Tests. The amount of material released includes material in the leachate solution and material adsorbed onto the vessel wall.

	MCC-1		
	Leachate, mg	Acid strip, mg	NL
Al	$1.5 \pm 1.0 \times 10^{-4}$	$5.9 \pm 3.7 \times 10^{-4}$	$1.2 \pm 0.9 \times 10^{-1}$
Ca	$3.4 \pm 0.83 \times 10^{-3}$	bdl*	$3.3 \pm 1.2 \times 10^{-1}$
Ti	bdl	bdl	bdl
Zr	$1.2 \pm 1.2 \times 10^{-4}$	$4.1 \pm 5.2 \times 10^{-4}$	$6.5 \pm 12 \times 10^{-3}$
Gd	$8.7 \pm 9.4 \times 10^{-4}$	$1.0 \pm 1.4 \times 10^{-4}$	$9.3 \pm 13 \times 10^{-2}$
Pu	$1.0 \pm 1.1 \times 10^{-5}$	$4.7 \pm 2.0 \times 10^{-5}$	$1.1 \pm 2.1 \times 10^{-2}$
Ba	$3.1 \pm 2.3 \times 10^{-3}$	$8.3 \pm 5.9 \times 10^{-5}$	na**
	PCT-A		
	Leachate, mg	Acid strip, mg	NL
Al	$5.6 \pm 1.1 \times 10^{-3}$	$1.2 \pm 0.5 \times 10^{-3}$	$1.1 \pm 0.2 \times 10^{-2}$
Ca	$1.9 \pm 1.1 \times 10^{-2}$	$1.4 \pm 1.2 \times 10^{-3}$	$3.1 \pm .0.4 \times 10^{-2}$
Ti	$2.1 \pm 0.6 \times 10^{-5}$	$2.9 \pm 1.9 \times 10^{-4}$	$1.2 \pm 6.7 \times 10^{-4}$
Zr	bdl	bdl	bdl
Gd	bdl	$4.3 \pm 2.7 \times 10^{-4}$	$4.8 \pm 3.0 \times 10^{-4}$
Pu	bdl	$4.0 \pm 1.2 \times 10^{-4}$	$2.2 \pm 0.7 \times 10^{-4}$
Ba	$1.6 \pm 0.0 \times 10^{-1}$	$4.6 \pm 1.4 \times 10^{-3}$	na

*Below detection limit.

**Normalized mass loss could not be calculated for this element.

particularly susceptible to corrosion. Their release is dependent on the total solid surface area. The total release of Pu and Gd in PCT-A and MCC-1 are about the same.

Long-Term Results

Figure 1 and Table 3 show the results of a series of PCT-B tests conducted at 90°C and an S/V of about 12,000 m^{-1}. The amount of an element released (y-axis in Figure 1) includes the material in the leachate and the acid strip solutions. The uncertainty in these concentrations is about ±15%, based on previous studies, and error bars in Figure 1 lie within the printed symbols. Trends in these data as a function of time can be interpreted in terms of processes affecting the corrosion of the ceramic.

Three of the elements (Al, Ca, and Ba) show decreasing release as a function of time. Such a decrease can be explained as a result of precipitation of solid phases. Small particles of BaO have been observed on the surface of glass particles reacted

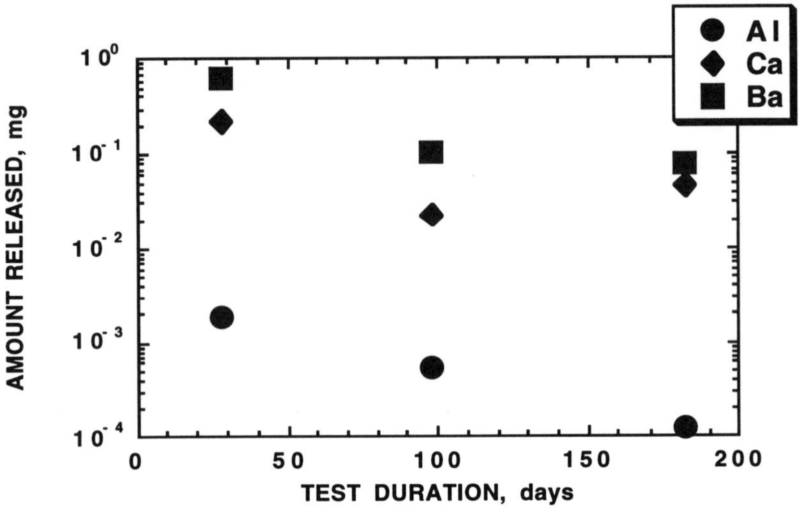

Al
Ca
Ba
10^0
10^-1
10^-2
10^-3
10^-4
AMOUNT RELEASED, mg
0
50
100
150
200
TEST DURATION, days

(a)

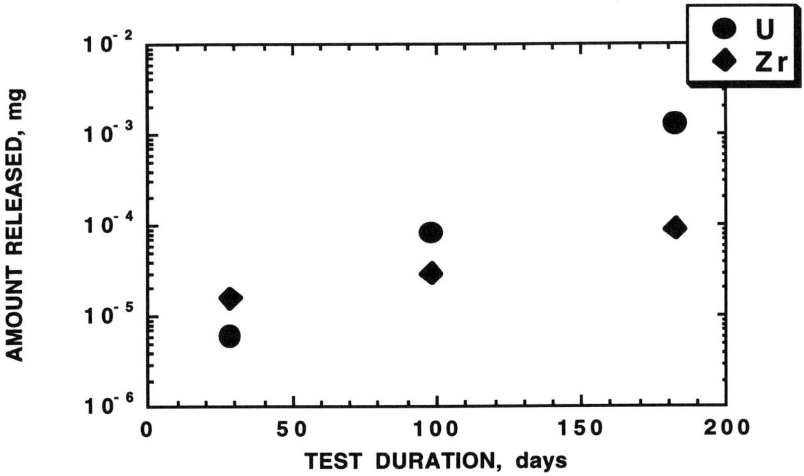

U
Zr
10^-2
10^-3
10^-4
10^-5
10^-6
AMOUNT RELEASED, mg
0
50
100
150
200
TEST DURATION, days

(b)

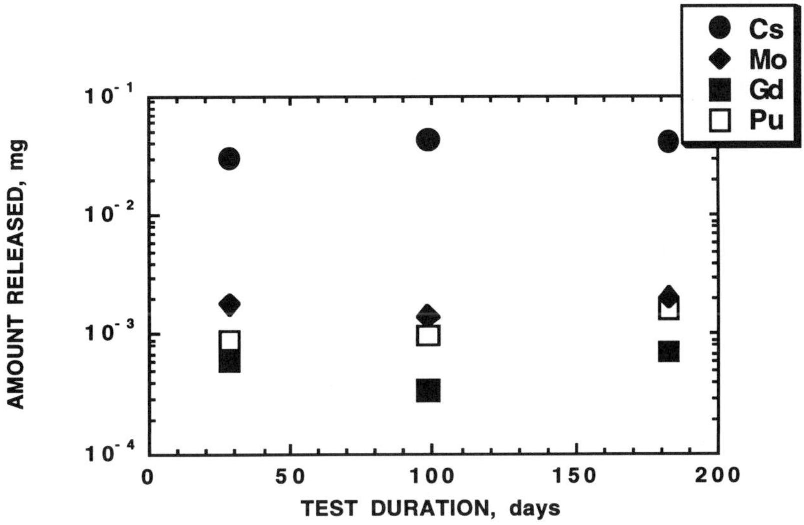

(c)

Figure 1. Results from PCT-B (90°C, 12,000 m^{-1}) for the indicated elements as a function of time. The amount of each element released includes what is present in the leachates and what is adsorbed on the vessel walls. Note that the y-axis is logarithmic.

for 98 days. This explains the decrease in the amount of Ba released with time (Figure 1a). Particles of $CaCO_3$ were detected with AEM in the leachate from 182 day PCT-B. A small amount of the leachate filtered through a carbon grid (pore size, 5 nm) was examined. Formation of these Ca-rich particles may explain the decrease in Ca concentrations during PCT-B (Figure 1a).

The amount of U and Zr released from the ceramic continues to increase in tests with durations up to 182 days (Figure 1b). This shows that these elements are still being released into solution up to 182 days.

The amounts of Cs, Gd, Mo, and Pu released into solution remain quite constant between 28 and 182 days (Figure 1c). This pattern would result if a highly soluble minor phase(s), containing these elements dissolved in the first 28 days of reaction. Alternatively, such a pattern would result if the release of these elements from the ceramic, under these conditions, is solubility controlled and that the maximum solubilites was reached before 28 days. This seems

Table 3. Measured elemental compositions (in mg/L) of solutions from the long-term PCT-B tests, and the total amount of each element released from the ceramic during the test (in mg).

	PSN-4, 28 days			PSN-5, 98 days		
Element	Leach[a]	AS[b]	mg[c]	Leach	AS	mg
Al	0.080	0.064	0.0019	<0.050	0.031	0.000030
Ca	22	0.083	0.22	2.2	<0.20	0.019
Ba	63	0.22	0.63	8.6	1.1	0.10
U	<0.000030	0.00040	0.0000056	0.0070	0.00050	0.000080
Zr	0.00050	0.00070	0.000020	0.0020	0.00050	0.000030
Cs	3.0	<0.0002	0.030	4.3	0.0030	0.043
Mo	0.18	<0.00050	0.0018	0.14	<0.0008	0.0014
Gd	<0.00020	0.037	0.00062	0.0010	0.019	0.00034
Pu	0.0010	0.051	0.00087	0.016	0.048	0.0010

	PSN-6, 128 days		
Element	Leach	AS	mg
Al	<0.30	0.070	<0.0020
Ca	4.4	0.14	0.046
Ba	6.8	0.62	0.078
U	0.13	0.00090	0.0014
Zr	0.0070	0.0020	0.000090
Cs	4.1	0.0070	0.041
Mo	0.20	0.0005	0.0020
Gd	0.019	0.031	0.00072
Pu	0.040	0.076	0.0017

[a]Concentration of element in mg/L of leachate.
[b]Concentration of element in mg/L of acid strip solution.
[c]Total amount of each element released during the test in mg.

unlikely given the high solubility of Cs and Mo. Calculations are underway to estimate the solubility limits of these elements in these solutions. In either case, the data show that the release of these elements is quite slow after 28 days.

Possible Marker Elements

It is often desirable to define a marker element that accurately describes the extent of corrosion of a material. Such a marker element must be released from the material at a maximum rate and must not be incorporated into any alteration phases. Previous studies have used Ca as a marker element for the corrosion of titanate ceramics in short-term static tests [2-5]. Our data show that Ca is incorporated into alteration phases in long-term PCT (182-day, 90°C, 12,000 m^{-1}), and that the release of Ca from the ceramic decreases with time between 28 and 182 days. In addition, previous characterization studies have shown that Ca is present in several distinct phases in the ceramic [10]. Based on these observations, we view Ca as a poor marker element. The relatively low solubility of Ti and Zr compounds [13] makes them likely to be involved in precipitation reactions, particularly during long-term or accelerated tests. This presents a serious difficulty for using either of these elements as markers. Therefore, the major elements present in the ceramic are all inappropriate for use as corrosion marker elements. The presence of most of the minor constituents in multiple phases within the ceramic makes them inappropriate as well. A highly soluble element, which is rarely incorporated into alteration phases (such as B) might be incorporated into the zirconolite structure to serve specifically as a marker element.

CONCLUSIONS

Data from this ongoing study show that interpretation of static test results are complicated by the heterogeneous nature of the ceramic material. The release of each element should be considered separately. We believe that the release of any one element depends on the phases in which it occurs, and the solubility of the element in the leachate solution. In addition, the release of Ba, and Ca in some long-term tests appears to be affected by the formation of alteration phases.

The corrosion behavior of this material observed in this study is consistent with previous studies. A relatively large amount of material, primarily Ba, Ca, and Al, is initially released from grain boundaries and from highly soluble phases. These processes are reflected in the results of short-term tests (MCC-1 and PCT-A).

Data presented here shows that the corrosion of this ceramic in short- and long-term static tests releases only small amounts of Gd and Pu. Therefore, based on the limited data presented in this study, this ceramic appears to be an effective immobilization material for Pu.

ACKNOWLEDGMENTS

This work was supported by the U.S. Department of Energy, Office of Materials Disposition, under contract W-31-109-ENG-38. Assistance was provided by J. Emery, T. DiSanto, L. Hafenrichter, and M. Surchik. AJB would like to dedicate this work to Thelma "Nana" Will.

REFERENCES

[1]Department of Energy, *Record of Decision for the Storage and Disposition of Weapons-Usable Fissile Materials; Final Programmatic Environmental Impact Statement*, January 14, 1997.

[2]E.R. Vance, C.J. Ball, R.A. Day, K.L. Smith, M.G. Blackford, B.D. Begg, and P.Angel, "Actinide and Rare Earth Incorporation into Zirconolite," J. Alloys Comp. **213/214**, 406-409 (1994).

[3]V.M. Oversby and W.E. Ringwood, "Leaching Studies on Synroc at 95° and 200°C," Rad. Waste Mgmt. **2**, 223-237 (1982).

[4]G.R. Lumpkin, K.P. Hart, P.J. McGlinn, T.E. Payne, R. Giere, C.T. Williams, "Retention of Actinides in Natural Pyrochlores and Zirconolites," Radiochim. Acta. 469-474 (1995).

[5]P.J. McGlinn, K.P. Hart, E.H. Loi, and E.R. Vance, "pH Dependence of the Aqueous Dissolution Rates of Perovskite and Zirconolite at 90°C," Mater. Res. Soc. Symp. Proc. (1996).

[6]A.G. Solomah and H.J. Matzke, "Leaching Studies of SYNROC Crystalline Ceramic Waste Forms," Mater. Res. Soc. Symp. Proc. **127**, 241-248 (1989).

[7]J.A. Cooper, D.R. Couens, R.A. Lewis, S. Myhra, R.G. Segall, R.St.C. Smart, P.S. Turner, and T.J. White, "Microstructural characterization of SYNROC C and E by electron microscopy," J. Am. Ceram. Soc. **68**, 64-70 (1985).

[8]K.L. Smith, G.R. Lumpkin, M.G. Blackford, R.A. Day, and K.P. Hart, "The Durability of SYNROC," J. Nucl. Mater. **190**, 287-294 (1992).

[9]D.K. Pham, F.B. Neall, S. Myhra, R.St.C. Smart, and P.S. Turner, "Dissolution Mechanisms of $CaTiO_3$ - Solution Analysis, Surface Analysis and Electron Microscope Studies - Implications for SYNROC," Mater. Res. Soc. Symp. Proc. **127**, 231- 240 (1989).

[10]E.C. Buck, B. Ebbinghaus, A.J. Bakel, J.A. Fortner, D.B. Chamberlain, and J.K. Bates, "Characterization of a Pu-Bearing Zirconolite-Rich Synroc," Mater. Res. Soc. Symp. Proc. (1996).

[11]ASTM, "Test Method for Static Leaching of Monolithic Waste Forms for Disposal of Radioactive Waste, "**ASTM Standard C1220-92**, ASTM Philadelphia, PA. (1992).

[12]ASTM, "Standard Tests Methods for Determining Chemical Durability of Nuclear Waste Glasses: The Product Consistency Test (PCT), **ASTM Standard C1285-94**, ASTM, Philadelphia PA (1994).

[13]H.F. Wu, C.C. Lin, and P.Y. Shen, "Structure and Dissolution of CaO-ZrO_2-TiO_2-Al_2O_3-B_2O_3-SiO_2 Glass," J. Non. Cryst. Sol. **209**, 76-86 (1997).

SAFETY CONCERNS IN DISPOSAL OF THE NATION'S THORIUM NITRATE STOCKPILE*

L. A. Harris,** D. P. Hoffmann, R. C. Melton, R. J. Stevenson, W. H. Hermes, L. V. Gibson, and F. R. VanRyn
East Tennessee Technology Park
Oak Ridge, TN 37831-7274

ABSTRACT

Disposal of the Defense Logistics Agency thorium nitrate hydrate (ThNH) stockpile is currently under consideration. This paper describes a series of low-temperature (0–650°C) tests performed with ThNH only and with mixtures of ThNH and $Ca(OH)_2$ and ThNH and plastic to identify potential safety problems. Results from these tests reveal no significant reactions from ThNH only and or from ThNH mixed with $Ca(OH)_2$. In contrast, ThNH and plastic react exothermically and form a mushroom-like growth in the center of the mixture. The growth is preceded by fire and billowing smoke reminiscent of volcanic action. This observation is important in that it identifies a need for caution in attempting to convert whole drums of ThNH.

INTRODUCTION

The Defense National Stockpile Center managed by the Defense Logistics Agency (DLA) currently stores 2355 metric tons of thorium nitrate hydrate (ThNH) in several facilities. The ThNH is reportedly present as the tetrahydrate and is contained in plastic bags within fiberboard barrels, which are subsequently enclosed in steel drums.[1] Variations in the storage configurations consist of the addition of $Ca(OH)_2$ as an overpack on the ThNH within the steel drums. Other modifications consisting of bricks and/or tar as pedestals for the steel drums emplaced in larger steel drums are found at one of the sites for a limited number of containers.

A diminishing need for breeder reactor fuel and the desire to mitigate any potential environmental hazards are the primary reasons for converting this stockpile to a

---

*Work sponsored by the Defense Logistics Agency, Department of Defense, under Interagency Agreement 1872-H070-A1 with the Department of Energy and performed by Lockheed Martin Energy Systems, Inc., under contract DE-AC05-84OR21400.

**Member, American Ceramic Society.

nonoxidizing form. Due to the health and safety concerns associated with handling this material, one of the program objectives was to determine the feasibility of whole drum processing by plasma heating. Preliminary tests of surrogate nitrates (magnesium nitrate and calcium nitrate) showed that the interaction with polyethylene can produce exothermic reactions.[2]

The above observations suggested a need for determining whether a similar exothermic reaction occurs when heating ThNH and plastic. Unlike the surrogate material, ThNH must not be released into the environment because it presents an unacceptable radiological hazard. In the present paper the results of low-temperature (0–650°C) testing of ThNH with $Ca(OH)_2$ and with polyethylene are presented and discussed.

EXPERIMENTAL

Sample Selection

Samples were obtained from a 35-gallon steel drum (Figure 1) originating from the DLA storage facility at Curtis Bay, Maryland, and transported to the East Tennessee Technology Park (ETTP) at Oak Ridge, Tennessee. Constituents from within the drum (e.g., ThNH, polyethylene from the inner and outer dense plastic bags, and fiberboard from the inner plastic drum lid) were collected and placed into small steel canisters for storage. The hygroscopic nature of the ThNH required that the canister be kept in a drierite-filled container. Although $Ca(OH)_2$ has been reported as a constituent in these drums, none was present in the drum at ETTP. However, to account for the possibility that $Ca(OH)_2$ was inadvertently removed from th drum prior to its shipment to ETTP, reagent grade $Ca(OH)_2$ was used in tests where $Ca(OH)_2$ was required.

Sample Testing

Heating tests were performed using a low-temperature (0–650°C) furnace system as shown in Figure 2. The furnace consists of a nichrome heating coil wound around the walls of a ceramic cup fashioned from a fire brick. The sample chamber is made up of interchangeable glass jars; this scheme permitted initial and final sample weighing without removing the samples from the jars. Temperature data were collected from between the ceramic liner and jar (furnace temperature) and from within the jar (sample temperatures) as a function of time for heating rates of 6 and 25°C/min.

The furnace system was enclosed within a plexiglass box to contain any ejected sample and to allow direct observation of the sample and temperature recordings during heating. This arrangement allowed for videotaping of each test and its corresponding temperature data. Visual observations of changes in color and/or

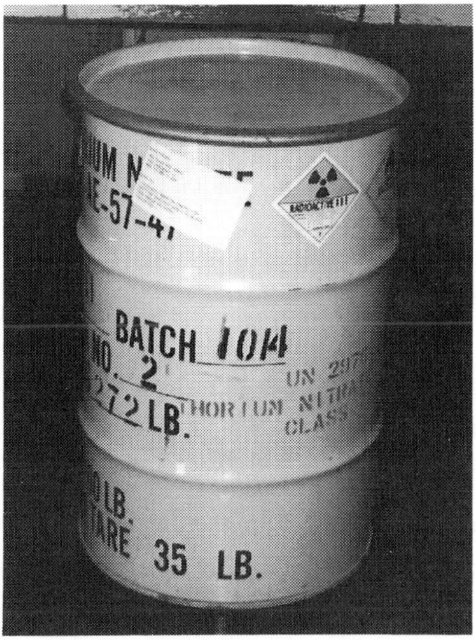

Fig. 1. Photograph of Maryland drum containing ThNH, plastic, and fiberboard.

Fig. 2. Two views of experimental degas furnace developed for ThNH testing.

appearance of the samples (e.g., tan vapors as an indicator of denitration) were also noted.

The plexiglass box contained an outlet that was connected to an off-gas collection system used in part to evaluate particulate transport properties during the heating process and to ensure complete constituent collection. Air from the enclosure was pulled through a Teflon filter exhibiting at least 99.95% efficiency (<0.05%

penetration) on 0.3-μm dioctylphthalate smoke particles followed by three impingers connected in series with leak-free connectors. The first impinger contained at least 200 mL of dilute sodium hydroxide solution, the second was initially empty, and the third contained at least 200 g of color-indicating silica gel. A commercial stationary source-sampling train was used to control and meter the air removal from the enclosure. A water manometer was used to verify the slightly negative pressure inside the enclosure relative to surroundings. The filters were recovered for characterization of collected material.[3]

Sample Characterization

Characterization of the feed material, the reaction products from the low-temperature heating tests, and particulate captured on the off-gas system HEPA filters was performed with the aid of X-ray diffraction (XRD), scanning electron microscopy (SEM), and energy-dispersive X-ray spectroscopy (EDS). XRD was particularly important in that it served to identify the phases in the test products, whereas SEM and EDS analyses were more useful in the examination of carry-over particles embedded on the Teflon filters. In addition, XRD peak profiles were used to compute approximate crystallite size for the thoria product powders.

RESULTS

Prior to discussion of the results from this study, it should be noted that the composition of the drum material was reported as $Th(NO_3)_4 \cdot 4H_2O$. However, weight-loss data obtained in the course of the present study suggested that the actual composition of the drum material is $Th(NO_3)_4 \cdot 5H_2O$. In spite of these observations, some ambiguity still exists, as the XRD pattern from the stockpile nitrate does not fit any reported ThNH compound.[4]

ThNH

ThNH was heated to 600°C at rates of 6° and 25°C/min to establish a baseline for the thermal treatment of the DLA ThNH stockpile and to aid in assessing the effects from mixing other drum constituents with ThNH. Visually observed changes concomitant with thermocouple readings from the furnace and sample reveal the various reactions that occur on heating ThNH. For example, it was observed that the sample started to boil at ~165°C, going to a liquid phase with tan vapors emanating from the furnace vial. By the time the sample had reached ~226°C, the reaction appeared to be close to completion, rendering the sample a solid white cake. These observations appear to be in good agreement with literature data on differential thermal analyses results from the thermal treatment of ThNH.[5,6,7]

The identity of the final product was obtained by means of XRD analyses, which revealed broad diffraction maxima belonging exclusively to ThO_2. These diffraction maxima were used to compute an approximate crystallite size of 10 nm for the thoria products. Figures 3 and 4 are SEM photomicrographs of the thoria taken at 3000x and 700x, respectively. These photomicrographs illustrate the nature of the thoria product, which contains particles ranging from submicrometer size to >50 μm. The morphology of the thoria suggests fracturing of a caked material, as evidenced in part by the particles' angularity.

Fig. 3. SEM photomicrograph of thoria particles at 3000x.

Fig. 4. SEM photomicrograph of thoria particles at 700x.

In consideration of future thermal processing of ThNH, additional heating tests at 6 and 25°C/min were performed to determine the lowest practical temperature for the transformation of ThNH to ThO_2. The tests were halted immediately following the last visible reaction. The final temperature ranged from 330 to 450°C for the slow and fast heating rates, respectively. A comparison of the XRD spectra for the present test and the earlier one heated to 600°C is presented in Figure 5. It can readily be seen that the lower-temperature test results in a diffraction pattern with increased peak breadths, indicating much smaller crystallite sizes and possibly incomplete conversion. Data from the weight loss analyses also suggest incomplete dewatering, with residual water adhering to the ThO_2 surfaces; these observations are consistent with the findings of Ferguson.[8]

ThNH and $Ca(OH)_2$

As stated in the introduction, $Ca(NO_3)_2 \cdot 6H_2O$(CaN) reacts exothermically with plastic. Because $Ca(OH)_2$ is reportedly present in these drums, it was necessary to determine whether reactions between ThNH and $Ca(OH)_2$ (on heating to 600°C) result in a transfer of nitrate to CaO and, consequently, an exothermic reaction if plastic is also present. To explore this hypothesis a set of tests using mixtures of ThNH and $Ca(OH)_2$ heated to 600°C were carried out. The results from these experiments are illustrated in Figure 6, which is an XRD pattern of the final product, clearly shows two phases, ThO_2 and $Ca(OH_2)$, without any indication of CaN. In addition, visual and thermocouple data revealed no unexpected reactions.

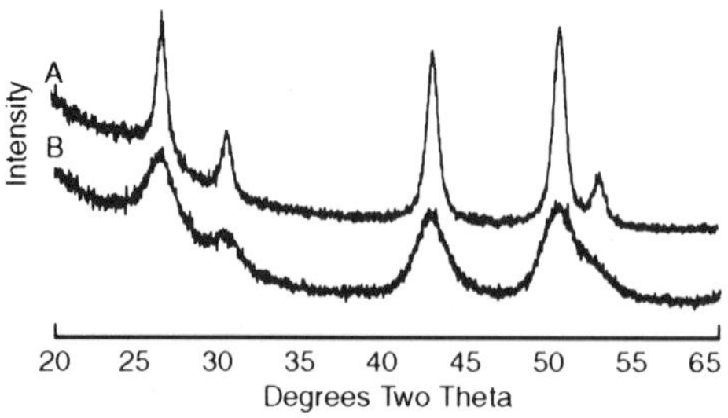

Fig. 5. Comparison of ThO_2 at (a) 600°C and (b) 350°C.

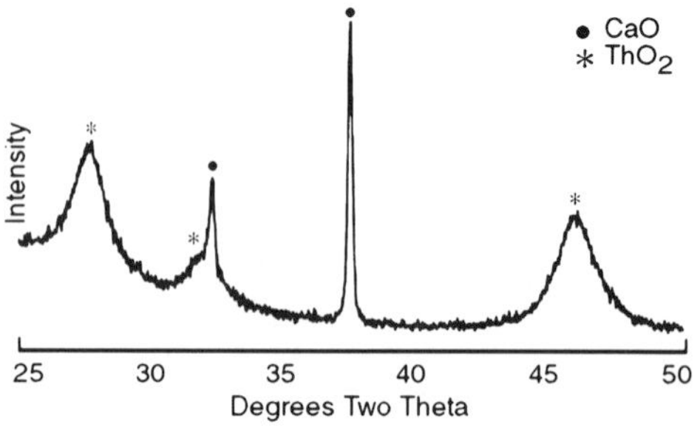

Fig. 6. XRD pattern of products from heating ThNH and $Ca(OH)_2$ mixture to 600°C.

ThNH and Plastic

The final series of tests, and probably the most significant, pertain to reactions between ThNH and plastic. Results from these tests may well affect the treatability pathway for the stockpile. Heating tests using a mixture of ThNH and plastic were performed with some variations in material configurations and heating protocols. (The ratio of plastic to ThNH was based on that ratio within the drums, namely, 0.15 to 1.00.) For example, two tests consisted of a physical mixture of ThNH and plastic and were carried out at heating rates of 6 and 25°C/min, respectively, whereas in a third test, performed at a heating rate of 6°C/min, the sample was held at a temperature of 145°C for 10 minutes before being heated to 600°C. The final test was performed with plastic in the bottom of the vial to which ThNH was added, in order to simulate ThNH in plastic bags.

General observations common to all of these samples during heating were melting followed by encirclement of a light-colored core of material by a dark (tan) outer rim and the emanation of tan vapors over a broad range of temperatures. Continued heating of the samples resulted in events as shown in Figure 7. In frame 2, a profusion of smoke is seen, but flame occurring in the vial is not capture in this frame. The billowing of smoke continues, and in frame 3 an exothermic reaction takes place, as the sample temperature reads 320°C, compared to a furnace temperature of 223°C. The reaction proceeds as shown in frame 4, where the center core is raised above the surrounding outer rim (see arrow), analogous to volcanic action. Further heating of the samples sees the collapse of the white core and some less dramatic occurrences with the cessation of the smoke. Some samples exhibited some flashing even at temperatures near 577°C, which may be due to incompletely converted ThNH nodules.

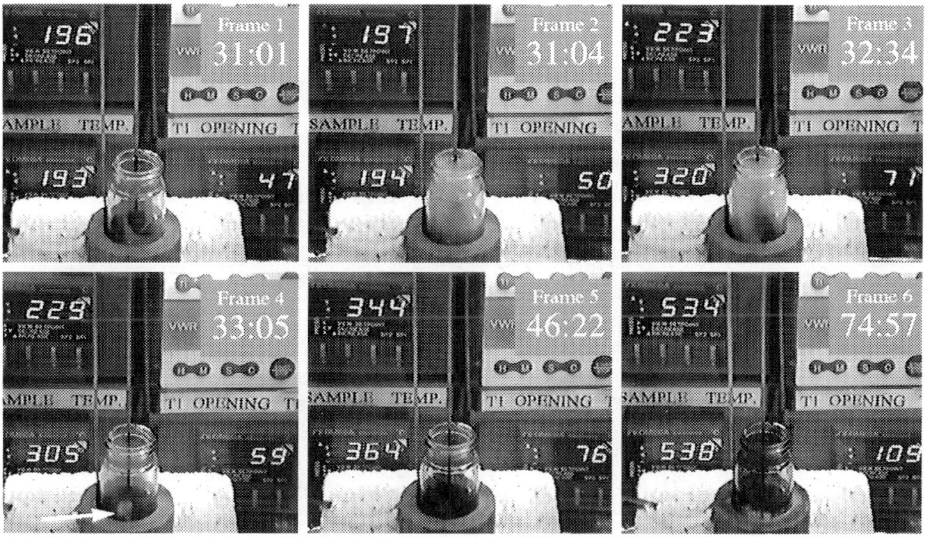

Fig. 7. Videorecorded exothermic reaction for of ThNH-plastic mixture heated at ~6°C/min. Frames 2 and 3 show smoke profusion; frame 4 shows white center (see arrow) risen above material. Frame 6 shows second reaction resulting in vapor filling vial.

The off-gas system was briefly discussed in the section on sampling, and it was noted that the Teflon filters would be recovered and analyzed for particle carryover. Table I compares results of alpha monitoring of Teflon filters from heat treatment tests involving ThNH only and ThNH plus plastic. The marked increase in alpha counts for the ThNH-plus-plastic tests are due to exothermic reactions resulting in particles being ejected from the reacting materials. Figures 8 and 9 are an SEM and an EDS analysis of a particle captured in the Teflon filter. The thorium and oxygen peaks in Figure 9 are from a ThO_2 particle, whereas the fluorine and carbon peaks are from Teflon, with chromium and copper peaks due to sample preparation. Further analyses suggest a size range of 2 to 15 μm for the particles.

CONCLUSIONS

Several significant conclusions drawn from these tests are listed below.

- With no other material present, ThNH decomposes to ThO_2 with no significant exothermic reaction.
- Mixtures of ThNH and $Ca(OH)_2$ decompose to form ThO_2 and CaO without any observable exothermic reaction.

Table I. Alpha readings from test meters

Test	Alpha reading (counts/min)[a]
ThNh only, heated at 6 °C/min	–0.02
ThNH only, heated at 25 °C/min	–0.02
ThNH and plastic, heated at 6 °C/min	0.58
ThNH and plastic, heated at 25 °C/min	0.60

[a]All values are corrected for background.

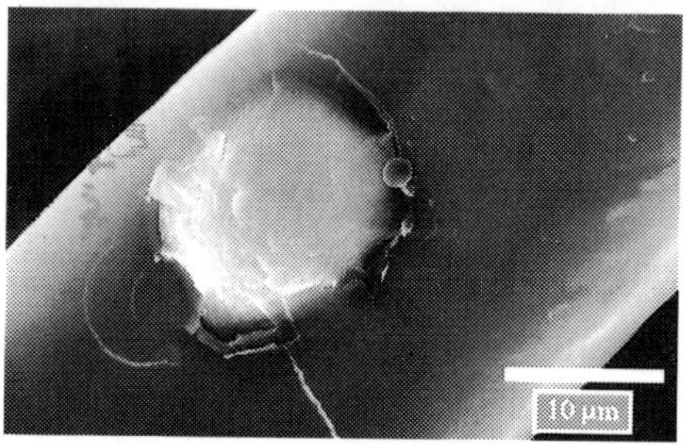

Fig. 8. SEM photomicrograph of thorium-rich particle embedded in Teflon filter fiber.

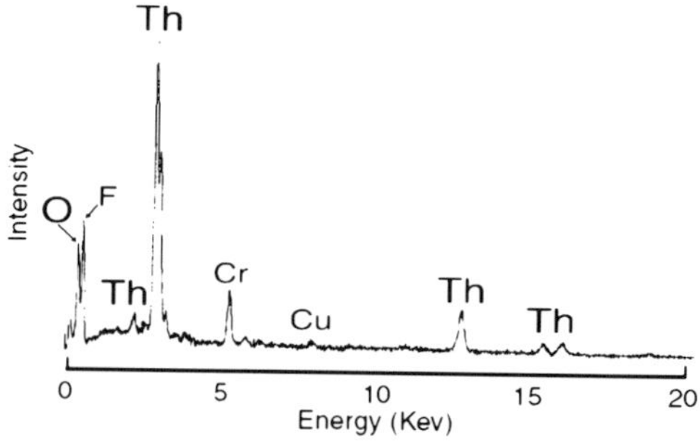

Fig. 9. EDS data for thorium-rich particle shown in Fig. 8.

- ThNH and plastic react exothermically and form a mushroom-like growth in the center of the mixture. The growth is preceded by flame and billowing smoke reminiscent of volcanic action. This demonstrates that an exothermic reaction occurs when ThNH and plastic are heated together. Thus it is necessary to separate plastic from ThNH in processing ThNH.

- A minimum temperature of 350 °C was established for the conversion of ThNH to ThO_2. However, the reaction may be incomplete in that some water may be sorbed on the ThO_2.
- Thoria carryover particles captured on the surface of the filters range in size from 2 to 15 μm. Much smaller particles (submicrometer) may be present in the filter cross-hatching.
- Although faster heating rates made reactions end at higher temperatures than did slower heating rates, the resulting reaction products were equivalent.

ACKNOWLEDGMENTS

The authors wish to acknowledge Mr. F. Kevin Reilly of DLA for his support in this study.

REFERENCES

[1]F.K. Reilly, Defense Logistics Agency, personal communication to F.R. Van Ryn, Environmental Management Division, Oak Ridge K-25 Site (East Tennesse Technology Park), Oak Ridge, Tennessee, November 25, 1994.

[2]D. Laird, Science Ventures, Inc., personal communication to S.L. Camacho, Plasma Technology Corp., September 15, 1995.

[3]W.H. Hermes, et al., *Defense Logistics Agency Thorium Nitrate Surrogate Test Results Report*, K/EM-409, prepared by the East Tennessee Technology Park Environmental Management Division of Lockheed Martin Energy Systems, Inc., for the Defense Logistics Agency, Oak Ridge, Tennessee, February 1996.

[4]*Powder Diffraction File*, JCPDS International Center for Diffraction Data, Philadelphia, 1989.

[5]R.N. Tiwari and D.N. Sinha, "Studies on the Thermal Decomposition of Thorium Nitrate and Thorium Hydroxide," *Indian Chemical Journal*, **14** [7] 25–28 (1980).

[6]D.R. Moorehed and E.R. McCartney, "The Formation of Thoria by Decomposition of Nitrate, Oxalate, and Hydroxide," *Journal of the Australian Ceramic Society*, **12** [2] 27–33 (1976).

[7]H. Hayashi and T. Watanake, "Properties of Thoria Obtained by Thermal Decomposition of Thorium Salts," *Nippon Kagaku Karshi*, **4** 858–60 (1973).

[8]D.E. Ferguson, "Sol-Gel Technology in the Nuclear Reactor Fuel Cycle"; pp. 37–78 in *Progress in Nuclear Energy Series III: Process Chemistry*, Vol. 4, Permagon Press, Hungary, 1970.

Low-Level Radioactive and Mixed Waste Vitrification

HIGH-TEMPERATURE MELTER RUNS ON MIXED WASTES FROM LOS ALAMOS

K.S. Matlack, I.S.Muller, H. Hojaji, R.K. Mohr and I.L. Pegg, Vitreous State Laboratory, The Catholic University of America, Washington, DC 20064.

ABSTRACT

An entire stream of mixed wastes (i.e., radioactive and hazardous wastes) from Los Alamos National Laboratory was vitrified using a High-Temperature Joule-Heated Melter system that is installed at Catholic University. The waste stream, which consisted primarily of nitrates and perchlorates, was converted into a glass product containing over 5.5 wt.% of both UO_2 and ThO_2. All glass samples taken throughout the run passed the TCLP test by a considerable margin. Extensive processing and emissions data were collected throughout the run. Prior to the actual radioactive run, additional melter emissions data were obtained by feeding the melter a surrogate waste spiked with Ag, As, Cd, Se, Sb, Ni, Cu, Cs and Co. The extent of melter emissions was evaluated as a function of temperature (1150-1350°C), percent cold cap, and glass pool air sparging rate.

INTRODUCTION

The Department of Energy (DOE) has the responsibility for the environmentally acceptable dispositioning of a wide spectrum of waste streams that have been generated over the course of decades of operations of a vast complex of sites across the US. The chemical and physical diversity of these streams will likely necessitate the implementation of several different treatment technologies. So-called "mixed wastes" are especially troublesome since they contain both radioactive and hazardous constituents which therefore necessitate treatment to meet both NRC and EPA standards for disposal. DOE has estimated the complex-wide inventory of mixed wastes to be over 165,000 m^3, with another 27,000 m^3 projected to be generated over the next five years[1]. Vitrification, in a variety of different forms (joule-heated, plasma, combustion, microwave, etc.), is being evaluated for treatment of several mixed waste streams. Joule-heated vitrification is currently being used for the treatment of high-level nuclear wastes at West Valley and at the Savannah River Site and for low-level mixed wastes at the M-Area vitrification facility at Savannah River[2]. All three of these systems employ "low-temperature" (below about 1200°C) joule-heated vitrification processes, with the temperature limit being imposed by the use of Inconel 690 electrodes. This paper presents the results from the treatment of mixed wastes from Los Alamos National Laboratory (LANL) in a new high-temperature joule-heated melter developed at

Catholic University[3,4]. In this work, an entire LANL mixed waste stream (LA-W923, "Mixed Waste Oxidizers") was shipped to Catholic University where the wastes and all of the primary containers were vitrified using the high-temperature melter system. Thus, in addition to obtaining valuable process performance data, an important LANL waste treatment need was addressed and a volume reduced, leach resistant, non-hazardous glass product was returned to the site.

The LA-W923 waste stream consisted of a variety of oxidizing compounds of heavy metals, uranium, and thorium, including nitrates and perchlorates. The EPA waste codes included D001, D003, D005, and D011, due to the reactive nature of these materials. Since none were listed wastes, the glass product was a non-hazardous, low-level waste because the reactive characteristic was removed by vitrification and the product passed the TCLP requirements. The LA-W923 was stored in 19 containers at LANL and was consolidated into five 55-gallon drums for shipment to Catholic University.

MELTER RUNS

The High-Temperature Joule-Heated Melter (HT-JHM) system at Catholic University has been described previously[3,4] and has been used extensively for a wide variety of feed compositions over the past two years. The system employs composite molybdenum-conducting ceramic electrode systems that permit operation up to about 1500°C and have a tolerance to a wide variety of glass chemistries. The melter is supported by a glovebox feed system and an off-gas treatment train consisting of a film cooler, quencher, scrubber, baghouse and HEPA filter. The melter was fed surrogate LANL mixed wastes from Sept.16 to 25, 1996 and actual LANL waste on Sept. 26 and 27, 1996. The surrogate run was used to turn over the glass inventory in the melter to the composition range expected for the LANL wastes and to identify preferred operating parameter ranges to be used for the radioactive runs. In addition, the surrogate feed was spiked with selected hazardous constituents in order to obtain additional melter emissions data on the effects of temperature, bubbling rate (i.e., air sparging of the glass pool), and cold-cap.

Surrogate Wastes

The surrogate feed for these tests was based on projected formulations for the actual wastes but with thorium and uranium replaced by zirconium and cerium on a molar basis. In addition, the feeds were spiked with Ag, As, Cd, Se, Sb, Ni, Cu, Cs and Co in order to obtain additional emissions data. The surrogate feed was composed of an aqueous slurry of oxides and carbonates resulting in a target glass composition in weight percent of the following: Al_2O_3 - 6.00, B_2O_3 - 7.36, CaO - 12.06, CeO_2 - 5.42, Cr_2O_3 - 0.49, Fe_2O_3 - 7.25, K_2O - 0.26, Li_2O - 1.93, MgO - 0.35, Na_2O - 8.40, NiO - 0.48, SiO_2 - 43.07, ZrO_2 - 4.90, AgO - 0.12, As_2O_3 - 0.12, CdO - 0.12, SeO_2 - 0.12, Sb_2O_3 - 0.12, CuO - 0.48, Cs_2O - 0.48, and CoO - 0.48.

Over 130 liters of the surrogate material was fed at an average feed rate of between 3.1 and 4.8 liters/hr, producing 58.2 kg of glass. Based on data obtained from the surrogate runs, a nominal glass temperature of 1250°C and a moderate bubbling rate of 15 scfh were selected for use with the radioactive feed. The typical power demand under these conditions was approximately 13 kW.

The variation of element volatility with respect to various melter conditions was determined during the surrogate melter runs by comparing elemental fluxes into and out of the melter. The elemental flux into the melter was determined from the feed composition and the feed rate while the flux out of the melter was determined by particulate and gas emission sampling. A list of the melter conditions used during each sampling period and the elemental percentages of feed emitted are given in Table I. The two primary variables that were set for each test were the glass temperature and bubbling rate. For each combination of these variables, the feed rate was increased to the point that the cold cap was essentially complete and this was taken as the maximum feed rate for that set of conditions. Since the extent of the cold cap can also affect the amount of emissions from the melter, the effect of this variable was investigated by reducing the feed rate from its cold-cap limited maximum value while holding the temperature and bubbling rate constant.

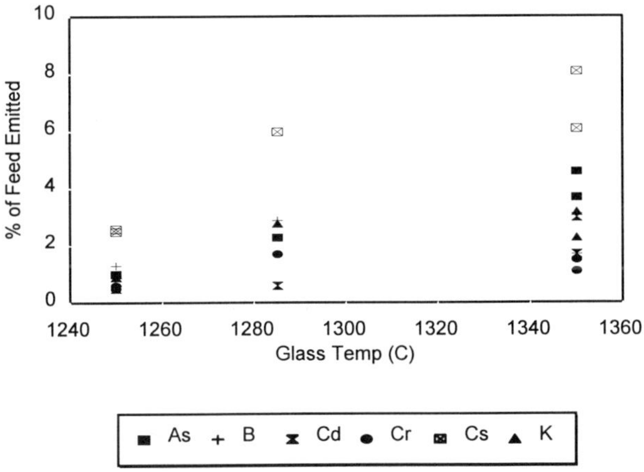

Figure 1. Effect of glass temperature on emissions.

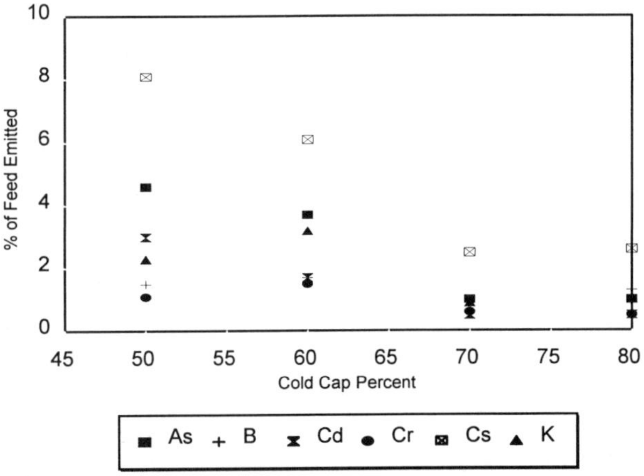

Figure 2. Effect of cold cap on emissions.

Particulate emissions increased with increasing temperature (Figure 1) and decreasing cold cap (Figure 2). Conversely, bubbling rate did not have an impact on particle emissions but did appear to increase submicron boron emissions. Data from the first sample taken (included in Table I) are omitted from Figure 2 since the cold cap was highly variable during that measurement. Overall, retention in the melter was excellent, with only 0.1 % of the solid mass exiting the melter at 1250°C and only 0.3 % at higher temperatures. The influence of cold cap is particularly striking considering that the data include two different temperatures.

Table I. Summary of Melter Emissions During Los Alamos Surrogate Runs

Date	9/18/96	9/19/96	9/20/96	9/24/96	9/25/96
Sampling Start Time	1600	1600	1502	1515	1436
Sampling End Time	1700	1630	1532	1545	1506
Glass Temp (°C)	1273-1311	1350	1250	1350	1250
Bubbling Rate (scfh)	6	5	5	20	20
Cold cap (%)	15-70	50	70	40-80	70-90
Feed Rate (l/hr)	3.2	4.8	3.7	5.3	5.0
Sample Vol. (dscf)	39.349	33.911	34.708	34.949	34.163
Moisture (%)	11.4	5.7	3.4	5.5	5.8
Particle Loading (mg/dscf)	5.3	4.0	0.9	4.2	1.4
Particulate Flux (mg/min)	186	280	63	294	98
Solids Emitted as Particles (%)	0.29	0.29	0.09	0.28	0.10
Sub-particulate B Emitted (%)	2.63	0.75	0.56	1.16	1.25
Ag emitted (%)	0.10	0.10	0.35	0.16	0.37
As emitted (%)	2.34	4.59	0.96	3.74	0.95
Cd emitted (%)	0.60	2.98	0.46	1.66	0.45
Ce emitted (%)	0.02	< 0.01	< 0.01	< 0.01	< 0.01
Co emitted (%)	0.02	0.02	0.02	0.03	< 0.01
Cr emitted (%)	1.68	1.06	0.58	1.51	0.48
Cs emitted (%)	6.01	8.08	2.45	6.08	2.58
Cu emitted (%)	0.16	0.10	0.05	0.07	0.04
K emitted (%)	2.84	2.29	0.86	3.23	0.95
Li emitted (%)	0.25	0.21	0.06	0.24	0.07
Ni emitted (%)	0.03	0.03	0.03	0.03	0.02
Sb emitted (%)	0.09	0.07	0.06	0.09	0.02
Se emitted (%)	44.98	17.81	8.89	21.56	9.49
Zr emitted (%)	0.01	0.01	0.01	< 0.01	0.01

The relative overall volatility of the elements observed in this study were as follows: Se > Cs > As > K > Cd > Ag, Li > Sb > Cu > Ni >> Co, Ce, Zr. This series covers a range of three orders of magnitude from the most to the least volatile. Several elements were deliberately omitted from the above ranking: Na, Mo, Cr, and B. Boron, although volatile, was not included because it is emitted primarily as a gas or submicron particles and therefore does not respond to melter conditions in the same manner as the other constituents. Chromium and molybdenum are constituents of the melter refractories and electrode materials and therefore the presence of these elements in the exhaust stream cannot be attributed solely to the feed. Molybdenum, for example is not present in the feed and yet its mass flux from the melter is similar to that of cesium. Sodium cannot be accurately measured in the impinger solutions; however its volatility based on particles alone is somewhere between potassium and lithium. Although this same suite of spiked elements was not investigated in other studies, the overall volatility trends agree with other relevant work[3-6].

Another method for determining elemental volatility is to measure the amount of each element that accumulates in the quencher, scrubber, and baghouse during feeding. In the three time intervals, the direct off-gas sampling data indicate that significantly more material left the melter than was measured in these solutions. This disparity can be attributed to elements precipitating from the basic solutions or the elements being inefficiently removed from the exhaust stream during wet scrubbing. Table II compares average values for the various contaminant retentions in various localities throughout the off-gas system to the direct air sampling. All the cadmium and approximately two-thirds of the cesium and arsenic are trapped on the baghouse coating medium (diatomaceous earth (DE)) presumably due to the fine particulate nature of the emissions. Silver and uranium readily precipitate from solution and are subsequently removed on the quencher bag filter. The nickel observed on the bag filter and the DE probably originated from the corrosion of the quencher tank and not from the feed. The selenium emissions measured in the melter exhaust were significantly higher than can be accounted for in the quencher and scrubber materials and the baghouse DE.

Los Alamos Radioactive Run

The continuous radioactive run with LANL feed was successful in vitrifying all the radioactive chemicals as well as their glass storage bottles. The magnesium perchlorate portion of the waste was mixed up as a separate feed stream from the rest of the waste; the feeding of these two streams was then alternated throughout the run. Overall, the feeds processed smoothly even at the highest feed rates. Sixty-eight liters of the primary waste feed slurry along with 14 liters of magnesium perchlorate waste were fed over 27 hours of operation resulting in an overall feed rate slightly greater than 3 liters per hour. Average feed rates of 3.6 and 5.25 liters per hour were achieved during slurry and perchlorate feeding, respectively. During this run, 56.7 kg of glass were produced resulting in an overall production rate of 50 kg of glass per day. The waste that was received in five 30-gallon drums was returned as a durable glass in only one drum giving an overall volume reduction of approximately five.

Table II. Comparison of Direct Off-Gas Sampling with Accumulations in Off-Gas Treatment Stages. (Average Percent)

	Source	Off-Gas Sampling	Quencher + Scrubber Solutions	DE	Quencher Filter
Ag	Surrogate and Radioactive Feeds	0.2*	0.3	<0.1	1.90
As	Surrogate Feed	2.5	0.9	1.90	<0.1
Cd	Surrogate Feed	1.2	< 0.1	1.30	<0.1
Cs	Surrogate Feeds	5.0	1.5	3.00	<0.1
Ni	Surrogate Feed and Melter System	< 0.1	< 0.1	0.20	2.40
Se	Surrogate Feed	20.5	6.2	1.00	<0.1
U	Radioactive Feed	0.1	< 0.1	<0.1	0.30

* - An unquantifiable amount of AgCl precipitate was observed in off-gas samples taken from the radioactive run and there this number is known to be an under estimate.

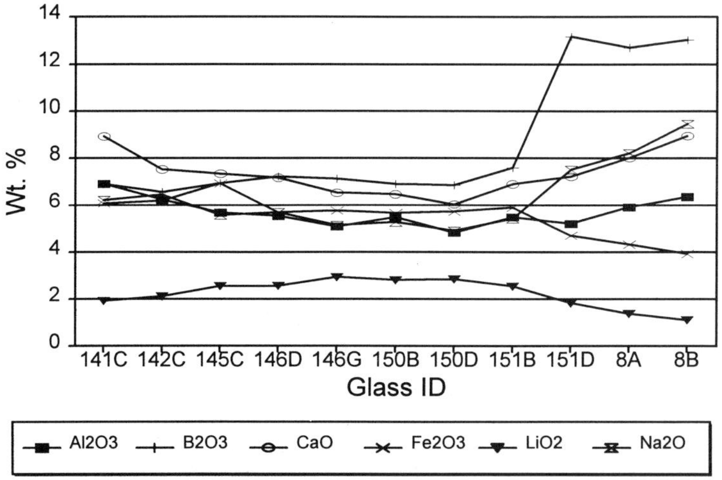

Figure 3. Glass product composition (samples in chronological order).

The glass produced during the radioactive runs approached the target composition as expected The bulk composition of the glasses did not change much over the course of the runs as a result of feed composition being very similar to the glass pool (Figure 3). The turnover of the glass pool is best indicated by the changes in uranium and thorium in the glass, as shown in Figure 4; note that glass samples after 151B were taken during the "flush-out" period, hence the drop in U and Th. All of the glasses tested successfully passed the EPA TCLP leach resistant requirements. Melter operating conditions were held relatively

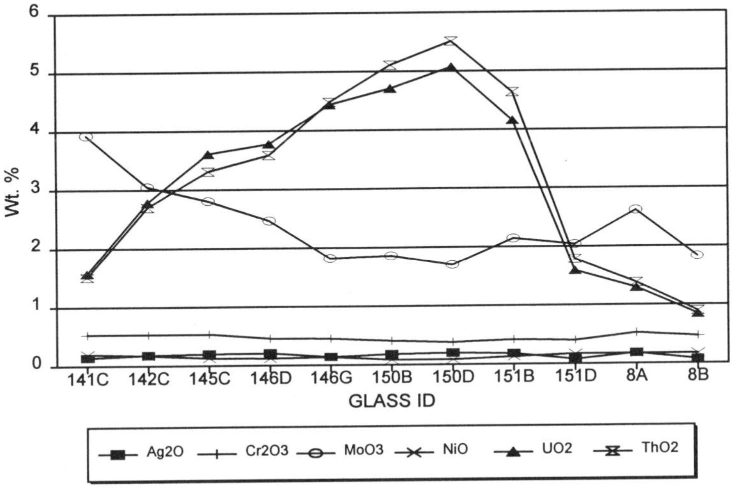

Figure 4. Glass product composition (samples in chronological order)..

constant throughout the melter runs. This waste could have been processed at lower temperatures in a joule-heated melter with Inconel electrodes instead of molybdenum, however production rates and waste loading would have decreased. Approximately three kilowatt hours were required for each kilogram of glass produced due, in large part, to the small size of this system and the significant water content of the feeds.

Four exhaust air samples were taken during the radioactive run; three from the melter and one down-stream from the HEPA filter. The sampling conditions are provided in Table III. All three of the melter exhaust samples provided data for calculating the elemental fluxes exiting the melter. During the first two melter exhaust samples, the elemental feed rates into the melter were known, which permitted the calculation of elemental retention in the melter. The post-HEPA exhaust sample taken at the end of the run showed that no measurable amounts of radionuclides or other harmful contaminants escaped the melter off-gas system; all elements in the final exhaust were at concentrations of less than 1 ug/dscf with the exception of boron.

Throughout the run, the uranium flux from the melter was no more than 0.1 percent of that fed into the melter. Boron emissions were similar to those of the surrogate run. Chlorine fluxes increased when chloride salts and perchlorate solutions were fed into the melter, as would be expected. Detailed interpretation of the melter exhaust data for the radioactive run is complicated by the differences in feeding conditions during the three

Table III. Melter Emission Sampling Summary for Los Alamos Radioactive Run

Date		9/26/96	9/26/96	9/27/96	9/27/96
Sampling Time		1632-1652	2321-0021	1200-1230	1710-1810
Sampling Site		Melter Exhaust	Melter Exhaust	Melter Exhaust	HEPA Exhaust
Conditions		Feeding Batch #1	Feeding Batch #2, Idling and Feeding $MgClO_4$	Feeding Batch #5	Feeding Batch #5
Sample Size (dscf)		22.498	67.053	32.368	52.735
Moisture		4.6%	3.8%	5.5%	2.3%
Particle Load (mg/dscf)		3.1	NA	3.3	NA
Elemental Fluxes (mg/min)	B	19.05	23.78	11.06	All elements at concentrations less than 1 ug/dscf with the exception of boron.
	Ba	0.07	0.08	0.05	
	Cr	4.70	7.34	3.43	
	Cu	2.78	7.94	1.82	
	K	3.65	9.68	5.16	
	Li	14.60	86.59	10.94	
	Mn	32.55	27.98	13.16	
	Mo	31.74	70.68	37.57	
	Ni	0.62	12.93	0.49	
	Sr	0.43	0.27	0.43	
	Zr	0.51	0.37	0.33	
	U	2.74	2.13	1.13	
	Cs	13.40	10.52	2.60	
	Cl	< 1 HCl 9.3 particle	575 total	98 HCl 80.0 particle	< 1 HCl

sampling periods. Also, the feed data were incomplete for the third sample. However, the elemental fluxes for the third sample are generally within a factor of two of the other two samples. The first sample tends to have higher fluxes due to the concentrated nature of the feed while the second sample tends to be lower due to idling time and the feeding of magnesium perchlorate. Particulate loadings between the first and third samples differ by less than 10%, illustrating the overall similarity of the exhaust streams. The second sample, which includes both idling and magnesium perchlorate feeding time, provides a reasonable composite of the emissions conditions over the entire run.

CONCLUSIONS

An entire mixed waste stream was volume reduced and rendered non-hazardous by vitrification and is now in a form suitable for disposal as simply low-level radioactive waste. In addition, valuable processing and off-gas composition data were obtained in this study over a range of conditions. The melter emissions were primarily as fine particulates and a ranking of the observed elemental volatilities showed Se, Cs, and As to have the highest volatilities; 99.9 wt% of the uranium fed was retained in the glass melt, with the remainder captured in the off-gas treatment system. Melter emissions increased with increasing melt temperature and decreasing cold cap coverage, as expected, but the glass bubbling rate showed no discernible effect. The waste was processed at a glass production rate of 50 kg/day and an overall volume reduction of a factor of five was obtained. All of the glass product samples easily passed the EPA TCLP leachability requirements. The magnesium perchlorate waste processed easily with no adverse effects other than increased chloride emissions into the off-gas treatment system.

ACKNOWLEDGEMENTS

The authors would like to thank Troy Eshleman and John Kelley (GTS Duratek); James Nunz (DOE-LAAO); and Kapil Goyal and Chris Duey (LANL) whose efforts were essential to the success of this project, and M.C. Paul for manuscript preparation. This work was supported by the Department of Energy Office of Environmental Management.

REFERENCES

[1]Department of Energy, Mixed Waste Inventory Report, February, 1997.

[2]B.W. Bowan and M.M. Brandys, "Development of a Vitrification Process to Solidify Savannah River M-Area Wastes," Ceramic Transactions, **61** 213 (1995).

[3]K.S. Matlack, I.S. Muller, H. Hojaji, R.K. Mohr, M. Brandys, E. Wang, I.L. Pegg, P.B. Macedo, "Vitrification of Actual Mixed Wastes in a New High-Temperature Melter," Ceramic Transactions **72** 11 (1996).

[4]S.S Fu, K.S. Matlack, R.K. Mohr, W. Luo, E. Wang, H. Hojaji, M. Leontiev, M. Brandys, I.L. Pegg, P.B. Macedo, " Minimelter Runs of Mixed Wastes High in Lead, Barium, Phosphorous and Sulphur," Ceramic Transactions **72** 27 (1996).

[5]T.J. Overcamp, J.M. Harden, C.A. Cicero, D.F. Bickford, "Offgas Emissions from Slurry-Fed Joule-Heated Melters," Ceramic Transactions **72** 261 (1996).

[6]K.S. Matlack, H. Hojaji, S.S. Fu, I.L. Pegg, P.B. Macedo, "Vitrification of Radioactive Fernald Wastes with the Continuously Fed DuraMelter 100," Ceramic Transactions **72** 221 (1996).

RADIONUCLIDE UPTAKE BY WASTE GLASS MELTER REFRACTORIES

A.C. Buechele, S.S. Fu, Q. Yan, and I.L. Pegg, Vitreous State Laboratory, The Catholic University of America, Washington, D.C. 20064

ABSTRACT

Ceramic refractories are widely used in the construction of melters for the vitrification of radioactive and hazardous wastes. For radioactive wastes, and particularly for low activity wastes, the potential for absorption and concentration of radionuclides by the refractories can be an issue in terms of projecting shielding requirements and maintenance planning (e.g., contact vs. remote). In addition, decommissioning and disposal options can also be impacted. Tests have been conducted on a variety of refractory materials that are used, or could be used, in the construction of waste glass melters. The data show instances in which reactions occur that concentrate radionuclides. It appears that one of the phases responsible for such concentration is a partially cesium-substituted form of leucite (α-(Cs, K)$AlSi_2O_6$). The conditions under which this phase forms and the potential impact in terms of melter operation or decommissioning are discussed.

INTRODUCTION

Vitrification is the method of choice for stabilization of an increasing number and variety of waste streams. A major element of concern in many radioactive and mixed waste streams is cesium due to its high specific activity and mobility. During the melting phase of the vitrification process, the potential exists for cesium from the waste to react with melter components both in direct contact with the glass pool and in the plenum area above the melt. If reaction does occur and if it leads to concentration of cesium in these components, the result will be increased radiation levels in and around the melter leading to increased shielding and/or remote operations requirements and complicating change-out, maintenance and ultimate disposal operations. A study to identify possible interactions with a few of the more common melter refractories was, therefore, deemed useful and complementary to several previous corrosion studies conducted on melter refractories at The Vitreous State Laboratory (VSL) of The Catholic University of America (CUA) [1-3].

TEST METHODOLOGY

Typical melter refractories were exposed to Cs-containing melts in a variety of ways. The extent of Cs incorporation was determined using a JEOL JSM-35C scanning electron microscope (SEM) equipped with a Noran 55/5600 energy and image analysis system incorporating energy dispersive x-ray spectroscopy (EDS).

Four refractories were selected for study: Carborundum K-3 (~59% Al_2O_3, ~27% Cr_2O_3, ~6% Fe_2O_3, ~6% MgO, ~4% other), Corhart ER-2161 (~27% each Al_2O_3, Cr_2O_3 and ZrO_2; ~15% SiO_2; ~4% other), Carborundum Monofrax H (Sodium beta-alumina with <1% total impurities, mainly SiO_2, CaO, and Fe_2O_3) and North American Refractories Zirmul (70% Al_2O_3, 10.2% SiO_2, 19.5% ZrO_2 mainly as zircon, 0.3% other). Carborundum K-3 has become the glass contact refractory of choice for the US DOE High Level Waste (HLW) vitrification programs, being currently in use in both the West Valley and Defense Waste Processing Facility (DWPF) melters, as well as in all existing DuraMelters™ built by GTS Duratek. Corhart ER-2161 has been identified as a potential alternative to K-3 in tests conducted at CUA/VSL over the past three years. Carborundum Monofrax H and North American Refractories Zirmul are commonly used in the plenum areas and to back up the primary refractories in melter systems.

A coupon of K-3 exposed to Cs-containing glass during a simulated Hanford Low Activity Waste (LAW) vitrification run conducted at VSL in a DuraMelter™ 100 system was sectioned and examined as the first step in this study. The test run had been conducted as part of the Westinghouse Hanford Company (WHC) Melter Vendor Test Program using a Double Shell Slurry Feed (DSSF) simulant somewhat similar to Hanford Low Level Waste (LLW) [4]. An accelerated test employing molten Cs_2SO_4 was conducted to assess the gross effects that could be expected in refractory interaction with cesium under extreme conditions. Immersion tests were made in which coupons were exposed to a glass composition similar to that used in the Hanford LLW Melter Vendor tests at VSL, but enriched in cesium to accelerate effects. Finally, tests were made also using the Cs-enriched glass in specially constructed test vessels designed to simulate actual melter conditions to some degree.

RESULTS

Figure 1 shows the microstructure of K-3 which was used in a majority of the tests. Variability of grain morphology and size is quite evident. The material is a fused cast ceramic consisting of two main phases: about 30-35% complex spinel containing Fe, Al, Mg, Cr and O (lighter grains in Figure 1), and about 60-65% solid solution phase containing mainly Cr_2O_3 and Al_2O_3 (darker grains). There is also about 1% of an intergranular silicaceous glassy phase, less than 0.5% iron metal and about 4% apparent porosity with some columnar pores extending several millimeters in length.

Figure 1. Typical unaltered microstructure of K-3.

Simulated LAW Vitrification Test Coupon

The coupon designated D1C-003V from the Hanford LLW Melter Vendor test was sectioned and examined in the SEM using EDS to look for Cs. We estimate the lowest detectable level of Cs_2O in our instrument to be about 0.2 wt%. Independent analyses of glass samples produced during this test showed about 0.1 wt% Cs_2O; while we did not expect to detect Cs_2O in the glass adhering to the coupon, we did expect to be able to detect concentrations in the K-3 at or above 0.2 wt%. In fact, no detectable concentration of Cs was found anywhere in the coupon. Other components of the test glass were found to have penetrated the coupon via the intergranular phase to the depth of sectioning (~10 mm). The coupon had been exposed to the test glass for a total of 49 hours at about 1150°C; apparently, therfore, Cs does not grossly concentrate in K-3 within this time under realistic processing conditions.

K-3 in Molten Cs_2SO_4

An accelerated test was performed by partially immersing a coupon of K-3 in molten Cs_2SO_4 contained in a quartz crucible at 1020°C for one hour. The coupon was then removed, sectioned after dissolving residual Cs_2SO_4 in water, and examined using SEM/EDS. Penetration occurred by reaction with the grain boundary vitreous phase. Penetration depended strongly on the microstructure of the exposed K-3 surface and on whether the surface was immersed or not. Figure 2 is an SEM micrograph of a cross section through the immersed portion of the K-3 coupon, and Figure 3 through the top, unimmersed portion. The white intergranular regions are highly enriched in cesium. Penetration of the immersed coupon ranged from 50 μm in surface areas dominated by the spinel phase with very few grain boundaries, to 100-200 μm in fine grain areas with very narrow boundaries, to 400-500 μm in large grain areas with broad boundaries. At the top of the coupon, where exchange could occur only from the vapor phase or from wick-up material, penetration was only about 50 μm in fine grain and 100 μm in large grain areas. The composition of the grain boundary phase was approximately correct for a Cs-substituted version of leucite ($CsAlSi_2O_6$), although it could be amorphous. Leucite (normally $KAlSi_2O_6$) crystallizes geologically in silica poor melt environments [5], and the high alumina content of the intergranular vitreous phase makes it relatively poor in silica. Although this test represents extreme conditions, it suggests that over a long time under more realistic operation conditions accumulation of Cs in the refractory may occur.

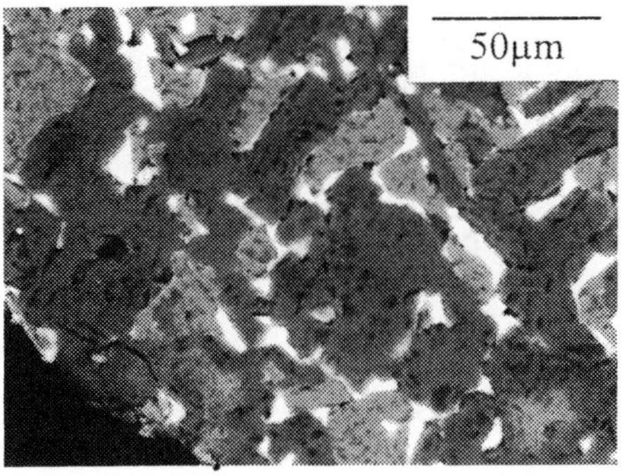

Figure 2. Immersed surface of K-3 coupon from molten Cs_2SO_4 test. White intergranular phase is high in Cs.

Figure 3. Top, unimmersed surface of K-3 coupon from molten Cs_2SO_4. Note less penetration than in Figure 2.

Tests with Cs-Containing Glass

As mentioned previously the glass used for other tests in this study was formulated to be similar to the glass used in the Hanford LLW Melter Vendor test, but enriched in Cs_2O to about 5 wt%; the composition of the glass, CsGlass2, is given in Table I. Chemicals were mixed and melted in a platinum crucible at 1150°C to produce a homogeneous glass. The liquidus of this glass was determined to be below 1050°C indicating that it is suitable for coupon tests at 1050°C and above. Total dissolution in acid was performed followed by inductively coupled mass spectrometry analysis (ICP-MS) on samples of the as-melted glass, and glasses melted and held in platinum crucibles at 1150°C for 5 and 10 hours to determine Cs loss due to volatilization. The measured Cs_2O contents were 4.9 wt%, 2.5 wt% and 1.8 wt%, respectively.

Three partial immersion tests at 1150°C were run using K-3 coupons and CsGlass2 for durations of 5 and 25 hours in platinum crucibles, and for 72 hours in a quartz crucible. It had been hoped that these tests would provide the basis for a quantitative measurement of Cs penetration versus time. Unfortunately, the variability of the microstructure of the K-3 made this impossible. For instance, measurable penetration of Cs was seen in the 5-hour coupon to a depth of 1000 μm, but in a fine-grained region of the 72-hour coupon the Cs concentration was unmeasurable at a depth of 500 μm. However, in another coarse-grain-size area of the same 72-hour coupon, measurable penetration of Cs occurred to at least 7 mm, the maximum depth of the prepared section. Local penetration in the 25-hour coupon was found up to the 7 mm width of the section prepared for SEM/EDS examination due to infiltration of the porosity present in the coupon. Nonetheless, several important properties of the interaction of the glass with K-3 were learned from these tests. In no case was any Cs found in the two major phases of the K-3, nor was the formation of any Cs-containing phase observed at any point within the coupon. The attack of the K-3 at the glass interface in the 25-hour coupon is typical of the glass/coupon interface in all three of these tests and is shown in Figure 4. The material at the lower right-hand corner of this micrograph, about 150 μm into the K-3 coupon, is representative of virgin K-3. Moving from this unmodified material toward the glass/K-3 interface, the collapse of the solid solution phase is progressively more evident as the Al_2O_3 from the solid solution is incorporated into the glass matrix leaving the Cr_2O_3 behind, and the spinel phase recrystallizes with glass infiltration as it is progressively modified toward a composition closer to chromite by loss of Al_2O_3 to the glass. About 25-50 μm into the coupon in the dissolution widened grain boundary regions, there are some dark crystals which are a solid-solution phase intermediate between nepheline ($NaAlSiO_4$) and kalsilite ($KalSiO_4$). The measured concentration of Cs_2O in the vitreous intergranular regions within about 100 μm of the interface is higher than in the glass outside the coupon, but lower than in the original CsGlass2. It is believed that this greater local concentration of Cs_2O is caused by uptake during the early stages of the test when the Cs_2O of the test glass is undiminished, and that it persists as the test progresses because of restricted transport through the grain boundaries. The cesium concentration in the intergranular vitreous phase falls off after this first 100 μm from the glass/K-3 interface, but potassium penetrates further.

Table I. Formulation of CsGlass2 Used for Cs Absorption Studies

Components	Wt.%
Al_2O_3	6.39
B_2O_3	5.76
CaO	7.02
CdO	0.29
Cr_2O_3	0.30
Fe_2O_3	6.22
K_2O	4.86
NaCl	1.21
Na_2O	15.16
Cs_2O	4.98
P_2O_5	0.80
SiO_2	40.31
ZrO_2	3.84
NaF	2.14
Total	99.29

Miniature Melter Tests

The vessels used for the final stage of the tests were "miniature mock-up melters"; their construction is illustrated in Figure 5. Two such vessels were constructed. The 1" diameter glass-containing cavities were bored 1-3/4" deep in cut blocks of K-3 and ER-2162. These blocks were cemented into welded cans of an Inconel cladding using K-3 cement. Half of the opening in each block was covered with a piece of Monofrax H and the other half with a piece of Zirmul as shown in Figure 5 in order to simultaneously test both refractories for absorption of volatilized cesium in the plenum space.

The miniature melters were filled roughly three-quarters full with approximately 29 g of CsGlass2, covered and placed in a furnace at 1150°C. Cracks developed in both the K-3 and the ER-2161 miniature melters, making it necessary to recharge them with glass each day. The additions over the course of the test to the K-3 miniature melter totaled 70 g and to the ER-2161 system, 175 g.

After a total of 7 days at 1150°C, the remaining glass was poured out of the miniature melters and each melter was sectioned vertically and subjected to SEM/EDS examination. There were obvious macroscopic cracks along both the K-3 walls and the ER-2161 walls, but most noticeably in the ER-2161 walls. Glass had diffused into these cracks and through the K-3 cement and had begun to corrode the Inconel cladding. The Cs_2O contents of the glass remaining in the K-3 and ER-2161 miniature melters were 3.5 and 3.6 wt%, respectively.

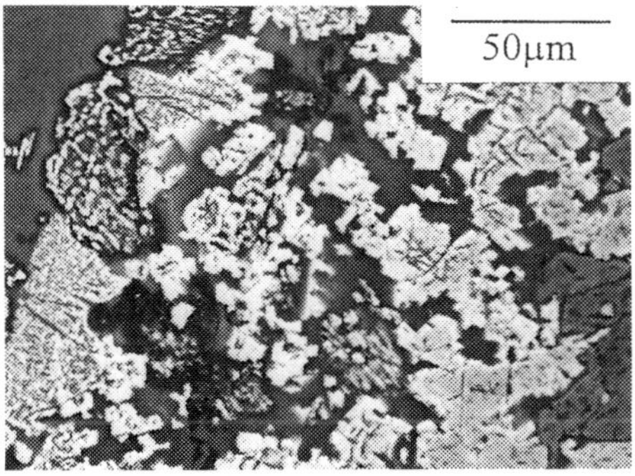

Figure 4. Glass/K-3 interface of 25-hour test coupon. Glass is at upper left corner.

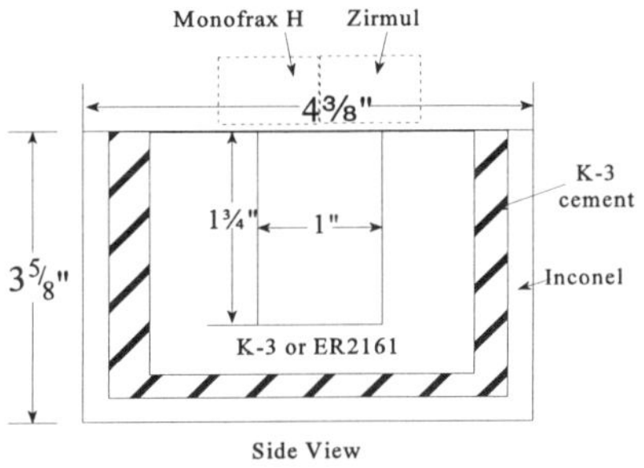

Figure 5. Schematic of miniature mock-up melter construction.

The interaction of the Cs-containing glass with the walls of the K-3 miniature melter was similar to that seen in the K-3 coupon test described above. The reaction region was about 200 μm deep. Crystals of nepheline/kalsilite ($(Na,K)AlSiO_4$) were found in the intergranular regions starting at about 100 μm into the K-3 brick. The nepheline crystals form a band about 50 μm wide, and occasionally in this band, Cs-enriched (up to 8 wt.% Cs_2O) areas 20-50 μm wide are observed. However, these areas appear to be vitreous and probably become enriched in Cs by segregation as the nepheline/kalsilite crystallizes. No evidence was found of the Cs-substituted leucite-type phase that was found in the case of K-3 exposed to Cs_2SO_4.

Both cesium and potassium provide good indications of the extent of glass diffusion into the refractories since neither of these are present in the virgin materials. SEM/EDS analysis of the K-3 refractory shows that the Cs diffused about 8 mm into the K-3 brick after 7 days at 1150°C. At 9 mm depth, no Cs was detectable by SEM/EDS (<0.2 wt.% Cs_2O), but the potassium was still detectable at about 1.6 wt.%.

The ER-2161 refractory is composed of about 20 wt% vitreous phase (constituting a substantial portion of the bulk rather than being just a minor intergranular phase), 27 wt% ZrO_2 and 53 wt% Cr_2O_3-Al_2O_3 solid solution. The grain structure of virgin ER-2161 is much more uniform than that of K-3, consisting of about 125 μm diameter solid solution grains and 50 μm diameter ZrO_2

grains, with many of the ZrO_2 grains elongated to 50 x 100 μm dimensions. A modification of the microstructure of the refractory developed to a depth of 1500-2000 μm in the ER-2161 miniature melter during one week at 1150°C. Figure 6 is an SEM micrograph showing the glass/refractory interface through about 1200 μm of this modified layer. Various levels of dissolution and reaction are seen in the components of the ER-2161, which define zones with distinct characteristics. Of greatest interest, however, are the phenomena which concentrate Cs. Concentration of Cs, probably due to segregation, is found in the vitreous phase proximate to nepheline/kalsilite crystallization which occurs at a depth of 600-700 μm into the coupon. About 100 μm deeper is a zone about 100 μm wide displaying elongated clusters of crystals of a Cs-rich phase which appears to be a partially Cs-substituted form of leucite ($(Cs,K)AlSi_2O_6$) containing about 23 wt% Cs_2O by EDS determination. These crystals are visible in Figure 6, but a closeup in Figure 7 clearly shows their faceted morphology. The vitreous phase in this region is quite high in Al_2O_3 (~25 wt%) and contains about 4 wt% Cs_2O. Although no further crystallization of the leucite phase is seen, the Cs_2O content of the vitreous phase continues to be at the 4-5 wt% level for the next 500 μm into the coupon and then drops rapidly to undetectable levels over another 500 μm. Beyond this point potassium penetration of the vitreous phase is the only detectable change in the ER-2161.

As described above, coupons of Monofrax H and Zirmul were used to cover both the K-3 and ER-2161 crucibles in these miniature melter experiments. In general, the coupons covering the ER-2161 crucible had less adherent glass and showed less darkening than those covering the K-3 melter. This may have been due to the fact that the level of glass in the ER-2161 miniature melter was on the average lower due to leakage, and therefore, the surface tension-induced wicking path to the cap was longer. Color changes in the refractories were clearly evident after the tests. The outer surface of the Monofrax H in both cases showed an overall aquamarine coloration with areas of magenta.

The caps to the miniature melters were sectioned and mounted for SEM/EDS observations. The Monofrax H is made up of crystalline sodium beta-alumina. Beta-alumina is known to readily exchange cations of radius close to that of Na from electrolyte melts [6]. However, no detectable cesium interaction with or diffusion into the Monofrax H cap on top of either miniature melter was

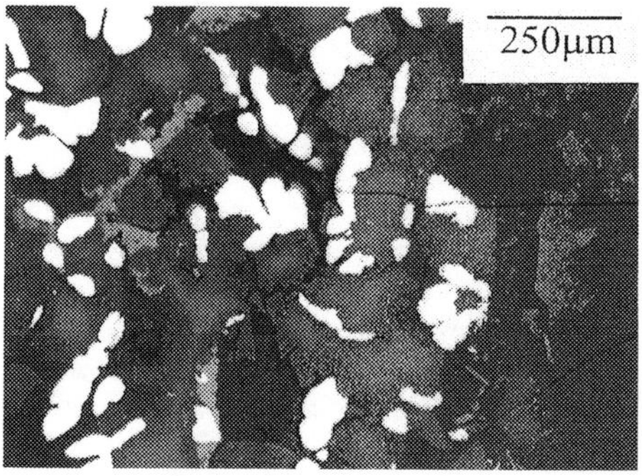

Figure 6. Glass/ER-2161 interface from miniature melter test.

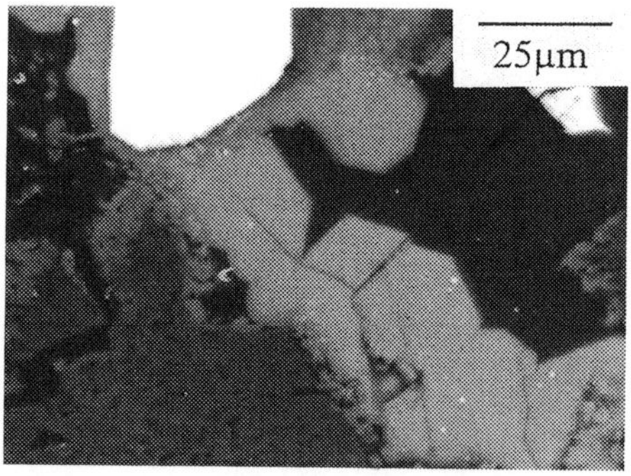

Figure 7. Partially Cs-substituted leucite crystals (light grey, faceted) in ER-2161.

found by SEM/EDS observations; this is probably due to the fact that the Cs^+ radius (1.65Å) is considerably greater than that of Na^+ (0.98Å). On the other hand, K (K^+ radius=1.33Å) from the glass partially exchanged with Na in the bulk of the Monofrax H up to the limit of the depth sampled (about 10 mm). About 1.2 wt.% K_2O was found in both Monofrax H covers at a depth of 9 mm. The dominant surface interaction of Monofrax H covering the K-3 miniature melter is the conversion of a few grains of the beta-alumina in Monofrax H into Cr-Al oxide and Cr-Fe-Al oxide up to a depth of about 200 μm. Detectable penetration of glass components into the Monofrax H from the K-3 miniature melter experiment occurred along grain boundaries for distances between 100 and 1000 μm. In contrast to the Monofrax H cap from the K-3 miniature melter, detectable chromium and iron interaction with the Monofrax H cap for the ER-2161 miniature melter was confined only to grain boundaries; however, the aquamarine/magenta coloration of both Monofrax H caps after one week at 1150°C suggests that chromium and iron have penetrated at low concentrations throughout the caps. Another difference between the two caps is that the Monofrax H cap on the ER-2161 miniature melter showed some nepheline/kalsilite ($(Na,K)AlSiO_4$) crystals in the adherent glass, whereas the adherent glass in the Monofrax H cap for the K-3 miniature melter was free of crystalline phases.

The structure of virgin Zirmul is comprised mainly of Al_2O_3 grains ranging in size from about 10-3000 μm in diameter cemented together by a mixture of zircon grains (50-200 μm diameter) and zirconia grains (1-8 μm diameter) in a glassy silicate phase of variable composition. The edge of each individual zircon grain is bordered by a reaction zone about 10 μm wide in which the zircon has converted to or is converting to zirconia particles in a silicaceous matrix. The data sheet from North American Refractories Co. for Zirmul indicates about 18 vol% open porosity as typical, and pores up to 1000 μm diameter are found throughout the material.

In contrast to the lack of cesium interaction with Monofrax H, cesium interacted extensively with Zirmul to form what is believed to be the same partially Cs-substituted leucite-type phase ($(Cs,K)AlSi_2O_6$) observed earlier, and it is found here as deep as 8 mm into the Zirmul refractory. The Cs content of these Cs-rich phases was found to be as high as 26 wt.% Cs_2O by EDS determination. These regions are amorphous in appearance, and our basis for suggesting that they are crystalline is the relative constancy of composition. The amorphous appearance is most likely due to crystallites smaller than the resolution of our SEM. These Cs-rich regions comprise at most 1 vol.% of the Zirmul refractory. The location of the Cs-rich phase within the Zirmul is usually in the vicinity of, or within the reaction zone around, the zircon grains. When found within the reaction zone, the phase resulting is Cs-rich (22-26 wt.% Cs_2O) toward the zircon grain core and Cs-poor (2-3 wt.% Cs_2O) at the outer edge with a rather sharp transition (within 5 μm) between Cs-poor and Cs-rich. The only difference found between the two Zirmul caps is the existence of some Cr_2O_3-Fe_2O_3-Al_2O_3 platelet crystals in the adherent glass near the Zirmul surface for the cap on top of the K-3 miniature melter, whereas no crystals were found in this region for the Zirmul cap on top of the ER-2161 melter.

DISCUSSION AND CONCLUSIONS

To summarize a few of the more important results:

- Rapid Cs penetration of K-3 is observed to occur with rates on the order of a few millimeters a day when exposed to a glass containing 5 wt% Cs_2O. Local enrichment of Cs is observed, probably by segregation during crystallization of secondary phases, but no Cs-containing crystalline phases are confirmed..
- Lower penetration rates on the order of a few tenths millimeter per day are observed in ER-2161, but greater local enrichment of Cs is observed within about 500 μm of surface due to crystallization of a Cs-substituted leucite phase.
- Uptake of Cs in melter plenum conditions is drastically lower for Monofrax H than for Zirmul.
- Cs penetration of refractories appears to occur exclusively via intergranular vitreous phases and infiltration of pores.

These tests have identified several mechanisms by which cesium may penetrate and concentrate in melter refractories. These tests use a glass containing 5 wt% Cs_2O, about three orders of magnitude greater than would be found under normal processing conditions, but they also represent a much shorter duration than normal melter operating lifetimes. The greater concentration of Cs_2O is employed to accelerate the effects of glass contact with the refractory in a short term test, but the possibility that at least some of the modifications observed may have a concentration threshold should not be discounted. There are two principal factors to consider in assessing the impact of these results in terms of the potential increase in radiation levels from the melter refractories over time during actual operations: The first is the ultimate bulk concentration of cesium that is possible in the refractory materials; the second is the rate of cesium penetration into the materials. The measured rates appear to be more than sufficient to achieve complete penetration of a typical contact refractory brick well within the expected melter life span (at a penetration rate of 1 mm/day cesium would completely penetrate a 6-inch thick brick in about 150 days). The ultimate bulk concentration of cesium that is possible in the refractory materials is more difficult to determine from the relatively short-term tests conducted in the present study. In all cases the *overall* concentrations of cesium in the regions of the refractories that had been penetrated were below those found in the glass melt itself, suggesting *no* build up of activity above that in the glass. However, there were clear indications of *local* enrichments of cesium (up to 26 wt% Cs_2O, five times that in the glass) in all of the refractories in which cesium was detected. Unfortunately, therefore, reactions *do* occur that result in the concentration of cesium, albeit locally, within the refractories. An important remaining question is then "how does the mass fraction of these regions of cesium enrichment increase with time?" At least over the time scales covered by the present experiments, the mass fraction of these regions was sufficiently small that the *overall* cesium concentrations in the refractories were below that of the glass melt. Sufficient concern is raised by the results of this study to merit further testing at longer times with glass containing more realistic levels of cesium.

ACKNOWLEDGMENTS

We wish to express our gratitude to M.C. Paul for assistance in manuscript preparation.

REFERENCES

[1]R.K. Mohr, Z.B. Djordjevic, E. Wang, A.C. Buechele, S.B. Xing, and W. Wagner, "High-Temperature Corrosion Resistance of Glass Refractories and Electrodes," *Ceramic Transactions, Volume 78: Corrosion of Materials by Molten Glass,* (Edited by G. Pecoraro, J. Marra, and J. Wenzel) The American Ceramic Society, Westerville, Ohio, 1996, pp. 155-170.

[2]E. Wang, R.K. Mohr, A.C. Buechele, and I.L. Pegg, "Current Density Effects on the Corrosion of Ceramic and Metallic Electrode Materials in Waste Glasses," *Scientific Basis for Nuclear Waste Management XIX*, (Edited by W.M. Murphy and D.A. Knecht) MRS, Pittsburgh, PA, **412**, 173-180 (1996).

[3]S.B. Xing, Y. Lin, R.K. Mohr, and I.L. Pegg, "Corrosion Resistance of Ceramic Refractories to Simulated Waste Glasses at High Temperatures," *Scientific Basis for Nuclear Waste Management XIX*, (Edited by W.M. Murphy and D.A. Knecht) MRS, Pittsburgh, PA, **412**, 181-188 (1996).

[4]"Phase I Hanford LLW Melter Vendor Tests for Westinghouse Hanford Company," Final Report, August 1995, Vitreous State Laboratory, Catholic University and GTS Duratek.

[5]W.A. Deer, R.A. Howie and J. Zussman, p. 302, *An Introduction to the Rock-Forming Minerals,* Longman Group, Ltd., London,1966

[6]Y.-F.Y. Yao and J.T. Kummer, "Ion Exchange Properties of and Rates of Ionic Diffusion in Beta-Alumina," *J. Inorg. Nucl. Chem.* **29**, 2453-2475 (1967).

PROCESSING OF OAK RIDGE B&C POND SLUDGE SURROGATE IN THE TRANSPORTABLE VITRIFICATION SYSTEM

John R. Zamecnik, David K. Peeler, Steven R. Young, Michael E. Smith
Westinghouse Savannah River Company
Savannah River Technology Center
Aiken, SC 29808

ABSTRACT

The Transportable Vitrification System (TVS) is a joule-heated, fully transportable melter system designed by the Westinghouse Savannah River Company Savannah River Technology Center (SRTC) in conjunction with EnVitCo, Inc. to demonstrate the treatment of low-level and mixed hazardous and radioactive wastes. The TVS is large enough to completely treat small to medium sized waste streams. An extended shakedown test of the TVS was completed in 1996 to demonstrate the processing of a surrogate wastewater sludge material. The surrogate waste was formulated to simulate the B&C Pond sludge waste from the Oak Ridge Reservation (ORR) K-25 Site. This surrogate contained several RCRA metals and a surrogate for depleted uranium.

The glass formulation used was soda-lime-silica (SLS), due to the high quantity of Ca in the waste. Glass forming materials were SiO_2, Li_2CO_3, and Na_2CO_3. While the melter was designed to produce 68 kg/hr of glass when slurry fed, improvements increased this rate to 86 kg/hr. Operational experience led to numerous improvements that increased the overall waste processing rate. During operation, the predicted glass viscosity was maintained between 44-62 poise. Glass produced was of acceptable durability compared to the Universal Treatment Standards for RCRA metals. The glass oxidation/reduction state was not found to influence the quality of the glass produced, even though redox increases due to the addition of organics were significant. The melter glass residence time was found to be approximately described by a well-mixed tank model with a plug flow time delay. The potential for devitrification of the glass melt during shutdown was minimized by flushing the melter with 1.5 melter turnovers of SLS glass.

INTRODUCTION[1,2]

Vitrification has been declared to be the "Best Demonstrated Available Technology" for heavy metals and high-level radioactive waste by the EPA. In the past, high capital costs have prevented vitrification from being considered as a viable option for treating low-level and mixed wastes (LLMW). By using a low cost, transportable unit, these up front capital costs can be significantly lowered and can result in lower life cycle costs (as compared to other existing technologies) for certain waste streams, especially if the resultant waste glass can be delisted from a LLMW to a low-level waste. The TVS was built for this reason. It is a large scale, fully integrated, vitrification system designed to treat LLMW in the form of soils, wastewater treatment sludges, incinerator ash, and similar material.

While the TVS is primarily intended for waste demonstration activities, it is large enough to completely treat waste streams of up to 500 000 kg. The TVS was procured by SRTC per a Department of Energy (DOE) request and was designed by SRTC in conjunction with EnVitCo, Inc., a division of Toledo Engineering. Fabrication of the TVS was performed by the Dreicor subsidiary of EnVitCo in Erwin, Tennessee. Acceptance testing was performed by EnVitCo and SRTC personnel, and then the TVS was shipped to Clemson, S.C. for more extensive testing by SRTC and Clemson personnel.

The first stream tentatively chosen to be treated in the TVS was B&C Pond Sludge from the Oak Ridge Reservation K-25 site, which is a sludge recovered from several wastewater ponds. Scoping tests on a smaller melter were first done at the Clemson University Environmental Systems Engineering Department's DOE/Industry Center for Vitrification Research.[3] This smaller scale melter of similar design to the TVS Melter was used to successfully treat surrogate B&C waste. Upon completion of these tests, SRTC personnel then ran extended tests of the TVS at Clemson using the same B&C surrogate waste composition. These tests produced about 11 500 kg of surrogate B&C waste glass. After these tests were completed, minor equipment modifications were made, and the TVS was then shipped to ORR. Additional tests using surrogate B&C waste were also done at ORR to verify system integrity after reassembly and to train ORR operations personnel. Startup with mixed waste is scheduled to begin in July 1997 with a Central Neutralization Facility (CNF) sludge/B&C sludge mixture.

SYSTEM DESCRIPTION

The TVS consists of four modules which contain all the equipment required to perform vitrification. It was designed to be disassembled into containers which can be shipped on standard trailers. Fourteen trailers are needed to ship the entire TVS. Disassembly, transport, and reassembly takes about 4-6 weeks. The TVS requires a process area of about 25 m by 30 m. A schematic diagram of the TVS is shown in Figure 1.

In the waste and additives module, liquid or dry wastes and dry glass forming materials are added to the 900 L mix tank to prepare the feed for the melter. These materials are mixed to the proper proportions by weight; the mix tank weight is monitored by a set of load cells. Dry wastes are added directly to the mix tank via a screw feeder. Slurry or other aqueous wastes can be added via a liquid addition port using an external, customer supplied pumping system. Glass forming additives are delivered to the TVS in bulk bags which are hung over small hoppers. Metering screw feeders transport the additives from the hoppers to the mix tank. The feed slurry mixture is then sampled prior to transfer to the feed tank to verify the targeted composition. If analyses show the feed composition does not meet specifications, the feed can be remediated with additional additives.

The feed tank is similar in design to the mix tank. The slurry feed is recirculated through a loop which passes into the melter module, where a side stream is drawn off to feed the melter. The design production rate for the TVS is 68 kg/hr of glass produced when slurry fed. The waste and additives module is enclosed in a sealed container, which provides ventilation through HEPA filters. This system also maintains the module at a slight negative pressure to prevent inadvertent leakage of contaminated material. An airlock is provided for workers to dress out before entering the portion of the module containing radioactive materials. Surfaces inside the container are designed to allow easy decontamination.

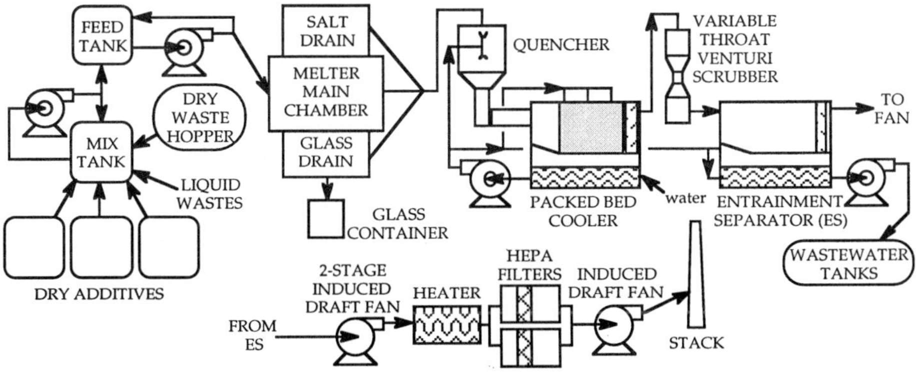

Figure 1. Transportable Vitrification System

The melter module contains the joule-heated glass melter and auxiliary equipment. The melter is divided into three refractory lined chambers. Slurried feed is introduced into the main, central chamber of the melter. Heat input is provided by molybdenum electrodes. The melter operates with glass temperatures in the range of 1100 to 1400°C. Control of the glass temperature is achieved indirectly by controlling the power to the joule heating electrodes. A propane gas burner supplies additional energy input into the main melt chamber plenum. The propane burner is used during cold startup of the melter to heat the solid glass in the melter to a temperature sufficient for joule heating to begin (ca. 700°C). The TVS melter is a "cold top" melter, wherein the "cold cap" of feed slurry on top of the glass insulates the plenum from the heat of the glass. The propane burner is also used during normal operation to provide additional energy input into the plenum to aid in the evaporation of water, and to maintain the plenum and offgas temperatures so that water does not condense before entering the offgas quencher. Operation with a cold cap also limits the loss of volatile metals from the glass. Those metals that are not incorporated in the glass collect at the bottom of the central chamber where they can be drained off as needed from a bottom drain (metals drain).

Two side chambers are provided in the melter; the glass drain chamber contains a mechanical plug-type glass drain. Separate joule and plenum resistance heaters are provided to maintain the correct temperature in the glass drain chamber. Continuous glass production rates of 50 to 150 kg/hr can be maintained, depending on the type of feed used and the percent moisture in the feed. Glass is poured into either drums or specially designed stainless steel containers. The second side chamber contains a salt skimmer to remove molten salts from the glass surface. Accumulation of molten salts is not expected with most of the waste feeds planned for the TVS. This chamber also has separate plenum heaters and a drain.

The offgas module was designed by Andersen 2000 under contract to EnVitCo. Offgases from each of the three chambers of the melter are conveyed to the offgas system through a refractory lined pipe. Flow through the offgas system and melter vacuum is maintained by two induced-draft fans in this module. The other components of the module are the quencher, packed bed cooler, variable throat venturi, entrainment separator, heater, HEPA filters, and a 50 foot stack. The quencher cools the offgas by evaporative cooling and removes the coarser particulates. The packed bed cooler further cools the gas and removes some additional particulates from the gas stream; fresh water is added to the packed bed cooler. The variable throat venturi, which uses recirculated pH adjusted water, is designed to remove 95% of the particulates greater than 1 micrometer and the entrainment separator removes water droplets. A constant blowdown of water from the entrainment separator is used to maintain the solids content in the water below a maximum level. A reheater prevents condensation before the offgas enters the HEPA filters, which are designed to remove the remaining particulate to a total removal of 99.99%. The second fan maintains the HEPA filters at a slight negative pressure and sends the offgas into the stack. Sampling ports are provided at the stack for performing gas and particulate measurements in accordance with Environmental Protection Agency (EPA) methods. Continuous sampling for radioactive particulates is done by isokinetic sampling.

The final module is the control and services module which contains the melter power supplies and a control room. Most of the process is monitored and controlled from this module via a programmable logic controller interface and surveillance cameras.

SURROGATE WASTE FORMULATION

The actual ORR B&C Pond Sludge is a sludge material that contains both waste sludge and excavated soil. RCRA metal components found in this waste include Cr, Ni, and Pb. The main radioactive component is depleted uranium. The surrogate was prepared from bulk technical grade chemicals, Indian red pottery clay, and a "metals spike" mixture (metal chlorides, nitrates, and acetates in HCl). The bulk chemical components were mixed in a large tumbler with the Indian red pottery clay. This clay was used since it is similar in composition to the excavated soil. The metals spike mixture contained minor component metals (Cr, Cu, Ni, Pb, Ce) in an acid solution. Cerium was the surrogate for uranium.

Table I shows target, calculated, and analyzed surrogate compositions. The calculated composition is the composition determined from a material balance on the masses of the components batched into the process. The calculated results are the average of the results for 24 batches of feed; the analytical results are the average of the analyses of nine of the batches from the mix tank. Small quantities of Ba and Zn (impurities in the clay), and Cl^-, NO_3^-, and acetate (from the metals spike solution) were also present. These data show that the analyzed composition of actual surrogate batches made matched the target values reasonably well. The most significant differences occur in the minor components; also, the presence of Li in the surrogate analyses is the result of residual material in the mix tank from mixing the previous batch of surrogate with the glass formers.

The glass formulation for this waste was a soda-lime-silica (SLS) glass, which is the best glass choice for wastes containing significant amounts of calcium since it maximizes waste loading by taking advantage of the glass forming potential of the waste.[4] Borosilicate glasses were not chosen due to the large immiscibility gap in the CaO-B_2O_3-

SiO_2 system. The glass forming additives used were silica, lithium carbonate, and sodium carbonate. Table II shows the targeted, calculated, and analyzed composition of the melter feed. The data show that the batching of the surrogate and glass forming materials into the TVS was also accurate, with the exception of Al, for which the actual concentrations were higher than the target. There are slight discrepancies when comparing the feed analyses with the glass analyses; the glass analyses for Li, Mn, Na and P were all somewhat different than the feed analyses, but were comparable to the target concentrations, indicating that there may be an analytical bias when analyzing these components in the feed. Again, the differences in the minor components were generally greater than the differences in the major components.

Table I. Target, Calculated, and Analyzed Composition of ORR B&C Surrogate

	Element Basis			Oxide Basis		
Element or Anion	Calculated (wt %)	Analyzed (wt %)	Oxide	Target (wt %)	Calculated (wt %)	Analyzed (wt %)
Al	2.350	2.437	Al_2O_3	5.400	5.356	5.842
Ca	13.15	11.92	CaO	23.97	22.19	21.16
Ce	0.091	0.106	Ce_2O_3	0.134	0.128	0.158
Cr	0.028	0.036	Cr_2O_3	0.045	0.049	0.067
Cu	0.027	0.029	CuO	0.047	0.041	0.046
Fe	4.647	4.339	Fe_2O_3	8.330	8.013	7.870
K	1.339	1.325	K_2O	2.060	1.945	2.024
Li	0.000	0.213	Li_2O	0.000	0.000	0.581
Mg	0.418	0.449	MgO	0.814	0.836	0.945
Mn	0.396	0.139	MnO_2	0.777	0.756	0.279
Na	0.260	0.837	Na_2O	0.246	0.422	1.431
Ni	0.172	0.163	NiO	0.276	0.265	0.264
P	0.334	0.327	P_2O_5	0.995	0.922	0.950
Pb	0.021	0.013	PbO	0.028	0.027	0.018
Si	22.62	21.28	SiO_2	56.21	58.36	57.77
Ti	0.339	0.278	TiO_2	0.667	0.681	0.589
$CO_3^=$	19.63	15.79				

The major anion present in the B&C sludge was carbonate ($CO_3^=$). Therefore, the calcium component of the surrogate was added as the carbonate. The lithium and sodium glass formers used were also carbonates. About 20% of the carbonates in the surrogate sludge and 35% in the melter feed were found to decompose prior to sampling. The acidic metals spike solution may have accounted for a small amount of this decomposition. The accuracy of the measured carbonate compositions was indirectly verified by measurements of the amount of CO and CO_2 in the offgas from the melter.

The total solids content of the melter feeds ranged from 39 to 61 wt%, while the density ranged from 1.3 to 1.6 kg/L. The processibility of these feeds depended greatly on the type of glass formers used. Several different types of glass formers were used, ranging from very fine (precipitated silica) to very coarse (sodium carbonate). Augering of the very fine precipitated silica was so difficult and the feed slurry made was so viscous that coarser silica was eventually substituted.

Table II. Target, Calculated, and Analyzed Composition of Melter Feed

Element or Anion	Element Basis Calculated (wt %)	Element Basis Analyzed (wt %)	Oxide	Oxide Basis Target (wt %)	Oxide Basis Calculated (wt %)	Oxide Basis Analyzed (wt %)	Glass @ Three Melter Turnovers
Al	1.138	1.450	Al_2O_3	2.700	2.724	3.580	3.716
Ca	6.370	6.018	CaO	11.99	11.29	11.00	11.60
Ce	0.044	0.042	Ce_2O_3	0.070	0.065	0.065	0.079
Cr	0.014	0.019	Cr_2O_3	0.020	0.025	0.036	0.042
Cu	0.013	0.013	CuO	0.020	0.021	0.021	0.018
Fe	2.251	2.168	Fe_2O_3	4.170	4.076	4.050	3.822
K	0.649	0.720	K_2O	1.030	0.989	1.132	1.038
Li	2.644	2.881	Li_2O	7.500	7.209	8.103	7.308
Mg	0.203	0.251	MgO	0.410	0.425	0.544	0.519
Mn	0.192	0.069	MnO_2	0.390	0.384	0.142	0.374
Na	4.706	4.255	Na_2O	7.620	8.034	7.493	7.946
Ni	0.084	0.079	NiO	0.140	0.135	0.131	0.122
P	0.162	0.229	P_2O_5	0.500	0.469	0.687	0.451
Pb	0.010	<0.010	PbO	0.010	0.014	0.014	<0.010
Si	23.55	22.42	SiO_2	63.11	63.79	62.66	62.62
Ti	0.164	0.157	TiO_2	0.330	0.347	0.342	0.355
$CO_3^{=}$	26.92	17.51					

The particle size of the glass formers can affect the rate of melting in the melter and also the amount of waste that can be incorporated into a given glass composition. Smaller particle sizes for glass formers usually result in higher possible waste loadings and melt rates; however, the overall processing time suffered due to difficulties with augering the very fine precipitated silica and with highly viscous feed slurries that were difficult to pump. Therefore, a compromise between waste loading and feed processibility was needed. Use of a coarser silica improved augering and feed viscosity greatly, such that the overall processing rate increased. Additional improvement in the reliability of all of the glass former augers was also made by inserting solid cores into the augers.

MELTER OPERATION

The design glass production for the TVS melter is 68 kg/hr when slurry fed with an approximately 50 wt% slurry. Initially, a maximum overall glass production rate of only 50 kg/hr could be achieved due to lower than design melt rate. The vacuum in the melter could only be maintained at about 25-75 Pa due to a large amount of air inleakage (475 m^3/hr) into the main melter chamber. Modifications done between the surrogate runs at Clemson and at Oak Ridge reduced this inleakage to about 135 m^3/hr, increased the operating vacuum to 250-375 Pa, and resulted in an increase in the average plenum temperature during feeding of about 200 °C. The bottom line was an increase in the attainable steady state glass production rate from 50 kg/hr to 86 kg/hr.[5]

Since the melter was started up with an inventory of SLS glass, the composition and physical properties of the glass in the melter changed as the surrogate waste was introduced. The most significant impact on operation is the change in the glass viscosity as the less viscous B&C surrogate glass formed displaces the SLS startup glass. Since viscosity is a parameter of critical importance to both glass pouring and refractory life, the melter operating temperature was adjusted periodically during the transition from SLS to the B&C surrogate glass. The composition and viscosity of the glass in the melter was estimated by treating the melter as a well-mixed tank and using viscosity-composition correlations developed by the SRTC. Table III summarizes the predicted viscosity as a function of melter turnovers of B&C surrogate glass. Viscosities between 20 and 100 poise are considered to be acceptable for pouring; higher viscosities in this range are better for refractory wear. Viscosities in brackets [] in Table III show the approximate temperature and the corresponding viscosities during the transition; viscosities in braces {} show temperatures that would have resulted in viscosities outside the acceptable range.

Table III. Predicted Glass Viscosity (Poise) During the Transition from SLS to B&C Glass

Melter Turnovers	1150 °C	1200 °C	1250 °C	1300 °C	1350 °C
0	{172}	119	84	61	[46]
1/8	{143}	98	70	[51]	38
1/4	{120}	83	[59]	43	32
3/8	{103}	71	[50]	37	27
1/2	89	[62]	44	32	24
5/8	79	[54]	39	28	21
3/4	71	[49]	35	25	{19}
7/8	64	[44]	31	23	{17}
1	[59]	40	29	21	{15}

Crystallization or devitrification in the melter can result in accumulation of a viscous layer on the bottom of the melter which could interfere with glass pouring or lead to localized corrosion. To mitigate the adverse effects of crystallization, the melter was maintained at a temperature of at least 100°C above the liquidus temperature predicted as a function of composition. The refractory was inspected following the demonstration and only minor wear was observed.

The maximum steady-state glass production rate was found to be limited by the maximum melt rate (86 kg/hr glass) that could be achieved in the melter. The maximum instantaneous rate of glass pouring achieved was about 365 kg/hr. However, the glass pouring rate should be limited to reduce the potential for formation of devitrified glass in the glass container due to slow cooldown. Pouring at up to 140 kg/hr showed no evidence of devitrification.

Joule heated glass melters are often assumed to be well mixed, such that their residence time distribution can be described by the continuously stirred tank reactor (CSTR), or well-mixed tank model. Several residence time distribution models for the concentrations of iron and lithium, which were not present in the SLS glass cullet used to initially charge the melter, were fit to the TVS data. Figure 2 shows the concentrations of Li as a function of melter mass turnover.

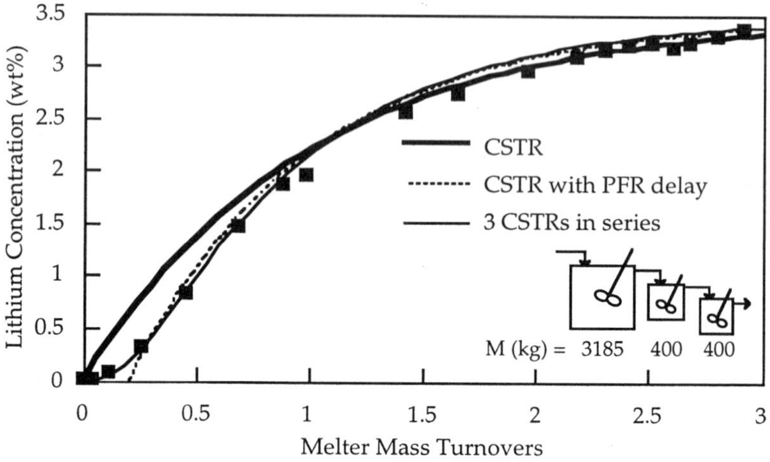

Figure 2. Lithium Concentration in Melter

The CSTR model was found to not fit the data as well as models with a plug flow reactor (PFR) delay time or with three reactors in series. The physical layout of the melter is consistent with these models. The melter consists of the large main chamber, which is likely to behave like a CSTR. The glass from the main chamber passes through a throat block to the glass drain chamber, from which it is poured. Figure 3a shows a simple interpretation of the possible glass flow patterns.

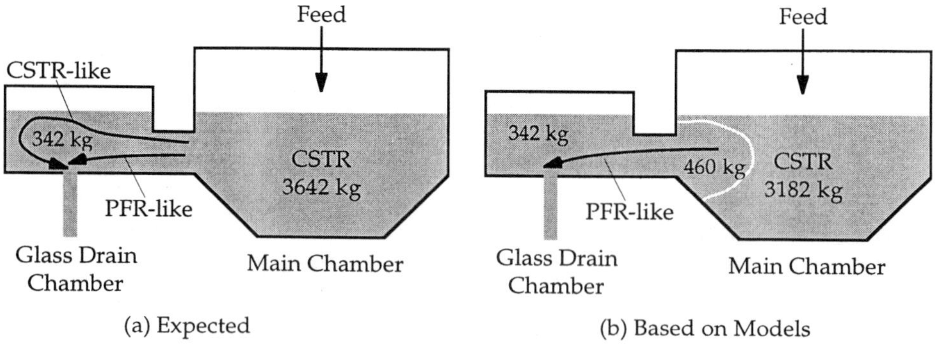

Figure 3. Residence Time Behavior of the TVS Melter (not to scale)

The flow through the glass drain chamber would be expected to be somewhere between a PFR and a CSTR. This chamber contains joule electrodes that would tend to promote some convective mixing, but the lack of a cold cap and the short residence time significantly limits the actual convection. The best model (3 CSTRs in series) has one large CSTR (main chamber) followed by two smaller equal sized CSTRs in series that approximate a non-ideal PFR with substantial back mixing,[6] as shown in Figure 2 for the Li data. One would expect that the predicted mass of the main chamber CSTR would be equal to the actual mass in the main chamber; however, the best fit of the data gives about 460 kg less than the actual amount, which may indicate that the region of the main chamber near the passage to the glass drain also has PFR behavior, as shown in Figure 3b. Figure 4 shows data for most of the other elements and fitted curves for the CSTR with PFR delay model. The fitted curves use the CSTR and PFR volume determined from the Li data.

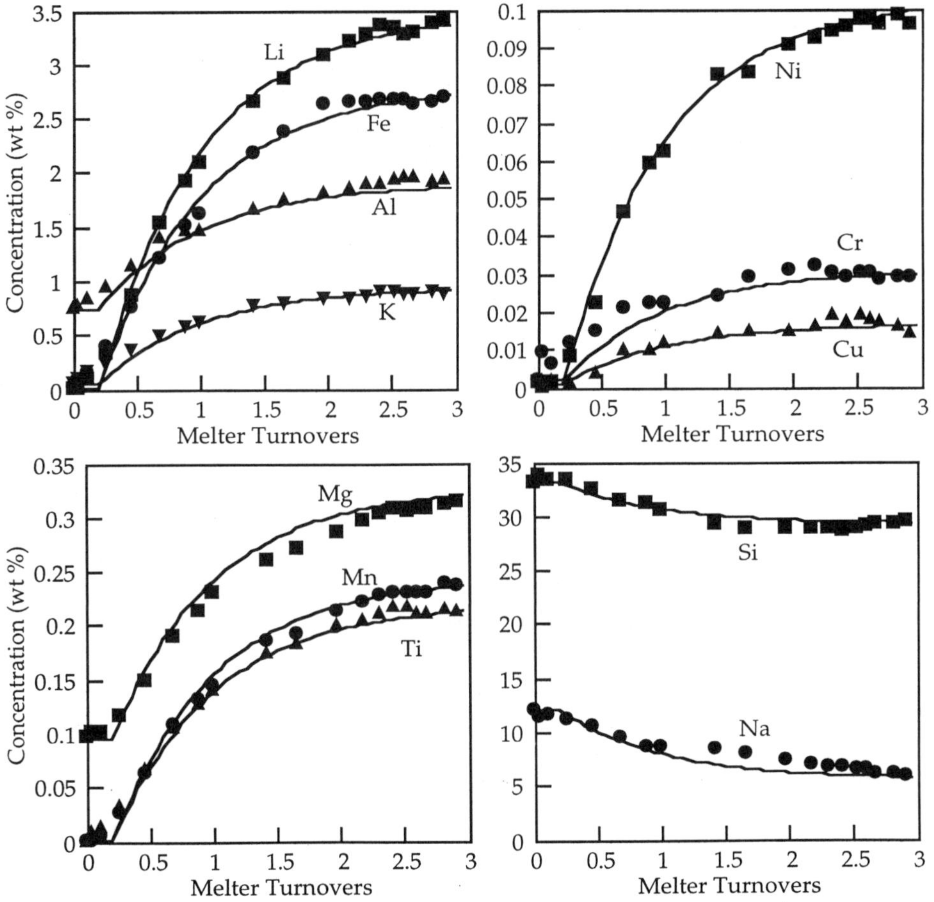

Figure 4. Melter Glass Composition Fit by CSTR-PFR Model

Although the melter does not behave exactly as a CSTR, the concentrations of all components at three melter turnovers are very similar to the concentrations that would be seen with true CSTR behavior and the CSTR model is essentially as accurate as the more complicated models at greater than one melter turnover.

The durability of the glasses produced was also of concern, as the B&C Pond sludge is a listed waste as defined in the Land Disposal Restrictions[7] (LDRs) for hazardous waste. Specifically, it is listed as F006 waste, which is for wastewater sludges from electroplating operations. The constituents of concern in F006 wastes are Cd, Cr, Ni, Pb, Ag, and cyanides; of these, B&C Pond sludge contains Cr, Ni, and traces of Pb. This wastewater sludge also contains the RCRA metals Ba and Zn. Glass pour grab samples and samples from the tops of glass containers were taken and their durability measured with the EPA Toxicity Characteristic Leaching Procedure (TCLP).[8] Based on the TCLP analyses of all the samples, the glass produced was judged to have acceptable durability as defined by the EPA Universal Treatment Standards (UTS).[9,10] Table IV summarizes the TCLP results and compares them to the UTS limits. The metals placed in the B&C Ponds sludge surrogate that are addressed by the LDRs were Cr, Ni, and Pb. Although not intentionally added to the surrogate, trace amounts of Ba and Zn, present as impurities in the Indian red pottery clay, gave measurable TCLP responses.

Table IV. TCLP Results for Glasses Produced

	Measured Values (mg/L)		
Metal	Minimum	Maximum	UTS Limits
Ba	<0.01	0.29	7.6
Cr	<0.02	<0.02	0.86
Ni	<0.01	<0.01	5.0
Pb	<0.10	<0.10	0.37
Zn	0.17	0.36	5.3

Crystallization in the TVS melter upon melter shutdown was also a concern. Devitrification of the remaining melt pool upon shutdown could cause severe problems with melter restart. To evaluate the devitrification potential of the melt pool, representative glass samples were obtained from the glass poured as the melter was transitioned from the startup inventory of SLS glass to the glass containing the B&C surrogate. Samples were taken at approximately 0, 0.5, 1.0, 1.5, 2.0, and 3.0 melter turnovers.

These samples were subjected in the laboratory to a thermal cooldown profile that approximated the cooldown data recorded from the TVS during a cooldown with SLS glass. The profile used was based on the temperature at approximately the center of the main melter chamber, so this profile is near the slowest cooldown rate, and thus is conservative. X-ray diffraction analysis on the as-received (pre-heat treatment) glasses showed no indication of devitrification.

Results of the thermally heat treated glasses showed that the high SLS glass (initial inventory) was "thermally stable" with respect to devitrification when the melter cooling curve was applied while the high B&C glass at three melter turnovers partially devitrified. The majority of the crystallization associated with this last sample was surface devitrification. At least three crystalline products were identified by x-ray diffraction: sodium aluminum silicate (nepheline), lithium silicate, and calcium silicate (wollastonite). With the exception of the glass at 0.5 melter turnovers, all intermediate glasses were x-ray amorphous after heat treatment. Nepheline was detected in the 0.5 melter turnover sample after heat treatment , but semi-quantitative analyses suggested an extremely small quantity.

Based on these results, it was recommended that the melter be flushed with at least one melter turnover of SLS glass to avoid or minimize devitrification in the melter. The melter was then flushed with 1.5 melter turnovers of SLS glass before shutdown, and was then successfully restarted at Oak Ridge.

Samples of the glass product poured from the melter were characterized for their oxidation/reduction (redox) state. Since the mixing in the melter can be approximately described as a well mixed tank, the glass poured from the melter is considered to be a fairly representative sample of the bulk contents of the melter. The redox state of the glass was monitored by measuring the ratio of Fe^{+2} to the total iron ($\Sigma Fe = Fe^{+2} + Fe^{+3}$), or $Fe^{+2}/\Sigma Fe$. Figure 5 shows the redox ratio as the melter glass inventory transitioned from SLS to B&C and the later parts of the demonstration when organics were added to three batches.

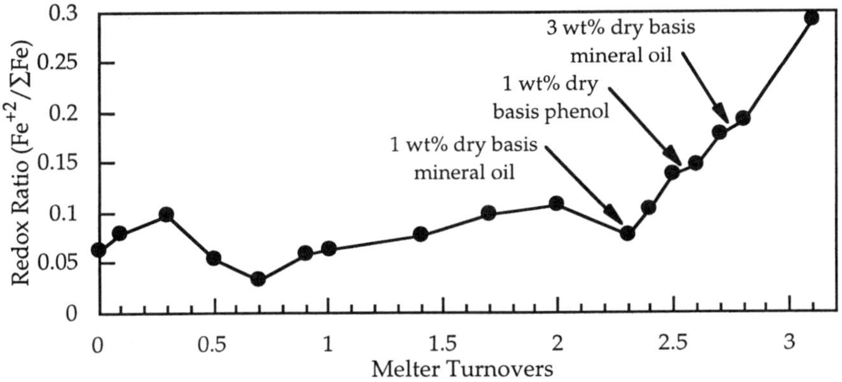

Figure 5. Redox of Glass Produced

For three batches, 1 wt% dry basis petroleum oil, 3 wt% dry basis phenol, and 3 wt% dry basis petroleum oil were added to the feed at approximately 2.4, 2.5 and 2.8 melter turnovers, respectively. The petroleum oil used was a hydrotreated heavy naphthalenic grade lubricating oil with an approximate molecular weight of 480 and a boiling point of 300 °C. These organics were added to estimate combustion efficiency in the melter. As Figure 4 shows, the transition from SLS glass to the B&C surrogate glass did not have a significant effect on the redox state of the glass. The addition of the organics, as expected, had a considerable effect. Each 240 gallon feed batch is approximately equivalent to 1/8 of a melter turnover. The three organic additions had the cumulative effect of raising the redox ratio from 0.08 to 0.29. Although the glass became more reduced with the addition of the organics, there was no impact processibility or product quality. It should be noted that if these levels of organics had been added throughout the demonstration, the redox ratio would have likely increased to unacceptable levels.

REFERENCES

[1] J. C. Whitehouse, C. M. Jantzen, F. R. Van Ryn, D. H. Davis, "Design and Fabrication of a Transportable Vitrification System for Mixed Waste Processing," Proceedings of the Third Biennial Mixed Waste Symposium, Baltimore, MD (August 1995).

[2] J. C. Whitehouse, C. M. Jantzen, D. F. Bickford, A. L. Kielpinski, B. D. Helton, F. R. Van Ryn, "Development of a Transportable Vitrification System for Mixed Waste," Proceedings of Waste Management '95, Tucson, AZ (February 1995).

[3] C. A. Cicero, T. J. Overcamp, D. L. Erich, "Vitrification Demonstration with Surrogate Oak Ridge Reservation K-25 B&C Pond Sludge," Proceedings of the International Symposium on the Environmental Issues and Waste Management Technologies in the Ceramic and Nuclear Industry, 1996.

[4] C. M. Jantzen, D. K. Peeler, C. A. Cicero, "Vitrification of Ion-Exchange (IEX) Resins: Advantages and Technical Challenges," Proceedings: Vitrification of Low-Level Waste - The Process and Potential (December 1995).

[5] J. R. Zamecnik, S. R. Young, E. K. Hansen, J. C. Whitehouse, "Design, Operation, and Evaluation of the Transportable Vitrification System," to be presented at the 1997 International Conference on Incineration and Thermal Treatment Technologies, Oakland, CA, May 12-16, 1997.

[6] C. G. Hill, An Introduction to Chemical Engineering Kinetics and Reactor Design, Wiley, New York, 1977, p. 405.

[7] "Land Disposal Restrictions Phase II - Universal Treatment Standards , and Treatment Standards for Organic Toxicity Characteristic Wastes and Newly Listed Wastes," Federal Register, Vol. 59, No. 180, Sept. 19, 1994, p. 47982.

[8] Toxicity Characteristic Leaching Procedure, Test Method 1311 in "Test Methods for Evaluating Solid Waste, Physical/Chemical Methods," EPA Publication SW-846.

[9] "Universal Treatment Standards", Code of Federal Regulations, 40 CFR 268.48.

[10] "Applicability of Treatment Standards - Treatment Standards for Hazardous Wastes", Code of Federal Regulations, 40 CFR 268.40.

VITRIFICATION OF SAVANNAH RIVER SITE SOIL IN A DC ARC MELTER

Thomas J. Overcamp, Douglas M. Cash, Stewart J. Griner, and Matthew P. Speer
Department of Environmental Systems Engineering
Clemson University
342 Computer Court
Anderson, SC 29625-6510

ABSTRACT

A series of eight trials was conducted on two Savannah River Site soils that were spiked with metal compounds and radionuclides surrogates. Soil A had an alumina content of 4.1%, and Soil B had an alumina content of 0.7%. Batch materials, prepared with these spiked soils and calcium hydroxide and lithium carbonate, were fed to the DC arc melter. Some of the trials were conducted with air fed to the melter's plenum to promote oxidation. Others were conducted with nitrogen fed to the melter to promote reducing conditions. Half of the tests also had small amounts of wood shavings added to simulate organic debris. The trials were conducted with semi-batch operations. There was incremental feeding during the run, but the melter was cooled prior to removing the wasteform.

The wasteforms from soil A tests were glasses or glassy ceramics. Due to the long cooling period, the wasteforms for soil B tests completely devitrified. Overall material balance and the recovery of elements within the melter were determined. The durability of the wasteforms was assessed with both the ASTM product consistency test and the EPA toxicity characteristic leaching procedure.

INTRODUCTION

The U.S. Department of Energy (DOE) must treat large volumes of buried wastes and contaminated soils that are legacies of World War II and the Cold War nuclear weapons programs. Many of these are mixed wastes in that they are both Resource Conservation and Recovery Act (RCRA) hazardous wastes and radioactive wastes under the Atomic Energy Act. These wastes must be treated to meet the RCRA land disposal restrictions (LDR) before disposal.

Due to varied organic and metals content, many buried wastes are difficult to treat successfully in conventional joule-heated vitrification systems. Organic materials can promote strongly reducing conditions in a system designed to produce an oxidized glass. Metal objects may be difficult to melt, promote reducing conditions, accelerate corrosion of the refractory and electrodes, and short the melter.

The graphite electrode DC arc melter is a potential solution to treating these wastes. It uses a DC arc to produce a high-temperature plasma. Radiation from the plasma as well as joule heating can melt most materials including soils and metals. Since the reduced metals and the slag or glass from the soil oxides form distinct phases, the melter can be used to separate and recover bulk metals from wastes. In addition, high temperatures can destroy organic materials and contaminants. The DC arc melter is normally operated under a reducing atmosphere to produce a combustible byproduct, but it can also be operated with an oxidizing atmosphere by providing air or oxygen.

This paper presents the results of the analysis of the wasteform from experimental tests of a DC arc melter to treat two Savannah River soils spiked with metal salts, surrogates for radionuclides, and organic debris. The melter was operated under either oxidizing or reducing atmospheres.

BACKGROUND

Surma *et al.*[1] conducted tests in an Electro-Pyrolysis, Inc., (EPI) Mark I melter on simulated Idaho National Engineering Laboratory (INEL) buried wastes. The Mark I melter, the predecessor to the melter used in this research, was operated in batch mode. In two tests with cesium-contaminated INEL soils, they reported particulate concentrations of 6.5 and 11.4 g/m^3 and cesium emissions of 0.9 and 1.9% for the submerged arc and exposed arc runs, respectively.

Soelberg *et al.*[2] processed simulated INEL buried wastes with the U.S. Bureau of Mines AC arc melter. This melter is a full-scale, 3 phase arc furnace capable of processing up to 700 kg of contaminated soil per hour. They tested a simulated waste containing approximately 53% INEL soil, 14% calcium, 10% iron, 7% solid combustible materials, 7% organic materials, 2% nitrate salts, and trace levels of various metals including nickel, copper, cerium, cesium, and lead. In four runs, they reported particulate concentrations ranging from 23 to 237 g/m^3 and over 80% emission of the lead.

Wenger *et al.*[3] developed a thermodynamic model for a DC arc melter and compared predictions of this model to data from Surma *et al.*[1]

METHODS AND MATERIALS

DC Arc Melter

The EPI Mark 1-W DC arc melter was purchased by Westinghouse Savannah River Company (WSRC) and installed at the Vitrification Research Laboratory at Clemson University. The Mark I-W melter is the water-cooled successor to the Mark I melter used by Surma *et al.*[1]

The melter, shown in Figure 1, is housed in a water-cooled, stainless-steel vessel containing two graphite crucibles. The lower half of the vessel, which holds the crucibles, can be lowered and rolled outward to allow for feeding and emptying of the melter. The outer crucible is permanently installed in this vessel. Refractory between the outer graphite crucible and the water-cooled vessel decreased the heat transfer from the crucible. An inner, removable graphite crucible has a working capacity of approximately 30 L and a mass of about 50 kg. A graphite disk holding a starting electrode was placed in the bottom of the crucible. This electrode and the graphite crucible were the positive electrode. A negative, graphite electrode was mounted on a movable carriage that could be raised or lowered with a screw-driven mechanism housed above the melter.

Power was supplied to the melter at a maximum output of 133 V at 750 A. The current was controlled by a manual potentiometer. The voltage was controlled by the electrode separation.

The melter had several purge gas systems. One system continuously purged the upper electrode assembly with nitrogen to keep it free from oxygen. Another system was designed to continuously purge the region between the outer crucible and the stainless-steel vessel. In addition, there was a nitrogen purge through the lens assembly of a video camera. Finally, there was a purge system that fed either nitrogen or air into the plenum of the melt chamber.

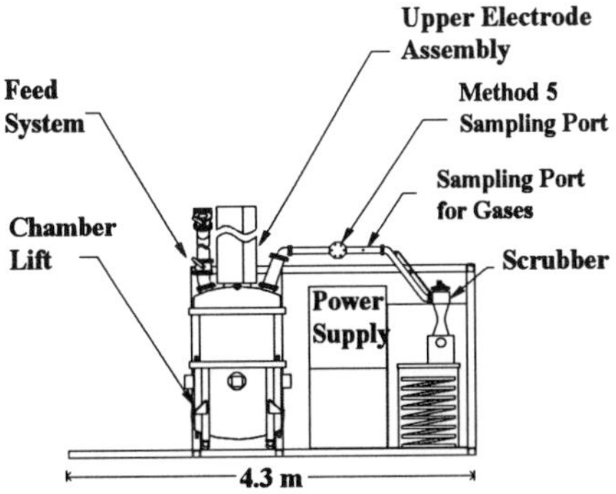

Figure 1. EPI Mark 1-W DC Arc Melter.

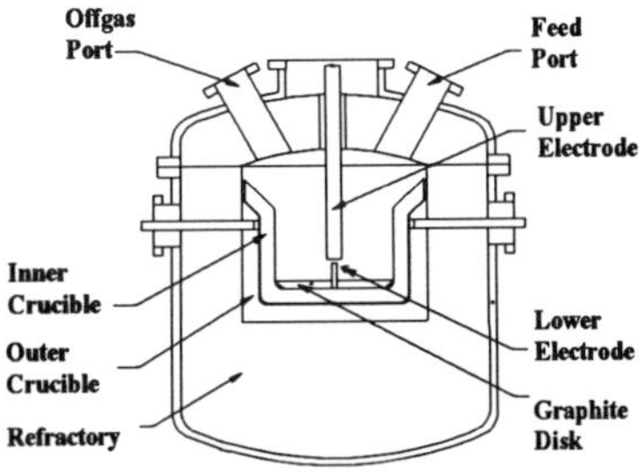

Figure 2. Schematic of the chamber.

Feed System

The melter was designed for batch operations. It was modified to introduce waste materials incrementally during the run. A feed system was assembled with two stainless-steel ball valves in a section of stainless-steel pipe that was connected to the melter. The feed system was designed so that it could be isolated from the melt chamber with a ball valve, filled with feed materials, sealed and purged. Then the lower ball valve could be opened to charge the melter.

Simulated Wastes and Feed Materials

The composition of the two Savannah River Site soils are given in Table 1. Unlike the INEL soil used in tests by Surma *et al.*[1] and Soelberg *et al.*[2], these soils had low alkali metals and alkaline earth fractions. Type A soil, a sand with a small amount of clay, had a 4.3% by mass alumina content, a 1.7% by mass iron oxide content, and a 93% by mass silica content. Type B, a sandy soil, had a 0.7% by mass alumina content, 0.6% by mass iron oxide content, and a 98% by mass silica content. Due to their high silica content, these soils have high melting temperatures.

The composition of the various batch materials is shown in Table 2. To aid in processing, calcium hydroxide and lithium carbonate were added to lower the melt temperatures of the feed materials and modify the viscosity of the melt. Metals were added to the soil to simulate heavy metal contaminants. To avoid producing RCRA characteristic wastes, the primary metals added were copper acetate, iron oxide, and nickel nitrate. A very small amount of lead acetate was added to determine if lead could be detected in the particulate emissions. To simulate radionuclides, nonradioactive cesium nitrate and cerium carbonate were added. Cerium is frequently used as a surrogate for uranium and plutonium in mixed-waste vitrification demonstrations. Cedar shaving pet bedding material was added to half of the runs to simulate organic debris.

In each test, the inner crucible was lined with a layer of Grafoil,™ which is a flexible graphite material. Several Grafoil disks and powdered graphite were used to protect the removable graphite disk at the bottom of the crucible. Approximately 10 kg of feed material was placed in the crucible at the start of the run. An additional 40 kg was fed in 3.75 L batches. A new batch was added when it was observed that the cold cap was nearly melted.

Table 1. Oxide Composition of the Savannah River Site Soils.

Oxide	Soil A, % Mass	Soil B, % Mass
Al_2O_3	4.130	0.733
B_2O_3	<0.014	<0.014
BaO	<0.006	<0.006
CaO	0.094	0.021
CdO	<0.006	<0.006
Cr_2O_3	<0.016	<0.015
Fe_2O_3	1.730	0.580
LaO	<0.004	<0.004
MgO	0.014	<0.002
MnO	<0.004	<0.004
Na_2O	0.342	0.337
NiO	<0.034	<0.034
P_2O_5	0.077	0.060
PbO	<0.035	<0.034
SiO_2	93.017	97.977
SrO	<0.003	<0.003
TiO_2	0.236	0.147
ZnO	<0.004	<0.004
ZrO_2	0.056	0.020

The particulate sampling was done with the EPA Method 5 train in the horizontal pipe downstream of the melter. Gaseous samples were pulled from this pipe, conditioned, and analyzed for permanent gases by gas chromatography and NO/NO_x with a chemiluminescent analyzer. The gas chromatographs were operated on 2–3 min cycle times. The chemiluminescent analyzer was manually switched between NO and NO_x every minute.

Experimental Design

An eight test experimental plan, shown in Table 3, was proposed. In this plan, two Savannah River Site (SRS) soils were contaminated with metal compounds and surrogates for radionuclides. Organic material was added to the soil materials in four tests. Four tests were conducted with a reducing atmosphere in the plenum, and four were conducted under more oxidizing conditions with air fed to the plenum.

During all tests 14 L/min of nitrogen was supplied to purge the upper electrode assembly. There was also 14 L/min of nitrogen supplied through the camera lens. The nitrogen purge beneath the outer crucible was not operated. For the tests with air, 140 L/min was supplied to the main chamber. For the tests with nitrogen only purge, only 85 L/min could be supplied.

Analyses

The glass or crystalline wasteform was sampled by taking a single sample from the middle of each wasteform at a point approximately halfway between the center and the outer wall and halfway between the top and bottom of the wasteform. Metal pellets were also sampled. At SRTC, these wasteform and soil samples were digested and analyzed by inductively coupled plasma emission spectroscopy (ICPES) or by atomic absorption (AA). Redox analyses were also performed on the glassy or crystalline samples. Both the RCRA Toxicity Characteristic Leaching

Procedure (TCLP) and the ASTM Product Consistency Test (PCT) were performed on the glassy or crystalline samples. In addition, samples of the particulate matter were chemically analyzed at SRTC by ICPES or AA. XRD analyses were performed on the glassy or crystalline wasteform and particulate samples to determine if crystalline phases were present.

Table 2. Batch Composition on a Calcined Basis.

Constituent	Level 1 Mass %	Level 2 Mass %
Soil (Type A or B)	79.270	76.544
Modifiers		
CaO from $Ca(OH)_2$	14.400	14.400
Li_2O from Li_2CO_3	3.600	3.600
Metal Additives		
CuO from $Cu(C_2H_3O_2){\cdot}H_2O$	0.500	1.000
Fe_2O_3	1.000	2.000
NiO from $Ni(NO_3){\cdot}H_2O$	0.500	1.000
PbO from $Pb(C_2H_3O_2){\cdot}H_2O$	0.004	0.004
Radionuclide Surrogates		
Ce_2O_3 from $Ce_2(CO_3){\cdot}H_2O$	0.500	1.000
Cs_2O from $CsNO_3$	0.226	0.452
Total	100.000	100.000
Debris		
Wood chips (dry basis)	0.034	0.000

Further details are given in three Clemson University theses. Cash[4] presents information on the batch loading and results on the wasteform including crystallinity, mass recoveries, and TCLP results for the tests. Griner[5] discusses the particulate sampling and analyses and the results. Speer[6] presents information on the gaseous emissions analysis system and the results for each test. Overcamp *et al.*[7] discussed the data on gaseous and particulate emissions.

Table 3. Test Matrix.

Trial	Soil	Mixed-Waste Simulant		Direct Purge	
		Debris	Metals	Gas	Flow, L/min
1	A	Yes	1	Air	140
2	A	No	2	Air	140
3	A	Yes	1	Nitrogen	85
4	A	No	2	Nitrogen	85
5	B	Yes	1	Air	140
6	B	No	2	Air	140
7	B	Yes	1	Nitrogen	85
8	B	No	2	Nitrogen	85

RESULTS

Melter Performance

The eight tests were successfully conducted. The average total power consumed per mass of batch ranged from 1.43 to 1.68 kWh/(kg feed). The power consumed while feeding was 0.57 to 0.72 kWh/(kg feed). The total power consumed per mass of product ranged from 1.97 to 2.31 kWh/(kg wasteform).

The consumption of the upper graphite electrode of power ranged from 0.0012 to 0.0036 kg/kWh. There was no measurable loss of mass in the inner graphite crucible for the runs with nitrogen plenum purge. The inner crucible lost about 0.5 kg per test using an air plenum purge. The latter loss was most evident above the melt line.

Gaseous Emissions

Table 4 gives a summary of the gaseous emission data. Due to both the incremental feeding and the cycle times of the analyzers, the values ranged widely. The high CO_2 emissions in all tests were largely due to the decomposition of carbonates. Oxygen concentrations were much lower for the tests with the nitrogen purge. Carbon monoxide was measured in all tests with the higher values for the tests with reducing conditions and debris. Methane was measured in all tests with higher methane content in the tests with debris. The nitric oxide and nitrogen dioxide concentrations in all tests ranged up to 10,000 ppm_v. The highest levels occurred immediately after charging a new feed sample. The majority of these emissions were felt to be due to the decomposition of the nickel and cesium nitrates added to the feed materials.[6,7]

Table 4. Summary of Gaseous Emissions.

Test	Condition	H_2 %	O_2 %	CO %	CO_2 %	CH_4 ppm_v	NO ppm_v	NO_2 ppm_v
1	Oxidizing Debris	0–5	0–10	0–10	10–25	0–5,000	0–6,000	0–6,000
2	Oxidizing	ND*	ND	0–2	ND	0–1,500	2,000–5,000	2,000–5,000
3	Reducing Debris	>6	<1	6–16	3–15	1,000–7,000	100–4,000	ND
4	Reducing	ND	<1	1–7	ND	200–1,000	0–10,000	0–10,000
5	Oxidizing Debris	0–5	0–7	0–10	8–25	0–7,000	0–3,000	ND
6	Oxidizing	0–1	5–10	0–2	12–20	0–2,000	1,000–10,000	1,000–10,000
7	Reducing Debris	>6	<1	10–20	5–10	1,000–10,000	0–1,000	ND
8	Reducing	0–6	<1	0–4	5–15	1,000–4,000	0-10,000	ND

*Not determined.

Particulate Emissions

The gas flow from the melter was relatively low, but the particulate emissions were high, ranging from 32 to 145 g/m^3. The particulate matter consisted primarily of vitreous silica and glass-forming modifiers resulting from the volatilization of the melt and entrainment of feed.[7]

Wasteform

Since the wasteform cooled slowly over a period of hours, test conditions promoted crystallization. Except for the likely contamination of the samples with stainless steel in milling the samples, XRD analysis showed that the samples from tests 2 and 4 were glasses. These tests were conducted using Soil A without debris. The samples from tests 1 and 3, which used Soil A with debris, had partially devitrified to form crystobalite crystals within a predominately amorphous phase. Analyses from tests 5–8, which used Soil B, clearly indicated crystalline materials. These were identified as crystobalite, lithium silicate, and pseudowollastonite, which is a calcium silicate. If the ceramic wasteforms were poured from the melter and cooled quickly, they may have formed amorphous glasses.

The major difference between the soils is their alumina content. Soil A has a 4.13% alumina content, whereas Soil B had a 0.73% alumina content. The additional alumina in Soil A may have increased the melt's viscosity and discouraged crystallization.

The overall material balance for the tests is given in Table 5. The overall mass balance closure was very good.

Table 5. Overall Material Balance for Each Test.

Component	Mass, kg							
	1	2	3	4	5	6	7	8
Glass/Crystal-line Phase	35.660	35.600	35.22	36.62	34.70	35.32	34.38	35.96
Metal Phase	0.308	0.254	0.363	0.197	0.545	0.207	0.109	0.234
Batch	2.772	3.103	0.313	1.10	2.320	2.306	1.816	1.586
Emissions	2.02	0.81	1.20	1.03	2.40	1.76	4.13	1.12
Total	40.76	39.77	37.10	38.95	39.96	39.59	40.44	38.90
Closure	101.9%	99.4%	92.8%	97.2%	99.9%	98.1%	101.1%	97.7%

Since no reduced metals were added to the feed, the metal pellets recovered formed due to the reducing conditions in the melt. The only metals found in the analyses were copper, iron, nickel, and silicon. There was a overall difference in fraction of some elements partitioned in the metal phase in Soil A versus Soil B. For copper, the fraction of added copper in the metal ranged from 3.8 to 11.3% for Soil A and from 0.6 to 4.7% for Soil B. For nickel, the fraction of the added nickel in the metal ranged from 24.5 to 40.1% for Soil A and from 7.0 to 22.0% for Soil B. For iron, the fraction of both the added iron and the iron in the soil ranged from 7.3 to 22.6% for Soil A and from 4.5 to 11.0% for Soil B. Because the tests with debris also had low metals addition and the runs without debris had high metals addition, no obvious trend relating the fraction partitioned to the metal phase was found with respect to debris or metals loading. The mass fraction of silicon in the metal ranged from 2.9 to 15.6% for the tests without debris and from 8.2 to 33.3% for the tests with debris.

The overall elemental recovery in the glassy or ceramic phase and the metal phase is given in Table 6. These values are based on the analyses of one metal and one glassy or crystalline sample. Excluding test 5, the recovery of silicon ranged from 88.1% to 98.8%. The recovery of other elements varied considerable due to offgas emissions, analytical errors, or the heterogeneity of the wasteform. The recovery of cerium in the wasteform ranged from 55.3 to 77.0%. The recovery of cesium ranged from 65.7 to 82.8%.

Table 6. Elemental Recovery in Both the Glassy or Crystalline Wasteform and the Metal Phase.

Test	Al %	Ca %	Ce %	Cs %	Cu %	Fe %	Li %	Ni %	Pb %	Si %
1	136.7	87.8	77.0	76.2	39.4	96.3	87.1	60.2	>6.9	98.9
2	139.2	94.1	59.2	73.9	55.7	98.4	91.1	60.5	>7.8	99.6
3	129.9	81.9	72.3	68.5	39.4	101.5	81.8	55.6	>0.8	92.2
4	5.8	88.3	70.1	82.8	58.3	104.7	89.8	62.0	>2.8	95.2
5	114.7	58.2	87.6	67.3	21.0	60.	65.2	22.4	>5.8	68.0
6	141.7	83.1	59.4	78.1	46.9	93.5	82.5	53.4	>5.8	96.1
7	152.1	82.3	55.3	65.7	42.9	87.4	12.6	56.9	>4.5	94.9
8	125.3	69.7	66.3	76.7	44.0	81.9	74.2	55.9	>4.0	88.1

The Fe^{2+}/Fe^{3+} ratio of the glassy or ceramic sample was determined using a colorimetric method. The values ranged from 4.2 to 17.2, which indicates that the glass or ceramic was very reduced. There was no apparent correlation with either oxygen or nitrogen plenum purge or the presence of debris.

The volume decrease was significant because of a decrease in porosity and destruction of organic materials. In comparing the initial volume of the waste to the final volume of the wasteform, there was a 55 to 68% decrease.

Wasteform Durability

The durability was evaluated using both the ASTM product consistency test and the RCRA toxicity characteristic leaching procedure. The normalized PCT data are given in Table 7 along with reference values for the Environmental Assessment (EA) glass. The final pH for tests 1–4 ranged from 10.17 to 10.74. The final pH for tests 5–8 ranged from 11.60 to 11.85. The relatively high values for aluminum in test 4 and for lithium in test 7 were largely due to anomalously low values of the respective element found in the chemical analysis of the glassy or crystalline wasteform. Except for calcium, the PCT values for tests 1–4, which produced glassy material, have lower values than those for tests 5–8, which produced a ceramic material. For calcium, the values in all tests were low, but those tests 5–8 were lower than those in tests 1–4. The normalized release for silicon for all tests is less than the EA glass reference value. The normalized release for lithium in tests 1–4 is much less than the EA reference value, but the values for tests 5–8 are higher than those for EA glass.

The TCLP data are given in Table 8. All nickel values were below the RCRA's Land Disposal Requirement (LDR). Since nickel was the only metal analyzed with any regulatory significance, the others values represent the relative release in a slightly acidic aqueous environment. Except for nickel and copper, the TCLP values are much lower for samples from tests 1–4 than for tests 5–8. For copper, the values for tests 1–4 were somewhat lower than those for tests 5–8.

If the wasteform could have been poured from the melter and cooled more quickly to form amorphous glassed, the PCT and TCLP data for the soil B tests may have been lower.

Table 7. PCT Data on the Glassy or Crystalline Wasteform for the Eight Tests.

Element	Normalize Release, g/L								
	1	2	3	4	5	6	7	8	EA
Al	0.073	0.054	0.066	2.005	2.375	1.115	1.288	1.645	NA*
Ca	0.097	0.099	0.098	0.086	0.014	0.010	0.013	0.011	NA
Cu	NC**	NC	0.020	0.002	0.940	0.074	0.352	0.110	NA
Fe	NC	0.001	0.002	0.001	0.022	0.070	0.011	0.065	NA
Li	0.987	0.258	0.504	0.285	10.66	11.015	81.38	12.297	9.565
Ni	NC	NC	NC	NC	0.038	0.024	0.069	1.467	NA
Si	0.147	0.106	0.116	0.108	0.922	1.173	0.833	1.170	3.922
pH	10.74	10.18	10.43	10.74	11.76	11.65	11.60	11.64	11.85

*Not applicable

**Not calculated because one or more of the triplicate analyses were below detection limits.

Table 8. TCLP Data of the Glassy or Crystalline Wasteform for the Eight Tests.

Element	TCLP, mg/L								
	1	2	3	4	5	6	7	8	LDR
Ca	4.41	2.11	3.04	2.34	270.94	203.37	187.84	249.64	NA*
Cu	0.65	0.36	0.51	0.62	0.70	1.26	2.43	1.26	NA
Li	0.70	0.23	0.38	0.31	35.36	32.63	23.49	31.50	NA
Ni	0.81	0.15	0.31	0.23	<0.096	0.40	<0.096	0.75	5.0
Si	4.63	4.09	3.09	9.27	119.57	95.83	85.04	117.89	NA

*Not applicable

CONCLUSIONS

The DC arc melter was able to melt the feed materials successfully and form a durable wasteform. Due to the slow cooling in the melter and the composition of the batch, the four batches with Soil B crystallized. The two batches with Soil A without debris formed glasses. The two batches with organic debris had some crystallization within a glassy phase.

Metal pellets, consisting mainly of iron, nickel, silicon, and copper, were found at the bottom of the wasteform. On the average, 24% of the added nickel was recovered in the pellets with higher values for Soil A than for Soil B.

The gas analyses reported on the concentrations of oxygen, hydrogen, carbon monoxide, carbon dioxide, hydrogen, methane, nitric oxide, and nitrogen dioxide. Since large amounts of carbonates were in the feed materials, the carbon dioxide concentration was high in all runs.

The overall mass balance on an oxide basis was very good. Between 92.1 and 101.9% of the mass was recovered in the glassy or crystalline wasteform, metal pellets, any unmelted batch, and the particulate emissions. The total recovery of silicon and calcium was about 90%. The overall recovery of some added metals was not as good. About 40% of the copper, 60% of the nickel, and 65% of cesium were recovered in the wasteform.

The glassy wasteforms had lower PCT values and generally had lower TCLP values than the crystallized wasteforms from Soil B. Except for the high PCT values for lithium, all glassy or crystalline wasteforms performed much better than the Environmental Assessment glass. The nickel TCLP data for all wasteforms were less than the RCRA LDR limits.

Crystallization of the some of the wasteforms was, in part, due to the slow cooling of the melter after the run. In a production system with a means to pour the molten products, glassy wasteforms with better durability may result for these compositions.

ACKNOWLEDGMENTS

We thank Connie Cicero-Herman, James Marra, Ray Schumacher, and Kenneth Wittle for their guidance and encouragement. The analyses of samples reported were done by the Savannah River Technology Center. The melter was on loan from the Westinghouse Savannah River Company. Electro-Pyrolysis, Inc., supplied some of the electrodes used in these tests, and the Savannah River Technology Center supplied the cesium nitrate.

REFERENCES

[1]J.E. Surma, C.J. Freeman, T.D. Powell, D.R. Cohn, R.A. Hamilton, D.L. Smatlak, P. Thomas, C.H. Titus, J.K. Wittle, P.P. Woskov, *Evaluation of the Graphite Electrode DC Arc Furnace for the Treatment of INEL Buried Wastes*, PNL-8525, Pacific Northwest Laboratory, Richland, WA, 1993.

[2]N.R. Soelberg, A.G. Chambers, G.L. Anderson, W.K. O'Connor, L.L. Oden, and P.C. Turner, *Graphite Electrode Arc Melter Demonstration: Phase 2 Test Results*, INEL-95/0502, Idaho National Engineering Laboratory, Idaho Falls, ID.

[3]A. Wenger, B. Farouk, and K. Wittle, "Modeling of Thermal Treatment of Hazardous Solid Wastes in A DC Arc Melter, *J. Air & Waste Management Assoc.*, **46**[12] 1162–1170 (1996).

[4]D.M. Cash, *Stabilization of Simulated Mixed Waste Contaminated Savannah River Soils via Vitrification in a DC Arc Melter*, M.S. Thesis, Clemson University, Clemson, SC, 1996.

[5]S.J. Griner, *Particulate Emissions from a Graphite Electrode DC Arc Melter*, M.S. Thesis, Clemson University, Clemson, SC, 1996.

[6]M.P. Speer, *Gaseous Emissions from a Graphite Electrode DC Arc Melter*, M.S. Thesis, Clemson University, Clemson, SC, 1996.

[7]T. J. Overcamp, M. P. Speer, S. J. Griner, and D. M. Cash, "Gaseous And Particulate Emissions from a DC Arc Melter," Paper 97-WP106.06, 90th Annual Meeting of the Air & Waste Management Association, June 8-13, 1997, Toronto, Ontario, Canada.

SOLUBILITIES AND REDOX EFFECTS OF TRANSITION METALS IN AN ALUMINO-SILICATE WASTE-GLASS

A.C. Buechele, C.-W. Kim, I.S. Muller, H. Gan, I.L. Pegg, P.B. Macedo, The Catholic University of America, Washington, D.C. 20064

ABSTRACT

Transition metals are common components of many waste streams, and an understanding of their effects on glass properties is essential to successful vitrification efforts. We have made an extensive study of the influence of transition metal oxide content and redox state on the properties of a multicomponent lime-aluminosilicate glass system adaptable to waste processing at 1450°C. Previously reported results have presented data on Time-Temperature-Transformation (TTT) behavior, kinetics of crystal growth, solubility and enthalpies of solution. Isothermal pseudoternary projections at 1450°C of the equilibrium phase field involving the transition metal oxides of Cr, Fe, Ni, and Zn have been mapped and have revealed several interesting phenomena involving crystallization, solid solution in spinel phases, and redox effects. Questions raised in the course of this study have now been addressed in additional experiments and analysis, and the important results of the research have been assembled and corrected, where necessary, in this summary.

INTRODUCTION

The basis for this project as stated in previous work [1,2] is the desire to understand specific aspects of the behavior of a system that could be used for the blending and vitrification of multiple waste streams which contain a large component of contaminated soil. Strategic blending of waste streams so that they themselves constitute the majority of the resources necessary for vitrification is an effective way to reduce the overall cost of a project by maximizing waste loading and minimizing additives required (Minimum Additive Waste Stabilization (MAWS) approach.) Glass formulations used in this study were inspired by work done to apply the MAWS philosophy to waste streams at the DOE Oak Ridge site [3]. The alumina-silica content of soil from this site dictates a rather high processing temperature in the 1250-1450°C range. A simplified version of a glass made for the Oak Ridge study was designed and used as a base glass whose properties are here studied as transition metal-oxide concentrations and their redox states are varied. These two latter parameters play a central role in establishing waste loading and acceptable glass properties. The four transition metals, Fe, Ni, Cr, and Zn, were chosen for this project; iron is an especially common component in many waste streams and soils, and the other three elements have toxic release limits and also occur frequently in waste streams.

METHOD

Initial melts to mix components and produce starting glasses were made in fused quartz crucibles.

Solubility studies were conducted by fusing mixtures, which represented the base composition, or higher-iron modifications thereof, with an amount of the oxide being tested in excess of its solubility limit. Mixing and sampling were performed at regular intervals, and the fully vitreous portions of the sampled glass were analyzed in the SEM using EDS microbeam techniques.

Enthalpies of solution were determined from plots of saturation mole fractions vs. reciprocal absolute temperature.

Pseudoternary compositions were formulated so that the sum of the three principle oxides was 56.61 wt% and all of the other oxides were held constant in base glass proportions; they were melted at 1450°C, quenched in air, prepared and observed in the SEM. Phases present were identified on the basis of morphology, composition as determined by EDS microbeam analysis and additional information provided by XRD.

Redox study glasses were melted in fused quartz crucibles and bubbled through an alumina tube with appropriate reducing gasses.

Heat treatments for TTT work were carried out using Pt-Rh crucibles at the required temperature after a one hour premelt at 1450°C to clear any nuclei remaining in the glass from the cooling cycle of the initial melt. At the end of the heat treatment cycle glasses were quenched in air.

A procedure similar to that used for TTT studies was employed to produce specimens for determination of crystal growth kinetics.

More complete details of the experimental methods used will be found in our two previous papers [1,2].

Table I. Solubility Glass Formulas.

Oxide wt%	**KG-BASE**	**KG-4**	**KG-23**	**KG-24**	**KG-25**	**KG-36**	**KG-37**	**KG-86**
Al_2O_3	10.47	9.42	8.95	6.52	4.71	9.42	8.95	4.71
CaO	16.94	15.25	14.48	10.53	7.62	15.25	14.48	7.62
Cr_2O_3	0.00	**10.00**	0.00	0.00	0.00	**4.17**	0.00	0.00
Fe_2O_3	3.52	3.17	3.01	**40.00**	1.58	**9.00**	**6.54**	**8.58**
K_2O	1.98	1.78	1.69	1.23	0.88	1.78	1.69	0.88
MgO	2.30	2.07	1.97	1.43	1.03	2.07	1.97	1.03
MnO_2	0.25	0.23	0.21	0.16	0.11	0.23	0.21	0.11
Na_2O	10.58	9.52	9.04	6.58	4.76	9.52	9.04	4.76
NiO	0.00	0.00	**14.53**	0.00	0.00	0.00	**11.00**	0.00
P_2O_5	0.39	0.35	0.33	0.25	0.17	0.35	0.33	0.17
SiO_2	53.09	47.78	45.38	33.01	23.86	47.78	45.38	23.86
TiO_2	0.48	0.43	0.41	0.30	0.22	0.43	0.41	0.22
ZnO	0.00	0.00	0.00	0.00	**55.06**	0.00	0.00	**48.06**

RESULTS

Solubility and Enthalpy

Glass formulations for solubility studies are listed in Table I along with the base glass composition. These formulations are made up with an amount of the oxide of interest in excess of its solubility limit in the balance of the formulation over the test temperature range of 1350-1450°C. This set of compositions was designed to measure solubilities of Fe_2O_3, Cr_2O_3, NiO and ZnO in the base glass, and of the latter three oxides in higher-iron variants of the base glass. The results of the solubility studies are summarized in Table II. It can be readily seen that the solubilities of Cr_2O_3, NiO and ZnO are decreased by increasing the Fe_2O_3 content of the base formulation. This is due to the tendency of the system to form spinels when more iron is available. At a sufficient iron concentration, the equilibrium phase is no longer the pure oxide but a mixed spinel. The iron content of the base glass is such that the equilibrium phases in contact with the melt when an excess of a pure oxide is present can be said to be essentially the pure oxide: Fe_2O_3, Cr_2O_3, NiO or ZnO, but this is not the case in the higher-iron variants of the base glass where a spinel is the equilibrium phase. It should be noted that the solubilities in Table II represent the minimum at the temperatures of measurement because of the finite cooling rate achieved by "dipping" samples out of the crucible using a quartz rod and allowing them to drop on a graphite block at room temperature to cool. The crystallization patterns in some of the air quenched specimens suggested some rapid and extensive crystallization occurring during cooling. Such crystallization extracts iron from the surrounding glass so that the measured concentrations in the quenched specimens are lower than the equilibrium concentrations at the melt temperature. Despite this limitation, the data do represent the maximum concentrations that will remain in the glass under these cooling conditions.

Table II. Maximum Solubility Determinations.

	Solubility (wt%) at:		
Oxide/Glass	**1350°C**	**1400°C**	**1450°C**
Fe_2O_3/Base	23.1	26.7	30.7
Cr_2O_3/Base	1.31	1.59	1.85
Cr_2O_3/KG-36	0.95	1.03	1.12
NiO/Base	8.9	10.9	13.1
NiO/KG-37	6.8	7.8	8.8
ZnO/Base	31.9	39.9	49.0
ZnO/KG-86	24.8	29.7	36.8

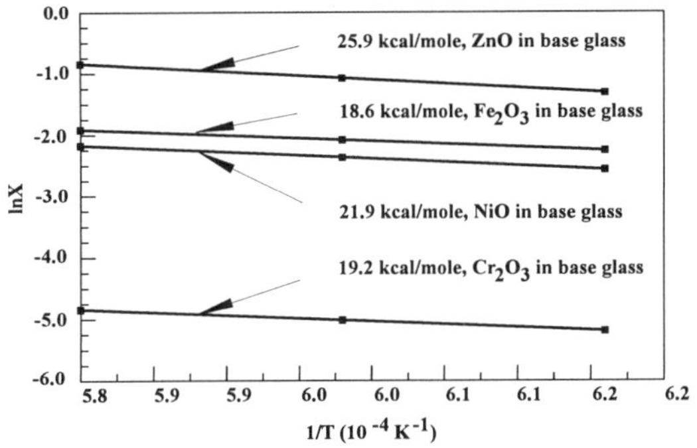

Figure 1. Plots for enthalpy determinations.

Enthalpies of solution for the pure oxides in the base composition were determined by calculating the mole fraction X of each oxide in the glass at saturation and plotting ln X vs. 1/T as shown in Figure 1. By making the assumption that the activity coefficient of each dissolved species remains constant over the relatively small temperature range employed here, the slope of each curve can be equated to the enthalpy of solution by application of the van't Hoff isochore [4]. The determined enthalpies of solution for Cr_2O_3, NiO, Fe_2O_3, and ZnO in the base glass are 19.2, 21.9, 18.6, and 25.9 kcal $mole^{-1}$, respectively. These values represent slight corrections of our previous work [1,2].The enthalpies are all positive, indicating that the dissolution reactions are endothermic. Taking an uncertainty of about 14% relative into account, we can say that the enthalpies measured for Cr_2O_3, Fe_2O_3, and NiO are comparable to each other and slightly lower than for ZnO. A comparison of these solution enthalpy values with the corresponding enthalpies of fusion can shed light on the character of the dissolution process. Normally, the value for heat of fusion found in tabulated data will be specified at the melting point under standard conditions. However, if the specific heat capacities of the reactant and product in a reaction are known as a function of temperature along with the enthalpy of transformation at one temperature, the enthalpy of transformation at any other temperature can be calculated by applying Kirchhoff's equation [4]:

$$\Delta H(T_2) = \Delta H(T_1) + \int_{T_1}^{T_2} \Delta C_P(T)\, dT \qquad (1)$$

where $\Delta H(T_2)$ is the enthalpy change at the temperature (T_2) of interest, $\Delta H(T_1)$ the enthalpy change at T_1, $\Delta C_P(T)$ the difference of the heat capacities of the individual products and reactants at the temperature T. In our case $\Delta H(T_1)$ is the heat of fusion of the oxide of interest at its melting point. The necessary thermodynamic parameters to do the calculations were taken from the *Handbook of Chemistry and Physics* (75th ed.) [5] and the *TAPP* database [6]. The temperature variation of C_P for the solid was approximated as linear and the integration in equation (1) done geometrically. The calculated value of the enthalpy of fusion for Cr_2O_3 at 1400°C is 27.1 kcal $mole^{-1}$ which is larger than the value for enthalpy of solution 19.2 kcal $mole^{-1}$ which we measure at saturation in the base glass in the temperature range of 1350°C to 1450°C. This indicates that the amount of heat required to dissolve Cr_2O_3 in the base glass at 1400°C is less than that required to simply liquefy the oxide at that temperature. The enthalpy of solution, ΔH_s, is the sum of an enthalpy of fusion, ΔH_f, and an enthalpy of mixing, ΔH_m:

$$\Delta H_s = \Delta H_f + \Delta H_m \qquad (2)$$

Therefore, the enthalpy mixing for Cr_2O_3 in the base glass is negative, that is, the mixing process is exothermic. This characterizes the interaction between solute and solvent as attractive. Using the same method, the enthalpy of fusion for ZnO at 1400°C was calculated to be 12.7 kcal $mole^{-1}$. This is smaller than our measured value for enthalpy of solution, 25.9 kcal $mole^{-1}$ indicating that the enthalpy of mixing is positive, that is, the mixing process is endothermic. This characterizes the interaction between solute and solvent as repulsive. These characterizations of the solvent-solute interactions for Cr_2O_3 and ZnO are exactly the opposite of what would be expected on the basis of the solubilities. The necessary values to do similar calculations for Fe_2O_3 and NiO could not be found in the sources we consulted.

Pseudoternary Diagrams

Although preliminary diagrams were published in our previous papers [1,2], a few additional data points have been acquired to better define the phase boundaries, and errors made in plotting the concentrations on the two original diagrams for Cr_2O_3-Fe_2O_3-SiO_2 and NiO-Fe_2O_3-SiO_2 have been corrected. This also brings together the results for ZnO-, NiO-, and Cr_2O_3- Fe_2O_3-SiO_2 in one place. The diagrams are presented as Figures 2-4. The crystal phase observed in Figure 2

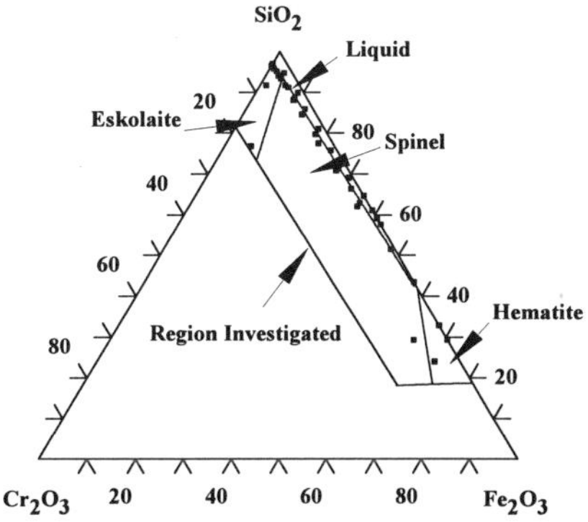

Figure 2. Pseudoternary at 1450°C for Cr_2O_3-SiO_2-Fe_2O_3

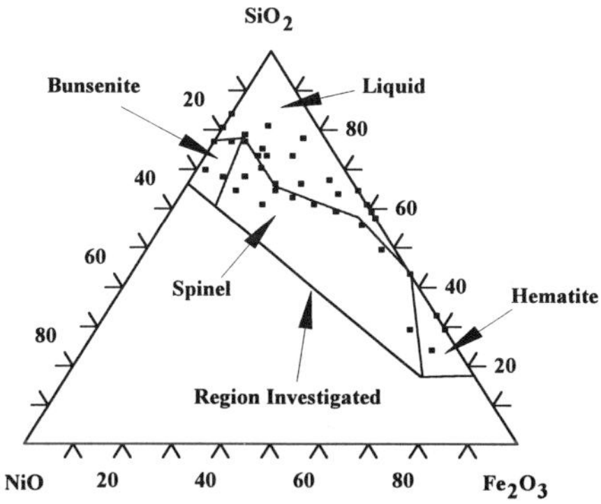

Figure 3. Pseudoternary at 1450°C for NiO-SiO_2-Fe_2O_3.

in the intermediate Cr_2O_3-Fe_2O_3 area is a spinel, predominantly chromite ($FeCr_2O_4$), which requires iron in the +2 state. It is suggested that the presence of Fe^{2+} is due to a mutual interaction between Cr^{3+} and Fe^{3+} in which some of the Cr^{3+} acts as an agent to reduce a fraction of the Fe^{3+} to Fe^{2+} while being itself oxidized to Cr^{6+}. The chromite contains small amounts of Mg, Al, and Ca in solid solution. In Figure 3, trevorite spinel ($FeNiFeO_4$) is observed in the intermediate NiO-Fe_2O_3 area. Again solid solution exists in the trevorite in that minor amounts of Mg^{2+} and Ca^{2+} may replace Ni^{2+}, and some Al^{3+} may replace Fe^{3+}. In Figure 4, spinel with a rather wide range of solid solution appears in the intermediate ZnO-Fe_2O_3 region. The make-up of the spinel extends from a composition intermediate between gahnite ($ZnAl_2O_4$), hercynite ($FeAl_2O_4$) and spinel ($MgAl_2O_4$) at the high ZnO end to a composition intermediate between franklinite ($FeZnFeO_4$), maghemite (γ-Fe_2O_3) and spinel at the high Fe_2O_3 end.

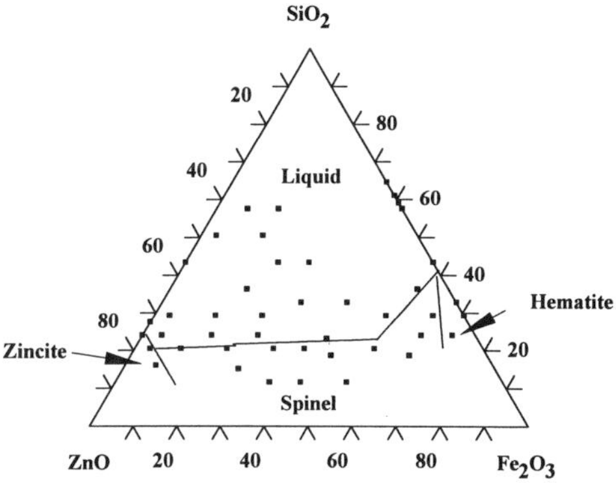

Figure 4. Pseudoternary at 1450°C for ZnO-SiO_2-Fe_2O_3.

Redox Effects

In our previous work [1] we noted that eskolaite precipitation is favored at intermediate iron redox states in chromium-containing glass (KG-0, composition given in reference 1). The EDS spectrum of the eskolaite always shows the presence of some Fe. No Fe or Cr metal is observed to form in any of the Cr-containing glasses at the levels of redox achieved in our reduction procedure. To help elucidate the possible mechanism for the crystallization of eskolaite at intermediate redox states in the glass, we will make use of data provided by Schreiber and Hockman [7] in a paper in which they compare redox chemistry in some experimental glass compositions developed for nuclear waste glass research at Savannah River Laboratories and at West Valley. The compositions considered there are borosilicates. There is no question that the values of electrochemical potentials for the redox couples of multivalent elements do vary with overall melt composition, but Schreiber and Hockman state that the order of the redox potentials in the electrochemical series is relatively independent of composition. They also show by direct measurements, which were made at 1150°C on three base glasses systematically doped with 1 wt% of several multivalent elements, that the redox chemistry is extremely similar in these systems despite significant compositional differences. The three systems characterized in their paper are Savannah River bases SRL-131 and SRL-165, and West Valley base WV-205. While acknowledging that our base composition is quite different from any of these three, and that the temperature of our experiments is 1450°C, we feel it is instructive to determine some values for redox couples of Cr and Fe using Schreiber and Hockman's curve for SRL-131 as a qualitative guide to what is happening in our Cr-containing glass. Table III lists the values determined at the oxygen fugacities of the gases used in our experiments for Cr^{3+},Cr^{6+}; Cr^{2+},Cr^{3+} and Fe^{2+},Fe^{3+} in SRL-131 at 1150°C. It can be seen that, initially, a substantial percentage of the Cr is present in the glass as Cr^{6+}. At $\log fO_2 = -2.8$, Cr^{3+} constitutes 98% of the Cr, and the percentage of Fe^{2+} has increased to 10%. Our glass, in which eskolaite precipitates at intermediate redox states, already contains a concentration of Cr_2O_3 near the solubility limit. The combination of an increase in both Fe^{2+} and Cr^{3+} would favor the crystallization of chromite which could then act as a nucleating agent to precipitate more Cr^{3+} as eskolaite. This would explain the small amount of Fe always observed in the eskolaite precipitating in the reduced glasses. In further support of this hypothesis, we note that Schreiber and Haskin [8] observed that at high Cr^{3+} concentrations in basaltic melts electron paramagnetic resonance (EPR) spectra show the presence of a β-peak attributed to the formation of exchange coupled Cr^{3+} ion pairs in which each Cr^{3+} ion is individually in octahedral coordination, and the pairs in turn are believed to be related to the creation of submicroscopic spinel-like aggregates near Cr^{3+} saturation. Basaltic melts bear some resemblance to our base glass, the main differences being that basaltic melts are always lower in Na and generally much higher in Fe. The precipitation of chromite-eskolaite can be expected to persist until the oxygen fugacity decreases to the point that Cr^{2+} begins to constitute a significant proportion of the Cr in the melt, and comparison with SRL-131 again shows that this might reasonably be the situation expected to prevail at higher redox states in our base glass. It has been observed that in basaltic glass melts Cr^{2+} is significantly more soluble than Cr^{3+} [9], and so as more of the Cr is converted to Cr^{2+} the melt becomes undersaturated in Cr^{3+} favoring dissolution of eskolaite.

Table III. Redox Couples in SRL-131 (based on Schreiber & Hockman [8]).

Gas	$\log fO_2$	Cr^{3+},Cr^{6+}	Fe^{2+},Fe^{3+}	Cr^{2+},Cr^{3+}
Air	-0.7	~65%	~4%	*
CO_2	-2.8	~98%	~10%	*
CO/CO_2 1:1	-8.25	*	~75%	~4%
CO/CO_2 10:1	-10.25	*	~89%	~14%
CO	-12	*	~95%	~25%

*Cr^{6+} and Cr^{2+} do not coexist in the glass.

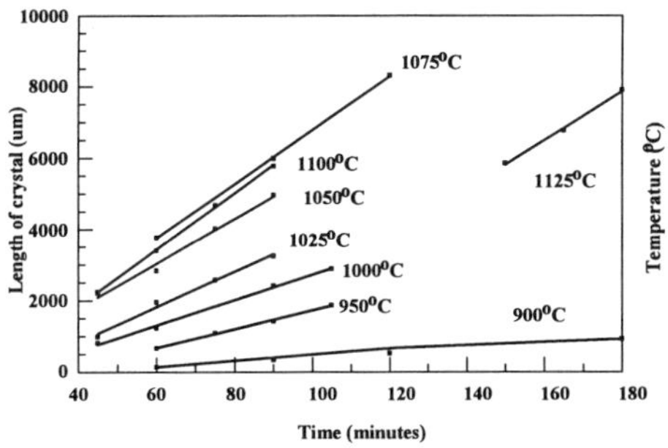

Figure 5. Crystal growth curves for KG-82.

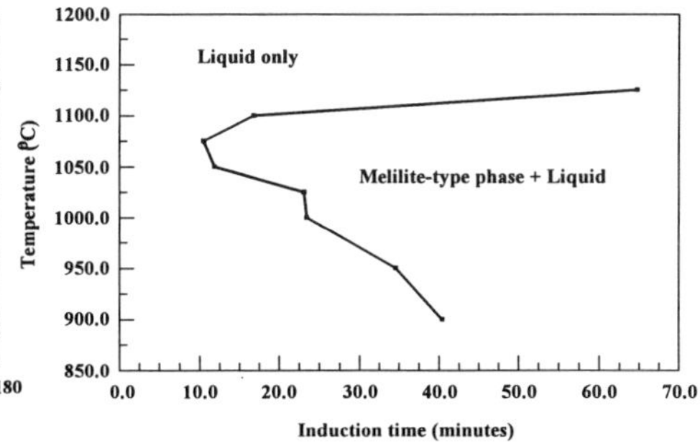

Figure 6. Induction times as determined from layer growth intercepts.

Crystallization and Time-Temperature-Transformation Behavior

Time-Temperature-Transformation plots were determined for several compositions and presented in our previous work [1,2]. One Zn-containing glass in particular (KG-82) developed only a melilite-type crystal phase over the entire heat treatment time-temperature matrix and was well adapted to the study of crystal layer growth rate. The results of the growth study were presented in a previous paper [2]. Figure 5 shows the results of the growth measurements plotting layer thickness vs. time. By extrapolating these curves back to zero thickness, a time intercept is determined that represents the incubation or induction time for the onset of crystal growth. If these times are plotted vs. temperature as is done in Figure 6, the result adds resolution to the nose of the TTT curve. The nose portion of the original TTT curve determined by a heat treatment matrix is replotted on a linear time scale in Figure 7. A comparison of the two shows good agreement between the two sets of data, and the improved resolution provided by the layer-growth intercept data.

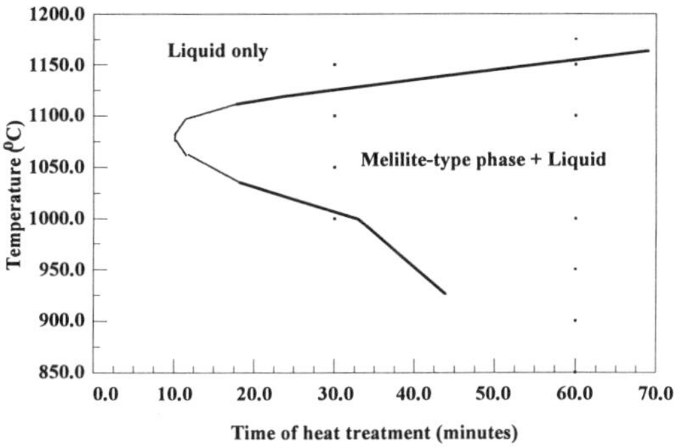

Figure 7. Nose of TTT curve for KG-82 determined by heat treatment.

CONCLUSIONS

In this paper we have summarized, elaborated and, where necessary, corrected the more important results of our previous research on this glass system. We have used the enthalpies of solution and tabulated data for ZnO and Cr_2O_3 to determine enthalpies of mixing and of fusion for these oxides in our base glass at 1400°C. The mixing enthalpy is positive for ZnO and negative for Cr_2O_3, which is the opposite of what would be expected on the basis of solubility data. A further study of the nature of the mixing process would be helpful in the understanding of this result. A maximum in the amount of precipitated eskolaite in a chromium-containing glass, KG-0, at intermediate redox states has been accounted for on the basis of minor chromite formation acting to promote nucleation and differences in the solubilities of Cr^{6+}, Cr^{3+}, and Cr^{2+} in the glass.

Finally, induction times have been determined from our previous crystal layer growth studies in the zinc-containing glass KG-82 and shown to be consistent with our previous TTT data from heat treatments.

ACKNOWLEDGMENTS

The authors would like to thank M.C. Paul for assistance in manuscript preparation.

REFERENCES

[1]C.-W. Kim, I.S. Muller, A.C. Buechele, I.L. Pegg, P.B. Macedo, "Redox State and Transition Metal Concentration Effects on Microstructure of Simulated Simplified Glasses for Waste Immobilization," in *Scientific Basis for Nuclear Waste Management XIX*, Materials Research Society Symposium Proceedings, Materials Research Society, Pittsburgh, PA, **412**, 149-156 (1996).

[2]C.-W. Kim, A.C. Buechele, and I.S. Muller, "Crystallization and Redox Effects in Waste Vitrification," *Ceramic Transactions, Volume 72*: Environmental Issues and Waste Management Technologies in the Ceramic and Nuclear Industries II, (Edited by V. Jain and D. Peeler) The American Ceramic Society, Westerville, OH, 1996, pp. 457-466.

[3]I.S. Muller, H. Gan, S.-T. Lai, A.C. Buechele, and I.L. Pegg, *Development of a Vitrification Composition Envelope to Support Complex-Wide Application of MAWS Technology* final report submitted to Argonne National Laboratory, (1995).

[4]P.W. Atkins, *Physical Chemistry, 2nd Ed.*, W.H. Freeman and Co., San Francisco, 1982, pp. 110, 268.

[5]CRC *Handbook of Chemistry and Physics*, 75th Ed., Edited by R.C. Weast, M.J. Astle and W.H. Beyer, CRC Press Inc., Boca Raton, FL, 1995.

[6]TAPP database, Version 2.2, E.S. Microware, Inc., Hamilton, OH, 1996.

[7]H.D. Schreiber and A.L. Hockman, "Redox Chemistry in Candidate Glasses for Nuclear Waste Immobilization," *J. Am. Ceram. Soc.*, **70** [8], 591-594, 1987.

[8]H.D. Schreiber and L.A. Haskin, "Chromium in Basalts: Experimental Determination of Redox States and Partitioning Among Synthetic Silicate Phases," *Pro. Lunar Sci. Conf. 7th*, 1221-1259, 1976.

[9]P.L. Roeder and I. Reynolds, "Crystallization of Chromite and Chromium Solubility in Basaltic Melt," *J. Petrology*, **32**, 909-934, 1991.

VITRIFICATION OF CHEMICAL, METAL, AND PESTICIDE CONTAMINATED SAVANNAH RIVER SITE SOIL

Connie A. Cicero-Herman
Westinghouse Savannah River Company
Savannah River Technology Center
Aiken, SC 29808

R. Kevin Poole and Russell J. Hubbard
Clemson University
Clemson Environmental Research Laboratory
Anderson, SC 29634

ABSTRACT

Treatability studies were performed to determine the suitability of vitrification for the treatment of Chemical, Metal, and Pesticide (CMP) Pit waste from the Savannah River Site (SRS). Waste from the CMP Pits was excavated as part of a remedial action and consists primarily of SRS soil. The treatability studies were performed at the Clemson University's Environmental Systems Engineering (CUESE) department in conjunction with the Savannah River Technology Center (SRTC). The intent of the studies was to determine the necessary glass additives for producing durable glass, the waste loading limits due to the contaminants present, and the processing parameters necessary. The treatability studies were successfully completed and homogeneous and durable glass was produced. Characterization of the waste and glass products will be discussed in this paper.

INTRODUCTION

The Department of Energy (DOE) has been operating nuclear material production sites since the late 1940's. During this time, extremely large quantities of radioactive and/or hazardous wastes have been accumulated. In some cases, these wastes were stored or disposed in earthen disposal pits. In the past, this was an acceptable means of disposal, but direct disposal into earthen pits without liners or barriers is no longer acceptable. Disposal pits are now facing extensive remediation because of the concerns with material migration to the surrounding environment or drinking water sources. Remediation is often an economic problem because of the large volumes of soil that have to be removed from the pit to ensure that all of the contaminants have been removed. Even after removal, the soil and contaminants still need to be stabilized.

One such disposal pit area was remediated at the SRS in the 1980s. The pit area consisted of seven disposal pits known as the CMP Pits. From 1971 through 1979, the CMP Pits were used for disposal of selected non-radioactive wastes (mostly volatile organic solvents and pesticides) generated at the SRS. The CMP Pit waste is not classified as a mixed waste because it is believed to be non-radioactive; however, in order to treat the material in a non-radioactive facility, the waste would first have to be screened for radioactivity.

Approximately three million pounds of waste were excavated during the remediation, which included the containerized and uncontainerized wastes and surrounding contaminated soil.[1] Originally, the CMP Pit waste was stored in B-25 boxes; however, these boxes degraded over time and, in the late 1980s, the waste was repackaged in B-12 boxes. Today, the B-12 boxes are intact and are being stored at the SRS, but because of concerns with eventual degradation of these boxes, methods are now being evaluated to determine a stabilization or disposal method for this waste.

The SRTC was requested by DOE-SR to determine the suitability of vitrification treatment for the CMP Pit wastes. Vitrification was believed to offer advantages over other treatment methods because of the volume reduction possible and the ability to destroy organics. In addition, radioactive vitrification facilities exist which would be able to process this waste, so the material would not have to be analyzed for radioactive content. Another advantage of vitrification is that some melter technologies provide the capability to process debris waste, so the waste would not have to be segregated to remove the drums, cans, and other debris that is mixed in with the waste.

To determine whether vitrification treatment is suitable for a particular waste stream, bench-scale vitrification studies are usually performed to determine optimum vitrification processing parameters (melt temperature) and necessary glass additives. These tests are usually performed on surrogate waste first and then the results can be applied in bench-scale treatment of the actual wastes. For the CMP Pit waste studies, additional objectives existed to determine one general glass composition suitable for all of the wastes and to determine if pre-treatment was necessary.

SURROGATE (NON-CONTAMINATED) SOIL STUDIES

Bench-scale vitrification studies with SRS soil were previously performed at the SRTC. A durable and homogeneous glass composition was developed for processing at 1350°C. At lower melting temperatures, it was found that the silica in the soil did not fully react. Therefore, the maximum amount of soil that could be incorporated in the glass matrix was assumed to be 60 wt%.[2] This loading was mainly bounded by the high temperatures required to melt the refractory soil.

An issue for the CMP Pit waste studies which was not present in the earlier soil studies was the organics present in the CMP Pit wastes. The amount of organics that can be safely vitrified in glass is limited due to the combustive nature of organics when exposed to high temperatures. Therefore, the waste loading could be limited by the amount of chemicals and pesticides present in the CMP Pit wastes. Another uncertainty in these studies was the types of organics present. Limits had to be defined as to how much of these contaminants can be processed in the given glass composition. The redox ratio, Fe(2+)/Fe(3+), of the glass becomes an important issue when organics are present in the waste stream. The redox ratio of glass must be controlled during melter operations to ensure that foaming or precipitation of metals does not occur. The presence of organics tends to cause more reduced glasses, which could lead to precipitation of metals or metal compounds which settle to the bottom of the melter.

The non-contaminated soil sample (or surrogate waste) obtained for this study was excavated from the SRS. The particular sample is characterized as the "clay" soil, as opposed to the "sandy" soil because of the slightly higher Al_2O_3 and Fe_2O_3 content. A representative composition is given in Table I, as is the composition of the soil used in the earlier studies. The soil for this study was similar in composition to the soil obtained for earlier studies. Therefore, glass formulations determined from the previous bench-scale studies could be applied to the soil to be used in the surrogate studies.

Table I. SRS Soil Characterization (Wt%)

Oxide	This Study	Earlier Study[2]
Al_2O_3	4.310	3.534
CaO	0.094	0.146
Fe_2O_3	1.730	0.961
Na_2O	0.342	0.033
SiO_2	93.017	94.429
TiO_2	0.236	0.460

The optimum glass composition from the previous soil studies was used as a starting point for these studies. This composition consisted of 60 wt% soil, 24 wt% Na_2CO_3, and 16 wt% $CaCO_3$.[2] This composition required a melting temperature of 1350°C, therefore additional compositions had to be tested to lower the melting temperature. The lower melting temperature would make the CMP Pit wastes more processable in the existing waste vitrification facilities. To lower the temperature, other fluxing agents and glass forming additives were used.

The bench-scale vitrification studies were performed at CUESE because of restrictive requirements imposed when performing treatability studies with listed wastes at SRTC. Since the studies with actual waste had to be performed at CUESE, the surrogate studies were also conducted there to be consistent and to avoid introducing additional errors for later comparisons.

Initial attempts to produce homogeneous glass using the previously determined glass composition and slight modifications to this composition were unsuccessful in reducing the glass melting temperature. Attempts were then made to tailor the glass composition to match glass compositions from bench-scale studies with Oak Ridge K-25 B&C Pond sludge. Two glass compositions were selected from these studies and were tested. These formulations successfully produced glass at lower melting temperatures, but appeared to be very viscous. To produce a more processable or flowable glass, seven additional glass compositions were formulated based on the two compositions. The targeted glass compositions are shown in Table II.

Table II. Glass Compositions Tested On SRS Soil (Wt%)

Batch ID	Soil	Na_2O	CaO	Fe_2O_3	Li_2O
1	65	9.8	9.1	6.3	9.8
2	65	5.6	14.7	9.1	5.6
3	65	11.15	5.0	7.7	11.15
4	65	8.65	10.0	7.7	8.65
5	65	8.0	10.0	9.0	8.0
6	65	11.0	7.0	6.0	11.0
7	65	6.75	12.5	9.0	6.75
8	70	10.5	9.0	0.0	10.5
9	62	8.5	11.0	10.0	8.5

All batches were melted in high purity (99.8%) alumina crucibles at 1150°C in a Deltec furnace for four hours. At the end of the four hours, they were removed and tested for pourability. All produced dark brown glasses, with the exception of #8 which was light green due to no Fe_2O_3 being present, and were visually homogeneous.

The optimum compositions were determined to be batch #5, #7, and #9 based on appearance and ease of pouring of the final product. The theoretical compositions of these three glasses based on the soil composition are given in Table III. These compositions were considered for further studies with the actual wastes.

Table III. Theoretical Glass Compositions for Non-Contaminated Soil Glasses (Wt%)

Oxide	#5	#7	#9
Al_2O_3	2.805	2.805	2.675
CaO	10.071	12.591	11.040
Fe_2O_3	10.156	10.121	11.068
Li_2O	7.980	6.738	8.512
Na_2O	8.203	6.960	8.724
SiO_2	60.528	60.528	57.735
TiO_2	0.153	0.153	0.146

Further characterization of glasses #5 and #9 was performed and included redox state determination by the colorimetric method and durability as measured by the Product Consistency Test (PCT). The Fe(2+)/ Fe(3+) ratio was determined since the redox is a very important factor for glass processing, especially when organic compounds are present. The ratios were 0.033 and 0.022 for #5 and #9, respectively. These glasses were oxidized and had similar redox ratios.

The PCT (ASTM C1285-94) is a crushed glass leach test that measures the releases of glass elements in 90°C ASTM Type I water over a period of seven days.[3] Glasses are tested in triplicate, the results are averaged, and then normalized for the elemental glass content. The results are given in Table IV based on the theoretical compositions. Since no acceptance criteria have been established for hazardous waste glasses, the durabilities were compared against the High Level Waste (HLW) Environmental Assessment (EA) glass results[4]. The normalized PCT results for the EA glass are also contained in Table IV. The measured leachate pHs are listed in Table IV, since this provides a secondary indication of glass durability.

Table IV. PCT Results (g/L) for Non-Contaminated Soil Glasses

Glass ID	B	Si	Na	Li	pH
5	0.00	0.148	0.883	1.209	11.28
9	0.00	0.126	0.597	0.866	11.12
EA	16.695	3.922	13.346	9.565	11.91

From the results presented, it appeared that the higher waste loading decreased the durability of the glass. This was inconsistent with what would be hypothesized based on the fact that the #5 glass should have contained more SiO_2 and less Na_2O and Li_2O, which should have resulted in lower elemental releases for glass #5 based on the effects of these oxides on durability. Based strictly on the PCT results, glass #9 was a better glass because of the slightly lower normalized releases and measured pH. However, both glasses were significantly more durable than the EA glass and should be considered acceptable from a durability perspective.

CMP PIT WASTE STUDIES

Waste Composition

To perform these studies, six samples of the CMP Pit waste from six of the pits were obtained from the Solid Waste Department (SWD) of the SRS. As mentioned earlier, seven pits were in the pit area known as the CMP Pits, but no B-12 boxes of the seventh pit were found when the six samples were removed. Estimates as to the contaminants present were also provided. Historically, contaminants found included 1,4-dioxane, endrin, toxaphene, dichlorodiphenyltrichloroethane (DDT), 2,4-dichlorophenoxyacetic acid (2,4-D), 2,4,5-trichlorophenoxypropionic acid (2,4,5-TP), toluene, perclene, petroleum distillates, methylene chloride, triclene, freon 11, chlordane, lindane, and methoxychlor.[1] For the six samples, characterizations performed when the wastes were packaged indicated the types and amounts of contaminants shown in Table V.

Table V. CMP Pit Waste Contaminants[1]

Sample	B-25 #	Waste Contaminant
1	91984-21	82 ppm toluene, 273 ppm perchloroethylene
2	100584-24	377 ppm 1,4 dioxane
3	91084-13	3 ppm endrin, 26 ppm toxaphene, 19 ppm DDT
4	100584-06	1 ppm 2,4-D, 116 ppm 2,4,5-TP
5	110884-32	494 ppm perchloroethylene
6	92784-09	215 ppm 2,4,5-TP

Visually, the waste samples did not look very different from sample to sample, other than the size of the soil particles. Some of the samples contained other material besides soil such as pieces of glass and rocks. Two of the samples had a very strong smell which was attributed to the organic compounds present.

Before any testing or studies could be performed with the waste samples, the samples were screened to determine if any radioactive contaminants were present. All of the results fell below the allowable limits for shipping samples offsite as non-radioactive.

The six samples were analyzed by General Engineering Laboratories (GEL) using the Toxicity Characteristic Leaching Procedure (TCLP) to determine the leachable hazardous contaminants. Several EPA methods were used to determine the Resource Conservation and Recovery Act

(RCRA) metals, pesticides and PCBs, BNA compounds, volatile organics, and herbicides that leached from the waste samples.

Table VI gives the leachable contaminants detected by GEL for the waste samples. RCRA metals screened but not detected included arsenic, lead, mercury, and selenium. The concentration of these metals were less than the analytical detection limits of 0.0421, 0.0364, 0.00015, and 0.0841 mg/L, respectively. Several pesticides and PCBs were screened. These compounds were not present above the detection limits, which were 0.00125 mg/L for heptaclor and heptaclor epoxide, 0.000625 mg/L for gamma-BHC (lindane), 0.00125 mg/L for endrin, 0.0125 mg/L for methoxychlor, 0.0250 mg/L for toxaphene, and 0.00625 mg/L for chlordane. Testing for the BNA compounds included pyridine, hexachloroethane, nitrobenzene, 2,4,6-trichlorophenol, 2,4,5-trichlorophenol, 2,4-dinitrotoluene, hexachlorobenzene, hexachlorobutadiene, pentachlorophenol, m,p-cresol, and o-cresol. None of these were found above the detection limits, which were 0.100 mg/L for all compounds for sample 1, 0.0500 mg/L for all compounds for sample 2 - 5, and 0.125 mg/L for all compounds for sample 6. The volatile organics evaluated included 1,1-dichloroethylene, vinyl chloride, chloroform, 1,2-dichloroethane, 1,4-dichlorobenzene, 2-butanone, carbon tetrachloride, trichloroethylene, benzene, tetrachloroethylene, and chlorobenzene. 2-butanone and trichloroethylene were the only organics found, with the rest below the detection limits of 10.0 mg/L for 1,1-dichloroethylene and 0.010 mg/L for all others. Herbicides evaluated were 2,4,5-TP and 2,4-D. 2,4,5-TP was detected in several of the samples, whereas 2,4-D was not found in any above the detection limit of 0.025 mg/L for samples 1 and 6, 0.00500 mg/L for samples 2 - 4, and 0.100 mg/L for sample 5.

Table VI. TCLP Results from CMP Pit Waste Samples (mg/L)

Contaminant	1	2	3	4	5	6
Ag	0.0119	0.0116	0.0094	0.0124	0.00787	0.0127
Ba	0.322	0.562	0.623	0.448	0.100	0.466
Cd	0.0103	0.0030	0.00589	0.000250	0.000150	0.00752
Cr	0.00645	0.00568	0.00461	0.00588	0.00231	0.00781
2-Butanone	0.78	<0.0200	<0.0200	<0.0200	<0.0200	<0.0200
Trichloroethylene	0.005	<0.0100	<0.0100	0.00350	<0.0100	<0.0100
2,4,5-TP	0.385	<0.00250	<0.00250	<0.00250	2.74	0.0900

As shown in Table VI, the concentrations of RCRA metals leached from the samples were very small. Barium was consistently the highest metal leached. None of the results were above the Universal Treatment Standards (UTS) for Land Disposal Restrictions (LDR) for non-wastewaters (0.30 mg/L for Ag, 7.6 mg/L for Ba, 0.19 mg/L for Cd, and 0.86 mg/L for Cr). Although trichloroethylene was found in two of the samples, it was below the analytical detection limit in both instances. The results above the detection limits for 2-Butanone and 2,4,5-TP would qualify these samples as hazardous for these constituents.

The GEL results were only used to provide an estimate of the leachable contaminants in the waste; not actual estimates of the concentration of the contaminants in the waste. The leached amounts for the waste could then be compared to the final wasteform leach results to determine the suitability of the treatment process.

The bulk density and loss on ignition of the six samples were measured so accurate volume reduction calculations could be obtained. The bulk density is important for determining the true volume of waste being treated, while the loss on ignition (which was performed at 1050°C) gives an estimate to the amount of water and non-oxide anions present in the waste. The density varied from 0.802 to 1.180 g/cc, while the loss on ignition varied from 4.63% to 18.63%. Sample 6 had the highest density which was consistent with the lowest loss on ignition. This sample appeared to be the least wet and had the smallest sized particles.

Vitrification Studies

In order to determine the suitability of vitrification for all of the CMP Pit samples, the optimum glass compositions determined for the non-contaminated soil had to be tested on all of the samples. To do this, each of the pit samples was thoroughly mixed with the proper amounts and types of glass forming additives. Each mixture was placed in a high purity alumina crucible and melted in a Deltec furnace at 1150°C for four hours using an air atmosphere. Batches of composition #9 were heated quicker in the furnace, but were maintained at temperature for six hours to ensure combustion and complete homogenization. All samples were removed after four or six hours at temperature, tested for pourability, and then cooled rapidly. The targeted compositions corresponded to compositions #5, #7, and #9 given above and were, respectively, as follows:

- 65 wt% waste, 8 wt% Na_2O, 10 wt% CaO, 9 wt% Fe_2O_3, and 8 wt% Li_2O
- 65 wt% waste, 6.75 wt% Na_2O, 12.5 wt% CaO, 9 wt% Fe_2O_3, and 6.75 wt% Li_2O
- 62 wt% waste, 8.5 wt% Na_2O, 11 wt% CaO, 10 wt% Fe_2O_3, and 8.5 wt% Li_2O

All three compositions produced glass for each waste sample, with no apparent variation from pit sample to pit sample. However, after the data from CUESE was reviewed, it was determined that the actual batch compositions used to make the waste glass samples were slightly off target. This occurred because the loss on ignition was accounted for when batching the waste samples and not the surrogate samples. Therefore, the waste loadings for the waste glasses were actually greater than targeted. The actual batch compositions were roughly as follows for composition #5, #7, and #9, respectively:

- 68.5 wt% waste, 7.2 wt% Na_2O, 9 wt% CaO, 8.1 wt% Fe_2O_3, and 7.2 wt% Li_2O
- 68.2 wt% waste, 6.1 wt% Na_2O, 11.3 wt% CaO, 8.2 wt% Fe_2O_3, and 6.1 wt% Li_2O
- 66.4 wt% waste, 7.5 wt% Na_2O, 9.7 wt% CaO, 8.9 wt% Fe_2O_3, and 7.5 wt% Li_2O

The glasses were black-brown in appearance and seemed to have a viscosity that would allow for easy pouring from a melter. Some of the glasses made from composition #7 had some unreacted material on the surface. This material was determined by Scanning Electron Microscopy with Energy Dispersive Spectroscopy to be mostly unreacted silica with some small quantities of Al, Fe, and Na.

Of the three compositions, glasses made from composition #5 and #9 seemed to be better in appearance as far as homogeneity and the remnants on the crucible were concerned. The estimated composition of the as batched glasses is shown in Table VII based on the soil composition given in Table I. Only glasses of composition #5 and #9 were considered for further characterization. They were characterized for redox state, TCLP leaching, and PCT performance.

Table VII. Estimated Chemical Composition of CMP Waste Glasses (Wt%)

Oxide	#5 Glasses	#9 Glasses
Al_2O_3	2.956	2.865
CaO	9.064	9.763
Fe_2O_3	9.285	10.051
Li_2O	7.201	7.499
Na_2O	7.436	7.727
SiO_2	63.788	61.832
TiO_2	0.161	0.156

The redox, Fe(2+)/Fe(3+), ratio for the waste glasses was determined. This number helps to determine the effect of the organic compounds on the vitrification capability, since organics tend to increase the redox ratio in waste glasses. The Fe(2+)/Fe(3+) values for the waste glasses are given in Table VIII, along with the base Fe(2+)/Fe(3+) values for the non-contaminated soil glasses.

Table VIII. Fe(2+)/Fe(3+) Ratio for CMP Waste Glasses

Glass Comp.	Waste Sample	Redox Ratio	Glass Comp.	Waste Sample	Redox Ratio
#5	1	0.080	#9	1	0.052
#5	2	0.404	#9	2	0.022
#5	3	0.134	#9	3	0.023
#5	4	0.102	#9	4	0.021
#5	5	0.116	#9	5	0.019
#5	6	0.092	#9	6	0.022
Base 5	N/A	0.033	Base 9	N/A	0.022

Overall, the glasses made with composition #5 had higher redox ratios, which would be consistent with expectations because of the greater amounts of organic compounds present in these batch formulations. When the values are compared to the non-contaminated soil glasses, the redox ratios for the glasses made using composition #5 were very high while the redox ratios for the glasses made using composition #9 were close to the base composition value. Part of the reason for this difference, besides the higher contaminants present, might be differences in melting of the glasses. Since the glasses made with composition #9 were heated differently in the furnace and allowed to react for longer times, the organics would have had more of an opportunity to be oxidized and the redox ratio would correspondingly have decreased.

The TCLP was used to determine RCRA metal performance and to verify the organic destruction. The TCLP was performed by WESTON using standard TCLP methods, with the only exception being the much smaller quantity of sample available for testing. For the TCLP testing, the leaching behavior of the RCRA metals, semi-volatile organics, pesticides, herbicides, and volatile organics were determined. Overall, the TCLP results indicated that the glasses were very durable. RCRA metal leaching was very low, with Ba being the only metal found above detection limits. Ba leaching ranged from the detection limit of 0.023 to 0.0472 mg/L, which was much better than the UTS for LDR for non-wastewaters limit of 7.6 mg/L and lower than the values reported for the unstabilized waste. Arsenic, cadmium, chromium, mercury, lead, and silver were not detected above the detection limits, which were 1.11, 0.029, 0.061, 0.0007, 0.56, and 0.033 mg/L, respectively.

No semi-volatile organics, pesticides, or herbicides were detected in the TCLP extractions. The detection limits were 0.10 mg/L for all semi-volatile organics except for 2,4,5-trichlorophenol and pentachlorophenol, which were 0.50 mg/L. For the pesticides and herbicides, the detection limits were 2.5 μg/L for heptachlor, alpha-chlordane, gamma-chlordane, gamma-BHC, and heptachlor epoxide, 5.0 μg/L for endrin, 25 μg/L for methoxychlor, 250 μg/L for toxaphene, 100 μg/L for 2,4-D, and 50 μg/L for 2,4,5-TP. Two of the waste samples had indicated some 2,4,5-TP leaching when subjected to the TCLP on the waste only. The glasses made with the samples had 2,4,5-TP glass leaching TCLP results that were lower compared to the waste samples. Thus, the organics were either destroyed during processing or were contained in the glass matrix.

For the volatile organics, only one sample had a reportable result. The detection limit for most of the volatile organics was 0.050 mg/L, except for vinyl chloride and 2-butanone, which were 0.10 mg/L. In the glass made from waste sample 5 using glass composition #5, the volatile organics levels were 0.012 mg/L for trichloroethane and 0.031 mg/L for tetrachloroethane, which were both less than the detection limit. This result was very surprising since vitrification processes have been shown to destroy the organics present in waste. In addition, the quantities of trichlorothylene, which can be detected as these two organics, found to leach from the waste samples were all below 0.01 mg/L, so it does not seem possible that the quantity leached from the glass would be higher than what was originally found in the waste. Since the concentrations reported were below the analytical detection limit, it is possible that an error in reporting may have occurred.

The PCT[3] was also performed to determine the durability of the glass matrix as compared to radioactive glasses. The results were normalized based on the chemical compositions given in

Table VII. The results are contained in Table IX. These normalized results were compared to the EA glass results[4], since no criteria yet exists for hazardous waste glasses. The base glass PCT results were included in the table, so the effect of the contaminants on the durability could be analyzed.

Table IX. Normalized PCT Results (g/L) for CMP Waste Glasses

Glass #	Waste #	Si	Na	Li	pH
5	1	0.106	0.550	0.764	11.42
5	2	0.119	0.611	0.846	11.30
5	3	0.105	0.609	0.839	11.40
5	4	0.083	0.465	0.656	11.01
5	5	0.090	0.509	0.720	11.15
5	6	0.087	0.479	0.689	11.16
Base 5	N/A	0.148	0.883	1.209	11.28
9	1	0.116	0.706	0.934	11.43
9	2	0.113	0.750	1.033	11.37
9	3	0.106	0.707	0.934	11.37
9	4	0.114	0.850	1.081	11.45
9	5	0.125	0.861	1.135	11.32
9	6	0.129	0.873	1.157	11.08
Base 9	N/A	0.126	0.597	0.866	11.12
EA[4]	N/A	3.922	13.346	9.565	11.91

The normalized releases for the glasses were very low compared to the EA glass. For the waste glasses made with composition #5, the durability did not appear to be adversely affected by the particular contaminants present since there was not much difference from the base glass PCT results. The PCT results for waste glasses made of composition #9 were a little higher for Na and Li than the base glass composition. Not much difference existed between the individual samples of the same glass composition, so the contaminants present did not appear to greatly affect PCT performance. A comparison of the PCT results for the two batch compositions indicates that the glasses made with batch composition #5 had better durability. This was as expected since these glasses should have contained less glass modifying additives (soda and lithia in particular) and more silica which would tend to give a better glass durability.

When compared to soda-lime-silica glasses used by the glass industry for such things as bottle glass, the Si PCT results for the waste glasses and for the glasses made from the non-contaminated soil are much lower. The Na PCT results, on the other hand, are approximately the same, with releases from the waste glasses made with composition #5 being slightly lower.

The volume reductions for the waste glasses were calculated assuming a glass density of 2.6 g/cc, which is typical for soda-lime-silica waste glasses. The calculated volume reductions varied from 33.3% to 60.5% for glasses of composition #5 and from 30.2% to 59.1% for glasses of composition #9. Glasses made from waste sample 4 had the highest volume reduction.

Based mostly on the PCT results for the waste glasses made from composition #5, composition #5 should be considered the optimum composition for processing CMP Pit wastes. Although this glass had slightly higher redox ratios, the values were within acceptable processing limits for joule-heated melters.

CONCLUSIONS

Crucible studies have shown that CMP Pit wastes can be incorporated into a durable glass wasteform. Waste loadings of up to 68.5 wt% were possible, which represented up to a 60.5% volume reduction. The optimum glass composition was 68.5 wt% soil, 7.2 wt% Na_2O, 9 wt% CaO, 8.1 wt% Fe_2O_3, and 7.2 wt% Li_2O. This glass easily melted at 1150°C and poured easily during testing. Leach resistance of the glass products was exceptionally good as indicated by the TCLP and PCT results. TCLP results indicated that the RCRA metals and hazardous organic compounds did not leach above allowable limits from the glass. For the PCT, no significant quantities of the glass structure components were leached from the glass product.

ACKNOWLEDGMENTS

Funding for the vitrification work was provided by the DOE - Office of Technology Development under the auspices of the Mixed Waste Focus Area and Technical Task Plan SR1-6-MW-42. Funding for the analyses of the soil samples was provided by the SWD of SRS. The efforts of all personnel involved with this project, especially John McCullough of the SWD and Bernard Nora of the Technical Services Division, are greatly appreciated.

REFERENCES

1. "Preliminary Report: Site Remediation and Closure CMP Pits Savannah River Plant", Woodward-Clyde Consultants, July 1985.

2. C.A. Cicero, "Bench-Scale Studies with Mercury Contaminated SRS Soil", WSRC-TR-95-00413, Westinghouse Savannah River Company, May 8, 1996.

3. ASTM C 1285-94, "Standard Test Methods for Determining Chemical Durability of Nuclear Waste Glasses: The Product Consistency Test (PCT)", February 1995.

4. C.M. Jantzen, N.E. Bibler, D.C. Beam, C.L. Crawford, M.A. Pickett, "Characterization of the Defense Waste Processing Facility (DWPF) Environmental Assessment (EA) Glass Standard Reference Material," WSRC-TR-92-346, Rev. 1, Westinghouse Savannah River Company, June 1, 1993.

THE EFFECT OF BARIUM, PHOSPHORUS, AND LEAD COMPOSITIONAL VARIATIONS ON WASTE GLASS PROPERTIES

Wen Luo and Sabrina S. Fu, Vitreous State Laboratory, The Catholic University of America, Washington, DC 20064.

ABSTRACT

Phosphorus, barium, and lead are components prevalent in a number of hazardous and mixed waste streams. The effect of these three components on waste glass properties was examined by systematic addition of P_2O_5, BaO, and PbO to a base glass consisting of Al_2O_3-B_2O_3-CaO-Fe_2O_3-K_2O-MgO-Na_2O-SiO_2. The chemical durability of these glasses was examined using the EPA TCLP test. The melt viscosity, melt conductivity, and liquidus temperatures of these glasses were also measured. Phosphorus was shown to affect the Pb leach rate both positively and negatively, by as much as 300 relative%, depending upon the composition of the glass matrix. Phosphorus had a similar dual role in its ability to both increase and decrease melt viscosity. Barium, phosphorus, and lead had little effect on melt conductivity, as expected, and phosphorus tended to increase the liquidus temperature of the glass.

INTRODUCTION

Barium, phosphorus, and lead are prevalent in hazardous and radioactive waste streams. For example, Hanford tank wastes may contain up to 13 wt.% BaO, 11 wt.% P_2O_5, and 3 wt.% PbO and the combined silo wastes at the Fernald site contain 7 wt.% PbO, 3 wt.% BaO, and 5 wt.% P_2O_5. These three components are often considered undesirable due to their negative effect on chemical durability. However, crucible studies of surrogate Fernald CRU-4 wastes showed that greater than 70 wt.% waste loading could be achieved and still pass the EPA Toxicity Characteristics Leaching Procedure (TCLP) for Pb and Ba by at least a factor of two, despite the unusually high contents of Pb, Ba, and P in these waste glasses[1]. Hence, systematic studies on the effect of phosphorus, lead, and barium were performed to examine their individual and combined effects on waste glass properties. This paper shows that phosphorus can have a beneficial effect on the chemical durability of the waste glass as determined by the EPA TCLP test; this increase in chemical durability at room temperature is accompanied by an increase in melt viscosity at temperatures of 1100-1300°C.

EXPERIMENTAL PROCEDURE

A series of glasses were melted beginning with the base glass whose composition is shown in Table I(a). Systematic additions up to 2.0 mole% BaO, 2.2 mole% P_2O_5, and 5.5 mole% PbO were made to this base glass as shown in Table I(b). The compositions of the resulting glasses were confirmed by direct-current plasma atomic emission spectroscopy (DCP-AES) analyses after complete acid dissolutions.

Table I(a). Base Glass (Glass #1) Composition (Mole%).

composition	Al_2O_3	B_2O_3	CaO	Fe_2O_3	K_2O	MgO	Na_2O	SiO_2
mole%	7.8	13.7	9.9	2.5	0.8	9.9	6.4	49.0

Glasses were made by mixing chemicals in a plastic container, then melted in quartz crucibles at 1250-1350°C for an hour with stirring after half an hour. The resulting glass melts were poured onto graphite plates and cooled in air.

Table I(b). The Addition in Mole% BaO, P_2O_5 and PbO to the Base Glass and Their Normalized Lead and Barium Leach Rates.

Glass #	BaO (mole%)	P_2O_5 (mole%)	PbO (mole%)	Normalized lead leach rate (10^{-4}/day)	Normalized barium leach rate (10^{-4}/day)
1	0	0	0	---	---
2	0	0	1.4	11.6	---
3	1.0	1.1	1.4	28.4	37.3
4	0	0	2.7	11.6	---
5	2.0	2.2	2.7	10.3	16.3
6	0	0	5.5	12.7	---
7	1.0	1.1	2.7	18.6	25.9
8	1.0	1.1	5.5	14.5	23.7
9	2.0	2.2	1.4	7.3	15.1
10	2.0	2.2	5.5	10.6	16.6
11	2.0	2.2	0	---	14.9
12	1.0	1.1	0	---	24.5
13	0	1.1	5.5	17.7	---
14	0	2.2	5.5	11.5	---
15	1.0	0	5.5	16.9	25.7
16	2.0	0	5.5	15.8	21.5

--- indicates that concentration in leachate is below detection limit.

All the glasses, subjected to the US Environmental Protection Agency's toxicity characteristic leaching procedure[2] (EPA TCLP), were crushed with mortar and pestle until they passed through a 9.5 mm standard sieve. Each glass sample was combined with the appropriate volume of leachant (acetic acid/sodium acetate buffer of pH 4.93±0.05) to produce a surface area to volume ratio at about 20 m^{-1}. The glass and leachant were sealed tightly in a wide-mouth plastic bottle, and rotated by an agitator at 30±2 rpm for 18 hours at an ambient temperature of 23±2°C. The resulting leachate was then analyzed by DCP-AES. The normalized leach rate (NLR) of each element was calculated by dividing the weight of each element in the leachate by the weight in the glass, and multiplied by 24/18 to obtain the normalized leach rate per day.

The melt viscosity, η, was calculated from the determination of torque at various rotation speeds using a calibrated spindle attached to a Brookfield viscometer. Measurements were performed at 1100-1300°C, and the data interpolated to standard temperatures using the Volger-Fulcher equation: $\ln\eta = A/(T-T_0)+B$, where A, B, and T_0 are fitting parameters. The viscometer was calibrated at room temperature using standard oils of known viscosity, and then checked at 1000-1250°C using a National Institute of Standards and Technology (NIST) standard reference glass (SRM711). The precision and accuracy of the viscosity measurements are estimated to be within ±10 relative%.

The electrical conductivity of the glass, σ, was determined by measuring the resistance of the glass melt as a function of frequency using a calibrated platinum/rhodium electrode probe attached to a Hewlett-Packard model 4194A impedance/gain phase analyzer. Measurements were performed over similar temperature ranges to those employed for the melt viscosity measurements. The results were extrapolated to zero frequency to obtain the DC conductivity. The conductivity data were then interpolated to standard temperatures using the Arrhenius equation: $\ln\sigma = A/T + B$, where A and B are fitting parameters. Estimated uncertainties in the conductivity measurements are ±10-20 relative% for conductivity measurements greater than 0.1 S/cm, but 30-40 relative% for measurements below 0.1 S/cm.

Liquidus temperatures were estimated by examination of the volume% crystals produced after various heat treatments. Heat treatments were made at 50°C intervals, and the heat-treated glasses examined by optical microscopy and/or SEM/EDX. The liquidus temperature was estimated as the temperature between the highest heat-treatment temperature at which crystals appeared and the lowest heat-treatment temperature at which no crystals could be found; since the heat-treatments are performed every 50°C, the estimated uncertainty for each liquidus temperature is ± 25°C.

RESULTS AND DISCUSSION

TCLP Leachate Results

Table I(a) provides the mole% composition of the base glass used in this study, and Table I(b) shows the mole% of BaO, P_2O_5, and PbO added to this base glass for the formation of the remaining glasses, along with their normalized lead and barium leach rates. Figure 1 shows the normalized leach rates (NLR) for all of the glass components with the exception of sodium (this element cannot be analyzed due to the sodium in the sodium acetate/acetic acid buffer used for the TCLP leach test). The normalized leach rates of Pb, Mg, Ca, and B all increase with the first 1 mole% addition of P_2O_5, and then decrease with the second 1 mole% addition of P_2O_5. The normalized leach rates of Si, Al, and Fe all remain constant within a 10 relative% experimental error, and the normalized leach rate of phosphorus increases when 1 mole% P_2O_5 is added to a glass containing 1 mole% P_2O_5. This suggests that phosphorus may be bound differently from the other cations.

The addition of BaO slightly increases the NLR of Pb, Ca, Mg, B, Fe, Al, and Si (15-50 relative%) upon the first 1 mole% BaO addition. The second 1 mole% BaO addition produces only a 5-16 relative% change from the first 1 mole% BaO addition. Barium has been shown to increase the leachability of glasses on other glass systems[3].

In addition to the effect of P_2O_5 alone and the effect of BaO alone, the effect of the two components combined on the chemical durability of the glass was examined. The normalized lead leach rate as a function of BaO addition, P_2O_5 addition, and

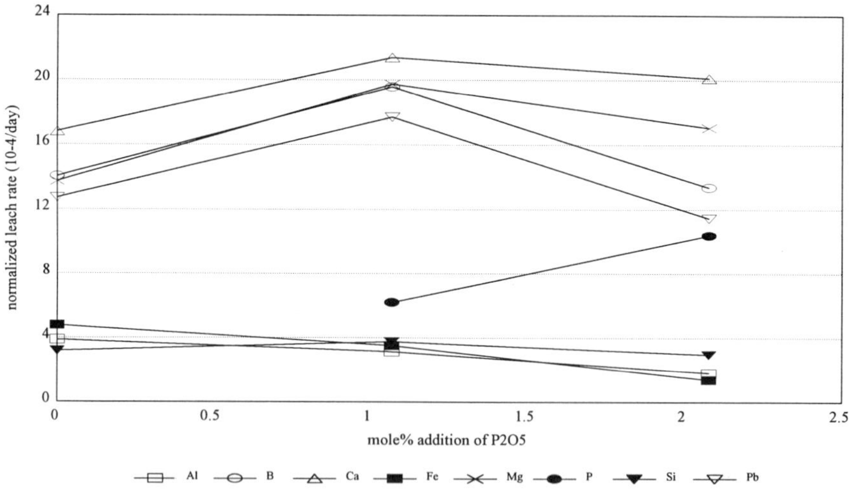

Figure 1. Effect of P_2O_5 addition on the normalized leach rates of glass components (glasses with no BaO and 5.5 mole% PbO addition).

P_2O_5 and BaO addition is shown in Figure 2. The addition of BaO tends to increase the normalized Pb leach rate while P_2O_5 addition, in the absence and presence of BaO, initially increases the Pb NLR, and then decreases it. Note also that the lead normalized leach rate appears to be higher in the absence of BaO at the 1 mole% P_2O_5 addition level. Similar trends appear for Ca, B, and Mg NLR as for the Pb NLR.

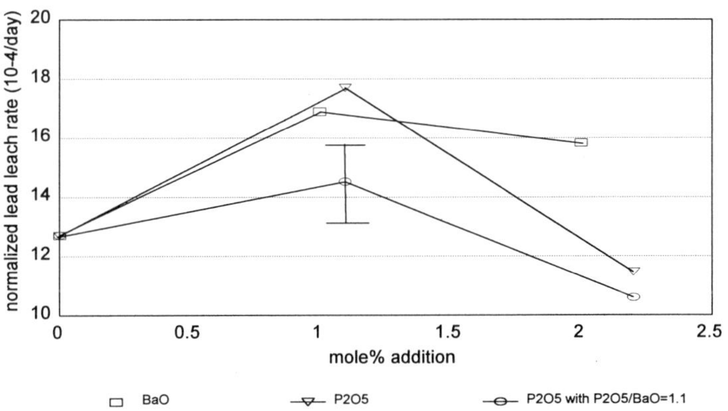

Figure 2. Effect of P_2O_5 and BaO additions on the normalized lead leach rate (glasses with 5.5 mole% PbO addition).

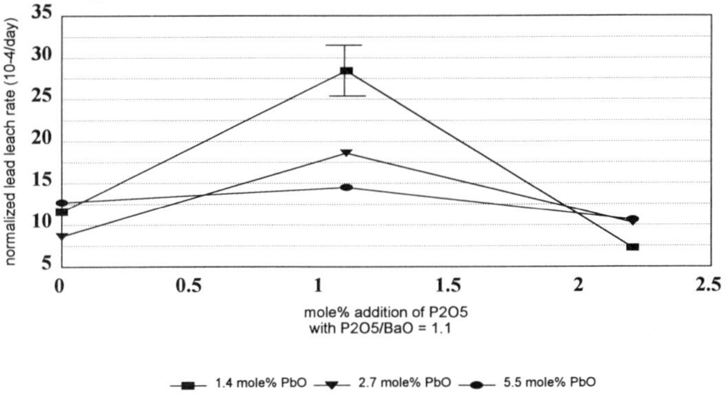

Figure 3. Effect of P_2O_5 additions (with $P_2O_5/BaO = 1.1$) on the normalized lead leach rate at various PbO levels in glass.

The lead NLR depends not only on the P and Ba contents of the glass, but also on the lead content of the glass. Figure 3 shows the Pb NLR as a function of P_2O_5 addition in the presence of BaO at three different PbO contents. The maximum in leach rate obtained with P_2O_5 addition increases as the PbO content decreases from 5.5 mole% PbO to 1.4 mole% PbO addition to the base glass. The greatest changes in Pb NLR as a function of P_2O_5 and BaO addition occur at 1.4 mole% PbO or 5.0 wt.% PbO added to the base glass. Table II shows the actual Pb TCLP leachate concentrations in ppm for glasses with systematic additions of P_2O_5 and BaO; the amount of Pb TCLP leachate concentration for a glass with 2.2 mole% P_2O_5 and 2.0 mole% BaO is actually more than a factor of three *less* than when there is half of that P_2O_5 and BaO concentration in the glass matrix. This shows that P_2O_5 and BaO can actually decrease the Pb TCLP leachate concentration.

Table II. Lead TCLP leachate concentration for glasses with 5.0 wt.% (1.4 mole%) PbO addition. This shows that 1 mole% P_2O_5 and 1 mole% BaO can be added to Glass #3 to decrease its lead TCLP leachate concentration by more than a factor of three.

Glass #	mole% P_2O_5 and BaO addition to the glass	lead TCLP leachate concentration (ppm)
2	no P_2O_5 and BaO addition	1.7
3	1.1 mole% P_2O_5 and 1.0 mole% BaO addition	3.8
9	2.2 mole% P_2O_5 and 2.0 mole% BaO addition	1.2

We have seen that barium decreases the chemical durability of the base glass slightly while P_2O_5 with and without BaO can either decrease or increase, by up to a factor of three, the amount of Pb leached in the TCLP test. Examination of the normalized leach rates of all the available glass components showed that the only NLR to increase when P_2O_5 was increased from 1 mole% to 2 mole% addition was the P NLR (see Figure 1). This suggests that phosphorus may be in a different structural unit than the remaining glass components. A number of studies on the role of P_2O_5 in silicate glasses have shown that the interaction of P_2O_5 with the glass matrix is dependent upon the cations and anions in the glass matrix as well as the amount of P_2O_5 added to the glass matrix[4-9]. Since those structural studies were performed on two to five component glass systems, it is difficult to extrapolate those structural changes to this eleven component glass system. The eleven component glass system examined in this paper contains K_2O and Na_2O, which should interact strongly with the P_2O_5, thus polymerizing the silicate matrix. However, Al_2O_3 has been shown to shift the distribution of alkali from the phosphate phase to the silicate phase[7,9], the structural role of boron has been shown to change with the amount of phosphorus in the glass matrix[8], and the role of CaO and Fe_2O_3 will also complicate the prediction of the extent of silicate polymerization. In addition, the effect of any structural changes on the lead and barium leach rates will be dependent upon the type of bonds that Pb^{2+} and Ba^{2+} will form in this eleven component glass system; for example, one would expect that if the addition of phosphorus induced Pb^{2+} and Ba^{2+} to bond to a more polymerized silicate glass matrix, the normalized leach rates of these ions would decrease. This may be the case when P_2O_5 addition, in the absence and presence of BaO, is increased from 1 mole% addition to 2 mole% addition, but this cannot explain the changes in NLR for Pb when the P_2O_5 addition is increased from zero to 1 mole%. We hope that further studies on multicomponent glass systems of this type will clarify the reason(s) for the dual role of P_2O_5 observed in these leach studies.

Phosphorus has been shown to decrease the chemical durability of borosilicate glasses[10], but it can increase the chemical durability of glasses by the formation of lead phospho-silicate crystals on the surface[11,12]. This possibility was examined by SEM/EDX analysis of these glasses before and after TCLP tests. No crystals were found[13] in any of the glasses examined. Another possible explanation for these changes in NLR with P_2O_5 addition is phase separation; results of this were presented in another paper [13] and showed that phase separation does not account for the changes in leach rates with P_2O_5 addition. A third possible explanation for the changes in leach rates shown in Figure 3 is structural changes in the glass matrix due to P_2O_5 addition. We will see in the next subsection that structural changes suggested by the changes in TCLP leachate concentrations at room temperature are also suggested by melt viscosity measurements at 1100-1300°C.

Melt Viscosity And Melt Conductivity Results

Table III shows the melt viscosity and melt conductivity at 1200°C for the 16 glasses used in this study. Typically, measurements were made from 1100-1300°C, but only the data at 1200°C will be used to compare the effect of compositional changes on melt viscosity and melt conductivity. Similar trends are observed at all temperatures measured. The melt conductivity at 1200°C varied from 0.03 to 0.06 S/cm, just slightly outside the estimated 30-40 relative% error for measurements below 0.1 S/cm. Since the alkali content of these sixteen glasses varied by only 10 relative%, the melt conductivity of these glasses was not expected to vary greatly, as was observed.

The melt viscosity at 1200°C ranged from 52 Poise to 191 Poise; in this viscosity range, the reproducibility of the viscosity measurements is about 10 relative%. Figure 4 shows that the effect of P_2O_5 in the absence and presence of BaO is to produce a minimum in the melt viscosity at about 1 mole% P_2O_5 addition (this minimum becomes more pronounced at lower melt temperatures). The effect of BaO is to decrease the melt viscosity, in agreement with other studies[14]. It is interesting to note that the addition of BaO resulted in a slight increase in Pb TCLP leachate and a slight decrease in melt viscosity while the addition of P_2O_5, both in the presence and absence of BaO, resulted in a minimum in melt viscosity and a maximum in Pb TCLP leachate concentration at 1 mole% P_2O_5 addition. Glass formers tend to increase melt viscosity and decrease leachate concentrations, while glass modifiers tend to decrease melt viscosity and increase leachate concentrations. Hence, both the melt viscosity data and the TCLP leachate concentrations suggest that P_2O_5 can behave both as a glass modifier and as a glass former.

Table III. The Addition in Mole% BaO, P_2O_5 and PbO to the Base Glass and Their Viscosities and Conductivities at 1200°C.

Glass #	BaO (mole%)	P_2O_5 (mole%)	PbO (mole%)	Viscosity at 1200°C (poise)	Conductivity at 1200°C (S/cm)
1	0	0	0	115	0.04
2	0	0	1.4	101	0.04
3	1.0	1.1	1.4	98	0.04
4	0	0	2.7	95	0.04
5	2.0	2.2	2.7	88	0.03
6	0	0	5.5	72	0.04
7	1.0	1.1	2.7	71	0.04
8	1.0	1.1	5.5	52	0.05
9	2.0	2.2	1.4	127	0.04
10	2.0	2.2	5.5	73	0.04
11	2.0	2.2	0	191	0.04
12	1.0	1.1	0	122	0.03
13	0	1.1	5.5	63	0.05
14	0	2.2	5.5	72	0.06
15	1.0	0	5.5	63	0.03
16	2.0	0	5.5	57	0.03

Liquidus Temperature Results

All sixteen glasses examined had a liquidus temperature of 975±25°C except Glass#5, #9, #10,#11, and #14; Glass#5 has a liquidus temperature of 1125°C, Glass#10 has a liquidus temperature of 1075°C, and the remaining three glasses have a liquidus temperature of 1025°C. Hence, all of the glasses with 2 mole% P_2O_5 addition to the base glass have an increased liquidus temperature of 50-150°C when compared to glasses with less P_2O_5. Typical processing temperatures for waste glasses are in the range of 1150-1250°C, and most melters require a liquidus temperature about 100°C below that of the operating temperature. All of the glasses with the exception of Glass#5 and Glass#10 have a processable liquidus temperature for operating at 1150-1250°C. Hence, although the addition of 1 mole% P_2O_5 to glasses containing 1 mole% P_2O_5 does increase the liquidus temperature (while decreasing the Pb TCLP leachate concentrations), most of the glasses will still be processable at the typical operating temperatures of 1150°C. For example, if an additional 1.1 mole% P_2O_5 and 1 mole% BaO are added to Glass#3, the normalized lead TCLP leachate concentration is reduced by a factor of three while the liquidus temperature is increased from 975°C to 1025°C; the resulting glass melt is still processable at 1150°C, but the resulting glass product at room temperature has a factor of three higher chemical durability as defined by the EPA TCLP test.

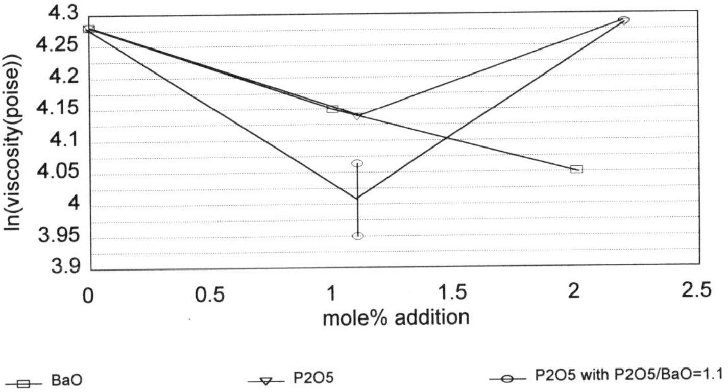

Figure 4. Effect of mole% addition (with 5.5 mole% PbO addition) on glass viscosities at 1200°C.

CONCLUSIONS

We have seen that 1 mole% P_2O_5 addition, in the presence and absence of BaO, can either decrease or increase the lead leach rate by up to a factor of three; these data suggest that glasses containing more P_2O_5 may be more durable. This ability of P_2O_5 to either stabilize or destabilize the glass matrix with respect to leaching was consistent with the melt viscosity changes induced by P_2O_5 additions; the minimum produced in melt viscosity measurements coincided with the maximum produced in Pb leach rates. Additions of P_2O_5, BaO, and PbO produced small changes in melt conductivity as expected. The stabilization effect of increasing the P_2O_5 addition from 1 mole% to 2 mole% was accompanied typically by a 50°C increase in liquidus temperature.

ACKNOWLEDGMENTS

The authors would like to thank Fluor Daniels Fernald for partial financial support; M.C. Paul for manuscript preparation; Dr. Shouxiang Hu, Dr. Shantao Lai, Chufen Feng, Li Su, and Hannelore Rittinger for TCLP results; Dr. Hao Gan, Wei Guo, and Wenyu Chen for most of the viscosity and conductivity measurements.

REFERENCES

[1]S.S. Fu, K.S. Matlack, R.K. Mohr, M.C. Paul, I.L. Pegg, and P.B. Macedo, "Vitrification Testing for Fernald CRU-4 Silo Wastes," Final Report to Fluor Daniel Fernald, May 1996, Revised September 1996.

[2]EPA, SW-846 method 1311, Toxicity Characteristic Leaching Procedure.

[3]C.B. Azzoni, G.L. Del Nero, A. Krajewski, and A. Ravaglioli, "A Diffusive Model of Pb^{2+} Release by Lead-Ceramic Glazes," *J. Mater. Sci.*, **16**, 1081-1087 (1981).

[4]F.J. Ryerson and P.C. Hess, "The Role of P_2O_5 in Silicate Melts," *Geochim. Cosmochim. Acta*, **44** 611-624 (1980).

[5]R. Dupree, D. Holland, M.G. Mortuza, J.A. Collins, and M.W.G. Lockyer, "An MAS NMR Study of Network-Cation Coordination in Phosphosilicate Glasses," *J. Non. Cryst. Solids*, **106** 403-407 (1988).

[6]C. Nelson and D.R. Tallant, "Raman Studies of Sodium Silicate Glasses with Low Phosphate Contents," *Physics and Chemistry of Glasses*, **25** [2] 31-38 (1984).

[7]H. Gan and P.C. Hess, "Phosphate Speciation in Potassium Aluminosilicate Glasses," *American Mineralogist*, **77** 495-506 (1992).

[8]H. Gan, P.C. Hess, and R.J. Kirkpatrick, "Phosphorus and Boron Speciation in K_2O-B_2O_3-SiO_2-P_2O_5 Glasses," *Geochimica et Cosmochimica Acta*, **58** [21] 4633-4647 (1994).

[9]D.R. Tallant and C. Nelson, "Raman Investigation of Glass Structure in the Na_2O-SiO_2-P_2O_5-Al_2O_3 System," *Physics and Chemistry of Glasses*, **27** 75-79 (1986).

[10]E.C. Bloor, "Glaze Composition, Glass Structural Theory and its Application to Glazes," *Trans. Br. Ceram. Soc.*, **55** 631-660 (1956).

[11]R.L. Lehman, "Lead-Ion stability in Soda-Lime Lead Silicate Glasses," *J. Am. Ceram. Soc.*, **75** [8] 2194-2199 (1992).

[12]R.L. Lehman, V.A. Greenhut, "Surface Crystal Formation During Acid Corrosion of Phosphate-Doped Lead silicate Glass," *J. Am. Ceram. Soc.*, **65** [9] 410-414 (1982).

[13]W. Luo, A.C. Buechele, and S.S. Fu, "The influence of phosphorus, barium, and lead compositional variation on the chemical durability of a multi-component glass system," *J. Am. Ceram. Soc.*, in press.

[14]D.R. Uhlmann & N.J. Kreidl, "Glass Science and Technology, Volume 3, Viscosity and Relaxation"; Academic Press, London, United Kingdom, 1986.

Low-Level Radioactive and Mixed Waste Vitrification

CERAMIC CONTAINERS FOR RADIOACTIVE WASTE

Anatoly E. Rokhvarger
Nucon Systems, Inc.
Syosset, NY 11791

INTRODUCTION

The major reason for the chemical corrosion of concrete, glass, and carbon steel materials is their susceptibility to hydrogen forms of other dissolved chemical ingredients and/or material-body interaction with (or dissolution in) water or steam when the water steam and hydrogen forms are permanently present in a surplus concentration in any Natural ambiance, including the low level of the conventional humidity of the Yucca Mountain tunnel. Moreover, surplus hydrogen molecules are present within the sealed (welded or bolted) metal canister of the waste package. This corrosion should be dramatically increased under combined radiation, heat and aggressive chemical impacts, such as happened in welding or bolting areas of even stainless steel canisters. Additionally, any glass is thermodynamically unstable material and there will appear micro-cracks that will incidentally induce a waste leakage [1-7].

Thus, it is necessary to exclude concrete, glass and carbon steel as materials for permanent capsulation of nuclear waste because any additional waste treatment or additional material improvement cannot change the chemistry of these materials and their products having more or less corrosion in any natural ambiance.

However, according to an existing paradigm of the nuclear waste management technology, which follows a major methods of the chemical engineering, glass or concrete are used as diluting agents for waste immobilization or incorporation of radioactive materials within an increased volume of mentioned above concrete or glass in order to decrease a critical concentration and poisoning impact of the highest radioactivity. All spoiled by the same amount of the decreased concentration of radioactive waste glass and concrete packages are the second hand waste products with a dramatically increased volume, and these packages must be finally stored within an additional shell or barrier in order to insulate and constitute the diluted nuclear waste.

It should be understandable that nuclear or radioactive products cannot lose their poisoning matter after any chemical and mechanical treatment, such as happened with other chemical hazards. Therefore, chemical treatments and existing clean-up methods do not applicable to nuclear products, and these techniques can not obtain complete, safe and permanent isolation of radioactive products.

Fortunately, there are known natural stone minerals and simulated them artificial ceramic materials that are thermodynamically and chemically stable for a geological period of time, such as 10^6 years. These Natural minerals are successfully isolating people and all creatures of the Earth from Natural's radioactivity.

Thousands of years ago man developed a ceramic technology for the conversion of the plastic clay or kaolin mass into stone-like shaped products of traditional silicate ceramics. More recent times have seen the development of multi-layer and fiber-content ceramic composites and advance oxide, carbide and nitride ceramics from non-plastic powder compositions.

It is possible now to control and vary both the composition and properties of the ceramic products in order to obtain all necessary refractory, chemical and nuclear resistance and retention/attenuation properties in combination with the structural/mechanical properties of the shaped item. Extensive study of ceramic behavior in a radioactive environment has been conducted in former USSR, The United States and United Kingdom (see a list of the provided references). These studies have found ceramic compositions that keep structural stability under highest radioactive impact [8, 9]. Meantime, it may be also designed material compositions that will shield and completely attenuate particular radioactive emanation [10-14].

We developed a new scientific and engineering approach to produce large sized and thick-walled multi-purpose ceramic containers (MPC). It is possible to do using a high temperature microwave assisted ceramic sintering process Since microwave energy is absorbed simultaneously throughout the work-piece (microwave heating acts internally within the bulk), microwave energy should provide uniform heating of the entire ceramic unit. As a consequence, microwave energy eliminates thermal gradients which produce stresses and cracking during heating, as may occur in conventional processing by external heating. Microwave assisted (hybrid) ceramic heating presents a unique opportunity for ceramic technology to bring to the market new large (and specially shaped) products with special properties and at reasonable cost. A second advantage of microwave ceramic processing is the possibility of internally joining and sealing ceramic materials because of the transparency of the thick-wall dielectric materials to microwave energy.

In general, we developed an advanced ceramic container technology for direct loading of dry radioactive waste at the production site, interim (temporary) storage, transportation, and permanent disposal of nuclear waste either directly at waste production site or in an underground repository, for example, at the Yucca Mountain underground tunnel loop.

The developed technology will be practicable and cost effective for all types (high, medium and low levels) of dry solid nuclear waste and many types of other hazardous materials.

BRIEF CHARACTERISTICS OF THE DEVELOPED TECHNOLOGY

A developed technology unites scientific discoveries, pioneering high tech accomplishments, engineering inventions and know-how from three fields:
(i) nuclear waste management
(ii) materials science
(iii) ceramic engineering, employing innovative microwave technique.

The project work should bring on the national and international market unique engineering techniques, equipment and products for the inexpensive and safe isolation, transportation and permanent disposal of nuclear and hazardous waste in initial solid conditions, including high level radioactive waste [15 - 25].

The special challenge of this project is the opportunity to employ new processing methods, using advanced technique, to solve the critical problem of the nuclear waste management. We developed a multi-purpose ceramic container (MPC) as the waste package for aggressive and hazardous/poisonous radioactive materials. The MPC will be completely sealed and will have permanent integrity in order to respond to all safety, engineering-manufacture, and economic requirements from the waste production site to the millennial waste repository.

We will produce and use a set of MPC modules with different forms, size and capacities (from $0.1m^3$ to $10m^3$) from different ceramic materials depending on waste production conditions, the shape of solid waste, waste radiochemistry and in-site production yield, as well as waste handling, loading, transportation and repository disposal techniques. The developed technology keeps as more as possible an existing waste management technology and will include:
(i) loading of nuclear waste product in the container
(ii) covering and sealing of the container (packaging process)
(iii) temporary storage, which could be at the waste production site
(iv)transportation of MPC from site to site or to ultimate repository, using conventional trucking and rails
(v) permanent storage of MPC within an underground repository
(vi) backfilling of MPC in the repository using other mineral materials.

During the packaging process and repository disposal, the MPC must provide a complete barrier for radioactive and chemical hazards and will have:

(i) high compressive strength in combination with material creep and

fracture resistance in case of transportation collision, lithographic stress or rock collapse at an underground repository
(ii) electrical-chemical corrosion resistance in the presence of other elements, radioactive materials and solvents; internal (due to radioactive reactions) and external continuous heating up to 1000°C of volcanic geological cataclysm
(iii) structural stability, permanent integrity and gas-tightness of body under high and low level of radioactive in combination with thermal enhanced chemical and water corrosion impacts.

The developed technology is a method of effective ceramic production and a safe storage of nuclear and hazardous waste in multi-purpose ceramic containers of small, medium and large capacities. The containers consist of ceramic cup-vessels and lids (covers) that are joined and sealed for full integrity as a perpetual barrier and attenuation shield for the stored nuclear and hazardous dry materials that cause nuclear and hazardous physical and chemical effects such as radioactive and chemical corrosion and structural and thermal impacts to the containers.

The developed multi-purpose ceramic containers will be used for on-site loading, transportation, and temporary or permanent storage of high, medium and low level dry nuclear waste in initial conditions, including untreated nuclear products, such as dismantled atomic weapons or nuclear fuel rods. These containers may be also used for secondary containerization of previously produced waste packages or their parts, including immobilized glass and concrete blocks or other dry waste products, such as chemical and medical solid and dry products, diagnostic ampoules and devices, such as irradiated parts of X-ray apparatuses.

These containers are produced from inexpensive ceramics using developed low cost, effective ceramic technology. The technology applies multi-faceted optimization of the versatile engineering system which includes structural design of the ceramic containers and their manufacturing, assembling and use for waste loading, transportation and permanent deposit in any natural repository. The container development includes:

a) determination of the matter and size and radiation and hazardous properties of the particular or typical containment product;
b) selection and formulation of ceramic materials composition which obtains a durable service of ceramics as mechanical container shells for the necessary period of time, for example, for million years;
c) determination of the necessary shielding and attenuation properties of the ceramic compositions, choice of the bulk materials mixture and determination of the shell thickness, including a composition and thickness of the interim bulk layer;
d) structural design of the ceramic construction and especially outside ceramic shell, applying probabilistic approach of the finite analysis to fully satisfy transportation and repository requirements;

e) design of the container as a customized product which should be suitable for waste loading in production site, container covering and sealing, its transportation to repository and repository setting;
f) ceramic material formulation, forming, drying and firing of ceramic cup-vessels and lid-covers which cause R&D optimization work, employing a several technological inventions and advanced Design of Multi-Factor Experiments and Response Surface Analysis methodology;
g) development, design, and erection of the invented forming and thermal aggregates (pre-dryers, dryers and kilns using hybrid microwave-radiative heating of ceramics) with a set of accompanied facilities, which make possible manufacturing of the unusual large and thick-walled ceramic vessels and lids;
h) development, design and production of the microwave heating system which provides on-site joining and sealing of the shell parts of the ceramic containers.

The project contemplates design of versatile (waste adjustable) construction of containers which will meet environmental and industrial requirements, such as:

a) existing customer needs for on-site waste loading, container covering and sealing, their safe transportation and storage for inexpensive costs and
b) design and construction of the pilot plant for development and manufacturing of customer-suitable containers and technique for their use.

Ceramic production technique should include high-tech microwave assisted ceramic processing which will decrease production costs, intensify thermal processing and improve quality of the thermal treatment (drying and firing) of thick-wall large-size ceramic vessels and lids and will make possible their integration. In order to improve process and quality control of the microwave ceramic processing, it is intended to employ a few advanced chemical and physical methods. The project employs several innovations of the materials science and ceramic engineering, such as: - alumina based ceramic composition with glass-fiber reinforcement (for outer shell), a few eutectic, reological and sintering stimulation additives

- high density grain-fraction composition and a special order of material mixing
- original forming technique including extrusion improvement
- combination of vibration-compacting forming and pre-drying systems
- hybrid heated pre-dryers, dryers and kilns for large and thick-walled products
- specially developed hydraulic gas/air circulation system for dryers and kilns
- dryer and kiln cars with a special construction to provide optimized heat treatment of ceramic vessels or lids of small and medium capacities
- unique construction of microwave furnace system for joining of container parts
- special performing of the both joining ends of container vessels and lids
- other advanced methods and techniques for material and product quality control to insure container integrity.

TECHNOLOGICAL, ENVIRONMENTAL AND COST BENEFITS

It is possible to consider a major technological advantage of the developed nuclear waste management technology in comparison with glass immobilization technology which is now a strategic concern of the U.S. Department of Energy. An existing glass immobilization technology includes:

(i) various nuclear waste treatment, including glass immobilization

(ii) carbon steel canister production and handling

(iii) glass block loading, MPC transportation, and a settlement of the waste package into a repository.

The scheme below introduces a principle flow chart of the developed ceramic containerization technology in comparison with a flow chart of the glass immobilization technique.

Current Technology - The existing flow chart: 1* --- 2* --- 3 --- 4 --- 5*--- 6*
NUCON's Technology - The proposed flow chart: 1 --------------------------- 6

where:1 = solid nuclear waste production site
2 = nuclear waste collection and temporary storage
3 = radioactive material initial treatment
4 = borosilicate glass melting and waste immobilization processes
5 = interim handling and storage in additional metal canisters
6 = permanent storage at repository site, such as Yucca Mountain repository
------ = transportation process
* = technological points where it would be possible to use multi-purpose ceramic containers either to contribute or supersede glass immobilization technique.

NUCON's technology [15 -18, 23-25] does not need stages 2 - 5 of the above scheme and therefore provides a simplified and beneficial method: direct and permanent capsulation of the solid radioactive waste in ceramic containers.

This major technological advantage produces tremendous economical benefit. Additionally, universal multi-purpose ceramic container may be used to capsulate preliminary treated nuclear waste if it is inserted in borosilicate glass. NUCON' technology will be able to prevent chemical corrosion and radioactive leakage which occasionally occurs after short period of time if glass waste immobilization and a metal canister technique are employed. Nevertheless, waste saturated blocks of glass would be loaded in ceramic containers which will be able to sure obtain completely safe perpetual storage of radioactive waste.

It is possible to calculate comparative economic benefits of multi-purpose ceramic containers and related waste management technology for the American economics. According to "The American Ceramic Society Bulletin", Vol.74, No.8,

August 1995, p.11, the cost of borosilicate glass (BSG), which incorporates 10%-20% of radioactive waste, is $\$600/ft^3$.

To determine full cost of the glass immobilization technology, the cost of BSG should be as minimum doubled in consideration of the necessity to treat and handle some solid parts of high level nuclear waste and mill them into powder which is poured into glass-melt-basin.

For example, for a waste-glass block with a considered today capacity of $9.5m^3 = 339.28ft^3$ and $\$600/ft^3$, the cost of the glass-waste material should be \$203,571 x 2 = \$407,142. This amount does not include the costs of the metal canister, handling and packaging processes.

A concentration of radionuclides that could be encapsulated in a ceramic container would be in 5-10 times higher than within borosilicate glass. This can decrease the container capacity from $9.5m^3$. Indeed, $1.5m^3$ ceramic container with a useful capacity of $1.0m^3$ may contain of ten times increased concentration of radionuclides in comparison with glass-waste-product. Consequently, it is decreased in the same ten times an amount of mass of the solid waste product or its volume. If this ceramic container supersedes $9.5m^3$ glass-waste package, the saving, which would accompany the use of a $1.5m^3$ ceramic container, should be \$407,142.

While this project is especially designed to obtain permanent environmental safety, a significant cost benefits should be achieved as well - Table I.

Table I. Comparison of the known and proposed nuclear waste disposal techniques

Disposal characteristics	Immobilization medium*			Containers from	
	glass	concrete	zircon ceramics	steel **	ceramics ***
1. Thermodynamic stability	-	+	+	+	+
2. Combined chemical hazards, water and thermal corrosion resistance	-	-	+	-	+
3. Nuclear resistance	-	-	+	+	+
4. 100% prevention of the radioactive leakage during first 50 years	-	-	-	-	+
5. Designed durability, years	300	100	10^5	100	10^6
6. Production cost estimation in comparison with ceramic container cost	100*	10*	1,000*	5-7	1

Remarks: + = ***positive response;*** - = ***negative response;***

* = containment costs consider costs of preliminary treatment and other conversions of high level nuclear waste
** = stainless steel container construction cannot assure 100% environment protection even for few hundred years due to corrosion process in the areas of the welding seams or bolt joining
*** = developed ceramic container construction.

Some additional cost benefits should be expected for the container manufacturer, such as Nucon Systems, Inc. The capital costs and operating expenditures of the proposed microwave supported dryer and kiln should be less than that for traditional dryers and kilns because of the elimination of thick refractory insulation shells, reduction of kiln dimensions, significant decreases of the sintering temperature in the range of 1300°C - 1350°C, and an increase of the thermal efficiency of the thermal processes. Additionally, microwave assisted ceramic firing has the potential to be 15 - 30 times faster than, and to require only 10 - 20% of the energy used by conventional firing process since ceramics can absorb more than 95% of the delivered microwave energy.

The use of cost beneficial ceramic technology will allow NUCON to produce advanced ceramic containers that will be thermodynamically and chemically stable for geological periods of time in such hostile conditions as high temperature, chemical and water impact, high compressive strength and radioactive emanation. It will be the major environmental and economical benefits for the U.S.

CONTAINER PRODUCTION PLAN

We developed a time-schedule work program which includes:
a) physical-chemical and applied engineering R&D, employing multi-faceted and multi-factor modeling and optimization of technological parameters
b) production and testing of ceramics and container prototypes
c) mechanical equipment choice and accommodation
d) pilot plant design and erection
e) design and construction of thermal equipment
f) manufacturing of ceramic components and assembly of ceramic containers
g) design and manufacturing of a few microwave joining systems
h) design of the industrial plant to produce containers of small, medium and large capacities.

A container with medium capacity of $1.5m^3$ ("1.5") would have 1,600mm height and 1,100mm diameter of inner room space, and container with large capacity of $5.5m^3$ ("5.5") would have 5,300mm height and the same 1,100mm diameter of the inner room space.

It makes possible to provide perpetual containerization of the depleted nuclear fuel rods of the research reactors in containers "1.5" and spend fuel rods of commercial nuclear power reactors in containers "5.5". Additionally, preliminary dried radioactive sludge of the nuclear power plant pools may be loaded in all types of the containers.

There are considered two stages of the five year project:

a) Three year plan of technology development and production of the containers with small and medium capacity up to 1.5m^3 using facilities of the erected pilot plant. During third year of the company operation, it is planned R&D works for design of the industrial plant.

b) Two years plan for the fourth and fifth years of the company operation which includes further R&D to produce containers with large capacity, using specially erected industrial plant. It is also planned a parallel production of the medium and small containers, employing pilot plant facilities.

CONCLUSION

While a history of the nuclear waste management industry has already fifty five years, still now we have a real chance to reach ultimate goal of the nuclear waste management when this project will be realized. Developed beneficial method overcomes an existing chemical engineering paradigm of the nuclear waste management science and technology, i.e., any waste treatment, such as extracting, dissolving and mixing, will be unnecessary.

A special version of the nuclear waste management technology, including an original ceramic container construction, and an innovative ceramic container production technology (altogether it would be abbreviated as CCT) have been invented and developed. To realize CCT, a R&D plan and projects of the pilot and production plants have been also developed.

CCT should obtain perpetually safe and durable solution of the nuclear waste problem for relatively inexpensive expenditures and with maximum possibility to use an existing nuclear waste management methods and techniques. It should be significant environmental and economical breakthrough.

CCT does not have analogy in an existing practice and it is an alternative for all known temporary and expensive methods and techniques of the nuclear waste management. New containerization approach should obtain a big market and profit for container manufacturers.

CCT is a tremendous opportunity for both electrical energy manufacturers and world-wide economics since it takes away a major limitation factor for the further development and operation of nuclear power energetic units, opening new perspectives for nuclear energy plant industry.

References

[1] Ewing,R.C. and Lutze,W., "Materials Science of Radioactive Waste Forms", Materials Research Society Bulletin, 12, Vol. XIX, (1994), p.16-18

[2] "Radioactive Waste Forms for the Future", Edited by W.Lutze and R.C.Ewing, North-Holland, Amsterdam, (1988), 778 p.

[3] Grambow,B. "Borosilicate Glass: Future Research Requirements or "What We Don't Know", Materials Research Society Bulletin, 12, Vol.XIX, (1994), p.20-23

[4] Vernaz,E.U. and Dussossoy,J.L., "Current State of Knowledge of Nuclear Waste Glass Corrosion Mechanism", The Case of R7T7 Glass, Appl. Geochem. Suppl. Issue No.1 (1992)

[5] Glasser,F.P. and Atkins,M., "Cements in Radioactive Waste Disposal", Materials Research Society Bulletin, 12, Vol.XIX, (1994), p.33-38

[6] Atkins,M., Damidot,D., and Glasser,F.P. in "Scientific Basis for Nuclear Waste Management", Mater. Res. Soc. Symp. Proc.333, Pittsburgh, PA, 1994 p.315

[7] Vance, E.R., "Synroc: A Suitable Waste Form for Actinides", Materials.Research Society Bulletin., 12, Vol. XIX, (1994), p.28-32

[8] Clinard, F.W., "The Effect of Irradiation-Induced Defects on Fusion Reactor Ceramics", Crist. Latt. Def. And Amorph. Mat., 1987, Vol. 14, pp. 241-254

[9] Sickafus, K.E., "The Remarkable Radiation Resistance of the Oxide Spinel" Internet: http://andromeda.mt.ac.uk/sem25sep.htm (1997)

[10] Lehman,R., Engineering materials Handbook, Vol.4: Ceramics and Glasses; ASM International, Materials Park, Ohio. 1992

[11] Roy,A.K., Fish,R.L., and McCright,R.D.,"Waste Packge Materials Selection Process", in Proc. 5th High Level Radioactive Waste Management Conference, American Nuclear Society, 1994, p.993

[12] Ewing, R.C., Weber,W.J. and Clinard,F.W.(1995) Jr., Radiation Effects in Nuclear Waste Forms for High Level Radioactive Waste, Progress in Nuclear Energy, Vol.29, No.2, 63-127

[13] John R. Lamarsh, Introduction to Nuclear Engineering, Addison-Wesley, 1983. 689 pp.

[14] Dresner, Lawrence, Principles of Radiation Protection Engineering, McGraw-Hill Book Company, Inc., New York, 1965. 451 pp.

[15] Rokhvarger, A."Urgent Principles and Prospects for Industrial Application of Microwave Ceramic Technology".First World Congress on Microwave Processing, Jan.,1997,Orlando,Fl., Abstr.

[16] Rokhvarger, A.E., "Recent Breakthrough and Modern Challenge of Ceramic Kilns", Abstracts, p. 199, (1996) Annual Meeting of The American Ceramic Society, Indianapolis, In.

[17] Rokhvarger, A.E. and Haber, N., "Development of Large Ceramic Container Technology for Nuclear Waste Disposal", Ceramic Trans.: Env. Issues and Waste Man. Techn.; Ed by V.Jain and R.Palmer, Vol.61, Am.Cer.Soc., Westerville, OH, pp. 345-356, (1995)

[18] Rokhvarger, A.E., "Use of Combustion and Incineration Wastes for Structural Clay Products", Ceramic Transactions, Volume 45: Environmental and Waste Management Issues in the Ceramic Industry II, 125-134 (1994)

[19] H.Larker (Sweden), Method of Containing Spent Nuclear Fuel or High Level Nuclear Fuel Waste. U.S. Patent 4,209,420; June 24, 1980

[20] B. Albert and S. Carpentier (France), Method for Embedding and Storing Dangerous Materials, such as Radioactive Materials in Monolitic Container. U.S. Patent 4,726,916; Feb. 23, 1988

[21] E.J.Ramm and A.E.Ringwood (Australia), Encapsulation of Waste Materials. U.S. Patent 4,834,917; May 30, 1989

[22] G. D. Crove (USA), Ceramic Container. U.S. Patent 5,302,565; April12, 1994

[23-25] A.E. Rokhvarger and A.B. Khizh (USA); Three patents are pending simultaneously from U.S., France, and Japan Patent Offices; January 1996.

IMMOBILIZATION OF ^{99}Tc IN LOW-TEMPERATURE PHOSPHATE CERAMIC WASTE FORMS

D. Singh, V. Mandalika, A. Wagh, R. Strain, and M. Tlustochowicz
Energy Technology Division, Argonne National Laboratory, Argonne, IL 60439

ABSTRACT

Radionuclides such as ^{99}Tc are by-products of fission reactions in high-level wastes. Technetium poses a serious environmental threat because it is easily oxidized into its highly leachable pertechnetate form. Magnesium potassium phosphate ceramics have been developed to treat ^{99}Tc that has been separated and eluted from simulated high-level tank wastes by sorption processes. Dense and hard ceramic waste forms were fabricated by acid-base reactions between mixtures of magnesium oxide powders and wastes, and acid phosphate solutions. Standard leaching tests, such as ANS 16.1 and the Product Consistency Test, were conducted on the final waste forms to establish their performance. The fate of the contaminants in the final waste forms was established with scanning electron microscopy techniques. In addition, stability of the waste forms in aqueous environments was evaluated by long-term water immersion tests.

INTRODUCTION

Technetium-99 is a long-half-life fission product isotope that is present in U.S. Department of Energy (DOE) high-level waste (HLW) along with other volatile fission products such as ^{137}Cs and ^{90}Sr. Over a wide range of oxidizing conditions, the predominant form of Tc is the pertechnetate anion (TcO_4^-), which is highly soluble in water and thus readily mobile in the environment. Because of its long half-life (2.13 x 10^5 yrs) and high solubility, immobilization of ^{99}Tc is of great concern to DOE. Conventional thermal treatments for the stabilization of fission-product (Tc, Cs) wastes generates volatiles that subsequently must be stabilized before disposal. Hence, we need technologies that stabilize these waste streams by nonthermal stabilization processes. To this end, we are developing and demonstrating a low-temperature stabilization technology based on chemically bonded phosphate ceramics (CBPCs).

Chemically bonded phosphate ceramics are inorganic materials that can be fabricated inexpensively at relatively low temperatures by acid-base reactions between an inorganic oxide/hydroxide powder and an acid-phosphate solution [1,2]. The resulting phosphate ceramics are extremely insoluble in aqueous media and exhibit excellent long-term durability. Using CBPC technology, we have successfully demonstrated superior stabilization of heavy metals such as Pb, Cd, Cr, Hg, Ni, and radioactive contaminants like ^{238}U [3].

In this work, magnesium potassium phosphate ceramics (MKPCs) were developed for the stabilization of Tc-containing waste streams generated by a complexation-elution process developed at Los Alamos National Laboratory (LANL). Performance of the phosphate ceramic waste forms was established with various leaching, strength, and durability tests. Investigations were carried out by incorporating rhenium (Re) as a surrogate (chemical but nonradioactive equivalent) for ^{99}Tc, as well as ^{99}Tc in the LANL waste. The results of this study show that MKPCs have several advantages that make them ideal candidates for the treatment of waste streams that contain ^{99}Tc.

WASTE STREAM COMPOSITION

As part of the DOE's Efficient Separations and Processing Crosscutting Program, several separation technologies are being developed to partition volatile fission products from HLW to reduce the overall volatiles and radioactivity levels in the wastes. These partitioning efforts also minimize waste handling operations. In this regard, LANL has developed a complexation-elution process to separate ^{99}Tc from HLW such as the Hanford supernatant process. However, this separated waste must be immobilized before disposal. The expected composition of the Tc waste that is obtained from the complexation-elution process is 1 M NaOH, 1 M ethylene-diamine, 0.005 M Sn (II), and 0.00005 M ^{99}Tc. Surrogate waste with this composition was prepared by using Re as a nonradioactive surrogate for Tc. Also, waste forms were fabricated from the ^{99}Tc-containing waste from LANL's complexation-elution process.

FABRICATION OF MKPC WASTE FORMS

Magnesium potassium phosphate ceramic waste forms were fabricated by incorporating simulated formulations of the Tc waste stream obtained from LANL's complexation-elution process [4]. The magnesium potassium phosphate matrix is fabricated by reacting calcined magnesium oxide powder with an acid-phosphate solution. Phosphate ceramic waste forms were fabricated by mixing a surrogate waste stream with the magnesium potassium phosphate binder. Reducing agents such as $SnCl_2$ were added to the waste form formulation to keep the Re/Tc in its reduced state. The binder was continuously added to the waste solution, and the mixture was vigorously mixed to a desired consistency and then transferred to a mold and allowed to set. The mix hardened in 2 h to form hard, dense monoliths. After curing for >7 days, the monolithic waste forms were removed from the molds and subjected to various performance tests. Waste forms were fabricated so that the waste loading in the final waste forms was 36%. Concentration of the contaminants in the final waste forms ranged from 5 to 225 ppm.

RESULTS AND DISCUSSION

Physical and Mechanical Properties of Final Waste Forms

The physical properties most relevant to the final waste form are density, porosity, and compression strength. Therefore, these properties were measured on the MKPC waste forms fabricated from surrogate formulations of LANL's elution wastes. The density of the samples was measured by determining the mass and geometric volume; the porosity, by water immersion; and the compression strength, with an Instron machine in the compression mode. The density and open porosity of the waste forms were 1.8 g/cm^3 and 4%, respectively. The average compression strength of the MKPC ceramic waste forms was 4320 ± 954 psi. The land disposal requirement for the final waste forms is 500 psi. When compared with the required value, the strengths of the phosphate ceramic waste forms are much higher, indicating that MKPCs form superior waste forms. Overall, the waste forms are dense and hard materials, with very low open porosity and high strength.

X-ray Diffraction Analysis

To identify the crystalline phases in the MKPC waste forms, X-ray diffraction (XRD) using CuK_α radiation were performed. Figure 1 shows the XRD pattern of the MKPC waste form with 36% waste loading. Phosphate ceramic waste forms are highly crystalline. The prominent peaks in Fig. 1 represent those of the magnesium potassium phosphate system, which is the binding phase in the waste forms. Magnesium potassium phosphate is a highly insoluble phase (solubility in water is $\approx 10^{-9}$ mol/L); hence, it is desirable in the final waste forms. Also present in Fig. 1 are peaks of unreacted magnesium oxide, which acts as a filler material.

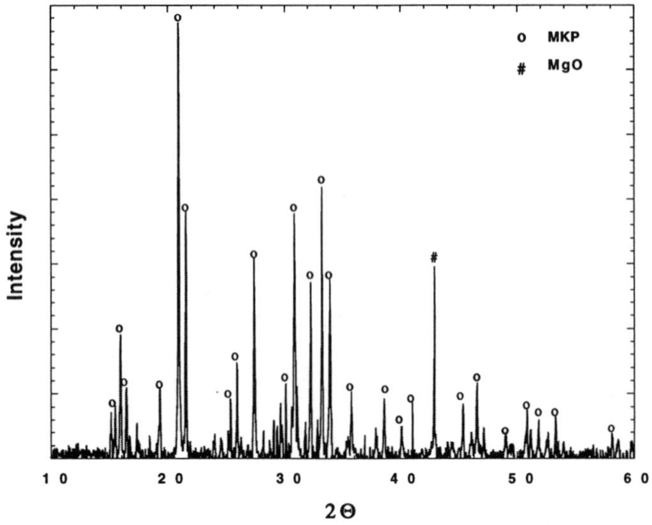

Fig. 1. X-ray diffraction pattern of magnesium potassium phosphate ceramic (MKPC) waste form with 36% waste loading.

Microstructural Studies

The microstructure of the phosphate ceramic waste forms was investigated by scanning electron microscopy (SEM). Figure 2 is an SEM photomicrograph of a fractured surface of a magnesium phosphate ceramic waste form with 36% waste loading. The photomicrograph shows a very dense, crystalline structure, with a small number of pores. The crystalline phase is the magnesium potassium phosphate binding matrix. The waste form microstructures were further analyzed by X-ray mapping of the contaminant in the waste form. The inset in Fig. 2, which shows the distribution of Re within the area of the photomicrograph, indicates that Re is evenly distributed. This implies that the distribution of the contaminant is homogeneous and the contaminant is not segregated into localized regions. In addition, contaminants are thoroughly encapsulated by the crystalline binding matrix of the MKPC. This makes the contaminant less leachable and accounts for the extremely low leachability of Re during leach testing of final waste forms.

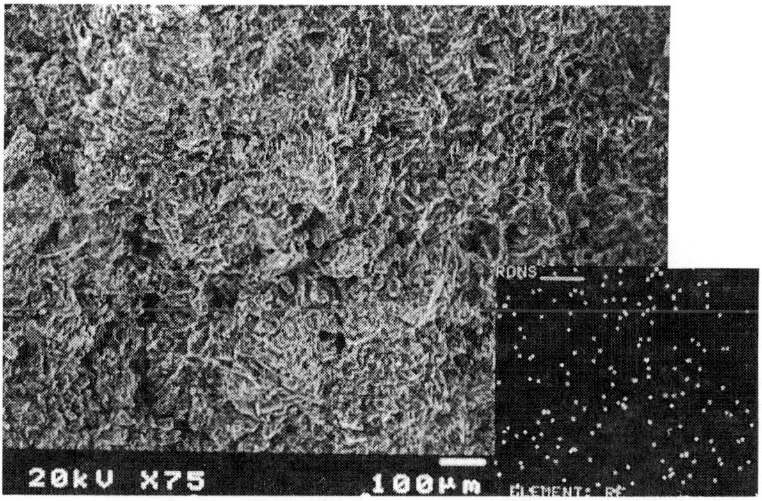

Fig. 2. Photomicrograph of fractured surface of MKP waste form with 36% waste loading; inset shows distribution of Re within the area of the photomicrograph.

Leaching Studies

To establish the effectiveness of the MKPC waste forms in stabilizing the contaminants, leaching studies were performed on the fabricated final waste forms according to American Nuclear Society's ANS 16.1 and Product Consistency Test (PCT).

MKPC waste forms with 36% elution waste loading were subjected to ANS 16.1 leach testing for 90 days under oxidizing conditions, in accordance with the procedure given in Ref. 6. This testing was conducted on waste forms made with Re (surrogate for Tc), and on wastes that contained ^{99}Tc. In this procedure, lechate solutions are periodically changed to simulate continuous flow of water. At the end of the 90-day test, parameters that are obtained are the effective diffusivity and leachability index of the contaminant. The effective diffusivity is given by

$$D_e = \pi[(a_n/A_0) / (\Delta t)_n]^2 (V/S)^2 (T) \quad (1)$$

where

a_n = Amount of contaminant leached from sample during leaching interval n,

A_0 = Total amount of contaminant in sample at tbeginning of leach test,

$(\Delta t)_n$	=	$t_n - t_{n-1}$, duration of nth leaching interval, s,
D_e	=	Effective diffusivity, cm^2/s,
V	=	Volume of sample, cm^3,
S	=	Surface area of sample, cm^2, and
T	=	Mean time of leaching interval, s,

The logarithm of the average effective diffusivity (cm^2/s) is defined as leachability index

$$L_i = 1/n\ \Sigma[\log(\beta/D_e)]_n \tag{2}$$

where β is a defined constant (1 cm^2/s) and D_e is the effective diffusivity of the contaminant. Larger values of the leachability index are desirable because they represent lower values of contaminant diffusivity. Moreover, based on Godbee's model [7], a cumulative fraction of the contaminant species in the leachant solution after a particular time t can be expressed by

$$F = \Sigma a_n/A_0 = [2.\ D_e/\pi]^{0.5}.\ t^{0.5}\ (S/V) \tag{3}$$

Figure 3 shows the cumulative fraction of the contaminants (Re or ^{99}Tc) plotted against the square root of the leaching time. A straight line was not obtained for the first four days, indicating that the removal of the contaminant in the first four days was likely to be by washoff from the surface and the inner pore walls of the specimen. After four days, a linear relationship between the cumulative fraction leached and the square root of the leach time was obtained, indicating that diffusion was the rate-limiting mechanism. At present, it is not clear why the samples with higher contaminant loadings showed a lower cumulative leached fraction. It is possible that the amount of contaminant released depends on the rate at which oxygen diffuses into the specimen and oxidizes the reduced contaminant species. Thus, the diffusion of oxygen is probably the rate-limiting step and, hence, even though the contaminant concentration in the waste form is high, only a limited amount is oxidized and released into the lechant.

Effective diffusivity and leachability index values were obtained from the ANS 16.1 standard equations given above. The results are presented in Table I. Diffusivities of the contaminants are extremely low, indicating excellent stabilization of the contaminants in the phosphate matrix. The determined leachability indices

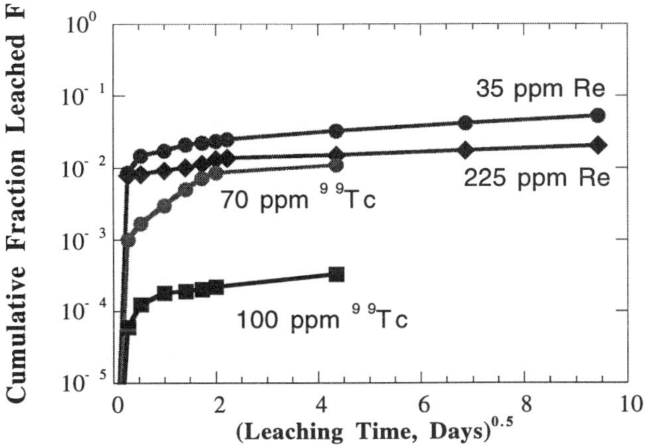

Fig. 3. ANS 16.1 test results, showing square root of leach time vs. cumulative fraction contaminant in leachate.

Table I. Results of ANS 16.1 test on MKP waste forms with 36% waste loading

Contaminant Conc., ppm	Effective Diffusivity, cm^2/s	Leachability Index
Re: 35	6.55×10^{-11}	10.2
Re: 225	6.39×10^{-12}	11.2
Tc: 70	2.13×10^{-11}	10.8
Tc: 100	2.36×10^{-14}	14.2

ranged from 10 for specimens with 35 ppm Re loading to as high as 14 for specimens with 100 ppm Tc loading. These figures of merit are significantly higher than the Nuclear Regulatory Commission's passing criterion of 6.

A static leaching test, such as the PCT [8], was performed to evaluate the chemical durability of MKPC waste forms at temperatures <100°C for various surface-area-to-volume (S/V) ratios. The PCT is designed to evaluate chemical durability of crushed homogeneous and vitrified glasses. To allow a comparison of leach test results of waste forms with differing compositions, results are given in terms of normalized elemental loss and normalized elemental leach rate in both tests. These tests simulate the worst-case leaching conditions. Product Consistency Tests were carried out at 90°C for seven days in teflon containers on >1 g from -100 to +200 mesh crushed waste form particles. Deionized water was the leachant and the S/V ratio in this test was held constant at 3000 m^{-1}. These leaching tests were performed under oxidizing conditions.

Table II summarizes the PCT leaching rates observed with waste form samples fabricated with 35 and 225 ppm Re loading. Normalized leaching rates of Re after a seven-day test period are extremely low ($\approx 10^{-4}$ g/cm^2-d) for both of the waste forms. These values are as good as those observed in high-temperature technologies applied to materials like borosilicate glass, vitrified glass, etc. The low leach rate values observed with phosphate ceramic waste forms in PCTs are attributed to the reduction of Re to its lower oxidation state. It should be noted that maintenance of low-oxygen conditions, to prevent oxidation and thus to reduce the mobility of Re, significantly improves the leaching rates of the contaminant in phosphate ceramic waste forms. We are currently performing a PCT on waste forms that were fabricated with the actual Tc waste.

Table II. Test results of product consistency

Waste Form	Re conc., ppm	Leach Rate, g/m^2-d
MKPC + 36% Eluted Waste	35	1.26×10^{-3}
MKPC + 36% Eluted Waste	225	4.90×10^{-4}

Water Immersion Tests on Final WasteForms

Immersion studies were initiated to study the durability of the final waste forms in an aqueous environment. Three waste form specimens were immersed in deionized water for 90 days. One of the specimens was periodically removed, dried, and weighed to determine weight change over the 90-day period. The other two specimens were not disturbed. The procedure we followed was similar to that outlined in ANS 16.1. Figure 4 shows the weight change in the waste form containing 36% LANL-eluted waste loading. The percent change in weight is

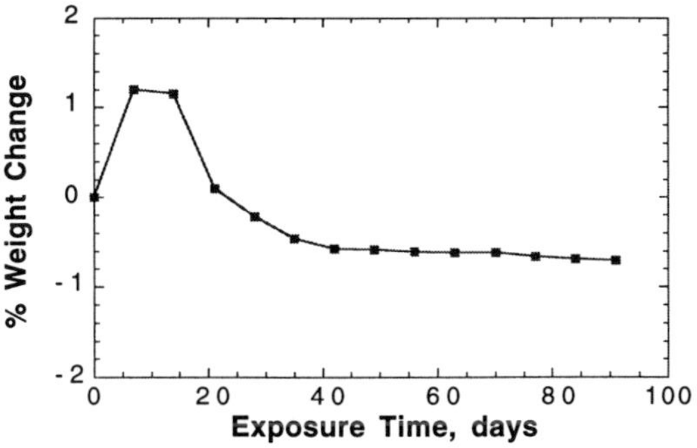

Fig. 4. Percent weight change of waste form during 90-day water immersion test.

computed with respect to the weight of the waste form before immersion study. Initially, there was an increase in weight. After three weeks of immersion, the weight of the sample started to decrease, and from five weeks onwards, it remained nearly constant. At the end of 90 days, there was an ≈0.5% net decrease in weight. However, specimens that were not disturbed for 90 days showed a 0.2 wt.% increase in weight.

X-ray diffraction analyses (Fig. 5) of as-fabricated and long-term-water-immersed waste forms provide insight into the initial gain and subsequent loss in weight of water-immersed specimens. All of the peaks of the magnesium potassium phosphate binding matrix that were present in the XRD plot of the as-fabricated waste forms were also present in the water-immersed samples. This finding indicates that the matrix of the fabricated waste forms is not being dissolved even after long-term exposure to water and demonstrates the insolubility of the phosphate binding phase in water, a highly desirable characteristic for the durability of the final waste forms in aqueous environments. Furthermore, it was observed that the relative intensity of the magnesium oxide peak at 42° has decreased significantly in the water-exposed samples. It can be inferred from this observation that the initial weight gain might be due to the hydration of magnesium oxide to mganesium hyroxide. The weight loss in the following weeks is probably due to the release of these reaction products, along with some unreacted MgO, into the water. The increase in the pH of the deionized water from 5.5 to 8 after 210 days

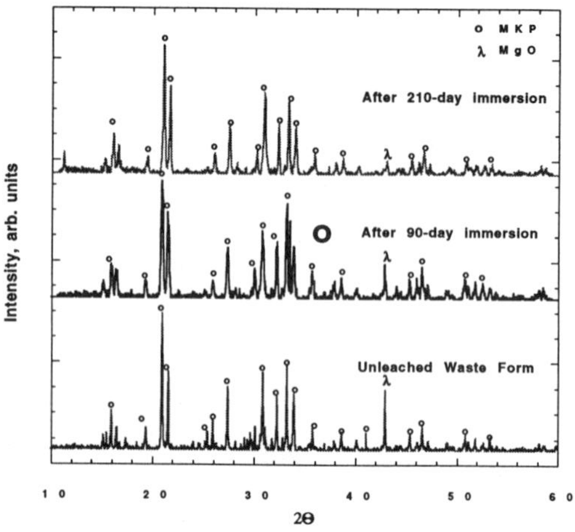

Fig. 5. XRD plots of as-fabricated and long-term-water-immersed MKPC waste forms with 36% waste loading.

of immersion suggests the release of the unreacted magnesium oxide from the water-immersed specimens. Over all, there was a very small decrease (≈0.5%) in the weight of the specimens after 90 days of immersion in water. We believe that this small weight loss can be further reduced by optimizing the processing conditions so that the residual MgO content in the final waste form is minimized.

In addition to monitoring the weight change of the waste forms, we immersed three specimens in water and tested them for compression strength at the end of 210 days. Figure 6 shows the strength of the as-fabricated final waste forms and of the phosphate ceramic waste forms after a 210-day water immersion. The compressive strength of the waste forms is ≈3000 psi, which is significantly higher than the regulatory requirement of 500 psi, even after a 210-day exposure to an aqueous environment. The observed weight change and compression strength results clearly indicate that MKPC waste forms have the potential to form highly insoluble and durable waste forms.

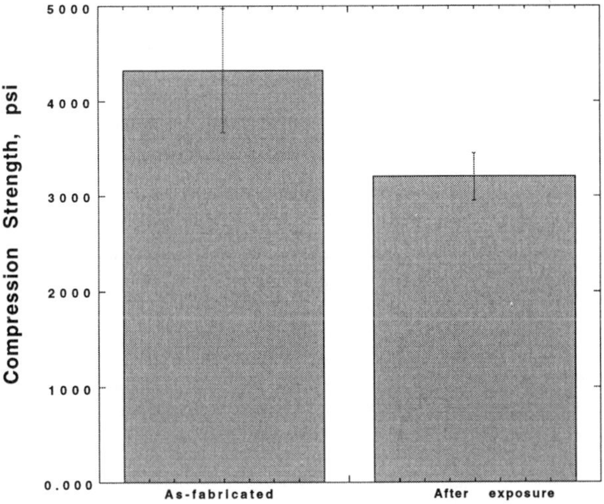

Fig. 6. Average strengths of as-fabricated MKPC waste forms and after 210-day immersion in water.

SUMMARY

Magnesium potassium phosphate ceramics were used to stabilize separated ^{99}Tc waste stream from LANL's complexation-elution process. Waste forms were produced with low open porosity, and hence low permeability, and high compression strength. Preliminary investigations were carried out by incorporating Re in the phosphate matrix as a nonradioactive surrogate for ^{99}Tc. The waste forms were subjected to various leaching tests to demonstrate retention of the contaminants in the surrogate wastes. Magnesium potassium phosphates are extremely effective in stabilizing Re, the surrogate of Tc. A Product Consistency Test conducted on MKPC-LANL-eluted waste forms showed leaching rates of Re of $\approx 10^{-3}$-10^{-4} g/m^2-d. Similarly, ANS 16.1 tests on the waste forms indicated leachability indices from 10-14 for ^{99}Tc. Ninety-day water immersion tests showed an $\approx$0.5% change in sample weight, indicating no apparent dissolution of the matrix material. Overall, MKPCs have several advantages that make them candidates for the treatment of waste streams that contain ^{99}Tc.

Acknowledgments

This work has been supported by the U.S. Department of Energy, Office of Science and Technology (EM-50), as part of the Efficient Separations and Processing Crosscutting Program, under Contract W-31-109-Eng-38.

References

[1]W. D. Kingery, "Fundamental Studies of Phosphate Bonding in Refractories: II. Cold-bonding Properties," *Journal of the American Ceramic Society*, **33** 242-247 (1950).

[2]D. M. Roy, "New Strong Cement Materials: Chemically Bonded Ceramics," *Science*, **235** 651-658 (1987).

[3]D. Singh, A. S. Wagh, J. Cunnane, and J. Mayberry, "Chemically Bonded Phosphate Ceramics for Low-Level Mixed-Waste Stabilization," *Journal of Environmental Science and Health*, A32(2), 527-541 (1997).

[4]N. Schroeder et al., "Annual Report on Technetium Partitioning for the Hanford Tank Waste Remediation System: Anion Exchange Studies for Partitioning Tc from Synthetic DSSF and DSS Simulants and Actual Hanford Wastes Using Reillex-HPQ Resin," LA-UR-95-4440, Los Alamos National Laboratory (1995).

[5]H. V. A. Briscoe et al., "The Reduction of Potassium Per-rhenate," *Journal of Chemical Society*, **1** 666-669 (1931).

[6]American Nuclear Society, American National Standard Measurement of the Leachability in Solidified Low-level Radioactive Wastes by a Short-Term Test Procedure, Method ANSI/ANS 16.1 - 1986, American Nuclear Society (1986).

[7]H. W. Godbee et al., "Application of Mass Transport Theory to the Leaching of Radionuclides from Waste Solids," *Nuclear and Chemical Waste Management,* **1** 29-35 (1980).

[8]Standard Test Methods for Determining Chemical Durability of Nuclear Waste Glasses: The Product Consistency Test (PCT), ASTM Designation: C 1285-94 (1994).

PERFORMANCE MODELING APPLIED TO THE TREATMENT AND DISPOSAL OF A MIXED WASTE AT THE SAVANNAH RIVER SITE

J. B. Pickett, C. M. Jantzen, and J. R. Cook
Westinghouse Savannah River Co., Aiken, SC 29808

A. R. Whited
Dames and Moore, Orchard Park, NY 14127

R. A. Fjeld
Clemson University, Clemson, SC 29643-0919

ABSTRACT

Performance modeling for Low Level Mixed Waste disposal was conducted using the measured leach rates from a number of vitrified waste formulations. The objective of the study was to determine if the improved durability of a vitrified mixed waste would allow trench disposal at the Savannah River Site (SRS).

Leaching data were compiled from twenty-nine diverse reference glasses, encompassing a wide range of exposed glass surface area to leachant volume ratios (SA/V), and various leachant solutions; all of which had been leached at 90°C, using the MCC-1 or PCT procedures (ASTM Procedures C1220-92 and C1285-94, respectively). The normalized leach rates were scaled to the ambient disposal temperature of 25°C, and compared to the allowable leach rate of uranium - which would meet the performance assessment requirements. The results indicated that a glass of above average durability (vs. the reference glasses) would meet the uranium leaching concentration for direct SRS trench disposal.

INTRODUCTION

Vitrification, the conversion of source materials into a solid amorphous glass matrix, has emerged as a viable treatment technology for the immobilization of Low-Level Radioactive Waste (LLW) and mixed waste (LLW/hazardous waste). Vitrification has been selected for the High Level Radioactive Waste (HLW) stored at the Department of Energy (DOE) West Valley, NY; Savannah River Site, SC; and Hanford Site, WA sites. The U.S. Environmental Protection Agency (EPA) has declared vitrification to be the "Best Developed Available Technology" for the treatment of the DOE HLW/mixed wastes [1]. The advantages of vitrification include immobilization of radioactive and hazardous constituents, substantial volume reduction, and a final wasteform with high chemical/physical durability.

Over the past two decades, substantial research has been conducted to vitrify the legacy of High Level Radioactive (HLW) waste remaining from the United States weapons production programs. More recently, vitrification has been proposed for the treatment of Low Level Radioactive (LLW) and mixed (radioactive/hazardous) wastes. Vitrification has been selected by the DOE to treat the low level fraction of the high level wastes stored at the Hanford reservation, to treat some of the mixed wastes stored at the Idaho National Engineering Laboratory site, and a mixed waste at Savannah River Site (SRS).

Final disposal of the vitrified wastes will depend on whether the final wasteform remains a hazardous waste by the EPA hazardous waste management Resource Conservation and Recovery Act (RCRA) regulations, or whether the waste can be de-characterized or "delisted" as hazardous. The SRS waste being vitrified is a RCRA listed F-006 wastewater treatment sludge from electroplating operations. The sludge was generated during the production of depleted uranium "targets" in the SRS "M-Area", for the eventual production of plutonium-239. The treated waste will remain a "listed" F-006 waste after treatment, and would therefore have to be disposed to a RCRA Subtitle C disposal facility. Disposal of the vitrified waste at the SRS is the preferred option, but the SRS low level waste disposal facilities are not permitted Subtitle C facilities, and no such facilities are anticipated at the SRS. Therefore, the SRS has submitted a "Delisting Petition" to the Region IV EPA for the vitrified M-Area sludge [2].

In order for low-level radioactive wastes to be disposed at DOE sites, the DOE Order 5820.2A mandates that a site-specific radiological Performance Assessment (PA) must be conducted. The final waste and disposal facility must comply with the performance objectives of the order. The time period for which compliance is required is not specified in the order, but it is generally assumed to that the performance objectives be met for a minimum of 10,000 years. [3] The key performance objectives for the SRS E-Area disposal are:

- Protect public health and safety in accordance with other applicable DOE orders (other than 5820.2A),
- Radioactive dose to public is less than 25 mrem/yr; all pathways,
- Prevent dose to inadvertent intruder to:
  - less than 100 mrem/yr for continuous exposure, or
  - less than 500 mrem for a single acute exposure, and
- Protect groundwater resources consistent with federal, state, and local requirements - usually assumed to be less than 4 mrem/yr from radionuclides or the Maximum Contaminant Limits (MCLs) for chemical constituents from water ingestion.

The performance assessment assumes that the integrity of the wasteform, the wasteform container, and/or the vault structure itself will degrade within the 10,000 year period. Therefore, only naturally occurring geologic phenomena are used to calculate the affect on the environment or potential human uses (or intrusion) from the disposed materials.

For example, the affect of a disposed waste on the groundwater, and its potential use as a human drinking water supply, is a function of the chemical mobility of the waste in the natural environment, and the distance and attenuation of the waste in the soil column between the waste and the water table. In the case of the M-Area plating line sludge, the final vitrified waste form will contain quantities of uranium isotopes (from the depleted uranium) which would exceed the SRS E-vault or trench Performance Assessment disposal limits. However, as stated above, the durability of the vitrified wasteform is not included in the performance assessment. The retention of the constituents in the glass matrix will inhibit the migration of the uranium isotopes to the drinking water supply. If the durability of the glass were included in the performance assessment calculations, the authors believe that the final M-Area product could meet the performance objectives.

The objective of this presentation is to demonstrate that current laboratory leaching procedures can be used to "predict" the long-term leaching performance of vitrified wastes, and that specifically, the vitrified M-Area waste should meet the performance criteria with respect to groundwater protection for the SRS low level radioactive trench disposal. The Curies of the various uranium isotopes present in the M-Area sludge are shown in Table I.

Table I Curies/Kilograms of Uranium Isotopes In M-Area Sludge

Isotope	Ci	kg
U-234	4.23	0.69
U-235	0.19	86.1
U-236	0.13	2.0
U-238	9.55	28,550

SRS PERFORMANCE ASSESSMENT GROUNDWATER MODELING

Previous performance assessments for the SRS have shown that the limiting pathway for public exposure is via transport to a drinking water weli and subsequent ingestion [3]. The modeling of the M-Area waste considered both the radiological and chemical toxicity of uranium. The limits used were:

- radiological toxicity, 4 mrem/yr from uranium (equivalent to 45 μgm/L), &
- chemical toxicity, a maximum proposed drinking water limit of 20 μgm/L (20 ppb) uranium, due to heavy metal (kidney) toxicity [4].

The groundwater modeling code used for the SRS EAV performance assessment [3] and the 1987 SRS Groundwater Protection Environmental Impact Statement [5,6,7] is the "PATHRAE" code. This code was designed and accepted by the EPA for modeling the land disposal of LLW waste [8]. The PATHRAE code is based on one-dimensional advection through the unsaturated soil zone, and is primarily affected by water infiltration rate, fraction of saturation, porosity, soil density, and contaminant sorption coefficients. Saturated zone transport is a one-dimensional advection and 2 dimensional dispersion as a primary function of groundwater velocity, dispersion coefficients, aquifer porosity and soil density and contaminant sorption coefficients. The rate of infiltration to groundwater is assumed constant over the entire period, in this case 10,000 years. The leach rate from the inventory is assumed constant, except for radiological decay and solubility influences. The modeling conducted in this presentation for the disposal of the vitrified M-Area waste at the SRS E-Area Low Level Waste disposal utilized the following waste and site specific parameters and disposal site assumptions. It was assumed that the waste would be distributed over the entire area of the waste trenches, to minimize a localized uranium concentration.

Table II. Parameters for PATHRAE Modeling of SRS E-Area Disposal Site

Parameter	Value	Source (Ref. #)
Trench Dimensions (m)	200 x 48	EAV-PA (3)
Downgradient Drinking Well Distance(m)	100	EAV-PA (3)
Annual Rainfall Infiltration (cm/yr)	40	EAV-PA (3)
Unsaturated Zone Depth (m)	9	EAV-PA (3)
Uranium Solubility (moles U/L)	2.0×10^{-6}	EAV-PA (3)
Drinking Water Uptake (L/yr)	730	EPA (4)
Uranium K_d (ml/g)	35	Sheppard (10)
Volume of Wasteform (m^3)	380	Pickett (2)
Density of Wasteform (kg/m^3)	2600	Pickett (2)
SRS Specific Hydrogeologic Factors	-	EAV(3) and Ryan (9)

The PATHRAE code calculates contaminant leaching, and subsequent groundwater concentration at the drinking water well, based on a constant fractional leach rate Δ_L (yr^{-1}). This Δ_L is multiplied by the initial inventory of each isotope I_0 (Ci) to obtain a preliminary value for the annual activity release in Ci/yr. The preliminary value is corrected for isotopic radioactive decay and solubility limitations, and corrected to human dose equivalents in millirem/year (mrem/yr). The significant decay products for the M-Area uranium inventory, in a 10,000 year period, are U-238 and U-234. At a fractional release rate of

6.0 x 10^{-6}/yr, U-238 causes 2.5 mrem/yr dose, U-234 causes a 1.25 mrem/yr, and all other radionuclides provide a dose of 0.25 mrem/yr, for a total of 4.0 mrem/year. Since the major isotope present is U-238 (Table I), this translates to a concentration of 45 μg U/L (45 ppb). The fractional release (Δ_L) resulting in a uranium concentration of 20 μg/L (20 ppb) is 2.65 x 10^{-6}/yr. This fractional release from the E-Area trenches, of 2.65% of the uranium inventory over 10,000 years, is the release rate which would meet the health based drinking water goal.

GLASS CORROSION

Corrosion Mechanisms And Glass Dissolution

Waste glass corrosion involves a series of complex processes which are highly dependent on environmental conditions and initial glass composition. The model most widely accepted at present involves three primary mechanisms (11):

- diffusion controlled ion-exchange (diffusion of water into the glass, and exchange of hydrogen for soluble cations),
- dissolution of the glass matrix, (complete breakdown of the glass structure via hydrolysis reactions), and
- formation of thermodynamically stable products from the glass constituents (crystalline phases on the glass surface, or precipitation of amorphous phases).

The diffusion controlled ion-exchange dominates the early stages of glass dissolution, while glass matrix dissolution dominates the latter stages (12). The surface layers may reduce leaching by forming a protective barrier [13], or could cause leaching to increase by altering the solution chemistry [11,14].

While these three corrosion mechanisms control the mechanism by which glass dissolves, solubility considerations determine the dissolution rate. The dissolution rate is highest initially, when the chemical potential gradient between the glass and the solution is the greatest [15]. As soluble constituents are released, the driving concentration gradient is decreased, and the corrosion rate declines. In most leaching environments, highly soluble elements such as sodium, lithium, and boron are almost completely released into solution. Components such as silicon, aluminum, and iron become supersaturated as leachant chemistry changes, and can be re-precipitated as surface layers. If the vitrified wasteform is subjected to limited infiltration of groundwater over long time periods, then the contacting solution may experience saturation with respect to all but the most soluble components. Solution saturation, especially with respect to the primary glass former, silicon, causes the dissolution rate to decline dramatically [11,16].

Most long-term leaching data suggest that the overall glass corrosion rate should decrease with time[12]. Test conducted with high Surface Area (SA) to Volume (V) ratios, at durations long enough to reach silicon saturation, generally show that dissolution rates approach a low, steady-state value [17].

Glass Leach Rate Calculations

An analytical model that reliably predicts contaminant leaching from glass in a disposal environment has yet to be developed. However, a substantial amount of empirical leaching date has been generated in the last two decades, during the development of durable glass wasteforms for the stabilization of radioactive and mixed wastes. Standard leaching test protocols report the concentration of the elemental glass components (i.e., silicon, boron, sodium) observed in the leaching solution, at various times. The leach solution concentration is normalized to based on the concentration of each component in the initial glass. The Normalized Concentration NC_i is defined as:

$$NC_i = \frac{C_i}{F_i} \text{ where,} \tag{1}$$

NC_i = normalized concentration of the element i in the leachant (g/m^3, or ppm),
C_i = concentration of element i in leachant (g/m^3), and
F_i = mass fraction of element i in glass (g_i/g_{glass})

If leaching data exist for two different times, at otherwise identical conditions, an approximation of the Normalized Leach Rate (NLR), at average time, can be calculated.

$$NLR_i = \frac{(NC_i)_n - (NC_i)_{n-1}}{(SA/V)\,(t_n - t_{n-1})} \tag{2}$$

where;
NLR_i = normalized leach rate for element i ($g/m^2 \cdot day$)
SA = exposed glass surface (m^2), and
V = leachant volume (m^3)
$t_n - t_{n-1}$ = delta time (days)

This approximation is poor during the early stages of glass dissolution, when the rate is affected strongly by the large differences in glass vs. solution concentrations. However, over long time periods (months to years), and in situations where the surface area is large compared to the volume of leachant, the NLR_i will remain relatively constant. High SA/V conditions are expected in disposal sites where groundwater infiltration is limited due to climatic or geological conditions, and the surface are of the waste is significant.

Boron has been selected as a good indicator of the overall leach rate of a glass, since it is a key component of boro-silicate glasses, is among the most soluble of glass constituents, and does not tend to saturate the solution or form secondary precipitation products. NLR_i can be approximated as NLR_{boron}.

If both the NLR_i and the effective surface area are be assumed to be relatively constant over a delta period of time t_n to t_{n-1}, then the total mass release (M_{Total}) can be calculated by:

$$M_{Total} = (NLR_{boron})\ g/m^2{\cdot}day\ (SA)\ m^2\ (delta\ time)\ days \qquad (3)$$

The mass release of less soluble glass components, such as uranium, will be a fraction of the boron release. A conservative assumption for these calculations is $NLR_{uranium} = NLR_{boron}$. Therefore, the total mass release of uranium at time t can be found as:

$$M_{Total\ uranium\ release} = (NLR_{boron})\ g/m^2{\cdot}day\ (SA)\ m^2\ (delta\ time)\ days \qquad (4)$$

then the fractional release = $\dfrac{M_T}{M_{initial}\ g} = \dfrac{(NLR_{boron})\ g/m^2{\cdot}day\ (SA)\ m^2\ (days)}{M_{initial}\ g}$ (5)

PATHRAE vs. GLASS LEACH RATE MODELING

PATHRAE's use of a constant fractional leach rate Δ_L is therefore similar to the constant approximation of glass leaching based on boron release and surface area to volume of leachant ratio. I. e. the PATHRAE fractional release is:

$$\Delta_L\ (yr^{-1})\ ,\ \text{fractional release/year} = \frac{M_T\ g\ _{released\ by\ time\ t}}{M_{initial}\ g} /\ \text{time (years)} \qquad (6)$$

Substituting the average fractional release from the leaching model; Eq. 5, gives

$$\Delta_L(yr^{-1}) = \frac{(NLR_{boron}\ g/m^2{\cdot}day)\ (SA,\ m^2)\ (delta\ time,\ days)/years}{Mass\ _{initial}\ g} \qquad (7)$$

In the specific case of the modeling of the release of uranium from a vitrified waste in an E-Area trench, the PATHRAE Δ_L can be related to the approximate glass leaching. The final volume of vitrified waste is assumed to be ~100,000 gallons, or 380 m^3. Since the density of the final glass is ~2.6 x 10^6 g/m^3, then the final mass will be ~ 9.9 x 10^8 g.

The glass will be prepared as small glass ovoids, approximately 13 mm in diameter. The volume (V) of a single ovoid = $2/3\pi r^3$ = 5.75 x 10^{-7} m^3. The surface area of a single ovoid = $3\pi r^2$ = 3.98 x 10^{-4} m^2. The number of ovoids can be estimated by dividing the total volume V by the volume per ovoid = 380 m^3/ 5.75 x 10^{-7} m^3. = ~6.6 x 10^8 ovoids. The surface area (SA) of the vitrified waste will be ~6.6 x 10^8 x 3.98 x 10^{-4} m^2 = 2.6 x 10^5 m^2.

To relate Δ_L to the leach rate calculations for this case, substitute the M-Area parameters in equation 7;

$$\Delta_L(yr^{-1}) = \frac{(NLR_{boron}\ g/m^2{\cdot}day)\ (SA,\ m^2)\ (delta\ time,\ days)/\ yr}{Mass_{\ initial}\ g} \qquad (7)$$

or, rearranging,

$$NLR_{boron}\ g/m^2{\cdot}yr = \frac{\Delta_L\ yr^{-1}\ (Mass_{\ initial}\ g)}{(SA,\ m^2)} \qquad (8)$$

NLR_{boron} = Average bulk corrosion rate, $g/m^2{\cdot}yr$
SA = initial effective area = $2.6 \times 10^5\ m^2$,
M_{glass} = initial mass of glass = 9.9×10^8 g

$$NLR_{boron}\ g/m^2{\cdot}yr = \frac{\Delta_L\ yr^{-1}\ (9.9 \times 10^8\ g)}{(2.6 \times 10^5\ m^2)} = \Delta_L\ yr^{-1}\ 3.8 \times 10^3\ g/m^2 \qquad (9)$$

Assuming $NLR_{uranium} = NLR_{\ boron}$, then

$$NLR_{uranium} = \sim 3800\ \Delta_L\ yr^{-1} g/m^2 \qquad (10)$$

This is a direct relationship between the leaching of the M-Area glass beads and the PATHRAE fractional leach rate parameter. The PATHRAE code was used previously to calculate a fractional release rate $\Delta_L = 2.65 \times 10^{-6}\ yr^{-1}$, giving a drinking water concentration of 20 ppb. Substituting this into equation 10 gives:

$$NLR_u = (3800\ g/m^2)(2.65 \times 10^{-6}\ yr^{-1}) = 0.010\ g/m^2{\cdot}yr = 2.7 \times 10^{-5}\ g/m^2{\cdot}day \quad (11)$$

Therefore, the normalized average bulk glass uranium release rate ($NLR_{uranium}$) that would meet the 20 ppb drinking water standard is $\sim 2.7 \times 10^{-5}\ g/m^2{\cdot}day$.

ACCEPTABILITY OF M-AREA GLASS LEACH RATE

Now that a normalized uranium leach rate has been calculated for the vitrified M-Area waste, it can be determined whether the actual waste will met this level. Three methods can be used:

1) comparison to natural glasses such as obsidian and tektites,
2) comparison to other glasses using standard leaching procedures, and
3) long term leaching studies with actual M-Area glasses.

Natural Glasses

The first method is appropriate to obtain a rough estimate of whether the M-Area waste glass can meet the 2.7 x 10^{-5} g/m^2·day leach rate, under ambient conditions. Jantzen has compared the durability of natural glasses (10^5 to 10^6 years) to ancient man-made glasses (10^3 years)[18] and found that these glasses bounded the expected durability of nuclear waste glasses. Advanced waste glasses should approach 10^6 years in a repository environment. Since the PATHRAE code calculations indicated that a release of 2.65% per 10,000 years would meet the 20 ppb drinking water limit, a total release would occur in 3.8 x10^5 years. This is consistent with the durability of nuclear waste glasses.

Standard Leaching Test Results

A number of long term static leach test results were collected from the literature. Twenty-nine studies, from seven different references, were utilized to form a reference database [20-26]. All the tests were conducted at 90°C, with a variety of leachants and SA/V ratios. The times ranged from 84 to 920 days, at a minimum SA/V of 340 m^{-1}. An average long term normalized leach rate for boron was determined from each of the 29 studies. Using the Arrhenius equation, the $NLR_{boron\ 90°C}$ data was then corrected to an ambient temperature of 25°C. An activation energy of 75 kJ/mole for the Arrhenius calculation was selected from the average of the range activation energies of 60-90 kJ presented in Reference 19. The temperature corrected leaching data ranged from 1.0 x 10^{-6} to 1.6 x 10^{-4} g/m^2·day. A lognormal probability plot (Figure 1) showed that seventy % of the glasses from the 29 different reference glasses would satisfy the calculated leach rate of 2.7 x 10^{-5} g/m^2·day needed to meet the 20 ppb groundwater concentration.

CONCLUSION

The allowable leach rate needed to meet the drinking water standard for uranium at a groundwater well downgradient of an E-Vault trench is well within the durability attained by modern nuclear waste glass formulations. Empirical leaching studies to confirm that actual M-Area glasses will meet this standard are now underway.

ACKNOWLEDGMENTS

This presentation was prepared in connection with work conducted under DOE Contract No. DE-ACO9-89SR18035, and a S. C. Universities Research and Education Foundation under Department of Energy contract DEFCO993SR-18262. The DOE's Mixed Waste Focus Area, EM-50 also contributed to the funding. The work discussed in this paper is based primarily on work conducted by A. Ryan Whited to prepare a thesis for presentation to the Graduate School of Clemson University, in partial fulfillment of a Master's Degree.

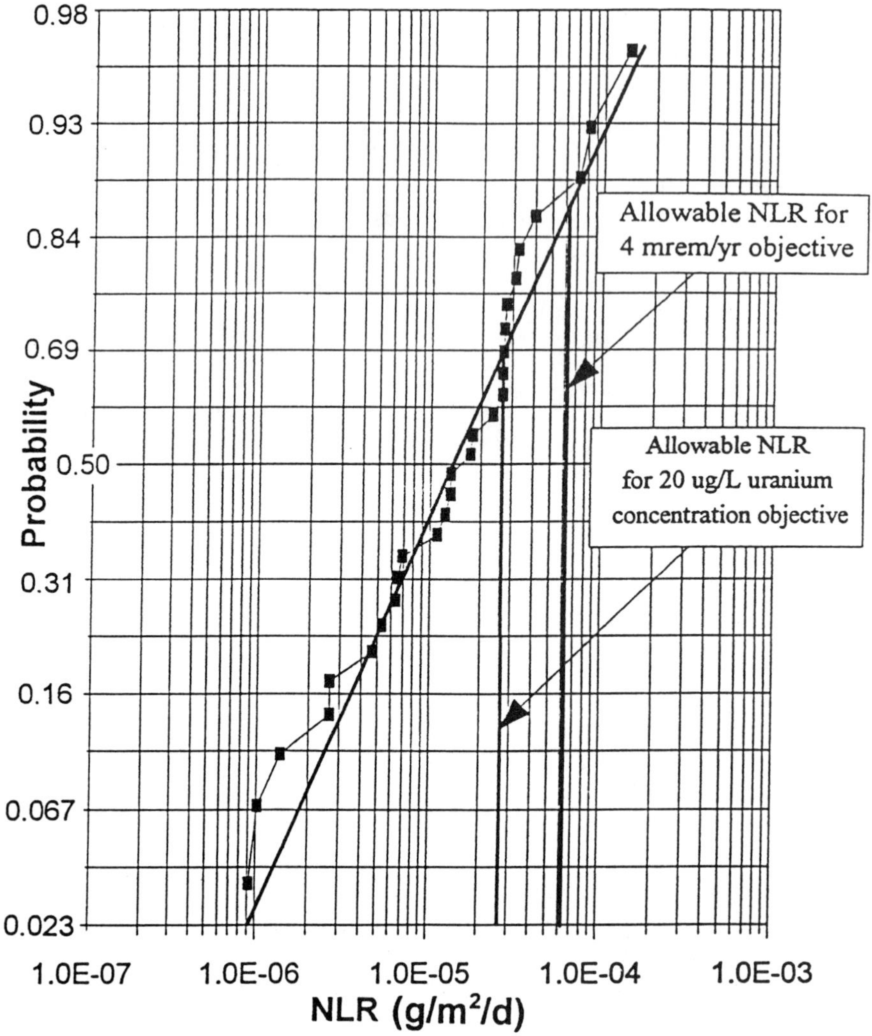

Lognormal probability plot for long-term leach rates with allowable leach rates from performance assessment calculations.

REFERENCES

1. U. S. EPA, "Land Disposal Restrictions for Third Scheduled Wastes, Final Rule", *55 Federal Register 22627* (1990).

2. J. B. Pickett, "Upfront Delisting Petition for Vitrified M-Area Plating Line Wastes", WSRC-TR-96-0244, U. S. Department of Energy, Savannah River Site, Aiken, SC 29898, 1996.

3."Radiological Performance Assessment for the E-Vaults Disposal Facility", WSRC-RP-94-218, Rev. 0, WSRC, Aiken, SC 29898.

4. U. S. EPA, "40 CFR Parts 141 and 142, National Primary Drinking Water Regulations; Radionuclides, Proposed Rule", *56 Federal Register 33050 (1991*).

5. U. S. DOE, "Final Environmental Impact Statement, Waste Management Activities for Groundwater Protection, Savannah River Site, Aiken, SC", DOE/EIS-0120, (1987).

6. B. B. Looney, C. M. King, and D. E. Stephenson, "Environmental Information Document, Quality Assurance Program for Environmental Assessment of Savannah River Plant Waste Sites", DPST-86-725, E. I. Dupont de Nemours & Co., Savannah River Laboratory, Aiken, SC 29808 (1986).

7. D. E. Stephenson, C. M. King, B. B. Looney, and M. W. Grant, "Environmental Information Document, Methodology for Predictive Modeling of Environmental Transport and Health Effects for Waste Sites at the Savannah River Plant ", DPST-86-710, E. I. Dupont de Nemours & Co., Savannah River Laboratory, Aiken, SC 29808 (1986).

8. G. B. Merrell, V. C. Rogers, K. K. Nielson, and M. W. Grant, "The PATHRAE Performance Assessment Code for the Land Disposal of Radioactive Wastes", Report to the U. S. EPA, RAE-8469/3, Rogers and Associates Engineering Corp., Salt Lake City, UT (1985).

9. A. Ryan Whited, "Vitrified Waste Form Performance Modeling Applied to the Disposal of a Mixed Waste Sludge at the Savannah River Site", A Masters Degree Thesis Presented to the Graduate School of Clemson University, Clemson SC 29643 (1996).

10. M. I. Sheppard and D. H. Thebault, "Default Soil/Liquid Partition Coefficients, K_ds, for 4 Major Soil Types: A Compendium", *Health Physics* **59** [4], 471 (1990).

11. W. L. Bourcier "Critical Review of Glass Performance Modeling", ANL-94/17, Argonne National Laboratory, Chicago, IL. (1994).

12. J. C. Cunnane and J. M. Allison, "High Level Waste Glass Compendium; What is Tells Us Concerning the Durability of Borosilicate Waste Glass", *Mat. Res. Soc. Symp. Proceedings*, **333**, 3 (1994).

13. X. Feng, "Surface Layer Effects on Waste Corrosion", *Mat. Res. Soc. Symp. Proceedings*, **333**, 55 (1994).

14. W. L. Ebert, J. K. Bates, E. C. Buck, and C. R. Bradley, "Accelerated Glass Reaction Under PCT Conditions", *Mat. Res. Soc. Symp. Proceedings*, **294**, 569 (1993).

15. A. J. G. Ellison, J. J. Mazer, and W. L. Ebert, "Effect of Glass Composition on Waste Form Durability: A Critical Review", ANL-94/28, Argonne National Laboratory, Chicago, IL. (1994).

16. B. Grambow, "Remaining Uncertainties in Predicting Long Term Performance of Nuclear Waste Glass From Experiments", *Mat. Res. Soc. Symp. Proceedings*, **333**, 167 (1994).

17. L. R. Pederson, C. Q. Buckwalter, and G. L. McVay, "The Effects of Surface Area to Solution Volume on Waste Glass Leaching", Nuclear Technology, 62, 151 (1983).

18. C. M. Jantzen, "Prediction of Nuclear Waste Glass Durability from Natural Analogs", *Adv. Ceram.* **20**, *Nuclear Waste Management II*, D. E Clark, (Ed), American Ceramic Society, Westerville, OH (1986).

19. W. B. White, "Dissolution Mechanisms of Nuclear Waste Glasses: A Critical Review", *Adv. Ceram.* **20**, *Nuclear Waste Management II*, D. E Clark, (Ed), American Ceramic Society, Westerville, OH (1986).

20. K. Lemmens and P. V. Iseghem, "The Long-Term Dissolution Behavior of the Pamela Borosilicate Glass SM527-Application of SA/V as Accelerating Parameter", *Mat. Res. Symp. Proc.*, 257, **47** (1992).

21. S. Fillet, E. Vernaz, J. L. Nogues, and N. Jacquet-Francillion, "Corrosion Rate of Nuclear Glasses in Saturated Media", *Adv. Ceram.*, **20**, *Nuclear Waste Management II*, D. E. Clark (Ed.), Am. Ceram.. Soc., Westerville, OH (1986)

22. E. Vernaz, T. Advocat, and J. L. Dussossoy, "Effects of SA/V Ratio on the Long-Term Corrosion Kinetics of R7-T7 Glass", *Ceram. Trans.*, **9**, *Nuclear Waste Management III*, J. G. Moore, (Ed.), Westerville, OH (1990).

23. W. L. Ebert and J. K. Bates, "A Comparison of Glass Reaction at High and Low Glass Surface/Solution Volume", *Nuclear Technology*, **104**, 372, (1993).

24. X. Feng, J. K. Bates, E,. C. Buck, C. R. Bradley, and M. Gong, "Long-Term Comparison of Dissolution Behavior Between Fully Radioactive and Simulated Nuclear Waste Glasses", *Nuclear Technology*, **104**, 193, (1993).

25. J. K. Bates, *et. al.*, "ANL Technical Support Program for DOE Environmental Restoration and Waste Management, Annual Report, October 1992-September 1993", ANL-94/19, Argonne National Lab., Chicago, IL (1994).

26. W. G. Ramsey, "Glass Dissolution of the System Na_2O-B_2O_3-SiO_2-Al_2O_3-Fe_2O_3-CaO", Doctoral Dissertation, Clemson Un., Clemson, SC (1995)

WASTE VITRIFICATION PROJECTS THROUGHOUT THE U.S. INITIATED AT THE SAVANNAH RIVER SITE

C.M. Jantzen, J.C. Whitehouse, M.E. Smith, J.B. Pickett, and D.K. Peeler
Westinghouse Savannah River Company
Aiken, South Carolina 29803

ABSTRACT

Technologies are being developed by the US Department of Energy's (DOE) Savannah River Site (SRS) to convert low-level and mixed (radioactive and hazardous) wastes, radioactively contaminated asbestos containing material (ACM), and plutonium to solid stabilized waste forms for permanent disposal. The low-level and mixed wastes include wastewater treatment sludges, spent filter aids, incinerator ash, incinerator blowdown, cementitious waste forms in need of remediation and radioactive mill tailings from throughout the DOE complex. Disposal of ion-exchange (IEX) resins and sludges from commercial nuclear reactors is being examined jointly by SRS and the Electric Power Research Institute (EPRI) via a Cooperative Research and Development Agreement (CRADA). Disposal of weapons grade and scrap plutonium from the DOE complex for long-term storage and disposition is also being actively pursued. In addition, SRS is developing technologies to recycle (1) SRS americium (Am) and curium (Cm) wastes for reuse in the Oak Ridge Reservation (ORR) medical source program and (2) non-radioactive asbestos containing materials (ACM). The technologies center around vitrification via electric melting, either Joule† heated or bushing melters, which melt the waste/recycle material with glass formers at elevated temperatures. Vitrification of this wide variety of wastes into glass is an attractive option because it atomistically bonds the hazardous and radioactive species in a solid glassy matrix, and reduces waste volume by up to 97%. The large volume reductions allow for large associated savings in interim storage, shipping, and/or long term disposal costs. To date, the results of these studies have led to two vendor privatizations and two "field-scale" demonstrations of Joule heated melting of actual mixed waste streams in the DOE complex.

INTRODUCTION

The US DOE Savannah River Site (SRS), operated by Westinghouse Savannah River Company (WSRC), is currently investigating vitrification of various low-level and mixed wastes throughout the DOE complex and the commercial sector. Development of "cradle-to-grave" vitrification processes have been initiated at SRS for a wide variety of wastes, which include, but are not limited to, the following:

† Joule-heated melters vitrify waste in a refractory-lined vessel containing diametrically opposed electrodes. The electrodes are used to heat the glass by passing an electric current through the material. The process is called Joule heating.

- spent filter aids from waste water treatment
- waste sludges
- combinations of spent filter aids from waste water treatment and sludges
- combinations of supernate and waste sludges
- incinerator ash, incinerator off-gas blowdown
- combinations of incinerator ash and off-gas blowdown
- cement formulations in need of remediation
- ion exchange resins and zeolites
- soils, geologic material/media
- asbestos containing material (ACM) or inorganic fiber filter media
- radioactive materials including transuranic (TRU), plutonium (Pu), and other actinide wastes, e.g. Am and Cm

Mixed low level waste (MLLW)[1] in any of the above categories must meet the regulatory requirements imposed on hazardous waste by the Resource Conservation and Recovery Act (RCRA) and the regulatory requirements imposed on radioactive species governed by the US Department of Energy (DOE) orders or Atomic Energy Act (AEA) regulations. The need to provide MLLW treatment has been driven by the RCRA Land Disposal Restrictions (LDR) that require the treatment of the existing MLLW stockpiles. As of 1992, the MLLW waste volumes were ~250,000 m^3 and projected to increase to 1,200,000 m^3 in by 1997 [1]. A schedule for DOE to come into compliance with RCRA was mandated by the passage of the Federal Facilities Compliance Act (FFCAct) of 1992. Large volumes of MLLW must, therefore, be converted to a solid stabilized waste form for permanent disposal.

Seventy six percent of the existing mixed wastes in the DOE complex are candidates for electric and/or Joule heated vitrification [1]. Several RCRA listed mixed (hazardous and radioactive) wastewater sludges at SRS [2-4] and ORR [5-6] have been identified as the first candidates for demonstration of Joule heated vitrification. Five vitrification treatability studies with real wastewater sludges, including sludges with up to 30 wt % organic, have been completed at the laboratory "proof-of-principle" scale.

Three vitrification "proof-of-principle" studies of simulated RCRA wastes have also been initiated at SRS. These include incinerator ash and blowdown from SRS [2,7], waste sludge from Rocky Flats admixed with Portland cement, and sludge from Los Alamos National Laboratory. "Proof-of-principle" vitrification studies of non-RCRA wastes have also been initiated by SRS. These include vitrification of mill tailings from the Fernald Environmental Management Project (FEMP) K-65 site, "clean" as well as radioactively contaminated asbestos containing material (ACM) from the DOE complex [9], some ion-exchange resins and sludges from commercial nuclear reactors [8], recycle of SRS americium and curium wastes to ORR for medical

---

1 waste that contains source, special nuclear, or byproduct material subject to regulation under the Atomic Energy Act and hazardous waste species subject to regulation under the Resource Conservation and Recovery Act waste as defined in 40 CFR 261 (U.S. Code Title 42, Section 2011); if the hazardous waste species are on the U.S. Environmental Protection Agency (EPA) list of hazardous wastes as outlined in 40 CFR261.31,.32,.33, then the waste is considered a "listed" MLLW.

applications [10-11], and vitrification of weapons grade and scrap plutonium [10-11] from the DOE complex.

RATIONALE FOR VITRIFICATION

Vitrification of wastes/recycle materials into glass is an attractive option because it atomistically bonds the hazardous and radioactive species in a solid glassy matrix. The waste forms produced are, therefore, very durable and environmentally stable over long time durations. Vitrification processes are flexible to process chemistry variations and can accommodate dry or wet waste/recycle materials. The Environmental Protection Agency (EPA) has declared vitrification the Best Demonstrated Available Technology (BDAT) for high-level radioactive waste [12] and produced a Handbook of Vitrification Technologies for Treatment of Hazardous and Radioactive Waste [13]. The DOE Office of Technology Development (OTD) has taken the position that mixed waste needs to be stabilized to the highest level reasonably possible to ensure that the resulting waste forms will meet both current and future regulatory specifications.

Vitrification produces large volume reductions, e.g. up to 97% [2] using cheap sources of glass formers, e.g. sand, soil, crushed scrap fluorescent bulbs, crushed reagent bottles, etc. A new generation of high throughput Joule heated melters, available from the commercial glass industry, allow for rapid vitrification of large volumes of MLLW waste. These vitrification units have been designed by SRS and EnVitco to be compact enough to be transportable, e.g. the SRS Transportable Vitrification System (TVS) [14,15]. This enables the Joule heated melter to be transported from waste site to waste site. Bushing melters with high throughput, as used in the commercial glass industry, are robust and compact enough to handle high throughput vitrification of TRU waste in glove box applications. Large reductions in volume minimize long-term storage costs, compact melter technology minimizes costs, making vitrification cost effective on a life cycle basis compared to other stabilization technologies.

PROTOCOL FOR DEVELOPMENT OF VITRIFICATION PROCESSES

Vitrification processes are being developed for commercialization and demonstration within the DOE complex. For each vitrification process the following "cradle-to-grave" protocol is followed:

- Analyze wastes
- Surrogate "proof-of-principle" laboratory scale studies (optional if actual waste is readily available)
- Actual waste "proof-of-principle" laboratory scale studies
- Surrogate "pilot-scale" demonstration (optional if actual waste is readily available)
- Actual waste "pilot-scale" demonstration
- Production scale ("field-scale or full-scale") testing of melter with surrogate waste (necessary for initial check-out of equipment, otherwise optional)

- Actual waste processing, e.g. "field-scale or full-scale" demonstration, processing, and/or commercialization

Throughout "proof-of-principle" laboratory scale testing, a systems approach to glass formulation and process optimization is applied. The systems approach simultaneously evaluates product performance and processing considerations [16-17]. Parameters affecting the product performance, such as chemical durability, are optimized relative to processing considerations such as volatility of hazardous species, melt viscosity, melt corrosivity, electrical resistivity, etc. The process/product models developed for HLLW and MLLW vitrification, allow this optimization [6-7] to be based on melter feed composition. "Proof-of-Principle" laboratory scale crucible testing is often performed with surrogates (1) in order to optimize glass product performance and processing considerations and/or (2) if the amount of waste available is limited. "Proof-of-Principle" laboratory scale crucible studies are necessary with actual waste whether or not a surrogate is available. "Proof-of-Principle" laboratory scale crucible testing should evaluate the following parameters:

- waste loading
- melt temperatures
- varying types of alkali additives
- varying types of silica additives, e.g. Reactive Additive Stabilization Process (RASP)[2] using high surface area sources of silica such as various filter aids, perlite, precipitated silica, rice husk ash, etc. vs conventional vitrification with granular sand, soil, scrap glass from light bulbs reagent bottles, etc.
- melt line refractory corrosion
- general refractory corrosion
- electrode corrosion
- determination of glass homogeneity, e.g. crystallization and/or phase separation
- waste form performance evaluation using the Environmental Protection Agency (EPA) Toxic Characteristic Leaching Procedure (TCLP) and/or the Product Consistency Test (PCT) developed for HLLW and MLLW waste glass durability testing (ASTM C1285-94).
- predictability of process/product models

If the glasses are either characteristically hazardous or listed wastes under RCRA, the "Proof-of-Principle" crucible testing can also include "doping studies." In these studies, the optimized glass composition is "doped" with excess RCRA metals at 20x, 50x, and 100x their nominal composition in the waste (or to a minimum of 10,000 ppm, whichever is larger).

---

2 Reactive Additive Stabilization Process (RASP), U.S. Patent 5, 434,333Reactive high surface area silica, used as a waste form additive, was determined to greatly enhance the solubility and retention of hazardous, mixed and heavy metal species in glass [4-6]. Vitrification using this Reactive Additive Stabilization Process (RASP)* was found to increase the solubility and tolerance of Soda(Na_2O)- Lime(CaO)-Silica(SiO_2) glass (SLS) to atomistically bond waste species. Highly reactive silica lowers glassification temperatures, increases waste loadings which provides for large waste volume reductions, minimizes melt line corrosion, and produces EPA acceptable glasses.

"Proof-of-scale-up" testing is necessary in a pilot scale melter. "Pilot-scale" testing with actual waste allows the following parameters, which cannot be assessed in crucible scale testing, to be determined:

- data on actual vitrified RCRA listed waste forms for "Upfront" Delisting Petitions
- confirmation of the processability of the glass compositions optimized in the "proof-of-principle" studies
- determination of off-gas emissions as a function of melt temperature
- verification of melter behavior as a Continuously Stirred Tank Reactor (CSTR) to ensure that waste and glass formers are homogenized during melting
- demonstration of recycle of secondary waste condensate produced
- demonstration of decontamination of melter between melting of different RCRA listed waste types by flushing with non-radioactive glass cullet
- demonstration of lack of cross contamination between waste campaigns of different types of wastes
- predictability of process/product models
- demonstration of decontamination of the off-gas system/condensate tank
- evaluation of melter refractory and electrode corrosion

This is the same protocol that was used to develop the vitrification process for High-Level Liquid Waste (HLLW) vitrification at SRS and at West Valley Fuel Services. Although development of the process for vitrification of HLLW took ~25 years to develop, the process for the M-Area waste sludges took ~7 years while the development of the ORR waste sludges was completed in ~1 year. Figure 1 shows the status of the various vitrification projects initiated by SRS.

VITRIFICATION DEVELOPMENT MATRIX

	DWPF HLLW	SRS M-AREA	ORR WETF	ORR K25	ORR CPCF	LANL SLUDGE	RF SLUDGE	FERNALD MILL TAILINGS	SRS CIF ASH	SRS ASBESTOS	SRS Am/Cm	EPRI RESINS	SRS Pu
Analyze Waste	■	■	■	■	■	■	■	■	■	■	■	■	■
Surrogate "Proof of Principle"	■	■	■	■	■	■	■	■	■	■	■	□	■
Actual Waste "Proof of Principle"	■	■	■	■	■						□		□
Surrogate Pilot Scale Testing	■	■	■	■		■	■		■				□
Actual Waste Pilot Scale Testing	■	■	V	2/96									
Prod. Integrated Testing	■	[V]	V	□									
Waste Processing	2/96	V	V	5/96									

■ COMPLETED BY SAVANNAH RIVER SITE

□ IN PROGRESS BY SAVANNAH RIVER SITE

V VENDOR PRIVATIZED

Figure 1. Savannah River Site Vitrification Initiative Development Matrix.

VITRIFICATION INITIATIVES BY WASTE CATEGORY

A. RCRA LISTED WASTE SLUDGES (SOMETIMES ADMIXED WITH SPENT FILTER AIDS, SOILS, AND/OR CEMENTS)

SRS M-Area Sludge + Spent Filter Aid - 3,000,000 kgs

- Analyze wastes (SRS) - high SiO_2 (~45 wt%), Al_2O_3 (~20 wt% as $Al(OH)_3$), $NaNO_3$ (~20 wt%) RCRA listed F006 nickel plating line sludge mixed with spent filter aid, Ni at ~ 1.2 wt% is the primary hazardous constituent, while the prime radioactive constituent is ~4.2 wt% U [2]
- Actual waste "proof-of-principle" studies (SRS) - 44 glass formulations in alkali borosilicate (ABS)[3] and alkali-lime-silica (ALS)[3] systems; waste loadings between 70-90 wt%; melt temperatures between 1150-1350°C; varied composition of waste from high alkali to high silica; one to three glass forming additives; volume reductions of 86-88%; all glasses passed [2-4] the TCLP Land Disposal Restriction Universal Treatment Standards [UTS,18]
- Surrogate "pilot-scale" demonstration (SRS/Clemson DOE/Industry Waste Vitrification Center) - 6 glass formulations in sodium borosilicate (SBS) system; one glass forming additive; waste loadings between 70-95 wt%; melt temperatures 1150-1500°C, all glasses passed TCLP LDR/UTS limits [19]
- Actual waste "pilot-scale" demonstration (SRS) - 2 glass formulations in ABS[3] system; waste loadings of 80 wt%; processed 400 kgs of waste; all glasses passed TCLP LDR/UTS limits; TCLP and Multiple Extraction Procedure (MEP) data to be used for "Up-Front Delisting" Petition [20]; first Delisting Petition in the DOE complex for vitrified mixed waste
- Production/integrated "full-scale" testing (Duratek) - first commercial vitrification of MLLW in DOE complex; design and construction complete January 1996, simulant testing scheduled March, 1996 [21]
- Actual "full-scale" waste processing (Duratek) - fixed unit price treatment contract; all construction and operations costs borne by sub-contractor until waste treated to meet delisting standards; treatment of M-Area wastes scheduled to begin April 1996 [21]

ORR West End Treatment Facility (WETF); ~ 8,000,000 kgs

- Analyze wastes (ORR) - high CaO (60-75 wt% on a dry oxide basis from $CaCO_3$), and Fe_2O_3 (2-10 wt% from FeOOH) RCRA listed F001, F002, F005 (from treatment of solvent residues), and F006 (from treatment of

---

3 Lithium oxide was used preferentially over sodium as a glass forming flux additive and various silica sources were investigated since recent studies had shown that the known glass forming region in the SLS system could be expanded using reactive sources of SiO_2 and or reactive fluxes like Li_2O [2-4,7, U.S. Patent 5,434,333]. In addition, ORR was known to have a large surplus of suspect contaminated LiOH [24] which could be used as a glass formers during waste stabilization.

plating waste) wastes; nickel (~0.25 wt% is the primary hazardous species of concern while U at ~0.42 wt% and traces of Tc^{99} and TRU (Np^{237}, etc.) are the radioactive species of concern [22]

- Surrogate waste "proof-of-principle" studies (SRS) - 120 glass formulations in the ALS^3 system with Tank 7 surrogate waste [22]; waste loadings of 20-70 wt%, melt temperatures between 1150-1350°C; no more than three glass forming additives; severe melt line and general refractory corrosion at high waste loadings and high temperatures; sources of alkali and silica varied; glass viscosity vs temperature studied; all glasses passed TCLP LDR/UTS limits and PCT durability testing [5].
- Actual waste "proof-of-principle" studies (SRS/ORR) - ~30 glass formulations with Tank 8 waste, and ~30 glass formulations with Tank 13 waste in the ALS^3 system due to the large known immiscibility gap in the CaO-B_2O_3-SiO_2 system [23] where glasses are known to phase separate (form immiscible liquid phases); waste loadings between 20-70 wt%; melt temperatures between 1150-1350°C; no more than three glass forming additives; volume reductions of 73-87%; sources of alkali and silica varied; glasses doped at with excess RCRA metals at 20x, 50x, and 100x their nominal composition in the waste (or to a minimum of 10,000 ppm, whichever is larger); all glasses passed TCLP LDR/UTS
- Surrogate "pilot-scale" demonstration (SRS/Clemson DOE/Industry Waste Vitrification Center) - 2 glass formulations in ALS system; three glass forming additives; waste loadings 20-40 wt%; melt temperature 1050-1350°C; 20 wt% glass passed TCLP LDR/UTS limits [25]; 40 wt% glass crystallized in the canister but passed TCLP
- Actual waste "pilot-scale" demonstration - vendor privatized by ORR
- Production/integrated "full-scale" testing - vendor privatized by ORR
- Actual "full-scale" waste processing - vendor privatized by ORR

ORR K-25 B&C Pond Waste; ~ 16,000,000 kgs

- Analyze wastes (ORR) - high SiO_2 (wt%) and CaO (~25 wt% from $Ca(OH)_2$) sludge with Fe_2O_3 (~16 wt%) from admixed clay basin liner, RCRA listed mixed F006 wastes derived from plating line activities, Ag and Ni (~0.51 wt%) are primary hazardous components, ~0.30 wt% U is the primary radioactive constituent, trace concentrations of Tc^{99} [22]; the relative proportions of SiO_2, $Ca(OH)_2$ and Fe_2O_3 vary greatly from drum to drum since clean RCRA closure of the basins in 1988-89 involved intermixing pond sludge with dredged clay pond liner and some partially successful stabilization efforts with Portland cement.
- Surrogate waste "proof-of-principle" studies (SRS) - 120 crucible studies with surrogate formulations containing high Fe_2O_3, high SiO_2 and intermediate compositions in the ALS^3 system; waste loadings of 40-90 wt%, melt temperatures between 1150-1350°C; a maximum of three glass forming additives; sources of alkali and silica varied; general refractory corrosion studied, PO_4 solubility studied, glass viscosity vs. composition studied; all glasses passed TCLP LDR/UTS limits and PCT durability testing; crystallization and liquidus vs. composition studied.

- Actual waste "proof-of-principle" studies (SRS/ORR) - 70 glass formulations with waste from the rotary drier used in K25 B/C pond remediation efforts in 1991-92 in the ALS[3] system; waste loadings between 40-90 wt%; melt temperatures between 1150-1350°C; no more than three glass forming additives; volume reductions of 70-90%; sources of alkali and silica varied; glasses doped at with excess RCRA metals at 20x, 50x, and 100x their nominal composition in the waste (or to a minimum of 10,000 ppm, whichever is larger); all glasses passed TCLP LDR/UTS
- Surrogate "pilot-scale" demonstration (SRS/Clemson DOE/Industry Waste Vitrification Center) - high SiO_2 B/C simulant developed by SRS; 1 glass formulation in ALS[3] system; three glass forming additives; waste loading 50 wt%; melt temperature 1250°C; glass passed TCLP LDR/UTS limits and PCT testing
- Actual waste "pilot-scale" demonstration (SRS/RUST Federal Services[4])- planned February, 1996, 400 kgs of dried B/C Pond sludge
- Production/integrated "field-scale" testing (SRS) - high SiO_2 B/C simulant developed by SRS; SRS Transportable Vitrification System (TVS); waste loading 50 wt%; melt temperature 1150°C; in progress
- Actual "field-scale" waste processing (SRS) - planned April, 1996 at ORR K-25 site in the SRS TVS

ORR Central Pollution Control Facility (CPCF); ~ 186,200 kgs

- Analyze wastes (ORR) - There are three categories of CPCF wastes: oily, wet non-oily, and dry non-oily; oily RCRA listed F006 CPCF plating line sludges containing 20-30% organics and ~0.50 wt% U and 0.2 wt% Ni. The CPCF wastes are listed RCRA waste codes D007, F003, F007 and U210. The oily CPCF wastes studied are high in SiO_2 (~50% on a dry oxide basis), ~4 wt% CaO as $Ca(OH)_2$, ~12 wt% Fe_2O_3 from FeOOH, and ~30-40 wt% organics.
- Actual waste "proof-of-principle" studies (SRS/ORR) - 30 glass formulations with oily CPCF waste in the alkali-lime-silica (ALS)[3] system and 3 glass formulations in the ABS system; waste loadings between 70-90 wt%; melt temperatures between 1150-1350°C; three glass forming additives; volume reductions of 85-90% sources of alkali and silica varied; organics driven off with slowly with slow heat up ramps; all glasses in ALS system passed TCLP LDR/UTS; 3 glass formulations in ABS system formed immiscible phases between the waste and the glass formers due to the large known immiscibility gap in the $CaO\text{-}B_2O_3\text{-}SiO_2$ system [23]
- Actual waste "pilot-scale" demonstration (SRS/RUST Federal Services[4]) - under consideration if a pretreatment technique can be developed to destroy the 30-40% organics before vitrification (the maximum safe

---

4 Under a Cooperative Research and Development Agreement between Westinghouse Savannah River Co. and Rust Federal Services (RFS).

organic content for a Joule heated melter is <10 wt% organics); pretreatment options such as solvent extraction, wet oxidation, or incineration are being investigated.

ORR Central Neutralization Facility (CNF); ~ 900,000 kgs

- Analyze wastes (ORR) - CNF wastes are listed RCRA F and U waste codes resulting primarily from the treatment of ORR TSCA incinerator scrubber blowdown solution.
- Surrogate waste "proof-of-principle" studies (SRS) - planned during 1996
- Actual waste "proof-of-principle" studies (SRS/ORR) - planned during 1996
- Surrogate "pilot-scale" demonstration (SRS/Clemson DOE/Industry Waste Vitrification Center) - planned during 1996
- Actual waste "pilot-scale" demonstration (SRS/RUST Federal Services[4]) - planned during 1996, 600 kgs of wet sludge
- Actual "field-scale" waste processing (SRS) - planned 1996 at ORR K-25 with SRS TVS

Los Alamos National Laboratory (LANL) Liquid Waste Processing Plant; ~ 324,000 kgs

- Analyze wastes (ORR) - high CaO (50 wt% on an oxide basis) as $CaCO_3$ from processing and admixed Portland cement and gypsum, high SiO_2 (38 wt% from perlite and diatomaceous earth), and Fe_2O_3 (8 wt% from FeOOH) RCRA listed F001, F002, F005A (from treatment of solvent residues); U at ~0.23 wt% and traces of Pu^{239} and Am^{243} are the radioactive species of concern and the hazardous species of concern are not well documented except for Cd [22]
- Surrogate waste "proof-of-principle" studies (SRS) - 19 glass formulations of which 7 were non-crystalline (5 in the ABS system, 3 in the CaO-Al_2O_3-SiO_2 (CAS) system, 3 in the CaO-Fe_2O_3-SiO_2 (CFS) system, 3 in the SLS [26]); waste loadings of 25-75 wt%, melt temperatures between 1150-1500°C; two glass forming additives; severe crystallization noted in certain composition regions in all systems; least amount of crystallization in SLS glasses; glasses doped with Ba, Cd, Cr and Ni; all glasses passed TCLP LDR/UTS limits and PCT durability testing [26].
- Surrogate "pilot-scale" demonstration (SRS/Clemson DOE/Industry Waste Vitrification Center) - 1 glass formulation at 65 wt% loading in the CAS system at 1350°C, the glass was difficult to pour due to high viscosity, TCLP and PCT testing in progress

Rocky Flats By-Pass Sludge

- Analyze wastes (ORR) - high Fe_2O_3 (~36 wt% on a dry oxide basis from $Fe(OH)_3$), CaO (~25 wt% from $CaSO_4$), and Na_2O (~12 wt% as oxide from $NaNO_3$) leaving about 12 wt% NO_x and >20 wt% SO_4 ; some waste admixed with up to 30% Portland cement; RCRA listed F006 (nickel plating line waste), D006, D007, D008, D011 for Cd, Cr, Pb, Ag, and Ni hazardous species; Pu as primary radioactive species of concern [22]

- Surrogate waste "proof-of-principle" studies (SRS) - 10 glass formulations in the SBS system were melted but all crystallized and produced a sulfate layer which floated on the top of the melt; waste loadings of 25-75 wt% at melt temperatures between 950-1050°C did not get rid of the sulfate layer; charcoal was added to 65 and 75 wt% waste loaded glasses which deterred the formation of the sulfate layer and the crystallization. two glass forming additives plus charcoal; the homogeneous glasses passed TCLP LDR/UTS limits and PCT durability testing [26].
- Surrogate "pilot-scale" demonstration (SRS/Clemson DOE/Industry Waste Vitrification Center) - 1 glass formulation at 75 wt% loading in the SBS system; melt temperature of 1350°C caused

B. INCINERATOR WASTES (ASH AND/OR OFF-GAS BLOWDOWN)

SRS Consolidated Incinerator Facility (CIF); ~ 800 m^3/year blowdown and 124 m^3/year ash generation for 25+ years

- Simulate wastes (SRS) - high Na_2O (~65 wt% on a dry oxide basis in blowdown primarily from NaCl, ~50 wt% as oxide in a mixture of 68 wt% blowdown and 32 wt% bottom ash), high CaO (~35 wt% on a dry oxide basis in blowdown from ash carryover, ~50% as oxide in the reference mixture of blowdown previously given) [2,7]; RCRA listed F, D, and U wastes for all the RCRA metals of concern and Zn; radioactive constituents include Cr^{51}, Sr^{90}, Cs^{137}, traces of Pu
- Surrogate waste "proof-of-principle" studies (SRS) - 20 glass formulations in the ALS[3] system surrogate waste [2,7]; waste loadings of 30-50 wt%; melt temperatures between 1150-1250°C to avoid volatilization of hazardous species as chlorides; 94-97% volume reduction; one glass forming additive, SiO_2; sources of alkali and silica varied; all glasses passed TCLP LDR/UTS limits; pyrohydrolysis investigated to remove Cl as HCl plus steam [2,7].

C. ION EXCHANGE RESINS AND ZEOLITES

Electric Power Research Institute (EPRI) Cooperative Research and Development Agreement (CRADA) ~220,000 kgs Boiling Water Reactor (BWR) and 66,000 kgs Pressurized Water Reactor (PWR) Ion Exchange (IEX) resins per reactor per year

- Analyze wastes (SRS/EPRI) - samples of six ion exchange resins from EPRI undergoing wet chemical analysis for cationic and anionic species; undergoing Differential Thermal Analysis (DTA) with coupled mass spectrometry to identify inorganic and organic volatile components

- Surrogate waste "proof-of-principle" studies (SRS) - preliminary data indicates 50 wt% waste loading for PWR wastes which gives a 77% volume reduction and a 35 wt% waste loading for BWR wastes which gives a 66% volume reduction [8]
- Surrogate "pilot-scale" demonstration (SRS) - planned in late 1996

D. SOILS, GEOLOGIC MATERIAL/MEDIA

Fernald Environmental Management Project (FEMP) K-65 silos of depleted uranium (mill tailings from processing pitchblende ore) for atomic bomb development ~10,000,000 kgs (~10,000 metric tons)

- Analyze wastes [27] - Residues from processing pitchblende ores from 1949-1958, high in SiO_2 (~63 wt%), BaO (~6.5 wt%), Pb (~12.5 wt%), and Fe (~5 wt%).
- Surrogate waste "proof-of-principle" studies (SRS) - 2 glass formulations in the ALS^3 system; waste loadings of 80-90 wt%, melt temperature 1050°C; two glass forming additives; both glasses passed TCLP limits.

E. ASBESTOS AND/OR GLASS FIBER FILTERS (UNCONTAMINATED OR CONTAMINATED)

Decommissioning and Decontamination (D&D) throughout the DOE, DOD, and commercial sectors

- Analyze wastes (SRS) - analysis of asbestos coated pipe indicates that asbestos containing materials (ACM) are admixed with up to 50 wt% $MgCO_3$ and/or $CaSO_4$ as cementitious binder material
- Surrogate waste "proof-of-principle" studies (SRS) - use of patented [9] caustic dissolution process to remove ACM from adhering pipe; allows pipe or other adhering metal to be sold/recycled; 10 glass formulations of high Mg silicate glasses render ACM non-crystalline and non-hazardous; waste loadings of 30-60 wt%; melt temperatures between 1150-1350°C; volume reductions of 99% for asbestos covered pipe; non-contaminated glass can be sold for recycle

F. RADIOACTIVE MATERIALS INCLUDING TRANSURANIC (TRU), PLUTONIUM (Pu), AND OTHER ACTINIDE WASTES

Pu - 50,000 kgs (50 metric tons) of excess fissile materials from nuclear weapons and excess residues in the DOE complex left from the production of these weapons

- Analyze wastes (DOE Complex) - includes pit, metal, oxide, scrap, compounds, residue wastes and fuel targets (irradiated and unirradiated)
- Surrogate waste "proof-of-principle" studies (SRS) - borosilicate and iron phosphate glasses using ThO_2

- Actual waste "proof-of-principle" studies (SRS) - in progress with PuO_2 nitrates and oxides; waste loadings up to 13/15 wt% in borosilicate on a nitrate/oxide basis and 25 wt% in iron phosphate glass
- Surrogate waste "pilot-scale" demonstration (SRS) - "can-in-canister" demonstration completed using CeO_2 [11]

Am/Cm - 15,000 liters to be stabilized in glass for shipment to ORR for reuse as medical target sources [11].

- Analyze wastes (SRS) - dilute 4N nitric acid solution containing approximately 10.1 kgs Am and 2.7 kgs Cm
- Surrogate waste "proof-of-principle" studies (SRS) - 5-45 wt% oxides using La and Nd nitrates so that glass can be safely shipped to ORR as a solid for their Isotope Sales Program; SRS waste reclaimed as a source of revenue for DOE complex; full (100%) recovery of La and Nd from glass demonstrated by nitric acid extraction; >90% volume reduction
- Actual waste "proof-of-principle" studies (SRS) - in progress

CONCLUSIONS

Electric heated vitrification is a viable option for a large variety of wastes in the DOE complex.

REFERENCES

1. J.B. Berry, "Mixed Waste Integrated Program-Demonstration Technologies to Meet the Requirements of the Federal Facility Compliance Act, Spectrum '94 Nuclear and Hazardous Waste Management International Topical Meeting Proceedings, 249-254 (1994).

2. C.M. Jantzen, J.B. Pickett, and W.G. Ramsey, "Reactive Additive Stabilization Process (RASP) for Hazardous and Mixed Waste Vitrification,". Proceed. Second Inter. Symp. on Mixed Waste, A.A. Moghissi, R.K. Blauvelt, G.A. Benda, and N.E. Rothermich (Eds.), Am. Soc. Mech. Eng., p.4.2.1 to 4.2.13 (1993).

3. C.M. Jantzen, J.B. Pickett, J.B. and H,.L. Martin, H.L., "Reactive Additive Stabilization Process (RASP): Vitrification of Nickel Plating Line Wastes," Proceedings of the American Chemical Society Symposium on Emerging Technologies for Hazardous Waste Management (September, 1993).

4. C.M. Jantzen, J.B. Pickett, W.G. Ramsey and D.C. Beam, "Treatability Studies on Mixed (Radioactive and Hazardous) M-Area F006 Waste Sludge: Vitrification Via the Reactive Additive Stabilization Process (RASP)," Proceedings of Spectrum 94 Nuclear and Hazardous Waste Management International Topical Meeting, Am. Nuclear Soc., 737-742 (1994).

5. C.M. Jantzen, N.D. Hutson, T.M. Gilliam, A. Bleier, and R.D. Spence, "Vitrification Treatability Studies of Actual Waste Water Treatment Sludges," WM95.

6. C.M. Jantzen and T.M. Gilliam, "Technical Task Plan: SRTC/ORNL Joint Treatability Studies for Vitrification of Low-Level Mixed Wastes at the Oak Ridge Reservation, " U.S. DOE Report WSRC-TR-95-0330, Rev. 0, Westinghouse Savannah River Co., Aiken, SC (August, 1995).

7. C.M. Jantzen, D.F. Bickford, and W.G. Ramsey, "Reactive Additive Stabilization Process (RASP): Vitrification of Mixed Waste Incinerator Ash and Blowdown," Proceedings of the American Chemical Society Symposium on Emerging Technologies for Hazardous Waste Management (September, 1993).

8. C.M. Jantzen, D.K. Peeler, and C.A. Cicero, "Vitrification of Ion-Exchange (IEX) Resins: Advantages and Technical Challenges," DOE Report WSRC-MS-95-0518, Rev. 0, Westinghouse Savannah River Co., Aiken, SC (December, 1995).

9. C.M. Jantzen, "Method for Dissolution and Stabilization of Silica-Rich Fibers," patent S-72504 pending.

10. W.G. Ramsey, N.E. Bibler, and T.F. Meaker, "Compositions and Durabilities of Glasses for Immobilization of Plutonium and Uranium," Waste Management 95 (1995).

11. W.G. Ramsey, R.F. Schumacher, R.L. Minichan, R.H. Jones, B.J. Hardy, D.H. Miller, A.S. Choi, C.J. Coleman, and T.F. Meaker, "Vitrification of F-Area Americium/Curium: Feasibility Study and Preliminary Process Recommendation," U.S. DOE Report WSRC-TR-94-0211, Rev. 0, Westinghouse Savannah River Co., Aiken, SC (April, 1994).

12. Federal Register, "Land Disposal Restrictions for Third Third Scheduled Wastes, Final Rule," 55 FR22627 (June 1, 1990).

13. U.S. Environmental Protection Agency, "Handbook: Vitrification Technologies for Treatment of Hazardous and Radioactive Waste," EPA/625/R-92/002 (May, 1992).

14. J.C. Whitehouse, C.M. Jantzen, D.F. Bickford, A.L. Kielpinski and B.D. Helton, "Development of a Transportable Vitrification System for Mixed Waste," WM95.

15. J.C. Whitehouse, C.M. Jantzen, F.R. VanRyn, D.D. Davis, "Design and Fabrication of a Transportable Vitrification System for Mixed Waste Processing," Proceed. Third Inter. Symp. on Mixed Waste, A.A. Moghissi, B.R. Love, and R.K. Blauvelt (Eds.), Am. Soc. Mech. Eng., p.8.3.1 to 8.3.8 (1995).

16. C.M. Jantzen, "Systems Approach to Nuclear Waste Glass Development," J. Non-Crystalline Solids, 84, 215-225 (1986).

17. C.M. Jantzen, "Relationship of Glass Composition to Glass Viscosity, Resistivity, Liquidus Temperature, and Durability: First Principles Process-Product Models

for Vitrification of Nuclear Waste," Nuclear Waste Mgt. IV, G.G. Wicks, D.F. Bickford, and L.R. Bunnell (Eds.), Ceram. Trans., V. 23, Am. Ceram. Soc., Westerville, OH, 37-51 (1991).

18. Federal Register, "Land Disposal Restrictions Phase II-Universal Treatment Standards, and Treatment Standards for Organic Toxicity Characteristic Wastes and Newly Listed Wastes" 59 FR47982 (September 19, 1994).

19. D. M. Bennert, T.J. Overcamp, D.F. Bickford, C.M. Jantzen, and C.A. Cicero, "Demonstration of Vitrification Surrogate F006 Wastewater Treatment Sludges," Environmental and Waste Management Issues in the Ceramic Industry, II, D.F. Bickford, S. Bates, V. Jain, and G. Smith [Eds.], Ceramic Transactions, v.45, 59-72 (1994).

20. D. Poulous, J.B. Pickett, C.M. Jantzen, "Upfront Delisting of F006 Mixed Waste," WM95.

21. J.B. Pickett, J.C. Musall, Alan F. Hayes, and E.E. Campbell, "Novel Procurement Concepts Utilized to Award Subcontract for Vitrification of an F006 Mixed Waste Sludge," Proceedings of Spectrum 94 Nuclear and Hazardous Waste Management International Topical Meeting, Am. Nuclear Soc., 110-113 (1994).

22. W.D. Bostick, D.P. Hoffmann, R.J. Stevenson, A.A. Richmond and D.F. Bickford, "Surrogate Formulations for Thermal Treatment of Low-Level Mixed Waste, Part IV: Wastewater Treatment Sludges," USDOE Report DOE/MWIP-18, 34p (January, 1994).

23. M.B. Volf, "Chemical Approach tø Glass," Glass Science and Technology, 7, Elsevier, NY (1984).

24. E.B. Munday, "Investigation of Lithium Hydroxide Disposal Options with Cost Estimates," USDOE Report K/PS-1239, Martin Marietta Energy Systems, Inc. (April, 1987).

25. K.J. Hewlett, "Vitrification Demonstration of West End Treatment Facility Mixed Waste Sludge," Unpublished Masters Thesis, Clemson University, 90p (December, 1994).

26. C. A. Cicero, D.F. Bickford, M.K. Andrews, K.J. Hewlett, D.M. Bennert, and T.J. Overcamp, "Vitrification Studies with DOE Low Level Mixed Waste Wastewater Treatment Sludges," Waste Management 95.

27. R.A. Merrill and D.S. Janke "Results of Vitrifying Fernald OU-4 Wastes," Nuclear Waste Management 93, US DOE Report PNL-SA-21856, Battelle Pacific Northwest Laboratory, Richland, WA (1993).

CONTROLLING THE PRODUCTION OF VITRIFIED, HIGH-LEVEL WASTES WITHIN SPECIFIED COMPOSITION TOLERANCES

Vijay Jain*, Ronald A. Palmer, and Steven M. Barnes
West Valley Nuclear Services Company, Inc.
West Valley, NY 14171-0191

ABSTRACT

The immobilization of high-level radioactivewaste (HLW) into a stable and nondispersible form using vitrification was successfully initiated at the West Valley Demonstration Project (WVDP) in July 1996. Since then, over 40 batches of the HLW, each between 7.5 and 11.5 m^3, have been vitrified into a stable canistered waste-form. The HLW transferred from Tank 8D-2 is mixed with glass-forming chemicals and fed to a joule-heated melter at 1,150° C. The glass is then poured into stainless steel canisters. After the canisters have cooled, they are welded, decontaminated, and stored on site to await eventual disposal. To ensure process reliability and that the canistered waste-form complies with all Waste Acceptance Product Specifications (WAPS) requirements, the batch-to-batch variability of the transferred waste, the variability associated with the sampling and analysis, and the differences observed between the feed and the glass shard samples were analyzed and compared with the WAPS. The vitrification process is discussed based on the process data collected to date and an assessment is provided of the canistered waste-form properties relative to the WAPS requirements.

*Now with Southwest Research Institute, San Antonio, TX.

INTRODUCTION

The high-level radioactive waste (HLW) at the West Valley Demonstration Project (WVDP) was generated by commercial reprocessing of 640 tons of spent reactor fuel. At the inception of the WVDP, two waste streams were stored in underground tanks. The majority, 2,200 m^3, was neutralized PUREX waste. Another 40 m^3 of partially processed THOREX waste was also stored in a separate tank. The PUREX waste was washed to minimize the sodium and sulfate salts to be vitrified. The original supernatant and wash solutions were decontaminated using a zeolite ion-exchange process to remove the cesium. The effluent was encapsulated in cement. Prior to the start of radioactive operations, the THOREX and zeolite wastes were blended with the washed PUREX sludge in Tank 8D-2. The volume of the waste at the start of radioactive operations was approximately 1,000 m^3.

PROCESS OVERVIEW

The vitrification process can be separated into three phases: tailoring the waste slurry by adding glass-forming chemicals to achieve the desired composition, melting the waste slurry and casting the glass into canisters, and closing the canistered waste-forms. The overall process is shown in Figure 1 and each phase is discussed below.

The initial step in the vitrification process is the pumping of a batch of HLW slurry from Tank 8D-2 to the concentrator feed makeup tank (CFMT) in the Vitrification Facility (VF). Approximately 100 batch transfers will be required to process the HLW inventory. At the CFMT, the waste is sampled and concentrated by evaporating water from the slurry. Chemicals are added to the CFMT to adjust the waste slurry to the target glass composition based on the results of the waste sample chemical analysis. The target glass composition and its acceptable process region are shown in Table 1. Following confirmation that the desired slurry chemistry has been achieved and that the feed is adjusted to the target reduction/oxidation (redox) ratio by the addition of sugar, the feed is transferred to the melter feed hold tank (MFHT).

The second phase is initiated by pumping the slurry from the MFHT to the melter. At the melter, the remaining water is evaporated from the slurry and the waste fuses with the glass-forming chemicals thereby producing the target waste-form composition. Molten glass is periodically poured from the melter into 3 m tall by 0.6 m diameter stainless steel canisters. After the glass solidifies, the canisters are routed for closure, decontamination, and on-site storage.

In the final process stage, the canisters are closed by welding a lid to the canister. The welding technique is an automated, autogenous, pulsed-gas tungsten arc welding process.[1] Following a visual quality verification inspection of the closure weld, the canisters are decontaminated using an acidic, cerium-based, chemical oxidation procedure.[2] After decontamination, the canistered waste-forms are stored on site until they can be shipped to a federal repository for storage or disposal.

The following sections discuss efficiency and control of the process to produce a consistent glass waste-form.

PROCESS CONTROL AND PERFORMANCE

The process starts with a transfer of a known amount of waste from Tank 8D-2 to the CFMT. The HLW transfer is made on the top of the existing heel (about 1,000 liters) and the recycled solution from the aqueous process off-gas scrubber. Figure 2 shows the Sr-90 and Cs-137 activity in the transferred HLW to the CFMT. Sr-90 and Cs-137, along with their daughters Y-90 and Ba-137m, are the major radioactive species in the waste. They account for more than 98% of the total activity.

The amount of waste transferred for each batch depends on the available volume in the CFMT for a new transfer and the expected volume of the glass-former chemical addition. Figure 3 shows the percentage of waste present in the glass. Figure 4 shows the amounts of the major and minor components (as oxides) in the waste. The composition of the HLW transfers for the first 18 batches have been very consistent. Both Figures 3 and 4 show a step change in the amount of waste transferred and the composition of the waste. This change is attributed to the improved suspension of solids in Tank 8D-2 upon replacement of one of the failed, long-shafted mixing pumps. The observed step change in the HLW composition did not result in any change in the production of the target glass composition because the required adjustments to the composition were made by the addition of glass formers.

The amount of HLW that could be transferred or the maximum glass that could be produced from a waste transfer depends upon the concentration potential of the wastes in the CFMT. The concentration potential is enhanced in the CFMT by the addition of sodium metasilicate as a deflocculating agent.[3] Efforts have been made to maximize the concentration of the HLW in the CFMT since the beginning of radioactive operations. Figure 5 shows the amount of glass produced from each HLW transfer. The glass loading is defined as the amount of glass produced per liter of feed. So far, a maximum glass loading of 530 g/l has been attained in the CFMT without any processing problems. This has enabled enhancing the batch size from an initial 6,000 kg of glass (3.0 canisters) to 10,000 kg of glass (5.0 canisters) per batch of HLW transfer without any slurry handling equipment problems. The improved batch size is a direct result of the WVDP's ability to concentrate and process concentrated feed through the vitrification process lines and pumps.

Based on the chemical analysis of the CFMT contents, the total amount of glass-forming chemicals required to produce the WVDP target glass composition is calculated. The glass formers are prepared in a nonradioactive facility and transferred to the CFMT.

The waste-feed slurry is the final mix of the HLW and glass-former slurries. The mix contains all chemical species needed to produce the target waste-form. The chemical analysis of this slurry is the key hold point for approval to transform the final slurry mix into glass. The Waste Form Qualification Report (WQR)[4] requires that for the feed batch to be of acceptable composition for making the waste form, it should fall within the processing region shown in Table 1. The final compositions of all the batches processed to date were within the processing region of the target glass composition. Even though the composition of all the batches fell within the acceptance range, the composition of batches 11, 12, 13, 28 and 30 were outside the acceptance region at the first chemical analysis. The WQR states that if the composition of the feed falls outside the acceptance region, a new set of samples will be taken and analyzed, and then the analyses will be grouped with the previous set to re-determine the chemical composition of the slurry. Furthermore, if the batch is still not acceptable the required shim chemicals will be added and the contents will be resampled and reanalyzed. Batches 11,12, 28, 29, 30 required chemical shimming to bring the composition within the acceptable range, while batch 13 was acceptable after reanalysis.

In addition, WQR Section 1.3 requires that before the final mix is approved for making glass, the mean of the normalized release rates for B, Na, and Li in the feed (as estimated by the prediction equations in reference 4) is at least two standard deviations below the mean normalized release rates for B, Na, and Li in the Environmental Assessment (EA) glass using the product consistency test (PCT). Figure 6 shows the estimated upper bound 95% confidence limit values for the normalized PCT release rates for B, Na, and Li in the feed, along with the lower bound 95% confidence limit values for the normalized release rates for the EA glass. The data indicate that the release rates are not only acceptable but are well below the lower limit of the EA glass.

The final step in feed preparation is to estimate the amount of reductant to be added to the feed. The reductant addition is necessary to provide proper redox conditions in the melter. The impacts of these conditions are discussed in detail in reference 5. The amount of sugar needed is calculated using the empirical relationship shown below:

$$IFO = NO_3(1-TS)/TOC$$

where IFO is the index of feed oxidation, NO_3 is the total nitrate concentration (ppm), TS is the fraction of total solids in the feed, and TOC is the total organic carbon concentration (ppm). A target IFO of 3.0 was used to estimate the amount of sugar required to process the feed and ensure that the Fe^{2+}/Fe^{3+} ratio remained between 0.01 and 0.20.

The sugared waste-feed slurry is transferred to the MFHT and mixed with the heel of the previous batch. The combined slurry is then fed to the melter. The feed rate depends upon the melting characteristics of the feed. Higher total solids and sugar concentrations promote higher feed rates, while higher water and nitrate concentrations decrease feed rates in the melter. An optimum balance is maintained between total solids, nitrates, and sugar to maximize the feed rate while maintaining the Fe^{2+}/Fe^{3+} ratio within the specified range. Figure 7 shows the three-canister moving average for the glass production rate. The data indicate that the glass production rate has steadily increased to 35 to 40kg/hr. This increase is attributed to the higher total solids in the feed.

Table 2 shows the average composition of feed batches 10 through 30 and the glass shards analyzed from randomly selected samples. The data indicate that the feed composition, as well as the glass composition, are within the target glass processing range.

SUMMARY

In summary, the WVDP has successfully demonstrated the conversion of HLW into a stable, borosilicate glass waste-form. As of July 1, 1997, 50% of the radionuclide inventory has been processed and all waste acceptance criteria have been met. The entire process is anticipated to be completed in June 1998. The canistered waste-form is stored in the Interim Storage Facility on site until a decision is made on a final HLW federal repository.

REFERENCES

1. D. K. Ploetz and J. J. May, "Unique Design Features of the West Valley Demonstration Project Vitrification Project," Proceedingsof the Waste Management '96 Symposia (1996).

2. T. F. Kocialski and E. F. Dinsmore, "Full-Scale Testing of the WVDP Canister Decontamination System," Proceedings of the Waste Management '94 Symposia (1994).

3. K. Firstenberg, W. Dabney and V. Jain, "The Role of Deflocculants During Concentration of the Simulated High Level Radioactive Wastes," Ceramic Transactions, Volume 61, (1995).

4. Waste Form Qualification Report, WVNS-186, Rev. 1, West Valley Nuclear Services Company, Inc., West Valley, NY (1996).

5. S. M. Barnes and V. Jain, "Vitrification Systems Testing to Support Radioactive Glass Production at the West Valley Demonstration Project," Proceedings of the Waste Management '96 Symposia (1996).

Table 1. WVDP Target Glass Composition

Oxide	Lower Process Limit (weight %)	Target Composition (weight %)	Upper Process Limit (weight %)
Al_2O_3	5.43	6.00	6.57
B_2O_3	10.96	12.89	14.82
CaO	0.36	0.48	0.55
Fe_2O_3	10.22	12.02	13.82
K_2O	4.37	5.00	5.63
Li_2O	3.25	3.71	4.17
MgO	0.76	0.89	1.02
MnO	0.70	0.82	0.94
Na_2O	7.00	8.00	9.00
P_2O_5	1.02	1.2	1.38
SiO_2	38.73	40.98	43.23
ThO_2	2.67	3.56	4.09
TiO_2	0.68	0.80	0.92
UO_3	0.47	0.63	0.72
ZrO_2	1.12	1.32	1.52
Others*	na	1.7	na

*Includes BaO 0.16%, Ce_2O_3 0.31%, CoO 0.02%, Cr_2O_3 0.14%, Cs_2O 0.08%, CuO 0.03%, La_2O_3 0.04, MoO_3 0.04%, Nd_2O_3 0.14%, NiO 0.25%, PdO 0.03%, Pr_6O_{21} 0.04%, Rh_2O_3 0.02%, RuO_2 0.08%, SO_3 0.23%, Sm_2O_3 0.03%, SrO 0.02%, Y_2O_3 0.02%, and ZnO 0.02%.

Table 2. Comparison of Feed and Glass Analysis with WVDP Target Glass Composition

Oxide	Lower Process Limit (weight %)	Feed Analysis (weight %)	Glass Analysis (weight %)	Upper Process Limit (weight %)
Al_2O_3	5.43	6.13	5.91	6.57
B_2O_3	10.96	13.38	12.86	14.82
CaO	0.36	0.48	0.42	0.55
Fe_2O_3	10.22	12.13	11.90	13.82
K_2O	4.37	4.96	4.97	5.63
Li_2O	3.25	3.69	3.65	4.17
MgO	0.76	0.89	0.80	1.02
MnO	0.70	0.82	0.79	0.94
Na_2O	7.00	8.15	7.80	9.00
P_2O_5	1.02	1.22	1.28	1.38
SiO_2	38.73	40.82	42.22	43.23
ThO_2	2.67	3.54	2.88	4.09
TiO_2	0.68	0.77	0.80	0.92
UO_3	0.47	bdl	bdl	0.72
ZrO_2	1.12	1.34	1.32	1.52
Others*	na	1.69		na

*Includes BaO 0.16%, Ce_2O_3 0.31%, CoO 0.02%, Cr_2O_3 0.14%, Cs_2O 0.08%, CuO 0.03%, La_2O_3 0.04, MoO_3 0.04%, Nd_2O_3 0.14%, NiO 0.25%, PdO 0.03%, Pr_6O_{21} 0.04%, Rh_2O_3 0.02%, RuO_2 0.08%, SO_3 0.23%, Sm_2O_3 0.03%, SrO 0.02%, Y_2O_3 0.02%, and ZnO 0.02%

bdl (below detection level)

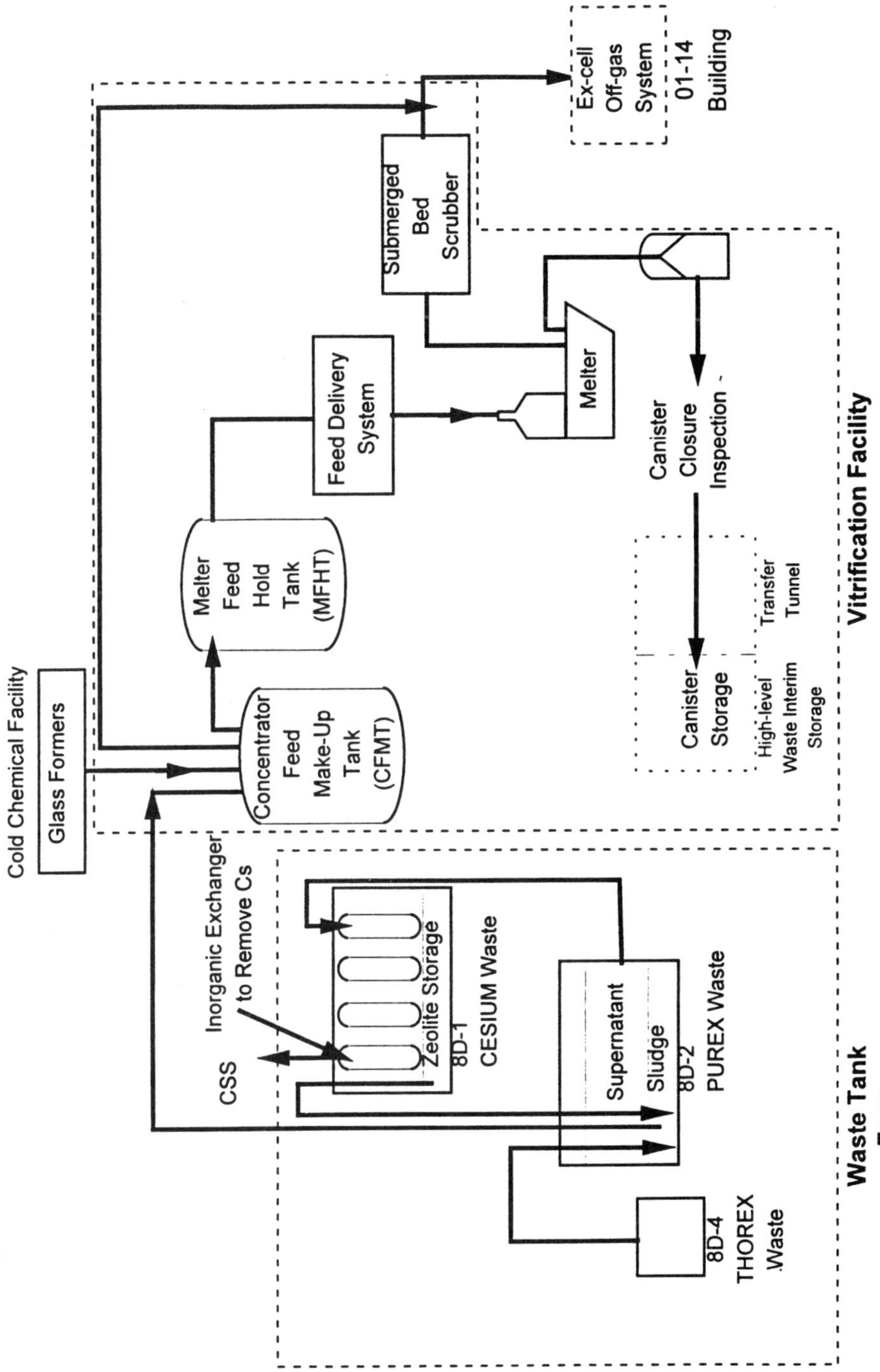

Figure 1. WVDP process flowsheet

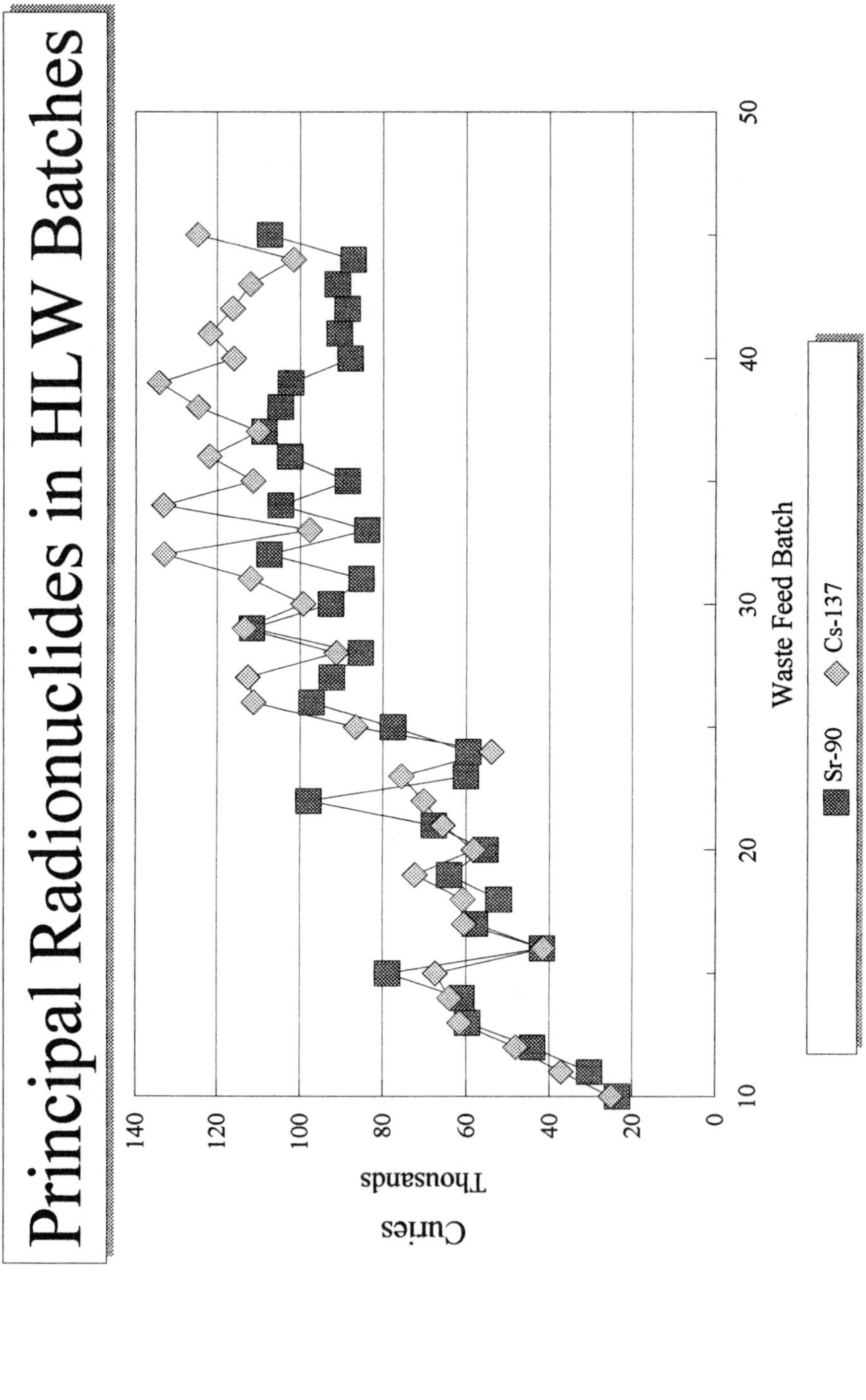

Figure 2. Analyzed Sr-90 and Cs-137 activity in HLW batches

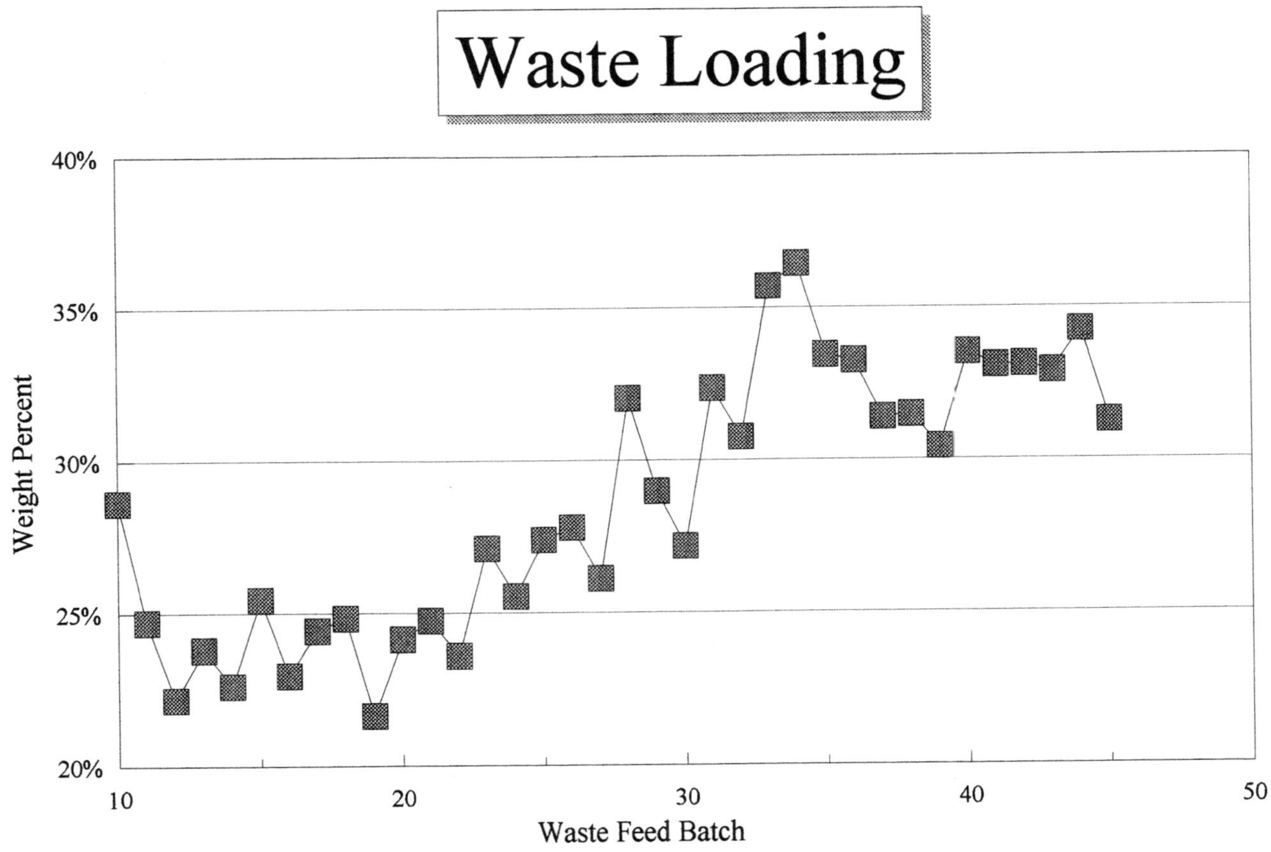

Figure 3. Amount of waste present in glass in each HLW transfer

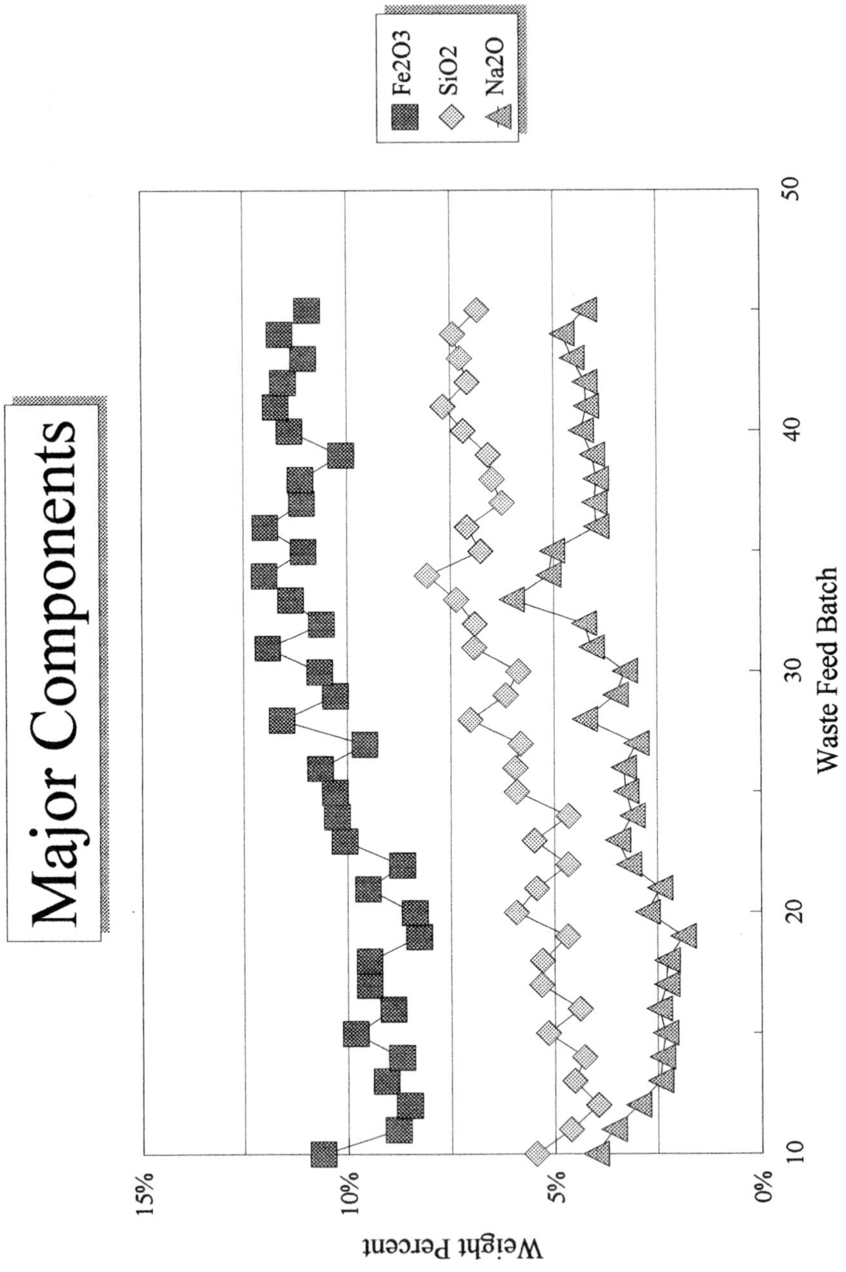

Figure 4a. Major components in HLW transfers

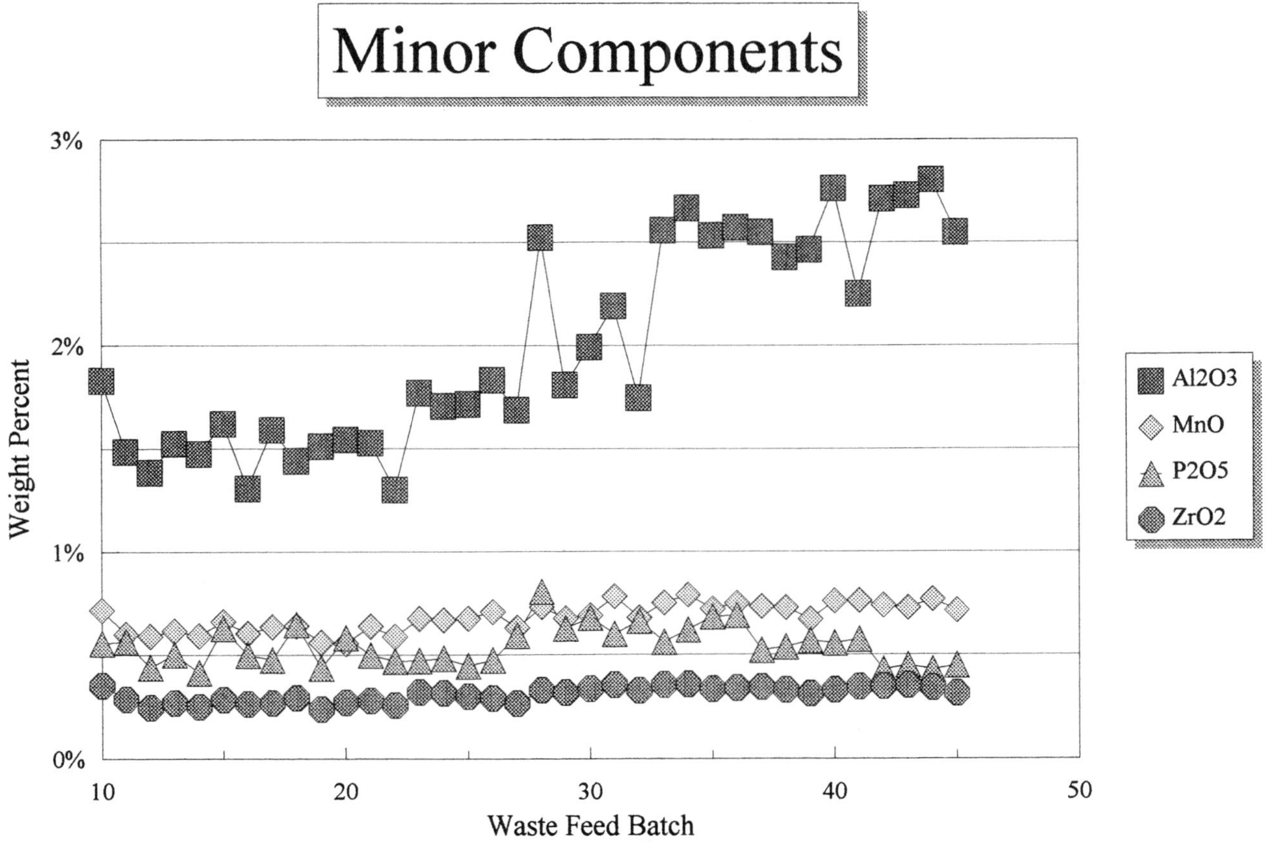

Figure 4b. Minor components in HLW transfers

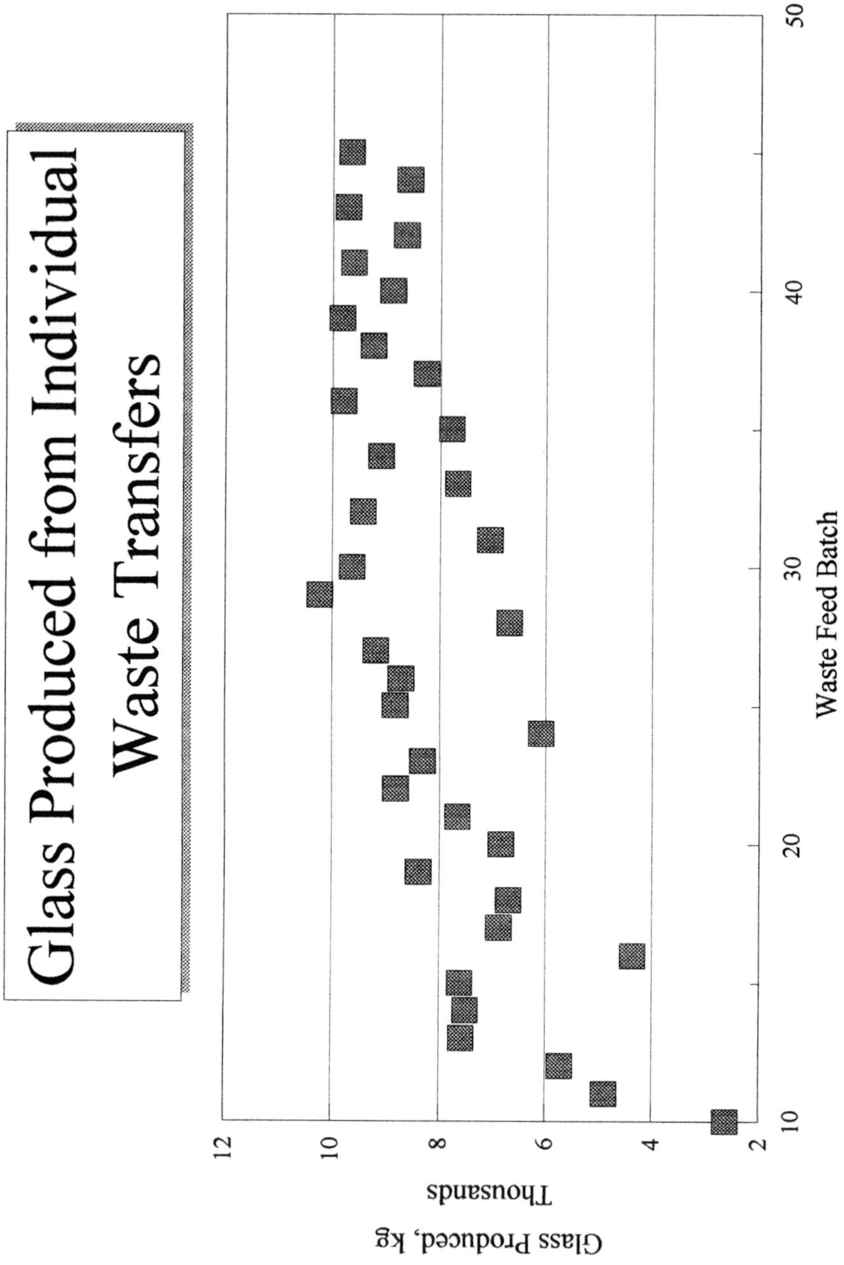

Figure 5. Amount of glass produced in each HLW batch

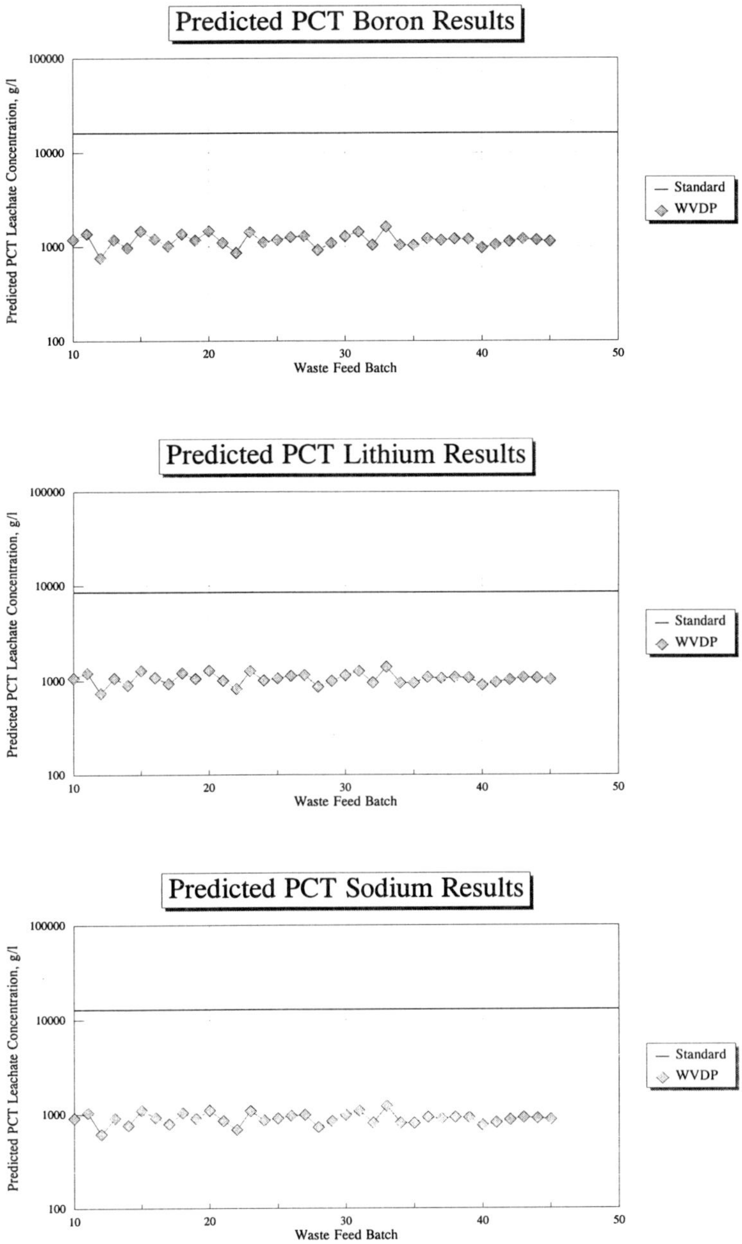

Figure 6. PCT performance of the WVDP feed batches

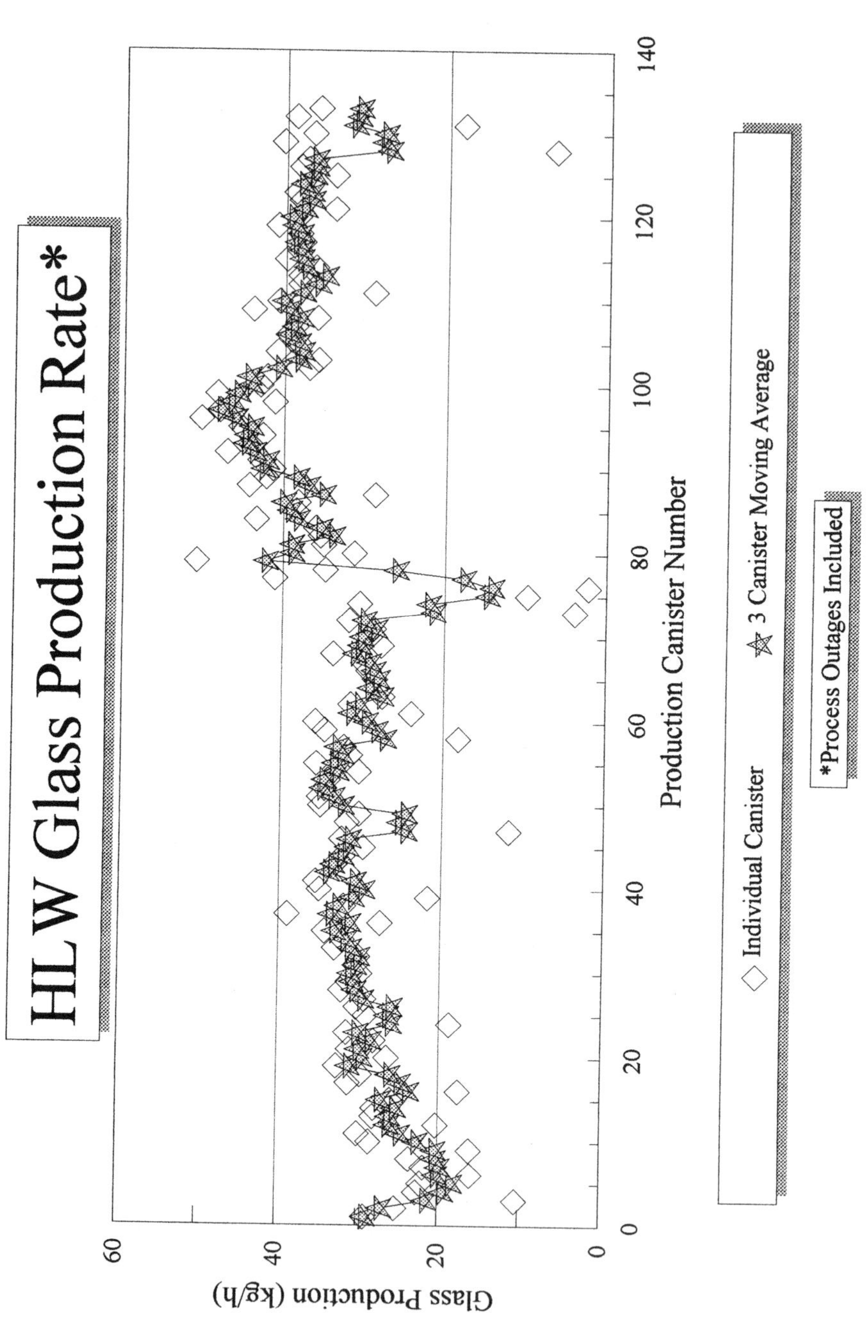

Figure 7. HLW glass production rate

KEYWORD AND AUTHOR INDEX